Standard Normal Distribution: $P(Z \leq z)$ (*continued*)

z	P	z	P	z	P	z	P	z	P
0.39	.6517	0.94	.8264	1.49	.9319	2.04	.9793	2.59	.9952
0.40	.6554	0.95	.8289	1.50	.9332	2.05	.9798	2.60	.9953
0.41	.6591	0.96	.8315	1.51	.9345	2.06	.9803	2.61	.9955
0.42	.6628	0.97	.8340	1.52	.9357	2.07	.9808	2.62	.9956
0.43	.6664	0.98	.8365	1.53	.9370	2.08	.9812	2.63	.9957
0.44	.6700	0.99	.8389	1.54	.9382	2.09	.9817	2.64	.9959
0.45	.6736	1.00	.8413	1.55	.9394	2.10	.9821	2.65	.9960
0.46	.6772	1.01	.8438	1.56	.9406	2.11	.9826	2.66	.9961
0.47	.6808	1.02	.8461	1.57	.9418	2.12	.9830	2.67	.9962
0.48	.6844	1.03	.8485	1.58	.9429	2.13	.9834	2.68	.9963
0.49	.6879	1.04	.8508	1.59	.9441	2.14	.9838	2.69	.9964
0.50	.6915	1.05	.8531	1.60	.9452	2.15	.9842	2.70	.9965
0.51	.6950	1.06	.8554	1.61	.9463	2.16	.9846	2.71	.9966
0.52	.6985	1.07	.8577	1.62	.9474	2.17	.9850	2.72	.9967
0.53	.7019	1.08	.8599	1.63	.9484	2.18	.9854	2.73	.9968
0.54	.7054	1.09	.8621	1.64	.9495	2.19	.9857	2.74	.9969
0.55	.7088	1.10	.8643	1.65	.9505	2.20	.9861	2.75	.9970
0.56	.7123	1.11	.8665	1.66	.9515	2.21	.9864	2.76	.9971
0.57	.7157	1.12	.8686	1.67	.9525	2.22	.9868	2.77	.9972
0.58	.7190	1.13	.8708	1.68	.9535	2.23	.9871	2.78	.9973
0.59	.7224	1.14	.8729	1.69	.9545	2.24	.9875	2.79	.9974
0.60	.7257	1.15	.8749	1.70	.9554	2.25	.9878	2.80	.9974
0.61	.7291	1.16	.8770	1.71	.9564	2.26	.9881	2.81	.9975
0.62	.7324	1.17	.8790	1.72	.9573	2.27	.9884	2.82	.9976
0.63	.7357	1.18	.8810	1.73	.9582	2.28	.9887	2.83	.9977
0.64	.7389	1.19	.8830	1.74	.9591	2.29	.9890	2.84	.9977
0.65	.7422	1.20	.8849	1.75	.9599	2.30	.9893	2.85	.9978
0.66	.7454	1.21	.8869	1.76	.9608	2.31	.9896	2.86	.9979
0.67	.7486	1.22	.8888	1.77	.9616	2.32	.9898	2.87	.9979
0.68	.7517	1.23	.8907	1.78	.9625	2.33	.9901	2.88	.9980
0.69	.7549	1.24	.8925	1.79	.9633	2.34	.9904	2.89	.9981
0.70	.7580	1.25	.8943	1.80	.9641	2.35	.9906	2.90	.9981
0.71	.7611	1.26	.8962	1.81	.9649	2.36	.9909	2.91	.9982
0.72	.7642	1.27	.8980	1.82	.9656	2.37	.9911	2.92	.9982
0.73	.7673	1.28	.8997	1.83	.9664	2.38	.9913	2.93	.9983
0.74	.7703	1.29	.9015	1.84	.9671	2.39	.9916	2.94	.9984
0.75	.7734	1.30	.9032	1.85	.9678	2.40	.9918	2.95	.9984
0.76	.7764	1.31	.9049	1.86	.9686	2.41	.9920	2.96	.9985
0.77	.7793	1.32	.9066	1.87	.9693	2.42	.9922	2.97	.9985
0.78	.7823	1.33	.9082	1.88	.9699	2.43	.9925	2.98	.9986
0.79	.7852	1.34	.9099	1.89	.9706	2.44	.9927	2.99	.9986
0.80	.7881	1.35	.9115	1.90	.9713	2.45	.9929	3.00	.9987
0.81	.7910	1.36	.9131	1.91	.9719	2.46	.9931	3.01	.9987
0.82	.7939	1.37	.9147	1.92	.9726	2.47	.9932	3.02	.9987
0.83	.7967	1.38	.9162	1.93	.9732	2.48	.9934	3.03	.9988
0.84	.7995	1.39	.9177	1.94	.9738	2.49	.9936	3.04	.9988
0.85	.8023	1.40	.9192	1.95	.9744	2.50	.9938	3.05	.9989
0.86	.8051	1.41	.9207	1.96	.9750	2.51	.9940	3.06	.9989
0.87	.8078	1.42	.9222	1.97	.9756	2.52	.9941	3.07	.9989
0.88	.8106	1.43	.9236	1.98	.9761	2.53	.9943	3.08	.9990
0.89	.8133	1.44	.9251	1.99	.9767	2.54	.9945	3.09	.9990
0.90	.8159	1.45	.9265	2.00	.9772	2.55	.9946		
0.91	.8186	1.46	.9279	2.01	.9778	2.56	.9948		
0.92	.8212	1.47	.9292	2.02	.9783	2.57	.9949		
0.93	.8238	1.48	.9306	2.03	.9788	2.58	.9951		

Critical Values of z for

Significance Level	Two tails	Lower tail	Upper tail
0.10	±1.65	−1.28	+1.28
0.05	±1.96	−1.65	+1.65
0.01	±2.58	−2.33	+2.33

Basic Statistics in Business and Economics

FOURTH EDITION

George W. Summers
University of Arizona

William S. Peters
University of New Mexico

Charles P. Armstrong
University of Rhode Island

Wadsworth Publishing Company
Belmont, California
A Division of Wadsworth, Inc.

Students: There is extra help available . . .

This text has been written to make your study of statistics as simple and rewarding as possible. But even so there may be concepts or techniques that will be difficult for you. Or you may want a little "insurance" for success as mid-terms and finals come up. If so, you will want to buy a copy of the *Student Supplement for Basic Statistics in Business and Economics*. It's described in the preface of this book. Your campus bookstore either has it in stock or will order it for you.

Statistics Editor: Stephanie Surfus
Production: Greg Hubit Bookworks

Printed in the United States of America
2 3 4 5 6 7 8 9 10—89 88 87 86

0-534-04299-6

Library of Congress Cataloging in Publication Data

Summers, George William.
Basic statistics in business and economics.

Includes index.
1. Social sciences—Statistical methods.
2. Statistics. 3. Commercial statistics.
I. Peters, William Stanley, 1925 joint author.
II. Armstrong, Charles P., joint author. III. Title.
HA29.S83 1985 519.5 84-23426
ISBN 0-534-04299-6

CONTENTS IN BRIEF

CONTENTS

PREFACE

Basic Statistics in Business and Economics, Fourth Edition, is intended primarily for business students taking beginning statistics courses. These courses may be as short as one quarter or as long as an academic year. Chapter sequences can be designed to fit either extreme.

Our main emphasis is on presenting statistical concepts in an applied context, developing the underlying statistical methods through verbal explanation and example. Careful attention is paid to necessary assumptions and to the consequences of violating them, and to interpretation of results. Knowledge of algebra is assumed, but concepts and procedures are explained with minimal use of mathematics. Our objective is to promote appreciation for underlying concepts without overmathematizing or oversimplifying.

Concrete applications from functional areas of business, industry, and government are used extensively in text examples and exercises. Many of these applications are based on situations reported by managers of statistical departments at a number of companies, who responded to the authors' requests for real-world uses of statistics. In addition, we have employed examples from the business literature. Sources are cited where appropriate. A brief list of selected applications follows.

Accounting

Comparison of operating ratios with control group (Exercise 6-24, page 171)

Sampling audit of accounts payable (Exercise 9-30, page 288)

Estimation of the proportion of delinquent charge accounts (Exercise 8-6, page 230)

Advertising

Hardware buying; cooperative regression study of sales results from three media (page 455)

Newspaper readership (page 121)

Response to TV spot ads (Exercise 10-17, page 317)

Media market penetration (Exercise 4-38, page 116)

Economics

Recession and balance of trade assessments (Exercise 4-18, page 106)

Distribution of number of employees per firm (Exercise 3-18, page 65)

Relationship of collective bargaining law to movement in wage offers (Exercise 11-10, page 355)

Finance, Insurance, and Real Estate

Selling price, floor area, age of single-family residences—a regression study (page 436)

Mergers and stock price changes (Exercise 4-23, page 107)

Life insurance premiums and actuarial probabilities (page 133)

Stock portfolio performance (Exercise 4-10, page 98)

General Management

W.R. Grace: major sources of net sales 1950, 1962, 1977 (Figure 2-5, page 23)

Ages of chief executive officers of larger corporations (Exercise 3-14, page 61)

Functional areas of business from which chief executives came (Exercise 2-23, page 48)

Marketing

Connecticut Peak-load Pricing Study: electricity consumption change for large users (page 354)

Packaging effects on sales (Exercise 16-6, page 524)

Effects of point-of-sale displays (page 402)

Effect of message on telephone sales appointment rate (Exercise 11-14, page 364)

Effectiveness of endorsements (page 496)

Operations Management

Machine maintenance cost versus age (Exercise 10-23, page 324)

Multiple-head machine control (page 492)

Product development cost ratios (Exercise 3-32, page 82)

Warehouse space control (Exercise 14-1, page 446)

Acceptance sampling by attributes (Excercise 6-26, page 176)

Product spoilage for two shifts (Exercise 11-4, page 349)

Personnel

Estimation of employee absenteeism (Exercise 8-42, page 252)

Factors influencing job performance (Exercise 14-4, page 447)

Recruiting professional athletes (Exercise 16-7, page 524)

Public Administration

Growth policies of county commissioners (Exercise 11-12, page 356)

Pedestrian fatalities from motor vehicles (Exercise 8-44, page 253)

Probability of arrest versus police response time (Exercise 10-3, page 303)

The changes made for this edition are not major, but are designed to advance our objectives of making the study of statistics interesting and accessible.

Exercise sets have been edited, winnowed, and updated where needed. In Chapters 1 through 10 especially, more drill exercises have been included as the first group in a set.

Further materials emphasizing the use of computers in statistics have been introduced. These include end-of-chapter sections on interpreting computer output and extended sets of computer exercises. The computer exercises draw from data files on real estate transactions, 50 leading utilities' and 50 leading retailers' operating data, a company personnel file, and baseball team statistics.

Optional sections have been included on exploratory data analysis. These include Stem and Leaf Displays in Chapter 2, Boxplots in Chapter 3, Robust Estimation in Chapter 8, and Resistant Regression in Chapter 10.

Chapters 8 and 9 on estimation and hypothesis tests for means and proportions have been streamlined. Material employing the finite population multiplier has been moved to optional end-of-chapter sections. Summarizing tests by means of P-values is covered.

Chapter 15, Two-Way Analysis of Variance, is new. This chapter includes the use of randomized blocks, analysis of factorial experiments, use of regression programs in ANOVA data processing, and two extended case studies.

The book is divided into five parts: Descriptive Statistics, Background for Statistical Inference, Basic Statistical Inference, Further Topics in Statistical Inference, and Additional Topics. The first three parts form a sequence that we recommend for a one-semester course. Optional topics can be omitted in shorter courses; some instructors may wish to include Inferences from Two Samples (Chapter 11) in such a course.

The additional topics in Chapters 11 through 20 provide ample material for a second semester. Instructors may wish to emphasize the multiple regression and analysis of variance chapters (13, 14, and 15), analysis of frequencies (chi-square) and nonparametric statistics (12 and 16), time series, index numbers, and forecasting (17 and 18), and (Bayesian) decision making (19 and 20), or some combination of these.

A *Student Supplement* furnishes additional learning and practice aids, including programmed learning modules, self-correcting exercises with solutions, and practice examinations.

We wish to express our appreciation to the organizations and firms that responded to our requests for real applications of statistical methods. We want to thank Stephanie Surfus of Wadsworth Publishing Company and Greg Hubit Bookworks for their helpful and expert editing and production efforts. We wish to thank Jay M. Bennett from Analytics and John A. Flueck of Temple University for furnishing the baseball data file which is referenced in their article "An Evaluation of Major League Baseball Offensive Performance Models," *The American Statistician*, February, 1983. Thanks are also due to the Academic Computer Center at the University of Rhode Island and its Director, David Clayton. Special thanks are due to Mike Shaughnessy and Roger Greenall for their advice on computer applications. We also want to acknowledge the many helpful suggestions made by colleagues, students, and reviewers. The reviewers for this Fourth Edition were Charles Barrett, Habib Bazyari, William Beaty, Bruce Bowerman, P. L. Claypool, Douglas C. Darran, Frank Kelly, Ron Koot, Leonard Presby, John Stack, and William Stein. Finally, we want to thank the authors and publishers who gave us permission to reprint figures and tables. Their permissions are acknowledged in footnotes.

George W. Summers
William S. Peters
Charles P. Armstrong

Basic Statistics in Business and Economics

FOURTH EDITION

PART ONE

Descriptive Statistics

1

STATISTICS AND DATA

1.1 The Nature of Statistics

The word *statistics* has two major meanings. In ordinary speech, statistics means any collection of numerical facts. The business section of a daily newspaper contains security market statistics, and the sports section reports statistics on sports events. Libraries devote entire sections to statistics from government, industry, and commerce. We shall use *data* to refer to such a collection of numbers and reserve *statistics* for its second meaning: a professional field that seeks ways to gather and interpret data.

> *Statistics* is a discipline concerned with developing and applying effective techniques for collecting, organizing, and analyzing data and for presenting conclusions based on those data.

Branches of Statistics

The field of statistics can be divided into two branches, descriptive statistics and statistical inference. In our definition of statistics, *descriptive statistics* covers organizing and presenting data, while *statistical inference* involves analyzing data.

The objective of descriptive statistics is to bring out important features and patterns in collections of data. Some type of summarizing technique is especially helpful with large collections of data to make the data intelligible. Imagine a listing of the ages of people living in a city. Until devices such as tables, charts, and averages capture the pattern and basic characteristics of the data, the data are difficult to use and comparisons with similar collections for other cities are virtually impossible.

Descriptive techniques not only summarize data and make them more manageable, they also furnish guides to action. For instance, in one large firm, executives use cumulative year-to-date sales graphs to locate trouble spots. Figure 1-1 presents typical graphs for two sales regions in the firm. The brown line traces year-to-date sales by month for the last year. The black line shows cumulative sales through August of the current year. The graph clearly shows that region A is doing much better than last year, whereas region B is doing considerably worse. Investigation may show that depletion of the sales force is responsible for the poor showing. If so, recruiting, training, and motivation can be focused on region B where they are needed.

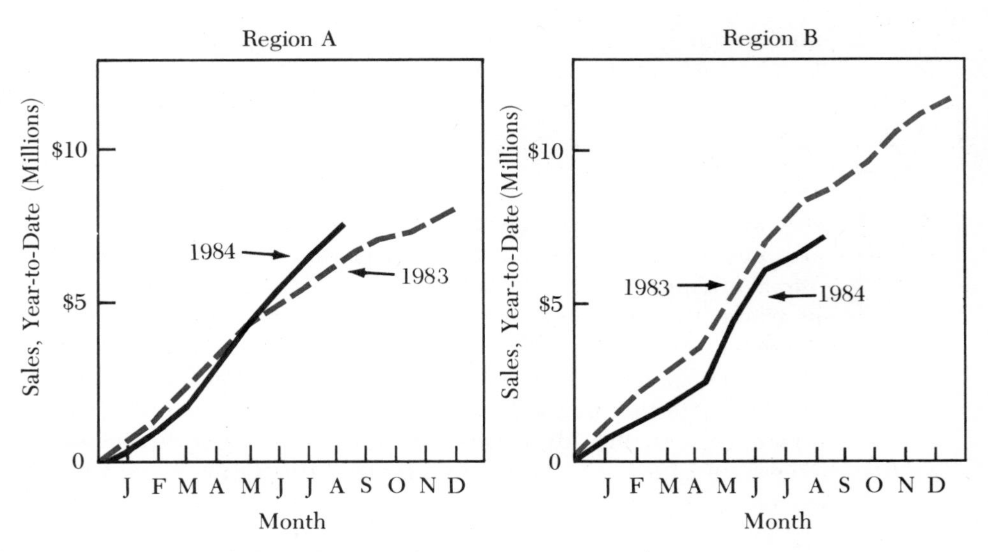

Figure 1-1 Regional Year-to-Date Sales

The objective of statistical inference is to use a subset of a set of data to draw dependable conclusions about the entire set. The entire set is called a *statistical population;* the subset is called a *sample*. Statistical inference discovers and applies procedures for gathering and analyzing samples that will produce safe generalizations about the populations from which the samples come. For example, a company developing a new drug could use a properly drawn sample of patients to estimate what percentage of potential users will suffer unwanted side effects. Any manufacturer involved in such an expensive development program must have such information to prevent going into full production with products that would later prove unacceptable.

The Impact of Statistics

Rapid transportation, rapid communication, and electronic computers have greatly increased the use of descriptive and inferential statistics. Despite the fact that the United States census now produces many times as many descriptive statistics as did earlier censuses, results are available much sooner. Organizations can buy data as computer input and use that input in any way they wish. Computer suppliers have inexpensive program packages for deriving descriptive summaries from many thousands of observations in a few minutes. Combining these resources, analysts in marketing, banking, politics, and real estate, for example, can construct up-to-date profiles of their constituencies, which can be very helpful in planning. As another example, vast inventories of stocks in large manufacturing, wholesaling, and retailing firms are kept correct to the minute on central computers linked to all branches in the firm. Such tasks as planning purchases and setting manufacturing or delivery schedules can be based on accurate, current data summarized according to the need of the moment with appropriate descriptive techniques.

Sampling procedures developed in statistics are now used in science, business, and government. Important guidelines for public policy come from frequent sampling of the labor force, the physical volume of production, the time pattern of prices, the size and composition of the population, automobile traffic and accident rates, fires and crimes, migration patterns, weights and measures, and myriad other facets of a complex society.

Modern agricultural crop productivity owes a great deal to statistically guided sampling experiments on the relative effectiveness of different types and combinations of factors such as fertilizers, moisture, soil chemicals, climate, and seed stock. Animal husbandry has been similarly improved by statistical comparisons of different genetic strains, feeding regimens, disease treatments, and so forth.

In industry, quality control of output has become nearly synonymous with that part of statistical inference known by the same name. Estimating the demand for products, determining the characteristics of buyers, and evaluating the effects of product features and advertising are all based on statistical techniques. Buyers use acceptance sampling to decide whether a supplier's wares meet specifications. Personnel managers statistically analyze application forms to guide them in selecting new employees. Auditors also use statistical sampling procedures.

This compilation only begins to suggest the increased use of statistical techniques during the last 75 years. If the present trends are any indication, the next 75 years should see us relying even more heavily on statistics than we do today.

1.2 The Nature of Data

To be sure that the data we collect will be relevant, we need to establish the general objectives of the study, specify the objects or elements to be observed and the relevant properties of these elements, and develop appropriate techniques for measuring the varying amounts of a property present in the elements. The first two of these steps are so closely related that they can be considered together. Their outcomes often appear as the statement of objectives in a study report.

Relevance

We shall first define the term *population*, which often occurs in discussions about the objectives of a study. The term is used in two senses: The collection of all elements of interest is called a population, as is the collection of numbers representing measurements of a property of each element. The following definitions distinguish the two usages.

> A *population* is the entire set of elements we want to study; a *statistical population* is the collection of numbers that would result if a property of each element in the entire set were measured.

It is common practice to use the word *population* in both senses and let the context make clear which meaning is intended. Observe also that we sometimes want information about several statistical populations from the same population of elements.

An important aspect of determining the objectives in a statistical investigation is to focus on what we want to learn about which population, or collection of people, organizations, or events. For example, suppose a refrigerator manufacturer wants to determine which of three new designs potential customers like best. A logical population might be the actual purchasers of refrigerators. However, because the decision to buy a refrigerator is usually made by several members of a household, households may provide better data than individuals.

Next, we must specify the property to be observed. The manufacturer wants to determine which of the three designs is preferred. But what exactly is to be determined? What data will be most helpful, the one design preferred most, or a level of preference for all three designs? If the manufacturer wants a level of preference for all three designs, then, at the very least, preferences must be ranked. How are the members of a household to produce a single set of preferences? Perhaps having each household reach consensus by its usual means for making major purchases would lead to the most dependable result.

Measurement

The third step in producing relevant data is developing appropriate techniques for measuring varying amounts of a property present in the elements to be observed. We first select one of four scales to use as a measuring scale. Then we apply the chosen scale to a property to define a variable. Finally, we observe the variables to produce data. Each of these three parts of the third step will be described in turn.

Measurement Scales Before a property or characteristic can be measured for each element of the population, one must decide how the measurements are to be made. In the refrigerator preference study, one approach would be to present each of the three possible pairs of designs, one at a time. The number 1 could be recorded for the preferred design and 0 for the other design in each pair. This is an example of a nominal scale. *Nominal scales use numbers to classify elements.* As another example, respondents in a survey might be classified according to their places of residence: The number 1 could stand for metropolitan areas, 2 for smaller urban areas, and 3 for rural areas.

Another approach in the refrigerator study would be to rank the designs. Each household could assign the number 1 to the preferred design, 2 to the next preferred, and 3 to the least preferred. This is an example of an ordinal scale. *Ordinal scales use numbers to rank elements.* With an ordinal scale, no constant distance between consecutive rank order numbers is implied. The ranks assigned to the refrigerator designs do not tell us *how much* better a household likes the highest-ranked design than the second, or the second than the third.

Interval scales are measures that allow us to compare distances between any two measurements. The classic example of an interval scale is temperature. If, as in the Celsius scale, the difference between the freezing point and boiling point of water is divided into 100 intervals, the intervals, or degrees Celsius, represent equal amounts of heat. The zero point does not represent absence of heat. But given three temperatures, 12°, 14°, and 18°, the difference between 18° and 14° represents twice as much heat as the difference between 14° and 12°.

The fourth and final type of measuring scale is the ratio scale. We use this kind of scale to record credit balances, to count the number of children born to each of several couples, to measure the lengths of several boards, or to determine the weights of several people. In these cases, and whenever we use a ratio scale, we interpret zero to mean that the element being observed has none of the characteristics being measured. *Ratio scales, along with interval scales, use numbers to put elements in order and measure the distance between them. In addition, ratio scales use the number zero to represent the absence of the characteristic being measured.*

The type of scale selected to measure a characteristic affects what can be done in subsequent steps of a statistical project. The degree of sophistication in the use of our number system increases as we go from nominal to ordinal, interval, and, finally, to ratio scales. Usually the more sophisticated the scale, the more elaborate the analyses that can be applied to the data.

Variables A measuring scale applied to a characteristic produces a variable. For example, credit balances measured in dollars and cents could be the variable of interest to a department store credit manager. *A variable represents any numerical value a given measuring scale can produce in a given application.* In the refrigerator study, if all three designs are presented at once, the variable is the rank of a given design as it is shown in successive households.

Variables can be either continuous or discrete. A variable such as the number of children born to each of several couples is discrete. *A discrete variable can assume values only at a countable number of points on its measuring scale.* On the other hand, a person's weight is a continuous variable; a person can weigh 112 pounds, 112.3 pounds, 112.3278 pounds, and so on to any desired degree of accuracy. *A continuous variable can assume any value on its measuring scale.* In practice, of course, there is a limit to the accuracy with which a continuous variable can be measured. For instance, length can be measured only to the nearest eighth of an inch with most yardsticks. When the length of a piece of cloth measured by such a yardstick is 18 in., it may, in fact, be any length from $17^{15}/_{16}$ to $18^{1}/_{16}$ in.

Data We have already referred to data as numerical facts. Now that we have become familiar with the concepts of measuring scales and variables, we can view data in a different light. When we use a measuring scale to determine the quantities of a specified characteristic present in a set of elements, the numbers produced are data. When we use a tape measure calibrated in sixteenths of an inch to measure the lengths of several boards, the numbers we record are data. *Data are observed values of a variable.*

Data can be either qualitative or quantitative. *Qualitative data* are produced by nominal scales and *quantitative data* are produced by ordinal, interval, or ratio scales.

Another important way of classifying a collection of observations is as time-series or cross-section data. *Time-series data* arise from measuring a characteristic of the same element at different times. For example, if the element is a retail shop, then a sequence of total daily sales constitutes a time series. *Cross-section data* arise from measuring a characteristic of several elements at the same time, as, for example, the credit balances of department store accounts on a particular date.

Figure 1-2 summarizes our discussion in this section. Households within some geographic area constitute the population; water usage is the characteristic of interest in each household. Gallons per week is the ratio scale selected to measure the characteristic. Together, the characteristic and the measuring scale define the variable to be observed: water consumption in gallons per week. The resulting observations are the data. In this case they are quantitative data. If we observe the pattern of water consumption by each household over a period of time, the result will also be time-series data. If all households are observed only once and at essentially the same time, the result will be cross-section data.

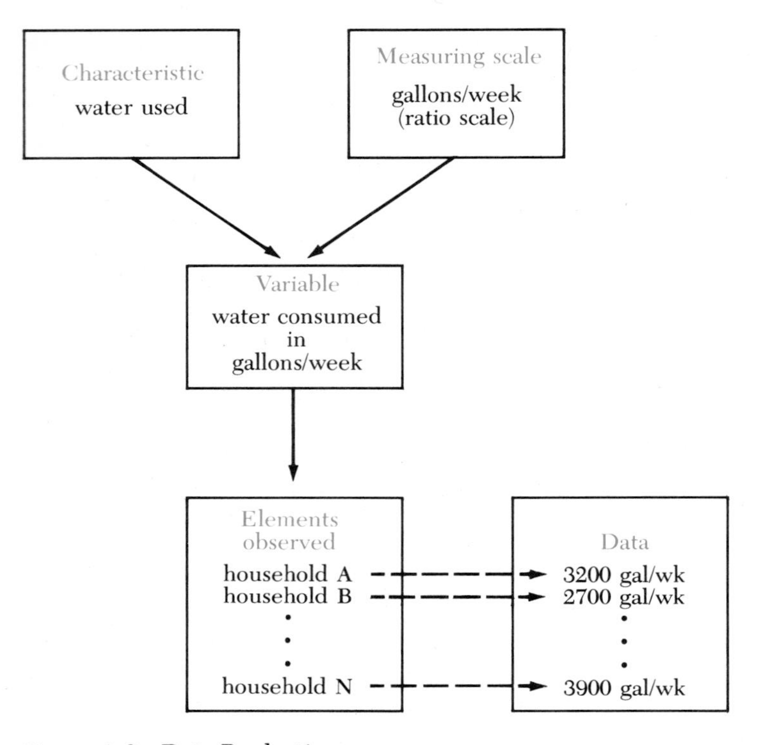

Figure 1-2 Data Production

When the objectives for a statistical project have been established and the data that will satisfy these objectives have been determined, the next step is to collect the data. The following section gives a brief description of some important considerations in data collection.

Exercises

1-1 *Personnel morale.* The personnel manager of an organization is studying employee attitudes. One item on the questionnaire is the following.
"I feel that I am performing a valuable service for society when I do my job well." Circle the number that most closely agrees with your feeling.

Strongly Agree	Agree	Undecided	Disagree	Strongly Disagree
1	2	3	4	5

a. For this study, describe the elements to be observed, the characteristic about which the item asks, the type of measuring scale, and the variable.
b. Is the variable discrete or continuous?
c. Are the data qualitative or quantitative? Cross-section or time-series?

1-2 *Automobile accident control.* Every month a statistical clerk in the city traffic engineering department must record the number of accidents that occurred during the month at each street intersection.

a. What is the population of elements being observed? What is the statistical population?
b. Define the variable. Is it continuous or discrete?
c. Which type of measuring scale is being used? Justify your choice.

1-3 *Investment portfolio pricing.* An investor owns 220 stocks, for which daily opening prices are given in the newspaper. The investor has recorded these daily quotations for the past 2 years, along with the earnings and price-earning ratios.

a. Briefly describe a study in which the investor would use some of these observations as cross-section data.
b. Briefly describe another study in which the investor would use some of these observations as time-series data.

1-4 *Utility rate setting.* A peak-load pricing test for residential electric power consumption was completed recently in Connecticut. The stated purpose was "to determine customer response to a time- and price-differentiated rate structure."*

a. Define the population of elements you would expect to be of interest in this study.
b. What two characteristics of each element constitute the central focus for the study?
c. Describe some alternative ways to measure consumption and discuss advantages and disadvantages of these alternatives.

1.3 Data Collection

Clearly defined, carefully collected data are the foundation of any soundly constructed statistical project. The degree to which the data are relevant to the objectives of the project establishes one upper limit of the quality of the result. Another upper limit is established by the extent to which the data are representative of the statistical population on which the project is focused.

Sources of Data

A central purpose in listing objectives for a project is to identify the variables that should be studied. This identification process is heavily influenced by alternative sources of data. Sources of data are of two types: secondary and primary.

Secondary sources of data are collections of published and unpublished results from previous projects. General references such as the *Survey of Current Business* and the *Monthly Labor Review* are examples of such sources. Unpublished reports on file in firms, trade associations, government agencies, and educational institutions are also available on many topics.

---

*Connecticut Public Utilities Control Authority and others, *Final Report — Connecticut Peak Load Pricing Test*, Hartford, Connecticut, May 1977.

Some precautions should be observed before accepting any secondary data for a project. The definitions of variables, populations, and elements used in the data must be appropriate to the project. For example, such terms as *household* and *family* are subject to several interpretations, so they must be defined in a similar manner in the data and in the project under consideration.

Data collection processes used for secondary sources also need to be examined. Reviewing questions asked of respondents may bring out facts that would invalidate the results of a project. Asking people who never work with tools which of four power saws they would choose for the home could result in answers based on appearance of the product rather than on its usefulness or reliability. A review of processes used to collect the data may reveal practices that yield nonrepresentative results. Mail questionnaires, for example, typically generate only about 30% to 40% response rates. Those with strong opinions in either direction on the subject at issue tend to be overrepresented. Callbacks on nonrespondents are necessary to correct this failing.

The censuses of the United States are very important as sources of secondary data for people in business, industry, and government. Information from the 1980 census is compared with that from the 1970 census in Table 1-1. The "100 percent items" were recorded for every individual and household, using a "short form." Additional items, called *sample items*, were included in a "long form," distributed to a random sample of individuals and households. One in 6 people were sampled in all places with more than 5000 population and 1 in 2 were sampled in places with less than 5000 population.

One important change in the 1980 census was the introduction of the "householder." The householder is the person (or one of the persons) who owns or rents a housing unit. The term *householder* took the place of the previously used *head of household*, who was always a male in households with a husband and wife. The first census item, *household relationship*, is the relation of each person in the household to the householder, including roomer, boarder, partner, or roommate for persons not related to the householder.

Among the new items included in the 1980 census were energy-related questions about car pools, travel time to work, and the number of light trucks and vans owned. Questions about the usual hours worked per week and number of weeks spent looking for work were two new employment-related questions.

When data of sufficient quality are not available in secondary sources, part of the project must be devoted to obtaining these data directly. All or part of the population must be approached and the relevant variables must be measured. The resulting observations are called *primary data*.

Collecting Primary Data

To observe the variables being studied, investigators can use either a census or a sample.

> A *census* is a complete enumeration for all elements in a population; a *sample* is an enumeration for any subset of such elements.

Suppose the credit manager in the department store mentioned earlier wants to know the average weekly balance last year for accounts 2 or more years old. A census can be conducted by having a computer find the weekly balances for every account at least 2 years old and calculate the average. These weekly balances comprise the entire statistical population of interest. Alternatively, a computer can be programmed to select only 25% of such accounts and to find the desired balances for this subset. This portion of all balances constitutes a sample.

Care should be taken to make the sample representative of important characteristics of the population from which it comes, such as the average balance of the credit accounts. In practice, this is accomplished by random sampling (defined on page 14). Even so, the representation will almost certainly not be perfect. For instance, the average balance of the credit accounts in the sample just described cannot be expected to *exactly* equal the average balance for all accounts in the population. The better the sample, the closer this correspondence can be expected to be. But for any sample short of the entire population, we must expect some discrepancy. This difference is called a *sampling error*.

Since sampling errors are inevitable, why are samples so often used in preference to censuses? Although it may seem foolish to deliberately select a data collection technique that could produce sampling error, there are reasons why sampling is so prevalent.

Reasons for Sampling

Destructive measurement is one situation in which sampling is the only practical choice. As an example, to determine how great an impact a type of automobile bumper can sustain, the manufacturer will hit the bumpers hard enough to destroy them. The population of interest is all bumpers of that type and a census would entail destroying all of them. Clearly sampling is the only sensible alternative.

A second case in which sampling is favored is that in which there is a large, widespread population. Collecting data on all automobiles in the United States by examining each automobile would be prohibitively expensive and time consuming.

A third reason sampling is preferred is that, if the sample is selected by random sampling, the expected sampling error can be reduced to any desired amount by increasing the number of observations in the sample. Seldom must sampling error be reduced to zero, however. For instance, if the refrigerator manufacturer finds that design B is preferred by 14% more people in the sample than the design of second choice and that the sampling error is practically certain to be no more than 4%, then sufficient knowledge has been gained to select the most preferred design. Such results are possible because statisticians have found the relationships between sample size and the amount of sampling error to be expected. Hence we can decide how close our estimate of the value of a variable must be and then determine in advance how large a sample to take to limit our expected maximum error to that amount.

Random Sampling

In order for us to predict the maximum expected sampling error in a given instance, two requirements must be met. First, nonsampling errors must be avoided. Such

Table 1-1 Content of the 1980 and 1970 Censuses

Questions in 1980	*Source in 1970*		
	100 percent item	*15 percent sample item*	*5 percent sample item*
100 PERCENT ITEMS			
Household relationship	x		
Sex	x		
Race[1]	x		
Age	x		
Marital status	x		
Spanish/Hispanic origin or descent			x
Number of units at address	x		
Complete plumbing facilities	x		
Number of rooms	x		
Tenure (whether the unit is owned or rented)	x		
Condominium identification[2]	x		
Value of home (for owner-occupied units and condominiums)[3]	x		
Rent (for renter-occupied units)	x		
Vacant for rent, for sale; and period of vacancy	x		
SAMPLE ITEMS			
School enrollment		x	
Educational attainment		x	x
State or foreign country of birth		x	x
Citizenship and year of immigration			x
Current language and English proficiency[4]		x	
Ancestry[5]			
Place of residence 5 years ago		x	
Activity 5 years ago		x	x
Veteran status and period of service		x	
Presence of disability or handicap[6]			x
Children ever born		x	x
Marital history			x
Employment status last week		x	x
Hours worked last week[7]		x	x

[1] In 1980, race categories were expanded to separately identify Vietnamese, Asian, Indian, Guamanian, Samoan, Eskimo, and Aleuts.

[2] New question to get separate identification of housing units which are condominiums. In 1970, these units were only identified if they were owner-occupied.

[3] Revised to include condominiums as well as one-family houses.

[4] In 1970, this question referred to one's native tongue.

[5] New question, replacing 1970 question on parents' country of birth.

[6] Expanded to inquire about transportation disability as well as work-related disability.

[7] Revised in 1980 to get actual hours rather than hour intervals.

Table 1-1 Content of the 1980 and 1970 Censuses (*continued*)

Questions in 1980	*Source in 1970*		
	100 percent item	*15 percent sample item*	*5 percent sample item*
Place of work		x	
Travel time to work			
Means of transportation to work		x	
Persons in car pool			
Year last worked		x	x
Industry, occupation, and class of worker		x	x
Weeks worked last year[8]		x	x
Usual hours worked per week last year			
Weeks looking for work in 1979			
Amount of income by source and total income in 1979[9]		x	x
Type of unit[10]		x	x
Stories in building and presence of elevator			x
Year built		x	x
Year moved into this house		x	
Acreage and crop sales		x	x
Source of water[11]		x	
Sewage disposal		x	
Heating equipment[12]		x	x
Fuels used for house heating, water heating, and cooking			x
Costs of utilities and fuels[13]		x	x
Complete kitchen facilities	x		
Number of bedrooms			x
Number of bathrooms		x	
Telephone	x		
Air conditioning		x	
Number of automobiles		x	
Number of light trucks and vans			
Homeowner shelter costs for mortgage, real estate taxes, and hazard insurance			

[8] Revised in 1980 to get actual weeks worked rather than week intervals.

[9] Revised in 1980 to ask for property income separately and get total money income.

[10] Expanded classification in 1980 to identify housing units such as boats, vans, or tents.

[11] Revised in 1980 to distinguish between drilled versus dug wells.

[12] Expanded classification in 1980 to separately identify heat pumps.

[13] Modified in 1980 to collect utility cost data from homeowners as well as from renters.

Note: No entry under "Source in 1970" indicates that question was not included in the 1970 census. Items included in the 1970 census but not included in 1980 are not shown.

Source: *Monthly Labor Review*, September, 1979.

errors include sampling the wrong population, making observations in a biased manner, or any other occurrence that would produce the wrong result even if the entire population were enumerated. Second, one of several forms of random sampling must be used. Random sampling is not haphazard and has an exact meaning in the field of statistics.

> A *random sample* results if every combination of n elements in the population has the same chance of being selected, where n is the sample size.

Suppose we want to select a random sample of 4 students from a class of 35. We can write the names of all 35 on separate slips of paper, mix them well, and draw 4. If the mixing is very thorough, this technique works fairly well. In Chapter 7 we shall learn how to use a table of random numbers to do an even better job.

While statisticians have found the relation between sample size and the amount of sampling error to be expected, this relationship holds only if the sample is a random sample. A random sample allows us to study relationships between sample results and the populations from which they come by means of the laws of probability. No such mechanism exists to connect sample results and their parent populations if any other type of sampling is done. It is often erroneously assumed that a large sample will be representative and useful for estimating population characteristics, even though care was not exercised to make sure it is a random sample.

To compare what can happen with random and nonrandom sampling, consider a case in which a city manager has been asked to sample the opinions of residential property owners on a proposal to increase bus service. The increase will entail a tax subsidy by property owners.

A large nonrandom sample can be gathered relatively cheaply by putting interviewers in the town's largest shopping center. Passersby could then be asked whether or not they own property in the city, and, if so, whether they feel bus service should be increased.

Alternatively, a current list of all property owners can be prepared and a random sample can be selected from the list. Interviewers can be sent to the addresses selected, where they can ask for the property owner by name and interview those available. Follow-up procedures can be used for those not at home. Telephone interviews can be made with absentee property owners.

It is clear that the random sampling procedure will cost more per interview than the nonrandom method. On the other hand, the random sampling procedure does not rely on persons to correctly report whether they own property. It is also clear that the first method of sampling gives absentee and shut-in property owners virtually no chance of being included in the sample; this is only one obvious way in which such an approach poorly represents the population.

With the random sampling approach, a given property owner has the same chance as any other to be included in the sample. For reasonably large random samples, various classes of property owners can be expected to contribute nearly the same proportion of elements to the sample as they represent in the population.

Hence opinions in the sample can be expected to fairly represent those of the population.

Once relevant data have been collected, they must be organized and summarized in preparation for analysis. In the next two chapters, we shall consider ways to organize and summarize data.

Exercises

1-5 *Censuses and samples.* Classify the following as either a census or a sample and explain your answer.

a. The annual physical count of stock in inventory for a retail hardware store.

b. Recording on the first of every month the price of a 5-lb chuck roast at every meat market in town.

c. The body temperatures of all the children in a family.

1-6 *Medical association survey.* A computer tape contains the current file of all members of a leading national medical association. Several characteristics, such as age, address, medical specialty, and sex, are recorded for each member. A random selection process has been programmed into the computer and used to select a sample of 5000 members from the tape. Describe how the sample might be compared with the population as a comprehensive check on the selection process.

1-7 *University faculty-staff survey.* University authorities want a survey to determine faculty and staff willingness to support a recreational sports complex on the campus. There are 9000 people in the population. A 1-page questionnaire is the survey instrument to be used. One proposal is to include the questionnaire in everyone's pay envelope with instructions to turn in the completed form to supervisors. This was turned down in favor of a separate mailing of the questionnaire and a return envelope to a random sample of 900 faculty and staff members. The mailing was followed up with a reminder phone call and then an interview phone call if necessary. What are some reasons for preferring the sampling approach?

1-8 *Sampling and nonsampling errors.* Classify each of the following as a sampling or nonsampling error and explain your choice.

a. Mistakes in keypunching questionnaire responses into cards to be used as computer input.

b. Inability to interview a family selected as an element in a random sample.

c. The proportion of freshmen in a random sample of all students selected from a current file on a computer is 0.32; the proportion of freshmen in the student body is 0.35.

d. The average age of students in an undergraduate statistics class is 1.2 years less than the average age of all undergraduate students at the university.

Summary

As a professional field, *statistics* is concerned with ways to collect, organize, and analyze data and with ways to present the resulting conclusions. *Descriptive statis-*

tics is the branch of the field that covers organizing and presenting both data and conclusions. *Inferential statistics* is the branch that covers data analysis.

Useful data result from observing applicable variables for entities selected as being relevant to the objectives of a study. A *variable* is the combination of a characteristic of an entity and a suitable scale to measure that characteristic.

Primarily for reasons of economy, data are often collected only for a sample of the population being studied. *Random* (probability) *sampling* provides a way to control sampling error by determining sample size before sampling is carried out.

Supplementary Exercises

1-9 *Establishing sales territories.* An insurance company is setting up life insurance sales operations in your state. Describe some characteristics of the population that might be available from the census and other sources that would help determine how many agents should be assigned and where they should be located.

1-10 *Price change effects.* When students register at the beginning of an academic term, the university collects personal information such as local address, home address, age, sex, marital status, academic schedule, work schedule, and financial sources. The university authorities are considering a tuition increase for out-of-state students. What are some uses that might be made of the personal information described in assessing the impact of such an increase?

1-11 *State quality inspection.* Samples of a new product sent by mail or given out in a supermarket usually consist of 1 unit per consumer. On the other hand, a sample for purposes of statistical inference almost always contains more than 1 unit. Why would a state inspector want several units of a new product in the sample that will be used to check the advertised net weight of the product in the packages?

1-12 *Sports management.* What data on players might help the manager of a baseball team decide on the batting order for the team?

1-13 *Automobile accident study.* Classify each of the following as an element or an observation and explain your answer.

a. An automobile accident.
b. The number of accidents at a given intersection during last January.
c. The state of the weather at the time of an accident classified as one of the following: clear, cloudy, rain, snow.

1-14 *Manufacturing quality control.* Once each hour a random sample of 12 light bulbs is selected from the assembly line delivering this type of bulb. The number of bulbs in each sample that will not light is divided by 12.

a. What elements are being observed?
b. What characteristic is being observed?
c. What type of measuring scale is being used?
d. What is the variable?
e. Is the variable continuous or discrete?

1-15 *City unemployment survey.* In a large city, an unemployment survey is to be conducted by interviewing a random sample of heads of households. Two sample sizes are being considered: 500 households and 8000 households. Discuss the relative importance of sampling and nonsampling errors for these two sample sizes.

1-16 *Census for a historical association.* A nationally prominent historical association is considering a book containing a biography of every surviving member of the U.S. armed forces with service in Europe during World War I (1914 – 1918). Discuss how these people might be located and the biographical information obtained. Consider the balance of cost of getting biographies by various means against the quality of the biography and the percentage of all veterans reached. Also consider what secondary sources might be useful.

Computer Exercises

1 Source: Personnel data file

a. List the first fifty observations and identify the type of measurement scale for each of the following variables: ID, salary, management level, participation in the tax shelter program, department, years employed.

b. Which of the variables in part (a) produce qualitative data? Quantitative data?

2 Source: Utilities data file

a. Compute income per employee (dollars per person) for each firm.

b. Produce a list of firms ranked by income per employee. What are the five firms with the most income per employee? The five with the least income per employee?

c. Repeat part (b) using assets as the variable. Are the lists the same?

d. Sort the data by industry type. Within each type, sort the data by income per employee. Which industry type appears to have the most income per employee? The least?

2

ORGANIZING AND PRESENTING DATA

In Chapter 1 we discussed the nature of statistics, the nature of data, and collecting data. In this chapter and in Chapter 3, we will discuss ways to organize collections of data. This chapter considers ways to present collections of data so that the general pattern is easy to see. The next chapter describes how to find and use averages and other descriptive measures to summarize characteristics of data collections.

We pointed out in Chapter 1 that data and variables can be either qualitative or quantitative. Recall that qualitative variables have nominal scales. For instance, peoples' opinions could be recorded as Yes, No, or Other with assigned values of 0, 1, and 2, respectively. Quantitative variables, on the other hand, are either discrete, such as the number of children per family, or continuous, such as diameters of pipes, and are measured on ordinal, interval, or ratio scales. We will first discuss organizing and presenting collections of qualitative data and then turn to collections of quantitative data.

2.1 Qualitative Variables

Qualitative data result from measuring qualitative variables, which are variables with nominal classes. Such variables have two or more classes. For example, a variable representing whether or not each person in a sample is male or female has two classes. Qualitative variables of this type occur so often that they have been given a special name. They are called *binary variables*. This is the first type of qualitative variable we will discuss. Then we will cover qualitative variables with more than two classes.

Binary Variables

As we have said, a qualitative variable with exactly two classes is a *binary variable*. Light bulbs either good or defective or voters either favoring or not favoring one of the candidates in an election are examples of such variables.

Binary variables occur frequently in business and industry, and summarizing techniques for such variables are often required. A vertical bar chart is the device we shall use for binary variables.

Suppose an inspector in a plant manufacturing light bulbs has selected a random sample of twenty 100-watt bulbs from the production line. Each bulb is tested. Of the bulbs, 18, or 90%, are good, and 2, or 10%, are defective.

The vertical bar chart on the left in Figure 2-1 shows the *number* of observations in each of the two states. On the right, the *percentage* of observations in each

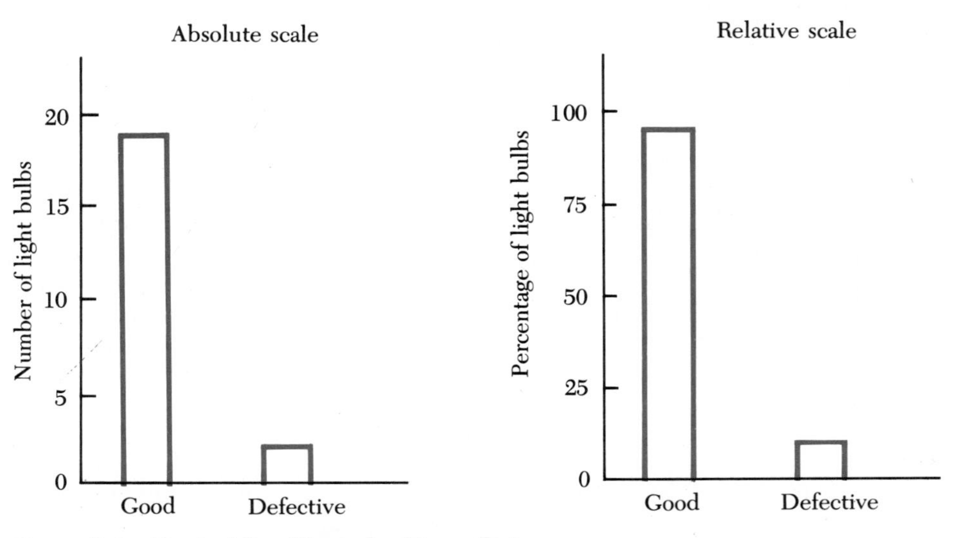

Figure 2-1 Vertical Bar Charts for Binary Data

state is shown. It would have been equally appropriate to use 0.90 and 0.10 rather than percentages in the latter case. A scale using the number of observations is an *absolute* scale, and one using percentages or proportions is a *relative* scale.

It helps to use relative scales to compare two or more collections of data containing different numbers of observations, since relative scales adjust for these differences. Consider the observations on the operating condition of the three samples of light bulbs in the top half of Table 2-1. The original observations have been summarized by using an absolute scale. But the different sample sizes make it difficult to compare the three samples. The bottom half of Table 2-1 shows the data from the top half written as decimal proportions of their respective totals. It is now clear that useful output is greatest for the sample of 100-watt bulbs, less for 150-watt bulbs, and least for 75-watt bulbs.

A quicker impression of the differences among the three light bulb samples is gained from a graphical display. Figure 2-2 uses a bar chart, with a relative scale on the vertical axis, to show the data from the bottom half of Table 2-1. It is immediately clear that output quality runs from best for the 100-watt sample to worst for the 75-watt sample.

Qualitative Variables with More than Two Classes

Several situations involving qualitative variables with more than two classes arise that require special forms of graphical treatment. One situation is the presentation of data for a single qualitative variable. Another is the comparison of two or more sets of data for such a variable. The last situation we shall discuss is the presentation of data when the variable represents geographical location.

A graphical device often used to present observations of a single qualitative variable with more than two classes is the *horizontal bar chart*.

Table 2-1 Operating Condition for Three Samples of Light Bulbs

	Number of Bulbs		
	Good	*Bad*	*Total*
75-watt bulbs	38	12	50
100-watt bulbs	18	2	20
150-watt bulbs	20	5	25
	Proportion of Bulbs		
	Good	*Bad*	*Total*
75-watt bulbs	0.76	0.24	1.00
100-watt bulbs	0.90	0.10	1.00
150-watt bulbs	0.80	0.20	1.00

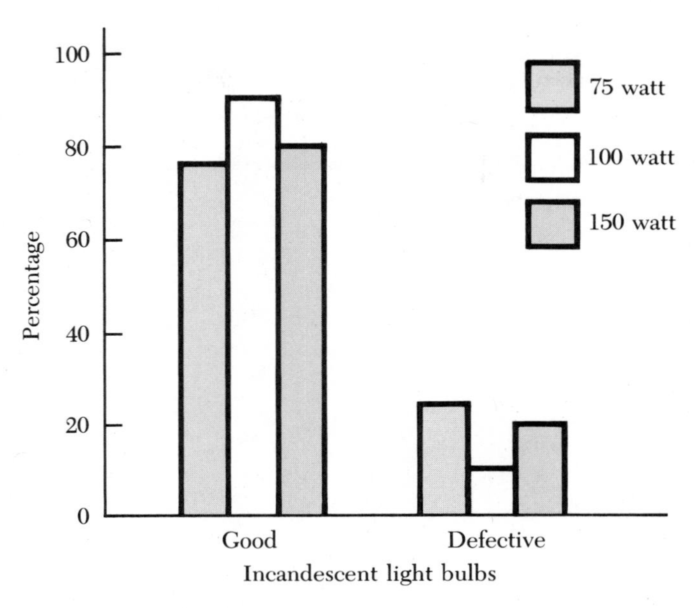

Figure 2-2 Comparison of Three Binary Variables on a Common Relative Scale

Example

The manager of a supermarket is reviewing the sales records of half-gallon cartons of milk. Sales by brand for the past month are as follows.

Brand	*Number of Cartons*
Health Farms	215
Pure	122
Golden Guernsey	64
Other (4 brands)	89

Figure 2-3 shows the horizontal bar chart for sales for each of the three major brands of milk and an "other" category for the combined sales of the minor brands. Health Farms milk is by far the largest seller, Pure milk sales are somewhat over half those of the leading brand, and Golden Guernsey milk sales are not quite a third. Together, sales of the minor brands are greater than those for Golden Guernsey but less than for Pure.

As is customary for horizontal bar charts, the classes are arranged in descending order of importance in Figure 2-3. Furthermore, space is left between adjacent bars. If there is a class combining several minor classes, it is also customary to place it last even though it may be somewhat larger than those immediately above it.

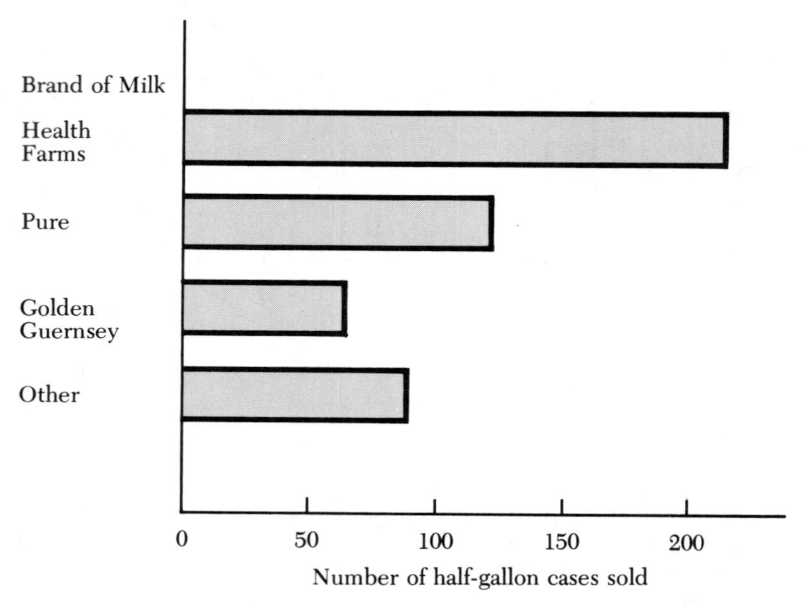

Figure 2-3 Horizontal Bar Chart

Often two or more sets of observations of a qualitative variable with several classes must be compared. *Pie charts* are effective for such a comparison. Suppose the supermarket manager just mentioned wants to compare the percentage of total milk sales for each brand last month with the percentages for the same brands during the same month in a nearby market. Figure 2-4 contains the pie charts from which the comparisons can be made. It is apparent that Health Farms and the minor brands have larger shares of sales at Market A (our manager), while Pure and Golden Guernsey have less.

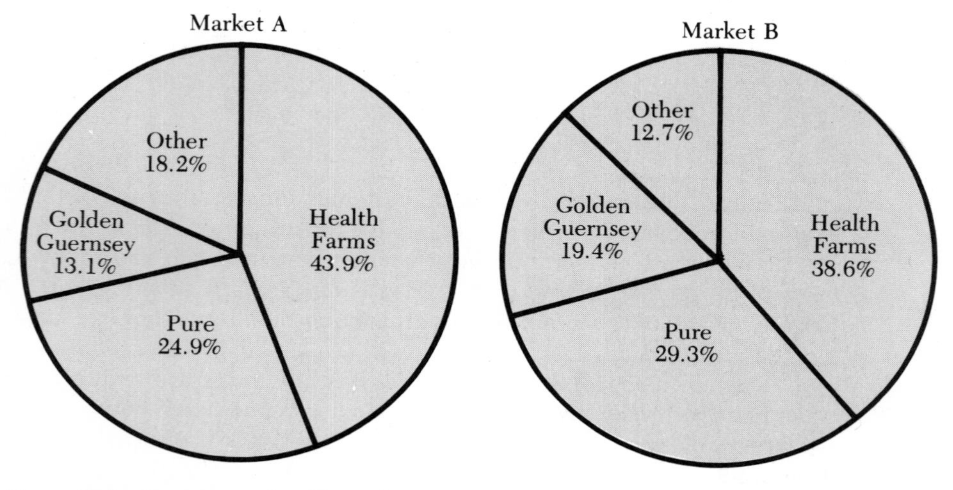

Figure 2-4 Percentage Pie Charts for Comparison

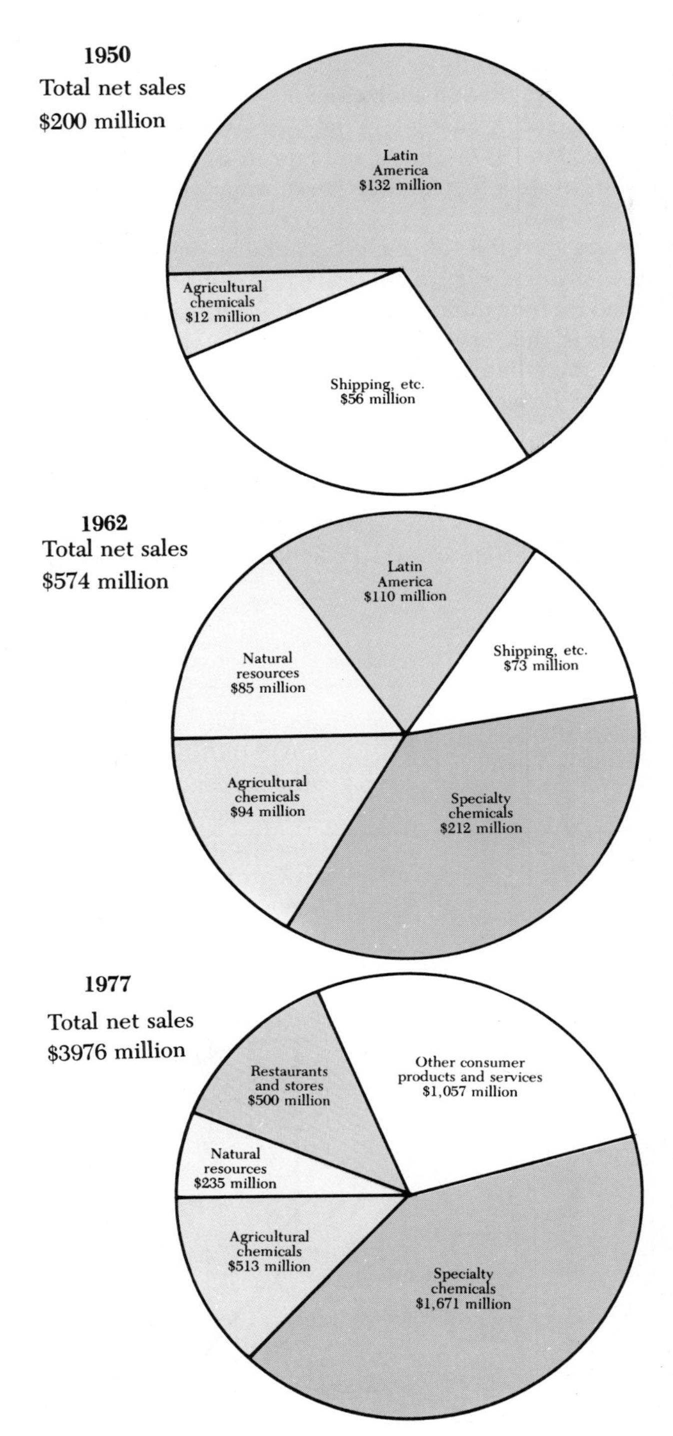

There's Always Something New at W. R. Grace

These charts show three stages in the company's transformation under the grandson of the founder. Serious change started in 1950, five years after Peter Grace was named president. Midway through his tenure in 1962, he had firmly established chemicals as the core of the company and made the first acquisitions of consumer businesses. By 1977, not a single piece of the old W. R. Grace was left.

Most of Grace's business in 1950 was in Latin America. Operations included sugar refining, paint, textiles, paper, wholesaling, and commodity trading. The "Shipping, etc." category takes in the Grace Line and some banking and insurance. The foray into chemicals began in 1950. Two early acquisitions that proved to be especially successful were Davison Chemical Corp., a producer of agricultural chemicals and catalytic agents used in oil refining, and Dewey & Almy Chemical Co., a manufacturer of specialty products, such as sealing compounds for containers. Since 1962, growth in chemicals has been largely internal.

The most dramatic shifts during the fifteen years from 1962 to 1977 were the withdrawal from Latin America and shipping and the push into consumer businesses and natural resources. The consumer thrust began with food companies in Europe, but by the late 1960's Grace was already selling some of them off and moving in new directions, primarily in the U.S. Most of the acquisitions in the last few years have been chains of restaurants or retail stores, which the company groups together as "specialty merchandising" ("Restaurants and stores" in the chart).

There is also a big, miscellaneous category of other consumer products and services, composed of some of the remnants of the company's fifteen-year buying spree, including two cocoa companies, automotive-parts manufacturers, and an Italian pasta maker. The natural-resources segment is made up of no fewer than ninety-three acquisitions, mostly in oil, coal, natural gas, and energy services.

Peter Grace says that the pattern for future growth is now set, and that the company will not be adding any new slices to the pie. But he does expect some change in the shares that the slices contribute to total company profits: chemicals will provide relatively less in years ahead and consumer businesses and natural resources will provide relatively more.

Figure 2-5 Application of Pie Charts at W.R. Grace
Source: Fortune, *May 8, 1978, p. 119, with permission.*

For effective comparison, the circles should be of equal size and the same class labels and order — if possible — should be used in each chart.

An application of pie charts illustrating changes in the business composition of W. R. Grace and Company from 1950 to 1977 is shown in Figure 2-5. The pie charts and accompanying description constitute a highly effective summary of the article in which they appear.

Modern computers offer several possibilities for geographical representation when geographical location is the variable. Figure 2-6 is a two-dimensional outline map of the individual states shaded to show motor vehicle thefts. The computer not only draws the outline but also fills in the varying degrees of shading that indicate ranges on the relative scale. It is immediately obvious that Alaska, Hawaii, and portions of the Northeast and West have higher theft rates than the rest of the country.

An even more spectacular way to illustrate geodata by means of computer graphics is the three-dimensional presentation in Figure 2-7. The 1975 population density for each political subdivision in Rhode Island is shown as a height above the geographical base. The heavy concentration around Providence is dramatically evident.

Exercises

2-1 *Opinion poll.* Two weeks before an election, an opinion polling firm surveyed registered voters to determine how many favor Walters, the candidate who hired the firm.

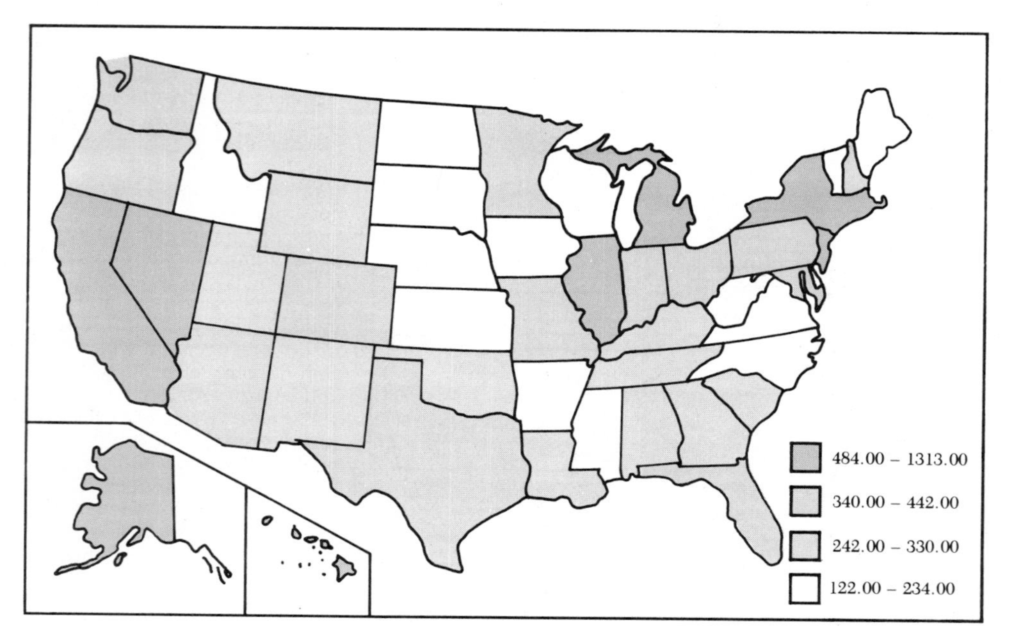

Figure 2-6 Motor Vehicle Thefts per 100,000 Population, 1976

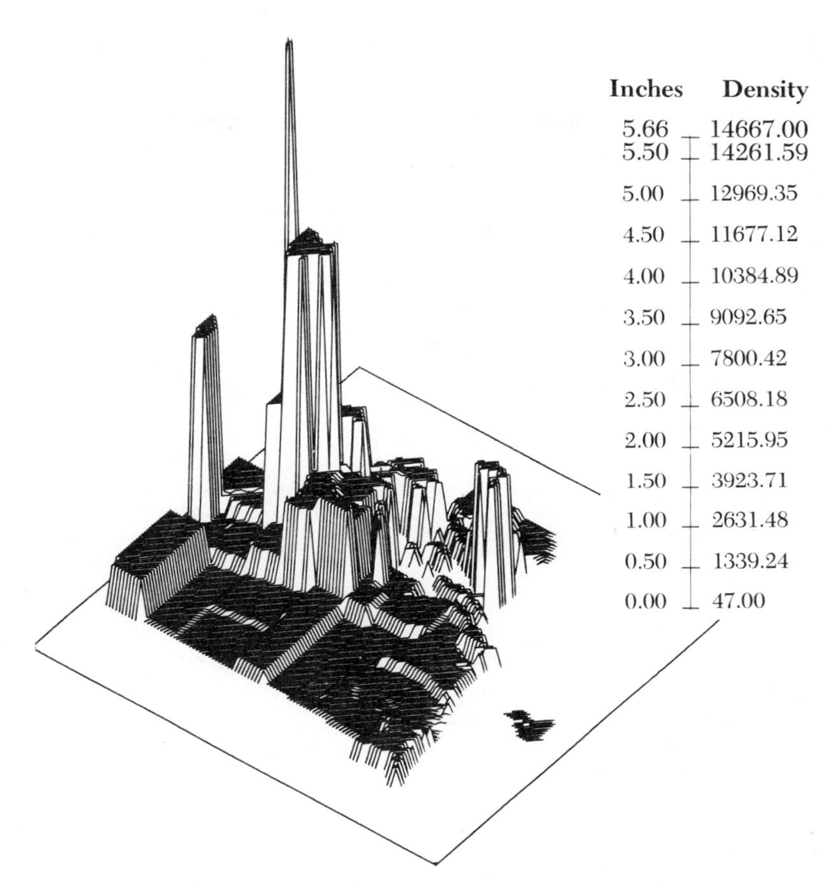

Figure 2-7 Population Density in Rhode Island, 1975

The survey has shown that 758 of the 1723 voters interviewed favor Walters, 620 do not favor him, and the remainder have not expressed an opinion. Prepare a bar chart showing the number of voters favoring Walters and the number of voters not favoring him. Prepare a second bar chart showing the proportion of voters with an opinion who favor Walters and the proportion who do not. Is Walters favored by a majority of interviewed voters who have an opinion?

2-2 *Personnel training.* A large firm has a training program on company procedures that all new employees must attend. New employees are assigned to one of three instructors. A common examination is given to every class and is graded as either pass or fail. On the last examination, 16 students taught by Anthony passed and 4 failed. Seventeen taught by Michaels passed and 6 failed. Twenty-seven taught by Roberts passed and 10 failed. Using the same relative scale, draw a bar chart to compare results for the different instructors.

2-3 *Composition of retail sales.* A retail ready-to-wear store reported the following sales last year: notions, $50,700; men's clothing, $125,400; women's clothing, $93,200; other, $34,600.

a. Construct a horizontal bar chart to illustrate these data.
b. Interpret the chart you prepared in (a).

2-4 *Composition of inventory.* An automobile tire distributor has counted the tires in stock that will fit 14-inch wheels. The result is: polyester, 56; glass-belted bias, 43; glass-belted radial, 39; steel-belted radial, 17.

a. Why is a horizontal bar chart appropriate for illustrating these data?
b. Construct a horizontal bar chart for this inventory and interpret the chart.

2-5 *Consumption patterns.** The following data show average household expenditures for moderate income urban families in the United States by expenditure categories.

	1960–61	*1972–73*
Food	$1234	$1664
Housing	1433	2604
Clothing	553	647
Medical Care	340	528
Transportation	770	1768
Other	723	1071
Total	$5054	$8282

a. Construct pie charts for the two periods showing the percentage of total expenditures in each category.
b. From the pie charts, determine the major changes in consumption patterns and describe them.

2-6 *Scheduled airline service.* Recent passengers on a leading commercial airline, picked at random, were asked to state the most satisfactory and least satisfactory single feature of their flight. These comments were classified as follows.

Category	*Satisfaction*	
	Most	*Least*
Check-in service	283	92
Flight schedules	296	120
Meal quality	116	350
In-flight service	412	111
Baggage handling	77	212
None	103	37
Total	1287	922

**Monthly Labor Review*, September, 1977.

a. Make a table that shows the proportion of the commendations and of the complaints in each service category.
b. Construct two pie charts to illustrate the relative frequencies of complaints and commendations in each service category.
c. Which types of service do you think may deserve some remedial attention and why?

2.2 Frequency Distributions

When the data collection is fairly large, an important technique for showing the pattern in a set of observations of a quantitative variable is the *frequency distribution*, which can be either a table or a graph. We first discuss frequency distributions for discrete variables in which each value of the variable is treated as a class. Then we shall talk about both discrete and continuous variables with a range of values used for each class.

Distributions with Single-Value Classes

When data come from a discrete variable such as the number of children per family, each successive value of the variable can be used as a class for a frequency distribution provided there are enough observations to produce a smooth pattern.

Example

A marketing research firm has sent interviewers to a random sample of 50 households. One of the observations interviewers are to make is the number of portable television sets in each household. The collection of 50 observations is listed below. Even for this modest number of observations it is difficult to get any idea of pattern from the list itself.

1	2	1	0	3	0	2	4	2	1
3	4	2	0	2	1	3	1	3	2
1	0	3	2	2	0	2	0	0	1
5	1	2	1	3	1	0	5	3	1
0	1	2	4	2	1	4	1	0	3

We begin organizing the data by noting that we are working with a discrete variable for which only small whole number values are possible: The variable ranges from 0 through 5 with several instances of each value. We can list these six values in a column and tally the number of times each value appears, as shown in Table 2-2. The tally marks for each value have been counted and recorded in the right column. Statisticians call these counts *frequencies*.

In Table 2-3, the values of the variable are given in the left column. The frequencies and relative frequencies are given to the right. The first two columns are

Table 2-2 Tally and Count of Number of Portable Television Sets per Household

Number of Sets	*Tally*	*Count*
0	~~////~~ ~~////~~	10
1	~~////~~ ~~////~~ ////	14
2	~~////~~ ~~////~~ //	12
3	~~////~~ ///	8
4	////	4
5	//	2

an example of a frequency distribution. The first and third columns constitute a *relative* frequency distribution.

> When a collection of data is summarized by showing the frequencies over the range of values of the variable, the result is called a *frequency distribution*.

Note that many households in Table 2-3 have no portable television set and that most have two or fewer.

Table 2-3 Frequency Distribution and Relative Frequency Distribution for Portable Television Sets

Receivers per Household	*Number of Households*	*Proportion of Households*
0	10	.20
1	14	.28
2	12	.24
3	8	.16
4	4	.08
5	2	.04
	50	1.00

A frequency distribution with single-value classes can be displayed in tabular form, as we have just shown. Such a distribution can also be displayed graphically, using a *frequency diagram*. The frequency diagram for our present example is shown in Figure 2-8. In a frequency diagram, the frequencies are plotted on the

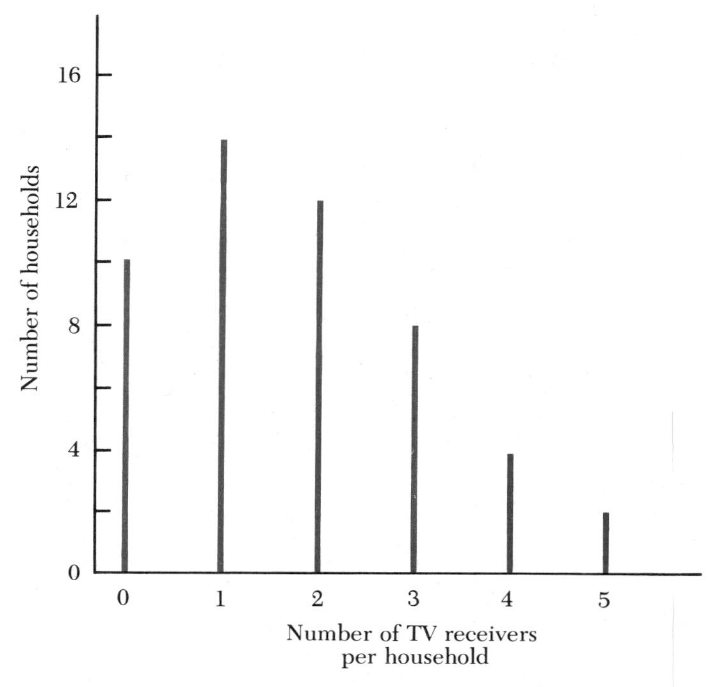

Figure 2-8 Frequency Diagram

vertical axis and the values of the variable are plotted on the horizontal axis. To display the relative frequency distribution, we would change the vertical scale to proportions.

Notice how the the frequency diagram illustrates the data. There appears to be about two receivers per household, on the average. The diagram shows a relatively high frequency for zero sets, a rapid rise to a maximum at one set per household, and then a drop in the number of households as the number of sets increases. This type of description is one important benefit from making a frequency distribution.

Distributions with Class Intervals

Sometimes there are not enough observations of a discrete variable to produce a reasonably smooth pattern when each value is treated as a separate class. When this is the case, several adjacent values can be combined to make up an interval for each class. This practice will put enough observations into all or most of the classes to smooth the pattern. Similarly, when the variable is continuous, class intervals are required. To illustrate techniques involving class intervals, we will use an example based on a continuous variable.

Example

An inspector for a flour milling company has weighed the contents of 80 bags supposedly containing 100 pounds intended for commercial customers. The inspector recorded these observations to the nearest hundredth of a pound. The list of these results is given in Table 2-4.

Table 2-4 Content Weights of 80 Bags of Flour (Pounds)

100.02	99.85	99.85	100.58
99.92	100.03	99.63	100.51
100.10	100.04	99.83	100.18
99.77	99.81	100.11	100.13
100.50	99.68	99.73	100.04
100.58	99.93	99.52	100.22
99.84	100.00	100.03	99.91
100.33	100.11	99.71	99.70
100.15	100.19	100.24	100.23
100.03	99.11	99.17	99.41
100.17	99.69	99.76	99.49
100.32	99.65	99.99	100.40
100.06	100.33	99.96	100.26
100.21	100.39	100.34	100.47
99.97	100.15	99.98	100.15
100.35	99.70	100.08	99.49
100.17	99.50	100.41	100.61
100.04	99.98	99.55	100.77
99.76	100.54	100.10	99.93
99.93	100.11	99.65	100.07

Since we can again make little sense of the raw data, we begin constructing a frequency distribution table for the 80 observations by grouping the data into weight classes. The general practice is to use from 4 to 20 classes, and, if possible, to make the class intervals equal in size. There are no hard and fast rules for determining the number of classes in advance, but if there is an average of at least 10 observations per class, a smooth pattern usually results. The following general rule holds.

> For a frequency distribution, the *estimated number of classes* is found by dividing the number of observations by 10. If the estimate exceeds 20, use just 20 classes. If the estimate is less than 4, use 4.

For our example, the estimate is 8 classes for the 80 observations.

Next, we need to determine the length of each class interval for the variable in question. Taken together, the class intervals must cover the range from the smallest to the largest observation in the set. Furthermore, we have already decided to make the intervals of equal length and we have already estimated the number of classes. Hence, the following guideline is reasonable.

For a frequency distribution, the *length of the class intervals* is found by dividing the range of the data by the estimated number of classes and adjusting to a convenient length.

In our example, the largest observation is 100.77 lb and the smallest is 99.11 lb. The range is the difference, 1.66 lb. The range divided by the estimated number of classes is $1.66 \div 8$, which gives 0.2075 lb for the length of each class interval. An awkward number such as this is not unusual for a first estimate. The remedy is to choose a convenient interval length close to the first estimate, such as 0.20 lb. We keep the same number of decimal places as the data have; thus each observation can be put in only one class.

Rather than begin our first class with the smallest observation, 99.11, we select the more convenient 99.10 lb. The first class begins at 99.10 lb, the second begins at 99.30 lb, and so on. The intervals end at 99.29 lb, 99.49 lb, and so forth. To finish, we tally, count, and record the frequencies in each class. Table 2-5 is the result. Observe that, because we shortened the class interval to 0.20 pounds, it took 9 classes rather than 8 to cover the entire range.

In frequency distributions, the stated ends of the class intervals are called *class limits*. In Table 2-5, the class limits are 99.10 to 99.29, 99.30 to 99.49, and so on. We must realize, however, that the actual measurements could have been rounded to the number of places indicated by the class limits. For instance, actual weights from

Table 2-5 Frequency Distribution of Weights for 80 Bags of Flour

Content Weight (Pounds)	*Number of Bags*
99.10– 99.29	2
99.30– 99.49	3
99.50– 99.69	8
99.70– 99.89	12
99.90–100.09	21
100.10–100.29	18
100.30–100.49	9
100.50–100.69	6
100.70–100.89	1
	80

99.095 lb up to 99.295 lb would be included in the class with limits of 99.10 to 99.29. The actual bounding values for a class are called the *class boundaries*.

Another characteristic of class intervals that is often useful is the class *midpoint*. As the name implies, the midpoint is the average of the class limits. It is also the average of the class boundaries.

The Histogram A graphic device for illustrating frequency distributions with many-value classes is the *histogram*. Figure 2-9 shows a histogram for the frequency distribution in Table 2-5. Class intervals of equal length are shown as adjacent bars of equal width. The bars are drawn vertically to a height equal to the class frequency. The variable always appears on the horizontal axis and the frequencies on the vertical axis. The break in the scale at the left of the horizontal axis indicates missing numbers and avoids useless empty space.

The Frequency Polygon An alternative to the histogram as a graphic device for illustrating frequency distributions with many-value classes is the *frequency polygon*. The frequency polygon for the weight distribution of Table 2-5 is shown in Figure 2-10. A point is plotted above the midpoint of each class at a height equal to the class frequency. A class midpoint can be found by averaging the upper and lower limits for the class in question. These points are connected with straight lines. At the ends of the distribution, straight lines are drawn to connect the midpoints of the last occupied classes with the midpoints of the adjacent empty classes to close the figure.

Both the histogram and the frequency polygon emphasize the peaking of weights in the immediate vicinity of 100 lb and the rapid tailing off as we move in

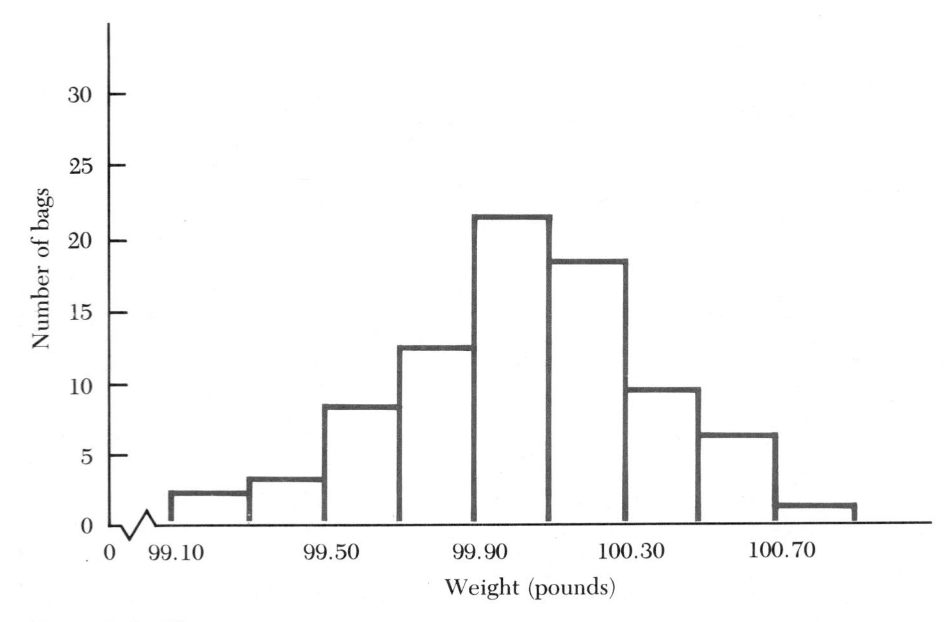

Figure 2-9 Histogram

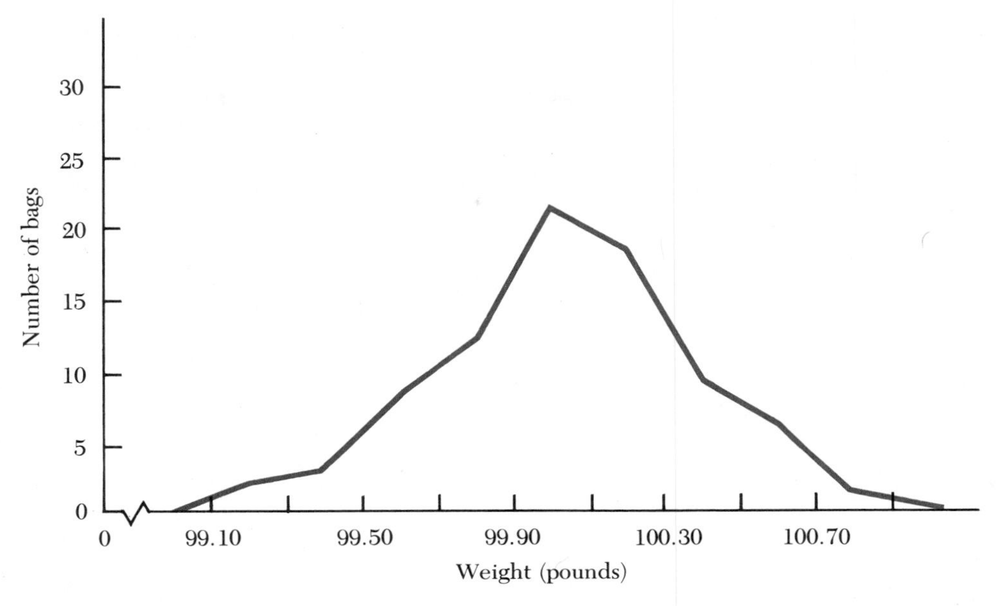

Figure 2-10 Frequency Polygon

either direction from 100 lb. The tails extend about equally far above and below 100 lb to give an impression of symmetry. All observations lie within 0.9 lb of the advertised 100-lb content weight.

Special Considerations In preparing frequency distributions, three situations arise often enough to merit comment. The first is concerned with the effect of choosing the wrong number of classes, the second with relative frequency distributions, and the third with sets of data that contain a few observations, called *outliers*, that are well separated from the main body.

Selecting the right number of classes for a frequency distribution produces a smooth pattern such as the one shown in the histogram and frequency polygon in Figures 2-9 and 2-10, respectively. Typically, the frequency count rises to a single peak and then descends. Both too few and too many classes disguise the underlying pattern. At one extreme, using a single class reduces any frequency distribution to a rectangle. At the other extreme, too many classes create a sawtooth effect, from which it is difficult to extract the desired pattern.

Often we wish to compare the patterns in two or more sets of data. Relative-frequency distributions for each set offer an effective way to do so. Any frequency distribution table is converted to a relative-frequency distribution table by changing the class frequencies to percentages or proportions of the total number of observations.

When outliers are present at one end of a data collection, they can sometimes be put in a single class of sufficient length to include them all. In other instances, however, the distance to the farthest observation is so great that the final class must be open-ended; for example, we can use "$100,000 and over" for an income distribution. Both approaches violate the principle of equal class intervals, but this is sometimes necessary. The frequency distribution tables for such cases present no special problems. Histograms and frequency polygons do require special treatment, however, to prevent a false visual impression. For histograms, the principle is to make the *areas*, rather than the heights, of all bars proportional to the frequencies they represent. The equal-interval classes are treated in the manner described earlier. If the final class interval is four times as long as the others, that class is given a height equal to only one-fourth of the frequencies included. For a frequency polygon, the terminal point in the final longer class is placed above the class midpoint at the same height as the final class for the histogram. Insofar as possible, the same principle is applied for an open-ended class. In graphical displays, the class is given very little height to convey the impression that only a few observations are included.

Exercises

2-7 *Auto parking control.* The following data represent the number of parking tickets given out on each of several weekdays in a downtown area.

42	47	46	35	43	39
38	40	50	37	43	37
47	47	44	49	41	34
38	41	36	42	38	38
58	34	32	42	49	52

a. Construct a frequency distribution table and frequency diagram with single-value classes.
b. Using a class interval of 4, group the data into classes and prepare a frequency distribution table and histogram.
c. Compare the results from (a) and (b). Which approach do you recommend and why?

2-8 *Travel advertising.* An advertising manager wants a frequency distribution for the number of travel advertisements that have appeared in the latest issues of 60 different magazines. The observations are arranged in numerical order for convenience.

14	26	31	35	39	43
15	26	32	35	39	43
18	26	33	35	39	43
19	27	33	35	39	43
19	28	33	36	39	44
21	28	33	36	40	44
22	28	34	37	41	44
23	29	34	38	42	46
25	29	34	38	42	49
26	30	35	38	42	52

a. From a visual inspection of the data, determine whether single-value or many-value classes would be better. What are the reasons for your choice?
b. Begin with 14 and use class intervals of 5 to construct a frequency distribution table and histogram.

c. Repeat (b) beginning with 13. Which of the two histograms do you prefer? Why?
d. Do the number of class intervals for (b) and (c) conform to the guidelines given for constructing frequency distributions in the text?

2-9 *Checking product length.* A machine is set to cut plastic stock 13 in. long so that 1-foot rulers can be made from the output. A sample of 120 pieces is chosen to check the setting. Each piece is measured to the nearest hundredth of an inch. The observations are as follows.

12.84	12.95	12.98	13.00	13.02	13.05
12.89	12.96	12.98	13.00	13.02	13.05
12.90	12.96	12.98	13.00	13.02	13.05
12.90	12.96	12.98	13.00	13.02	13.06
12.91	12.96	12.98	13.00	13.02	13.06
12.92	12.96	12.99	13.00	13.03	13.06
12.92	12.96	12.99	13.01	13.03	13.06
12.92	12.96	12.99	13.01	13.03	13.06
12.92	12.97	12.99	13.01	13.03	13.06
12.93	12.97	12.99	13.01	13.03	13.06
12.93	12.97	12.99	13.01	13.03	13.07
12.93	12.97	13.00	13.01	13.03	13.07
12.94	12.97	13.00	13.01	13.03	13.07
12.94	12.98	13.00	13.01	13.04	13.07
12.95	12.98	13.00	13.01	13.04	13.08
12.95	12.98	13.00	13.01	13.04	13.08
12.95	12.98	13.00	13.01	13.04	13.08
12.95	12.98	13.00	13.01	13.05	13.08
12.95	12.98	13.00	13.02	13.05	13.08
12.95	12.98	13.00	13.02	13.05	13.11

a. What do the guidelines in the text indicate for the number of class intervals?
b. According to the guidelines, why is 0.02 in. acceptable for the first estimate of the length of class intervals?
c. Begin with 12.84 and use class intervals of length 0.02 to construct a frequency distribution table and histogram.
d. Begin with 12.84 and use class intervals of length 0.03 to construct a frequency distribution table and histogram. Is this better than the result in (c)? Why?

2-10 *Automobile repair service time.* A customer relations specialist has collected the following data on the number of minutes (to the nearest tenth) that people have to wait for repairs they were told were minor and would not take long.

14.2	20.2	47.8
14.4	20.6	55.9
15.1	21.8	56.4
16.4	25.3	57.0
16.9	26.8	61.4
17.1	29.0	62.7
17.5	31.0	63.6
17.9	37.0	68.3
18.3	37.7	82.5
18.7	46.0	103.2

a. Are outliers present in this collection?

b. Use 14.0 as the lower limit of the first class and an interval of 10 to prepare a frequency distribution table and polygon for these data.
c. Can you think of anything in this type of business that may explain the pattern in (b)?

2-11 *Ages of entrepreneurs.** A recent study of business entrepreneurs produced the following frequency distributions.

Age	*Minority Entrepreneurs*	*Nonminority Entrepreneurs*
Under 30	11	—
30–34	19	1
35–39	33	6
40–44	50	10
45–49	39	18
50–54	37	24
55–59	11	12
60–64	10	4
65–69	7	5
70 & over	3	2
	220	82

Both of these distributions are to be plotted on the same set of axes to facilitate comparison.
a. What graphical device is the best choice? Why?
b. Use the device selected in (a) to illustrate both distributions. Describe the differences in patterns for the two collections of ages.

2.3 Frequency Curves

Throughout the previous section, we pointed out that one reason for constructing a frequency distribution is to approximate the underlying pattern in a collection of data. This section first describes how a sequence of better approximations can be obtained and proceeds to the ultimate realization of the pattern, the *frequency curve.* Then we shall present some of the more important frequency curves encountered in business and industry.

For the flour-bag example illustrated in Figure 2-9, suppose we take a random sample of 320 observations, four times as large as the original sample. The larger sample should still have practically all of its observations in the range 99.10 to 100.89 lb because this is a basic characteristic of the filling process.

*E. Gomulka, *American Journal of Small Business,* July 1977.

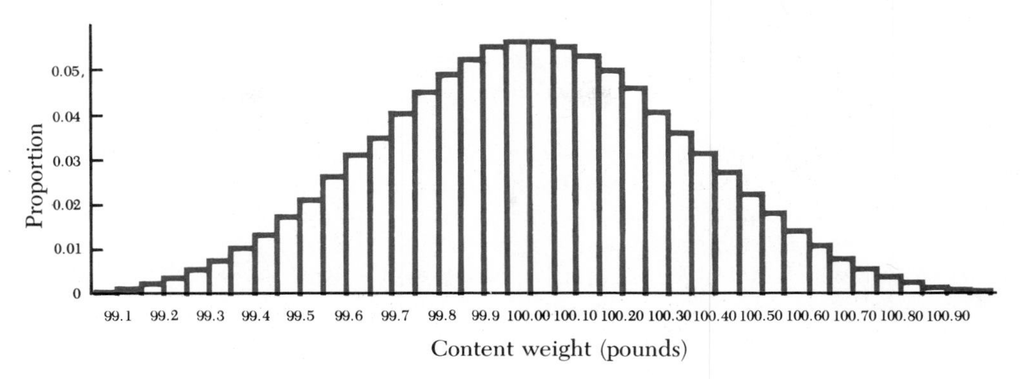

Figure 2-11 Expected Relative Frequency Histogram for 320 Bags of Flour Advertised as 100-Pound Bags

In preparing to construct a relative frequency histogram for the larger sample, we recall the guideline that said 20 classes are the most we need. Nevertheless, 320 observations are enough to support about 32 classes, so we shall use approximately this many classes for our current purpose. Since the range is essentially the same as for the smaller sample, four times as many classes will produce much shorter class intervals than we had before. Figure 2-11 shows the expected relative frequency histogram for a sample of 320 observations. Note that the range is only slightly greater than for the sample of 80. Also note that the class interval is only one-fourth as long.

Suppose we were to gather larger and larger samples from the population of fill weights and construct a relative frequency histogram for each. As the samples grow larger, the class intervals can be made shorter. As this process continues, the histograms will approach the smooth curve shown in Figure 2-12. This is the *frequency curve* of the underlying statistical population from which our random samples came. This is an example of the pattern we seek when we construct frequency distributions, histograms, and polygons from samples.

In the present case, the population pattern is a symmetrical bell-shaped curve that statisticians call a *normal* curve. The term *normal* is simply the name given a

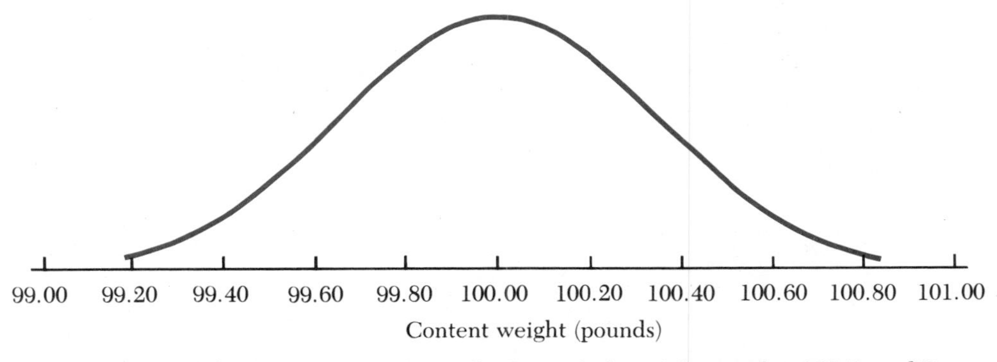

Figure 2-12 Relative Frequency Curve for Bags of Flour Advertised as 100-Pound Bags

curve of the shape illustrated and has no implication that this is the standard shape for curves. In fact, frequency curves for populations encountered in statistics assume several other shapes. Some of these appear in Figure 2-13. The curve at the top left sometimes occurs when waiting times—such as waiting times for customers to be served by bank tellers—are being studied. The curve at top right is typical of money distributions such as sales and incomes. Both of these curves have longer tails on the right, the positive direction on the axis, than on the left. They are said to have *positive skewness*. The curve at lower left could represent grades on an algebra test in which 100 is perfect and 70 is the lowest passing mark. This curve is *negatively skewed*. The curve at lower right is called a *uniform distribution*. The one shown here is for random numbers generated by a computer.

2.4 Cumulative Frequency Distributions

On the one hand, frequency distributions tell us how many observations have a single value or lie within a short interval. On the other hand, cumulative frequency distributions tell us how many observations lie below a certain value.

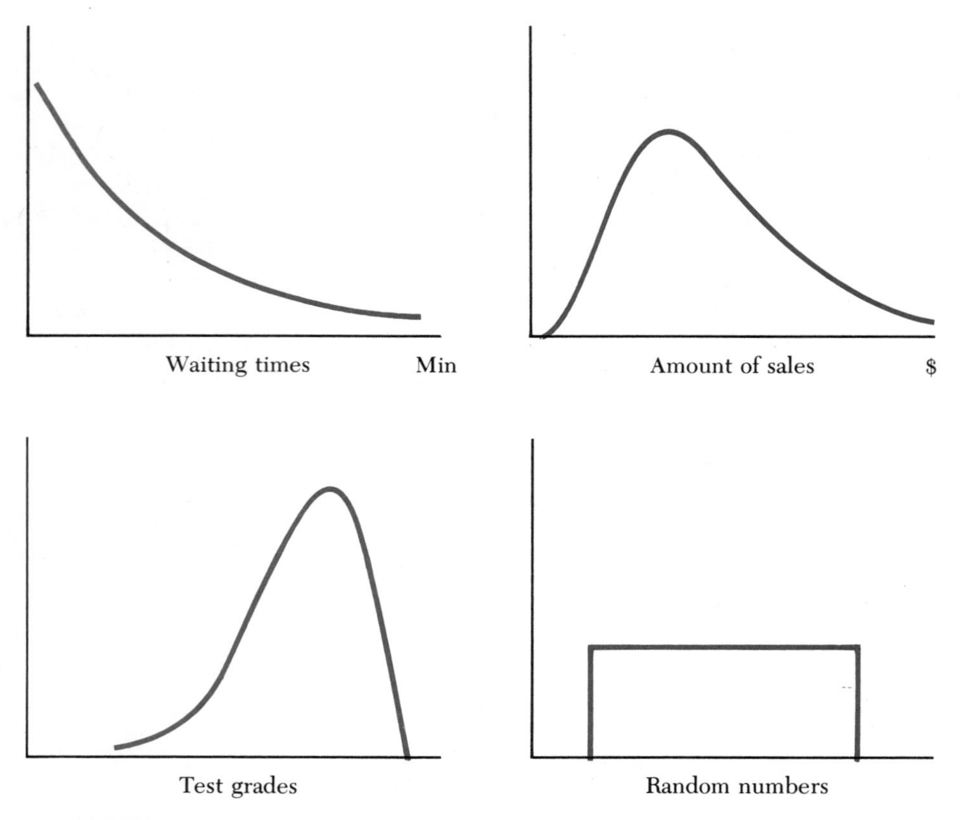

Figure 2-13 Some Common Frequency Curves

Single-Value Classes

Earlier we discussed a frequency distribution for the number of portable television sets per household. This frequency distribution is repeated in the left and center columns of Table 2-6. The right column shows the cumulative frequencies and how to calculate them.

The *cumulative frequency* is the number of observations less than or equal to a given value of the variable. For example, 44 households have 3 or fewer television sets and 36 have 2 sets or fewer.

Table 2-6 Finding Cumulative Frequencies for a Cumulative Frequency Distribution

Television Sets per Household	*Number of Households*	*Cumulative Frequency*
0	10	10
1	14	14 + 10 = 24
2	12	12 + 24 = 36
3	8	8 + 36 = 44
4	4	4 + 44 = 48
5	2	2 + 48 = 50

Graphical presentation of cumulative frequency distributions with single value classes is done with *step graphs*. Figure 2-14 clearly illustrates the origin of the name. This step graph shows the cumulative frequency distribution developed in Table 2-6. The changes in the heights of the individual steps are the frequencies in the classes, and the entire height from a value on the horizontal axis to a step is the cumulative frequency for that value of the variable.

Many-Value Classes

Constructing a cumulative frequency distribution table from a frequency distribution with class intervals rather than single-value classes presents no new problems, as the process is identical to that used with Table 2-6. The distribution of net weights of the 80 bags of flour we discussed earlier appears in the first two columns of Table 2-7. Column 2 gives the accumulated frequencies, beginning with the smallest value of the variable. This column tells us, for instance, that 25 bags weighed less than 99.90 lb. The cumulative relative frequency will be discussed shortly.

The graphical display for a cumulative frequency distribution with many-value classes is called an *ogive*. Figure 2-15 shows the ogive for the weight distribution of the flour bags. The cumulative frequency for each class (read on the vertical scale to the left) is plotted directly over the lower limit of the next class. Note that the 25 bags for the weight class 99.70 lb to 99.89 lb is plotted above 99.90 lb because not

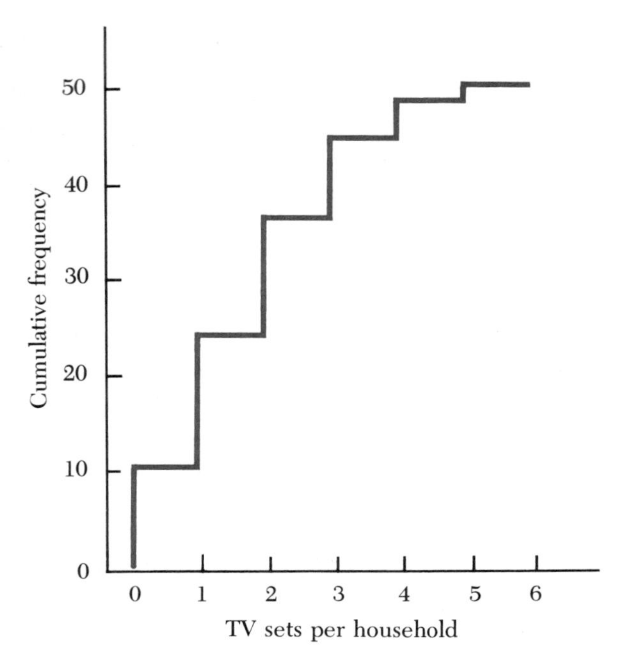

Figure 2-14 A Cumulative Frequency Step Graph

until we reach 99.90 lb are we sure that all 25 bags have been counted. For the same reason, 46 is plotted over 100.10 lb. The straight line connecting 25 and 46 implies that the weights in this class are distributed uniformly throughout the class, a reasonable assumption in view of our desire to show pattern rather than specific detail. Also note that the left end of the ogive begins just below the lower limit of the first class. This is the greatest weight that still is less than the weights of all the bags.

Variations

As was the case for frequency distributions, we may wish to compare cumulative distributions using relative frequencies. In a table, stating cumulative frequencies as percentages or proportions of the total number of observations in the complete data collection converts a cumulative frequency distribution to a cumulative *relative*-frequency distribution. Table 2-7 shows the cumulative relative frequency distribution for the flour data.

We defined the cumulative frequency as the number of observations less than the lower limit of the next class. This is the usual way of treating cumulative frequencies. On occasion, however, we may wish to reverse the direction in which frequencies are accumulated. We start with the observation of greatest, rather than least, value and accumulate from the top down to the bottom of the data collection. The tables and graphs for such cases show the cumulative frequency as the number of observations *greater than* a given value of the variable.

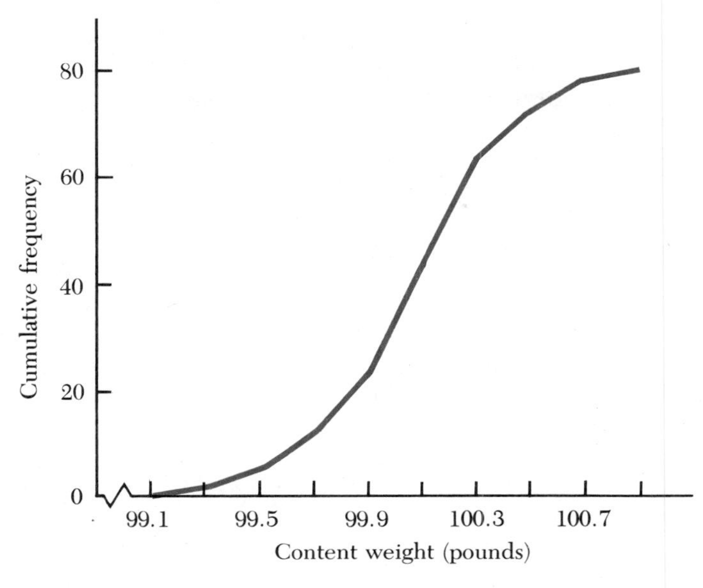

Figure 2-15 A Cumulative Frequency Distribution

Table 2-7 Cumulative Distributions for 80 Bags of Flour

Content Weight (Pounds)	*Cumulative Frequency*	*Cumulative Relative Frequency*
< 99.30	2	.0250
< 99.50	5	.0625
< 99.70	13	.1625
< 99.90	25	.3125
< 100.10	46	.5750
< 100.30	64	.8000
< 100.50	73	.9125
< 100.70	79	.9875
< 100.90	80	1.0000

Exercises

2-12 *Frequency curves.* Sketch an appropriate frequency curve for each of the following distributions.

a. Scores made by a nationwide random sample of 800 high school seniors on a well-known college aptitude test.

b. The grade-point averages of 500 college seniors (on the basis of 4.0 for a perfect grade-point average) who have applied for admission to one of the nation's top ten law schools.
c. Gasoline mileages reported by owners of a specified make and model of a 1978 compact automobile.

2-13 *Frequency curves.* For each of the following collections of data, sketch the frequency curve you would expect to apply.
a. Tourists entering a state are stopped for inspection. Each driver is asked to estimate the number of miles the automobile has traveled in the previous 24 hours.
b. The ages of men with bald heads.
c. The money a coin vending machine owner collects each week from a soft-drink machine that typically experiences heavy demand between weekly trips to refill the drinks and collect the money.

2-14 *Typographical errors.* A book publishing firm is studying the performance of a typesetting firm. A random sample of 100 pages has resulted in the following distribution of typographical errors.

Number of Errors per Page	*Number of Pages*	*Number of Errors per Page*	*Number of Pages*
0	42	5	2
1	33	6	1
2	12	7	0
3	6	8	1
4	3		

a. Prepare a cumulative frequency distribution table from this distribution.
b. Prepare a step graph from the table constructed in (a).
c. From the step graph and table, determine how many pages had no more than 2 errors each.

2-15 *Warranty tire replacement.* An automobile tire dealer must replace new tires of a certain brand if customers discover flaws within 90 days. The number of tires the dealer has had to replace over the past three years is shown in the following distribution.

Tires Replaced per Month	*Number of Months*	*Tires Replaced per Month*	*Number of Months*
0	11	4	4
1	8	5	2
2	6	6	1
3	4		

Construct a step graph for this distribution. During how many months did the dealer have to replace 2 tires or less?

2-16 *Employee weekly earnings.* During a recent week the employees of a local plant earned the amounts shown in the following distribution.

Weekly Earnings	*Number of Employees*
$150.00–159.99	2
160.00–169.99	4
170.00–179.99	10
180.00–189.99	13
190.00–199.99	25
200.00–209.99	22
210.00–219.99	19
220.00–229.99	12
230.00–239.99	8
240.00–249.99	5

a. Prepare a cumulative frequency distribution table for this distribution.
b. Construct a cumulative frequency ogive.
c. How many employees earned less than $200 during the week in question?

2-17 *Retail shop daily sales.* The record of daily sales of a small retail shop during a recent period is summarized in the following frequency distribution.

Daily Sales	*Number of Days*
$ 90.00– 99.99	18
100.00–109.99	25
110.00–119.99	12
120.00–129.99	9
130.00–139.99	6
140.00–149.99	3
150.00–199.99	2

a. Develop a cumulative frequency distribution table for this distribution.
b. Construct a cumulative frequency ogive.
c. For how many days during the period of observation were sales less than $110?

2.5 Stem and Leaf Displays (Optional)

An alternative to the frequency distribution for bringing out the pattern in an exploratory analysis of a data collection is the stem and leaf display. To illustrate this device, we will use the data in Table 2-8. These are observations of 1975 weekly grocery bills in 35 owner-occupied single family homes.

Table 2-8 Weekly Grocery Bills for Families in Owner-occupied Dwellings

$49.60	$43.20	$50.00	$66.00	$57.30
45.20	48.20	53.90	52.60	60.20
52.80	51.90	67.10	65.40	71.60
36.30	33.10	58.20	55.20	55.00
57.00	55.40	52.60	72.50	58.80
60.80	57.00	56.40	60.70	63.10
59.80	43.50	52.90	71.20	62.10

We begin by breaking each observation into a set of leading digits and a set of trailing digits. The break point for the leading digits is the same number of places from the decimal in every observation. The set of leading digits for each number is called the *stem*, and the leftmost of the trailing digits is called the *leaf*. The first observation in Table 2-8 is $49.60. For our first try, we will use 4 as the stem and 9 as the leaf. Similarly, we will break every observation between the first and second digit to the left of the decimal point. We will prepare and examine the first stem and leaf display and then decide whether we need to revise our selection of stems to prepare a more satisfactory result.

The second step is to list the stems in ascending order to the left of a vertical line and write down each leaf to the right of the line following its stem. The resulting initial stem and leaf display is in Table 2-9. Note that we do not round the data. Instead, we truncate it by just ignoring any digits following the leaf for a given observation. By doing so we preserve the ability to check our work easily.

Table 2-9 First Stem and Leaf Display for Data in Table 2-8

Unit: 1 dollar; 3 | 6 represents $36

Stem	Leaf
3	63
4	95383
5	27915703826225758
6	07650032
7	211

The first display makes it apparent that the distribution is unimodal and nearly symmetrical. The display is somewhat overly concentrated, however. Perhaps it could be improved if there were more stems.

We can make a display with twice as many stems by letting 3L be the stem for the lower half of the range from $30 to $40, 3U be the stem for the upper half, and by proceeding similarly for the remaining stems. These changes lead to the second display, shown in Table 2-10. Here the result is more satisfactory. Now we see there definitely is slight skewness toward the smaller values. The longer tail runs from the central mass down to $33.

Table 2-10 Revised Stem and Leaf Display for the Data in Table 2-8

Unit: 1 dollar; 3L | 3 represents $33
3U | 6 represents $36

Stem	Leaves
3L	3
3U	6
4L	33
4U	958
5L	2103222
5U	7957865758
6L	00032
6U	765
7L	211

Another strength of the stem and leaf display is illustrated by the data in Table 2-11. These are annual payments to 40 employees of a firm. A few employees in the sample worked only part of the year.

Table 2-11 Annual Payments to 40 Employees

$42,512	$43,677	$35,600	$48,433
26,000	15,678	70,244	22,400
38,400	43,600	18,897	27,450
37,600	14,200	89,586	46,775
42,800	43,500	19,378	24,500
17,600	39,500	34,200	23,500
43,400	39,000	23,500	16,600
18,500	4,384	21,900	39,500
17,500	41,000	36,743	21,500
22,456	23,567	36,400	42,450

We note that most of the data in Table 2-11 lie between $15,000 and $50,000. If we use 1U, 2L, and so on as stems, we will have 7 rows for the main part of the data with an average of about 5 observations per row. Usually, an average of from 5 to 10 observations per row makes a satisfactory display. The stem and leaf display with the stems as described is presented in Table 2-12. This display presents a new feature in the form of 3 *outliers*, two of which are much larger than the data in the central mass and one of which is much smaller. Rather than show many empty rows between the central mass and the outliers, we label outliers as HI or LO and list them, separated by commas. Another feature of the display for the salary data is the marked bimodality of the distribution. A reasonable speculation is that the distribution associated with the larger mode consists of persons in managerial or similar positions and the lower distribution is for other employees. The two high outliers may be two chief executives, and the low outlier may be someone who worked only part of the year. It is preferable to establish that such observations are, in fact, subject to different influences than the others before classifying them as outliers. Rather frequently, investigation establishes that the influence is a recording error.

Table 2-12 Stem and Leaf Display for Data in Table 2-11

Unit: $10 thousand; 1L	*4 represents $14,000*
1U	*7 represents $17,000*
LO	04
1L	4
1U	7875896
2L	23312431
2U	67
3L	4
3U	87995669
4L	22333312
4U	86
HI	70,89

Our final example of stem and leaf techniques illustrates the weights of the contents for 80 sacks of flour as listed in Table 2-4. The display in Table 2-13 uses a third scheme for defining stems. Here, 99L applies to leaves 0 and 1, 99T to two and three, 99F to four and five, 99S to six and seven, 99H to 8 and 9, respectively. This scheme provides five rows per set of digits in a stem. The L and H scheme provided two rows per set, and using the set itself provides one row. By moving the next digit from the leaves to the stems we can increase the number of rows by a factor of 10 as compared with the last scheme mentioned. As we saw for the first example in this section, sometimes we must try more than one scheme to find a satisfactory display.

Table 2-13 shows that the weight distribution is unimodal with possibly a slightly longer tail toward the smaller weights. There is also a possibility that the two smallest observations are outliers, although the slight gap between them and the rest of the data may be only a sampling fluctuation. The boxplot technique to be discussed in Chapter 3 will help us to decide which is the case.

Table 2-13 Stem and Leaf Display for the Data in Table 2-4

Unit: 0.1 lb; 99L | 1 represents 99.1 lb

Stem	Leaves
99L	11
99T	
99F	555444
99S	776667677767
99H	989988998899999
100L	011010100001111100111010
100T	33233323222
100F	55545544
100S	67

As compared with preparing frequency distributions, making stem and leaf displays has several advantages. In the first place, selection of class intervals is simpler. Recall the rather elaborate procedure we had to go through to set up classes for the flour weight frequency distribution, and note how much simpler it was to choose stems for Table 2-13. Second, should our initial effort prove unsatisfactory, changing class intervals is considerably more difficult for frequency distributions than it is for stem and leaf displays. Third, by keeping the spaces allotted to leaves equal, a stem and leaf display turned 90° counterclockwise becomes a histogram for the data collection. Finally, although the general pattern is disclosed by the stem and leaf display, the actual numerical detail is preserved. With this detail, unusual features can become apparent. For example, the presence of only multiples of 4 would result from the common practice of counting pulse beats for only 15 seconds, multiplying by 4, and recording the product as beats per minute.

Exercises

2-18 Prepare a stem and leaf display for the data in Exercise 2-7. Comment on the features of the distribution as made evident by your display.

2-19 Prepare a stem and leaf display for the data in Exercise 2-8 and discuss the characteristics of the pattern in this collection.

2-20 For the data in Exercise 2-9, what will be the effect of using stems of 12 and 13 inches in a stem and leaf display? Try alternative stems until you find a satisfactory display. Then describe the characteristics of the distribution.

2-21 For the data in Table 7-1, page 189, make side-by-side stem and leaf displays for the grocery bills from owner-occupied homes and from renter-occupied homes. Use the leftmost digits without modification as stems. Comment on any differences in the two patterns.

2-22 For the weight data in Table 2-4, make a stem and leaf display which uses 99.1, 99.2, . . . ,100.7 as stems. Compare the outcome with the display in Table 2-13.

Summary

Several devices can be used to present qualitative cross-section data. *Vertical* and *horizontal bar charts* are effective. *Pie charts* are especially useful for comparing percentages. When the qualitative variable is location, *background maps* are excellent.

The *frequency distribution* is the primary device for presenting quantitative cross-section data. Both discrete and continuous variables can be presented with this device, in which the pattern of frequencies is portrayed along the scale of the variable. Often the frequencies must be grouped into an appropriate number of class intervals to accomplish the objective. Either a *histogram* or a *frequency polygon* can be used to illustrate a frequency distribution. A *frequency curve* is the underlying pattern approximated by a frequency distribution.

In *cumulative frequency distributions* the number of observations no greater than the upper boundary of each class is tabulated. Such distributions can also be constructed to show the number of observations no less than the lower class boundary. *Step graphs* and *ogives* are graphical devices for presenting cumulative distributions.

Supplementary Exercises

2-23 *Background of executives.** The business backgrounds of two samples of executives of U.S. corporations is given below for two different years.

Background	*Executives Earning $100,000 or More per Year (Percent)*	
	1971	*1976*
Administration	23	22
Marketing	23	15
Finance	21	25
Production	6	11
Engineering	7	9
Legal	12	12
Other	8	6
	100	100

*W. G. Brown and K. K. Motamedi, *California Management Review*, Winter 1977.

Construct two pie charts for these data and use the charts to describe the changes from 1971 to 1976.

2-24 *Electric lamp shipments.* The following data show electric lamp industry shipments for 2 years.

Electric Lamp Type	*Shipments in Millions of Dollars* 1954	1972
Large incandescent	136.7	380.0
Electric discharge	76.7	303.8
Miniature incandescent	38.3	170.3
Photo incandescent	38.0	165.9
Christmas tree lamps	12.4	20.1
Cold cathode fluorescent	3.0	1.3
Total	305.1	1041.4

Construct two pie charts for these data and use them to describe the changes in shipment pattern.

2-25 *Retail shop sales comparisons.* A firm operates two retail ready-to-wear shops in a certain city. Records of daily sales are given below.

Daily Sales	*Number of Days* Shop 1	Shop 2
$ 90.00–109.99	12	18
110.00–129.99	29	37
130.00–149.99	20	55
150.00–169.99	14	26
170.00–189.99	8	24
190.00–209.99	4	15
210 & over	3	5
Total	90	180

a. On the same set of axes, construct two relative-frequency polygons for the two distributions.
b. Describe the difference in pattern for the two polygons in (a). Why were relative-frequency polygons rather than frequency polygons selected?
c. Why were relative-frequency polygons rather than relative-frequency histograms selected?

2-26 *Cumulative relative ogives.* For the data in the previous exercise, construct two cumulative, *relative*-frequency ogives on the same set of axes and use them to describe the differences in the two distributions.

Computer Exercises

These exercises are intended to illustrate the variety of frequency tables and graphical devices which can be used to describe data. You are encouraged to display other variables using various devices. Often several techniques must be tried to find the most suitable. If your program doesn't have the graphical displays required in the exercise, obtain frequency tables or rank-ordered listings to assist in creating the charts by hand:

1 Source: Personnel data file
   a. With management level as the variable, construct a pie chart, a frequency diagram, a relative frequency diagram, and a cumulative frequency step graph.
   b. Use the results in (a) to answer the following:
      How many people are at level 1?
      How many people are at level 2 or below?
      What proportion of people is at or above level 3?
      What proportion is at level 4?
   c. With salary as the variable, plot one histogram using frequencies and another using relative frequencies. Place midpoints at each 1000 dollars. Under what circumstances would each be preferred?

2 Source: Utilities and retailers data sets
   a. Find relative frequency distributions for 1982 earnings per share for utilities and retailers. Place midpoints at each percent. Write a brief description of these distributions.
   b. Repeat (a) using assets as the variable. What problem arises?
   c. (Optional) Draw a stem and leaf diagram for utility assets. How does this diagram help explain what happened in (b)?

3

SUMMARY DESCRIPTIVE MEASURES

We began the last chapter by considering the problem of describing a large collection of data; we saw that tables and graphs can be used as effective summary devices for such collections. Creating a visual impression of the pattern made by the observations using these devices constitutes one approach to the problem.

A second approach to the problem of describing a set of data is to define general properties of data collections and then develop descriptive measures for these properties. This approach is discussed in this chapter.

The first property to be described is illustrated in Figure 3-1. Relative frequency curves for the dollar values of life insurance policies taken out by men and women are shown on the same graph. Considered as a whole, the distribution of policy amounts for men lies to the right of that for women. Otherwise, the distributions are identical. Generally speaking, this indicates that men own larger policies than women.

The two policy-amount distributions differ with regard to a property called *central location*. In the figure, the central location for men is greater than that for women. A way to measure this property is to use averages. In the two distributions shown, the average for men is greater than that for women. We shall shortly describe ways to find averages.

A second general property of data collections is illustrated in Figure 3-2. These are relative frequency curves for the actual weights of a set of 5-pound bags of salt and of a set of 5-pound bags of potatoes. Each weight has been checked on a very precise balance beam scale. The property by which these two distributions differ is called *dispersion*. The weights of the salt bags are much closer to 5 lb than are the weights of potato bags. Very small amounts of salt can be added to or taken away from a bag to correct its weight, but an entire potato must be added to or taken away from a bag to come closer to 5 lb. Hence, the potato bag weights are more widely scattered than the weights of the salt bags. Later in this chapter we shall discuss two important measures of dispersion.

Another general property is *skewness*, which we discussed in connection with Figure 2-13. Recall that distributions with longer tails extending to the right (in the positive direction) have *positive skewness*, while distributions with longer tails extending to the left have *negative skewness*. On the other hand, Figure 3-2 shows examples of distributions with no skewness. Such distributions are said to be *symmetrical* with respect to their central locations. The left and right halves of symmetrical distributions are mirror images of each other.

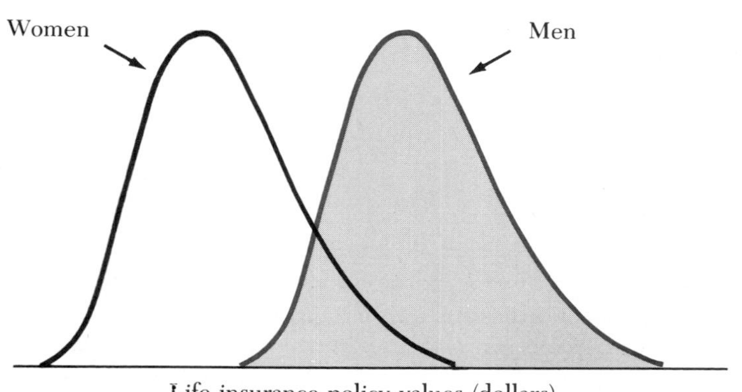

Figure 3-1 Frequency Curves Differing in Central Location

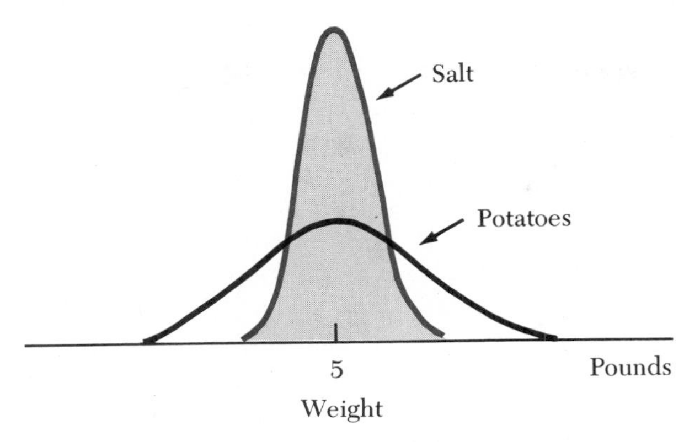

Figure 3-2 Frequency Curves Differing in Dispersion

3.1 Measures of Central Location

The three measures of central location we shall discuss are the *arithmetic mean*, the *median*, and the *mode*. The arithmetic mean is commonly known as the "average." The second measure, the median, is the middle observation in a data collection arranged by order of magnitude. For a collection with no two values equal, the number of observed values smaller than the median is equal to the number of observed values greater than the median. The third measure, the mode, is the value in the data collection that occurs most frequently. We are now ready to consider specific procedures for finding these measures.

3.2 The Arithmetic Mean

We have already said that the arithmetic mean is the familiar average for a data collection. More formally, we have the following.

> The *arithmetic mean* is the sum of the values in a collection of observations divided by the number of observations in the collection.

When there is no chance for confusion, the arithmetic mean is simply called *the mean*.

The Sample Mean

Suppose, by week, 7, 10, 4, 8, and 9 sedans were sold by a used car dealer in a random sample of 5 weeks. Then the mean is

$$\frac{7 + 10 + 4 + 8 + 9}{5} = \frac{38}{5}$$

or

7.6 sedans per week

Note that even though the observed values can be only whole numbers, the mean need not be.

Symbolically, one general equation for the sample mean is

$$\overline{X} = \frac{X_1 + X_2 + \cdots + X_n}{n}$$

For the above example, $X_1 = 7$, $X_2 = 10$, $X_3 = 4$, $X_4 = 8$, $X_5 = 9$, and $n = 5$, the number of observations in the sample. A somewhat more succinct representation is

$$\overline{X} = \frac{\sum_{i=1}^{n} X_i}{n}$$

where Σ (sigma) is the sign used to indicate summation and i is the *index* on X by which the values to be summed can be defined. In this book, summations will be understood to include every observation in the data collection being considered, so we can simplify the representation even more by deleting the indexing notation. Finally, then, we shall use the following equation for the *sample mean*:

$$\overline{X} = \frac{\Sigma X}{n} \tag{3-1}$$

As another example, suppose an operations specialist for a bank observed the time it took tellers to serve a random sample of 6 customers. The observations, in minutes, were 4.3, 2.6, 3.9, 1.7, 3.2, and 5.9. The mean of this sample is

$$\overline{X} = \frac{\Sigma X}{n} = \frac{4.3 + 2.6 + 3.9 + 1.7 + 3.2 + 5.9}{6} = \frac{21.6}{6} = 3.6$$

We can think of the mean as the constant value each observation must have to produce the same total as the actual observations. Six observations of 3.6 min will total 21.6 min, the same as the total of the actual observations.

Statistics and Parameters

The sample mean, $\overline{X}$, is the first example of a summary measure for a set of observations that make up a sample. Collectively, summary measures for samples are called *statistics*. Hence, in addition to being the sample mean, $\overline{X}$ is also a statistic.

Often we are interested in the population mean rather than the sample mean. If the population is accessible, we can find the mean by the same procedure we used for samples. Symbolically, however, we represent the population mean by μ (mu). In addition, we represent the number of observations in a population with N to distinguish it from the number of observations in a sample (n). In summary, the population mean for a finite population is

$$\mu = \frac{X_1 + X_2 + \cdots + X_N}{N} = \frac{\Sigma X}{N} \tag{3-2}$$

Suppose that last year the four executives of a small firm earned \$62,800, \$57,000, \$42,500, and \$42,500. These constitute the population of executive salaries for this firm last year. The mean executive salary is

$$\mu = \frac{\Sigma X}{N} = \frac{\$62{,}800 + \$57{,}000 + \$42{,}500 + \$42{,}500}{4} = \$51{,}200$$

The population mean, μ, is an example of a summary measure for a set of observations that make up a statistical population. Summary measures for statistical populations are called *parameters*. Hence, in addition to being the population mean, μ is also a parameter.

In practice, we often have neither time nor money to find the population mean. Imagine what a job it would be to find the mean salary of *all* executives in the United States for last year. Nonetheless, such a population mean is important for many purposes. In these circumstances, a random sample can be selected and the sample mean ($\overline{X}$) can be used to estimate the population mean (μ). In Part III, we begin with a chapter that discusses these estimation procedures. Even more generally, one of the chief uses for sample statistics is to estimate their counterpart parameters in the statistical populations from which the samples came. In Parts III and IV, we also cover estimation procedures for parameters other than the population mean.

The Mean from Frequency Distributions

For the most part, modern statistical data processing using computers calculates means of data collections in accordance with Equation 3-1. Consequently, calculation of the mean from frequency distributions is not as prevalent as it was. Nonetheless, there still is enough analysis done by hand with or without small calculators to necessitate knowing how to do so.

Our example for finding the mean from a frequency distribution will be Table 2-3 which exhibits the number of television sets per household for 50 households. The data are repeated in the first two columns of Table 3-1.

To find the mean of a data collection, be it either the original observations or a frequency distribution, we first must find the sum of the observations and the number of observations. For a distribution composed of single-value classes such as our example, the frequency count for a given class tells us how many repetitions of the X value occur. Hence the contribution to the grand total from that class is the X value multiplied by the frequency. For the third row from the top in Table 3-1, there are 12 households with 2 television sets per household. These households contribute a total of 24 television sets and 12 observations to the grand totals of television sets (88) and number of observations (50), respectively. Repeating this process for all classes and summing produces the grand totals for observations and sets. The mean is the total number of sets divided by the total number of households. In general, the sample mean from a frequency distribution is

$$\overline{X} = \frac{\Sigma fX}{n} \tag{3-3}$$

When we have only a frequency distribution with many-value classes, it is not immediately apparent what X value to assign to each of the frequencies in a given

Table 3-1 The Arithmetic Mean from a Frequency Distribution: Television Sets per Household

Sets per Household (X)	*Number of Households* (f)	*Class Totals* (fX)
0	10	0
1	14	14
2	12	24
3	8	24
4	4	16
5	2	10
	$n = \Sigma f =$ 50	$\Sigma fX =$ 88

$$\overline{X} = \frac{\Sigma fX}{n} = \frac{88}{50} = 1.76$$

class. The individual values are, of course, not available. In this case, we make the assumption that the values in the given class have a mean equal to the class midvalue. Then, granting this assumption, the contribution of the observations in this class to the grand total is the product of the class midpoint and the frequency count. In summary, we use Equation 3-3 by letting X be the midpoint for each class.

Exercises

3-1 *Sales distributions.* Two frequency distributions have been prepared for a small retail store. One is the distribution of daily sales and the other is the distribution of weekly sales. How would you expect these distributions to differ with respect to central location, dispersion, and skewness?

3-2 *Mileage distributions.* A car rental agency has prepared two different distributions of mileage per gallon of gasoline. One is for its fleet of compact sedans and the other is for its fleet of large and heavy luxury limousines. Compare these distributions with respect to differences in central location, dispersion, and skewness.

3-3 *Apartment rentals.* A random sample of the rents paid for 100 apartments in a community was collected. State the name of the measure of central location that applies to each of the following descriptions.

a. There are as many rentals less than \$192 as there are rentals greater than that amount.

b. More units rent for \$186 than for any other single amount.

c. A rental of \$197 is a measure of central location which, when multiplied by 100, equals the total of the observations in the sample.

3-4 *Accounts receivable.* The mean credit balance for the accounts receivable of a store is \$83. There are 410 accounts. Which of the following is a correct deduction from the information just given?

a. There are as many accounts with balances of $83 or less as there are with balances of $83 or more.
b. More customers have balances of $83 than of any other single amount.
c. The total of all accounts receivable is $34,030.

3-5 *Metropolitan bus traffic.* A city transit official rode the same bus route every weekday morning for one week and counted the number of passengers on board at 7:45 A.M. each day. The results were 32, 41, 35, 38, and 27 passengers.
a. What is the arithmetic mean number of passengers per day for this route and time during this week?
b. Use the mean to find the total number of passengers on board at 7:45 A.M. for the 5-day period. Check the result with the correct figure found from the original observations.

3-6 *Employee tardiness.* The timekeeper for a firm has records of the number of times that each employee was late for work last month. What is the mean number of times late for all employees?

Times Late	*Number of Employees*
0	52
1	27
2	13
3	6
4	3
6	1
9	1

3-7 *Typographical errors.* Find the mean number of errors per page made by the typesetting firm in Exercise 2-14.

3-8 *Percentage return on investment.* The frequency distribution for last year's return on investment for a random sample of 90 firms in the state is as follows.

Percentage Return	*Number of Firms*
0% and under 5%	25
5% and under 10%	32
10% and under 15%	20
15% and under 20%	9
20% and under 25%	4
	90

a. What is the arithmetic mean for this distribution?
b. Can we expect the result in (a) to equal the mean of the individual percentage returns for the 90 firms?

3-9 *Real estate listings.* Ace Realty has the following distributions of the number of new residential properties listed for sale by the company's agents.

This Year		*Last Year*	
Number of listings	*Number of weeks*	*Number of listings*	*Number of weeks*
0– 4	13	0– 4	10
5– 9	18	5– 9	24
10–14	7	10–14	5
15–19	2	15–19	1

a. Find the mean number of listings per week for the two distributions.
b. In view of the results in (a), how does this year's listing activity compare with last year's?

3.3 The Median

A second important measure of central location is the median. *The median is a value at the center of a set of observations when they are in numerical order.* For example, the number of messages received at a county recorder's office per hour on a weekday was 3, 5, 12, 18, 11, 6, and 8. When these are arranged from smallest to largest the result is

3 5 6 8 11 12 18

Median = 8

Because three values are less than 8 and three are greater, the central value is 8 messages. Hence 8 is the median.

When there is an even number of observations in the set, there is no central value. Suppose one more observation were made in the example just presented:

2 3 5 6 8 11 12 18

Median = 7

Any value between 6 and 8 will satisfy the definition of the median. By convention, we choose the value midway between the two nearest the middle. Hence the median is $(6 + 8)/2 = 7$ messages.

In general, we have the following definition.

For a collection of n observations arranged in order of size, the *median* is

$$Md = X_{(n+1)/2} \qquad \textbf{(3-4)}$$

In the first example, there are $n = 7$ observations in the set. Therefore, $Md = X_{(7+1)/2} = X_4$. With the observations ordered by size,

$$Md = X_4 = 8$$

In the second example, there are $n = 8$ observations, so

$$Md = X_{(8+1)/2} = X_{4.5} = 7 \text{ messages}$$

When several identical values appear at the center of a set of observations, the above definition still applies. The numbers of copies of a new novel sold per day by a certain bookstore were 22, 22, 22, 27, and 28. The central value is 22; this is designated as the median even though there are no smaller values.

The Median from Frequency Distributions

When data are given as a frequency distribution with single-value classes, the technique just described for use with Equation 3-4 still applies. In fact, in this type of frequency distribution, the data are already in order. Hence we need only count in from either end to the middle observation and that observation will be the median. For the now familiar data on number of television sets per household repeated in Table 3-2, application of Equation 3-4 identifies observation number

$$\frac{n+1}{2} = \frac{50+1}{2} = 25.5$$

as the median. Counting in toward the middle of the distribution from either end, we find that the median is one of the 2's in the third class of the distribution.

When a data collection is given only as a frequency distribution with many-value classes, the median often falls between the boundaries of a class. Because the locations of the observations within that class are unknown, we need a way to estimate their locations. In such an instance, we make the assumption that the data in the class are uniformly distributed, that there are equal intervals between these

Table 3-2 The Median from a Frequency Distribution: Television Sets per Household

Sets per Household (X)	*Number of Households* (f)	*Cumulative Frequency* (F)
0	10	10
1	14	24
2	12	36
3	8	44
4	4	48
5	2	50
Total	50	

$$Md = X_{(n+1)/2} = X_{25.5}$$

Cumulative frequency = 10 + 14 = 24; median is $X = 2$

observations. Then we find the proportional part of the class we need to get to the median observation and add this to the class boundary.

Table 3-3 shows the frequency distribution for content weights of 80 bags of flour first presented in the previous chapter. If the original data were available, the median would be observation 40.5, $(n + 1)/2$, in order of magnitude. From the column of cumulative frequencies we see that the median is in the class 99.90 to 100.09 lb. Because the median lies in this class we will call it the *median class*.

To position the estimated median within the median class, we use the following equation:

$$Md = L_{md} + \frac{c}{f_{md}}\left(\frac{n}{2} - F_p\right) \tag{3-5}$$

where Md = the median
L_{md} = the lower *boundary* of the median class
c = the median class interval; the difference between the lower and upper boundaries
f_{md} = the number of observations in the median class
n = the number of observations in the entire distribution
F_p = the cumulative frequencies in classes preceding the median class

Equation 3-5 divides the median class into a number of equal intervals, which, in turn, equals the number of frequencies in the median class. Then the equation positions the median at the center of the median interval.

Table 3-3 The Median from a Frequency Distribution with Many-value Classes: Content Weights of Flour Bags

Weight (pounds)	*Number of Bags (f)*	*Cumulative Frequency from Smallest Weight (F)*
99.10– 99.29	2	2
99.30– 99.49	3	5
99.50– 99.69	8	13
99.70– 99.89	12	25
99.90–100.09	21	46
100.10–100.29	18	64
100.30–100.49	9	73
100.50–100.69	6	79
100.70–100.89	1	80

$\frac{n+1}{2} = \frac{81}{2} = 40.5$; median class: 99.90–100.09 lb

$$Md = L_{md} + \frac{c}{f_{md}}\left(\frac{n}{2} - F_p\right)$$
$$= 99.895 + \frac{0.20}{21}\left(\frac{80}{2} - 25\right) = 99.895 + 0.143 = 100.04 \text{ lb}$$

Exercises

3-10 *Family size.* In 12 families, there are 3, 5, 2, 4, 3, 2, 2, 6, 3, 4, 2, and 5 people. What is the median number of people per family for this set of data?

3-11 *Number of ads.* An irate subscriber has just called the editor of the local newspaper to complain. The subscriber claims that even the first section is overly saturated with advertisements. There are 0, 3, 2, 5, 8, 1, 2, 4, and 3 ads on the pages of the first section. What is the median number of ads per page for this section of the paper? Would it be more appropriate to find the proportion of area on each page that is covered by advertising and then find the median proportion for the first section?

3-12 *Number of flaws.* A machine applies a plastic coating to 100-foot lengths of copper wire. A second machine checks each length for flaws in the coating. The distribution of flaws from a sample is as follows.

Number of Flaws	*Number of Lengths*	*Number of Flaws*	*Number of Lengths*
0	8	4	0
1	8	5	2
2	8	6	1
3	3		

What is the median number of flaws per length?

3-13 *Warranty tire replacement.* What is the median number of tires replaced per month for the distribution in Exercise 2-15 on page 42?

3-14 *Ages of executives.** Ages as of their nearest birthdays for chief executive officers of U.S. corporations who earned more than $100,000 annually in 1976 were reported as follows in a recent study.

Age	*Number of Executives*
30 and under 41	10
41 and under 46	30
46 and under 51	106
51 and under 56	158
56 and under 61	231
61 and under 66	172
66 and under 71	33
71 and under 81	14

What is the estimated median age of these executives?

*W. G. Brown and K. K. Motamedi, *California Management Review*, Winter 1977.

3-15 *Annual mileage.* The transportation department in a city recently made a study of the miles driven per year in private automobiles by a random sample of city residents. The distribution of the results is as follows.

Miles Driven	*Number of Residents*
0– 2,999	10
3,000– 5,999	28
6,000– 8,999	46
9,000–11,999	32
12,000–14,999	24
15,000–19,999	11
20,000 and over	7

What is the median number of miles driven per resident for this sample?

3-16 Suppose that the ages of executives in Exercise 3-14 had been reported as of their *last* birthday. How would this change the median calculation? Recalculate the median with this change in the data and compare your result with that from Exercise 3-14.

3.4 The Mode

Modes are values of a variable that identify peaks in the smoothed frequency curve. Many frequency curves, such as those in Figure 2-12 and all but the last in Figure 2-13, have just one peak. When this is the case the mode is the value of the variable beneath this peak.

For data reported as individual observations, *the mode is the value of the observed variable that occurs most frequently*. For example, over the past two weeks, the numbers of employees not reporting for work each day in a small plant are 1, 0, 0, 1, 2, 3, 2, 0, 0, and 1. The number 0 appears more than any other and is the mode. In this case the mode probably is reasonably representative of the typical number of daily absences over several months' time.

Now consider the actual weights of the random sample of 80 bags of flour given in Table 2-4. For this collection of ungrouped data, 99.93, 100.03, 100.11, and 100.15 each occur three times and no other value appears more often. Here the mode is not single-valued. Furthermore, we are likely to find sizable erratic shifts in the most frequently recurring values from one sample to the next. Several values in a data collection often satisfy the definition of the mode; also, the mode tends to be highly unstable among samples from the same statistical population. For these reasons, the mode is seldom used as a descriptive measure for ungrouped data.

For frequency distributions, however, the mode is an important measure of central location. The frequency distribution for the weights of the 80 bags of flour is shown in Table 3-3. The class from 99.90 to 100.09 pounds has the greatest number of frequencies, 21. Hence it is called the *modal class*. We must now position the mode within the modal class.

> For frequency distributions, the *mode* is defined to be the midpoint of the modal class; the modal class is the class with the most frequencies.*

The variable in the example is continuous, and the class boundaries are 99.895 lb and 100.095 lb. Hence the mode is $(99.895 + 100.095)/2 = 99.995$, or 100.00 lb.

The sample mode is an approximation of the mode of the statistical population from which the sample came. For our example, the relative frequency curve for the population was given in Figure 2-12. The highest point on the curve was directly over a weight of 100 lb. Since this is the value most likely to occur, it is the population mode. In this case, the sample mode and the population mode happened to be equal.

Multimodal Distributions

We defined the mode as the most frequently recurring value in a collection of data and as the value of the variable beneath the highest point on a frequency curve. Both of these definitions permit only one mode for a single collection of data. Yet cases such as that illustrated in Figure 3-3 occur. We now broaden our definition to include more than one peak of possibly different heights in a set of data.

When a statistical population has more than one mode, it is said to be *multimodal*. The relative frequency curve for personal incomes in a city might look like the one in Figure 3-3. Here there are three modes of different heights. The mode at \$8000 is attributable to unskilled laborers. The central mode at \$17,000 is for skilled workers receiving wages or salaries, and the mode at \$32,000 represents incomes of managerial and professional people.

Three distinct income populations are represented in Figure 3-3. When they are all included in the same data collection, they constitute a nonhomogeneous statistical population. Any single measure of central location calculated for the entire collection is not representative and tends to be misleading. Separating the data into three homogeneous income groups and describing each independently would be preferable.

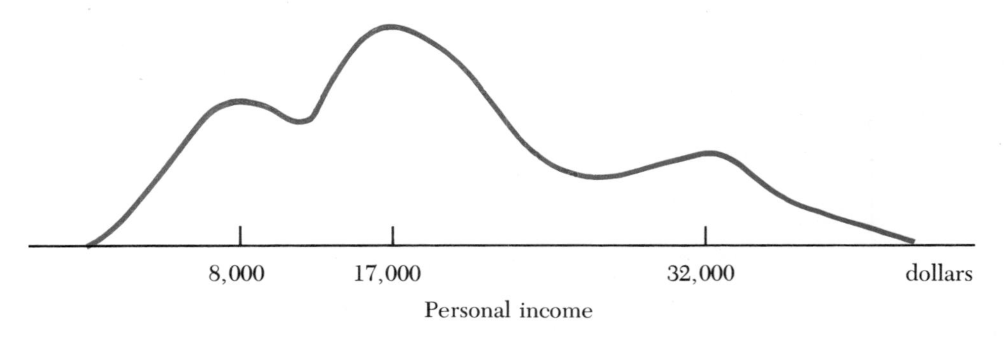

Figure 3-3 Multimodal Frequency Curve

* Interpolation formulas exist for the mode, but the midpoint is adequate for most purposes.

Whenever a set of data is judged to be large enough to produce a frequency distribution representative of the population, the presence of more than one mode usually indicates a lack of homogeneity in the population. We must then find a variable such as job level, sex, or geographical location that will classify the original observations into homogeneous groups and provide a logical explanation for the original lack of homogeneity. Then each group can be treated as an entity. If no such variable can be found, the existence, location, and relative magnitude of each mode should be noted.

3.5 Comparison of the Mean, Median, and Mode

Figure 3-4 shows the locations of the mean, median, and mode for three relative frequency curves. The curve on the left has marked negative skewness (it is skewed to the left). In the presence of such negative skewness, the mean is smaller than the median, which is, in turn, smaller than the mode. Typically the median is about a third of the distance from the mean to the mode.

For the curve on the right, everything is reversed relative to the curve on the left. Here there is considerable positive skewness. The long tail extends to the right. In such a case, the mean is greater than the median and the median is greater than the mode. Again, the distance from the mode to the median is about twice that from the median to the mean.

In symmetrical distributions, the left half is a mirror image of the right. When a symmetrical distribution has a single mode, the mode will be in the center of the distribution. Furthermore, the mean and the median will be equal to the mode. Such a frequency curve is shown in the center of Figure 3-4.

Figure 3-4 illustrates the fact that, of the three measures, the mean is the measure most affected by a few extreme observations in one tail of the distribution. The median is also affected, but much less so. The mode shows little effect. In the presence of appreciable skewness in either direction, the median often is used as the most representative measure of central location.

Even though it is affected by extreme values, the arithmetic mean is the most commonly used measure of central location. One major reason is that the mean has an exact mathematical definition and is subject to algebraic rules. For instance,

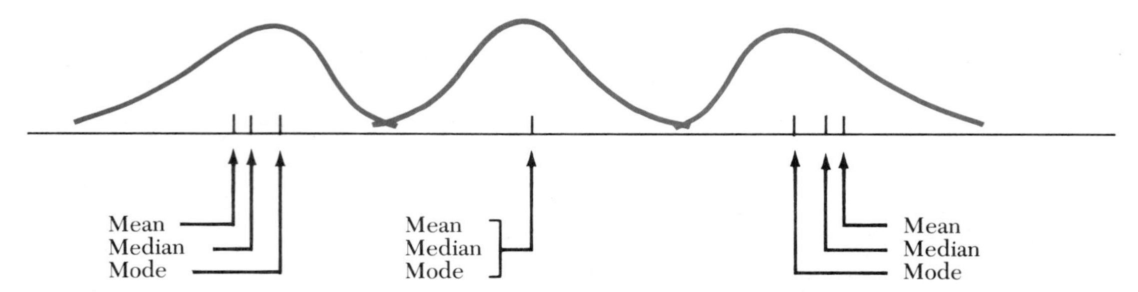

Figure 3-4 Effect of Skewness on Measures of Central Location

suppose that the central division of a company has 30 employees, for whom the mean overtime was 1.2 hours last week. The other division is composed of 12 sales employees for whom the mean overtime in the same week was 2.5 hours. Then the mean for all employees is

$$\overline{X} = \frac{30(1.2) + 12(2.5)}{30 + 12} = 1.57 \text{ hours}$$

The sum of the two products in the numerator, 30(1.2) + 12(2.5), is the total hours of overtime for the company, and the denominator, 30 + 12, is the number of employees for the entire company. The total hours of overtime divided by the total number of employees is the mean for the company. There is no way to combine similar information for the median or the mode to find the overall median or mode.

Exercises

3-17 *Equipment lifetime.* The distribution of lifetimes for one type of fluorescent tube in an office building is as follows.

Lifetime (h)	*Number of Tubes*
50 and under 150	35
150 and under 250	91
250 and under 350	58
350 and under 450	47
450 and under 550	21
550 and over	8

What is the mode for this distribution? Does there appear to be any skewness present? If so, in which direction? Would the median be a more representative measure of central location?

3-18 *Employees per firm.* The following distribution appeared in a recent publication.

Number of Employees	*Percent of Businesses*
1– 3	54
4– 7	19
8– 19	16
20– 49	7
50– 99	2
100–249	1.3
250–499	0.4
500 and up	0.3

a. What is the mode for this distribution?
b. Is the mode the best choice for a representative measure of central location in this case?

3-19 *Plant safety.* A person on the safety staff of a manufacturing plant reported that the mean number of accidents per week over the past year was 3.4 and that this was an increase from 3.2 from the previous year. The executive officer of the plant was puzzled by this result because a major safety campaign had been instituted during the past September. The officer requested the frequency distributions for accidents per week for the two years. The distribution for the first year showed a single mode at 3.1 accidents per week, along with positive skewness. The distribution for the past year showed a major mode at 3.3 accidents per week and a smaller mode at 2.8 accidents per week. Criticize the original report and explain how you would report the results.

3-20 *Student spending.* In a report concerning high school students, one result of a student survey stated that the mean amount spent per week was $22.42. The accompanying frequency distribution is reproduced below.

Weekly Spending	*Number of Students*
$ 0– 4.99	5
5.00– 9.99	12
10.00–14.99	19
15.00–19.99	13
20.00–24.99	10
25.00–29.99	14
30.00–34.99	16
35.00–39.99	9
40.00–44.99	4
45.00 and over	3

a. What is a likely explanation for the unusual pattern in the frequency distribution?
b. Criticize the representativeness of the reported mean and suggest a remedy.

3.6 Dispersion

In Figure 3-2, a relative frequency curve showing actual weights of 5-pound bags of salt was contrasted with a similar curve for 5-pound bags of potatoes. It was pointed out that very small amounts of salt could be used to correct the weight of a bag of salt, but only an entire potato could be used for the same purpose in the second case, so there was more variation in the weights of the bags of potatoes. We say that the latter distribution has more *dispersion* around its central location than the former.

Measuring central location is often very important, but there are many situations in which measuring dispersion or variability is just as important. Quality control for manufacturing operations usually involves periodic sampling to see that the process is performing acceptably on the average and that variability around the average is not excessive. For example, when shafts are being machined to fit inside half-inch sleeve bearings, the mean diameter must be just enough less than one-half inch to provide the necessary clearance. Furthermore, the variability around the mean must also be small. If an appreciable percentage of shafts are too large to fit their bearings, there will be expensive reworking or scrapping. On the other side, too many shafts with excessively small diameters will also result in excessive scrapping.

3.7 Measures of Dispersion Based on Deviations

In the most prevalent measures of dispersion, deviations of observations from the mean of the data collection, or values related to these deviations are averaged. The three such measures are the mean deviation, the variance, and the standard deviation.

The Mean Deviation

Let us consider once again the earnings last year of the four executives of a small firm. The observations are \$62,800, \$57,000, \$42,500, and \$42,500. We found earlier that the population mean for these observations is \$51,200 (μ = \$51,200). The deviations of the individual observations from the mean are:

$$\begin{aligned} \$62{,}800 - \$51{,}200 &= \$11{,}600 \\ 57{,}000 - 51{,}200 &= 5{,}800 \\ 42{,}500 - 51{,}200 &= -8{,}700 \\ 42{,}500 - 51{,}200 &= -8{,}700 \end{aligned}$$

Initially it may seem sensible to use the mean of these deviations as a measure of dispersion. But the sum of these deviations is zero. This is not a satisfactory method because, first, it is obvious that there is dispersion in this population. Only if all four observations were \$51,200 would we want our measure to be zero. Second, it can be shown that the sum of *any* such set of deviations is always zero.*

The problem with our first attempt was that negative deviations from the mean cancelled positive deviations. To prevent this, either of two approaches can be taken. The first approach is to ignore the algebraic signs of the deviations and average the absolute values.

---

* $\sum(X - \mu) = \sum X - N\mu = \sum X - N\left(\frac{\Sigma X}{N}\right) = \sum X - \sum X = 0$

The *mean deviation* of a statistical population is the mean of the absolute deviations of the observations from the population mean. In symbols,

$$\text{Mean Deviation} = \frac{\Sigma \mid X - \mu \mid}{N} \qquad (3\text{-}6)$$

where the vertical bars indicate absolute value.

For our example, the mean deviation is

$$\frac{11{,}600 + 5800 + 8700 + 8700}{4} = \$8700$$

To find the mean deviation for a sample, we use the sample mean, $\overline{X}$, in place of the population mean, and the sample size, n, in place of the number of observations in the population, N.

The mean deviation is readily understood and is satisfactory for descriptive purposes. It also introduces the concept of averaging deviations from the mean. Absolute deviations, however, are difficult to handle algebraically and, therefore, lead to subsequent complications. Consequently the mean deviation is not the most popular measure of dispersion for practical applications.

The Variance and Standard Deviation

The second way to prevent positive and negative deviations from the mean from cancelling each other is to square all the deviations before averaging.

The *variance* of a statistical population is the sum of the squares of the deviations of the observations from the population mean divided by the size of the population. In symbols,

$$\sigma^2 = \frac{\Sigma(X - \mu)^2}{N} \qquad (3\text{-}7)$$

For the example of the four executives, the variance is

$$\frac{11{,}600^2 + 5800^2 + (-8700)^2 + (-8700)^2}{4} = 79{,}895{,}000 \text{ dollars squared}$$

Note that squaring both positive and negative numbers produces only positive results.

The variance of our example, 79,895,000 dollars squared, shows that the variance presents difficulties when used as a descriptive measure of dispersion. The magnitude of the numerical result often differs greatly from the magnitudes of the original deviations. Furthermore, the unit of measure for the variance is the square of the unit of measure for the original observations (dollars squared versus dollars for our example).

To alleviate both of the difficulties with the variance, we take the positive square root of the variance and call it the *standard deviation*. Symbolically, *the population standard deviation* is

$$\sigma = \sqrt{\frac{\Sigma(X - \mu)^2}{N}} \tag{3-8}$$

For our example,

$$\sigma = \sqrt{79{,}895{,}000 \text{ dollars squared}}$$
$$= 8938.40 \text{ dollars}$$

The standard deviation has the same unit of measure as the data from which it comes and is usually reasonably close in magnitude to the deviations. It is the most often used descriptive measure for dispersion. Nonetheless, the variance plays a fundamental role in the theory underlying applied statistics.

Sample Variance and Standard Deviation

In the typical sampling situation, we know neither the mean nor the variance for the statistical population being sampled. Indeed, estimating these two population parameters often is a major purpose of sampling. In forming an estimate of the variance from a sample, we cannot use the population mean (μ) as a reference point when we do not know what it is. But we can substitute the sample mean ($\overline{X}$) and use it as the reference from which to get deviations. This leads us to the definition of the *sample variance*:

$$s^2 = \frac{\Sigma(X - \overline{X})^2}{n - 1} \tag{3-9}$$

Not only do we substitute the sample mean for the population mean, but we also divide by one less than the sample size, $n - 1$, rather than by the number of observations in the sample, n. We shall see in Chapter 8 that dividing by $n - 1$ makes the sample variance, s^2, a better estimate of the population variance, σ^2.

As was the case for statistical populations, the sample standard deviation is the positive square root of the sample variance. In symbols, the *sample standard deviation* is

$$s = \sqrt{\frac{\Sigma(X - \overline{X})^2}{n - 1}} \tag{3-10}$$

Let us find the sample variance and standard deviation for the number of cars coming to the drive-up windows of a bank. For a random sample of five business days, the numbers of cars that appeared were 128, 93, 106, 114, and 109. Calculation of the variance and standard deviation of this sample is shown in Table 3-4. Once again, we see from the second column that the sum of the deviations from the mean is zero. This serves as a check on our calculations. The sum of the deviations squared is 646 and division by one less than the sample size gives a variance of 161.5. The square root of this number is 12.71 cars, the sample standard deviation.

Table 3-4 Sample Variance and Standard Deviation for Ungrouped Data Based on Deviations from the Mean

Number of Cars	*Deviations*	*Deviations Squared*
128	128 − 110 = 18	324
93	93 − 110 = −17	289
106	106 − 110 = − 4	16
114	114 − 110 = 4	16
109	109 − 110 = − 1	1
550	0	646

$$\overline{X} = \frac{550}{5} = 110$$

$$s^2 = \frac{\Sigma(X - \overline{X})^2}{n - 1} = \frac{646}{4} = 161.5$$

$$s = \sqrt{161.5} = 12.71 \text{ cars}$$

Finding deviations from the mean, as in the last example, works well as long as the deviations are convenient whole numbers. But consider the following example. The number of messages received at a county recorder's office each hour of one day were 3, 5, 12, 18, 11, 6, and 7. The mean of these seven observations is 8.857, to three decimal places. With such a mean, the procedure outlined in Table 3-4 is computationally awkward. A more convenient procedure that does not involve finding deviations is to use an equation that is the algebraic equivalent of Equation 3-9.

For calculation, the *sample variance* is

$$s^2 = \frac{\Sigma X^2 - (\Sigma X)^2/n}{n - 1} \qquad \textbf{(3-11)}$$

Table 3-5 illustrates calculation of the sample variance and standard deviation using Equation 3-11. This approach avoids using deviations.

Variance and Standard Deviation from Frequency Distributions

We use much the same approach to calculate the variance and standard deviation from a frequency distribution as we used to find them immediately above. We calculate the sum of the observations and the sum of the squares of the observations. Then we proceed in accordance with Equation 3-12.

Table 3-5 Finding the Sample Variance and Standard Deviation Without Calculating Deviations: Number of Messages Received per Day for Seven Consecutive Days

Messages Received (X)	X^2
3	9
5	25
12	144
18	324
11	121
6	36
7	49
62	708

$$s^2 = \frac{\Sigma X^2 - (\Sigma X)^2/n}{n - 1} = \frac{708 - (62^2/7)}{7 - 1} = 26.48$$

and

$$s = \sqrt{26.48} = 5.15 \text{ messages}$$

The calculation procedure for the *sample variance* (s^2) from a frequency distribution is

$$s^2 = \frac{\Sigma f X^2 - (\Sigma f X)^2/n}{n - 1} \tag{3-12}$$

In Table 3-6, the data on television sets per household from Tables 2-3 and 3-1 are repeated along with the necessary calculations to find the variance and the standard deviation based on Equation 3-12.

The first three columns in Table 3-6 are identical to Table 3-1, where we found the mean. The values of X in the first column occur with the frequencies shown in the second column. Each X class contributes to the grand total the quantity shown for that class in the third column. Finally, the entries in the fourth column are the contributions to the sum from column 3 each multiplied by the value of the variable in column 1. Alternatively, the two households with 5 TV sets apiece contribute 10 sets to the grand total. Furthermore, X^2 is 25 for each of these two observations; so they contribute 50 to the sum of the squared observations.

When we have only a frequency distribution with many-value classes, we proceed as we did in calculating the arithmetic mean. We assume that every observation in a given class is equal to the midvalue of its class. This allows us to use Equation 3-12 if we make the X-values the midpoints of the successive classes.

Table 3-6 The Variance and Standard Deviation from a Frequency Distribution: Television Sets per Household

Sets per Household (X)	*Number of Households* (f)	*Class Totals* (fX)	*Class Sums of Squares* (fX^2)
0	10	0	0
1	14	14	14
2	12	24	48
3	8	24	72
4	4	16	64
5	2	10	50
	50	88	248

$$s^2 = \frac{\Sigma fX^2 - (\Sigma fX)^2/n}{n-1} = \frac{248 - 88^2/50}{50-1} = 1.9004;$$

$$s = 1.9004^{1/2} = 1.38$$

Exercises

3-21 *Shopping mall management.* As part of an annual review of tenant rental rates, the manager of a shopping mall has obtained the mean monthly net incomes for the mall's five shoe stores for last year. The data are

$22,500 $78,300 $47,700 $32,100 $28,900

a. What is the mean deviation for this collection of data?
b. Find the variance and standard deviation for this sample.

3-22 *Coding clerk errors.* A supervisor of people who write keypunch codes on product purchase reports from a sample of households is examining a sample of coding results for errors. The errors on a sample of 9 returns selected at random are

4 1 6 0 3 1 0 1 2

a. What is the mean deviation for this sample?
b. Find the variance and standard deviation for this sample.

3-23 *Number of flaws.* Find the variance and the standard deviation for the distribution of flaws in plastic-coated wire from Exercise 3-12.

3-24 *Employee tardiness.* Find the variance and standard deviation for the employee tardiness distribution from Exercise 3-6.

3-25 *Age of equipment.* A replacement study of the typewriters in a large law office includes the following distribution of their ages.

Age to Nearest Month	*Number of Typewriters*
25–29	4
30–34	6
35–39	6
40–44	2
45–49	2

Find the variance and standard deviation of this distribution.

3-26 *Daily low temperatures.* In a vegetable-growing region the distribution of daily low temperatures during May for the past several years has been determined.

Low Temperature (Degrees Fahrenheit)	*Number of Days*
20 to under 25	1
25 to under 30	7
30 to under 35	14
35 to under 40	28
40 to under 45	27
45 to under 50	13
50 to under 55	3

Find the variance and the standard deviation.

3.8 Descriptive Applications of the Standard Deviation

Normal Curve Guidelines

The first frequency curve we discussed in Section 2.3 was a symmetrical bell-shaped curve called the normal distribution. If we know the mean and standard deviation for such a distribution, the range within which a given percentage of the observations lie can be determined. For instance, 68.3% of the observations in a normal distribution are known to be within one standard deviation of the mean. Furthermore, 95.4% are within two standard deviations and 99.7% are within three. These percentages are illustrated in Figure 3-5.

The percentages in Figure 3-5, together with the mean and standard deviation, give us a very good mental picture of a given normal distribution. If we are told that the scores on a certain aptitude test for graduate study in business are normally distributed with a mean of 500 and a standard deviation of 100, we know that

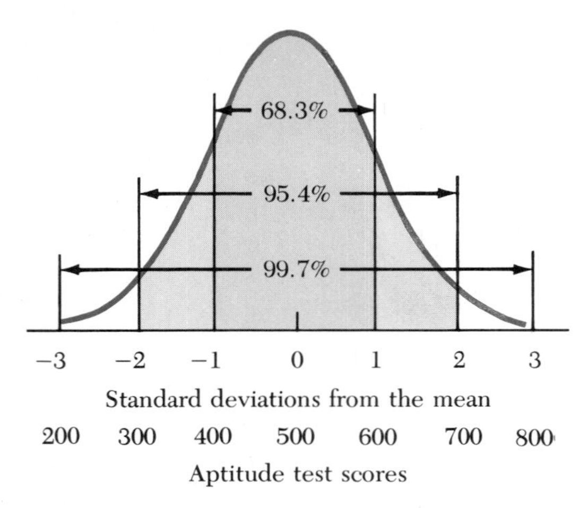

Figure 3-5 Percentage of Observations Within One, Two, and Three Standard Deviations of the Mean for Normal Distributions

practically all the scores (99.7%) lie in the range from 200 to 800 because 200 is three standard deviations below the mean and 800 is three standard deviations above the mean. We also know that 68.3% of the scores lie in the range from 400 to 600. A person who scored 600 on this test has a score which is equal to or greater than 84.15% of the scores. Fifty percent of the scores are below the mean (median) and 34.15% lie within the range from the mean to plus one standard deviation.

Even when a distribution is not quite normally distributed, the percentages in Figure 3-5 are good approximations. The weights of the 80 flour bags in Table 2-4 are not normally distributed. Nonetheless, the histogram in Figure 2-9 shows that the distribution has a single mode and is not seriously skewed. In such cases, the percentages in Figure 3-5 provide sound estimates. In this instance, the mean and standard deviation are 100.015 lb and 0.3285 lb, respectively. The range of plus or minus one standard deviation is 99.69 to 100.34 lb. There are 55 of the 80 observations (68.75%) in this interval. Furthermore, 77 of the 80 (96.25%) lie within two standard deviations and 100% lie within three.

Standard Scores

Another use for the standard deviation is in finding standard scores for observations.

> The *standard score* of an observation is the number of standard deviations that the observation lies from the mean. For a statistical population
>
> $$z = \frac{X - \mu}{\sigma} \tag{3-13}$$
>
> and for a sample
>
> $$z = \frac{X - \overline{X}}{s} \tag{3-14}$$

The person who scored 600 on the graduate business aptitude test mentioned earlier has a standard score of +1 because the score is 100 points (one standard deviation) above the mean of 500.

As we saw earlier, the person who has a standard score of +1 on the aptitude test for graduate study in business is above 84% of the people who took the test. Suppose this same person has taken an aptitude test for graduate study in law for which the mean is 300 and the standard deviation is 50. Suppose our subject's score on this test is 400. We then find that

$$z = \frac{X - \mu}{\sigma} = \frac{400 - 300}{50} = +2$$

Assuming that the percentages for a normal curve apply, a score two standard deviations above the mean is greater than 97.7% of the scores on the test.

Standard scores adjust for differences in units of measurement, means, and standard deviations between two distributions. If a grade school student has a standard score of +0.5 on the height distribution and −0.3 on the weight distribution that apply to the age and sex group for the student, we know the student is taller and thinner than average.

Coefficient of Variation

The coefficient of variation provides a way to measure the relative importance of variability or dispersion in sets of data. *The coefficient of variation for a statistical population* is

$$CV = \frac{\sigma}{\mu} \qquad \text{(3-15)}$$

and *for a sample* is

$$CV = \frac{s}{\overline{X}} \qquad \text{(3-16)}$$

The coefficient of variation is the ratio of the standard deviation to the mean. Typically, this ratio is expressed as a percentage.

Suppose the height distribution for the grade school students just mentioned has a mean of 60 in. and a standard deviation of 4 in. Then the coefficient of variation for this distribution is $4/60$, or 6.67%. Similarly, suppose the weight distribution has a mean of 100 lb and a standard deviation of 10 lb. The coefficient of variation for this distribution is $10/100$, or 10%. Relative to their means, the distribution of heights is less variable than that of weights.

3.9 Computer-Packaged Programs

During the past few years, several computer software packages have been developed for statistics. Calculations described in most chapters of this book are typical of those that can be carried out by these computer packages. An important feature of most programs is that they can be used by people with little or no computer programming experience.

Packaged programs are available for all of the major types of computers—the large mainframe computers, the intermediate minicomputers, and the microcomputers. At one extreme, mainframe programs can process very large data sets with extensive collections of integrated programs at high purchasing cost and high operating cost per unit time but with great speed. At the other extreme, microcomputer packages are relatively inexpensive, are limited to smaller data sets and less integrated processes, and sometimes require appreciable processing time. Microcomputer packages are, however, presently adequate for topics covered in this book and for data sets of one or two hundred observations and up to about a dozen variables. Furthermore, the current rates of development in microcomputer hardware and software are most rapid, and capabilities within the next few years are expected to greatly exceed those of today.

Figure 3-6 presents some output from a packaged program that is typical when the objective is statistical description of a data collection. The variables are the salaries of the employees, whether or not the employee participates in the tax deferral program, and employee job classification level. For the first two variables, salary and deferment, the descriptive measures reported have either been described in this chapter or are self-explanatory except for *std error of mean*. The standard error of the mean is found by dividing the standard deviation by the square root of N, the number of observations. Statistical uses of the standard error of the mean will be discussed beginning in the next part of the book.

The final variable in Figure 3-6, job level, has been processed by the computer-packaged program to produce a frequency distribution of job levels together with numerical summaries of the number of observations in each level.

3.10 Boxplots (Optional)

Boxplots are based on characteristics of the normal distribution, a bell-shaped distribution we discussed in connection with flour bag content weights in Figures 2-11 and 2-12. The normal distribution underlies much of the statistical inference we will discuss in Part Three, and that is one important reason for building the descriptive device known as the boxplot on this distribution.

We will first describe a procedure for constructing boxplots from stem and leaf diagrams. Then we will turn to comparing actual boxplots for our examples with those to be expected if the distributions were normally distributed.

The stem and leaf display for 1975 weekly grocery bills for 35 families as given in Table 2-10 is reproduced in Table 3-7(a) for convenience. To prepare the stem and leaf diagram for use in constructing a boxplot, we rearrange the leaves into ascending order within the stem classes, and we record a cumulative depth count of observations from each end of the distribution toward the middle. We stop this count at the stems adjacent to the class containing the median and simply record the number of observations within the median class. The result of this preparation appears in Table 3-7(b).

Given a stem and leaf display prepared as in Table 3-7(b), we next find the descriptive measures needed to construct the boxplot. These measures are the

```
VARIABLE     N         MEAN      STANDARD       MINIMUM       MAXIMUM     STD ERROR           SUM      VARIANCE
                                DEVIATION         VALUE         VALUE       OF MEAN

SALARY     672  54893.508929  9925.9768792  33447.000000  92011.000000  382.90287103  36888438.000  98525017.007
DEFER      672      0.797619     0.4020740      0.000000      1.000000    0.01551034       536.000         0.162


FREQUENCY BAR CHART

LEVEL                                                                                FREQ    CUM.  PERCENT     CUM.
                                                                                             FREQ           PERCENT
        :
LEVEL1  :**                                                                            11      11     1.64     1.64
        :
LEVEL2  :**************************************                                       154     165    22.92    24.55
        :
LEVEL3  :**********************************************************                   233     398    34.67    59.23
        :
LEVEL4  :********************************************************************         274     672    40.77   100.00
        :
         ----+----+----+----+----+----+----+----+----+----+----+----+----+----
            20   40   60   80  100  120  140  160  180  200  220  240  260

                                       FREQUENCY
```

Figure 3-6 Typical Summary Descriptive Output from a Computer-Packaged Program

Table 3-7 Stem and Leaf Display: Weekly Grocery Bills for 35 Families

(a)

Unit: 1 dollar — *3L | 3 represents $33*; *3U | 6 represents $36*

Stem	Leaf
3L	3
3U	6
4L	33
4U	958
5L	2103222
5U	7957865758
6L	00032
6U	765
7L	211

(b)

Depth	*Stem*	*Leaf*
1	3L	3
2	3U	6
4	4L	33
7	4U	589
14	5L	0122223
(10)	5U	5556777889
11	6L	00023
6	6U	567
3	7L	112

median, the first and third quartiles, and the lower and upper adjacent values. We will identify these with the symbols *MD*, *Q*1, *Q*3, *LAV*, and *UAV*, respectively. To find the adjacent values, we must first determine the two inner fences (lower and upper), although these do not appear in the boxplot. Similarly, to distinguish between mild and extreme outliers, we must find the two outer fences. We will identifiy the four fences as *LOF*, *LIF*, *UIF*, and *UOF*.

In Table 3-8, the top three rows show the calculation of the median and the quartiles for the grocery expense display in Table 3-7(b). The median is found as described earlier in this chapter. A quartile is just the median of its half of the distribution. We found the depth of the median to be 18. For convenience we say that the lower half of the distribution has eighteen observations, as does the upper

Table 3-8 Finding the Descriptive Measures for a Boxplot

Measure	*Depth*	*Value*
Median (MD)	$(35 + 1)/2 = 18$	\$56
First Quartile $Q1$	$(18 + 1)/2 = 9.5$	\$51.5
Third Quartile $Q3$	same as for $Q1$	\$60

Measure	*Procedure*	*Value*
Interquartile Range: IQR:	$Q3 - Q1 = 60 - 51.5$	\$ 8.5
Fence Interval: K:	$1.5(IQR) = 1.5(8.5)$	\$12.75
Lower Inner Fence: LIF:	$Q1 - K = 51.5 - 12.75$	\$38.75
Lower Adjacent Value: LAV:	Smallest value $\geq LIF$	\$43
Upper Inner Fence: UIF:	$Q3 + K = 60 + 12.75$	\$72.75
Upper Adjacent Value: UAV:	Largest value $\leq UIF$	\$72
Outliers: Less than LAV:		\$33, \$36
Greater than UAV:		None
Lower Outer Fence: LOF:	$Q1 - 2K = 51.5 - 25.5$	\$26
Extreme Outliers:		None

half. Then we use the familiar median-depth formula to find the depth of the quartiles. Had the depth of the median ended in 0.5, we would have dropped that fraction before finding the depth of the quartiles.

In the lower two-thirds of Table 3-8, procedures are given for finding the fences and the adjacent values. The fence interval is always 1.5 times the interquartile range. This interval is subtracted from the first quartile to find the lower inner fence and added to the third quartile to find the upper inner fence. The adjacent values are the smallest and largest values inside the lower inner fence and upper inner fence, respectively. Outer fences are found by going another fence interval beyond the respective inner fences. Mild outliers lie between inner and outer fences. Extreme outliers lie beyond outer fences.

Having completed Table 3-8, we now have the information needed to construct the boxplot. As illustrated in Figure 3-7, construction begins by laying out a suitable scale. Here we have chosen to make the scale horizontal and to position it below the boxplot. The box itself extends from the first to the third quartile. The median is indicated by a line across the box at the correct scale location. At either end of the

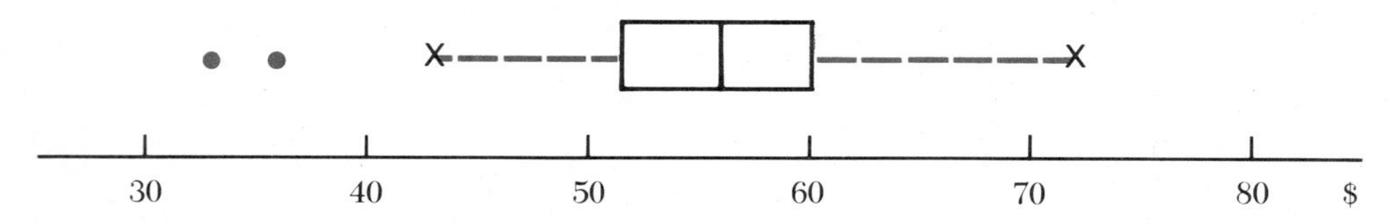

Figure 3-7 Boxplot for the Weekly Grocery Expense Data in Table 3-6

box are the whiskers. A given whisker extends from the end of the box to the adjacent value associated with that whisker. For instance, the left whisker in the figure extends from the box to $43, the lower adjacent value. The two mild outliers are shown as solid dots. Had there been an extreme outlier, it would have been shown with a different symbol such as a circle.

As would be expected for data from a normal distribution, the median in Figure 3-8 is essentially equidistant from the ends of the box. This indicates no skewness in the central half of the distribution, a condition not unusual even when the parent distribution for the data has unequally long tails.

A second characteristic of the grocery expense boxplot that is typical of random samples from normally distributed populations is the lengths of the whiskers. The factor 1.5 used in calculating the fence interval in Table 3-8 makes the fence interval 1.5 times as long as the box in the boxplot. Typically, for samples from normal distributions, finding the fence interval this way makes each whisker very nearly the same length as the box. Here the whiskers are nearly the same length as the box.

The third feature of the boxplot in Figure 3-7, the two mild outliers, is not typical of data from a normal distribution. The probability of getting an observation farther from the mean than the inner fences is 7 in 1000. This casts suspicion on these two observations and should lead us to review the data collection process which produced them. If nothing untoward is found, however, a case can be made for including them as representing a possible long thin tail at the left end of the parent distribution.

We have constructed a boxplot for the grocery expense data which is typical for data from a normal distribution in all but one respect. The two mild outliers on the lower tail suggest a thin tail to the left in the parent distribution.

Next we should take a look at a set of data for which construction of a boxplot is *not* appropriate. The 40 annual payments to employees of a certain firm were listed in Table 2-11. They were presented in a stem and leaf display in Table 2-12. There it became apparent that there was no single central location. Instead the distribution was seen to have two modes or peaks. All normal distributions have a single mode. Since boxplots are based on characteristics of normal distributions, the presence of more than one mode in a sample rules out the proper use of a boxplot. On the other hand, skewness is not a characteristic of a normal distribution either. Yet we do make boxplots for samples with skewed stem and leaf displays. From such boxplots we can determine whether the skewness is primarily a characteristic of the tails or if, indeed, the central portion of the distribution also seems to be skewed.

In summary, the stem and leaf display allows us to find out if a data collection is homogeneous, that is, if the collection has a single mode. If so, the boxplot then displays the central location and dispersion of the data collection, the degree of skewness in the central portion, and the lengths of the tails. Finally, it performs the important function of identifying and visually locating any outliers which may be present.

Exercises

3-27 Based on the stem and leaf display for the content weights of 80 bags of flour in Table 2-13, state why a boxplot is appropriate. Make the boxplot. Compare it with what would be expected from data from a normal distribution. Does the median have the value expected for the process? Did the two smallest values turn out to be outliers?

3-28 Would making a boxplot from the stem and leaf display to be constructed in Exercise 2-22 be likely to provide more information than a boxplot constructed from the stem and leaf display in Table 2-13? Why or why not?

3-29 Based on the most suitable stem and leaf display prepared for Exercise 2-20, prepare a boxplot for the data and evaluate the characteristics of the distribution as compared with those to be expected for data from a normal distribution. Is there any basis to question the applicability of a boxplot to this distribution?

3-30 Use the stem and leaf displays prepared in Exercise 2-21 to construct boxplots for each distribution. Evaluate the results with respect to apparent differences in central location, dispersion, skewness, and presence of outliers. Is there any indication that either distribution has more than one mode? If there were, what would you recommend?

Summary

Central location, *dispersion*, and *skewness* are three general characteristics of data collections. Important measures of central location are the arithmetic mean, the median, and the mode. The *arithmetic mean* is the measure commonly known as the "average." It is the sum of the observations in a set divided by the number of such observations. The *median* is a value at the center of a set of observations when they are ordered by magnitude. The *mode* is the value that occurs most frequently in a data collection.

Dispersion (scatter, spread) of a data collection can be measured by the mean deviation, the variance, and the standard deviation. The *mean deviation* is the average of the absolute deviations of the observations from their mean. The *variance* is the average of the squares of the deviations from their mean, where a statistical population is being considered. For samples, one less than the number of observations is used as the divisor. The *standard deviation* is the square root of the variance. Of these, the standard deviation is the most frequently used for descriptive purposes.

Supplementary Exercises

3-31 *Ages of chief executives.** The ages (to last birthday) at which 754 chief executives of U.S. corporations assumed their positions are given in the following table.

Age	*No.*	*Age*	*No.*	*Age*	*No.*	*Age*	*No.*	*Age*	*No.*	*Age*	*No.*
20	1	30	1	40	10	50	38	60	11	70	1
21	1	31	8	41	16	51	44	61	4	71	0
22	0	32	4	42	23	52	42	62	8	72	1
23	2	33	2	43	30	53	39	63	3		
24	0	34	4	44	29	54	46	64	2		
25	0	35	8	45	36	55	39	65	2		
26	1	36	4	46	38	56	36	66	1		
27	2	37	10	47	33	57	30	67	1		
28	3	38	8	48	33	58	21	68	0		
29	5	39	13	49	38	59	22	69	0		

a. Find the mean, median, and mode. Do these measures indicate the presence of skewness? If so, in which direction?
b. Find the variance and the standard deviation.
c. Prepare a frequency distribution with 5-year age classes (20 and under 25, and so on). Estimate the mean, median, and mode from this distribution and compare results with those in (a). Which mode is to be preferred? Why?
d. Find the sample variance and standard deviation for the distribution in (c). Compare these results with those in (b).

3-32 *Product development costs.*† The ratios of actual to projected costs for 49 new product developments in a pharmaceutical firm are given in the following distribution.

Cost Ratio	*Number of Products*
0–0.50	1
0.51–1.00	5
1.01–1.50	18
1.51–2.00	6
2.01–2.50	6
2.51–3.00	10
3.01–3.50	3

a. Calculate the mean, median, and modal cost ratio for this distribution.

*W. G. Brown and K. K. Motamedi, *California Management Review*, Winter 1977.
†J. E. Schnee, *Journal of Business*, July 1972.

b. Do you think it fair for the management to present any of these figures to the board of directors as a representative cost ratio for development projects? Explain.

3-33 *Automobile purchasers.* An advertising campaign is being designed for a new automobile. The age distribution of a sample of purchasers of the previous model is obtained as a basis for designing the advertisement. The distribution is as follows.

Age Last Birthday	*Number of Purchasers*
under 25	8
25 and under 35	11
35 and under 45	23
45 and under 55	16
55 and under 65	9
65 and over	4

a. Find the mean age of buyers in this sample (use 20 and 69 as midvalues for the open end classes).
b. What is the sample standard deviation?
c. What is the range of ages within one standard deviation of the mean? Estimate the percentage of ages in this distribution that lie within this range and compare the result with the normal curve guideline.

3-34 *Waiting times.* The number of minutes after their appointment times each of a random sample of 64 patients had to wait to be served in a major local health facility were observed as follows.

Waiting Time (min)	*Number of Patients*
0 and under 4	10
4 and under 8	17
8 and under 12	16
12 and under 16	11
16 and under 20	7
20 and under 24	3

a. What are the mean and standard deviation from this sample?
b. Estimate the percentage of observations within two standard deviations of the mean for this sample. Compare results with the normal curve guideline.

3-35 *Futures market.* Over a three-month period futures for a certain commodity sold for an average of \$3.02 per bushel with a standard deviation of 47 cents. Another commodity sold for an average of \$5.14 a bushel with a standard deviation of 63 cents. Which commodity had the greater relative variability?

3-36 *Sales stability.* The standard deviations for the monthly sales of two businesses over a five-year period are $4670 and $13,108. Given this information, a statistics student said that it shows the sales of the first firm are more stable than those of the second firm. Is this conclusion justified without qualification?

3-37 *IQ test results.* A government job application procedure requires the applicant to take IQ tests in verbal and mathematical reasoning. The verbal test scores have a mean of 100 and a standard deviation of 10, and the mathematical test scores have a mean of 32 and a standard deviation of 6. An applicant received scores of 115 on the verbal test and 30 on the mathematics test. Compare the applicant's performance to that of the group that was used to get the means and standard deviations.

3-38 *Measurement errors.* At a certain base, the electronic instrument used to measure the distance to a weather satellite has a mean error of zero and a standard deviation of 150 ft for errors. Estimate the percentage of such measurements that will yield standard scores greater than -1.

Computer Exercises

1 Source: Real estate data file

a. Obtain a frequency distribution of purchase prices with 10 class intervals. Find the mean and standard deviation of the purchase prices.

b. On a plot of the frequency distribution of purchase prices, locate the mean. Also, locate points representing one, two, and three standard deviations below and above the mean. Locate a purchase price of $80,000 and express that value as a z-score.

c. Obtain a cross-tabulation of incomes of buyers versus whether the buyer is a new resident of the community. Use six classes for income. Construct relative frequency distributions for the incomes of each group of buyers. Plot the two relative frequency distributions on the same graph. Write a brief comparison of incomes of new and old residents.

d. Find the means and standard deviations of the incomes of old and new residents. Do these values support the comparisons you made in (c)? Explain.

2 Source: Utilities data file

a. Using assets as the variable, find the mean, median, mode, variance, standard deviation, and coefficient of variation.

b. Delete GTE and AT&T and repeat (a). Which measures are changed the most? Which the least? Why?

3 Source: Utilities and retailers data files

a. For retailers and for utilities, find the mean and standard deviation for 1982 earnings per share.

b. Use the results from (a) and from Computer Exercise 2(a) to describe how the distributions differ.

4 Source: Utilities data file without GTE and AT&T

a. Find the relative frequency diagrams for assets and income per employee.

b. Which distribution is more like a normal (bell-shaped) distribution?

c. For the distribution in (b) that is more nearly normal, find the standard scores. What proportion of the data are within one standard deviation of the mean? Within two standard deviations?

5 Source: Utilities data file
   a. Find the mean income. Find the mean incomes by industry type (gas, electric, communications).
   b. Repeat (a) for income per employee.
   c. Using the results of the two previous parts, describe the utilities industry income and income per employee.

PART TWO

Background for Statistical Inference

4

PROBABILITY

Part One of this text was concerned with descriptive statistics. There we learned how to construct effective graphic and numerical summaries of the variables in a statistical study. The study might encompass an entire population of observational units or it might be limited to a sample of these units. Part Three of the text will be devoted to *statistical inference*. In statistical inference the objective is to reason from descriptive statistics of the sample to characteristics of the entire statistical population.

Example

A marketing research group conducted a study of the market potential for a portable sauna unit in a metropolitan market area. The study was based on a random sample of households in the market area. The study report included the following statement: "Our estimate of population mean income is $27,600 and the chances are 95 to 5 that the estimate is within $300 of the true mean household income in the market area."

The statement contains an inference, as well as a quantitative measure of the degree of uncertainty of the inference. The measure of uncertainty is expressed in the language of probability. In this chapter, which begins Part Two, we shall discuss and illustrate the basic concepts of probability on which statistical inference is based.

4.1 The Probability of an Event

Probability statements are common in everyday experience. The following expressions are examples of such statements.

The probability of heads in a coin toss is. . .

The probability that a voter prefers candidate A is. . .

The probability that this business will succeed is. . .

Each of these statements involves an event that is not certain. The degree of certainty that an event will occur, called the *probability* of the event, is a number between zero and 1. For example, an event with a probability of 0.9 has a high degree of certainty, while an event with a probability of 0.2 has a low degree of certainty. The second event is more *un*certain. If the probability of an event were 1, we would regard the event as certain to happen. If it were zero, we would regard the event as certain *not* to happen.

Sample Space

Each of the examples above involves a happening with alternative outcomes. We assume that a coin toss can result in only heads or tails. A voter will either prefer Candidate A, prefer some other candidate, or have no preference. A business will either succeed or not succeed, given a clear definition of success. The alternative outcomes of a happening are often referred to as the *sample space of an experiment*. In mathematical terms, probability is an assignment of numbers to the elements in a sample space in a manner such that: (1) the number assigned to any element (outcome) lies between zero and 1.0, and (2) the total (probability) assigned to all the elements in the sample space is 1.0. It should be clear in this language that we are talking about a set of outcomes that are *mutually exclusive*. That is, no two of the outcomes can occur in a single trial of the experiment.

How are probabilities established? There are several approaches to establishing probabilities of events, which we shall now discuss briefly.

Equally Likely Outcomes

A historical use of probability was in analyzing games of chance. Such games usually have a set of mutually exclusive basic outcomes that can be regarded as equally likely. A coin toss will result in either heads or tails; the ball in a roulette wheel will drop into one of 38 colored spaces; a playing card drawn from a standard deck will be one of 52 distinct cards. In a fair game, in which the coin is not weighted, the roulette wheel is evenly balanced, or the dealer shuffling cards is both thorough and honest, such outcomes are regarded as equally likely. Therefore, equal probabilities are assigned to each of the basic outcomes, or elements in the sample space. Thus, the probability of *heads* in a fair coin toss is $1/2$, as is the probability of *tails*.

Often our interest centers on events that are made up of several elements in a sample space. For example, we may want to bet on "red" in a roulette spin. We would be concerned with the event that the ball comes to rest in one or another of the 18 red slots on the board. Formally,

> An *event* is a subset of the elements in a sample space.

In the case of an "equally likely" assignment of probabilities to the elements, we can say:

> If an uncertain happening has $N(U)$ equally likely mutually exclusive outcomes and $N(A)$ of these satisfy the event A, then the probability of A is given as
>
> $$P(A) = \frac{N(A)}{N(U)} \qquad \textbf{(4-1)}$$

In roulette 18 of the 38 spaces are red. For the probability of red, A is the event that the ball comes to rest on red, $N(A)$ is 18, $N(U)$ is 38, and the probability of red is

$$P(\text{red}) = P(A) = \frac{N(A)}{N(U)} = \frac{18}{38} = \frac{9}{19}$$

This can also be expressed as 0.474, correct to three decimal places.

Relative Frequency

Many uncertain events involve a set of circumstances that can be repeated over and over. Tossing a coin or rolling a die can be done over and over in essentially the same manner. Selecting a voter at random from a registration list and finding the voter's opinion about a current public issue can be thought of as a trial that can be repeated again and again.

A second way of viewing probability is in terms of the *relative frequency* of the event, compared to the total number of trials, as trials involving the event are repeated over and over. Suppose an advertising agency is interested in selling a mailing list to a magazine publisher. The list contains 1 million names and addresses stored on two reels of magnetic tape. The computer is programmed to select names at random and check them against the publisher's subscriber lists. In the first 100 names checked, one subscriber is found; in the next 100 names three more subscribers are found. The computer continues to select and check names, with the following results.

Number of Trials	*Number of Subscribers*	*Relative Frequency of Subscribers*
100	1	0.01
200	4	0.02
300	7	0.023
400	9	0.0225
500	9	0.0180
1,000	24	0.0240
10,000	221	0.0221
100,000	2310	0.0231

As the number of trials increases, we expect better and better estimates of the probability. Indeed, if all the names on the list were checked, the observed relative frequency would correspond to the ratio $N(A)/N(U)$ of Equation 4-1, where $N(U)$ is the one million names and $N(A)$ is the number who subscribe to the client's magazine.

The relative frequency approach to probability is an empirical approach. To establish probabilities this way, many cases must be observed. This approach is used by insurance companies, who collect and analyze data on accidents, illnesses, and deaths to establish probabilities as a basis for setting premium payments for their policies.

The use of equally likely outcomes to establish probability is entirely consistent with a relative frequency interpretation. When we assign $1/2$ as the probability of *heads* for a coin toss, we expect that the relative frequency of *heads* will approach $1/2$ as the coin is repeatedly tossed.

Degree of Belief

The third approach to probability is *degree of belief*. We introduced the probability of an event as a measure of the degree of certainty that the event will occur. From an individual's standpoint, this is close to saying that probability is a measure of one's belief that the event will occur. In this context, equally likely outcomes or relative frequencies used in establishing probabilities are objective justifications for our degree of belief. If 18 slots on a roulette wheel containing 38 slots are red and we believe the wheel is fairly balanced, we have an objective basis for asserting that our degree of belief the next spin will be a "red" is $18/38 = 0.474$. If one knew that 37.6% of the voters in a registration list actually prefer candidate A, the degree of belief in the proposition that a randomly selected voter will prefer candidate A is also, objectively, 0.376.

In other situations, degree of belief cannot be so objectively assessed. Suppose a fast-food franchiser is trying to establish the probability that a new unit opening in Duluth will show a profit in its first year. Executive A knows that 80% of the units established last year showed a first-year profit. But because the economy is down this year, executive A assesses the probability at 0.70 rather than 0.80. Though not disagreeing about the state of the economy, executive B draws on knowledge of the competitive situation in Duluth to assess the probability at 0.75. Even though there are subjective, or judgmental, elements in the assessments, it may be useful for managers to quantify their beliefs in terms of probabilities.

Some probabilities will have a very objective basis, others will be mixed (like the example just given), and others will be very subjective. Assessing the probability of first-year profits for units of a franchise operation with no past record would be a case where subjectivity would be necessary. Subjectivity does not imply that the basis for a probability is ill-considered; it simply means that there is room for differences in belief about the proposition in question.

4.2 Marginal, Joint, and Conditional Probability

The elements in a sample space correspond to the units of observation in a statistical study. Events correspond to characteristics of these units that are of interest to us. Consider the following record of loans made by a bank in the first two years of a small-business loan program. The bank made loans to its own business clients, as well as to clients referred to it by the Small Business Administration. The loans have been classified as successes or failures, depending on whether they were repaid or had to be redeemed by the Small Business Administration under its guarantee.

		Own Customers C	*Referrals* R	*Total*
Success	S	312	262	574
Failure	F	51	20	71
Total		363	282	645

Suppose a loan is randomly selected from the 645 loans. The probability of each event in the table is the ratio of the number of times the event occurred to the total of 645 loans. We use C for the bank's own customers, R for referrals, S for success, and F for failure.

Marginal Probability

Several types of probabilities can be found from a table like the one given above. For example, we can focus on success and failure over the period and find

$$P(S) = 574/645 = 0.890 \quad \text{and} \quad P(F) = 71/645 = 0.110$$

On the other hand, if we are interested in customer loans and referrals, we can find

$$P(C) = 363/645 = 0.563 \quad \text{and} \quad P(R) = 282/645 = 0.437$$

If we were to choose a loan at random from the 645 loans, our belief that it would be successful (loan repaid) is 0.890, our belief that it was made to a customer is 0.563, and so on.

> A *marginal probability* is the probability of an event over the entire sample space.

The marginal probability of a success above was constructed without regard for whether the loan was a customer or a referral loan, and the marginal probability of a customer loan is constructed without regard to whether the loan was a success or a failure. The frequencies for constructing these probabilities will be found in the row or column totals of a cross-classified frequency table.

Joint Probability

The second type of probability is called a *joint probability*.

> For two events, a joint probability is the probability that both events will occur.

The frequencies in the main body of the table given above represent the joint occurrence of two characteristics. For example, the 312 loans in the upper left corner of the table were both successful and made to customers. If we select one of the 645 loans at random, the probability that it was a successful loan to a customer is $312/645 = 0.484$. We use *and* in referring to such probabilities. Symbolically,

$$P(S \text{ and } C) = 312/645 = 0.484$$

If we reduce the entire table of occurrences to relative frequencies, we have the following:

	C	R		
S	0.484	0.406	0.890	$= P(S)$
F	0.079	0.031	0.110	$= P(F)$
	0.563	0.437	1.000	
	$= P(C)$	$= P(R)$		

The joint probabilities are in the body of the table. The marginal probabilities appear as row and column totals in the margins of the table. Going across, the sums of the joint probabilities are the marginal probabilities in the column at the right; going down, the sums are the marginal probabilities in the row at the bottom.

Conditional Probability

Two questions naturally arise about the record of loans: What was the probability of success for customer loans? Also, what was the probability of success for a referral loan? Consider the first of these. The question asks for a restricted or *conditional* probability. We are concerned not with all loans (645) but only with customer loans (363).

> A *conditional probability* is the probability of an event in a subset of the elements in a sample space.

In our example, the probability of success among customer loans is computed from the original table of frequencies as

$$P(S \mid C) = {}^{312}/_{363} = 0.860$$

The vertical line in $P(S \mid C)$ means *conditional on*, or *given the event* that follows the vertical line. Thus we would read "the probability of success *given* a customer loan." Likewise, we find the probability of success given a referral loan as follows:

$$P(S \mid R) = {}^{262}/_{282} = 0.929$$

In calculating these conditional probabilities, we used the restricted sample space directly by finding the number of successes among customer loans and the number of successes among referral loans.

Conditional probabilities can also be found from appropriate joint and marginal probabilities. For example, the conditional probability of success among customer loans is

$$P(S \mid C) = \frac{P(S \text{ and } C)}{P(C)} = \frac{0.484}{0.563} = 0.860$$

If we reconstruct the probabilities in this expression, we have

$$P(S \text{ and } C) = {}^{312}/_{645} \quad \text{and} \quad P(C) = {}^{363}/_{645}$$

Their ratio gives

$$P(S \mid C) = \frac{{}^{312}/_{645}}{{}^{363}/_{645}} = \frac{312}{363}$$

Observe that the denominators of 645 loans cancel, leaving the denominator of 363 customer loans, which is the restricted sample space of concern to us.

It is useful to generalize the method for finding conditional probabilities from joint and marginal probabilities. In general, the probability of event A conditional on event B can be found by dividing the joint probability of A and B by the marginal probability of B.

$$P(A \mid B) = \frac{P(A \text{ and } B)}{P(B)} \tag{4-2}$$

To further illustrate conditional probabilities, consider the following joint probability table for occupied dwelling units in the United States in 1980.

Location	*Owner-occupied*	*Renter-occupied*	*Total*
Metropolitan, central city	0.147	0.150	0.297
Other metropolitan	0.272	0.112	0.384
Nonmetropolitan	0.237	0.082	0.319
Total	0.656	0.344	1.000

Restricting our attention first to the individual columns, we can construct the conditional probabilities for each location, given first owner-occupied units and then renter-occupied units. Each calculation is an application of Equation 4-2. For example,

$$P(\text{Nonmetropolitan} \mid \text{Owner-occupied}) = \frac{P(\text{Nonmetropolitan and Owner-occupied})}{P(\text{Owner-occupied})}$$

$$= \frac{0.237}{0.656} = 0.361$$

The results of all the calculations are given in the following table.

Conditional Probabilities of Location Given Occupant Status		
Location	*Owner*	*Renter*
Metropolitan, central city	0.224	0.436
Other metropolitan	0.415	0.326
Nonmetropolitan	0.361	0.238
Total	1.000	1.000

It is also possible to relate the joint probabilities to the marginal probabilities for each location. The result is the conditional probabilities for each type of occupant given, in turn, each location. The results are given in the following table.

Conditional Probabilities of Occupant Status Given Location			
Location	*Owner*	*Renter*	*Total*
Metropolitan, central city	0.495	0.505	1.000
Other metropolitan	0.708	0.292	1.000
Nonmetropolitan	0.743	0.257	1.000

From this table we see that if an occupied dwelling unit were selected at random from all units located in central cities of metropolitan areas, the probability that the unit is owner-occupied is 0.495. From the table given just before this one, we see that if an occupied dwelling unit were selected at random from all owner-occupied units, the probability that the unit is located in a central city of a metropolitan area is 0.224. Thus, owner-occupied units in metropolitan area central cities comprise 49.5% of all metropolitan area central city units and 22.4% of all owner-occupied units.

Exercises

4-1 *Assigning probabilities.* For each of the following events, would you use equally likely events, relative frequency, or degree of belief as the basis for assigning a probability? Discuss your choices.

a. The probability that a customer entering a store selling backpacking supplies is a female.

b. The probability that the New York Yankees will win the American League baseball pennant this year.

c. The probability that you will win a door prize that is awarded by a random draw from the names of those attending a social event.

4-2 *Weather forecasting.* Weather forecasters often give probabilities for various events. Two such events are (1) the probability of measurable precipitation on a day in June in your area and (2) the probability that it will rain tomorrow in your area.

a. Would you expect two different forecasters to give the same or different probabilities for (1)? For (2)? Explain.
b. Forecaster A has been forecasting for 500 days; on 200 of these days the forecaster gave the probability of rain tomorrow as 0.2. How could you check whether the forecasts are good?

4-3 *Interpreting probabilities.* Let A be the event that a company employee is male and B be the event that the employee is a college graduate. Interpret the following probabilities in the form *the percentage of* _____ *who are* _____.
a. $P(B)$
b. $P(A \mid B)$
c. $P(A \text{ and } B)$
d. $P(B \mid A)$

4-4 *Using symbols.* Let A be the event that a student at this college is a freshman, and B be the event that a student is a female. Write symbolic expressions for the following.
a. The probability that a student is a female.
b. The probability that a freshman student is a female.
c. The probability that a student is a female freshman.
d. The probability that a female student is a freshman.

4-5 *Odds and probability.* Odds are the ratio of chances in favor of an event to chances against the event, while probability expresses the ratio of chances for an event to total chances. Restate the following odds statements as probabilities.
a. The odds for a head on a toss of a coin are 50-50.
b. The odds for rain today are 1 to 4.
c. The odds against Navratilova beating Evert on clay are 5 to 3.

4-6 *Insurance statistics.* * Among 100,000,000 males, there were 68,500 accidental deaths in a recent year; of these, 32,500 involved motor vehicle accidents. Among an equal number of females, there were 27,900 accidental deaths; of these, 11,500 were from motor vehicle accidents. Of the deaths from motor vehicle accidents, 6510 occurred among 10,000,000 males aged 20 to 24, and 1790 occurred among 10,000,000 females aged 20 to 24. Assume random selection.
a. What is the probability of accidental death by motor vehicle among males? Among females?
b. What is the probability of accidental death by motor vehicle among males aged 20 to 24?
c. If a person aged 20 to 24 dies in a motor vehicle accident, what is the probability that the fatality occurred to a female?

4-7 *Passenger car drivers.* A study was made of 13,433 young drivers of passenger cars. Of the total, 4526 were driving subcompacts, 3581 were driving compacts, 3160 were driving intermediate models, and 2166 were driving larger models. The numbers using and not using seat belts were 3486 and 9947, respectively. If a young driver were selected at random from those observed, what is the probability that the driver was:
a. Driving an intermediate model automobile?
b. Driving a compact or subcompact automobile?
c. Using a seat belt?

*Adapted from *Statistical Bulletin*, Metropolitan Life Insurance Company, July–September 1978.

4-8 *Earth-escape launches.* The following is the record of earth-escape launches of U.S. spacecraft by time periods.

	1957–60 *I*	*1961–65* *II*	*1966–69* *III*
Success	2	11	18
Failure	8	3	1

Given random selection, state the following as ratios.

a. The probability of a successful launch.
b. The probability that a launch that was successful occurred during time period I.
c. The probability that a launch in time period I was successful.
d. The probability that a launch was both successful and occurred in time period I.

4-9 *Air force personnel.* The officers in the air force of a small nation are distributed as follows by age and rank.

			Age		
Rank	*21–30*	*31–40*	*41–50*	*Over 50*	*Total*
Lieutenant	154	132	36	4	326
Captain	54	292	112	20	478
Major		92	294	70	456
Lt. Colonel			132	96	228
Colonel			24	62	86
Total	208	516	598	252	1574

Assume random selection and:

a. Find the probability of a lieutenant.
b. Find the probability of a lieutenant, given that the officer is 30 years of age or less.
c. What is the most probable age bracket for a major? What is the most probable rank of an officer over 50 years of age?
d. Find the probability that a captain is aged 41 to 50.
e. Find the probability that an officer is a captain aged 41 to 50.

4-10 *Stock portfolio performance.* One hundred stocks in a portfolio were classed as either cyclical or growth stocks. Of 60 cyclical stocks, 45 had increased in price in the past month, while 15 of the growth stocks had increased in price. A stock is to be selected at random from the portfolio.

a. Construct a table showing four joint probabilities and two sets of marginal probabilities.
b. From the probabilities in the table determine (1) the probability that a cyclical stock has increased in price, and (2) the probability that a stock that has not increased in price is a cyclical stock.

4-11 *Business ventures.* The table gives the joint probabilities for success and failure of two new ventures under consideration by a business.

		Venture 2	
		Success	Failure
Venture 1	Success	0.08	0.12
	Failure	0.08	0.72

a. Given success in venture 1, what is the probability that venture 2 succeeds?
b. Given success in venture 2, what is the probability that venture 1 succeeds?

4.3 Probability Relations

To illustrate some important probability relations, we shall use an example of the place of residence in 1975 of persons (excluding those who moved here from abroad) living in the United States in 1980. The probabilities shown in Figure 4-1 are derived from the Current Population Survey, a large-scale continuing survey of the United States population conducted by the U.S. Bureau of the Census. Figure 4-1 shows that the probability of a person living in the same housing unit in a metropolitan area in 1975 and 1980 is 0.362, and the probability of a person moving within the same metropolitan area over the period is 0.210. The probability of a person moving to a different metropolitan area over the period is 0.073.

The probability that a person lived in the same housing unit in a nonmetropolitan area in 1975 and 1980 is 0.179, and the probability that a person moved but remained outside metropolitan areas is 0.109. The mutually exclusive categories are completed by considering those who moved between metropolitan and nonmetropolitan areas. The probability that a person moved from a metropolitan to a nonmetropolitan area is 0.037, and the probability that a person moved from a nonmetropolitan area to a metropolitan area is 0.030.

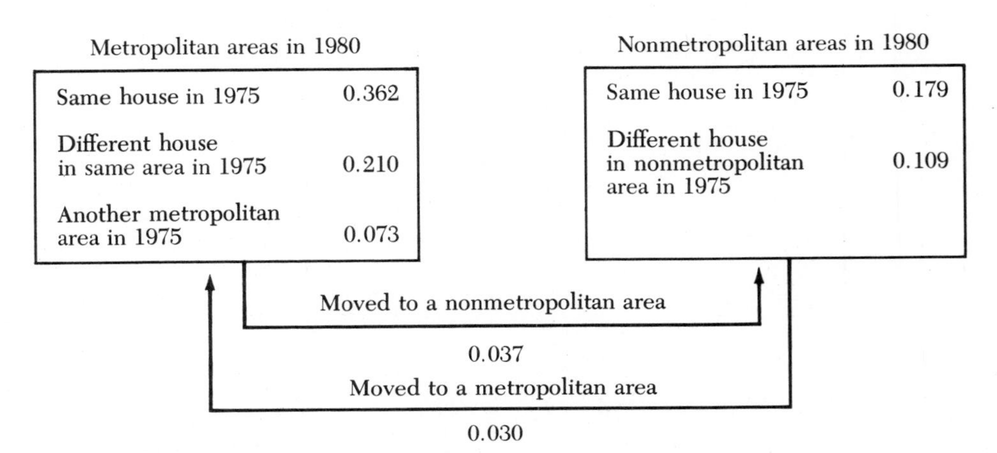

Figure 4-1 1975 Residence of 1980 Population Living in the United States in 1975

The Addition Relation for Mutually Exclusive Events

Suppose we want to know the probability that a person lived in the same metropolitan area in 1975 and 1980. Such persons could have not moved or could have moved within the same metropolitan area. The two categories—nonmovers and same-area movers—are mutually exclusive, so we can add the probabilities for the two different ways of satisfying the event *lived in the same metropolitan area in 1975 and 1980* to obtain the total probability for the event.

> When events A and B are mutually exclusive, we find the probability that either event A *or* event B occurs by adding the separate probabilities for event A and event B.
>
> $$P(A \text{ or } B) = P(A) + P(B) \qquad \textbf{(4-3)}$$
>
> where A and B are mutually exclusive.

If A is the event *metropolitan area nonmover* and B is the event *moved within the same metropolitan area*, no person can satisfy both of these events. Therefore if a person is randomly selected from the 1980 U.S. population, the probability that the person lived in the same metropolitan area in 1975 and 1980 is

$$P(A \text{ or } B) = 0.362 + 0.210 = 0.572$$

The Complement of an Event

Events that are always mutually exclusive and exhaustive (contain all the elements in the sample space) are an event and its complement. For any event A, the complement of A contains all events that do *not* satisfy A. It is customary to denote the complement of A by A'. Since the total probability assigned to a set of mutually exclusive and exhaustive events is 1.0, we can always get the probability for the complement of an event by

$$P(A') = 1 - P(A)$$

For our example, let A be the event *living in the same metropolitan area in 1975 and 1980*. Then,

$$P(A') = 1 - 0.572 = 0.428$$

The probability 0.428 is the probability that a person selected from the 1980 U.S. population *did not live in the same metropolitan area in 1975 and 1980.*

The Addition Relation for Nonmutually Exclusive Events

Suppose we are given that, for the population studied, the probability that a person lived in a metropolitan area in 1975 was 0.682 and that the probability that a person lived in a metropolitan area in 1980 was 0.675. We now want to find the probability that a person lived in a metropolitan area in 1975 *or* 1980. These two events are not mutually exclusive, because a person could have lived in a metropolitan area in *both* 1975 and 1980. The sum of the probabilities 0.682 and 0.675 counts such persons twice. To compensate for this, we subtract the elements that are counted twice. Using Figure 4-1 and the addition relation for mutually exclusive events, we see that the probability that a person lived in a metropolitan area in both years is

$$0.362 + 0.210 + 0.073 = 0.645$$

We can then calculate

$$P(\text{lived in metropolitan area in 1975 or 1980}) = 0.682 + 0.675 - 0.645 = 0.712$$

In general, when events are nonmutually exclusive, the probability that an observed event is either event A or event B is obtained by adding the separate probabilities $P(A)$ and $P(B)$ and subtracting the probability $P(A \text{ and } B)$.

$$P(A \text{ or } B) = P(A) + P(B) - P(A \text{ and } B) \qquad \textbf{(4-4)}$$

where A and B are not mutually exclusive.

Because Equation 4-3 dealt with mutually exclusive events, it did not have the term subtracted in Equation 4-4. If events A and B are mutually exclusive, the event A *and* B is a logical contradiction, or impossibility.

Figure 4-2 shows the relation of Equation 4-4 for our example. The events comprising A receive one kind of shading and those comprising B another. The event (A and B) receives both shadings. In a more general way, nonmutually exclusive events can be represented as overlapping circles. The overlap of the circles represents the joint occurrence of the two events. To obtain the probability that either event occurs, one must subtract the probability for the overlapping area from the sum of the probabilities for the two events.

We must be careful to test events to see if they are mutually exclusive. Suppose we find for a certain group of adults that 20% have been divorced and 12% have remarried. We must not conclude that 32% have been divorced or have remarried.

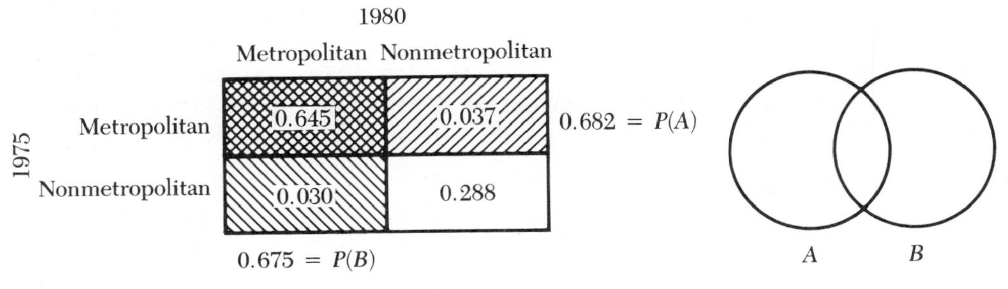

Figure 4-2 Nonmutually Exclusive Events

Since a person can have had both of these experiences, we would need to know what percentage have been both divorced and remarried. We would then subtract that percentage from 32 to find the percentage of persons who have been divorced or remarried.

Multiplicative Probability Relations

In our example involving small business loans, we examined marginal, joint, and conditional probabilities. We saw that a conditional probability could be obtained from appropriate joint and marginal probabilities or be determined using the underlying frequency table. Equation 4-2 summarized that probability relation.

A similar relation holds for finding a joint probability from appropriate marginal and conditional probabilities.

Example

Of the persons employed in a community, 40% work for forest-product industries that depend on timber from a national forest. If the national forest is incorporated into a proposed wilderness area, the forest-product industries will have to shut down. A survey of those industries was made and 70% of the employees said they would have to look for work outside the community. What would be the direct impact of the closings on the community?

In this example we have

$P(A)$ = probability that an employed person works in the forest products industry = 0.40

$P(B \mid A)$ = probability that a forest-products employee would leave the community to look for work = 0.70

$P(A \text{ and } B)$ = probability that an employed person works in forest products and would leave the community to look for work

If forest-product employees comprise 40% of total employment, and 70% of *these employees* would leave the community, then 70% of 40% or 28% of *total* employment would leave. In general

> The joint probability of events A and B can be obtained from the product of the marginal probability of event A and the conditional probability of event B given event A.
>
> $$P(A \text{ and } B) = P(A) * P(B \mid A) \qquad \textbf{(4-5)}$$

For our example,

$$P(A \text{ and } B) = 0.40 * 0.70 = 0.28$$

The direct impact of the closings would be that the community would lose 28% of its employment.

Independence

Consider the following cross-classification table derived from a census of 500 voters in a community who were asked the question: "Do you agree or disagree that the president is doing a good job?"

	Democrats	*Republicans*	*Total*
Agree	225	150	375
Disagree	75	50	125
Total	300	200	500

Now look at opinions among Democrats, among Republicans, and for all voters.

Probabilities of Agreement and Disagreement That President Is Doing a Good Job

	Among Democrats	*Among Republicans*	*Among All Voters*
Agree	0.75	0.75	0.75
Disagree	0.25	0.25	0.25
Total	1.00	1.00	1.00

The proportion of Democrats who agree that the president is doing a good job is the same as the proportion of Republicans who agree, and both are the same as the marginal probability of agreement. This means that if a voter is selected at random from the community, knowing the voter's political party does not change the probability of agreement. Such a situation is described by the term *independence*. In the current example, we say that opinion is independent of political party.

In our earlier small business loans example, success was *not* independent of the type of loan (customer versus referral). Recall the following table.

		Customer C	*Referral* R	*Total*
Success	S	312	262	574
Failure	F	51	20	71
Total		363	282	645

The probabilities of success and failure *depend* on whether the loans were to the bank's own customers or to SBA referrals. The failure rate was about twice as high for customer loans ($^{51}/_{363} = 0.140$) as for referrals ($^{20}/_{282} = 0.071$). The overall, or marginal, probability of failure was 0.110. Knowing whether a loan was a customer loan or a referral does change the probability of failure.

When independence occurs, a special multiplication relation among probabilities holds.

> Under independence, joint probabilities can be found by multiplying marginal probabilities for the constituent events.
>
> $$P(A \text{ and } B) = P(A) * P(B) \qquad \textbf{(4-6)}$$

Example

A fast-food franchise operation has determined that the probability of success for a new unit is 0.60. New units are located to serve different areas so that the success of any unit is independent of the success of other units. What is the probability that the next two units will both succeed?

If event A is success for the first unit and event B is success for the second unit, then

$$P(A \text{ and } B) = P(A) * P(B)$$
$$= 0.60(0.60) = 0.36$$

The concepts of independence and mutual exclusiveness are sometimes confused. Consider the two mutually exclusive events *getting a 1* and *getting a 6* on a throw of one die. Look at the following joint probability table.

	6	not 6	
1	$^0/_6$	$^1/_6$	$^1/_6$
not 1	$^1/_6$	$^4/_6$	$^5/_6$
	$^1/_6$	$^5/_6$	$^6/_6$

The marginal probability of a 1 is $^1/_6$. If a 6 occurs, a 1 cannot occur. The conditional probability of a 1 given a 6 is zero. Thus mutually exclusive events are never independent. Mutual exclusiveness involves complete dependence. The question of independence arises only in connection with nonmutually exclusive events.

Summary of Probability Relations

We can now summarize the probability relations developed in this section.

General Multiplication Law

$$P(A \text{ and } B) = P(A) * P(B \mid A)$$

Multiplication Law for Independent Events

$$P(A \text{ and } B) = P(A) * P(B)$$

General Addition Law

$$P(A \text{ or } B) = P(A) + P(B) - P(A \text{ and } B)$$

Addition Law for Mutually Exclusive Events

$$P(A \text{ or } B) = P(A) + P(B)$$

The definitions of independence and mutual exclusiveness are inherent in the differences between the general laws and the special cases. That is, A and B are independent if $P(B \mid A) = P(B)$, and A and B are mutually exclusive if they cannot occur together. Indeed, if all the probabilities are given, one way to check for independence is to see if the probabilities agree with the multiplication law for independent events.

Exercises

4-12 *Addition laws.* Indicate whether the procedure in each of the situations below is correct.

a. Adding the probabilities that an employee now with a department quits, retires, or transfers to another department within a year to find the probability of voluntary separation from a department within a year.
b. Adding the probability that a senior citizen graduated from college to the probability that a senior citizen graduated from high school to find the probability that a senior citizen received at least a high school education.
c. Adding the probability that your team wins the first game of a baseball doubleheader to the probability that it wins the second game to find the probability that your team wins one of the two games.

4-13 *Addition laws.* The probability that a retail pharmacist is under 30 years of age is 0.10, and the probability that he/she is over 60 years of age is 0.12. The probability that the retail pharmacist works for a chain store is 0.60.

a. What is the probability that a retail pharmacist is either under 30 or over 60 years of age?
b. If age of pharmacists is independent of store type, what is the probability that a retail pharmacist is under 30 years of age or works for a chain store?
c. If it is known that the percentage of pharmacists who are under 30 years of age and working for a chain store is 0.08, what is the probability that a pharmacist is under 30 years of age or works for a chain store?

4-14 *Multiplication laws.* Your name is in a hat along with three other guests attending a dinner. The host has three "door" prizes to award. What is the probability that you will not receive a prize if:

a. The prizes are awarded by making three draws from the hat in which the name drawn is replaced before the next draw?
b. The prizes are awarded by making three draws from the hat in which the name drawn is not put back in the hat?

4-15 *Product screening.* Of the new products proposed in a certain industry, only 30 percent survive an initial feasibility screening. Of these 60% are abandoned prior to prototype manufacture. Twenty-five percent subsequently survive marketing and product acceptance testing and are manufactured and brought to market. What is the probability that a new product idea will result in a product in the marketplace?

4-16 *Back-up systems.* A backpacker carries a butane lighter for lighting stoves and campfires. For back-up, "waterproof" matches are carried. The probability that the lighter will fail is 0.05, and the probability that the matches will fail is 0.10. If the events are independent, what is the probability that both systems will fail?

4-17 *Shirts 'n' Shorts.* Of 1000 shoppers entering the Shirts 'n' Shorts Boutique, 700 were interested in shirts and 600 were interested in shorts. Half the shoppers were interested in both shirts and shorts.

a. Make a cross-classified frequency table of interest or noninterest in shirts versus interest or noninterest in shorts.
b. Is interest in shirts independent of interest in shorts?

4-18 *Recession and balance of trade.* An economist assessed the probability of continued unfavorable balance of trade for the United States as 0.80, the probability of a recession within the next three years as 0.50, and the probability of both of these events as 0.40. Does the economist believe that occurrence of a recession is independent of an unfavorable balance of trade? Explain.

4-19 *Floods and tornados.* In a certain farming area an insurance company assessed the probability of flood during the coming year as 0.005 and the probability of a tornado as 0.001. If the events are independent, what is the probability of:

a. Both a tornado and a flood occurring?
b. Either a tornado or a flood occurring?
c. Neither a tornado nor a flood occurring?

4-20 *Mutual exclusiveness and independence.* Event B never occurs when event A occurs. Event C occurs half of the time when event A occurs.

a. Is event B independent of event A? Is event B mutually exclusive with event A?
b. Is event C mutually exclusive with event A? Under what condition would event C be independent of event A?

4-21 *Mutual exclusiveness and independence.* Jones never plays golf and tennis on the same day, but on half of the days when Jones plays tennis, Jones swims laps.

a. Is Jones' tennis playing independent of golfing? Is Jones' tennis playing mutually exclusive with golfing?
b. Is swimming laps mutually exclusive with playing tennis? Under what conditions would swimming laps be independent of playing tennis?

4-22 *Hot-air ballooning.* Ideal conditions for a hot-air balloon festival are fair weather and winds not exceeding 10 mph. During a certain season the probability of fair weather is

0.80, the probability of winds not exceeding 10 mph is 0.40, and the probability of either or both of these events is 0.90. What is the probability of ideal conditions for the festival?

4-23 *Gasco and Coalco.* A security analyst believes that Gasco stock can rise this month only if a merger with Coalco is announced. The analyst assesses the probability of the announcement as 0.30 and the probability of a rise in price, given the announcement, as 0.75. What is the probability that the stock rises this month?

4.4 Aids in Probability Calculations

In our work so far with multiplication and addition laws for probabilities, we have dealt with examples involving one law at a time. These fundamental laws can be applied to solve fairly involved probability problems, in which a complex event is a particular combination or sequence of simpler events. An example (which we will not solve) is a roller winning a game of craps. In craps, the roller wins if a 7 or 11 occurs on the first roll of two dice and loses if 2, 3, or 12 occurs. If any other number is rolled, that number is called *point*, and the roller wins if he or she subsequently rolls point before rolling a 7. If the probability of any number on a throw of one die is $1/6$, we can use only the addition and multiplication laws to find that the probability of a roller winning a game is 0.493.

In this section, we shall present some aids for finding probabilities when the situations become more complicated than those with which we have been dealing. All of these aids give ways to focus on a problem so that it can be more easily solved.

Listing Mutually Exclusive Ways

Though not always the most elegant way to solve a probability problem, listing the mutually exclusive ways in which a complex event can occur will usually help move us toward a solution. Starting this way helps us see how to get the unknown probabilities.

Example

Under international basketball rules, a player fouled in the act of shooting gets three attempts (if needed) to make two successful free throws. Assuming independent events, what is the probability that a 70% free-throw shooter at the foul line will score two points?

To list the mutually exclusive ways of scoring two points, we think of the sequence of free throws and realize that the shooter can (1) convert the first two throws, (2) fail the first and convert the next two, or (3) convert the first, fail the second, and convert the third. Listing these ways helps us to see that each one involves a sequence of events that are assumed to be independent. Using the multiplication law for independent events, the probability for each of these ways is

$$P(S_1 \text{ and } S_2) = 0.70(0.70) = 0.490$$
$$P(F_1 \text{ and } S_2 \text{ and } S_3) = 0.30(0.70)(0.70) = 0.147$$
$$P(S_1 \text{ and } F_2 \text{ and } S_3) = 0.70(0.30)(0.70) = 0.147$$

Because these are all the mutually exclusive ways that the shooter can score two points, we now can find the required probability

$$0.490 + 0.147 + 0.147 = 0.784$$

In other situations different laws might be used to find the probabilities for the mutually exclusive ways. The process of visualizing and listing these ways will help identify the needed laws.

Using Complementary Events

In some problems where we could use the approach of listing the mutually exclusive ways in which an event can occur, it may be quicker to think of the ways in which the event *cannot* occur. In the foul-shooting situation of the last example, suppose we want to find the probability that the shooter scores *at least one* basket. We could add the mutually exclusive ways of scoring one basket to the mutually exclusive ways of scoring two baskets already listed, work out the individual probabilities, and find the sum. On the other hand, there is only one way for the shooter *not* to score at least one basket, and that is to fail on all three attempts. Thus we want the complement of not scoring any baskets.

$$\begin{aligned} P(\text{at least one success}) &= 1 - P(F_1 \text{ and } F_2 \text{ and } F_3) \\ &= 1 - 0.30(0.30)(0.30) \\ &= 0.973 \end{aligned}$$

Using Several Approaches

Often more than one approach will yield the solution to a probability problem. Using the complement to find the probability of scoring at least one basket in the foul-shooting problem was an example. Another approach using mutually exclusive events is to realize that at least one basket will be scored if the shooter scores on the first try or, failing that, scores on the second try or, failing the first two, scores on the third try. This would lead to the following calculation.

$$\begin{aligned} & P(S_1) + P(F_1 \text{ and } S_2) + P(F_1 \text{ and } F_2 \text{ and } S_3) \\ &= 0.70 + 0.30(0.70) + 0.30(0.30)(0.70) \\ &= 0.70 + 0.21 + 0.063 \\ &= 0.973 \end{aligned}$$

Notice that the answers are the same with our two different approaches. Were they not, barring errors in arithmetic, we would have to go back and reexamine our thinking. We might find that the ways we thought were mutually exclusive are not, that we failed to make an exhaustive list of mutually exclusive ways, or that we had misapplied one of the probability laws. When the answers do check, our confidence in the solution is increased.

Probability Trees

Probability trees are useful in visualizing the mutually exclusive outcomes of a complex of events and in carrying out probability calculations involving such events.

Example

A sales representative for industrial resins is able to make advance appointments with 70% of the buyers she wishes to see. Eighty percent of these buyers will see the representative when she calls, but only 60% of buyers will see the representative if no prior appointment is made. The representative estimates the probability of selling a new product to a buyer who keeps a previous appointment at 0.50 and that of selling to a buyer seen without a previous appointment at 0.10. The representative wants to know (1) on what percentage of attempted calls a sale can be expected, and (2) whether it is worthwhile to call on buyers without appointments.

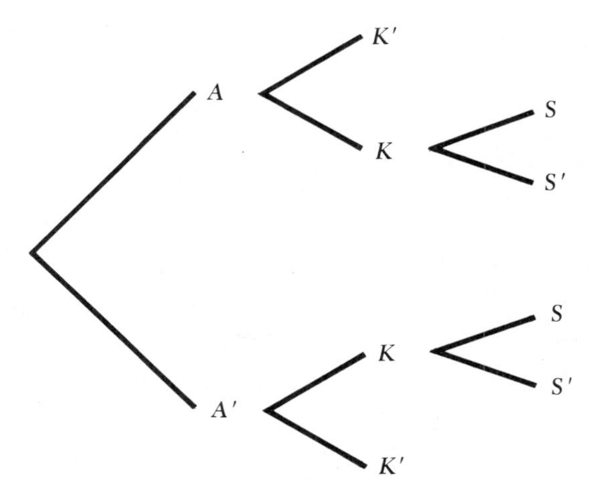

A: Appointment made $\quad$ A': Appointment not made
K: Able to see buyer $\quad$ K': Not able to see buyer
S: Sale made $\quad$ S': Sale not made

Figure 4-3 Tree Diagram for Salesperson's Problem

The probability tree for this problem is shown in Figure 4-3. The tree represents the sequence of events, and each path through the tree (from left to right) represents a mutually exclusive outcome of the sequence of events. For example, one sequence in Figure 4-3 is:

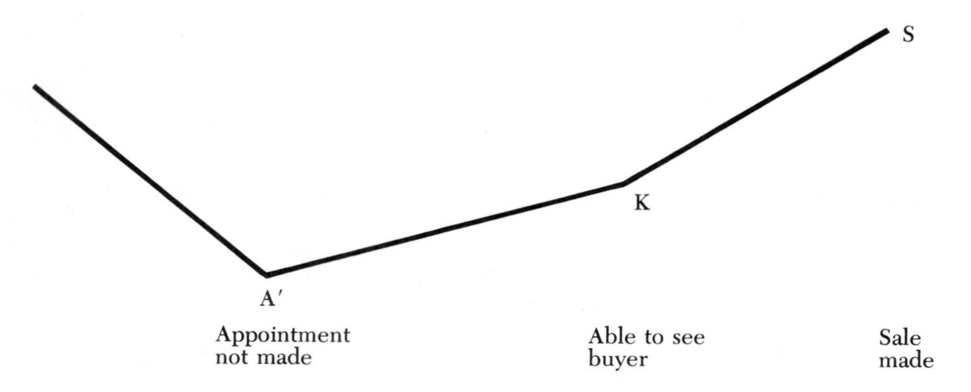

We put the (marginal) probabilities of the initial events on the leftmost branches. On the subsequent branches, we enter the probabilities of the subsequent events *conditional* on the earlier events. We can then find the probability for an event sequence from the probabilities on the path for that sequence. In Figure 4-4, we have put the salesperson's probabilities on the appropriate branches and then multiplied to obtain the probability for each path through the tree.

Once the tree is constructed it is easy to see that

$$\begin{aligned} P(\text{sale}) &= P(A \text{ and } K \text{ and } S) + P(A' \text{ and } K \text{ and } S) \\ &= 0.70(0.80)(0.50) + 0.30(0.60)(0.10) \\ &= 0.280 + 0.018 \\ &= .298 \end{aligned}$$

This answers the first question. To answer the salesperson's second question, we must find $P(S \mid A)$ and $P(S \mid A')$, the probability of a sale *given* an appointment and

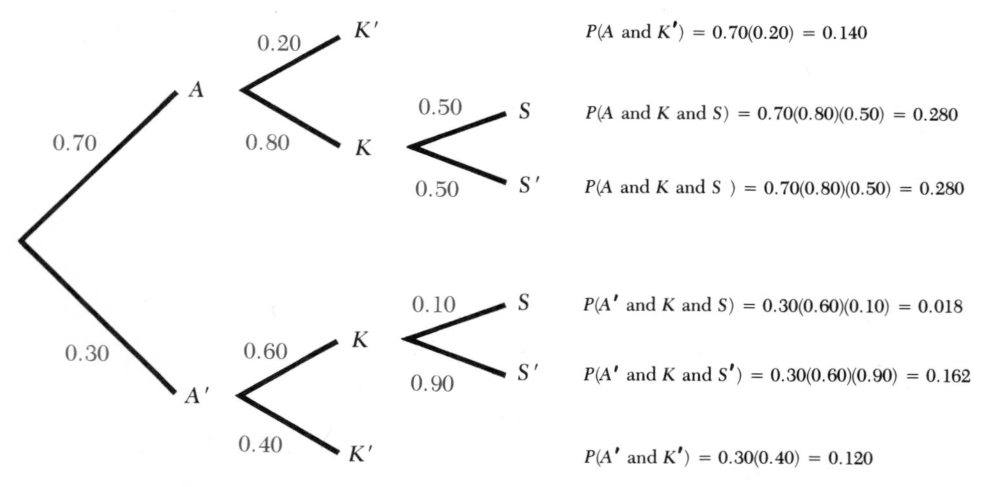

Figure 4-4 Probabilities of Mutually Exclusive Paths

the probability of a sale *not given* an appointment. With the aid of the tree diagram, we see these are

$$P(S \mid A) = \frac{P(S \text{ and } A)}{P(A)} = \frac{0.280}{0.700} = 0.40$$

$$P(S \mid A') = \frac{P(S \text{ and } A')}{P(A')} = \frac{0.018}{0.300} = 0.06$$

A call with an appointment has a greater probability of success (a sale) than a call without an appointment. In terms of making a sale, a prior appointment is worth more than six unannounced visits, and the salesperson might well question whether calling without an appointment is worth the time.

Cross-Classification Tables

We introduced joint events with a cross-classification table of outcomes for customer and referral loans. When a probability problem involves joint events, it is often useful to construct the underlying cross-classification table as a way of clarifying the problem.

Example

Suppose that 25% of the people in a community are under 21 years of age. A report on an influenza inoculation program indicates that 5% of the people were inoculated and that 60% of the inoculations were given to people under 21 years of age. What was the inoculation rate among adults?

Recognizing that two kinds of events—age status (youth and adult) and whether inoculated—are involved, we construct the cross-classification table and enter any known joint or marginal probabilities.

	Y	A	
I			0.05
N			0.95
	0.25	0.75	1.00

Sixty percent of the inoculations were given to persons under 21 years of age, that is, $P(Y \mid I) = 0.60$. Thus, for the joint event I and Y, we must have $P(I \text{ and } Y) = P(I) * P(Y \mid I) = 0.05(0.60) = 0.03$. Once this joint probability is found, the entire joint probability table can be completed, although not all the entries are needed to solve our immediate problem.

	Y	A	
I	0.03	0.02	0.05
N	0.22	0.73	0.95
	0.25	0.75	1.00

The problem asks for $P(I \mid A)$, which is $0.02/0.75 = 0.027$.

The cross-classification table is a helpful way to visualize the problem and to clarify what probabilities are given and what must be done to calculate a required probability.

Tree diagram equivalent The cross-classification problem just presented can be approached with a tree diagram. In this case the tree diagram (Figure 4-5) is used to portray the logical sequence of conditionality. Events shown later on the tree are conditional on events earlier in the tree, even though a time sequence may not be involved. In the inoculation problem the first branch on the tree is inoculated versus not inoculated because we recognize that the conditional probability given in the problem is the probability that a person is young given that he or she was inoculated, that is, 0.60.

We multiply the probabilities on the tree diagram to obtain $P(I \text{ and } Y) = 0.03$ and $P(I \text{ and } A) = 0.02$. The problem asks for $P(I \mid A)$, which is $P(I \text{ and } A)/P(A)$. We recognize that $P(A) = 1 - P(Y) = 1 - 0.25 = 0.75$. Then it follows that

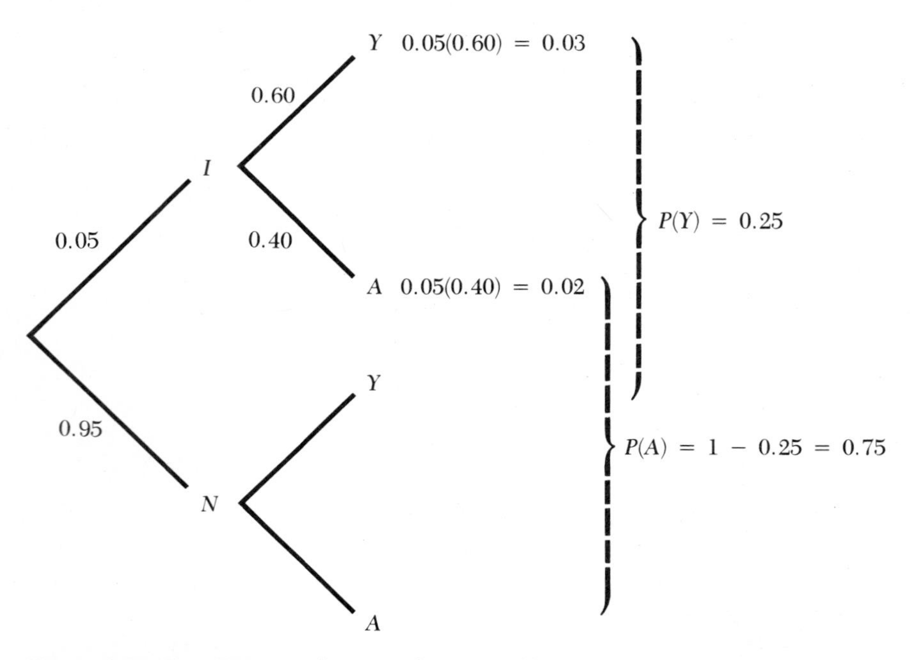

Figure 4-5 Tree Diagram for Inoculation Problem

$$P(I \mid A) = \frac{P(I \text{ and } A)}{P(A)} = \frac{0.02}{0.75} = 0.027$$

Exercises

4-24 *Job enrichment.* A personnel consultant stated that the probability that job enrichment would increase worker satisfaction was 0.75 if a worker were highly motivated and 0.40 otherwise. If only 30% of the workers in a firm are highly motivated, what is the marginal probability that job enrichment will increase a worker's satisfaction?

4-25 *Zoning and real estate.* A real estate dealer feels that the probability of selling a certain parcel is 0.60 if a desired zoning change could be obtained and 0.40 otherwise. If the probability of obtaining the desired zoning change is 0.70, what is the probability of selling the parcel?

4-26 *Drawing the duty.* To select 3 members to be responsible for setting up their sound system, the names of 6 members of the Rocky Mountain Mud Flats Band are placed in a hat and 3 names are drawn in succession without replacing the names drawn. Three of the 6 members are singers. What is the probability that exactly 2 singers will be selected?

4-27 *Insurance sales.* An insurance salesperson schedules 3 calls in an evening. If the probability of making a sale on any call is 0.20:

a. What is the probability of making 2 sales in an evening?
b. What is the probability of making at least 1 sale?

4-28 *Robbery detail.* The robbery division of a police department determined that the probability of apprehending a person holding up a convenience food outlet is 0.40 and the apprehension is independent of place and time of day. If four such robberies are committed in a day, what is the probability of apprehending all of the robbers? None of the robbers?

4-29 *Repeat sales calls.* A salesperson figures that the probability is 0.6 of making a sale on the first call, and that on each subsequent call on a prospect who has not been sold, the probability of selling drops by 0.1. If the salesperson is willing to make up to three calls on a prospect, what is the probability of making a sale?

4-30 *Sequential quality control.* A production inspector is checking quality control for two processes, A and B, where process B follows process A. The inspector knows that 11% of the production units going through the two processes are rejected after the completion of process B. The rejection rate for process A is 10%, but 80% of those units are corrected in process B. With the aid of a tree diagram, determine the rejection rate of process B for those production units that are acceptable after process A.

4-31 *Grading lumber.* In grading a certain kind of lumber, inspectors grade 10% of Number 1 boards as Number 2 and 20% of Number 2 boards as Number 1. If a lot to be graded is actually 60% Number 1 and 40% Number 2:

a. What is the marginal probability that a board will be misgraded?
b. Given that a board is misgraded, what is the probability that it is a Number 1 board?
c. Given that a board is graded number 2, what is the probability that it is really Number 1?

4-32 *Auto performance tests.* An experiment consists of subjecting a test model of a new automobile to a test for steering stability and a test for cornering ability. Let S and S' stand for passing and failing the steering test, respectively, and C and C' stand for passing and failing the cornering test, respectively. If $P(S \text{ and } C) = 0.50$, $P(S' \text{ and } C) = 0.30$, $P(S \text{ and } C') = 0.15$, and $P(S' \text{ and } C') = 0.05$, what is the probability that:

a. The car passes the steering test?
b. The car passes the cornering test?
c. The car passes at least one of the tests?

Summary

Probability is a measure of the degree of certainty that an event will occur or that a proposition is true. The probability measure, which can take on values from zero to 1.0, can be established in a particular instance by an assumption of equally likely outcomes, long-run relative frequency, or subjective degree of belief.

In dealing with compound events, we encounter *marginal, joint,* and *conditional probabilities.* The relations between marginal, joint, and conditional probabilities constitute a set of rules for calculating the probability of one event from the probabilities of other events.

In calculating probabilities for compound events, probability trees and cross-classification tables assist in distinguishing between marginal, joint, and conditional probabilities and in keeping track of the mutually exclusive ways in which an event can occur.

Supplementary Exercises

4-33 *British civil service exam.** After World War II, vacancies were filled in the administrative sections of the British civil and foreign services by a competitive examination. Some 30 years later there were 301 persons still in the services who had entered this way. The rank attained by persons who were normal entrants on the exam, borderliners (lowest acceptable) and fliers (exceptional) is shown below.

	Borderliners	*Normal*	*Fliers*	*Total*
Below Assistant Secretary	11	9	1	21
Assistant Secretary	49	45	3	97
Under Secretary	35	71	6	112
Deputy Secretary	8	30	10	48
Permanent Secretary	5	10	8	23
Total	108	165	28	301

*E. Amstey, *Journal of Occupational Psychology*, 50 (1977).

Find:

a. The probability that a flier attained the rank of at least Deputy Secretary.
b. The probability that a Permanent Secretary is a borderliner.
c. What is the most likely rank attained by a borderliner? A normal entrant? A flier?
d. What was the most likely examination status for persons who attained the rank of Permanent Secretary? Deputy Secretary? Under Secretary?

4-34 *Drilling record.** In a particular year in Oklahoma, 3105 prospective oil and/or gas wells were drilled, with the following results.

		Outcome		
Depth of well	*Oil well*	*Gas well*	*Dry hole*	*Total*
Under 2500 ft	636	215	417	1268
2500–4999 ft	347	223	438	1008
Over 5000 ft	277	252	300	829
Total	1260	690	1155	3105

a. Find the probability that a well was drilled to a depth over 5000 feet.
b. Find the probability that a well was a gas well.
c. Find the probability that a well was a gas well drilled to over 5000 feet.
d. If the outcome (type of well) was independent of depth of well, what would be the probability that a well was a gas well drilled to over 5000 feet?

4-35 *More addition laws.* In which of the following cases would you need the probability of *A and B* in order to find the probability of *A or B*?

a. *A* is the event that a job applicant is male and *B* is the event that a job applicant is a college graduate.
b. Job applicants are screened by the personnel department and sent on to operating departments only if they are not turned down by personnel. Let *A* be the event that a job applicant is turned down by the personnel department and *B* be the event that an applicant is turned down by an operating department.

4-36 *Mountain climbing.* A mountaineering expedition can use two different assault plans on a 25,000-foot peak. Under plan A, an advance base is established at 20,000 ft and the final assault is launched from there. The probability of successfully establishing the advance base is 0.90 and the probability of the final assault succeeding from there is 0.80. Under plan B, the probability of successfully establishing an advance base at 22,000 ft is 0.75, and the probability of a successful final assault from there is 0.95. Which plan has the greater probability of success?

4-37 *Sports attendance.* A class of 60 men and 40 women were polled on season attendance at varsity football and basketball games. Of 60 men, 24 had attended basketball and 10 had attended football games, while of the 40 women, 16 had been to a basketball game and 10 had seen a football game.

a. Is basketball attendance independent of sex?
b. Is football attendance independent of sex?
c. Why can't you tell if basketball attendance is independent of football attendance?

*American Petroleum Institute, 1976.

4-38 *TV market penetration.* An advertising research firm determined that 90% of the households in an area owned a television set, and that the probability that a household with TV had a set turned on between 6 P.M. and 7 P.M. on weekdays was 0.60. They estimate that a projected local news program would capture 40% of the listening audience during the 6 P.M. to 7 P.M. time slot. What is the projected market penetration, or percentage of households in the area that will be reached by the program?

4-39 *Airline punctuality.* An airline reported that 20% of its flights last year left on time and arrived late. Answer each question if 60% of all flights left on time and 50% of all flights arrived late, assuming that no flights leave early.

a. What is the probability of a late arrival given that a flight left on time?
b. What is the probability that a flight left on time given that it arrived late?
c. Is arriving late independent of leaving late? Explain.

4-40 *Equal opportunity.* In hiring persons for a certain position, 30% of qualified applicants were female. The probability that a qualified applicant was hired was 0.20.

a. If the events were independent, what is the probability that a qualified applicant was female and hired for the position?
b. How does the question of independence relate to discrimination and equal opportunity?

4-41 *Product differentiation.* A manufacturer of television sets trims sets in either black or grey and uses either a walnut or a maple stain. A consumer survey showed that 20% of consumers wanted a black trim with a walnut stain and that half of all consumers wanted a black trim or a walnut stain. If it has already been decided to trim 30% of the sets in black, what proportion of the sets should be given a walnut stain?

4-42 *Auto model sales.* A survey of automobile owners revealed that 50% had most recently purchased compact models. Among the purchasers of compact models, 60% had purchased compact models previously, and among noncompact buyers, 80% had purchased a noncompact previously.

a. What is the probability that an owner who had previously purchased a noncompact most recently purchased a noncompact?
b. What is the probability that a previous owner of a compact most recently purchased a noncompact?
c. What percent of the buyers' previous purchases were compacts?

4-43 *False signals.* A procedure has been designed that will detect 99% of the defective assemblies in a manufacturing process. However, the probability that the procedure will declare a good assembly defective is 0.05. If 2% of all assemblies are defective, what percentage of the assemblies signaled as defective by the process are really good?

4-44 *Competitive timber bids.* Deer Creek Timber Company is considering two upcoming competitive sales on standing timber. The management feels that the probability that its major competitor, Clearlake Lumber Company, will bid on the first sale is 0.60 and that the probability of Clearlake being the low bidder is 0.40. If Clearlake does not bid on the first sale or bids and is not the low bidder, Deer Creek assesses the probability of Clearlake bidding on the second sale at 0.80 and the probability of Clearlake being the low bidder at 0.50. On the other hand, if Clearlake wins the first sale the probability of Clearlake bidding on the second sale is assessed at 0.40 and the probability of Clearlake being low bidder is assessed at 0.30.

a. Construct a tree diagram for the sequence of events involving Clearlake.

b. What is the probability that Clearlake bids on both sales?
c. What is the probability that Clearlake bids on the second sale?
d. What is the probability that Clearlake wins the second sale?

4-45 *Embezzling funds.* A bank vice-president declared that the probability that the bank's funds were being embezzled was 0.70 and that the probability that an embezzlement would be discovered by the bank examiner was 0.90. Given that the bank examiner does not find the embezzlement and that the examiner would not discover embezzlement when none existed, what is the probability that the bank's funds are not being embezzled?

4-46 *Darwin and the seeds.** Charles Darwin once conducted an experiment in which he found that 64 out of 87 varieties of seed germinated after immersion in sea water for 28 days, and that, after being dried, 18 out of 94 distinct species with ripe fruit floated for 28 days or more. Given independence between floating and germinating capacity, what is the probability that a species would survive a sea voyage of 28 days and retain its power of germination (also noted by Darwin)?

4-47 *Demand for services.*† In a large survey of consumers, 30% indicated they would not carry their own bags onto an airliner, and 21% said they would not use a travelers' check machine at a bank. The percentage of consumers who would neither carry their own bags nor use the travelers' check machine was 9.5. Is the demand for baggage service independent of the demand for personal service in obtaining travelers' checks? Explain.

4-48 *Casino games.*‡ A large survey of Las Vegas casino players showed that the probability that a player who preferred slots was a repeat visitor was 0.65, and the probability that a player who preferred blackjack was a repeat visitor was 0.76. The probability that a player preferred slots was 0.39, and the probability that a player preferred blackjack was 0.34.
a. Find the probability that a player preferred slots and was a repeat visitor.
b. Find the probability that a player preferred blackjack and was a repeat visitor.
c. It is known that the marginal probability of a repeat visitor is 0.72. Find the probability that a repeat visitor preferred neither slots nor blackjack.

4-49 *Baseball fan.* A San Francisco Bay Area baseball fan's subjective probability for attending a San Francisco Giants game in the next month is 0.50, the fan's probability for attending an Oakland A's game is 0.30, and the probability for attending both is 0.20.
a. What is the probability for attending a Giants game, given that an A's game is attended?
b. What probability would you assess that the fan attended an A's game if you know that he or she attended a Giants game?

4-50 *Industrial accidents.* Two investigators examine the situation surrounding an accident resulting in hospitalization of a worker.

*E. Rubin, *The American Statistician*, October 1972.
†E. Langeard, J. E. G. Bateson, C. H. Lovelock, and P. E. Eiglier, *Marketing Science Institute*, 1981.
‡Lawrence Dandurand, *Journal of Travel Research*, XX, No. 4 (Spring, 1982).

a. Investigator Black decides that the probability that operator error was a contributing cause is 0.70, and the probability that inadequate lighting was a contributing cause was 0.60. If these events are independent, what is the probability that neither event was a contributing cause of the accident?
b. Investigator White declares that the probability that operator error caused the accident is 0.60, the probability that inadequate workplace conditions caused the accident is 0.30, and that these causes are mutually exclusive. What is the probability that neither of the conditions caused the accident?

4-51 *Home delivery route.* Sweet Home Dairy Products delivers milk to 40% of its customers daily. Thirty percent of the customers who take milk also take eggs. If 36% of all customers take eggs each day, what percentage of customers take neither eggs nor milk?

Computer Exercises

1 Source: Real estate data file
a. Obtain a cross-tabulation of whether the buyer is a new or old resident (in the community) versus whether the purchased home was new or old. For a randomly selected sale from these 100 home sales:
   i. What is the probability that the purchaser is a new resident of the community?
   ii. What is the probability that the home sold is a new home?
   iii. What is the probability that the sale is a new home to a new resident of the community?
   iv. What is the probability that the sale is a new home or a sale to a new resident of the community?
b. From the cross-tabulation obtained in (a), find:
   i. The marginal probability that a sale is to a new resident.
   ii. The probability that a sale is to a new resident, given that the sale is a new home.
c. From the probabilities in (a) and (b), show whether residence of purchaser is independent of new and old home sales.

2 Source: Personnel data file
a. Make a cross tabulation of management level versus degree type.
b. Use the results of (1) to answer these questions. All are based on the experiment: *select a personnel record at random.*
   i. What is the probability of selecting a person who has a graduate degree?
   ii. What is the probability of selecting a person who is at level 3?
   iii. What is the probability of selecting a person who has a graduate degree and is at level 3?
   iv. What is the probability of selecting a person who has a graduate degree or is at level 3?
   v. The person selected at random has a graduate degree. What is the probability that this person is at level 3?

vi. The person selected at random is at level 3. What is the probability that this person has a graduate degree?

c. Use the results from 2(a).

i. Draw relative frequency diagrams for each level and each degree type (six diagrams).

ii. Are management level and degree type independent?

5

DISCRETE PROBABILITY DISTRIBUTIONS

In Chapter 4 we discussed the probability of an event and relations between probabilities that allow us to find the probability of one event from probabilities for other events. In this chapter we concentrate on numerical values of a variable. We saw many such variables in the earlier chapters on descriptive statistics: Family size, ages of executives, and weights of bags of flour are some examples. Summarizing such events is the task of descriptive statistics. But the usual statistical investigation is based on sample data and is not complete with only a description of the sample data.

Inferences must be drawn about the underlying statistical population. For example, suppose a random sample of households in a community shows there are 1.8 television sets per household. What inferences are justified about television ownership in the entire community? In this chapter we shall develop some critical links between probability and statistical inference. The first link is the concept of a random variable, which we shall now discuss.

5.1 Random Variables

Consider the following investigations.

1. The advertising department of a city newspaper is concerned with how many people read each subscription copy of the Sunday paper because duplicate use is of interest to potential advertisers.
2. A public opinion polling firm wants to know if people think the president is doing a good job.
3. A sporting goods manufacturer wants to know the breaking strength of a new model of tennis racket.

In the first investigation, as with all statistical studies, there is an underlying population of units of observation. This population is all subscriptions to the Sunday paper. The characteristic of these units that is of concern to the investigator is the variable; in this case, it is the number of people reading each subscription copy. Suppose the advertising department decides to determine the number of people reading *last* Sunday's paper. Putting the units of observation and the variable of interest together, we have the following.

1. A list (or other representation) of all the units of observation, that is, subscriptions to the Sunday paper.
2. Attached to each subscription copy (unit of observation) is a value of the variable of interest, that is, the number of people who read it.

Suppose we select one subscription copy at random from the population of copies. Since we are uncertain which subscription will be selected, we do not know what value of the variable will be observed. Thus the numerical value, the number of people reading a (selected) subscription copy of last Sunday's newspaper, is an uncertain event.

> A variable whose numerical values are uncertain events is called a *random variable.*

Figure 5-1 shows the concept of a random variable. The subscribers are portrayed as elements, or points. Associated with each point is a value of the variable. For illustrative purposes, the figure just shows two of these—a subscription copy of the Sunday newspaper read by two persons and another copy read by four persons. Because it is uncertain what value of the variable will be observed, the variable in question is a *random variable.*

Mathematicians prefer to say that a *random variable* is a numerical function defined on the elements in a sample space. The elements in a sample space represent the possible outcomes of an experiment. In the example here the elements in the sample space are the subscribers and the experiment is selecting a subscription at random.

In Chapter 2, we dealt with two kinds of variables: discrete variables and continuous variables. The random variable in the newspaper subscription example is a *discrete* random variable. In the second investigation given at the beginning of this section, random selection of a citizen would produce a discrete random variable if we associated a zero with every person who thinks the president is not doing a good job and a one with every person who thinks the president is doing a good job. In the third example, the breaking strength of a racket randomly selected from a pilot production lot would be an uncertain quantity, in this case a *continuous* random variable.

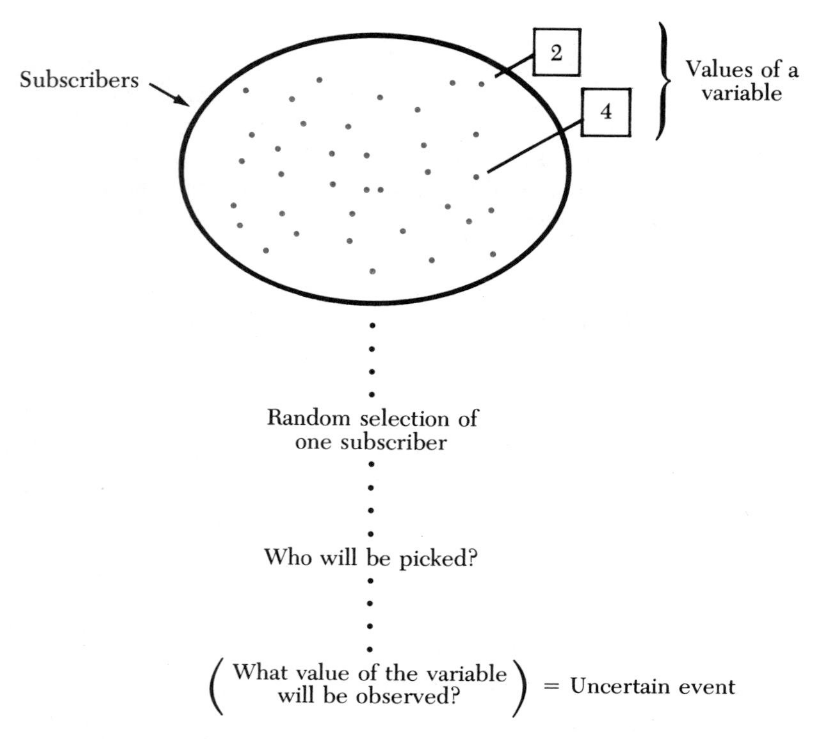

Figure 5-1 Random Variable

For either discrete or continuous variables, statisticians use uppercase letters to refer to the random variable in a general way and lowercase letters to refer to particular values of the random variable. In the newspaper example, we might write

$$X\colon x = 0,1,2,\ldots,k$$

and mean that the random variable X takes on discrete integer values (x) running from 0 to k. In the opinion polling example, the random variable (also discrete) is

$$X\colon x = 0,1$$

The third example, concerned with tennis racket breaking strengths, involves a continuous random variable. It could be expressed as

$$X\colon 0 \le x \le 400$$

which says that X is a continuous random variable that can take on positive values between zero and 400 (pounds breaking strength).

In this chapter we shall deal with discrete random variables, while the next chapter is concerned with continuous random variables. We have already discussed descriptive statistics for discrete and continuous variables. We shall now put these ideas into a probability context because the variables of concern are *random* variables.

5.2 Discrete Random Variables

A store that sells clock radios has advertised that each day during a special promotion, one purchaser of a radio will get the purchased radio free. The winner will be determined by a random drawing from the sales slips for the radios sold each day. During the first day of the promotion, the following 12 sales were made.

Slip 1	FM Regular, $42	Slip 7	FM Regular, $42
Slip 2	AM Regular, $36	Slip 8	AM Digital, $36
Slip 3	FM Regular, $42	Slip 9	AM Regular, $36
Slip 4	AM Regular, $36	Slip 10	FM Regular, $42
Slip 5	AM Regular, $36	Slip 11	AM Digital, $36
Slip 6	AM Digital, $36	Slip 12	FM Digital, $48

We are concerned with the cost of the promotion to the store, which is the amount paid by the winning customer. On the first day, the 12 sales shown are the possible units, from which one will be drawn at random. We can associate two numbers with each of these units. The first number is the probability that the sales slip is selected and the second is a value of the random variable.

Table 5-1 shows these numbers. We have reorganized the sales slips (units of observation) from the serial order in which they were listed earlier into an order based on the amount of the sale. The third column shows the selection probabilities, which were assigned on the basis of equally likely outcomes. Thus each unit of observation (sales slip) has the same probability ($^1/_{12}$) of selection. The fourth column

Table 5-1 Probability Distribution of a Random Variable

Unit of Observation	*Model*	*Selection Probability*	*Value of Random Variable*	x	$P(x)$
Slip 2	AM reg.	1/12	36		
Slip 4	AM reg.	1/12	36		
Slip 5	AM reg.	1/12	36		
Slip 9	AM reg.	1/12	36	36	7/12
Slip 6	AM dig.	1/12	36		
Slip 8	AM dig.	1/12	36		
Slip 11	AM dig.	1/12	36		
Slip 1	FM reg.	1/12	42		
Slip 3	FM reg.	1/12	42		
Slip 7	FM reg.	1/12	42	42	4/12
Slip 10	FM reg.	1/12	42		
Slip 12	FM dig.	1/12	48	48	1/12

shows the second number assigned to each slip, the price paid for the radio. It happens that the two different AM models sold for \$36, the four FM regular radios each sold for \$42, and the FM digital radios sold for \$48.

The final two columns show the possible values of the random variable and associate with each possible value its total probability. The probability that the promotion will cost the store \$36 on the first day is 7/12 because 7 of the 12 radios sold were priced at that figure. The fact that some of those radios were AM regular models and others were AM digital models may be important to the inventory records, but it does not directly affect the random variable of interest.

Probability Distribution

In summarizing a random variable, we generally show only the last two columns of Table 5-1. In our example, we would therefore show only the following.

x	$P(x)$
36	7/12
42	4/12
48	1/12

This list of values of the random variable along with the probabilities of these values is called the *probability distribution* of the random variable. Although the terminology is different, the distribution looks like a relative frequency distribution in Chapter 2. It is the same table that we would obtain if we were to construct the relative frequency distribution of prices paid for the 12 radios. The difference is in the random draw underlying the development that we have traced. This is where uncertainty is introduced and why we speak about a probability distribution rather than just a relative frequency distribution.

Just as we portrayed distributions graphically in Chapter 2, we can graph a random variable in terms of its probability distribution or its cumulative probability distribution. The first of these is analogous to the relative frequency distribution of Chapter 2, as we stated. The cumulative probability distribution is, in turn, analogous to the cumulative relative frequency distribution of Chapter 2.

The graphs for the promotion-cost random variable are shown in Figure 5-2. The *probability distribution* is the relation

$$P(X = x)$$

where X stands for the random variable and x stands for a particular value of the random variable. In our example, we have

$$P(X = 36) = 7/12$$

$$P(X = 42) = 4/12$$

$$P(X = 48) = 1/12$$

The *cumulative probability distribution* of the random variable is the relation

$$P(X \leq x)$$

where X is the random variable and x is a particular numerical value of the random variable; $P(X \leq x)$ stands for the probability that the random variable takes on a

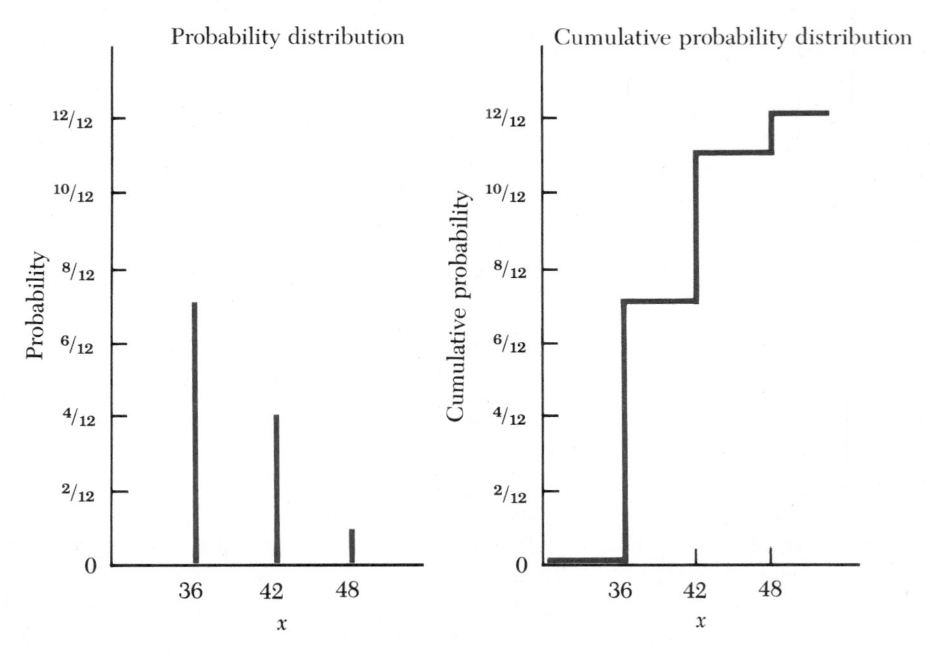

Figure 5-2 Graphs of a Random Variable

value *equal to or less than* x. In the promotion-cost example, the probability that the first day's promotion will cost \$36 or less is $7/12$, and the probability that the cost will be \$42 or less is $7/12 + 4/12 = 11/12$. It is a certainty that the cost will be \$48 or less. Thus

$$P(X \leq 36) = 7/12$$

$$P(X \leq 42) = 11/12$$

$$P(X \leq 48) = 12/12$$

The Sum of Two Random Variables

A random variable of interest is often the sum of two other random variables. Suppose the operator of a campground made an analysis of the extra days stayed by 200 groups that originally made reservations for specified lengths of stay. The following table shows the results of such an analysis.

Extra Days of Stay Classed by Days Originally Reserved by 200 Groups

Days Reserved	*Extra Days* 0	1	2	*Total*
1	36	72	12	120
2	30	18	12	60
3	14	4	2	20
Total	80	94	26	200

This is a frequency table for the joint events *days reserved* and *extra days*. If a party is selected at random, there is a probability distribution for days reserved, which comes from the marginal totals at the right. The probability distribution for extra days comes from the marginal totals at the bottom of the table. If we call days reserved X_R and extra days X_E, these probability distributions are as follows.

x_R	$P(x_R)$	x_E	$P(x_E)$
1	$120/200 = 0.60$	0	$80/200 = 0.40$
2	$60/200 = 0.30$	1	$94/200 = 0.47$
3	$20/200 = 0.10$	2	$26/200 = 0.13$
	1.00		1.00

Total length of stay is the sum of days reserved plus extra days. If we call total length of stay Y, then $Y = X_R + X_E$. The probability distribution of total length of stay can be obtained by listing the values of that random variable and finding the probability for each total length of stay from the cross-classification table. For example, a total stay of 2 days can result from 2 reserved and no extra days or 1 reserved and 1 extra day. The probability for $y = 2$ is, therefore, $(30 + 72)/200 = 0.51$. The probability distribution of total length of stay is given below.

y	$P(y)$
1	$\frac{36}{200} = 0.18$
2	$\frac{30 + 72}{200} = 0.51$
3	$\frac{14 + 18 + 12}{200} = 0.22$
4	$\frac{4 + 12}{200} = 0.08$
5	$\frac{2}{200} = 0.01$
	1.00

In some cases the joint events on which a random variable is defined are independent. When this is the case, the joint probabilities for the several events can be obtained from the products of marginal probabilities, and the probability distribution of the random variable can then be found from these joint probabilities.

Suppose a trucker owns two rigs, which are constantly on the road. The probabilities of weekly breakdowns for the two rigs are as follows.

Breakdowns	*Rig A*	*Rig B*
0	0.70	0.90
1	0.25	0.10
2	0.05	—
	1.00	1.00

If breakdowns of Rig A and Rig B are independent, then the probabilities of joint numbers of breakdowns can be found from products of the marginal probabilities just given. For example, the probability that Rig A has 1 breakdown and Rig B has 1 breakdown is $0.25(0.10) = 0.025$. This would lead to 2 total breakdowns, but so would 2 breakdowns for Rig A and no breakdowns for Rig B, which has a probability of $0.05(0.90) = 0.045$. The probability of a total of 2 breakdowns for the rigs owned by the trucker is, therefore, $0.025 + 0.045 = 0.070$. Introducing the symbols x_A and x_B for the number of breakdowns occuring to Rig A and Rig B, respectively, it is convenient to find the entire set of joint probabilities in the manner below.

	Rig A	*Rig B*		
x_A	$P(x_A)$	*0*	*1*	x_B
		0.90	0.10	$P(x_B)$
0	0.70	0.630	0.070	
1	0.25	0.225	0.025	
2	0.05	0.045	0.005	

Using this table, the probability distribution of Y, total breakdowns for the two rigs, can be found, as shown below.

Total Breakdowns $y = x_A + x_B$	*Probability* $P(y)$
0	0.630
1	$0.225 + 0.070 = 0.295$
2	$0.045 + 0.025 = 0.070$
3	0.005

Exercises

5-1 *Medical study.* A physician plans to sample hospital records of births in a large city in the past year to determine the weights and order of birth (first, second, third, . . . , child) of newborn infants.
a. What are the units of observation?
b. What are the random variables?
c. Are the random variables discrete or continuous?

5-2 *Weather statistics.* Ten weather stations reported minimum daily temperatures and the occurrence of overnight frost for a period of 14 days critical to the success of the peach crop in an area. No temperatures were reported below 25°F or above 38°F. If a station is selected at random to observe if frost occurred on the eighth day, the resulting random variable is X: $x = 0, 1$. Use symbols to define the random variables arising from each situation below.
a. A day is selected at random to observe the minimum temperature at the fifth station.
b. A station is selected at random to observe the number of days on which overnight frost occurred.
c. A day is selected at random to observe how many stations reported minimum temperatures below 32°F.

5-3 *Gasoline shortage indicators.* What are the units of observation connected with each of the following random variables?
a. The amount of fuel in an automobile gasoline tank in Los Angeles.

b. The number of states with an end-of-month shortage in supplies.
c. The number of days last month that a service station closed for lack of supply.

5-4 *Motel vacancies.* During a 20-week tourist season the number of vacancies at 6 P.M. on weekdays at the Canyonlands Motel was as follows.

Number of vacancies	0	1	2	3	4	5
Number of days	25	15	20	18	12	10

a. Find the relative frequency distribution of vacancies.
b. If two parties arrived after 6 P.M. on a weekday, what is the probability that they were accommodated?
c. How many parties could arrive after 6 P.M. before the probability is less than 0.25 that they could all be accommodated?

5-5 *Law partners.* The number of partners in law firms in a city has the following probability distribution.

Number of partners	2	3	4	5
Probability	0.40	0.30	0.20	0.10

a. Graph the distribution and the cumulative probability distribution of the random variable.
b. Find: $P(X \leq 3)$; $P(X \geq 3)$; $P(X < 3)$; $P(X > 3)$.

5-6 *Requests for service.* For nearly two years, Ace Landscaping Service has been managing the grounds for two industrial clients. The contracts call for maintenance services and additional services on request. Based on 100 weeks of experience, the following joint probability distribution of weekly requests from Clients A and B was formed.

Requests from A	*Requests from B* 0	1	2
0	0.20	0.10	0.05
1	0.05	0.20	0.05
2	0.05	0.10	0.20

a. Find the probability distribution of total requests.
b. Graph the distribution and the cumulative probability distribution of total requests.

5-7 *Replacement demand.* Weekly demand for replacement part 062 at a supply facility has been found to be equally likely for zero, one, and two demands, with zero probability for demand greater than two. Use a tree diagram to find the probability distribution of total demand in a two-week period, assuming demands from week to week are independent.

5-8 *Closing real estate deals.* A real estate agent found that the probability of closing no deals on Saturday was 0.6 and the probability of closing one and two deals was 0.2 each. On Sunday the probability of closing no deals was 0.5 and the probability of closing one deal was 0.5. Find the probability distribution of the number of deals closed in a weekend, assuming the outcomes for Saturday and Sunday are independent.

5-9 *Store promotion.* Suppose the store described in Section 5.2 decides to give a second prize by making a second draw from the sales slips remaining after the first draw. Find the probability distribution of the total cost of the promotion to the store.

5.3 Probability Distribution Measures

We have seen that probability distributions are a special kind of frequency distribution. Just as a frequency distribution gives the frequencies of values of a variable, a probability distribution gives the probabilities of values of a random variable. For this reason we should not be surprised to find that measures for probability distributions parallel those we encountered earlier as measures for frequency distributions. Central location and variability are the primary characteristics of the distribution of a random variable, and the mean and the standard deviation are the most important measures.

Mean

Table 5-2 shows the probability distribution for the store promotional drawing discussed earlier.

Table 5-2 Finding the Mean of a Discrete Random Variable

Probability Distribution x	$P(x)$	*Calculations for Mean* $x * P(x)$
36	$7/12$	$252/12$
42	$4/12$	$168/12$
48	$1/12$	$48/12$
		$\mu = 468/12 = 39$

We see that the mean is the sum of the products of values of the random variable weighted by their probabilities. In symbols,

$$\mu = \sum[x * P(x)] \tag{5-1}$$

In Table 5-2 we found μ as follows.

$$\mu = \sum[x * P(x)]$$
$$= \frac{468}{12} = 39$$

In this example we established probabilities on the basis of equally likely events. Notice that had we been working with the frequency distribution of sales slip dollar amounts, we would have used the technique of Chapter 3 to calculate the mean as:

x	f	fx
36	7	252
42	4	168
48	1	48
	12	468

$$\mu = \frac{\Sigma fx}{N} = \frac{468}{12}$$

$$= 39$$

Here the frequencies are *weights* for the values of the variable, and the mean is the sum of the products of values of the variable weighted by their frequencies after this sum is divided by the sum of the weights, which is the total of the frequencies. Probabilities are like relative frequencies; the sum of the probabilities is also 1.0. We get the same mean from weighting by frequencies and dividing the weighted sum by N as we do from weighting by probability (f/N).

Variance and Standard Deviation

Table 5-3 shows the probability distribution of costs for the first day of the radio promotion again, along with the calculations required for the variance and standard deviation.

Table 5-3 Finding the Variance and Standard Deviation of a Discrete Random Variable

Probability Distribution		*Calculations for Variance*	
x	$P(x)$	$(x - \mu)^2$	$(x - \mu)^2 * P(x)$
36	$^7/_{12}$	9	$^{63}/_{12}$
42	$^4/_{12}$	9	$^{36}/_{12}$
48	$^1/_{12}$	81	$^{81}/_{12}$
			$\sigma^2 = {}^{180}/_{12} = 15$

$$\sigma^2 = 15$$

$$\sigma = \sqrt{15} = 3.87$$

Again, the similarity to calculations from a frequency distribution with single-value classes should be apparent. The variance is the average of the squared deviations, which is a weighted average of the square of the deviation of each value of the random variable from the mean. The weights are now probabilities instead of frequencies. Since the sum of these weights is 1.0, there need be no division to obtain the average squared deviation. The formula is

$$\sigma^2 = \sum[(x - \mu)^2 * P(x)] \quad (5\text{-}2)$$

The standard deviation of a random variable is the square root of the variance.

$$\sigma = \sqrt{\sigma^2}$$

Expectation Notation

Expectation notation is often used in connection with random variables. The mean of a random variable is called its *expectation*; it is written $E(X)$, that is,

$$E(X) = \mu$$

To mathematicians, E is an "operator" just as the summation sign, Σ, is an operator. The summation sign indicates that we should sum the values that follow. For example, Σx is a direction to sum the values of the variable x. The notation $E(X)$ directs us to take the expected value of X, defined as $\Sigma[x * P(x)]$; this is, of course, the mean of X.

We have emphasized here and in Chapter 3 that population variance is the average squared deviation of a population from its mean. For a probability distribution, the variance is the *expected value* of the squared deviation, and can be written as

$$E[(X - \mu)^2] = \sum[(x - \mu)^2 * \mathrm{P}(x)] = \sigma^2$$

The variance of a random variable can also be expressed as the expected value of the squares of the variable minus the square of the expected value. That is,

$$\sigma^2 = E(X^2) - [E(X)]^2 \quad (5\text{-}3)$$

For our example in Table 5-3, where $E(X) = \mu = 39$, the variance can be obtained as follows.

x	$P(x)$	x^2	$x^2 * P(x)$
36	7/12	1296	756
42	4/12	1764	588
48	1/12	2304	192
			1536

$$E(X^2) = \sum[x^2 * P(x)] = 1536$$

$$\begin{aligned}\sigma^2 &= E(X^2) - [E(X)]^2\\ &= 1536 - (39)^2\\ &= 15\end{aligned}$$

Interpretation of Measures

The expected value and standard deviation of a random variable are descriptive measures of location and dispersion of the statistical populations represented. In the case of the clock radio promotional drawing, the population consisted of the 12 prices paid for clock radios by the day's customers.

When the random variable stems from repeated trials, the expected value can be interpreted as the long-run average for a large number of repetitions. For example, the annual risk of accidental death among males aged 20 to 24 in the United States is about 100 per 100,000 persons, which gives a probability of accidental death of 0.001. An insurance company proposes to market a \$20,000 accidental death policy for an annual premium of \$30. As the number of policies sold (repetitions) increases, the average return per policy sold can be expected to get closer and closer to:

$$\begin{aligned} E(X) &= 0.001(-\$19{,}970) + 0.999(\$30) \\ &= \quad -\$19.97 \quad + \quad \$29.97 \\ &= \$10.00 \end{aligned}$$

If it sells a large number of policies, the company can expect to have \$10 per policy available to meet the costs of doing business and to provide a profit or dividend.

On each policy written, the insurance company is taking a gamble on losing \$19,970 or gaining \$30. As we have seen, this gamble has a positive expected value. In other situations, gambles may not seem quite as repetitive or consistent as the insurance example, but expected value and standard deviation can still be used as guides. Suppose a small business investment company is considering investing in three ventures. The returns associated with failure and success of the ventures (in thousands of dollars) along with the probabilities for failure and success are as follows.

	Venture A		*Venture B*		*Venture C*	
Outcome	x	$P(x)$	x	$P(x)$	x	$P(x)$
Failure	−40	0.50	−150	0.20	−50	0.80
Success	60	0.50	50	0.80	250	0.20

The expected return for each of these ventures is 10 (thousand dollars). While each venture may be unique or not subject to repetition, the expectation can be used as a guide for action—in this case, a decision to invest in the ventures. The standard deviations of the random variables (or gambles) are quite different. The gamble on Venture A has a standard deviation of 50 (thousand dollars), while in Ventures B and C, the standard deviations are 80 and 120 (thousand dollars), respectively. Observe that in Venture A, the possible gain or loss is moderate, in Venture B there is a substantial possibility of a large loss, and in Venture C there is a fair chance of a very large gain. These last two ventures are riskier than Venture A, although in different ways. The standard deviations, as measures of dispersion of the random variables, reflect this difference in risk. The larger the standard deviation, the greater the risk.

Mini Case

In a promotion for the soft drink Sprite, The Coca-Cola Bottling Company of Chicago printed cash amounts on the caps of some of its 16-ounce returnable bottles of Sprite. In order to receive the cash amount, the consumer would peel back the cap liner and return the cap to the store. According to Coca-Cola's advertisement, the amounts which could be won and the chances of winning were:

Amount	*Chances of Winning*
$0.10	1 out of 4
0.25	1 out of 40
1.00	1 out of 400
10.00	1 out of 4000

Use this information to find the average cost per bottle for this promotion.

Exercises

5-10 *Mean and variance.*

a. Find the expected number of vacancies and the expected number of partners in Exercises 5-4 and 5-5.

b. Using Equation 5-2, find the variance of the probability distributions in Exercises 5-4 and 5-5.

c. Using Equation 5-3, find the variances in (b).

5-11 *Price-earnings ratios.* The probability distribution of price-earnings ratios of a large number of common stocks available through a certain exchange is approximated by the following table.

Price-earnings ratio	6	7	8	9	10
Probability	0.10	0.20	0.30	0.30	0.10

a. If stock is selected at random, what is the expected price-earnings ratio?

b. Give an interpretation of the expected value as a long-run average.

c. What is the probability of a price-earnings ratio of 9 or more? Give an interpretation of this probability as a long-run relative frequency.

5-12 *Truck-rig breakdowns.* Use the probability distribution of total weekly breakdowns for two trucking rigs (on page 127).

a. Find the expected number of total weekly breakdowns.

b. Interpret the expected value as a long-run average.

5-13 *Business ventures.* An investor is considering investing in both Venture A and Venture B (see page 133). Assuming independence, find the probability distribution of total return. Find the expected value and variance of total return.

5-14 *Expected values.* Interpret the following expected values as long-run averages.

a. The expected size of any party for dinner at the restaurant La Boheme is 2.8.
b. The expected number of defectives turned out per 100 stampings in the Ace Machine Shop is 0.7.
c. The expected number of days of sunshine at weather station Zebra in April is 3.0.

5-15 *Bonus clauses.* An agent for a professional baseball player finds that management will offer a bonus clause giving the player an extra $200,000 if attendance at home games exceeds 2 million for the season. Alternatively, management will offer the player a bonus of $50,000 if he hits 30 home runs or more in the season. The agent and the player agree that the probability of home attendance exceeding 2 million is 0.10 and that the probability of the player hitting at least 30 home runs is 0.30.

a. Find the expected value and the standard deviation of each of the propositions.
b. Which is the riskier proposition? Explain.
c. Which bonus clause would you recommend? Why?

5-16 *Construction bids.* A construction firm has decided to bid on a contract. If they bid so that the profit to the firm is $100,000, the probability of winning the contract is assessed at 0.20. On the other hand, shaving the profit to $50,000 would raise the probability of winning the contract to 0.50. If the firm does not win the contract the return is zero.

a. Find the expected value of the alternative bids.
b. Find the standard deviations of each of the money gambles.
c. What do the means and standard deviations tell about the alternatives?

5.4 Bernoulli Process and Binomial Distribution

Suppose we stand on a bridge over a highway in the country and observe the automobiles traveling in one direction. We record a 1 for each compact auto and a 0 for each noncompact auto. It is likely that we are involved in a process in which the probability of observing a compact auto remains constant throughout the traffic stream and in which observing that one auto is compact does not affect the probability that the next (or any other) auto is compact. Such a process is called a *Bernoulli process* after the Swiss mathematician James Bernoulli (1654–1705). In general, the following hold in a Bernoulli process.

1. The random variable is X: $x = 0, 1$. Often, the event for which $x = 1$ is called a *success* and the event for which $x = 0$ is called a *failure*.
2. The probability of a success, $P(X = 1)$, is constant over all trials.
3. The outcomes of trials are independent of one another.

When a process meets these requirements, the observations are sometimes called *Bernoulli trials*. Thus if we observed whether or not 40 automobiles were compact models, we could say that we conducted 40 Bernoulli trials if the situation met the second and third requirements above. Some other examples likely to qualify as Bernoulli trials are:

defective versus good products in a stream of output;

female versus male babies in a series of births at a clinic;

decisions to grant versus not to grant credit by a bank officer.

In the example of defective products, we would need a production process that had reached a "steady state" condition. If the stream of output included items produced while workers were learning their jobs, the probability of a defective item might not remain constant but would tend to decrease. In the example involving credit, a bank officer could make a series of decisions conforming to the requirement of constant probability but violating the condition of independence. Suppose, after a decision to grant credit, the officer applied stricter requirements to the next applicant and, after having disapproved an application, relaxed the requirements. The marginal probability of granting credit could remain constant, but the probability of granting credit would depend on the preceding decision. It is hard to imagine the sex of the babies in a series of births not constituting Bernoulli trials.

A Bernoulli process is also generated when repeated random draws are made from a population of 0's and 1's. Imagine a bin containing a large number of fuses, 10% of which are defective. The repeated trials constitute successive selections of a fuse from the bin and observing whether the fuse is good ($x = 1$) or defective ($x = 0$). If the successive draws, or trials, are made in such a way that the probability of selecting a good fuse remains constant over all trials, and if the outcomes of the trials are independent of one another, then the draws constitute trials of a Bernoulli process.

Bernoulli Random Variables

Bernoulli random variables are so common that a special set of symbols is used in referring to them. The probability distribution of a Bernoulli random variable is:

x	$P(x)$
0	$1 - \pi$
1	π

The Greek letter π is used for $P(X = 1)$, the probability of a success. The probability of a failure is, then, $1 - \pi$.

Suppose in our example involving compact autos that $\pi = 0.25$, the probability of observing a compact auto. The following table shows the calculation of the mean, or expected value, of the Bernoulli random variable.

x	$P(x)$	$x * P(x)$	$P(x)$	$x * P(x)$
0	0.75	0	$1 - \pi$	0
1	0.25	0.25	π	π
$\mu = \sum[x * P(x)] =$		0.25		$\mu = \pi$

Here we see that $\mu = 0.25$, and in the right-hand panel of the table we can see that μ will always equal π for a Bernoulli random variable.

In Table 5-4 the variance and standard deviation are calculated for our example. The right side of the table shows the calculation in terms of π, up to the point of weighting squared deviations by probabilities.

Table 5-4 Variance and Standard Deviation of a Bernoulli Random Variable

x	$P(x)$	$(x - \mu)^2$	$(x - \mu)^2 * P(x)$	x	$P(x)$	$(x - \mu)^2$	$(x - \mu)^2 * P(x)$
0	0.75	0.0625	0.046875	0	$1 - \pi$	π^2	$\pi^2(1 - \pi)$
1	0.25	0.5625	0.140625	1	π	$(1 - \pi)^2$	$(1 - \pi)^2\pi$
		$\sigma^2 = \sum[(x - \mu)^2 * P(x)]$					
		$= 0.1875$					
		$\sigma = \sqrt{0.1875} = 0.433$					

The variance is the sum of the weighted squared deviations in the right-hand column of Table 5-4. This sum is $\pi^2(1 - \pi) + (1 - \pi)^2\pi$. Factoring it yields $[\pi(1 - \pi)][\pi + (1 - \pi)] = \pi(1 - \pi)$. Knowing this, we could have found the variance originally from $\pi(1 - \pi)$. The standard deviation is, of course, the square root of the variance: $\sigma = \sqrt{\pi(1 - \pi)}$.

We now have the general result that for a Bernoulli random variable, the mean and standard deviation are

$$\mu = \pi \tag{5-4}$$

$$\sigma = \sqrt{\pi(1 - \pi)} \tag{5-5}$$

Figure 5-3 shows the probability distributions for three different Bernoulli random variables. Note that when π is less than 0.5, the distribution is skewed to the right; when π is greater than 0.5, the distribution is skewed to the left. Another feature that deserves emphasis is that π appears on both the vertical (probability) axis and the horizontal (variable) axis. This is because π is, in the first place, a probability; that is, $P(X = 1) = \pi$. In the second place, π is the expected value of the random variable; that is, $E(X) = \mu = \pi$.

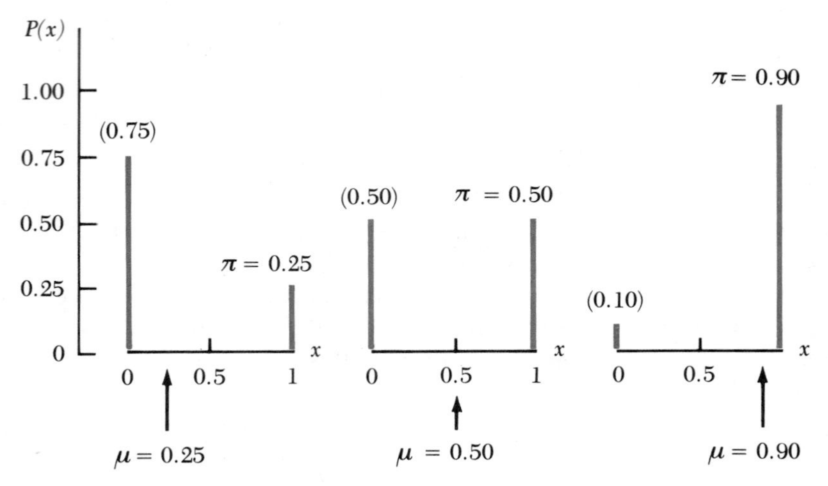

Figure 5-3 Three Bernoulli Distributions

The standard deviations of the three Bernoulli random variables are:

$\pi = 0.25$	$\pi = 0.50$	$\pi = 0.90$
$\sigma = \sqrt{\pi(1-\pi)}$	$\sigma = \sqrt{\pi(1-\pi)}$	$\sigma = \sqrt{\pi(1-\pi)}$
$= \sqrt{0.25(0.75)}$	$= \sqrt{0.50(0.50)}$	$= \sqrt{0.90(0.10)}$
$= 0.433$	$= 0.500$	$= 0.300$

Notice that the greatest variability exists when $\pi = 0.50$ and that the smallest variability is when π approaches zero or 1. Uncertainty is greatest when the probability of success is around 0.50 and least when the probability of success is near zero or 1.

The Binomial Probability Distribution

A single trial from a Bernoulli process would be pictured in tree diagram form as follows.

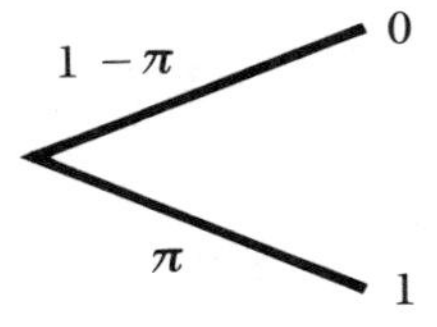

Figure 5-4 shows the tree diagram for possible outcomes of the sequence of trials when repeated trials are conducted from a Bernoulli process.

Suppose in our example we were concerned with the number of compacts in the next three autos passing by. Suppose also that π, the probability of a compact auto on any trial, is 0.25. The possible sequences of outcomes are shown in Figure 5-5.

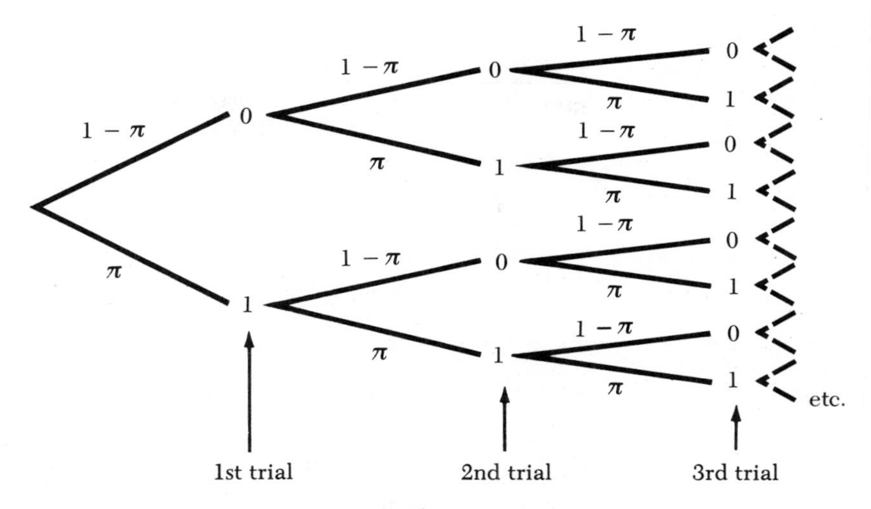

Figure 5-4 Repeated Trials of a Bernoulli Process

Our concern is to find the probability distribution of the number out of three autos that are compacts. Call this random variable R. Each route through the tree has an associated value of r, shown at the extreme right of the figure. The probabilities for each route have been calculated by finding the product of the individual probabilities lying along that path. For example, the sequence (0, 1, 0) leads to a value of $r = 1$ and has a probability of $9/64$.

We now add the probabilities for the different sequences leading to zero, 1, 2, and 3 compact autos. The result, shown in Table 5-5, is the probability distribution of the number of compacts in three trials from a Bernoulli process in which the probability of a compact auto on a single trial is $1/4$.

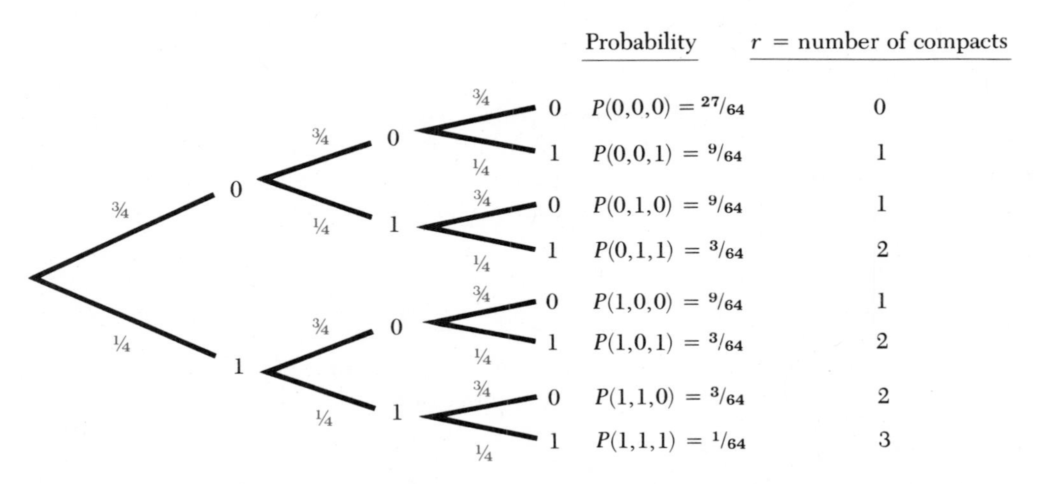

Figure 5-5 Outcomes for Three Trials of a Bernoulli Process with $\pi = 1/4$

Table 5-5 Binomial Probability Distribution for $n = 3$, $\pi = 1/4$

Number of Successes (Compact Autos) r	Probability $P(r)$
0	$27/64 = 0.4219$
1	$9/64 + 9/64 + 9/64 = 27/64 = 0.4219$
2	$3/64 + 3/64 + 3/64 = 9/64 = 0.1406$
3	$1/64 = 0.0156$

The *binomial distribution* is the probability distribution of the number of successes (r) encountered in a series of n Bernoulli trials.

Formula and Tables for Binomial Probabilities

It would be desirable to have a formula for the probability of r successes in n Bernoulli trials. We have seen that the multiplication law for independent events can be used to find the probability of any given sequence of successes and failures. For example, for the following sequence of 3 successes in 5 trials,

$$P(FSSSF) = (1 - \pi)(\pi)(\pi)(\pi)(1 - \pi)$$

The probability for another sequence with 3 successes is

$$P(SFSFS) = (\pi)(1 - \pi)(\pi)(1 - \pi)(\pi)$$

In both cases the factors can be collected to obtain $(\pi)^3(1 - \pi)^2$. In general, any sequence leading to r successes in n trials will have the probability

$$\pi^r(1 - \pi)^{n-r}$$

Since all sequences leading to r out of n successes have the same probability for a given r, we now need a rule for counting the number of equally likely sequences. For example, in how many different ways can 3 successes occur in 5 trials? Visualize the 5 trials as 5 empty boxes and ask in how many ways can 3 S's occur in the boxes.

Trial 1	Trial 2	Trial 3	Trial 4	Trial 5

The answer is called the *number of combinations* of 5 elements taken 3 at a time. The general formula for the number of combinations of n elements taken r at a time is

$$_nC_r = \frac{n!}{(n - r)!r!}$$

The symbol ! is called *factorial*, and $n!$ means $n(n-1)(n-2)\cdots(n-n+1)$. (We define 0! to be 1.) Thus for the number of combinations of 5 elements 3 at a time we have

$$
\begin{aligned}
{}_5C_3 &= \frac{5!}{(5-3)!3!} \\
&= \frac{5*4*3*2*1}{(2*1)(3*2*1)} \\
&= 10
\end{aligned}
$$

There are 10 ways that 3 successes (1's) could occur in a sequence of 5 trials. We can verify this by listing all the sequences as follows.

SSSFF	*SSFFS*	*SFFSS*
SSFSF	*SFSFS*	*FSFSS*
SFSSF	*FSSFS*	*FFSSS*
FSSSF		

We are now able to write the general binomial formula.

$$P(r) = {}_nC_r\pi^r(1-\pi)^{n-r} \tag{5-6}$$

The values we must know before we can apply the binomial formula are n, the number of trials; π, the probability of a success; and r, the specified number of successes. The conditions for Bernoulli trials must be met.

We shall apply the formula to find the same probabilities for observing r out of 3 compact autos that we obtained using the probability tree.

$$
\begin{aligned}
P(0) &= {}_3C_0(0.25)^0(0.75)^3 = (1)(1)(0.75)^3 = 0.4219 \\
P(1) &= {}_3C_1(0.25)^1(0.75)^2 = (3)(0.25)(0.5625) = 0.4219 \\
P(2) &= {}_3C_2(0.25)^2(0.75)^1 = (3)(0.0625)(0.75) = 0.1406 \\
P(3) &= {}_3C_3(0.25)^3(0.75)^0 = (1)(0.0156)(1) = 0.0156
\end{aligned}
$$

Because the binomial distribution is so commonly used, tables for binomial probabilities have been developed (see Appendix A-1). Given specified values of n and π, we read any desired $P(r)$ from the table. To check the values we used earlier, $n = 3$ and $\pi = 1/4$, or 0.25, we move down the column headed by $\pi = 0.25$ to the row for $n = 3$ and we read 0.4219 opposite $r = 0$.

n	r	π 0.25
.	.	.
.	.	.
.	.	.
3	0	0.4219
	1	0.4219
	2	0.1406
	3	0.0156

These values of $P(r)$ are the same as those we obtained from the probability tree and from the binomial formula. This gives three ways of finding binomial probabilities, each more convenient than the last. A complete way to refer to a binomial probability is to write

$$P(r \mid n, \pi)$$

For example, the probability 0.1406 above would be represented by

$$P(2 \mid 3, 0.25)$$

which is read "the probability of 2 successes in a binomial distribution in which n is 3 and π is 0.25."

5.5 Mean and Standard Deviation of a Binomial Distribution

The Bernoulli random variable and the binomial probability distribution are two very important random variables. The Bernoulli random variable refers to outcomes of a single trial from a Bernoulli process, while the binomial distribution refers to outcomes of multiple Bernoulli trials. The Bernoulli process is the source from which binomial distributions arise. In Section 5.4 we discussed the mean and standard deviation of Bernoulli random variables. In this section, we shall talk about these measures for binomial distributions.

Example

A landscaper is designing plantings for a new home. The client is particularly fond of azaleas and would like to have them included. The landscaper knows that in the local climate, azalea plants are given only a 30% chance of surviving the first year. Generally, if the plants survive the first year, they will continue to prosper. The landscaper is considering two plans, one calling for 5 azaleas and one for 10 azaleas. What should the landscaper tell the client about the possibilities of survival?

Consider first 5 azalea plantings. The survival of the plants may be assumed to represent 5 trials of a Bernoulli process with $\pi = 0.30$. The binomial distribution for the number of successes can be found in the binomial table (Appendix A-1). It is shown in Table 5-6, along with the calculations for finding the mean, or expected value, for the number of successes.

The mean, as before, is the sum of products of values of the variable weighted by their probabilities. It is probably not surprising that if individual plants are given a 30% chance of survival, the expected number of plants surviving out of 5 is 30% of 5, or 1.5 plants. The landscaper might tell the client that on the average, $1\frac{1}{2}$ out of 5 plants will survive. Because this statement suggests repeated plantings of five azaleas, it may not seem wholly satisfactory to the client; however, this is inherent in any statement involving probability.

Table 5-6 Finding the Mean of a Binomial Distribution

Binomial Distribution r	$P(r)$	*Calculation for Mean* $r * P(r)$
0	0.1681	0.0000
1	0.3602	0.3602
2	0.3087	0.6174
3	0.1323	0.3969
4	0.0284	0.1136
5	0.0024	0.0120
		$1.5001 = \mu$

As the above paragraph hinted, the mean of a binomial distribution can always be found from

$$\mu = n\pi \tag{5-7}$$

In our example,

$$\mu = 5(0.30)$$
$$= 1.5$$

Table 5-7 shows how to find the variance of a binomial distribution from the detailed listing of the distribution.

Table 5-7 Finding the Variance of a Binomial Distribution

Binomial Distribution r	$P(r)$	*Calculation for Variance* $(r - \mu)^2$	$(r - \mu)^2 * P(r)$
0	0.1681	2.25	0.3782
1	0.3602	0.25	0.0900
2	0.3087	0.25	0.0772
3	0.1323	2.25	0.2977
4	0.0284	6.25	0.1775
5	0.0024	12.25	0.0294
			$1.0500 = \sigma^2$

As we saw before, the variance is obtained by weighting the square of the deviation of each value from the mean by the probability of that value and adding the resulting products. As with the mean, however, it is not necessary to go through this calculation. The variance of a binomial distribution can be found from

$$\sigma^2 = n\pi(1 - \pi) \tag{5-8}$$

The standard deviation is the positive square root of the variance.

$$\sigma = \sqrt{n\pi(1 - \pi)} \tag{5-9}$$

For the landscaper's problem, $n = 5$ and $\pi = 0.30$. The variance is

$$\sigma^2 = 5(0.30)(1 - 0.30)$$
$$= 1.050$$

and the standard deviation is

$$\sigma = \sqrt{1.050}$$
$$= 1.025$$

We shall discuss this standard deviation again after looking at the binomial distribution for the landscaper's alternative plan of 10 azalea plants.

Now that we have convenient formulas for the mean and standard deviation of a binomial distribution, we can examine the probability distribution and the summary measures for the number of plants surviving out of 10. The binomial for $n = 10$ and $\pi = 0.30$ is given in Appendix A-1 and is reproduced below.

r	$P(r)$	r	$P(r)$
0	0.0282	6	0.0368
1	0.1211	7	0.0090
2	0.2335	8	0.0014
3	0.2668	9	0.0001
4	0.2001	10	0.0000
5	0.1029		

The mean of this binomial distribution is

$$\mu = n\pi$$
$$= 10(0.30)$$
$$= 3.0$$

and the standard deviation is

$$\sigma = \sqrt{n\pi(1 - \pi)}$$
$$= \sqrt{10(0.30)(1 - 0.30)}$$
$$= 1.449$$

If 10 plants are used, the expected number surviving is 3 and the standard deviation of the probability distribution of the number surviving is 1.449. The expected number of plants surviving is again 30% of the number planted. But in addition to the average, or expected value, the landscaper can give the client some idea of extremes and general variability. For example, the detailed binomials reveal that the probability of no plants surviving out of 5 is 0.1681, while the chance of no plants surviving out of ten is 0.0282. It should interest the client that the chances of complete failure are reduced by a factor of nearly six by doubling the number of plants.

Further indication of variability is afforded by noting that for 5 plants, the probability of 1 or 2 surviving is $0.3602 + 0.3087 = 0.6689$; and for 10 plants, the probability of 2, 3, or 4 surviving is $0.2335 + 0.2668 + 0.2001 = 0.7004$. Note that in both cases, these are the probabilities for the outcomes within one standard deviation of the mean.

$$\text{5 plants: } 1.5 \pm 1.025$$

$$\text{10 plants: } 3.0 \pm 1.449$$

In both cases, the total probability for outcomes within one standard deviation of the mean is not too far from the 68% mentioned in Chapter 3 for normal distributions. The binomial distributions we have been dealing with in the landscaper's problem are not normal; they are skewed to the right. In Chapter 6, which deals in detail with the normal distribution, we shall see that in some situations binomial distributions are indeed *very* close to the normal distribution pattern.

Exercises

5-17 *United Fund solicitation.* Two volunteers are working together as a team in soliciting the cooperation of firms that had not participated in previous years in the current United Fund drive. Would you expect successive outcomes of calls to conform to a Bernoulli process in each of the following situations? Explain.

a. The volunteers are an experienced worker and a novice and alternate in making the major presentation to successive firms.

b. The volunteers are both new at this work and work together in presentations to the firms, hoping to develop an effective team approach as time goes by.

c. The volunteers are experienced and have developed a team approach from previous campaigns, which they will use in the current drive.

5-18 *Errors in a flow of work.* Are the requirements for a Bernoulli process, constant underlying probability and independence, apparently met in the following situations? The event is an incorrect unit of work.

a. A trained punch-card operator works at a steady pace and is unaware of an error in a completed card until the work is checked later by another operator.

b. A computer terminal operator is making data entries using an interactive program that informs the operator about certain kinds of errors. When this happens, the operator becomes frustrated and tends to make additional errors for a brief period.

c. An electronics technician assembles parts under a microscope. The work is so exacting that management has been considering mandatory rest periods.

5-19 *Binomial formula.* Use the binomial formula to find the following probabilities. Check your answers by finding the same probabilities in the binomial table.

a. $P(R = 2 \mid n = 3, \pi = 0.60)$
b. $P(R = 0 \mid n = 2, \pi = 0.20)$
c. $P(R = 4 \mid n = 4, \pi = 0.90)$

5-20 *Binomial table.* Find the following probabilities by using the binomial table.

a. The probability of three or fewer successes in 8 Bernoulli trials with $\pi = 0.20$.
b. The probability of less than 8 successes in 10 Bernoulli trials with $\pi = 0.70$.
c. The probability of more than 2 successes in 12 Bernoulli trials with $\pi = 0.40$.

5-21 *Computer service interruptions.* The occurrence of an interruption in service during the 8 A.M. to 5 P.M. weekday shift at a computer facility is a Bernoulli process with $\pi = 0.20$. Find the probability of the following events.

a. On 3 of 5 weekdays, there will be an interruption.
b. There will be an interruption on Monday, Tuesday, and Wednesday, but not on Thursday or Friday during a week.
c. Five weekdays will go by without an interruption.

5-22 *Binomial distributions.* Plot the binomial probability distribution for $n = 4$ and $\pi = 0.20$ and for $n = 4$ and $\pi = 0.50$. Show the mean on your plot(s) and the distance on the variable scale of one standard deviation above and below the mean.

5-23 *Bernoulli process and binomial.* Occurrence of defective parts in a production stream is a Bernoulli process with $\pi = 0.10$.

a. Find the mean and standard deviation of the Bernoulli random variable.
b. Find the mean and standard deviation of the binomial distribution for 9 trials ($n = 9$) from the process.
c. Find the mean and standard deviation of the binomial distribution for 25 trials ($n = 25$) from the process.

5-24 *Finding qualified operators.* A textile firm has found from experience that only 20% of the people applying for a certain stitching-machine job are qualified for the work. Find the mean and standard deviation of the Bernoulli random variable and the binomial random variable in each case.

a. Sixteen persons are interviewed to find qualified persons.
b. Thirty-six persons are interviewed.
c. Sixty-four persons are interviewed.

5-25 *Product sampling.* Marketing representatives for a new laundry detergent have been instructed to leave a sample of the product at nine households, which have already been randomly selected, in each of a number of small areas in a large city. The representative must talk to the person who does the laundry in order to explain the product and obtain information about current product use. The Bernoulli random variable is the event that a laundry operator is at home when the representative calls. Find the expected value and standard deviation of the Bernoulli random variable and of the number out of nine households in which a laundry operator will be found at home in an area for which each given probability of finding a laundry operator at home holds.

a. 0.10 b. 0.50 c. 0.90

5-26 *Bernoulli process and sampling results.* Appearance of overdue accounts in accounts drawn randomly from a large population is a Bernoulli process with $\pi = 0.20$. What is the probability of finding no more than one delinquent account out of

a. 5 trials ($n = 5$) b. 10 trials ($n = 10$) c. 15 trials ($n = 15$)

5-27 *Standard deviations.* Find the standard deviation of Bernoulli processes having probabilities of success of .10, .20, and .50. Then find the standard deviations of the related binomial distributions given in the table below.

Standard Deviation	*Probability of Success (Bernoulli)* .10	.20	.50
Bernoulli process	____	____	____
Binomial ($n = 4$)	____	____	____
Binomial ($n = 9$)	____	____	____
Binomial ($n = 16$)	____	____	____

Summary

Uncertain events which are identified by values of a variable are called *random variables*. The probability distribution of a random variable can be studied by identifying its probability or its cumulative probability distribution, which were illustrated in the chapter for *discrete* random variables. Some discrete random variables are formed from the sum of two other random variables.

As with frequency distributions, the *probability distribution* of a discrete random variable can be summarized by its mean and standard deviation.

A *Bernoulli process* gives rise to a discrete random variable X: $x = 0, 1$. The mean and variance of a *Bernoulli random variable* depend only on the probability that $X = 1$. The probability distribution of successes ($x = 1$) for repeated trials of a Bernoulli process gives rise to the binomial probability distribution of the number of successes, R. The mean and variance of a binomial random variable depend on the probability of success for the underlying Bernoulli process and the number of trials (n) being considered.

Supplementary Exercises

5-28 *Student parking violations.* From considerable experience the campus parking authority has developed the following probability distribution of the number of tickets per semester issued to students with parking permits.

Number of tickets	0	1	2	3
Probability	0.50	0.10	0.30	0.10

a. Find the expected number of tickets and interpret the value as a long-run average.
b. Find the variance of the distribution and interpret it as a long-run average.

5-29 *More parking violations.* Answer the following questions if two permit holders (Exercise 5-28) are selected at random.
a. What is the probability that they each receive less than 2 tickets during a semester?
b. What is the probability that the total number of tickets issued to the pair is less than 2?
c. What is the probability that they receive 1 ticket each?

5-30 *Logging operations.* Grizzly Lumber Company is considering logging operations in a certain area in November. The probability that it will be too cold to operate on any given day is 0.30, and the probability that snow conditions will prevent operating is 0.40. Weather records indicate that these events are independent.
a. Out of 25 scheduled working days in November, on how many should Grizzly expect to be able to operate?
b. Would you expect the occurrence of days suitable for operations to constitute a Bernoulli process? Explain.

5-31 *To catch a thief.* A private detective, Arch Lewer, is approached by a widow whose jewels were stolen from her hotel room. The widow offers to pay Arch $5000 if he just recovers the gems, $2000 if he just apprehends the thief, $10,000 if he recovers the gems *and* apprehends the thief, and $1000 if he accomplishes neither. Asked by Arch if she would simplify the offer, the widow responds that if Arch would prefer it she would be willing to pay $4000 if he succeeds in any aspect of the case and $2000 if he should fail entirely. After listening to the details of the theft, Arch forms the following probabilities.

Apprehend Thief	*Recover Gems*	
	Yes	*No*
Yes	0.10	0.10
No	0.20	0.60

a. Plot the probability distributions for the two propositions, using the same scales for both graphs.
b. Find the expected value and standard deviation for each of the propositions.
c. Do the standard deviations reflect the comparative riskiness of the two offers? Explain.

5-32 *Damage awards.* A firm has a damage suit against a competitor for infringement of the firm's copyrighted brand label. The firm's attorney assesses the probability of obtaining an out-of-court settlement for $100,000 as 0.60. If no such settlement occurs, the attorney will go to court; there, the attorney assesses the probability of no award as 0.20, the probability of an award of $100,000 as 0.50, and the probability of an award of $200,000 as 0.30.
a. What is the expected award if the firm goes to court?
b. What is the expected settlement or award?

5-33 *Tournament prize money.* A professional tennis player is scheduled to play in an elimination tournament with 7 other players. There are 3 rounds in such an event, with the winners in each round moving on to the next round. The prize money in the forthcoming event is based on the number of rounds won, as follows.

Rounds won	0	1	2	3
Thousands of dollars	5	10	20	50

The prize money is not cumulative. For example, if a player wins the final (third round) match, the prize money is \$50,000. If the player loses in the finals, the prize money is \$20,000. This player believes that, for him, winning matches is a Bernoulli process with probability 0.60 for each match.

a. Work out the probability distribution of the random variable, the number of rounds won.
b. What is the player's expected prize money for the event?

5-34 *Exploratory oil wells.* Split Rock Drilling, Inc., plans to sink five exploratory wells at Black Desert. Consulting geologists give the probability that a well results in a discovery as 0.20 for each of the first 3 wells. If there is a find in the first 3 wells, the geologists' probability is 0.30 for each of the next 2 wells. Otherwise, the geologists give a probability of 0.10 for each of the last 2 wells.

a. Use binomial probabilities and a tree diagram to find the probability distribution of the number of discoveries.
b. What is the expected number of discoveries?
c. Does the distribution agree with the binomial distribution for $n = 5$ and $\pi = 0.20$? Why not?

5-35 *Weather forecasting services.* To compare the accuracy of two weather forecasting services, a local radio station selected past occasions when the forecasts of the services differed. Ten days were found when Precision Weather had given the probability of rain as 0.20 and Meteorology Associates had given the probability of rain as 0.50. On 3 of the days in question it did rain. What is the probability of the observed outcome in each case?

a. If Precision Weather was right about the probability of rain.
b. If Meteorology Associates was right about the probability of rain.

5-36 *Baseball hitters.* Assuming a Bernoulli process, compare the probabilities of a 0.400 hitter compared with a 0.250 hitter in baseball for each situation.

a. Getting four hits in 4 at bats in a game.
b. Getting no hits in 4 at bats in a game.
c. Getting the next hit in the fourth time at bat.

5-37 *Marketing premium apples.* From each incoming batch of apples, a farm-marketing cooperative samples a dozen for texture and taste. If 11 or 12 of the dozen meet a standard, the batch is used to fill cartons carrying the cooperative's label. These packages bring a premium price through direct sales to large urban department and food specialty stores. If 10 or fewer meet the standard, the batch is marketed through regular distribution channels.

a. Construct a graph that shows quality percentage from 50 to 95 at intervals of 5 percentage points on the horizontal scale and probability of "premium marketing treatment" on the vertical scale.
b. How could the cooperative use this graph to assure buyers about the quality of its premium product?

5-38 *Ceramics quality control.* A tribal enterprise turns out ceramic pots in small batches. Owing to variations in the character of native clays it is difficult to control the heat-resistant character of the pots. Therefore, 3 out of each batch of 43 pots fired are subsequently tested for heat resistance. If all 3 pots withstand the test, the remaining 40 in the batch are advertised and sold as "oven-proof."

a. Plot the probability of declaring a batch as oven-proof against the proportion in the batch that would withstand the test for the proportions 0.00, 0.10, 0.20, . . . , 0.80, 0.90, 1.00.

b. Discuss the protection that the testing scheme provides to customers of the enterprise.

5-39 *Installment delinquencies.* According to a Better Business Agency, insisting on a credit check will lower the incidence of delinquency on installment sales from 15% to 5%. An auditor notes that in a random sample of 14 closed installment sales, 2 had been delinquent. What is the probability of this outcome if credit checks were uniformly made? Not made?

5-40 *Federalist Papers.** In known writings of Alexander Hamilton the word "upon" occurs at least once in passages of 200 words with probability 0.45, while in 200 words written by James Madison the probability of encountering "upon" is 0.05. Federalist Paper 55, of unknown authorship, contains 10 passages of 200 words. The word "upon" is found in 2 of the ten passages.

a. What is the probability of this outcome in Hamilton's writing?

b. What is the probability of this outcome in Madison's writing?

c. Does the evidence suggest which man is more likely to be the author if one had to choose between the two? Why?

5-41 *Minicopier servicing policy.* Image Corp. has leased varying numbers of its mini-copiers to firms in its market area. The machines are thoroughly overhauled annually, but the probability that a machine will require interim monthly service is a Bernoulli process with $\pi = 0.20$. The frequency distribution of machines per firm is

Number of machines	1	2	3	4	5
Number of firms	10	40	30	20	10

a. For each number of machines find the probability that at least 1 out of that number of machines requires monthly service.

b. Find the expected number of firms in each size class that will need monthly service.

c. Image Corp. is considering having its servicepeople visit the firms monthly to check the machines and repair those requiring service. What is the probability that a visit will result in a need for repair service if (i) servicepeople visit all firms, and (ii) servicepeople visit only firms with 3 or more machines?

5.6 The Poisson Distribution (Optional)

A probability distribution that has found wide application in business and industry

*Adapted from F. Mosteller and D. L. Wallace, "Deciding Authorship," in *Statistics: A Guide to the Unknown*, edited by Judith M. Tanur, San Francisco: Holden-Day, 1972.

is the *Poisson distribution*, named after the French mathematician Siméon Poisson (1781–1840).

Example

We are standing on a railroad bridge overlooking an uninterrupted stretch of country highway. The traffic flow in one direction under the bridge averages 120 cars per hour during all normal daylight hours. The number of cars passing by in any minute is independent of the traffic in any other minute and the probability of 2 cars passing by in a small time interval is negligible compared to the probability of 1 car passing by. What is the probability distribution of the number of cars passing by in one minute?

In the example we are concerned with a unit of exposure (a minute) to the event *a car passing by*. The number of occurrences of the event is independent from one unit of exposure to another and there is a constant expected number of occurrences per unit of exposure. Further, we can think of a unit of exposure (time interval) small enough so that the probability of two occurrences of the event is negligible compared to the probability of one occurrence. When these three conditions prevail, the probability distribution of the number of occurrences per unit of exposure is given by

$$P(r) = e^{-\mu}\left(\frac{\mu^r}{r!}\right) \qquad \textbf{(5-10)}$$

where r is the number of occurrences per unit of exposure, μ is the expected number of occurrences per unit, and e is the constant approximated by 2.71828 (the base of the system of natural logarithms). For example, the probability of no cars passing by in a minute when the expected number (or average rate) is 2.0 per minute is

$$\begin{aligned} P(0) &= 2.71828^{-2.0}\left(\frac{2.0^0}{0!}\right) \\ &= \frac{1}{2.71828^2}\left(\frac{1}{1}\right) \\ &= 0.135 \end{aligned}$$

and the probability of 3 cars passing by in a minute is

$$\begin{aligned} P(3) &= 2.71828^{-2.0}\left(\frac{2.0^3}{3!}\right) \\ &= \frac{1}{2.71828^2}\left(\frac{8}{3 * 2 * 1}\right) \\ &= 0.180 \end{aligned}$$

Poisson processes can arise in connection with various kinds of units of exposure. Some additional examples where the unit of exposure is an interval of time are:

arrivals of customers at a service facility;

requests for replacement parts;

accidents at a plant.

Examples involving other units of exposure are:

lightning fires in grid units in a national forest during a summer season;

defects in square yards of carpet;

defects in linear feet of spool wire;

minor traffic violations in a year to individuals with otherwise good driving records.

Some examples that violate the Poisson process conditions might be enlightening. Traffic intensity on an uninterrupted stretch of highway is likely to be a Poisson process, but when traffic is congested and delayed intermittently by traffic lights, the independence assumption is violated. Location of convenience food stores in spatial units of a city would fail to be a Poisson process because the stores may be zoned out of strictly residential areas and may not locate close to one another. These factors lead to violation of both uniform expectation over units of exposure and independence of occurrence from unit to unit.

An extensive study of drivers' records in California found that the probability distribution of yearly numbers of accidents occurring to individual drivers was not Poisson distributed. However, when drivers were classified into groups by sex, marital status, age, area of residence, and prior accident history, the distributions of the numbers of accidents to individuals for 167 of 193 groups closely approximated Poisson probabilities. The fact that the expected number of accidents per person varied among subgroups of drivers violated the uniform expectation condition for a single Poisson process for all drivers.

Figure 5-6 shows Poisson distributions with means of 0.50, 2.0, and 10.0. In our example, these would be the probability distributions for the numbers of cars passing in 15 sec, 1 min, and 5 min. Notice that the first distribution, with a mean arrival rate of 0.50 cars per 15-second interval, is highly skewed to the right. The second distribution, where $\mu = 2.0$, is noticeably skewed to the right, but less skewed than the first. The distribution of arrivals in 5-minute intervals appears quite symmetrical. In general, as the expected number of occurrences per unit of exposure increases, the skewness of the Poisson distribution decreases.

A table for Poisson probabilities is given in Table A-7 of Appendix A. In Table 5-8, the distribution for $\mu = 2.0$ is repeated and the mean and variance for the distribution are calculated. Notice that while the table goes only to $r = 8$, there is some probability, however small, for $r = 9$, $r = 10$, and so on. The Poisson random variable can take on any positive integer value.

Notice that the mean, $\Sigma[r * P(r)]$, is calculated to be 1.994, rather than 2.000. This is due to rounding errors in the probabilities and to the fact that very small probabilities for more than 8 occurrences were left out. Without these factors, the mean would be exactly 2.0. We therefore used that value in figuring the squares of

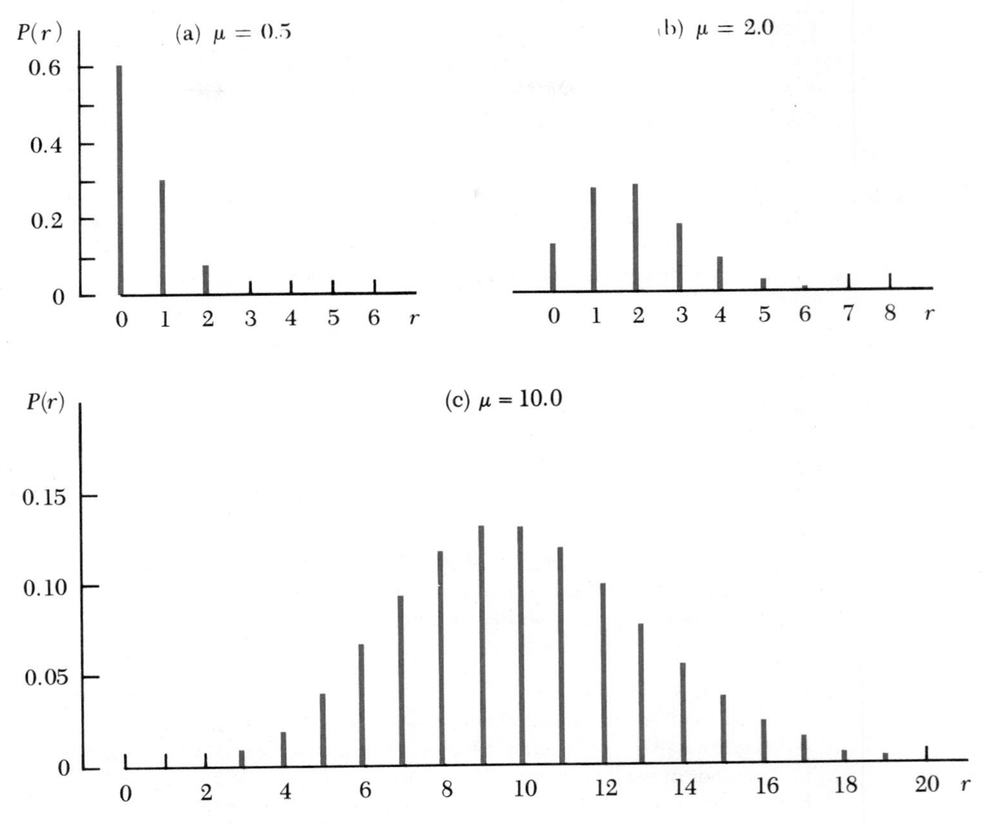

Figure 5-6 Poisson Distributions

Table 5-8 Mean and Variance of a Poisson Distribution

Probability Distribution		*Calculations*		
r	$P(r)$	$r * P(r)$	$(r-\mu)^2$	$(r-\mu)^2 * P(r)$
0	0.135	0.000	4	0.540
1	0.271	0.271	1	0.271
2	0.271	0.542	0	0.000
3	0.180	0.540	1	0.180
4	0.090	0.360	4	0.360
5	0.036	0.180	9	0.324
6	0.012	0.072	16	0.192
7	0.003	0.021	25	0.075
8	0.001	0.008	36	0.036
	0.999	1.994		1.978

the deviations in the variance calculation. Again, the variance is slightly small because the distribution was carried only to $r = 8$; for the complete distribution, the variance would be exactly 2.0. *For any Poisson distribution, the variance is equal to the mean.*

Exercises

5-42 *Poisson table.* Find the following Poisson probabilities from Table A-7, Appendix A.

a. $P(R \leq 5 \mid \mu = 3.0)$ b. $P(R > 3 \mid \mu = 3.0)$
c. $P(R = 8 \mid \mu = 6.4)$ d. $P(R \geq 8 \mid \mu = 6.4)$
e. $P(R < 6 \mid \mu = 6.4)$

5-43 *Demand for spare parts.* Demand for a spare part is a Poisson process with an expected demand of 2 per month.

a. What is the probability that demands in a month exceed 4?
b. What is the lowest level of demand that is exceeded with probability less than 0.01?

5-44 *Daily births.* A community of 100,000 projects a birth rate next year of 14.6 births per thousand population. Assume a Poisson process.

a. What is the probability of more than 7 births in a day?
b. What is the probability of no more than 1 birth in 2 days?

5-45 *Fatalities in coal mining.* From 1972 to 1976 in underground anthracite coal mining in the United States, there were 3 fatalities in 4.23 million worker-hours of operations. Experience from the early 1960's would have led us to expect 8.5 fatalities in the same number of worker-hours. What is the probability of 3 or fewer fatalities in 4.23 million worker-hours under conditions of the early 1960's?

5-46 *Supreme Court vacancies.* * Two vacancies on the Supreme Court were filled in 1971 by President Nixon after the retirement of Justices Black and Harlan. Historical studies show that since the inception of the Court in 1790, the number of appointments per year is Poisson distributed with a mean of very nearly 0.5. What is the probability of 2 or more appointments in a year?

5-47 *Fire alarms.*† Fire alarms during one year in Monterey, California, were found to be Poisson distributed with an average rate of 0.10 per hour.

a. What is the probability of more than 3 alarms in a 10-hour period?
b. What is the probability of more than 6 alarms in a 20-hour period?

5-48 *Highway repairs.* The probability distribution of the number of hot patches required per mile of a certain highway is Poisson distributed with a mean of 4.0. What is the probability that in a mile of highway

a. 6 patches will be required?
b. More than 10 patches will be required?
c. Less than 2 patches will be required?

---

*J. Kinney, *American Statistician*, October 1973.

†J. D. Finnerty, *Management Science*, July 1977.

5-49 *Pooling reserve stock.* Four depots where the demand for a certain spare part is 3 per period maintain a reserve stock level at the beginning of the period sufficient to make the probability less than 0.05 that the reserve stock will be exhausted during the period.

a. What is the stock level maintained?
b. If two depots pool their operations, the demand will be 6 per period. What stock level would be required to limit the probability of drawing stocks down to zero to 0.05?
c. Answer the question in (b) for pooling three depots. Four depots. Comment on the relation of stocks to demand supplied.

Computer Exercises

1 Source: Personnel data file. Consider this to be a population.

a. Which variables in this data file are variables which might qualify as variables in a Bernoulli process?
b. Draw the relative frequency diagrams for these variables.
c. Find the means and standard deviations for these variables using Equations 5-4 and 5-5.
d. Verify the results of (c) using the computer. (Note: you must make coding changes first.)

6

CONTINUOUS RANDOM VARIABLES AND THE NORMAL DISTRIBUTION

In this chapter, we introduce continuous probability distributions and the *normal distribution*, which is an important distribution in statistics. We first compare discrete and continuous random variables and show how to find a probability when the variable is continuous. Then we introduce the normal distribution and show how to solve various problems using the normal distribution. Later in the chapter, we shall see how the continuous normal probability distribution can be used to approximate the discrete binomial distribution in certain circumstances.

6.1 Continuous Random Variables

In the previous chapter, we treated only *discrete* random variables. The possible values of such a variable could be listed and probabilities could be assigned to each of the values. As an example, consider the following.

Discrete Random Variable Generator

A weighted, well-balanced spinner is mounted on a circular board. Around the sides of the board are slots numbered from 1 through 12. Dividing the slots are pieces of flexible spring steel. When the spinner is spun, the lead edge of the pointer strikes these flexible pieces, allowing the spinner to keep moving if it has enough momentum. Eventually the spinner slows so that it cannot pass the next spring barrier and comes to rest in one of the numbered slots.

Figure 6-1(a) shows the apparatus and the probability distribution associated with repeated trials, or spins. If the spinner is well balanced and the spring dividers are of equal flexibility, we would expect the long-run relative frequency of each of the 12 numerical outcomes to be the same, $1/12$ each.

Probability Density Function

Now consider another apparatus.

Continuous Random Variable Generator

A well-balanced spinner is mounted on a circular board. Around the edge of the board is a continuous circular scale whose major markings resemble the hour markers on a clock. The spinner is free to spin unimpeded until it comes to rest.

Figure 6-1(b) portrays this apparatus and the probability function that is appropriate to repeated spins, or trials. Unlike the "discrete" apparatus, we cannot

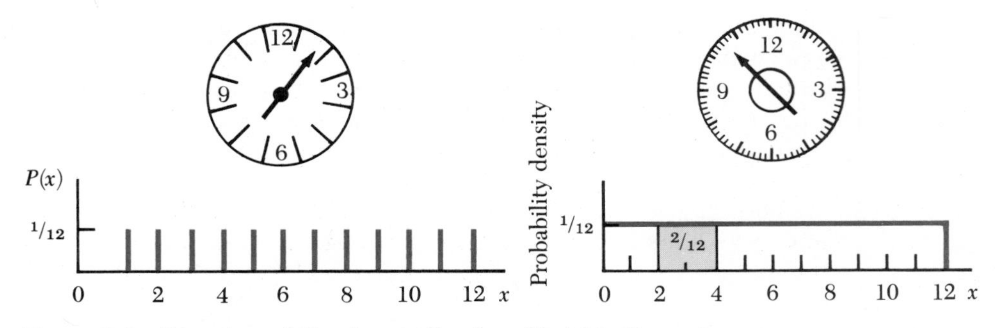

Figure 6-1 Discrete and Continuous Random Variable Generators

count the points at which the spinner may come to rest. However, we believe it is as likely to come to rest at any one "point" as another. That is, the pointer is as likely to come to rest at 2.0000 as at 8.6725, or at any other point that we might specify within the range of zero to 12.

In the graph of the probability distribution for a discrete variable, such as Figure 6-1(a), ordinates are drawn up to a vertical height corresponding to the probability at each value of the discrete variable. The sum of all these vertical ordinates is 1.0, the total probability. In the case of the continuous random variable generator, we would like to have a graph of the probability distribution of distances along the scale at which the spinner comes to rest. We assume the spinner is as likely to come to rest at any one point as another, but we cannot enumerate the points. However, we can use the assumption of equally likely probability to decide that $^1/_{12}$ of the total probability should be assigned to each of the 12 major intervals along the scale. In Figure 6-1(b), the shaded area between 2.0 and 4.0 on the variable scale is labeled $^2/_{12}$, which corresponds to our idea of what the probability should be for the spinner coming to rest between 2.0 and 4.0.

For a continuous variable, the appropriate representation of probability is a figure with a total enclosed *area* of 1.0. Probability is associated with any *interval* of the continuous variable and is represented by the proportion of the total area in the figure between the endpoints of the interval. The rectangular shape of the figure for the spinner outcomes reflects the assumption that it is equally likely that the spinner will come to rest anywhere along the continuous scale.

The base of the rectangle in Figure 6-1(b) is the range of the scale, 12.0 − 0.0 = 12.0. For the area under the rectangular figure to be 1.0, the height of the figure must be $^1/_{12}$. The height (at any value of x) of the figure representing a continuous probability distribution is called the *probability density* (at that point). The entire figure shows the relation between probability density and values of the random variable; it is called *the probability density function*.

Cumulative Distribution Function

Suppose the manager of a computer facility is interested in the time it takes the central processing unit to complete a certain class of job. The number of jobs of this kind submitted is very large and the computer records the times to the nearest hundredth of a second. In the manner of Chapter 2, an analyst could construct a frequency distribution of processing times using a large number of classes and, consequently, very small intervals. Reducing the frequencies to relative frequencies, the analyst could plot the relative frequency polygon and the cumulative "equal to or less than" frequency curve. Because there are many classes, the polygon and cumulative plots could easily be generalized into smooth curves. The result of this process is shown in Figure 6-2.

The manager wants to find the probability that a randomly selected job took between 5.00 and 7.50 sec to process. Without referring to the original data or tabulations, it would be easy to approximate the probability from the cumulative curve, or cumulative probability function. The height of the cumulative curve at 7.50 sec gives the probability of a processing time of 7.50 sec or less, and the height of

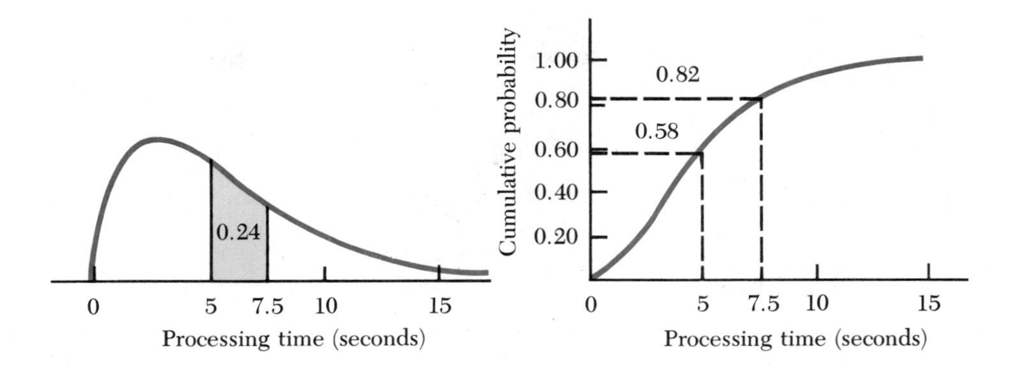

Figure 6-2 Probability Density and Cumulative Probability Functions

the curve at 5.00 sec gives the probability of a processing time of 5.00 sec or less. Subtracting the latter (0.58) from the former (0.82) gives the probability in the interval from 5.00 to 7.50 sec. In symbols,

$$\begin{aligned} P(X \leq 7.50) &= 0.82 \\ \underline{P(X \leq 5.00)} &= \underline{0.58} \\ P(5.00 \leq X \leq 7.50) &= 0.24 \end{aligned}$$

This probability is also indicated by the shaded area under the probability density function in Figure 6-2.

This section illustrated two concepts. First, when we speak of probabilities in connection with a continuous random variable, the probability must be for an *interval* rather than a discrete value of the variable. Second, the probability in an interval is found by using the cumulative probability function.

As with discrete probability distributions commonly used in statistical work, we shall find that tables exist for finding probabilities under commonly used continuous distributions. These tables, however, will correspond to *cumulative* relative frequency distributions.

6.2 The Normal Distribution

A continuous distribution of crucial importance in statistics is the *normal probability distribution*. It is important, first, because many observed variables are normally distributed, or approximately so. Second, the normal distribution plays a critical role in the methods of statistical inference. In this chapter we examine the normal distribution in the first of these uses. In Chapter 7 we begin to examine the second use.

Figure 6-3 shows a normal distribution of annual rainfall in an area. Like an example in Chapter 2, the distribution is symmetrical about its mean. This is a feature of all normal distributions. Because of this symmetry, the median rainfall and the most typical annual rainfall for the area are the same as the mean rainfall of

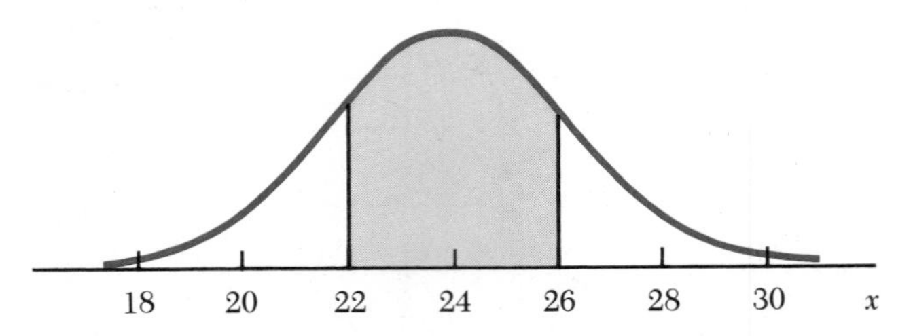

Figure 6-3 Normal Distribution of Annual Rainfall

24 in. Since areas under a continuous distribution curve are proportional to probabilities, we can see that the probability of an annual rainfall between 22 in. and 24 in. is the same as the probability of rainfall between 24 in. and 26 in., for example. In a normal distribution, the area between the mean and any distance above the mean is the same as the area between the mean and the same distance below the mean.

The normal density curve is bell-shaped; density decreases symmetrically as we consider x values deviating from the mean in either direction. The curve approaches but never reaches the x-axis. While the graph is not detailed enough to show it, there is some probability of rainfall below 17 in. and above 31 in. in the distribution in Figure 6-3. The equation for a normal density is

$$f(x) = \frac{1}{\sigma\sqrt{2\pi}} e^{-\frac{1}{2}\left(\frac{x-\mu}{\sigma}\right)^2}$$

where π is approximately 3.14159 and e is approximately 2.71828 and μ and σ are the mean and standard deviation of the random variable, respectively. Given the constants and a mean and standard deviation of 24 in. and 2 in. of annual rainfall, substitution of different rainfall values (x) in the normal density equation would produce the curve in Figure 6-3.

Figure 6-4 shows two normal distributions with different means and standard deviations. One is the rainfall distribution from Figure 6-3 and the other is an

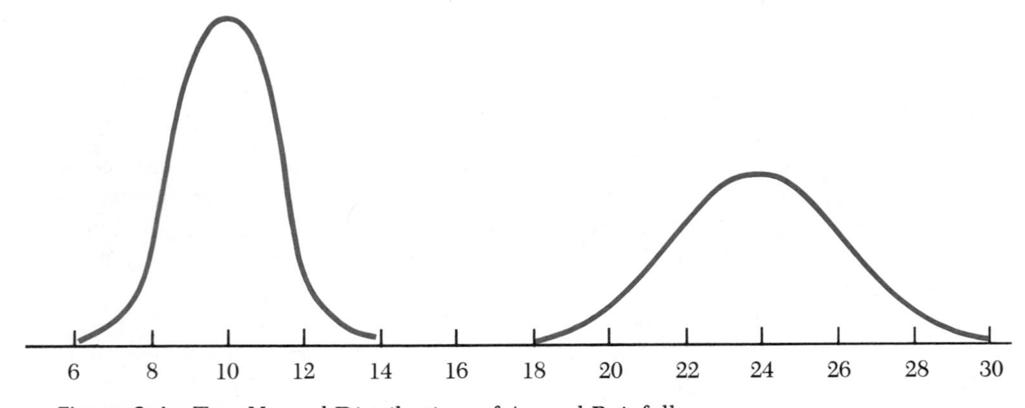

Figure 6-4 Two Normal Distributions of Annual Rainfall

annual rainfall distribution from a drier area. In the drier area, the mean is 10 in. and the standard deviation is 1 in., reflecting a smaller amount and less variability of rainfall than in the first area.

Examination of the equation for a normal density reveals that the density at any value of x depends on the exponent in the equation. The exponent involves the standard score from Chapter 3.

$$\frac{x-\mu}{\sigma} = z$$

This is the expression of x in terms of standard deviations from the mean. This should suggest that, in terms of standard scores, all normal distributions look and are the same. The effect of the change in variable scale to standard scores is shown for the two rainfall distributions in Figure 6-5. When they were plotted on the same inches-of-rainfall scale in Figure 6-4, their spreads were different. Now, in units of standard deviation, they are the same. In terms of z-scores, they are identical distributions. Since we can transform any distribution in this manner, we need only deal with one normal distribution, the *standard normal* distribution. It is the normal distribution with a mean of zero and a standard deviation of 1.0.

Areas of the Standard Normal Distribution

Table A-2 in Appendix A contains cumulative "equal to or less than" probabilities for the standard normal distribution. For any value of z, Table A-2 gives the cumulative probability up to that level. We call this a *lower-tail* probability. For example, at $z = -1.50$, we read 0.0668. Thus

$$P(Z \leq -1.50) = 0.0668$$

Graphically we have found the proportional area shown in Figure 6-6. Note that if we want $P(Z < -1.50)$, the answer will be the same, because there is no probability (area) at any exact value of a continuous variable.

Suppose we want the probability that Z exceeds −1.50. This is called an *upper-tail* probability. Since the total probability (area under the curve) is 1.0, we find the complement of the tabled lower-tail probability at $z = -1.50$.

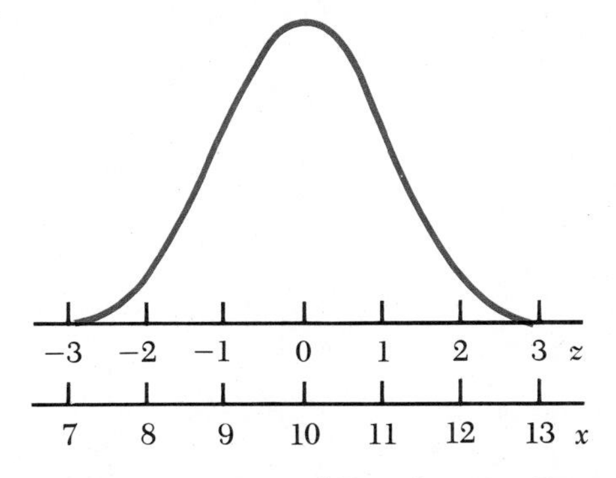

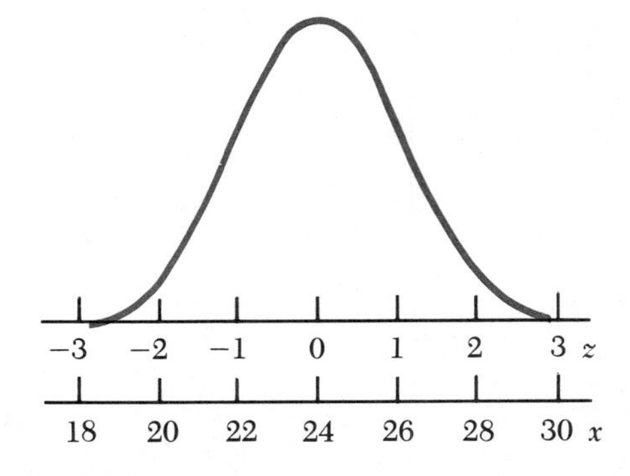

Figure 6-5 Normal Distributions Converted to Standard Scores

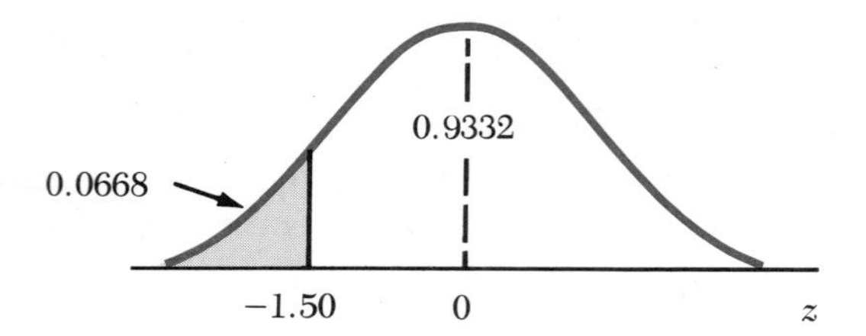

Figure 6-6 Lower- and Upper-Tail Standard Normal Probabilities

$$
\begin{aligned}
P(Z > -1.50) &= 1 - P(Z \leq -1.50) \\
&= 1 - 0.0668 \\
&= 0.9332
\end{aligned}
$$

If we want the probability that Z lies in the interval from −0.50 to 2.00, we use Table A-2 to find the probability associated with the upper limit of the interval and subtract from it the probability associated with the lower limit of the interval.

$$
\begin{aligned}
P(-0.50 \leq Z \leq 2.00) &= P(Z \leq 2.00) - P(Z \leq -0.50) \\
&= 0.9772 - 0.3085 \\
&= 0.6687
\end{aligned}
$$

This is shown graphically in Figure 6-7.

Percentiles of the Standard Normal Distribution

In Appendix Table A-2 we entered the table with a value of the standard normal deviate, z, and read the cumulative normal probability associated with that value, or $P(Z \leq z)$. In other situations we will encounter it is useful to find the value of z associated with a specified cumulative probability. While such values can be found using Table A-2, they can be found more directly from Appendix Table A-3. There, cumulative probabilities appear in thousandths in the left-hand column and associated z values in the right-hand column of each pair of columns. Some pairs are

P	z
.010	−2.33
.050	−1.65
.900	1.28
.980	2.05

The value −2.33 is the first percentile of the standard normal deviate. That is $P(Z \leq -2.33) = 0.01$. Similarly, −1.65 is the fifth percentile, 1.28 the 90th percentile, and 2.05 the 98th percentile of the standard normal deviate. In general, we will use

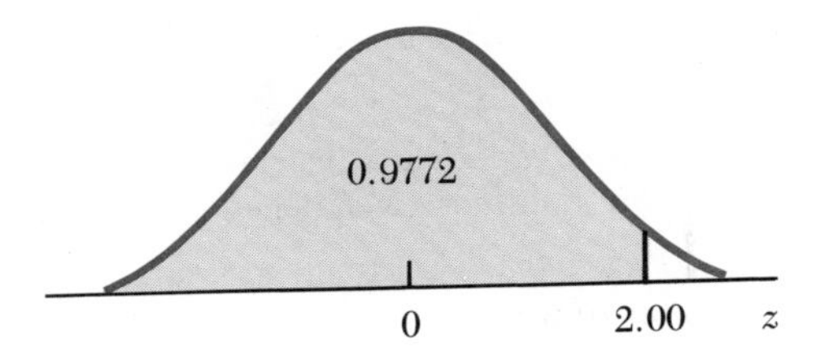

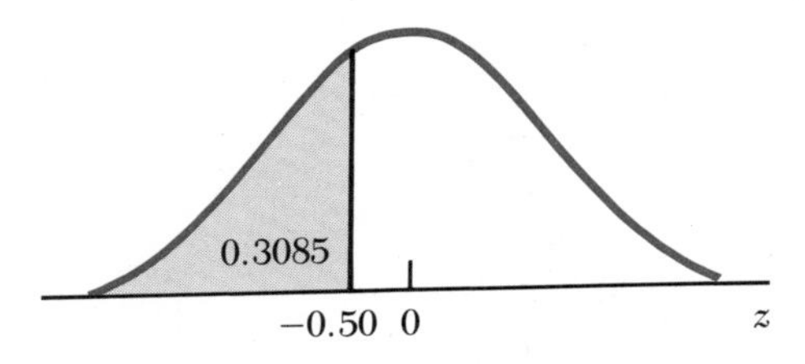

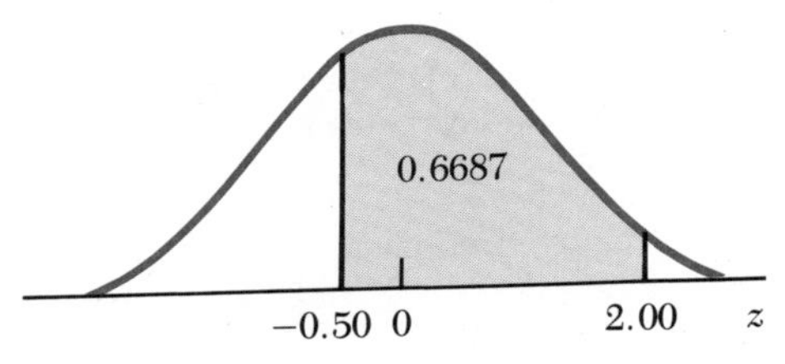

Figure 6-7 The Standard Normal Probability for a Closed Interval

the symbol z_p to denote the z value defined by an associated lower-tail cumulative probability. That is, for example, $z_{.01} = -2.33$ and $z_{.98} = 2.05$.

Exercises

6-1 *Ore truck loads.* A study was undertaken to find the percentages of recommended load weights at which ore trucks were operating. The relative frequency distribution is given below.

x	86–87.9	88–89.9	90–91.9	92–93.9	94–95.9	96–97.9	98–99.9	100–101.9	102–103.9	104–105.9	Total
f	.01	.03	.06	.10	.14	.16	.20	.17	.10	.03	1.00

a. Construct a frequency histogram for the data.
b. Shade the area corresponding to the following probabilities: $P(X \leq 90)$; $P(92 \leq X \leq 96)$; $P(X > 104)$.
c. Construct a separate frequency polygon for the data.

d. Shade the areas on the frequency polygon corresponding to probabilities in (b) above.

6-2 *Failure times.* The smoothed cumulative probability distribution of the hours that a knitting machine will operate without failure is given below.

x	0	1	2	5	10	20	30	40
$P(X \leq X)$	0.00	0.10	0.18	0.39	0.63	0.86	0.95	0.98

a. Graph the cumulative probability function as a smooth curve.
b. Approximate the following probabilities from your graph: $P(X \leq 25)$; $P(12.5 \leq X \leq 17.5)$; $P(X > 2.5)$.

6-3 *Mortgage interest rates.* In a particular year the banks in a southern city made 500 mortgage loans at the following rates.

Rate (%)	7.50–7.99	8.00–8.49	8.50–8.99	9.00–9.49	9.50–9.99	Total
Number of loans	6	22	97	271	104	500

a. Convert the data to a probability distribution.
b. Construct a frequency polygon for the probability distribution. Shade the area that represents the probability of an interest rate of less than 8.00; between 9.00 and 9.50.

6-4 *Time headways.** The cumulative relative frequency distribution of time headways (time between vehicles) on the Eisenhower Freeway in Chicago during certain hours is given below.

Headway (sec)	0.4	1.4	2.4	3.4	4.4	5.4	6.4	7.4	8.4	9.4
Cumulative %	0.3	11.4	30.3	50.5	61.5	71.5	78.3	83.1	88.0	91.0

a. Convert the distribution to frequency density and draw a frequency polygon for the distribution.
b. Shade the area corresponding to: $P(X \leq 3.4)$; $P(5.4 \leq X \leq 6.4)$.

6-5 *Notation.* Write symbolic expressions for the following.
a. The probability that a random variable exceeds 15.
b. The probability that the standard normal variate lies between zero and 1.7.
c. The fifteenth percentile of a random variable.
d. The 95th percentile of the standard normal variate.

6-6 *Probability notation.* Provide correct numerical values for the following.
a. $P(X > x_{.04})$ b. $P(Z \leq z_{.99})$
c. $P(x_{.03} \leq X \leq x_{.95})$ d. $P(Z \leq z_{.025} \text{ or } > z_{.975})$

*B.D. Greenshields and F.M. Weide, Statistics with Applications to Highway Traffic Analysis, *Eno Foundation*, 1978.

6-7 *Standard normal distribution.* Find the following for the standard normal distribution, using Appendix A-2.
a. $P(Z \leq 0.52)$ b. $P(Z \geq 2.50)$ c. $P(-0.50 \leq Z \leq 1.25)$

6-8 *Standard normal distribution.* Find the following for the standard normal distribution, using Appendix A-2.
a. $P(Z \leq 0.65)$ b. $P(Z > -0.50)$ c. $P(0.20 \leq Z \leq 1.20)$

6.3 Normal Distribution Applications

Now that we understand the standard normal table of areas, we are ready to apply it to normal distributions. We first return to the earlier rainfall distribution.

Example 1

Historical records of rainfall in an area have shown the distribution of annual rainfall to be normal with a mean of 24 in. and a standard deviation of 2 in. No time trend is present in the data. What is the probability of 22 or less inches of rainfall next year?

The key is to convert the value of x to a z-score. Then the standard normal table of areas can be used to find the desired probability. These steps can be used.

1. Problem: $P(X \leq 22)$
2. Standard score: $z = \dfrac{x - \mu}{\sigma}$

$$= \frac{22 - 24}{2} = -1.0$$

3. Probability: $P(Z \leq -1.0) = 0.1587$ (Appendix A-2)

The chances of rainfall not exceeding 22 in. next year are a little less than 16 in 100. This problem is shown in Figure 6-8 with the double scale of x and z.

Example 2

A large-scale study of waist measures of adult males revealed that the measurements are normally distributed with a mean of 36 in. and a standard deviation of 4 in. A buyer for men's slacks in a department store catering to the population at large is considering a line of slacks that are sized at 2-inch intervals. What proportion of the stock should be in size 38?

Assume that size 38 slacks have to cover waist sizes from exactly 37 inches to 39 inches. Knowing that the standard normal table of areas is cumulative, we plan to subtract the probability of a waist size of 37 or less from the probability of a waist

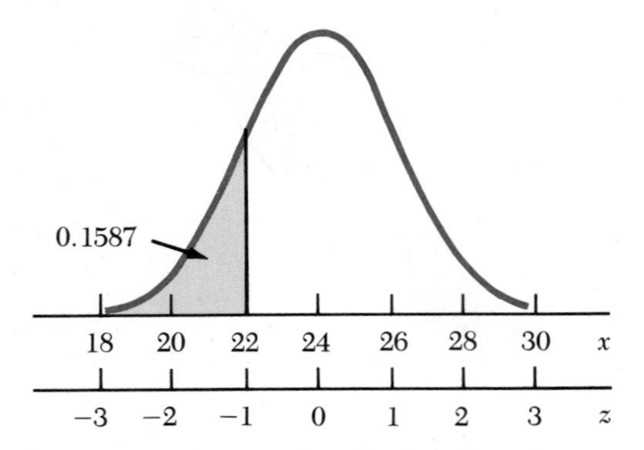

Figure 6-8 A Lower-Tail Normal Probability

size of 39 or less in order to find the probability in the interval 37 to 39. The problem can be set up as follows.

		x_{LOWER}	x_{UPPER}
1.	Problem:	$P(X \leq 37)$	$P(X \leq 39)$
2.	Standard score:	$z = \dfrac{37 - 36}{4}$	$z = \dfrac{39 - 36}{4}$
		$= 0.25$	$= 0.75$
3.	Probability:	$P(Z \leq 0.25) = 0.5987$	$P(Z \leq 0.75) = 0.7734$

The desired probability is

$$0.7734 - 0.5987 = 0.1747$$

About 17.5% of the stock should be in size 38 slacks. This problem is shown graphically in Figure 6-9.

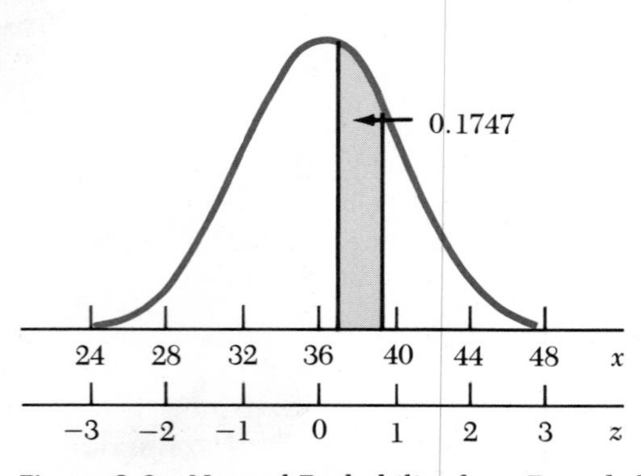

Figure 6-9 Normal Probability for a Bounded Interval

Example 3

Another appealing style of slacks is not made in sizes to fit waists larger than 43 inches. What percentage of the market is thus excluded?

This is an upper-tail problem in which we have to use the complement of a lower-tail probability given in a table.

1. Problem: $P(X > 43)$
2. Standard score: $z = \dfrac{43 - 36}{4}$

   $= 1.75$
3. Probability: $P(Z > 1.75) = 1.0 - P(Z \leq 1.75)$

   $= 1.0 - 0.9599 = 0.0401$

About 4% of the market have waist measures too large to be fitted by the style mentioned. This problem is illustrated in Figure 6-10.

Situations arise in working with normal distributions where we must find a value of the random variable with a particular associated cumulative probability.

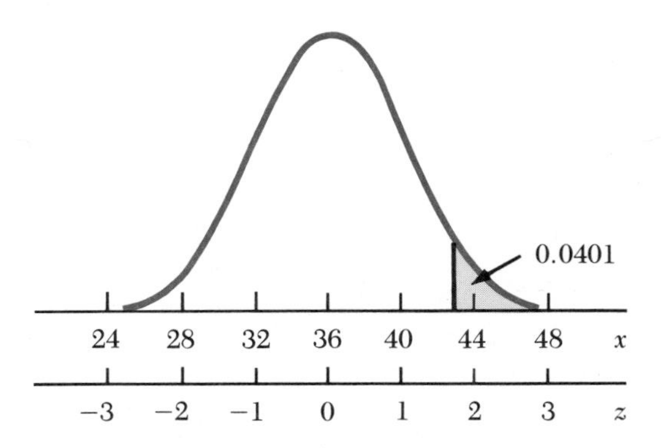

Figure 6-10 An Upper-Tail Normal Probability

Example 4

A study showed that monthly energy consumption in British thermal units (BTU's) per square foot in industrial buildings in a large city in February was normally distributed with a mean of 40 and a standard deviation of 5. The energy agency sponsoring the study wants to single out the lowest and highest 5% of energy-using buildings to find the factors contributing to low and high energy use. What monthly consumption levels should be used to classify the buildings?

Figure 6-11 shows the standard normal distribution. This is the graph of probability density of standard scores, z. The second horizontal axis shows corresponding energy consumption values. These reflect the relationship

$$x = \mu + z\sigma$$

In our case,

$$x = 40 + z(5)$$

For example, a z-score of +2.00 means an x-value of two standard deviations above the mean, which is at 40 + 2.00(5) = 50. The energy agency wants to find the consumption level that has an associated cumulative probability of 0.05—the fifth percentile of the energy consumption distribution. In Appendix Table A-3, we read that the 0.05 cumulative probability occurs at $z = -1.65$. We then substitute −1.65 in the following expression.

$$\begin{aligned} x_{0.05} &= \mu + z_{0.05}\sigma \\ &= 40 + (-1.65)(5) \\ &= 31.75 \text{ BTU/ft}^2 \end{aligned}$$

The 0.95 cumulative probability level, or 95th percentile of energy consumption, is found similarly from

$$\begin{aligned} x_{0.95} &= \mu + z_{0.95}\sigma \\ &= 40 + (1.65)(5) \\ &= 48.25 \text{ BTU/ft}^2 \end{aligned}$$

In another kind of situation, we know the standard deviation of a normal distribution and a particular cutoff point in terms of a percentile and need to find the mean of the distribution.

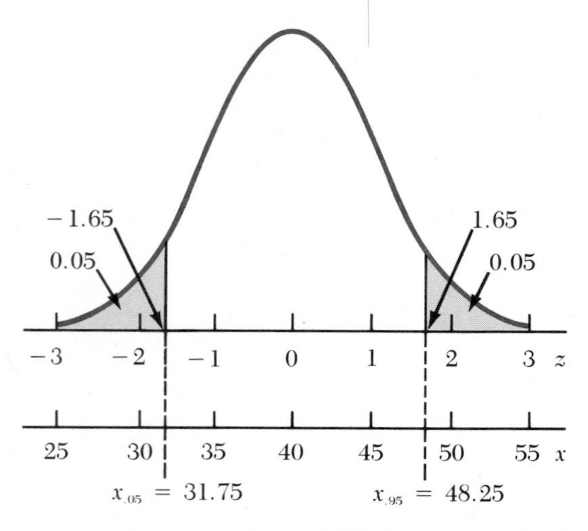

Figure 6-11 The 5th and 95th Percentiles of a Normal Distribution

Example 5

A cola bottler's filling machinery will fill "half-gallon" or 1892.5-milliliter containers to any mean fill setting, but the standard deviation of the fills will be 10 ml. The bottler's policy is to insure that 80% of the containers have at least a full half-gallon of cola. What mean fill setting should be maintained?

The mean of the normal distribution of fills is the fill setting we want to determine. The setting should be such that 1892.5 ml is at the 20th percentile of the distribution of fills. We solve for μ in $x = \mu + z\sigma$.

$$\mu = x - z\sigma$$

The 20th percentile of z is -0.84 and

$$\begin{aligned} \mu &= 1892.5 - (-0.84)(10) \\ &= 1900.9 \text{ ml} \end{aligned}$$

Figure 6-12 summarizes the problem.

One reason for emphasis on the normal distribution is that it will be found in later chapters to be important in statistical inference. While many observed variables will conform more or less to the normal distribution, we should be aware that many will not. Other continuous distribution models may be suitable for these variables. For example, some variables conform to a class of extreme right-skewed distributions called *exponential* distributions. Exponential distributions have a density function

$$f(x) = \mu e^{-(1/\mu)x}$$

where μ is the population mean. The cumulative exponential distribution function is

$$P(X \leq x) = 1 - e^{-(1/\mu)x}$$

The probability process that underlies the exponential distribution is frequently satisfied by variables involving the time or distance to the next occurrence of an

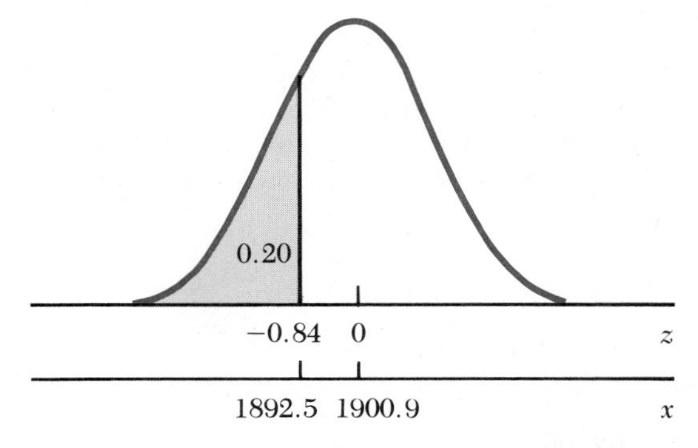

Figure 6-12 Solving for μ Given the 20th Percentile of a Normal Distribution with $\sigma = 10$

event. Examples are the time to next failure of equipment and time to next demand for service.

Exercises

6-9 *Normal percentiles.* Find the following percentiles of the standard normal distribution from Table A-3. Then find the cumulative probability associated with the z values from Table A-2.

a. $z_{.95}$ b. $z_{.99}$ c. $z_{.995}$ d. $z_{.999}$

6-10 *Cumulative normal.* Plot the cumulative standard normal distribution by finding the z values for the 0.01, 0.05, 0.10, 0.20, 0.30, 0.40, 0.50, 0.60, 0.70, 0.80, 0.90, 0.95, and 0.99 cumulative probabilities.

6-11 *Z values.* Weights of sacks of potatoes are normally distributed with a mean of 50 lb and a standard deviation of 2 lb. Convert each of the following probability statements into statements about the standardized normal variable, Z.

a. $P(X \leq 55)$
b. $P(47 \leq X \leq 53)$
c. $P(X < 46 \text{ or } X > 54)$

6-12 *Follow-on.* Find the probabilities in Exercise 6-11.

6-13 *Normal distribution.* In a normal distribution with a mean of 40 and a standard deviation of 10 find each of the following.

a. $P(X \leq 36)$
b. $P(32 \leq X \leq 44)$
c. $P(X < 25 \text{ or } X > 55)$
d. The 15th percentile of X.
e. The values that cut off the most extreme 5% of X values.
f. The range within which the middle 80% of X values lie.

6-14 *Typing speeds.* The times taken by applicants for clerical positions to type a standard passage were normally distributed with a mean of 140 sec and a standard deviation of 20 sec. Find the probability of each completion time.

a. Not exceeding 120 sec.
b. More than 155 sec.
c. Between 120 and 180 sec.

6-15 *Weights of concrete mix.* The weights of mixes of 4 ft^3 of a particular concrete mixed by a plant vary owing to the moisture content of the sand and the composition of the aggregate used. The weights are normally distributed with a mean of 600 lb and a standard deviation of 8 lb. What is the probability that 4 ft^3 of mix have each weight?

a. Less than 610 lb.
b. Less than 580 lb.
c. Between 590 and 610 lb.

6-16 *Percentiles of typing times.* In Exercise 6-14, what typing time satisfies each condition?

a. Cuts off the fastest 10% of the typists.
b. Cuts off the slowest 25% of the typists.

6-17 *Percentiles of mix weights.* In Exercise 6-15, what weight of 4 ft^3 of mix satisfies each condition?

a. Is exceeded with probability 0.20.
b. Is exceeded with probability 0.02.

6-18 *Full measure.* A machine fills bottles of wine with an average of 1.0 liters. The fills are normally distributed with a standard deviation of 0.04 liters.

a. Below what value will the shortest 10% of the fills lie?
b. If the bottler wants to guarantee that no more than 1% of the fills will be short of 1.0 liter, what mean fill will he have to provide?
c. An improved filling machine has a standard deviation of 0.02 liters. What percentage saving will this secure for the bottler in meeting the guarantee in (b)?

6-19 *Gasoline pumpage.* The amount of gasoline pumped at a service station each week is normally distributed with a mean of 20,000 gal and a standard deviation of 3600 gal.

a. What is the probability that the gallons pumped in a week is within 5000 of the expected value?
b. Find the weekly sales in gallons that cuts off the middle 50% of the distribution, that is, $x_{0.25}$ and $x_{0.75}$.

6-20 *Assembly times.* Completion times for an assembly operation among a group of workers are normally distributed with a mean of 80 sec and a standard deviation of 10 sec.

a. Find the completion times that cut off the fastest 2.5% and the slowest 2.5% of the times.
b. A particular worker had a completion time of 96 sec. What is the probability of a deviation this large or larger (in either direction) from the expected time?

6-21 *Drill press setting.* The actual sizes of holes drilled by a machine at any setting will vary slightly from the setting. In a given application, the differences are normally distributed with a mean of zero and a standard deviation of 0.2 mm. What should be the setting of the machine if holes are to be drilled that have a 0.99 probability of exceeding a size (diameter) of 25 mm?

6-22 *Bus arrivals.* A bus line has found that on certain intercity trips the difference between actual and planned arrival times is normally distributed with a mean of zero and a standard deviation of 10 min. A scheduled arrival time is, of course, an estimated time of arrival. If the company schedules an arrival time of 3 P.M. in a city, what should be each planned arrival time?

a. To make the probability 0.80 that the bus is not late.
b. If a 0.70 probability of a late bus is permitted.

6-23 *Dow Jones stock index.** Monthly percentage changes in the Dow Jones stock price index from 1949 to 1970 were normally distributed with a mean of +0.65 and a standard deviation of 3.5. Answer each question if a month is selected at random.

a. What is the probability that the index rose during the month?
b. What is the probability of an increase exceeding 5%?
c. What percentage change is exceeded with probability 0.05?

6-24 *National Bank of San Diego.*† The United States National Bank of San Diego was placed on the "serious problems" list of the Federal Deposit Insurance Corporation on November 7, 1972 and was declared insolvent by the comptroller of the currency on

*R.M. Fleming, *Journal of Business*, July 1973
†J. F. Sinkey, Jr., *Journal of Bank Research*, Spring 1975

October 8, 1973. Below are 1972 figures for certain operating ratios for the San Diego bank and a "control group" of 30 other California banks.

Ratios to Income (%)

	Total Expenses Control Banks			Occupancy Expenses Control Banks		
	μ	σ	USNB	μ	σ	USNB
1971	85.8	6.8	92.7	4.8	1.7	8.6
1972	85.8	7.6	101.2	4.8	2.0	8.1

Assume the control banks represent a standard of acceptable practice and that the distributions are normal. In each of the four cases, what is the probability that a bank operating under acceptable standards has a ratio as high as (or higher than) the San Diego bank?

6.4 Normal Approximation to the Binomial

When the sample size is adequately large, binomial distributions can be closely approximated by normal probabilities. Availability of the normal approximation means that cumbersome binomial tables for large values of n are not needed. Figure 6-13 shows binomial distributions for $\pi = 0.10$ and $\pi = 0.50$, with samples of 5, 10, 20, and 50. The graphs for $\pi = 0.10$ show decreasing skewness as n increases. Also present in the graphs is a definite suggestion of the normal distribution shape as n increases. The normal shape seems to appear sooner for $\pi = 0.50$ than for $\pi = 0.10$. Of course, binomial distributions are discrete, while the normal distribution is continuous. But as n is increased, the ordinates for the binomial become more numerous, which gives a suggestion of continuity.

As n increases, binomial distributions become closer and closer to the normal distribution shape and may be approximated by normal probabilities. A useful rule of thumb is that the normal distribution provides satisfactory approximations to binomial distributions as long as both $n\pi$ and $n(1 - \pi)$ equal or exceed 5. For example, if $\pi = 0.10$ we need to have $n \geq 50$ to use the normal approximation. If $\pi = 0.50$ then $n \geq 10$ would be sufficient.

We will illustrate the normal approximation to binomial probabilities with the binomial for $n = 16$ and $\pi = 0.50$. For this binomial

$$\mu = n\pi = 16(0.50) = 8.0$$

and

$$\sigma = \sqrt{n\pi(1 - \pi)} = \sqrt{16(0.50)(0.50)} = 2.0$$

To approximate the binomial probability that $r = 9$, for example, we need to convert the integer value, $r = 9$, to the continuous interval $8.5 \leq X \leq 9.5$. We then find $P(8.5 \leq X \leq 9.5)$ for a normal distribution with $\mu = 8.0$ and $\sigma = 2.0$.

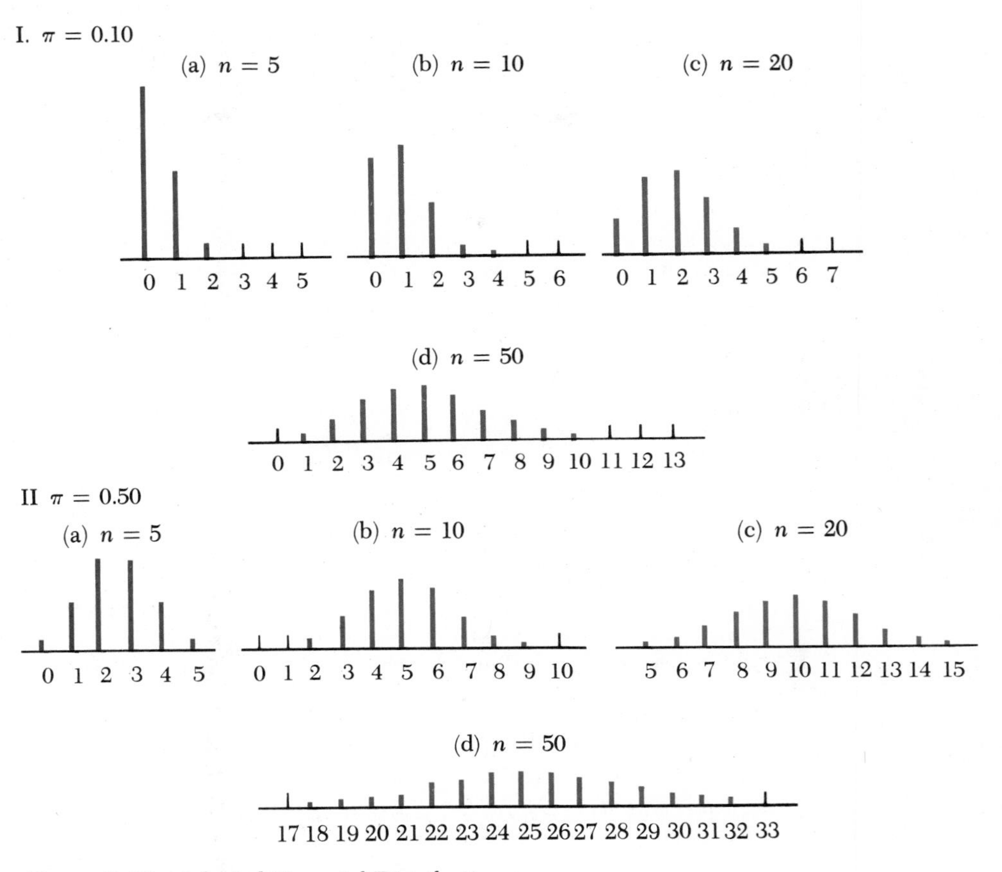

Figure 6-13 Selected Binomial Distributions

$$P(X \le 9.5) = P\left(Z \le \frac{9.5 - 8.0}{2.0}\right)$$

$$= P(Z \le 0.75) = 0.7734$$

$$P(X \le 8.5) = P\left(Z \le \frac{8.5 - 8.0}{2.0}\right)$$

$$= P(Z \le 0.25) = 0.5987$$

$$P(8.5 \le X \le 9.5) = 0.7734 - 0.5987$$

$$= 0.1747$$

This can be compared with the binomial probability for $r = 9$ given $n = 16$ and $\pi = 0.50$ in Appendix Table A-1 of 0.1746.

To find the normal approximation to a cumulative binomial probability, care must be taken about the limits. For example, let us approximate the binomial probability $P(R > 10 \mid n = 16, \ \pi = 0.50)$. This is the probability that r equals or exceeds 11, and we thus need the normal probability for the interval $x > 10.5$.

$$P(X > 10.5) = P\left(Z > \frac{10.5 - 8.0}{2.0}\right) = P(Z > 1.25)$$
$$= 1 - P(Z \leq 1.25) = 1 - 0.8944$$
$$= 0.1056$$

This approximates the sum of the binomial probabilities in Table A-1 for $r = 11$, 12, 13, 14, 15, and 16 for $n = 16$ and $\pi = 0.50$ of 0.1050.

Figure 6-14 provides a visual appreciation of the integer binomial variable versus the continuous normal variable. For the probability that r exceeds 10 we must include the "slices" of probability for integers of 11 and more, which begin at 10.5.

Examples

Some examples of normal approximations to binomial probabilities are carried out below.

Example 1

In the past, 10% of the items sold in the small-appliance department of a discount store have been returned. Of the most recent 100 items sold, 18 were returned. Is this many or more returns unusual in the light of past experience?

The binomial probability of 18 or more returns is approximated by the normal probability for $X \geq 17.5$ in a normal distribution with

$$\mu = n\pi = 100(0.10) = 10$$

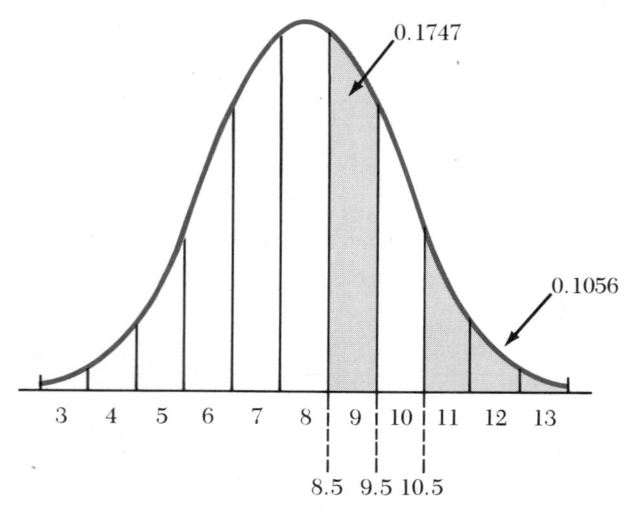

Figure 6-14 Normal Approximation to Binomial Probabilities

and

$$\sigma = \sqrt{n\pi(1-\pi)} = \sqrt{100(0.10)(1-0.10)} = 3$$

Then

$$P(X \geq 17.5) = P\left(Z \geq \frac{17.5-10}{3}\right)$$

$$= P(Z \geq 2.50)$$

$$= 1 - 0.9962$$

$$= 0.0038$$

The observed number of returns does seem unusual, providing that returns are a Bernoulli process. If the probability of a return normally goes up in connection with pre-Christmas sales, for example, the binomial model would not be applicable.

Example 2

A standard for Grade A pinto beans packed by a farm marketing cooperative is that no more than 5% of the beans be too soft. Otherwise beans received are processed and packed as "refries." A shipment received will be packed as refries if the number too soft in a sample of 200 falls in the upper 10% of the binomial distribution for $\pi = 0.05$. What is the smallest number of soft beans that will qualify a lot for processing as refries?

This is a "percentile" problem. The normal distribution that approximates a binomial with $\pi = 0.05$ and $n = 200$ is one with

$$\mu = n\pi = 200(0.05) = 10$$

and

$$\sigma = \sqrt{n\pi(1-\pi)} = \sqrt{200(0.05)(0.95)} = 3.08$$

In the normal distribution of X with this mean and standard deviation, the 90th percentile is at

$$x_{0.90} = 10 + 1.28(3.08)$$

$$= 13.94$$

The smallest integer that has an associated interval whose lower limit exceeds 13.94 is 15, with an associated lower limit of 14.5. Therefore, we classify a lot for refries if 15 or more out of 200 beans are too soft. The actual probability that 15 or more are too soft when $\pi = 0.05$ is

$$P(X \geq 14.5) = P\left(Z \geq \frac{14.5-10}{3.08}\right)$$

$$= P(Z \geq 1.46)$$

$$= 1 - 0.9279$$

$$= 0.0721$$

The entire "slice" of probability for the integer 15 falls in the upper 10% of the approximating normal distribution. While $x_{0.90}$ came to 13.94, "14 beans" is not entirely in the upper 10% of the distribution. Part of the interval from 13.5 to 14.5 is below the 90th percentile of x. The situation is sketched in Figure 6-15.

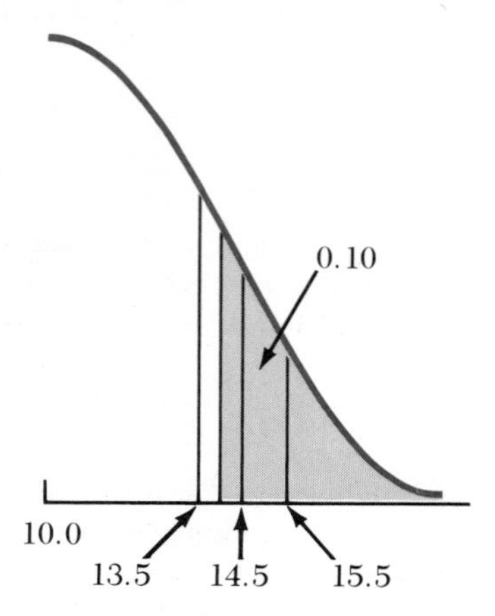

Figure 6-15 Normal Approximation to Binomial with $\pi = 0.05$, $n = 200$

Example 3

A customer purchased 100 hyacinth bulbs from a nursery. The bulbs were advertised as 87.5% sure to produce plants if planting directions were followed. The customer follows directions and 83 plants are produced. Does the customer have cause for complaint?

If the probability of 83 or fewer successes out of 100 trials of a Bernoulli process with $\pi = 0.875$ is very small, the customer would have a basis for complaining. In the normal approximation to this binomial,

$$\mu = n\pi = 100(0.875) = 87.5$$

and

$$\sigma = \sqrt{n\pi(1-\pi)} = \sqrt{100(0.875)(0.125)} = 3.31$$

For 83 or fewer successes, x_U is 83.5, and

$$P(X \leq 83.5) = P\left(Z \leq \frac{83.5 - 87.5}{3.31}\right)$$

$$= P(Z \leq -1.21)$$

$$= 0.1131$$

There would not seem to be a basis for complaint. What happened to this customer would happen to about one customer in nine if the advertisement were perfectly true.

Exercises

6-25 *Normal approximations.* Find the normal approximations to the following binomial probabilities and compare the results with the true binomial probabilities from Appendix Table A-2.

a. $P(R \geq 12 \mid n = 16, \pi = 0.5)$
b. $P(R = 6 \mid n = 16, \pi = 0.5)$
c. $P(R < 7 \mid n = 14, \pi = 0.6)$

6-26 *Acceptance sampling.* In an acceptance sampling procedure, a random sample of 144 parts was taken from each large lot of parts and lot quality declared unacceptable if 8 or more defective parts were found. What is the probability (normal approximation) under this scheme of each of the following?

a. Declaring a lot unacceptable when 4 percent of the parts are defective.
b. Declaring a lot unacceptable when 10 percent of the parts are defective.

6-27 *Melon cull.* A grower counted on having to cull out 20% of his melons due to inferior quality. A sample of 36 melons yields 10 below grade and another sample of 64 melons yields 15 below grade. Find the probability of as many or more culls if $\pi = 0.20$ for each case.

a. The first sample.
b. The second sample.
c. The two samples considered together.

6-28 *Student gripe.* On a true-false exam of 50 items, a student who got 35 items correct claimed that her real knowledge of the subject matter was "B" level, that is, 80% or better. If the questions were a random sample of the subject matter, what is the probability of a test score as bad or worse than 35 items correct if the student's real knowledge level was 80%?

6-29 *Gasket shipments.* Samples of 225 gaskets from each large production run are to be subjected to a watertightness test. A run is not to be shipped out when the probability of obtaining as many or more than the observed number of defectives from a 4% defective run is less than 0.05. What is the smallest number of defectives in the sample that will prevent shipment of a run?

6-30 *Allocating staff.* A metropolitan court system evaluated the probability at 0.10 that a person released on his or her own recognizance pending appearance in court would not appear. If 400 persons are so released in a month and the court wants to provide sufficient staff so that the probability of being able to check on nonappearances is 0.90, how many nonappearances should the court plan for? Assume a Bernoulli process.

6-31 *Males born at sea.* In 14 of the 20 years from 1871 to 1890, the number of males born at sea en route to the United States exceeded the number of females born en route. Use the normal approximation to the binomial to find the probability of an excess of males in this many or more years if the probability of an excess of male births in any year is 0.50.

Summary

We can define the probability of an outcome within an interval of the variable for a continuous random variable, but not the probability of an exact outcome. The *probability density function* of a continuous random variable represents the frequency curve of such probabilities as the interval being considered becomes smaller and smaller. The probability for any interval of the variable can then be represented by the proportion of total area under the probability density function that is within the interval.

The *normal distribution* is a particular continuous distribution model with widespread uses in applied statistics. The probability density of a normal distribution depends only on the mean and standard deviation of the random variable. A consequence of this is that, when reduced to standard scores, all normal distributions are the same. We can find probabilities under the *standard* normal distribution from tables of its cumulative distribution function. These tables are entered with values of the *standard normal deviate*, z. The tables can also be used to find values of a normal random variable associated with particular cumulative probabilities.

As n, the number of *Bernoulli trials* in a binomial distribution, becomes large, the binomial distribution becomes more "normal" in shape. We can use the continuous standard normal distribution to approximate probabilities for such discrete binomial distributions.

Supplementary Exercises

6-32 *Vehicle speeds.* A study of traffic speeds concluded that the speeds of vehicles passing a highway checkpoint were normally distributed with a mean of 52.5 mph and a standard deviation of 7.5 mph. What proportion of vehicles exceeded each speed limit?

a. The 55 mph speed limit.
b. A 65 mph citation limit.
c. A speed of 75 mph.

6-33 *Performance times.* A performance measurement team found that the time taken by employees to perform a certain assembly task was normally distributed with a mean of 78 sec and a standard deviation of 5 sec. What completion time is exceeded with each probability?

a. 0.20 b. 0.10 c. 0.01

6-34 *Moving times.* Household Transfer Company's estimate of the time required for a certain move is a normally distributed random variable with a mean of 20 h and a standard deviation of 1.5 h. What is the probability that the actual time will satisfy each condition?

a. Be within 2 h of the expected time.
b. Be more than 3 h from the expected time.
c. Be more than 2.5 h over the expected time.

6-35 *Determining a process mean.* The percentages (by weight) of the components of an animal feed in 50-pound bags can be controlled by a manufacturing process so that the percentages are normally distributed with a standard deviation of 0.8 percentage points. On the average, the process can deliver any desired or "target" percentage. What should the target percentages be in each case?

a. The manufacturer wants to be 99% sure that the percentage of protein is at least 20.
b. The manufacturer wants to be 99% sure that the percentage of moisture does not exceed 12.

6-36 *Birth and death months.** In a study of notable Americans, 42 out of 354 notables were found to have died in the same month that they were born.

a. Use the normal approximation to the binomial to find the probability of 42 or more deaths in the birth month if the true probability were $1/12$ for a Bernoulli process.
b. Would a low probability for (a) lead you to question the assumption of a probability of $1/12$?

6-37 *Pilot test-marketing.* Thirty-six homemakers are to be randomly selected for a comparison test of a manufacturer's "new and improved" furniture polish with the same manufacturer's current polish. The manufacturer is willing to put the new and improved polish into more extensive marketing tests if the number out of 36 who prefer it to the current polish falls in the upper 2.5% of the binomial distribution for $\pi = 0.50$ and $n = 36$. Use the normal approximation to the binomial distribution to answer each question.

a. What is the probability of 24 or more in the binomial distribution?
b. What is the probability of 25 or more in the binomial distribution?
c. What result of the homemaker test will result in more extensive marketing tests?

6-38 *Root infestation.* Sixteen out of 100 randomly selected plants in a large field were found to have a root infestation. Use the normal approximation to the binomial to find each probability.

a. The probability of 16 or more infested plants if the percentage of infested plants in the field is 10.
b. The probability of 16 or fewer infested plants if the percentage of infested plants in the field is 20.

6-39 *Drilling specifications.* The actual diameters of holes drilled by a machine are normally distributed with a standard deviation of 0.10 mm. The holes must accommodate a friction fitting which will not work if the holes are too large. What should be the specification (mean) diameter in order to limit the probability of a diameter in excess of 50 mm to 0.01?

6-40 *Predicting active-duty terms.*† Prior to receiving their commissions and entering four years of military obligation, cadets in the class of 1966 at the U.S. Military Academy completed a questionnaire; one of the questions asked for a rating on a scale from 1 to 100 of the cadet's confidence that he would continue in military service after his

---

*D. B. Phillips, "Deathday and Birthday: An Unexpected Connection," *Statistics: A Guide to the Unknown*, edited by Judith M. Tanur, Holden-Day, Inc., San Francisco, 1972, p. 60.

†R. P. Butler and C. F. Bridges, *Journal of Occupational Psychology*, 1978, p. 51.

obligatory term. The following table shows the results on this question for those who remained on active duty and those who had resigned as of seven years after graduation.

Class of 1966	n	μ	σ
Active duty	257	60	11.0
Resigned	139	45	12.2

Assume both distributions are continuously normal and find the probability of remaining on active duty after seven years for each group.

a. Cadets scoring below 45.
b. Cadets scoring between 45 and 60.
c. Cadets scoring over 60.

6.5 The Sum of Two Independent Normal Random Variables (Optional)

In Chapter 5 we encountered situations in which a random variable was defined by the joint outcomes of two other random variables. When the outcomes of the two variables were independent, we multiplied the marginal probabilities of the numerical outcomes of the one variable by those for the other to obtain the probabilities for the joint outcomes. We used accidents to two rigs operated by a trucking company to illustrate this idea. Once we obtained the joint probabilities of numbers of accidents to Rig A and Rig B, the probability distribution of total accidents to the two rigs was easily tabulated.

The equivalent procedure for continuous variables is beyond the scope of this text. However, certain useful results are known. They will be stated and illustrated here.

Result 1

The *expected value of the sum of two random variables* is the sum of their individual expected values.

Result 2

The *variance of the sum of two independent random variables* is the sum of their individual variances.

Result 3

A random variable that is the sum (or difference) of two independent normal random variables is a *normally distributed random variable*.

The first result applies in general—to discrete and continuous variables and whether or not the variables are independent. The second result about variances applies to discrete and continuous variables, but the variables must be independent. The third result is a statement about independent normal random variables.

When we are concerned with the probability distribution of the sum of two independent normal random variables, we can use these results to find particular probabilities.

Example

In order to grow a vegetable crop without supplemental irrigation in an area, it is important to have at least 8 in. of rain in April and May. Weather records indicate that April rainfall is normal with a mean of 5.4 in. and a standard deviation of 0.8 in., and May rainfall is normal with a mean of 3.8 in. and a standard deviation of 0.6 in. May rainfall is independent of April rainfall. What is the probability of adequate rainfall?

We let April rainfall be X and May rainfall be Y. We are concerned with total rainfall, $X + Y$. Result 1 says

$$\begin{aligned} E(X + Y) &= E(X) + E(Y) \\ &= 5.4 + 3.8 \\ &= 9.2 \end{aligned}$$

Result 2 says that

$$\begin{aligned} \text{Var}(X + Y) &= \text{Var}(X) + \text{Var}(Y) \\ &= (0.8)^2 + (0.6)^2 \\ &= 1.0 \end{aligned}$$

Result 3 tells us that the probability distribution of total rainfall is normal. From the other results, we know that the distribution has a mean (expected value) of 9.2 and a standard deviation of 1.0 inches. For the probability of at least 8 in., we have

$$\begin{aligned} z &= \frac{8.0 - 9.2}{1.0} \\ &= -1.20 \end{aligned}$$

and

$$P(Z \geq -1.2) = 0.8849$$

While this probability is high, the chance of insufficient rainfall may be large enough to warrant installation of standby irrigation capability.

Results 1, 2, and 3, used in the preceding example, apply to differences as well as to sums of independent normal random variables, since subtracting a value is the same as adding the value with a change in sign. That is,

$$X - Y = X + (-Y)$$

We can illustrate Results 1 and 2 for sums and differences using the example of truck breakdowns from page 127. The separate distributions were as follows.

x	0	1	2	*Mean*	*Variance*
$P(x_A)$	0.70	0.25	0.05	0.35	0.3275
$P(x_B)$	0.90	0.10		0.10	0.09

The probability distribution of $X_A + X_B$ under independence is:

$x_A + x_B$	0	1	2	3	*Mean*	*Variance*
$P(x_A + x_B)$	0.630	0.295	0.070	0.005	0.45	0.4175

The probability distribution of $X_A - X_B$ under independence is:

$x_A - x_B$	−1	0	1	2	*Mean*	*Variance*
$P(x_A - x_B)$	0.070	0.655	0.230	0.045	0.25	0.4175

The probability distribution of the sum was worked out on page 128. The probability distribution of the difference can be found by a similar method. A general proof of Result 2 requires relationships introduced in Chapter 10. You can gain an appreciation for the result, which is not intuitively evident, by working through the details of the above example.

For an example of the difference between a pair of independent normally distributed random variables, consider a grain dealer who makes a subjective probability forecast of receipts and another subjective forecast of shipments. The difference between these two random variables is the net change in the dealer's inventory. If the receipts and shipments forecasts are *independent* normally distributed random variables, then the change in inventory will be a normally distributed random variable. The expected change in inventory will be the *difference* between the expected receipts and expected shipments. The variance of the change in inventory will be the *sum* of the variances of receipts and shipments.

Exercises

6-41 *Egg 'n' Foam.** An analysis of alternative investment strategies for producing a new plastic package as a substitute for paper egg cartons produced normal distributions of prospective values of investment with the following means and standard deviations.

*D. B. Hertz and H. Thomas, *Sloan Management Review*, Winter, 1983, pp. 17–31.

Strategy	*Mean*	*Standard Deviation*
Superplant and fast investment	\$3.5 mil	\$ .8 mil
Superplant and slow investment	.7	.4
Small plant and fast investment	.9	1.4

a. What is the probability that investment value will be positive under each of the second two strategies?
b. Under the first strategy what is the investment value that has a probability of 0.95 of being exceeded?
c. If the first strategy is not available, how would you compare the second and third strategies?

6-42 *Release mechanism.* In order to release oceanographic measuring devices from beneath the sea, a release mechanism is used in which a steel knife edge corrodes one or more magnesium pins. In a particular instance, a research vessel has set a device in place with two pins for release. Corrosion time for the first pin is normal with a mean of 30 h and a standard deviation of 5 h; the corrosion time of the second pin is normally distributed with a mean of 20 h and a standard deviation of 3 h. Assume independence and find the earliest elapsed time at which the vessel should be scheduled to return if the captain wants the probability to be 0.90 that the device has been released.

6-43 *Cash planning.* The Western Mercantile Company currently has a balance of cash on hand of \$40,000. The treasurer has formed independent normal probability distributions for next month's cash receipts and disbursements as follows.

	Mean	*Standard Deviation*
Receipts	\$100,000	\$15,000
Disbursements	\$130,000	\$8,000

a. What is the probability of a negative cash balance, that is, that the company will be unable to meet anticipated disbursements, at the end of next month?
b. If the company takes out a \$20,000 short-term loan, what will the probability in (a) become?

6-44 *Consolidating inventories.* A company supplies liquid fertilizer from two depots located 500 mi apart in an agricultural area. Demand at Alpha Depot during the growing season is normally distributed with a mean of 5000 gal and a standard deviation of 400 gal, while at Beta Depot, the mean and standard deviation are 8000 and 600 gal, respectively. The distribution at Beta is independent of Alpha and is normal.

a. How many gallons must be stocked at Alpha to limit the probability of not meeting total demand to 0.01?
b. How many gallons must be stocked at Beta to provide the same level of service there?
c. The company is considering an overnight trucking service that would effectively (at a cost) pool the stocks at the two locations because orders at one location could be filled from stocks at the other. How many total gallons must be stocked under this plan in order to meet 99% of all demand?

6-45 *Probability forecasting.* Sales of potash produced in the United States depend on the effective demand for domestic and for foreign (export) sales. The Potash Institute has formed probability forecasts for next year as follows.

	Thousands of Short Tons	
	Mean	*Standard deviation*
Domestic	1500	100
Foreign	1000	80

Answer each question if the level of domestic sales anticipated is independent of the level of foreign sales anticipated.

a. What is the probability that total sales will be between 2300 and 2700 thousand short tons?
b. What range about the expected value has a 0.95 probability of including the actual total sales?

6-46 *Cost estimates.* An architect is asked to make a cost estimate for a city parks-and-recreation construction project that will be included in a future bond issue if approved by the voters. Raw materials prices, labor rates, and productivity in six months, when the project would be built, are uncertain. The architect's expectations concerning materials costs form a normal distribution with a mean of $1,000,000 and a standard deviation of $80,000, while labor costs are normally distributed with a mean of $500,000 and a standard deviation of $60,000. Expectations about the two cost components are independent. What total cost should the architect recommend for the bond issue in each case?

a. If the mayor is willing to run a 25% risk that the bond money would be insufficient.
b. If the mayor is willing to submit the proposition to the voters with a 60% chance that the bond money would be insufficient.

6-47 *Sums and differences.* The time to complete task A is normally distributed with a mean of 50 sec and a standard deviation of 8 sec. Times for task B are independent of A and are normally distributed with a mean of 40 sec and a standard deviation of 6 sec. Find the probability

a. That the total time for the two tasks is less than 100 sec.
b. That task B takes longer than task A; that is, time for A less time for B is negative.
c. That task A takes less than 40 sec and task B takes less than 30 sec.

6-48 *Investment analysis.** Over a period of 50 years the annual return from different investments is as given below.

Investment	*Mean*	*Standard Deviation*
Common stocks	11.6	22.4
Long-term corporate bonds	4.2	5.6
Long-term government bonds	3.5	5.8

*F. K. Reilly, *Investment Analysis and Portfolio Management*, Dryden Press, 1979.

Assume normality and independence in each of the following.

a. What is the probability of a positive return from a common stock?
b. What is the probability that a common stock returns more than a long-term corporate bond?
c. What is the probability that a long-term corporate bond returns more than a long-term government bond?

Computer Exercises

1 Source: Personnel data file.

a. Make a relative-frequency diagram of the variable salary. Does the distribution seem normal or skewed?
b. Repeat (a) for each management level. Give an explanation for the results.
c. Convert each salary to a standard score and make a relative-frequency diagram of the results.
d. For (c), what proportion of the values are within one standard deviation of the mean? Within two standard deviations? How do these values compare with the values predicted by normal curve theory?
e. What are the mean and standard deviation of any standardized distribution? Use the computer to verify this result for the standardized distribution found in (c).

7

SAMPLING METHODS AND SAMPLING DISTRIBUTIONS

Before introducing more topics, we shall briefly summarize what we have done and what lies ahead to clarify the role of the current chapter.

Descriptive Statistics Statistics deals with collections of numerical observations appropriate to a given investigation or problem. The most common summaries of quantitative characteristics are means and standard deviations, and the most common summaries of qualitative characteristics of units of observation are percentages (or proportions). Frequency distributions, especially of quantitative characteristics, are an important analytical and conceptual tool.

Probability We express the degree of belief in the occurrence of uncertain events with probability. We use the laws of probability to find the probability of complex events from the probabilities for simpler events that make up the complex event.

Random Variables We are often concerned with the numerical value associated with a randomly selected unit of observation. The Bernoulli random variable is associated with binary (0, 1) qualitative characteristics, and the normal random variable occurs in connection with many continuous quantitative characteristics.

Sampling and Inference It is often either impossible or impractical to collect data on all the units of observation relevant to a statistical study. Instead, we must study the summary characteristics of a sample of observations for what they tell us about the population. For example, we use the mean income of a sample to make an inference about population mean income or the percentage of defective products in a sample to make an inference about the percentage defective in the entire production lot. The sample mean is an uncertain numerical value because it depends on the particular collection of units of observation that are selected for the sample. We now want to study this random variable to learn how sample means behave with respect to the mean of the population from which they come.

In this chapter we shall see that the mean of a random sample from a specified population is related to the population mean by the laws of probability. Stated another way, the sample must be selected in a random manner in order for the probability laws that determine the relation between the sample mean and the population mean to hold. So we first discuss how to select a random sample.

7.1 Random Sampling

The map in Figure 7-1 shows all the occupied dwelling units in a hypothetical village called Georges Mills. Table 7-1 indicates whether each dwelling unit is owner- or renter-occupied and presents the most recent week's grocery bill for the occupant family. We shall use this list, which completely covers the universe of units of observation, to illustrate random sampling.

In Chapter 1, we defined a random sample as one that results from a process in which every element in the population has the same probability of selection on any given draw. Suppose we want a random sample of 12 households from the 60 households in Georges Mills. A random sample could be drawn either *with* or *without* replacement. Although a better method of "drawing" will be introduced shortly, suppose we were to number the households from 1 to 60 and place slips of paper with these numbers in an urn for drawing.

Sampling With Replacement A slip is drawn and the household recorded. Then the slip is returned to the urn. It would be possible for the same household to be included more than once in the sample.

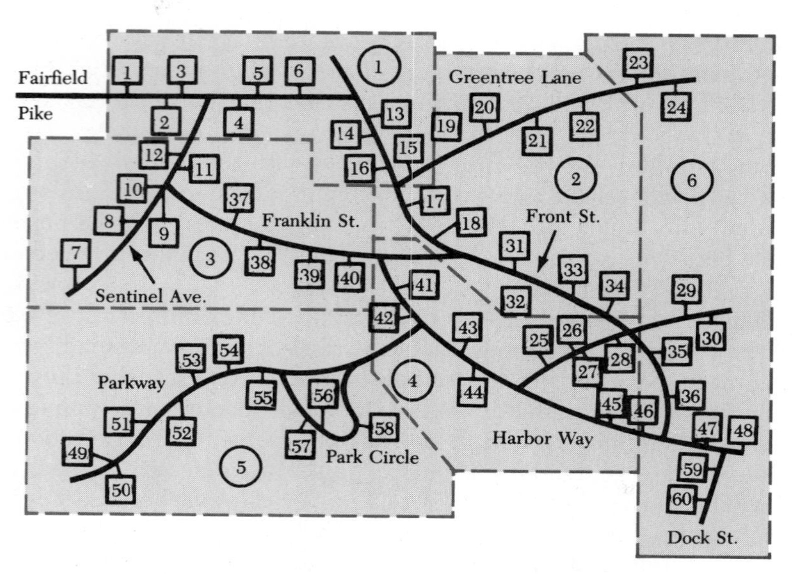

Figure 7-1 Map of Dwelling Units for Village of Georges Mills

Sampling Without Replacement Any slip (household) drawn is not returned to the urn before the next draw. It is guaranteed that a different unit of observation will be selected on each draw.

Most samples are actually drawn without replacement. There is no practical advantage to sampling with replacement. The simplest probability theory underlying sampling, however, applies to sampling with replacement. If this seems paradoxical, do not worry. In most situations the number of units selected for the sample is very small compared to the number of units in the population, and in those situations the theory of sampling with replacement is adequate. We shall show you the modifications anyway.

Random Number Tables

Table A-8 in Appendix A is a portion of a table of random rectangular numbers. It appears to be nothing more than a collection of the digits zero through 9 arranged in pairs with no discernible order, which is exactly the intention. If we machine a cylinder to have 10 balanced faces and label each face with one of the 10 digits, we can generate a table of random rectangular two-digit numbers by repeatedly rolling the cylinder twice and recording the two numbers that face up on each pair of rolls. The probability of each two-digit number occurring is $1/100$ and any two-digit number appears independently of the other two-digit numbers. A large table of such numbers approximates a population in which each two-digit number (00 through 99) appears independently and with equal probability. The numbers form a rectangular probability distribution.

Table 7-1 Dwelling Units with Ownership Status and Weekly Grocery Bill

Unit	*Bill*	*Unit*	*Bill*	*Unit*	*Bill*	*Unit*	*Bill*
1	$69.6	16	$79.8	31	$87.1	46	$75.2
2	65.2	17(R)	65.1	32	78.2	47(R)	72.2
3(R)*	61.5	18	63.2	33(R)	55.6	48	92.5
4(R)	48.0	19	68.2	34	72.6	49	80.7
5(R)	61.2	20(R)	51.6	35	76.4	50	91.2
6(R)	49.8	21(R)	55.8	36	72.9	51	77.3
7(R)	58.4	22(R)	46.0	37(R)	61.3	52	80.2
8	72.8	23	71.9	38(R)	55.1	53(R)	73.6
9(R)	79.1	24	53.1	39(R)	81.1	54	91.6
10(R)	51.3	25(R)	50.4	40(R)	68.8	55	75.0
11	56.3	26	75.4	41	86.0	56	78.8
12(R)	65.3	27	77.0	42	72.6	57(R)	67.5
13	77.0	28	63.5	43(R)	64.5	58(R)	69.0
14	80.8	29	70.0	44(R)	68.2	59	83.1
15(R)	50.4	30	73.9	45	85.4	60	82.1

* R indicates renter-occupied; all others are owner-occupied.

To use a random number table to draw a random sample (without replacement), we first number all the units of observation in the population. This has already been done on our dwelling unit map. Then we begin at an arbitrary starting point in a random number table and select our sample by reading successive random numbers. To select a sample of 12 units from 60 numbered units, two-digit random numbers are enough. We can decide that the random numbers 01 to 60 will correspond to the correspondingly numbered dwelling unit. Should we encounter the random numbers 00 or 61 through 99, we simply pass along to the next two-digit random number. Table 7-2 shows the sequence of random two-digit numbers found by starting in the first row of Table A-8. The third two-digit number, 73, does not select one of the 60 dwelling units. Notice that the thirteenth number, 35, has already been drawn (the ninth number). We ignore it and continue to read two-digit numbers until we have drawn 12 distinct numbers between 01 and 60. If sampling were with replacement, dwelling unit 35 would appear twice in the sample, and unit 29 would not have appeared at all.

The table below 7-2 (p. 190) gives the weekly grocery bill for each of the households selected in our sample, and the sample mean is calculated. The sample mean is

$$\bar{x} = \frac{\Sigma x}{n}$$

where x stands for the values of the sample observations and n is the sample size.

Table 7-2 Selection of a Random Sample of 12 Out of 60 Numbered Dwelling Units

Random Number	*Dwelling Unit Selected*	*Random Number*	*Dwelling Unit Selected*
10	10	67	—
09	9	35	—
73	—	48	48
25	25	76	—
33	33	80	—
76	—	95	—
52	52	90	—
01	1	91	—
35	35	17	17
86	—	39	39
34	34	29	29

Unit	x	*Unit*	x
10	\$51.3	35	\$76.4
9	79.1	34	72.6
25	50.4	48	92.5
33	55.6	17	65.1
52	80.2	39	81.1
1	69.6	29	70.0

$$\Sigma x = 843.9$$

$$\bar{x} = \frac{\Sigma x}{n} = \frac{843.9}{12} = \$70.32$$

The sample mean is, of course, a function of the particular set of households drawn in the sample. Many other equally likely combinations of households could have been selected as our sample; in fact, there are ${}_{60}C_{12} = 1.399 \times 10^{12}$ different combinations of 12 households that could have been selected from the 60 households in the community.

> A simple random sample from a finite population is *a sample selected in a manner that gives each possible sample combination an equal chance of being chosen.*

Exercises

7-1 *Sampling situations.* Comment on the relative ease of selecting a simple random sample in the following situations.
  a. A large insurance company wishes to sample its agents to determine their work patterns and satisfaction with company compensation and benefits.
  b. The Consumer Products Safety Commission wants a study of conditions surrounding bicycle accidents involving children 10 to 14 years of age.
  c. A regional forester wants to determine the percentage of mature trees in an area of 36 mi^2 that is being damaged by an insect infestation.

7-2 *Random sampling with replacement.* Use the table of random numbers to select a random sample, with replacement, of 10 households from the list of 60 households in Table 7-1. There are 100 rows in the random number table. Start with the row number corresponding to the last two digits of your social security number. Make a systematic record of your selection process, in the manner of Table 7-2.

7-3 *Random sampling without replacement.* Use the table of random numbers to select a random sample, without replacement, of 10 households from the list of 60 households in Table 7-1. Start with the row number corresponding to 100 minus the last two digits of your social security number. Make a systematic record of your selection process.

7-4 *Home sales.* Regard the 100 home sales in Appendix B-3 as the population of sales in a community over a limited period of time. Select a random sample of size 16, with replacement, from the population. Start with a row number in the random number table corresponding to 100 minus the last two digits of your social security number. Record the purchase price and whether the home was an old or new home.

7.2 Sampling Distribution of the Mean

In the preceding section, we saw how to select a random sample with the aid of a table of random numbers. We selected a random sample (without replacement) of 12 bills from a population of 60 grocery bills. The mean of the particular sample we selected was $70.32.

It is evident that we could have selected many other samples of size $n = 12$, depending on where we started in the random number table. Each of the many other possible samples would yield a specific value of the sample mean. In other words, the sample mean is a random variable. Our interest now centers on the behavior of this random variable—specifically, how the sample mean is related to the underlying population from which the sample is drawn.

A Computer-Simulated Example

In our example of grocery bills in Georges Mills, we could in principle study the behavior of sample means from samples of 12 households in two ways. We could do this only because Table 7-1 gives the listing of grocery bills for the entire population. The two ways are:

1. *Exact Probability Approach.* List all of the equally likely samples of 12 households and calculate the mean for each sample. Then, organize these means into a frequency distribution in the manner of Chapter 2. Given the number of possible sample combinations, this is not practical for our current example.
2. *Simulation Approach.* Select several hundred samples of 12 households by using the random number table. Then, calculate the sample means and construct a frequency distribution of them. This distribution will approximate the results of the first approach. We shall employ this method.

First, consider the population of grocery bills. The 60 bills in Table 1 have a mean of $69.79 and a standard deviation of $11.72. The frequency distribution of the 60 bills is listed below. It is unimodal with some skewness to the left.

Dollar Amount	*40–49.99*	*50–59.99*	*60–69.99*	*70–79.99*	*80–89.99*	*90–99.99*
Frequency	3	10	15	20	9	3

We used the computer to draw 1000 samples of size 12 (without replacement) from the population of grocery bills. The mean of each of the 1000 samples was calculated and the 1000 means were grouped into a frequency distribution. The result is shown graphically in Figure 7-2, along with the population distribution for comparison. The variable scale in Figure 7-2 accommodates both values of x from the population distribution and values of $\bar{x}$ from the frequency distribution of sample means.

Observe first from Figure 7-2 that the frequency distribution of sample means appears to concentrate around the population mean, μ. Second, note that the dispersion of the sample means is much less than the dispersion of the underlying population that was sampled. Finally, note that, while the underlying population was noticeably skewed, the distribution of sample means is essentially symmetrical.

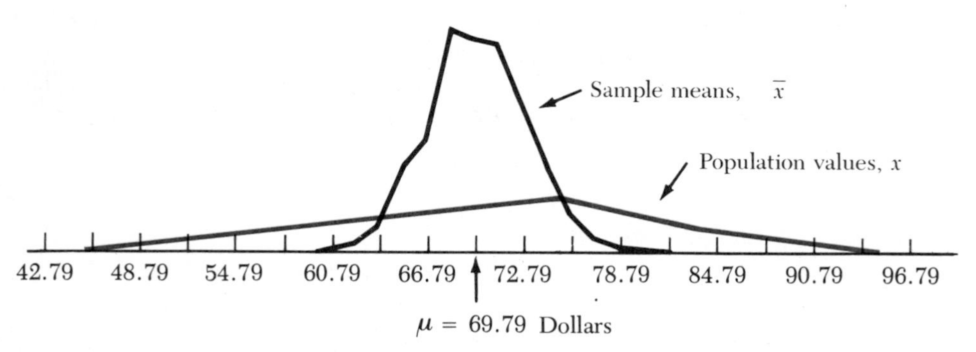

Figure 7-2 Population Distribution and Means of 1000 Samples

General Theorems

Our simulation has served to introduce the important concept of the *sampling distribution of the mean*. While in practice one generally has but one sample for consideration, the idea of the behavior of many similar samples from the same population is essential to statistical inference. The long-run relative frequency of means from repeated random samples of a given size from a given population is the probability distribution of the sample mean, a term often shortened to *the sampling distribution of the mean*. Like any other probability distribution, the sampling distribution of the mean has an expected value and a variance. Fortunately, these values can be obtained without actually drawing repeated samples. They are related to the population mean and variance in the following ways.

1. In random sampling with or without replacement, the expected value of the sample mean is equal to the population mean.

$$E(\bar{X}) = \mu \tag{7-1}$$

2. In random sampling without replacement, the standard deviation of the sampling distribution of the mean is

$$\sigma_{\bar{X}} = \frac{\sigma}{\sqrt{n}}\sqrt{\frac{N-n}{N-1}} \tag{7-2}$$

   where N is the number of units in the population and n is the sample size.

3. In random sampling with replacement, the standard deviation of the sampling distribution of the mean is

$$\sigma_{\bar{X}} = \frac{\sigma}{\sqrt{n}} \tag{7-3}$$

   Observe the difference between the formulas for the standard deviation of the sampling distribution in (2) and (3) above. The expression $\sqrt{(N-n)/(N-1)}$ is called the *finite population multiplier*. It applies to the sampling without replacement situation. But consider what happens when N, the number of elements in the population, is large in relation to the size of the sample, n. The finite population multiplier tends to approximate 1.0, and the standard deviation of the sample mean approaches the value given by the simpler formula in (3) for sampling with replacement. In most sampling situations, the sample size is small in relation to the population size. From here on we will emphasize the simpler formula for sampling with replacement. As a rule of thumb, most practitioners employ the finite multiplier only if a sample comprises 10% or more of the population elements.

4. If the population is normally distributed, the sampling distribution of the mean is normal for any sample size. If the population is not normally distributed, the sampling distribution of the mean approaches normality as the sample size increases.

Using the Sampling Distribution

Let us apply these theorems in our grocery bills example. First, the mean of the sampling distribution of the mean is

$$E(\overline{X}) = \mu = \$69.79$$

Then, the standard deviation of the sampling distribution of the mean is

$$\sigma_{\overline{X}} = \frac{\sigma}{\sqrt{n}}\sqrt{\frac{N-n}{N-1}} = \frac{\$11.72}{\sqrt{12}}\sqrt{\frac{60-12}{60-1}} = \$3.05$$

The sampling distribution in Figure 7-2 seems quite symmetrical and perhaps close to a normal distribution. In Figure 7-3 we show a normal distribution with a mean of \$69.79 and a standard deviation of \$3.05—the mean and standard deviation of the sampling distribution according to the formulas above. The actual distribution we obtained by computer simulation is plotted for comparison. Although there are some minor irregularities, the agreement between the simulated sampling distribution and the distribution according to the theorems is remarkably close.

Question

Earlier we selected one sample of size $n = 12$, without replacement. The mean of this sample was \$70.32. What is the probability of obtaining a sample mean this close (or closer) to the population mean of \$69.79?

The z-statistic, or standard score, for each value of a variable is the difference between the value and the mean of the values of the variable divided by the standard

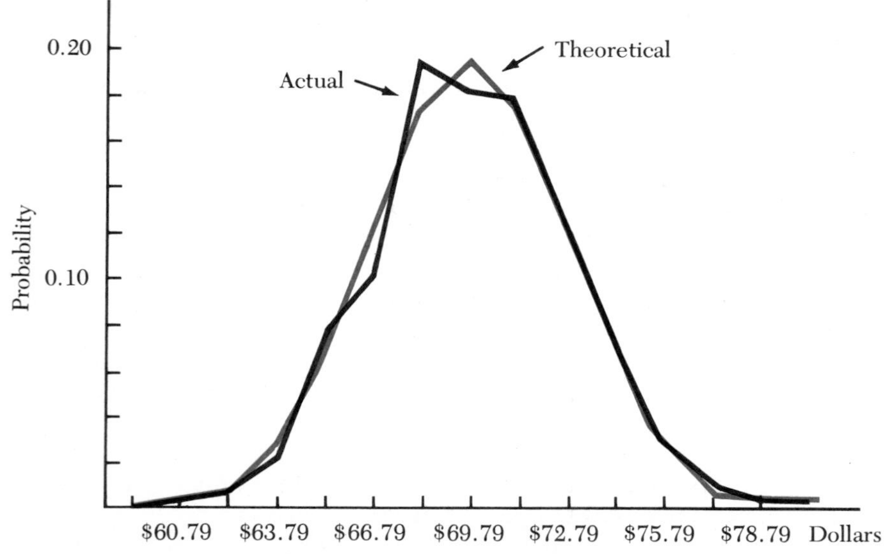

Figure 7-3 Actual and Theoretical Sampling Distributions

deviation of the values. When dealing with a sampling distribution of *means*, the value of the variable is a given sample mean, the mean of the distribution is the population mean, and the standard deviation of the random variable is the standard deviation of the sampling distribution of the mean.

$$z = \frac{\bar{x} - \mu}{\sigma_{\bar{X}}} \tag{7-4}$$

For our problem

$$z = \frac{70.32 - 69.79}{3.05}$$

$$= 0.17$$

The probability that a sample mean is as close to the population mean as ours is, then,

$$P(-0.17 \leq Z \leq 0.17) = 0.5675 - 0.4325$$

$$= 0.1350$$

The sample mean we obtained is rather close to the population mean. We would expect to get sample means closer to the population mean in only about 13.5% of a large number of samples selected in the manner of ours. This normal distribution problem is illustrated in Figure 7-4.

Exact-Probability Approach

When the probability distribution of a population is given, it is possible to derive the probability distribution of the sample mean directly. This approach is feasible when the population values are limited and the sample size is small. We will apply it here to an example of sampling with replacement.

Recall from Section 5.2 the merchandiser who was to make a random draw from sales slips of the day's sales of clock radios and refund to the winner the

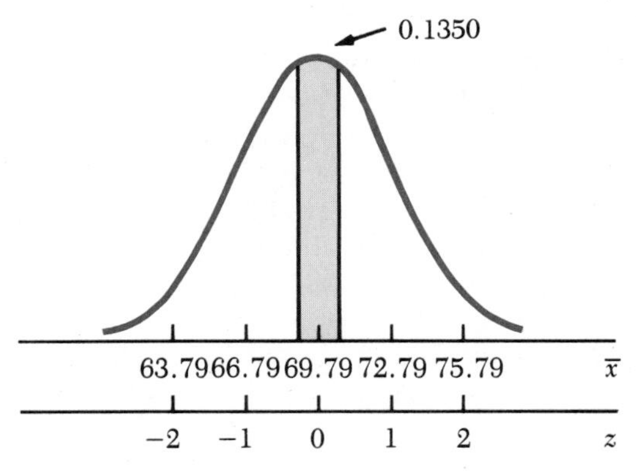

Figure 7-4 Sampling Distribution of Mean Grocery Bill, $n = 12$, Without Replacement

purchase price of the sales ticket drawn. In Chapter 5 we were concerned with the random variable, *cost of the first day's promotion*. This random variable had the distribution

x	$P(x)$
$36	7/12
$42	4/12
$48	1/12

which was also the relative frequency distribution of the sales ticket values for the radios sold on the first day. The mean and standard deviation of this population, calculated in Section 5.2, were $39 and $3.87.

Suppose now that the merchandiser plans two refunds, to be determined by two random draws, with replacement, from the 12 sales slips. We are concerned with the *average value of the refunds given*. Knowing the population distribution above, we can calculate the following probabilities.

$$P(\overline{X}=36) = P(36 \text{ and } 36) = (7/12)(7/12) = 49/144$$

$$P(\overline{X}=39) = P(36 \text{ and } 42) + P(42 \text{ and } 36)$$
$$= (7/12)(4/12) + (4/12)(7/12) = 56/144$$

$$P(\overline{X}=42) = P(36 \text{ and } 48) + P(48 \text{ and } 36)$$
$$+ P(42 \text{ and } 42) = (7/12)(1/12)$$
$$+ (1/12)(7/12) + (4/12)(4/12) = 30/144$$

$$P(\overline{X}=45) = P(42 \text{ and } 48) + P(48 \text{ and } 42)$$
$$= (4/12)(1/12) + (1/12)(4/12) = 8/144$$

$$P(\overline{X}=48) = P(48 \text{ and } 48) = (1/12)(1/12) = 1/144$$

$$144/144$$

The list of possible sample means together with their probabilities above is the sampling distribution of the mean for a sample of size $n = 2$, with replacement, from the sales ticket population. In Table 7-3 the sampling distribution is repeated and the mean and standard deviation of the sampling distribution calculated.

We have used Equations 5-1 and 5-3 with variable $\bar{x}$ rather than x. Because we have been able to derive the exact sampling distribution of the mean, we find in this case that the results agree exactly with Equations 7-1 and 7-3 presented earlier.

$$E(\overline{X}) = \mu = \$39$$

$$\sigma_{\overline{X}} = \frac{\sigma}{\sqrt{n}} = \frac{\$3.87}{\sqrt{2}} = \$2.74$$

Table 7-3 Mean and Standard Deviation of a Sampling Distribution

Sample Mean $\bar{x}$	$P(\bar{x})$	$\bar{x} * P(\bar{x})$	$\bar{x}^2 * P(\bar{x})$
36	$^{49}/_{144}$	$^{1764}/_{144}$	$^{63,504}/_{144}$
39	$^{56}/_{144}$	$^{2184}/_{144}$	$^{85,176}/_{144}$
42	$^{30}/_{144}$	$^{1260}/_{144}$	$^{52,920}/_{144}$
45	$^{8}/_{144}$	$^{360}/_{144}$	$^{16,200}/_{144}$
48	$^{1}/_{144}$	$^{48}/_{144}$	$^{2304}/_{144}$
		$^{5616}/_{144}$	$^{220,104}/_{144}$

$$\mu_{\bar{X}} = {}^{5616}/_{144} = 39$$

$$\sigma^2_{\bar{X}} = ({}^{220,104}/_{144}) - (39)^2 = 7.50$$

$$\sigma_{\bar{X}} = \sqrt{7.50} = 2.74$$

The Central Limit Theorem

Statement 4 on page 193 included the assertion that the probability distribution of the mean of a sample from a nonnormally distributed population approaches normality as the sample size increases. This statement is called the *central limit theorem*. It is an extremely important generalization.

We can extend our example of sampling from the clock radio sales distribution to illustrate the central limit theorem. Let the population distribution given earlier represent the relative frequencies of sales ticket values for a large number of daily sales—as might be the case for a mail-order house. The house is concerned with the average value of refunds. In the section just preceding we worked out the sampling distribution of the mean for two refunds, or $n = 2$. That distribution was quite skewed, reflecting the skewness of the underlying population.

Figure 7-5 shows the sampling distributions of the mean for samples of size 6, 12, and 24 from the sales ticket population. The mean of each of the sampling distributions is \$39, the population mean. The standard errors of the mean decrease with increasing sample size. They are, in turn, $\$3.87/\sqrt{6}$, $\$3.87/\sqrt{12}$, and $\$3.87/\sqrt{24}$, or \$1.58, \$1.12, and \$0.79. As the dispersions decrease, the skewness also diminishes, being noticeable but not marked in the sampling distribution for $n = 12$, and quite minor in the sampling distribution for $n = 24$.

The sample size required for the sampling distribution of the mean to be considered normal for practical purposes varies with the shape of the population distribution. A high degree of skewness, as in our example above, slows the approach to normality. A widely used rule is that it is safe to assume normality of the sampling distribution of the mean of a continuous variable if the sample size exceeds 30, no matter what the population distribution.

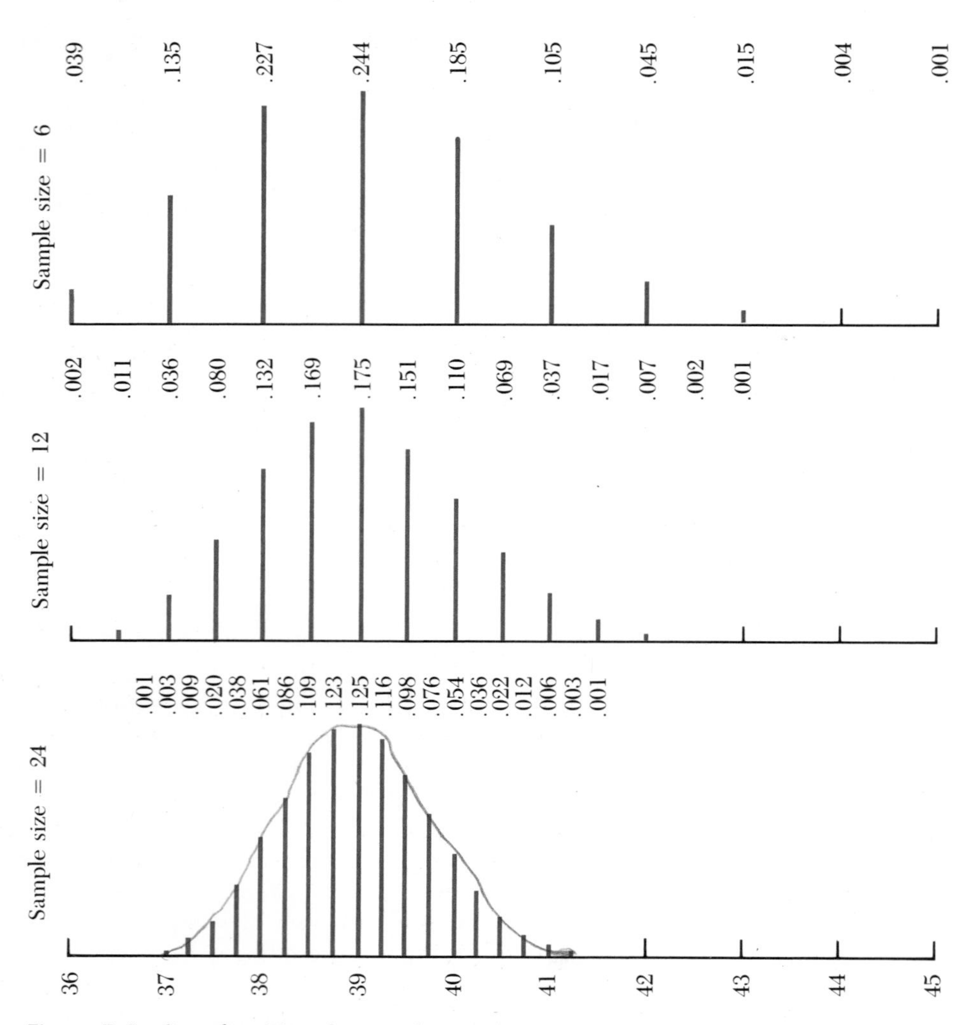

Figure 7-5 Sampling Distributions from Sales Slip Population

Exercises

7-5 *Georges Mills grocery bills.* Find the mean grocery bill for the sample of 10 households that you took in Exercise 7-2. Given the population mean of \$69.79 and standard deviation of \$11.72, assume a normal sampling distribution and find the probability of the mean of a sample of size $n = 10$ differing from the population mean by each of the following amounts.

a. No more than \$2.

b. No more than \$5.

c. No more than your sample mean.

7-6 *Sample mean(s) and random variable.* Refer to the Georges Mills grocery bill population in which the mean is \$69.79 and the standard deviation is \$11.72.

a. In Exercise 7-5, how many sample means did you actually observe?
b. Find the mean and standard deviation of the sampling distribution of the mean of a random sample (with replacement) of 25 household grocery bills.
c. Was it necessary to observe any sample means in order to answer (b)?
d. What is the "trial" underlying the random variable in (b)?

7-7 *Distributions.* The "equal to or less than" relative frequency distribution of Georges Mills grocery bills is as follows.

Dollar amount	45	50	55	60	65	70	75	80	85	90	95
Relative frequency	0.0	0.05	0.18	0.22	0.32	0.48	0.63	0.80	0.88	0.95	1.00

a. Plot the relative frequency population ogive given above.
b. Order the observations from your sample from Exercise 7-5 and plot a cumulative step graph on the same chart as (a).
c. Determine z-scores for the sampling distribution of the mean ($n = 10$, with replacement) at \$60, \$65, \$70, \$75, and \$80 and plot the "equal to or less than" ogive for the sampling distribution on the same graph as (a) and (b).
d. Which two of the three distributions that you have plotted look most alike? Why?

7-8 *Home sales.* Find the mean sales price for your sample of home prices in Exercise 7-4. Given that the population mean is \$76.96 and the population standard deviation is \$25.40, find the probability of a sample mean differing from the population mean by no more than your sample mean did.

7-9 *All possible samples.* An automobile accessories dealer announced that an equal number of batteries carrying 1-, 2-, and 3-year warranties was sold last month.
a. List the nine equally likely sequences of values for samples of size $n = 2$, with replacement, from the population. Calculate their means and present the sampling distribution of the mean.
b. List the 27 equally likely sequences for samples of size $n = 3$. Present the results in terms of the sampling distribution of the mean.
c. Prepare graphs of the sampling distributions for $n = 2$ and $n = 3$. Comment on the shapes revealed.

7-10 *Knitting machine breakdowns.* The probability distribution of monthly breakdowns for a certain type of knitting machine is as follows.

Breakdowns	0	1	2
Probability	0.50	0.30	0.20

a. The Fitzwell Hosiery Company operates two of these machines. Assume independence and find the probability distribution of average breakdowns per machine in a month.
b. Assuming independence of experience from month to month, use the results from (a) to find the probability distribution of average breakdowns per machine-month for two months of experience with the two machines.
c. The distribution of breakdowns for a single machine per month is highly skewed. Comment on the comparative skewness of the distributions of mean breakdowns per machine per month found in (a) and (b).

7-11 *Misconceptions.* Identify the misconceptions in each of the following statements.
a. The sampling distribution of the mean tells us how the sample observations are distributed about the sample mean.

b. The central limit theorem assures us that sampling distributions are always normal.
c. The central limit theorem assures us that if we take a large enough sample, the sample observations will be normally distributed.
d. Normally distributed sampling distributions stem only from normally distributed populations.

7-12 *Normal and nonnormal populations.* In one city, reading comprehension scores of students in sixth grade tend to be normally distributed. However, in twelfth grade, the scores tend to be substantially skewed to the right. Comment on the normality or lack of normality in the sampling distribution of each mean.
a. For a random sample of 5 sixth graders.
b. For a random sample of 5 twelfth graders.
c. For a random sample of 50 sixth graders.
d. For a random sample of 50 twelfth graders.

7.3 Sampling Distribution Examples

A brand of radial tires is manufactured to a specification that the tread, when tested on a machine, has an average equivalent life of 30 (thousand miles). When the manufacturing process is satisfactorily controlled, the distribution of tread life is normal with a mean of 30 (thousand) and a standard deviation of 3 (thousand miles).

Example 1

If 4 tires from a satisfactory production run are tested, what is the probability that the sample mean tread life will be less than 26.0?

If the distribution of tread lives is normal, the sampling distribution of the mean in samples of 4 tread lives will also be normal. The mean and standard deviation of $\overline{X}$ in samples of size 4 is

$$\mu_{\overline{X}} = \mu$$
$$= 30$$

$$\sigma_{\overline{X}} = \frac{\sigma}{\sqrt{n}}$$
$$= \frac{3}{\sqrt{4}}$$
$$= 1.50$$

The z-score is

$$z = \frac{\overline{x} - \mu}{\sigma_{\overline{X}}}$$

$$= \frac{26 - 30}{1.50}$$

$$= -2.67$$

The probability of a sample mean less than 26.0 is the probability of $Z \leq -2.67$, which is read from the normal table of areas as 0.0038. It would be very unusual for 4 tires from a satisfactory production lot to have an average tread life as low as 26,000 mi. Figure 7-6 shows the problem.

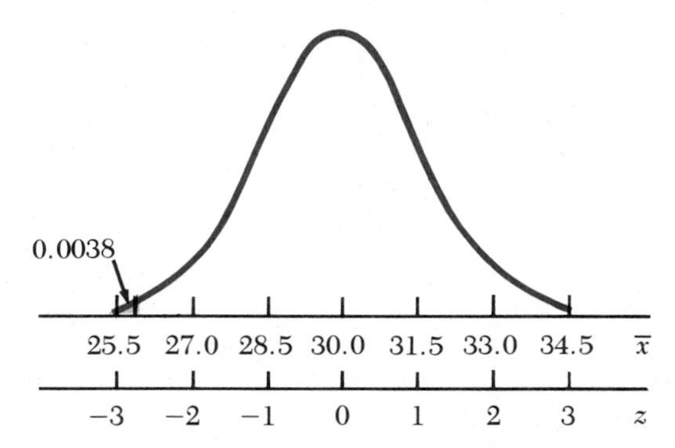

Figure 7-6 Sampling Distribution of Mean Tread Life, $n = 4$

Example 2

The company has decided to test 16 tires from each production lot. If the lot is satisfactory, what average tread life for the 16 tires has a 0.99 probability of being exceeded?

We want to find $\bar{x}_{0.01}$. We need to use

$$\bar{x}_{0.01} = \mu + z_{0.01}\sigma_{\bar{X}}$$

The first percentile of z is -2.33, and we find

$$\bar{x}_{0.01} = 30 + (-2.33)\sigma_{\bar{X}}$$

For samples of size 16, $\sigma_{\bar{X}} = 3/\sqrt{16} = 0.75$. Then $\bar{x}_{0.01}$ is

$$30 + (-2.33)(0.75) = 28.25$$

This problem is pictured in Figure 7-7.

The standard deviation of the sampling distribution of the mean is frequently called the *standard error* of the mean. Since the sampling distribution of the mean is centered on the population mean, the dispersion of the distribution of means measures the potential error in using the sample mean to estimate the population mean. The standard deviation of the sample mean is also the standard deviation of these *errors* of estimate, or, more briefly, the standard *error* of the (sample) mean.

You may still have an uneasy feeling about the deductive nature of sampling distributions. We must have a number for the population standard deviation before

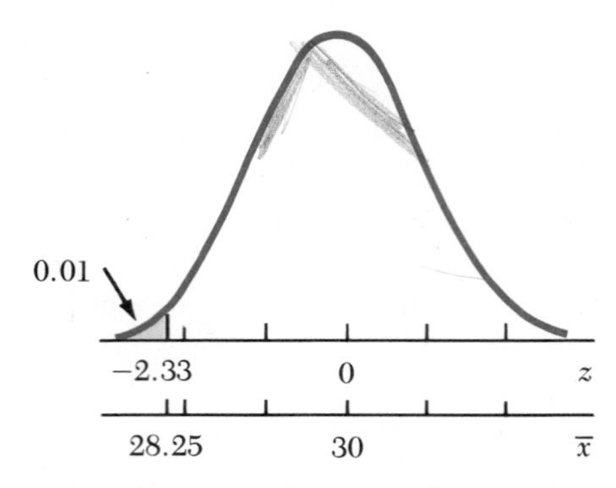

Figure 7-7 Sampling Distribution of Mean Tread Life, $n = 16$

we can deduce the standard deviation of a sampling distribution, and if we are to locate the sampling distribution on the scale of the variable, we must know the population mean. What good is all this when we do not know these population values? The answer is twofold.

1. The concepts, though deductive, are central to statistical thinking. They *are* employed in an essentially deductive way in statistical hypothesis testing, which begins in Chapter 9.
2. The methods of statistical inference that use only sample data also employ sampling distributions. The ideas are the same as we have been using here, but the population standard deviation does not have to be known in advance.

Exercises

7-13 *Assembly times.* The time taken for an assembly job is normally distributed with a mean of 80 sec and a standard deviation of 10 sec. Assuming independence, what is the probability that the average time for 4 assemblies does each of the following?
a. Exceeds 87.5 sec.
b. Fails to exceed 68.75 sec.
c. Lies between 75.0 and 87.5 sec.

7-14 *Flashlight lives.* Nine flashlight bulbs are randomly selected from a large lot of bulbs with lives normally distributed with a mean of 20 h and a standard deviation of 5 h. What is the probability that the mean life of the nine bulbs does each of the following?
a. Exceeds 22 h.
b. Is between 19 and 21 h.
c. Is less than 16 h.

7-15 For Exercises 7-13, find the following percentiles of the sample mean.
a. 10th b. 40th c. 70th

7-16 For Exercises 7-14, find the following percentiles of the sample mean.
a. 90th b. 95th c. 99th

7-17 *Short weights.* Actual weights of 32-ounce packages of ground meat sold at a supermarket are normally distributed with a mean of 32.1 oz and a standard deviation of 0.2 oz. What is the probability that you will receive less than the stated weight on the average if you buy each amount?
a. 1 package b. 4 packages c. 16 packages

7-18 *Stock price changes.* Over a period, the price changes of a universe of 500 common stocks were normally distributed with a mean of +4.0% and a standard deviation of 8.0 percentage points. What is the probability that the average price change would have been negative if you had selected each number of stocks at random?
a. 4 stocks b. 25 stocks

7-19 *Lottery revenues.* State lottery ticket sales per week in one city are normally distributed with a mean weekly revenue of $48,000 and a standard deviation of $2000. Answer each question, assuming independence of weekly outcomes.
a. Find the mean and standard deviation of the probability distribution of average weekly revenue for the next 10 weeks.
b. Find the 5th and 95th percentiles of average weekly revenue for the next 10 weeks.

7-20 *Cost control.* A financial manager in charge of a large number of profit centers has observed that monthly expenditures as a percentage of the budget are normally distributed with a mean of 100 and a standard deviation of 9. If month-to-month performance for a center were a random sample from this population, what is the probability that the average percentage of the budget for 12 months would exceed 105?

7-21 *Home price survey.* A realtors association in a large metropolitan county wants to conduct a monthly survey to determine the average purchase price for old homes in the city as a whole and in each of 50 community areas in the city. The distribution of purchase prices for the city as a whole is believed to be highly skewed to the right. The total sample size is to be 500.
a. Will the probability distribution of the sample mean be normal?
b. Will the probability distribution of the sample mean for a random sample of 10 purchase prices in a community area be normal? Discuss your answer.

7-22 *Two distributions.* A process produces integrated circuits with a mean leakage current of 10 microamps and a standard deviation of 3.0 microamps. The lives of the circuits in use have a mean of 500 h and a standard deviation of 400 h. Leakage currents are normally distributed and lives are highly skewed to the right.
a. Comment on the normality of the sampling distribution of the mean of each variable in samples of size $n = 4$; $n = 100$.
b. For a random sample of 100 circuits, what is the probability that the sample mean lies within 10% of the population mean for each variable?

7-23 *Food processing control.* The amounts of impurities in a food product are controlled in manufacturing so that the mean percentage of impurities (by weight) is 3% and the standard deviation is 3 percentage points. Such a distribution is highly skewed to the right.
a. Find the standard deviation of the sampling distribution of the mean in samples of size $n = 4$ and $n = 36$.

b. Can you find the 95th percentile of the sample mean by using the normal distribution when $n = 4$? When $n = 36$?
c. Find the 95th percentile of the sample mean for $n = 36$ and $n = 81$.

7-24 *Journey to work.* Suppose that the distribution of home-to-work trip lengths in a certain metropolitan area has a mean of 9.2 miles, a standard deviation of 7.5 miles, and is markedly skewed to the right.
a. Is the sampling distribution of the mean of a sample of size $n = 5$ normally distributed? A sample of size $n = 35$? Explain.
b. For a sample of size $n = 225$, what is the probability that the sample mean will be within 0.5 miles of the population mean? Within 1.0 miles?
c. Did you need the value of the population mean to answer (b)?

7.4 Sampling Distribution of the Proportion

Proportions are a special case of means. If all the elements in a population that possess a specified attribute are assigned the value 1 and all the elements not possessing the attribute are assigned the value 0, then the mean of the values assigned is the proportion of the population possessing the characteristic.

We have encountered π before. In Chapter 5, π stood for the expected value of a Bernoulli random variable. If draws are made from a finite population of households in such a way that the probability of observing, for example, a household living in a rented housing unit, is constant and independent of previous draws, the requirements of a Bernoulli process will be met. These requirements are exactly those achieved by random sampling with replacement.

We learned in Chapter 5 that the number of successes in n trials of a Bernoulli process follows the binomial probability distribution for the given n and π, the underlying probability of a success. The initial text example, given in Table 5-5, involved the number of compact autos observed out of three observations from a stream of compact and non-compact autos. That binomial distribution is repeated in Table 7-4 with one important addition.

Proportion of successes is just another way of identifying the binomial variable. An outcome of 1 success out of 3 can be referred to as a proportion of 0.33 in a sample of 3. If we want to talk about the sample *proportion* of successes rather than the sample *number* of successes, then we are talking about the mean of 0's and 1's in the sample.

$$p = \frac{\Sigma x}{n} = \bar{x}$$

Here we have introduced a new symbol, p, to stand for the sample proportion. The sample proportion is a statistic and the population proportion is a parameter. The probability distribution of the sample proportion follows the binomial distribution in the case of sampling with replacement. Its expected value and standard deviation are exactly $1/n$ times the expected value and standard deviation of the binomial distribution of *number* of successes.

Table 7-4 Binomial Probability Distribution for $n = 3$, $\pi = 0.25$ with Addition of Scale for Sample Proportion

Number of Successes	*Proportion of Successes*	*Probability*
0	0.00	0.42
1	0.33	0.42
2	0.67	0.14
3	1.00	0.02

$$\mu_P = \pi \tag{7-5}$$

$$\sigma_P = \sqrt{\frac{\pi(1-\pi)}{n}} \tag{7-6}$$

The standard deviation of sampling distribution of the proportion is also called *the standard error of the proportion.*

We saw also in Chapter 5 that as n increases binomial distributions approach the normal distribution form. The same holds then for the sampling distribution of the proportion. If the sample size is 100 or more and $n\pi$ and $n(1-\pi)$ are each at least 10, then probabilities under the sampling distribution of the proportion can be found from the standard normal distribution, where

$$z = \frac{p - \pi}{\sigma_P} \tag{7-7}$$

Example

A simple random sample of 225 voters is to be drawn, with replacement, from the list of registered voters in a community. If, in fact, 28% of the population of registered voters were receiving social security benefits, what is the probability that more than 32% of the voters in the sample will be found to be receiving benefits?

The normal approximation to the sampling distribution of the proportion is appropriate because (a) the sampling is at random with replacement, (b) the sample size exceeds 100, and (c) the value of $n\pi$ is $225(.28) = 63$, well above 10. The mean and standard deviation of the sampling distribution of p are

$$\mu_P = \pi = 0.28$$

$$\sigma_P = \sqrt{\frac{\pi(1-\pi)}{n}} = \sqrt{\frac{0.28(0.72)}{225}} = 0.030$$

The standard normal deviate for the sample proportion of 0.32 is

$$z = \frac{p - \pi}{\sigma_P} = \frac{0.32 - 0.28}{.030} = 1.33$$

and the probability required is

$$P(Z > 1.33) = 1 - 0.9082 = 0.0918$$

Notice in the example that the number of registered voters in the community was not mentioned. Because the sampling was at random with replacement, the requirements of a Bernoulli process were met. If sampling is without replacement, the standard error of the proportion should be calculated with the finite population multiplier. However, as was the case with means, if the sample number is no more than 10% of the number of population elements, the finite population multiplier can be ignored.

Exercises

7-25 *Related distributions.* Last year 20% of the people registering for conventions held at a city's convention center were women. Consider a random sample of 625 from that population.

a. What are the mean and standard deviation of the binary population of zeros and ones?

b. What are the mean and standard deviation of the binomial distribution of the number of women in the sample?

c. What are the mean and standard deviation of the sampling distribution of the proportion of women in the sample?

7-26 *Georges Mills sample.* Exercise 7-2 asked for a random sample, with replacement, of 10 households in Georges Mills.

a. How many of your sample of 10 households were renters? What proportion were renters?

b. What are the mean and standard deviation of the binary population of zeros (homeowners) and ones (renters)?

c. What are the mean and standard deviation of the sampling distribution of the proportion of renters in a random sample (with replacement) of 10 households?

7-27 *Home sales.* What was your sample proportion of new homes for your sample of home prices in Exercise 7-4? Given that the population proportion is 0.68, use the normal approximation to the binomial distribution of the number of new homes to find the probability that the sample proportion in a sample of size 16

a. differs from the expected number by more than your sample number.

b. differs from the expected number by as much as or more than your sample number did.

7-28 *Emissions testing.* In a long run of experience with compulsory auto emissions testing, the testing agency has found that 13% of the vehicles in Albuquerque, N.M. fail the test.

a. What are the mean and standard deviation of the binary population?

b. What are the mean and standard deviation of the sampling distribution of the proportion for a random sample of 100 vehicles? 1600 vehicles?

7-29 *Coupon prizes.* A store gives out coupons with each purchase. Ten percent of the coupons have winning numbers on them and can be used toward additional purchases at the store. Assume random selection and a customer who makes 100 purchases at the store over the next six months.

a. What are the mean and standard deviation of the probability distribution of the proportion of winning coupons?
b. Use the normal approximation to find the probability that the proportion of winning coupons will be 0.05 or less.

7-30 *Opinion polling.* Nationwide public opinion polls are frequently based on samples of around 2500 people. Assuming random sampling, what are the mean and standard deviation of the sampling distribution of the proportion holding an opinion in a sample of 2500 for each population proportion holding the opinion?
a. 0.50 b. 0.20 c. 0.10

7-31 *Sampling error.* Given the three population proportions in Exercise 7-30 and a random sample of size $n = 2500$, what is the probability that the sample proportion will differ by as much as or more than 0.01 from the population proportion in each case?

Summary

Random sampling from a finite population is accomplished by giving the same probability of selection to each unit of observation in the population at any given draw. In random sampling without replacement, units selected are not replaced in the population for succeeding draws; in sampling with replacement, they are replaced. A random sample of either type can be drawn by numbering the units of observation and then selecting the required number of units with the aid of a table of random numbers.

The *mean of a random sample* of n observations is a random variable with distribution determined by the laws of probability. The distribution is called the *sampling distribution of the mean*. The *expectation* and *variance* of the sampling distribution of the mean are related to the mean and variance of the underlying population.

For samples of size 30 or more from large populations, the sampling distribution of the mean of a continuous variable is approximately normal regardless of the shape of the distribution of the underlying population. Consequently, the probability that the sample mean will fall in a specified interval can be found from tables of the standard normal distribution after the specified value(s) of the sample mean have been expressed as standard scores.

Supplementary Exercises

The data below are for Exercises 7-32 and 7-33 and represent the number of new cars sold daily for 12 weeks by an automobile dealer.

Week	*Day*	*Sales*	*Week*	*Day*	*Sales*	*Week*	*Day*	*Sales*
1	M	4	5	M	5	9	M	6
	T	5		T	4		T	9
	W	6		W	5		W	9
	T	4		T	6		T	7
	F	7		F	9		F	8
	S	12		S	8		S	14
	S	9		S	7		S	12
	Average	6.7		Average	6.3		Average	9.3
2	M	3	6	M	3	10	M	8
	T	6		T	4		T	5
	W	1		W	6		W	7
	T	5		T	9		T	8
	F	4		F	8		F	10
	S	10		S	13		S	12
	S	4		S	11		S	9
	Average	4.7		Average	7.7		Average	8.4
3	M	5	7	M	8	11	M	8
	T	7		T	7		T	7
	W	3		W	2		W	9
	T	7		T	10		T	10
	F	6		F	2		F	10
	S	6		S	9		S	8
	S	10		S	9		S	13
	Average	6.3		Average	6.7		Average	9.3
4	M	4	8	M	8	12	M	7
	T	7		T	10		T	11
	W	2		W	3		W	10
	T	4		T	6		T	5
	F	8		F	6		F	4
	S	10		S	11		S	11
	S	7		S	13		S	15
	Average	6.0		Average	8.1		Average	9.0

The following data are given for the population:

Mean daily sales = 7.38
Standard deviation of daily sales = 3.02
Mean daily sales (weekdays) = 6.28
Standard deviation of daily sales (weekdays) = 2.42
Mean daily sales (weekends) = 10.12
Standard deviation of daily sales (weekends) = 2.62

7-32 *Random sampling with replacement.* Select a random sample (with replacement) of 21 daily sales from the car sales population. Find the mean of your sample.

a. Find the standard deviation of the sampling distribution of the mean.
b. Assuming a normal sampling distribution, find the probability that a sample mean differs from the population mean by more than your sample mean did.

7-33 *Random sampling without replacement.* Select a random sample (without replacement) of 21 daily sales from the car sales population. Find the mean of your sample.

a. Find the standard deviation of the sampling distribution.
b. Assuming a normal sampling distribution, find the probability that a sample mean differs from the population mean by more than your sample mean did.

7.5 Further Sampling Methods (Optional)

We have emphasized that, in order to have a quantitative measure of sampling error, sampling must be carried out in ways that allow us to use the laws of probability to derive the sampling distribution of a statistic like the sample mean. So far we have talked only about simple random sampling as a method that meets this requirement. Other methods, or variations of simple random sampling, will also provide measures of sampling error. Collectively, all these methods are called *probability sampling* procedures. This section provides an introduction to these methods, which are widely used in all kinds of statistical surveys.

Stratified Random Sampling

From the basic formula for the standard error of the sample mean,

$$\sigma_{\bar{X}} = \frac{\sigma}{\sqrt{n}}$$

we can see that, given a sample size, the standard error depends on the standard deviation of the population. If we are sampling a population with large variability, sampling error will be correspondingly large for a given sample size. If the standard deviation is small, then sampling error will be correspondingly small for a given sample size.

This feature of sampling error provides an opportunity to separate, or *stratify*, a population in advance of sampling into two or more subpopulations (called *strata*) that have smaller standard deviations than the population taken as a whole. The final effect, if successful, will be to reduce the sampling error for a given total sample size because it will be based on the smaller subpopulation standard deviations.

For stratified sampling to have the intended effect, the basis for stratification must be related to the variable(s) being studied in the survey. For example, in a sample survey of benefit payments to employees participating in a company-wide health insurance plan, we might consider stratification by sex. If average benefits paid to women differed substantially from average benefits paid to men, then the standard deviations for the separate groups would be smaller than the standard deviation for all participants considered as one group. In order to carry out the stratified selection, we would have to be able to identify whether each benefit payment record was for a male or a female employee.

To illustrate stratified sampling, let us return to the weekly grocery bills in Georges Mills. We have determined for each household whether the members of the household own or rent the dwelling unit. We think that homeowners may be more affluent as a group and spend more on groceries than renters. If this is true, there is something to gain from a stratified random sample. Table 7-5 shows that there are

35 owner-occupied households and 25 renter-occupied households. We want to consider a total sample of size $n = 12$. Since $35/60 = 7/12$ of the universe of households are owners, we plan to sample 7 households from the owner stratum. This leaves 5 households to be sampled from the renter stratum. The allocation of the total sample number to the strata duplicates the proportions of the strata numbers to the total in the population. We call such an allocation of sample numbers *proportionate stratified sampling*. Once the units of observation in the population have been divided into the strata and the sample sizes of the strata determined, all that remains is to select random samples of the required size from each stratum.

Table 7-5 lists the households by homeowner and renter strata and gives the weekly grocery bills, which of course would not be known. However, we shall use this advantage of complete knowledge about the population to examine the sampling distribution of the mean for a proportionate stratified random sample.

First, the true means and standard deviations for the populations of grocery bills for homeowners and renters are as follows.

	Mean	*Standard Deviation*	*N*
Homeowners	$75.90	$ 8.93	35
Renters	61.23	9.60	25
All households	69.79	11.72	60

We see that homeowners and renters do have different average grocery bills. The fact that the standard deviations for the separate groups are smaller than the overall standard deviation of $11.72 is a reflection of this difference.

The mean grocery bill for all 60 households is a weighted mean of the two strata means, where the weights are the proportions of total households in the strata. Using w for these weights, we have:

$$w_O = 35/60 = 0.5833 \qquad w_R = 25/60 = 0.4167$$

$$\begin{aligned}\mu &= w_O\mu_O + w_R\mu_R \\ &= 0.5833(\$75.90) + 0.4167(\$61.23) \\ &= \$69.79\end{aligned}$$

The expected value of the mean of a proportionate stratified random sample is the mean of the underlying population. The variance of the mean of a proportionate stratified random sample depends not on the overall variance, but on the variances of the observations *within* the strata. The formula is

$$\sigma_{\bar{X}}^2 = \frac{\sum_i (w_i \sigma_i^2)}{n} \tag{7-8}$$

where w_i are the individual strata weights and σ_i^2 are the individual strata variances. In our example, there were two strata, and

Table 7-5 Household Grocery Bills by Strata

Owner-Occupied				*Renter-Occupied*			
New Number	x	*New Number*	x	*New Number*	x	*New Number*	x
1	69.6	19	72.6	1	61.5	19	68.8
2	65.2	20	76.4	2	48.0	20	64.5
3	72.8	21	72.9	3	61.2	21	68.2
4	56.3	22	86.0	4	49.8	22	72.2
5	77.0	23	72.6	5	58.4	23	73.6
6	80.8	24	85.4	6	79.1	24	67.5
7	79.8	25	75.2	7	51.3	25	69.0
8	63.2	26	92.5	8	65.3		
9	68.2	27	80.7	9	50.4		
10	71.9	28	91.2	10	65.1		
11	53.1	29	77.3	11	51.6		
12	75.4	30	80.2	12	55.8		
13	77.0	31	91.6	13	46.0		
14	63.5	32	75.0	14	50.4		
15	70.0	33	78.8	15	55.6		
16	73.9	34	83.1	16	61.3		
17	87.1	35	82.1	17	55.1		
18	78.2			18	81.1		

$$\sigma^2_{\bar{X}} = \frac{0.5833(8.93)^2 + 0.4167(9.60)^2}{12}$$

$$= 7.08$$

so that

$$\sigma_{\bar{X}} = \sqrt{7.08} = \$2.66$$

If we were to take a simple random sample of $n = 12$ with replacement, the standard error would be

$$\sigma_{\bar{X}} = \frac{\sigma}{\sqrt{n}}$$

$$= \frac{\$11.72}{\sqrt{12}}$$

$$= \$3.38$$

The sampling error of the stratified sample mean is less than that for the simple random sample mean. This is because separating the population of grocery bills into those for homeowners and those for renters produced two subpopulations that were both more homogeneous in regard to weekly grocery expenditures than the population considered as a whole. By sampling these subpopulations separately, we are able to take advantage of this reduction in variability. Taking a predesignated number of homeowners and a predesignated number of renters for our sample makes it impossible for chance variation in the proportionate representation of owners and renters to contribute to sampling error. This is a source of sampling error in a simple random sample, but not in a stratified random sample.

Figure 7-8 shows the sampling distributions for the stratified and the simple random samples of size $n = 12$. Both have expected values equal to the population mean of \$69.79. The increased statistical efficiency of the stratified sample is shown by the smaller spread of the possible sample means about the population mean.

Cluster Sampling

The units of observation of concern in many surveys often come in convenient groupings, or *clusters*. Examples of clusters are pages in a directory of physicians, blocks of dwelling units in a city, and employers of wage and salaried employees in a labor market. Selection of ultimate units can be made by listing and selecting clusters rather than listing and selecting the ultimate units directly. Cluster samples may be single-stage or two-stage.

> A *single-stage cluster sample* is one in which all the ultimate units in each selected cluster are included in the sample. A *two-stage cluster sample* is one in which first clusters are selected and then a sample of units within each selected cluster is included in the sample.

We will illustrate single-stage cluster sampling with the example of the Georges Mills grocery bills. In Figure 7-1, the dwelling units are grouped into 6 areas, 10 dwelling units in each area. A sample of 20 grocery bills could be obtained by

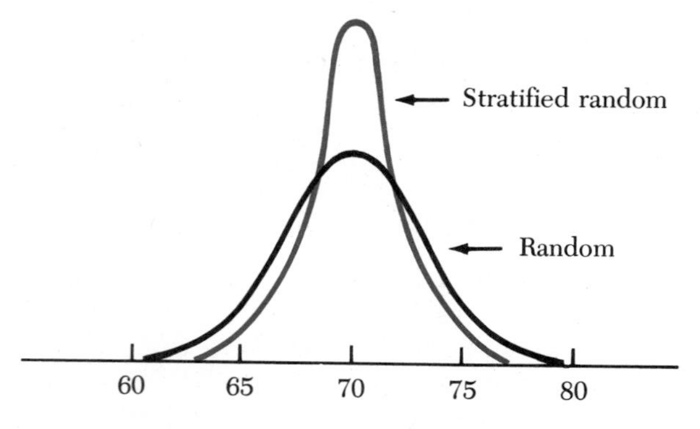

Figure 7-8 Sampling Distributions of Simple Random and Stratified Random Sample Means

selecting 2 areas (without replacement) and including the grocery bills for all 10 dwelling units in each of the selected 2 areas. This is a single-stage cluster sample.

Because we have the whole population at our disposal, we can examine this sampling procedure. The mean grocery bills for each of the 6 areas are:

Area 1	\$64.33	Area 4	\$71.82
Area 2	\$64.34	Area 5	\$78.49
Area 3	\$64.95	Area 6	\$74.81

In this single-stage cluster sample, 2 of these 6 means will be selected randomly without replacement. The sample mean grocery bill will be the average of the cluster means selected. Adapting Equations 7-1 and 7-2 to this context, the expected value and standard deviation of the mean of the cluster sample proposed are

$$E(\overline{X}) = \mu \text{ of cluster means} = \$69.79$$

$$\sigma_{\overline{X}} = \frac{\sigma \text{ of cluster means}}{\sqrt{k}}\sqrt{\frac{K-k}{K-1}}$$

where K is the total number of clusters and k is the number of clusters included in the sample. The standard error of the proposed cluster sample mean is

$$\sigma_{\overline{X}} = \frac{\$5.597}{\sqrt{2}}\sqrt{\frac{6-2}{6-1}} = \$3.54$$

where \$5.597 is the standard deviation of the population of 6 cluster means listed above. It is instructive to compare this standard error with the standard error of a random sample of 20 grocery bills selected without replacement. Recalling that the standard deviation of the 60 grocery bills is \$11.72, the latter standard error is

$$\sigma_{\overline{X}} = \frac{\sigma}{\sqrt{n}}\sqrt{\frac{N-n}{N-1}}$$

$$\sigma_{\overline{X}} = \frac{\$11.72}{\sqrt{20}}\sqrt{\frac{60-20}{60-1}} = \$2.16$$

The result here is typical. The standard error of the cluster sample mean is larger than the standard error of the mean of a random sample of an equal number of ultimate units. This is because convenient clusters of ultimate units will typically differ in systematic ways from each other. In a dwelling unit sample, socioeconomic differences among blocks and neighborhoods will prevail. Similarly, in our example involving wage rates, a cluster sample would have a larger standard error for mean wage rate of employees than would a random sample of an equal number of employees because of the differences in average wage rates paid by employers.

What we have illustrated about the lack of statistical efficiency of single-stage cluster samples is true of two-stage cluster samples as well. Cluster samples are used when random samples of ultimate units would be expensive per unit sampled or simply not possible. The relevant expenses are those required to list and select the population elements and collect the data. For a sample of householders in an urban area, a list of city blocks (clusters) can be constructed on a good street map. A list of ultimate householders or household units is difficult, if not impossible, to construct for a random sample of households. Once blocks are selected for a cluster sample, an interviewer on foot can list household units by proceeding around the block if such a list is needed for second-stage selection. Clustering interviews by blocks in household interviewing also tends to minimize interviewing costs because it lessens unproductive travel time between interviews. While less efficient statistically, a highly clustered sample might be used because — for the same dollar expenditure — it could produce greater reliability than a less highly clustered or even a random sample.

Two-stage cluster samples typically take one of two forms. In the first, clusters are selected with equal probability and then a fixed fraction of the ultimate units in each cluster is selected for the sample. Suppose a sports-equipment manufacturer wants to sample people in summer recreation programs sponsored by municipalities or school boards in order to measure a potential market. A list of such programs could be constructed with some effort, but it is unlikely that timely enrollment data could be obtained. The cluster design permits the manufacturer first to select programs at random. This requires no enrollment data. When interviewers visit the selected programs, they can obtain the local list of participants and randomly select and interview the required fraction of the participants.

When the number of ultimate units in each cluster is known, a commonly used two-stage cluster sampling method is to select clusters with probabilities proportional to cluster size and then to include in the sample the same number of ultimate units from each cluster. An advantage of this method is that the sample size can be controlled exactly. Where field work is required, it can be planned and scheduled more effectively because the amount of work at each location is the same.

Both of the methods just described insure that each ultimate unit of observation has an equal probability of inclusion in the sample. In the first case, both the cluster selection fraction and the fraction of ultimate units sampled within a cluster are uniform. The probability that an ultimate unit is selected is the product of these two selection fractions. In the second case, large clusters have a higher chance of inclusion, but this is exactly compensated for by the smaller fraction of ultimate units from a large cluster that will be selected, given that the cluster is selected.

When there is a choice of units to use as clusters, we select the units that will yield the most similar clusters. This is because less variability among clusters implies a smaller standard error of a cluster sample. By designing clusters that are as alike as possible, we minimize the sampling error.

In two-stage cluster sampling, if we include more clusters in the sample for a given sample size of ultimate units, the standard error of the sample mean will be smaller. For example, apprentice training programs run by local trade unions differ greatly in wages paid, length of training, and other features. If we include many local union programs (clusters) in a survey of pretraining and posttraining experience of apprentices, we have a better chance of representing the variety of existing programs in the sample.

Systematic Sampling

A selection procedure that is sometimes easier to carry out than simple random sampling is *systematic sampling*. Suppose there are 40,000 records in a credit file, but they are not numbered serially. To obtain a sample of 400 records it may be easier to draw every 100th record from the file rather than number all 40,000 records so that a random sample can be selected with the aid of a random number table. In the systematic sample, a random starting point within the first 100 records would be chosen and then every 100th record after that included in the sample.

> In a *systematic sample*, the elements in the population are ordered from 1 to N. A systematic sample of n elements is obtained by selecting a random starting point between 1 and k, where $k = N/n$, and selecting every kth element thereafter.

Systematic sampling is not equivalent to random sampling. If there is a periodic order to the elements that is related to the variables under study, a systematic sample may have a larger sampling error than simple random sampling. Imagine an apartment complex in which there are 6 units in each building—two 3-bedroom units, two 2-bedroom units, and two 1-bedroom units. Within each of 20 buildings in the complex, the units are numbered in that order. A systematic sample of 1/6 of the units from a list ordered by building and apartment number within buildings would contain units of only one size, with the size depending on the starting point. Systematic intervals (values of k) of 3, 9, or 12 would also produce samples that are nonrepresentative of the size distribution of the apartment units. In a survey dealing with leisure activities, which are known to differ by family size, the fact that family size composition could differ greatly from one systematic sample of the apartment units to another must be taken into account in calculating standard errors.

In some situations, a systematic sample will have a smaller standard error than a random sample of the same size. In the current example, suppose we made a list of the apartment units with all 3-bedroom units listed first, followed by the 2-bedroom units and the 1-bedroom units. A systematic sample would guarantee that the size distribution of all the apartment units would be very nearly proportionately represented. A systematic sample would then be very much like a proportionate stratified random sample. The sampling error for variables related to the size of the apartment unit would be smaller than for a random sample of the same size.

Exercises

7-34 *Stratified random sampling.* Select a stratified random sample of 15 weekdays and 6 weekend days (Saturday or Sunday) from the 84 days of sales history given in Exercise 7-32. Use sampling with replacement. Find the mean of your sample.

a. Find the standard deviation of the sampling distribution of the mean of such a stratified sample.

b. Assuming a normal sampling distribution, find the probability that a stratified sample mean differs from the population mean by more than your sample mean did.

7-35 *Comparison.* Compare your results from Exercise 7-34 with those from Exercise 7-32.

a. Which sampling method had the smaller standard error?

b. In which case was your sample mean closer to the true mean?
c. If your random sample mean was closer to the true mean, would that show that random sampling is better than stratified random sampling in this case? Explain.

7-36 *Sample of dwelling units.* The City Planning Department in a medium-sized community is designing a sample of households for a survey of city planning issues related to a proposed land-use plan. They have decided to select a sample of 400 dwelling units. The city is divided into 4 neighborhoods, and blocks within each neighborhood have been defined on city maps. The number of blocks and the estimated number of dwelling units in each neighborhood is as follows.

Neighborhood	*Number of Blocks*	*Number of Dwelling Units*	*Dwelling Units per Block*
Southside	200	8,000	40
Valley	100	3,000	30
Northeast	300	5,000	16.7
Westside	200	4,000	20
Total	800	20,000	

a. How many dwelling units should be drawn from each neighborhood for a proportionate stratified random sample of dwelling units?
b. A cluster sample of 80 blocks has been suggested, with the blocks to be allocated in proportion to total blocks by neighborhood. What fraction of the dwelling units in each selected block will be included in the sample?
c. If the cluster sample in (b) is followed through, will the expected number of dwelling units in the sample from each neighborhood be in proportion to total dwelling units? Demonstrate.
d. Compare the stratified random and cluster samples with respect to the work involved in listing and selecting units.

7-37 *Visitor survey.* The supervisor in a national park is designing a survey of visitors to the park who are brought by packaged bus tours during the four-month peak tourist season. Principal objectives of the survey will be to determine the anticipated and actual recreational activities at the park, the hometowns of the visitors, and readership of recreation and travel-related magazines among visitors. The tour schedule is known at the beginning of the season because the tours book blocks of rooms for each group at the major lodge in the park. The numbers and names in each group are not known until about two weeks before arrival. The groups contain an average of about 60 people, however. The tour groups are organized by international, national, and regionally based organizations. A sample of 600 visitors is planned. There are 300 tours scheduled for the season, and most stay two nights at the park.
a. Is it possible to select a simple random sample of 600 tour visitors? Explain.
b. A park ranger will be assigned to select and interview the visitors. Assuming tour groups will be used as clusters, discuss how the number of clusters selected might affect the ranger's ability to organize and carry out the work.
c. Discuss the advantages to the survey of stratifying the tours by type of tour organization.
d. What is statistically undesirable about just selecting 10 tours at random and interviewing everyone in those tours?

7-38 *Alternative clusters and strata.* Discuss the likely effectiveness of alternative clusters or strata in each of the following sample situations.

a. To estimate the frequency of different prescribed treatments in an outpatient clinic, a sample of patient visits will be drawn.
Strata: physicians versus week of visit.
Clusters: physicians versus week of visit.

b. There are 504 players on the rosters of 12 teams in a professional football league. A sample of around 50 players is desired to estimate average salary.
Strata: team versus offensive and defensive players versus starters, substitutes, and special teams.

c. In the wholesale used-automobile market, there tend to be about as many buyers as sellers. The sellers tend to specialize in particular models, while the buyers want an assortment of models to put on their lots. A sample of sales is desired monthly to estimate the average transaction price per vehicle.
Clusters: buyers versus sellers.

7-39 *Systematic sampling.* Along a major arterial street there are eight blocks to the mile, with major intersections every four blocks. Retail establishments have tended to concentrate near the major intersections. Comment on the following systematic sampling plans.

a. Sampling every fourth block to estimate the number of retail establishments per block.

b. Sampling every fifth block to estimate the number of retail establishments per block.

c. Sampling every tenth retail establishment to estimate the average number of off-street parking spaces per establishment.

7-40 *Cluster sampling situations.* Explain the advantages (if any) of cluster sampling in each of the following situations.

a. A nationwide gasoline distributor wants to select a sample of its service station operators for a personal interview study of attitudes of the operators toward company policies. The company has a computerized list of operators that can be sorted in various ways: by state, county, ZIP code areas, and so on.

b. A county medical society wants to survey patients of licensed physicians to determine the characteristics of private insurance coverage used to meet medical expenses. The society has no list of individual patients.

c. A manufacturer of lawnmowers wants to sample recent purchasers of the mowers to determine their satisfaction with the product. A telephone survey is planned. The source of names is warranty cards filled out by the buyers and mailed back to the manufacturer. Each card contains the name and address of the dealer from whom the mower was purchased.

7-41 *Stratified random sample.* Find the standard deviation of the sampling distribution of the mean of a stratified random sample of the grocery bills of 14 homeowners and 10 renters selected with replacement from the Georges Mills grocery-bill population.

7-42 *Haddock catch.** The catch of a commercial trawler landing haddock was sorted into four categories. Samples from each category were measured (lengths in cm) with the following results.

* J. A. Gulland, *Manual of Sampling and Statistical Methods for Fisheries Biology*, United Nations, 1966.

Category	*N*	*n*	*sum X*	*sum X * X*	*Mean*
Small	2,432	152	5,284	185,532	34.8
Small-med	1,656	92	3,817	158,953	41.5
Large-med	2,268	63	3,033	146,357	48.1
Large	665	35	2,027	118,169	57.9
Total	7,021	342	14,161	609,011	41.4

a. Find the estimated variance for each size category.
b. Use the estimated variances in Equation 7-8 to find the estimated standard error of the stratified sample mean.
c. Use the total line to find the estimated standard error of the sample mean if the strata were not taken into account.
d. Why does the stratified sample cut down so much on sampling error in this case?

Computer Exercises

1 Source: Personnel data file. Assume that this constitutes a population.
   a. Find the population mean and standard deviation for the variable salary.
   b. Suppose you were to draw a simple random sample of $n = 25$. Using the results of (a) and Equations 7-1 and 7-3, find the parameters of the sampling distribution of the mean.
   c. Draw five different random samples of $n = 25$. Find the sample means. For each sample mean, find the corresponding standardized score. How close is each sample mean to the population mean?
   d. Repeat (a) through (c) using the variable participation in the tax-shelter program. Should Equations 7-5 and 7-6 be used?

2 Use the variable degree (graduate or undergraduate) to define two strata. Use Equation 7-8 to compute the standard error of the mean for the variable salary for $n = 25$. How does this result compare with the one in Exercise 1(b) above? Give reasons for the result.

PART THREE

Basic Statistical Inference

8

ESTIMATION OF MEANS AND PROPORTIONS

As early as the first chapter, we pointed out that there are many situations in business and public affairs in which it is important to know the value of some population characteristic. But because we may be prevented by time or money constraints from taking the necessary census to determine the value, an estimate must be made from a sample. In statistical language, we say that we want to know the value of a certain parameter but must be satisfied with an estimate provided by a relevant statistic. For example, two of many important indicators of the level of economic activity in the United States are average weekly earnings in manufacturing and the proportion of the labor force that is unemployed. Business and government leaders want to know the values and changes in values of these two indicators several times each year. Sampling estimates provide the only practical way to keep track of such variables.

When we must estimate the value of a parameter such as average weekly earnings in manufacturing (μ) by using the results of a random sample, we are faced with a problem in *statistical inference*. If we plan to calculate and use the sample mean as our estimate of the population mean, we do this with the knowledge that any sample smaller than the entire population provides only partial information about the population mean. We know that the sample mean ($\overline{X}$) fluctuates from sample to sample and the particular value of $\overline{X}$ we get depends upon the random sample we happen to choose. The same situation faces us if we plan to use a sample proportion (p) to estimate the proportion of the nation's labor force that is unemployed (π). Only a census will provide the exact value of the parameter. A sample is based on incomplete information and the estimate it provides is necessarily accompanied by a degree of uncertainty. We learned in Chapter 7 how sample means ($\overline{X}$) are distributed in relation to the population mean (μ) and how sample proportions (p) are distributed in relation to the population proportion (π). Our task in this chapter is to use this knowledge to make the best possible estimate of the value of a parameter from a sample.

In general, *the objective in statistical inference is to obtain the best possible information about population characteristics from sample information*. One major branch of statistical inference is *statistical estimation—the use of a statistic to estimate the value of a relevant parameter*. This chapter will be concerned with estimation of a population mean or proportion. The other major branch of statistical inference is *hypothesis testing*. In the following chapter, we shall discuss testing hypotheses about a population mean or proportion. The remaining chapters in this section of the book will be devoted to other topics in statistical inference.

8.1 Point Estimation

One approach to the problem of finding the best sample estimate of a population characteristic is to restrict ourselves to finding the best single number we can get from a sample as an estimate of the value of the parameter in question. Given a random sample of weekly earnings in manufacturing, we want to calculate the best estimate we can get from these observations for the population mean weekly earnings. Such a single-number estimate is called a *point estimate*.

Estimators and Estimates

Intuitively, we chose to use the sample mean ($\overline{X}$) to estimate the population mean (μ) in our discussion of weekly earnings of manufacturing. But perhaps the sample median or a special statistic such as the average of the largest and smallest observations in a sample (the midrange) would produce an estimate that is in some sense better than the sample mean. Our goal is to choose a statistic that can be expected to yield an estimate closer to the value of the parameter than any other statistic. For guidance in making such a choice, statisticians turn to the sampling distributions of alternative statistics and examine these distributions with regard to the parameter to be estimated. The alternative statistics that might be chosen to estimate a particular parameter are called *estimators*. Within the constraint of the sample size we can afford, the sampling distributions of alternative estimators are considered before the sample is selected. Once we have decided upon an estimator, the sample can be selected and the procedure for calculating the estimator applied. The resulting summary descriptive measure is a single number that is an *estimate* of the parameter in question.

Before we discuss how sampling distributions of estimators can be used to compare the effectiveness of alternative estimators, let us use another example to summarize what we have said about estimators and estimates. Suppose a market researcher must estimate the current mean weekly income per household in a certain city and can afford a simple random sample of 80 households. Having obtained the observations of weekly income that constitute the data, how can the researcher best estimate the value of the population mean? Which of the sample mean, sample median, or sample midrange (to mention just three possibilities) can be expected to provide the best estimate? If the researcher, not knowing the answer to this question, arbitrarily elects to calculate the sample mean, finds the value to be \$198, and takes this number as the estimate of the population mean, then the sample mean ($\bar{X}$) is called the *estimator* of the population mean (μ). The resulting point *estimate* of the population mean is \$198.

Characteristics of Good Estimators

Instead of choosing an estimator arbitrarily, it is preferable to establish criteria that can be used to compare the effectiveness of alternative estimators so the best estimator can be chosen in a given situation. Such criteria have been developed by considering the sampling distributions of alternative estimators in relation to the parameter of interest. The criteria we will discuss are *unbiasedness*, *minimum variance (efficiency)*, and *consistency*.

Unbiasedness Figure 8-1 shows the sampling distributions of two estimators, S_1 and S_2, in relation to the parameter of interest, θ. The estimator S_1 has a sampling distribution with an expected value, or mean, equal to the parameter. Such an estimator is an *unbiased* estimator. Rarely will a sample produce a point estimate, s_1, for this estimator that equals the expected value, $E(S_1)$. But the mean of all possible values of s_1 is always equal to the value of the parameter θ for an unbiased estimator.

The estimator S_2 in Figure 8-1 has a sampling distribution with expected value (mean) to the left of θ. On the average, the point estimates—that is, values of s_2—produced by the estimator are too small; $E(S_2) < \theta$. If the expected value of an

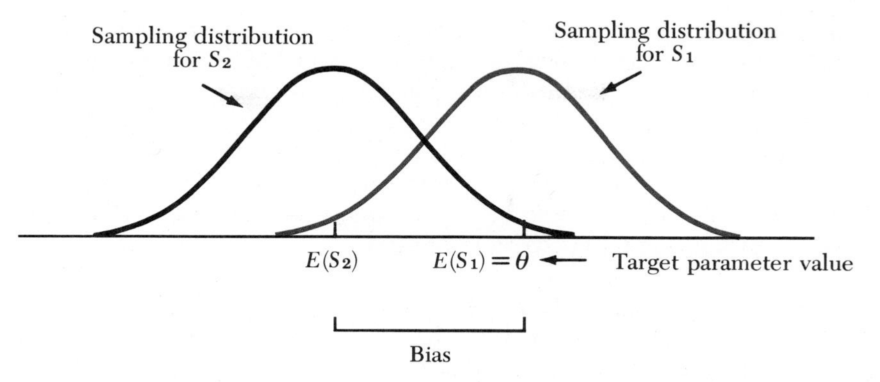

Figure 8-1 Sampling Distributions for an Unbiased (S_1) and a Biased (S_2) Estimator

estimator is not equal to the value of the parameter it estimates, then the estimator is called a *biased* estimator.

> If the expected value of an estimator equals the parameter to be estimated, the estimator is *unbiased*. If not, the estimator is *biased*. For a biased estimator, the amount of bias is the difference between its expected value and the parameter.

Suppose the sampling distributions of the two estimators in Figure 8-1 are identical in all respects except their expected values. Then the unbiased estimator S_1 can be expected to produce point estimates that are closer to the value of the parameter θ than those produced by the estimator S_2. If two estimators differ in no other respect, the unbiased estimator is preferable to the biased one.

One concern of this chapter is estimation of a population mean μ. In the last chapter, we learned that the sampling distribution of the sample mean has as its expected value the mean of the population from which the sample is to come.

> For any statistical population, the sample mean $\overline{X}$ is an unbiased estimator of the population mean. In symbols,
>
> $$\mu_{\overline{X}} = E(\overline{X}) = \mu \tag{8-1}$$

From the viewpoint of an unbiased estimator, the intuitive choice of the mean at the beginning of the chapter was justified.

As an estimator, how does the sample median fare with respect to the population mean? We know the median equals the mean for symmetrical populations. For such populations, it seems reasonable that the sample median should have the population mean as its expected value. This is the case. Hence, for symmetrical populations, the sample median is an unbiased estimator of the population mean. On the other hand, we know the median is not equal to the mean for skewed populations. It turns out that the sample median is a biased estimator of the population mean for skewed populations. Intuitively, of course, we would use the sample median to estimate the population median rather than the population mean. It can be shown that the sample median is an unbiased estimator of the population median, so intuition is a sound guide once again.

There are other estimators that could be discussed with respect to the criterion of being unbiased. We shall, however, delay this discussion until we have described two other criteria for choosing estimators.

Minimum Variance and Efficiency For a given sample size, consider the sampling distributions of two estimators, $\overline{X}$ and *Md*, in relation to the population mean μ of a normally distributed population. These two sampling distributions are illustrated in Figure 8-2. Because the parent population is symmetrical, both estimators are unbiased. The two are equally desirable when we want an unbiased estimator. Yet, as the figure shows, the possible values of the sample mean are less widely scattered around their central location than are those of the sample median. The variance of the sampling distribution of the sample median has been shown to be 57% larger than that of the sample mean for samples of a given size from normal populations. Because of this, the sample mean is an estimator with greater *efficiency* than the sample median in the situation described. Furthermore, in the case of normal populations, the sample mean has been proven to be a more efficient estimator of the population mean than any other estimator. For this reason, it is said to be ***absolutely efficient*** when the population is normal.

> For a given parameter and sample size, the more *efficient* of two unbiased estimators is the one with the sampling distribution that has the smaller variance. If an unbiased estimator has smaller variance than any other estimator, it is ***absolutely efficient***.

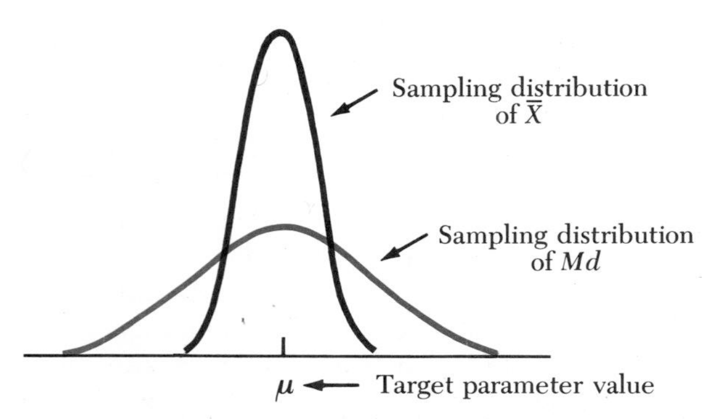

Figure 8-2 Sampling Distribution of an Estimator $\overline{X}$ That Is Efficient Relative to the Estimator *Md*

As an example of the foregoing, suppose a federal agency wants to estimate the mean IQ of this year's entrants into law colleges. A complete list of entrants' names is available, but their scores must be obtained. The decision has been made to select a random sample from the list to serve as the basis for the desired estimate. The distribution of IQ scores for the population is judged to be essentially normal. The statistician for the agency selects the mean IQ of the people in this sample to use as the point estimate of the mean IQ of the population. This choice is made because, as an estimator of the mean of a normal population, the sample mean is both unbiased and absolutely efficient. For the population mean, using the criteria of unbiasedness and minimum variance, no better estimate than the sample mean exists.

Consistency The final criterion for selection of an estimator that we shall discuss is the effect of an increase in sample size on the sampling distribution of that estimator. We realize intuitively that there is potentially more information in a large sample than in a small one. It seems reasonable that with larger samples, a higher proportion of the point estimates of a good estimator will be closer to the target parameter than is the case for smaller samples. An estimator with this property is called a *consistent* estimator.

Because the sample mean $\overline{X}$ is a consistent estimator of the population mean, we can use the sampling distribution of $\overline{X}$ to illustrate the behavior of a consistent estimator. Figure 8-3 shows sampling distributions of the sample mean for samples of size 10, 30, and 100. As we learned in the last chapter, the variance of the sampling distribution of $\overline{X}$ is σ^2/n, where n is the number of observations to be selected randomly, independently, and with replacement. We have learned in this chapter that the sample mean is an unbiased estimator of μ for any sample size. It follows that, as n grows large, the sampling distribution of $\overline{X}$ becomes more tightly packed around μ because σ^2/n becomes smaller and smaller.

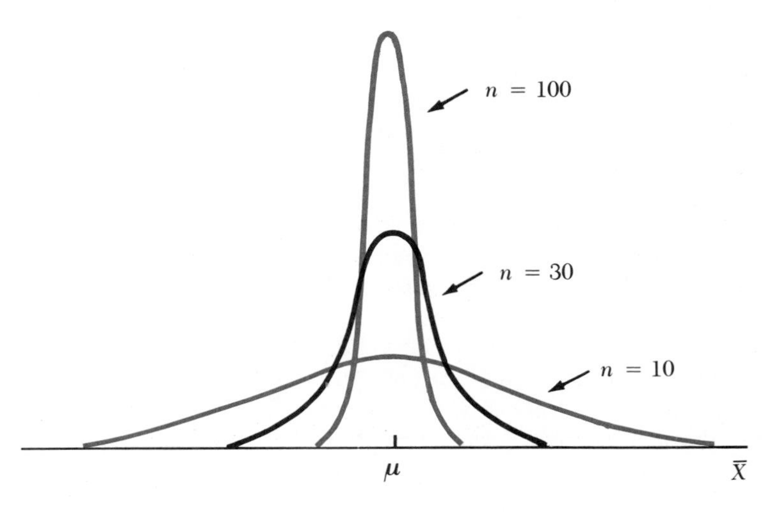

Figure 8-3 Consistency of $\overline{X}$ as an Estimator of μ

As sample size, n, increases, the sampling distribution of a *consistent* estimator becomes increasingly concentrated around the value of the parameter to be estimated.

As we mentioned, the estimator we used as an example of consistency is also unbiased. There are also biased estimators that are consistent; we shall discuss one such estimator shortly. There are also inefficient estimators that are consistent.

Although sampling error decreases as sample size increases, the cost of sampling also increases in practically all cases. A feature of a consistent estimator used in practical estimation situations is that sample size can be increased until further reduction in sampling error is judged not to be worth the cost increase of any larger sample.

Common Point Estimators

The central purpose of this chapter is to discuss estimation of two parameters—a population mean and a population proportion. We complete this section by describing point estimators commonly used for these two parameters. In addition, we shall discuss common point estimators for a population variance and standard deviation.

For a Population Mean As we have already indicated, the sample mean $\overline{X}$ is an unbiased and consistent estimator of μ for a population distribution of any shape and is absolutely efficient when the population is normally distributed. For large samples, the sample mean is also the most efficient estimator of μ even if the population is not normally distributed. The sample size is large enough when the central limit theorem has taken effect to the extent that the sampling distribution of the sample mean is essentially normal. A sample size of 30 should be adequate for practically all populations. In summary, except for small samples from some populations that are not normally distributed, the sample mean is the best estimator of μ with respect to the criteria discussed here (and other criteria covered in more advanced texts). Hence $\overline{X}$ is the most frequently used estimator of μ.

Let us return to our example of the market researcher estimating the mean income per household for all households in a certain city. The researcher has collected incomes from a random sample of 80 households. Income distributions are almost always skewed in a positive direction, and we should expect such skewness to be evident for this population of incomes. The marked positive skewness shown by the frequency polygon for the researcher's sample supports this assumption. Given this situation, what estimator of μ should be selected? We know from the discussion of the central limit theorem in the last chapter that the distribution of sample means from nonnormal populations such as this is essentially normal for samples as large as 80. Hence the sample mean is the efficient, unbiased, and consistent estimator of μ and the researcher is justified in using the sample mean of $198 as the point estimate of the population mean.

For a Population Proportion In Section 7.4 we discussed the sampling distribution of the sample proportion p, where p is the number of successes in the sample divided by the total number of observations in that sample. We saw there that p is an unbiased estimator of π, where π is the probability of success for a Bernoulli process. It has also been proven that p is a consistent and efficient estimator of π. Consequently, p is the estimator most commonly employed to estimate π. For example, if we want π, the proportion of a shipment of lamp bulbs that are defective, then the proportion p defective in a random sample is the best point estimator of that parameter.

For a Population Variance and Standard Deviation In Chapter 3, the population variance was defined as

$$\sigma^2 = \frac{\Sigma(X - \mu)^2}{N}$$

and the sample variance was defined as

$$s^2 = \frac{\Sigma(X - \overline{X})^2}{n - 1}$$

This is the only instance we have encountered in which the same procedure is not used to calculate both population and sample summary measures. We can now explain why. Suppose we were to use

$$v^2 = \frac{\Sigma(X - \overline{X})^2}{n}$$

as an estimator of σ^2. This is known to be a consistent and efficient estimator of σ^2, but it is biased. On the other hand, s^2 is unbiased and consistent, but it is not the most efficient estimator of σ^2. For reasonably large samples, however, the numerical values of the two estimates are essentially equal. The current preference is for the unbiased estimator (s^2) of the population variance (σ^2) at the expense of some loss of efficiency for small samples.

Neither square root (s or v) of the two variance estimators is a general unbiased estimator of the population standard deviation σ. Current practice is to use s, the square root of the unbiased estimator of σ^2, as the estimator of the population standard deviation.

Exercises

8-1 *Characteristics of estimators.* Distinguish between the following pairs of terms.
   a. Parameter and estimate.
   b. Estimator and estimate.

8-2 *Characteristics of estimators.* What is the difference between
   a. A biased and an unbiased estimator?
   b. A more efficient and a less efficient estimator for a given parameter?

8-3 *Bias.* Two estimators are being considered for estimating the variance of the dollar amounts on a year's collection of sales tickets in the cosmetic department of a large department store. The first estimator is the sum of squared deviations from the sample mean divided by $n - 1$. For the other estimate the same sum is divided by n.

a. Which of the estimators is biased?

b. Why is the estimator in (a) biased?

8-4 *Bias.* For the situation in the previous exercise, the expected value of the first estimator is 802.02 and the expected value of the second estimator is 721.82.

a. What is the population variance? Standard deviation?

b. How much bias is present in the biased estimator?

8-5 *Age and automobile preference.* In a national random sample, the sum of the ages of 1200 adults was 41,500 years and 744 of these adults said they preferred smaller, lighter cars with greater gasoline mileage than the most popular domestic models.

a. Use the best estimator to make a point estimate of the mean age of adults in the nation. Discuss why the estimator selected is best for this population.

b. Make a point estimate of the proportion of adults in the nation who prefer smaller, lighter automobiles. What is the best estimator that can be chosen for this case? Why?

8-6 *Delinquent charge accounts.* The credit manager for a large department store has selected a random sample of 200 customer charge accounts from the complete file of several thousand accounts. Forty-two of the accounts in the sample are delinquent and the total amount owed on delinquent customer accounts in this sample is \$5586.

a. Estimate the proportion of delinquent accounts in the complete file. Then sketch the Bernoulli population on the assumption that the point estimate is correct.

b. The manager wants a point estimate of the mean amount owed per delinquent account. Is this a random sample from the correct population for the stated purpose? What is likely to be the shape of the correct statistical population? Assume the sample described is from the correct population and find the point estimate of the desired mean. Justify the choice of estimator.

8-7 *Piston diameter control.* A firm makes pistons for a small gasoline engine. To work properly, the diameter of a piston must be between 2.98 and 3.02 in. If a high percentage of pistons coming from the production process is to be acceptable, the mean diameter of the production process must be very close to 3.00 in. and the variance must be very near 0.000036 in.^{2} (a standard deviation of 0.006 in.). The five pistons in a random sample have the following diameters: 3.006, 3.012, 3.004, 2.991, and 3.024 in.

a. Find the point estimate of the mean piston diameter being produced. Use the best estimator for this purpose.

b. Can we expect the sample median to be an unbiased estimator of the mean piston diameter? If so, why should it not be used here?

c. Use an unbiased estimator to find a point estimate of the population variance from this sample. Also find the point estimate of the standard deviation based on the unbiased estimator of variance.

d. Does it appear that the process is performing as desired?

8-8 *Variability of aptitude scores.* The director of admissions for a university has the entrance examination scores for a random sample of 145 newly admitted freshmen. For the nation, the standard deviation of scores on the examination is 50. The director thinks scores at the university in question are less variable than for the nation.

a. The sum of the squares of the deviations from the mean of the sample is 331,776. Use the unbiased estimator to find the point estimate of the population variance from this sample.
b. Based on the result in (a), find a point estimate of the population standard deviation.

8.2 Confidence Interval Estimation of a Population Mean

Rationale of Interval Estimation

Point estimates have a serious shortcoming: They do not reflect the different amounts of information on which they are based. Because we do not know how much confidence to place in such an estimate, we would like a method that overcomes this deficiency.

When we find a point estimate, such as a sample mean, we know it belongs to a sampling distribution of similar estimates from other possible samples. For a given large statistical population, suppose we have one sample mean from 16 observations and another from 400 observations. We know that the first mean belongs to a sampling distribution with a standard deviation (standard error) of $\sigma/\sqrt{16} = 0.25\sigma$, while the standard error of the second is only $\sigma/\sqrt{400} = 0.05\sigma$. In the majority of cases, the second sample mean will be much closer to the population mean than the first. Consequently we can place more confidence in the mean from the larger sample and we want the estimate to reflect that fact.

The classical method for attaching a confidence statement to a point estimate is called *confidence interval estimation*.

> *Confidence interval estimation* gives an interval estimate of the parameter and accompanies that estimate with a measure of confidence.

As we shall see, the measure of confidence is based on the standard error of the estimator used to form the interval estimate.

For the example of the sample means just given, the confidence interval estimate for the smaller sample might be 501 lb ± 10 lb, and, for the same level of confidence, the estimate for the larger sample might be 497 lb ± 2 lb. In this approach, the point estimate is placed at the center of an interval, the length of which depends on the standard error. The standard error for the first sample is five times that for the second. The larger the sample, the smaller the standard error and the shorter the interval.

Interval Estimation of μ with σ Known

In many, if not most, instances, the population mean must be estimated without knowing the population variance and standard deviation. We shall, however, begin with the case in which the population standard deviation is known because of its relative simplicity. For all but the next to last section of the chapter, we assume that the number of observations in the sample is but a small fraction (less than 10%) of the population.

Since confidence interval estimation is concerned with sampling error, we begin describing the method by referring to the relevant sampling distribution that includes such errors, the sampling distribution of $\overline{X}$. Figure 8-4 shows the sampling distribution of $\overline{X}$ for a fixed sample size. When the population is normal or when the sample size is large enough for the central limit theorem to take effect, the values of $\overline{X}$ calculated from simple random samples will essentially be normally distributed around μ. Furthermore, the standard error of the distribution is $\sigma/\sqrt{n}$.

The shaded portion of Figure 8-4 is the arbitrarily selected central portion covering 95% of the distribution. We already have learned that the standard normal deviate for the region is a z-value of 1.96. This is the region containing 95% of the sample means obtainable, and every sample mean in this region is within 1.96 standard errors of the population mean.

Before we select our single random sample of n observations, the probability is 0.95 that the mean ($\overline{X}$) of the sample will be in the shaded region, because we constructed the region to contain 95% of all possible sample means. Four possible sample means in the distribution are highlighted by arrows. The first three are in the shaded region and the fourth is not. Each of the four is a possible point estimate of μ.

Next we make each of the four possible point estimates in Figure 8-4 the center of an interval estimate equal in length to the interval containing the central 95% of the sampling distribution. In other words, we construct intervals of the form

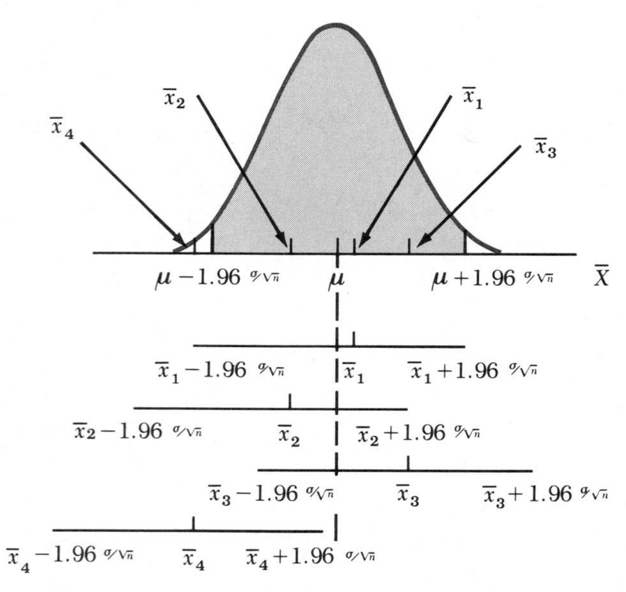

Figure 8-4 Sampling Distribution of $\overline{X}$ and Four 0.95 Confidence Interval Estimates of μ

$$\bar{x} \pm 1.96 \frac{\sigma}{\sqrt{n}}$$

The resulting four intervals are shown in the lower section of Figure 8-4. Note that the three intervals based on means from the shaded region contain μ, while the interval for the sample mean outside the region does not.

Before selecting a sample, we know that the probability of getting a sample mean from the shaded region is 0.95. Hence the probability is 0.95 that the interval we shall construct around that mean—called a *confidence interval*—will contain μ. Briefly, before the sample is selected, the probability is 0.95 that the confidence interval

$$\bar{X} \pm 1.96 \frac{\sigma}{\sqrt{n}}$$

will contain the population mean μ, where $\bar{X}$ is a random variable.

After the sample is selected and the value of the sample mean is known, the resulting interval

$$\bar{x} \pm 1.96 \frac{\sigma}{\sqrt{n}}$$

either does or does not contain μ, although we do not know which is the case. According to the classical theory of probability, the probability that a given confidence interval contains μ after sample selection is either zero or 1; we cannot say which. We can say that for every 100 such intervals constructed, 95 can be expected to contain μ. Instead of stating a probability, we can say that our *level of confidence* is 0.95 that an interval has been constructed that contains the population mean μ.

There is one final point. We used the central 95% as the shaded region in Figure 8-4 and the associated z-value of 1.96 in our confidence intervals. If we had chosen to use the central 99%, we would have replaced 1.96 with 2.58. The resulting intervals would have been somewhat longer, so a 0.99 level of confidence would have been appropriate. For a given sample, the larger the value of z, the longer the interval and the higher the applicable confidence level. Levels of 0.90 or more are customary.

Interval Estimation of μ with σ Unknown

Although we just described interval estimation for cases in which the population standard deviation is known, in the more typical situation the population standard deviation is not known.

When the population variance and standard deviation are not known, the sample from which the sample mean is found must also be used to estimate the unknown population standard deviation, σ. The sample estimate of σ is

$$s = \sqrt{\frac{\Sigma(X - \bar{X})^2}{n - 1}} \tag{8-2}$$

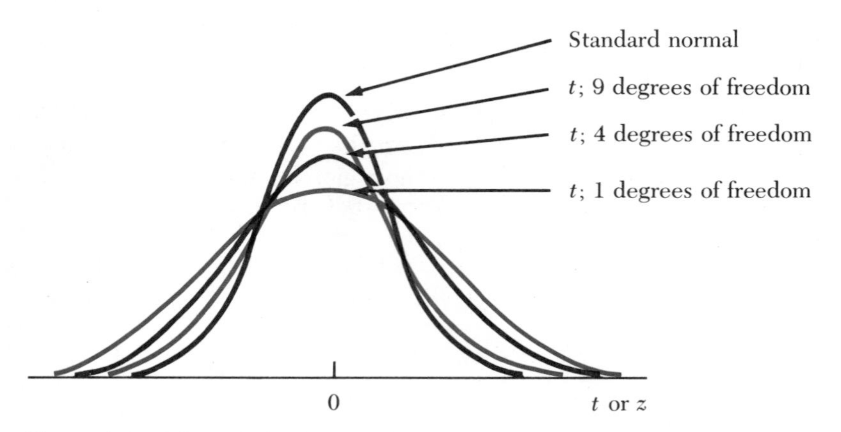

Figure 8-5 The Standard Normal Distribution and Three t Distributions

By calculating $\bar{x}$ and s from the sample, a procedure can be developed for making a confidence interval estimate of the population mean, even though the population standard deviation is not known. The procedure is based on the t statistic and its sampling distribution.

The *t* Distribution

The t statistic is

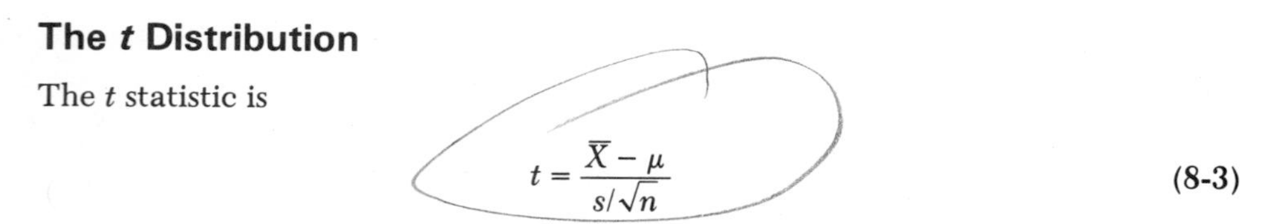

$$t = \frac{\overline{X} - \mu}{s/\sqrt{n}} \tag{8-3}$$

where s is the sample standard deviation as given in Equation 8-2. We say there are $n - 1$ *degrees of freedom* in the equation for s.

In 1908, W.S. Gossett* reported the general sampling distribution of t for samples from normally distributed populations. Distributions of the t statistic are not normally distributed, however. For samples of any fixed size, not only does the sample mean, $\overline{X}$, vary from one sample to the next, but so also does the sample standard deviation, s. Because there are two sources of sampling variability rather than one, the sampling variability of t is greater than that of z, the standard normal deviate.

Figure 8-5 shows several t distributions and the standard normal distribution. Three characteristics of t distributions stand out. In the first place, a t distribution has the same basic shape as the standard normal distribution: A t distribution is concentrated around a mean of zero and tails off symmetrically from its central location. In the second place, there is a family of t distributions rather than a single distribution. One t distribution is distinguished from another by the number of degrees of freedom on which each is based. In the third place, the degrees of freedom

*Gossett's employer prohibited him from using his own name on the report. Instead, Gossett gave "Student" as the author's name. As a result, the distribution is often called the Student t distribution.

increase as the sample size increases. As the degrees of freedom grow large without limit, the t distribution approaches the standard normal distribution as a limit.

Appendix Table A-4 presents commonly used upper-tail values of t for degrees of freedom in increments of 1 up to 100, increments of 2 up to 150, and increments of 100 up to 1000. The final row shows t-values when degrees of freedom have grown large without limit.

For a t distribution with 10 degrees of freedom, for example, the table shows that 0.75, or 75%, fall below a t-value of 0.6998. Seventy-five percent of the values of t from samples with 10 degrees of freedom are less than 0.6998. Similarly, 99% are less than 2.7638. In this same column, note that as the degrees of freedom increase, the value of t decreases from 2.7638—for 10 degrees of freedom—to 2.3263, when the degrees of freedom grow large without limit. As expected, this is the value of z on the standard normal distribution below which 99% of the distribution falls.

Interval Estimation of μ Based on *t*

When the mean of a normally distributed population is not known and is to be estimated with a confidence interval, we begin by collecting a random sample. Then we calculate the mean and standard deviation from the sample. Next, for the appropriate degrees of freedom, we select the value of t for which the proportion of the distribution in the range from $-t$ to $+t$ equals the desired confidence level. Finally, we use the expression

$$\bar{x} \pm t \frac{s}{\sqrt{n}}$$

to find the upper and lower limits of the interval estimate. In general, we have the following.

When the standard deviation of a normal population is not known, the $1 - \alpha$ *confidence interval estimate of the population mean*, μ, is

$$\bar{x} \pm t_{1-\alpha/2} \frac{s}{\sqrt{n}} \tag{8-4}$$

where $\bar{x}$ is the sample mean, s is the sample standard deviation, and t is selected from Appendix Table A-4 to make the central proportion of the distribution equal to $1 - \alpha$, the desired confidence level. The degrees of freedom for t are $n - 1$, one less than the sample size.

An example should help clarify the procedure.

Example

A beverage manufacturing firm must watch the output of its water softening process in order to control the hardness of the water for its product. Nine vials of water have been selected at random. The readings for calcium carbonate in parts per million (ppm) are 20.31, 21.13, 19.97, 19.40, 19.26, 19.62, 20.09, 19.66, and 18.20. Find the 0.99 confidence interval estimate of the mean hardness of the water.

The population from which these measurements came will be assumed to be normal. From the sample, we find the mean to be 19.73778 and the sum of squares of the observations to be 3511.3896. We will carry several decimal places to reduce the cumulative rounding error and then round off to two places when we are through calculating. The sample variance is

$$s^2 = \frac{\Sigma x^2 - n(\bar{x}^2)}{n-1} = \frac{3511.3896 - 9(19.737778^2)}{9-1}$$

$$= 0.6463346$$

From this, the sample standard deviation is 0.8039 ppm. For $n - 1 = 8$ degrees of freedom, Table A-4 shows that a t-value of 3.3554 will leave areas of 0.005 in the two tails and an area of 0.99 in the center. The resulting 0.99 confidence interval estimate of the mean hardness is

$$19.7378 \pm 3.3554 \frac{0.8039}{\sqrt{9}} = 19.7378 \pm 0.8991$$

or 18.84 to 20.64 ppm.

Interval Estimation for Nonnormal Populations

The confidence interval procedure described in conjunction with Equation 8-4 can often be applied even when the population is not precisely normal. If the population is not seriously skewed and if the sample size is not excessively small, the procedure based on t yields confidence intervals with very nearly the correct confidence coefficient. How large should a sample be to prevent serious errors when the population may not be normal? The answer depends upon how much the population in question departs from normality. A conservative rule of thumb is to require that samples from nonnormal populations contain at least 50 observations, and we shall use this as the required minimum.

We have just said that Equation 8-4 can be used for confidence interval estimates of the mean for nonnormal populations when the sample size is 50 or more. *The procedure is applied to all sample sizes of 50 or more, regardless of how large the sample is, because of the extensive t-table in Appendix A-4.* Some other statistics books replace t with the standard normal deviate z when the sample is large. This earlier practice came from having only very abbreviated t-tables available and from the fact that the t distribution approaches the standard normal distribution as sample size grows large. In summary then, interval estimates of the population mean when population variance is unknown will make use of the t distribution for any sample size from a normal population and for any sample of 50 or more from a nonnormal population.

Exercises

8-9 *σ known: normal population.* A food packing plant has a machine that fills cans with cream-style corn. The operator sets a dial to the desired mean fill, measured in ounces, for the size of can to be filled. Regardless of the setting, machine fills are normally distributed with a standard deviation of 0.31 oz. The operator has just set the dial to 18 oz for a run of several thousand cans. A random sample of eight fills is collected and is found to have a mean of 17.66 oz. Make a 0.99 confidence interval estimate of the population mean fill.

8-10 *σ known: normal population.* The mean of a random sample of 16 observations of the depth of a cabinet is 25.40 cm. Assume that the population of such measurements is normally distributed with a standard deviation of 0.3 cm. Find the 90% confidence interval estimate of the true depth of the cabinet.

8-11 *Values of t.* Answer the following questions for a t distribution based on 5 degrees of freedom.

a. What proportion is between $\pm$ 2.5706?
b. What proportion lies below 2.5706?
c. What proportion lies above zero?

8-12 *Probabilities for t.* What is the probability of getting a value from a t distribution that satisfies each condition?

a. Greater than 2.1604 from a sample containing 14 observations.
b. Greater than 2.0003 from a sample containing 61 observations.
c. Greater than 1.9639 from a sample containing 600 observations.

8-13 *σ unknown: normal population.* On a certain day, a random sample of 16 water meter readings is being used to estimate the mean water usage per meter for a neighborhood. The observations are 352, 382, 368, 287, 319, 257, 214, 289, 308, 356, 237, 307, 267, 335, 302, and 265 gallons. Find the 90% confidence interval estimate of the mean water usage for the neighborhood on this day.

8-14 *σ unknown: normal population.* A firm that installs large computers has introduced a new training course for installers. The first class of ten has just completed the course. Their scores are 68, 55, 57, 93, 91, 56, 64, 47, 27, and 48. Assuming that these ten people are a random sample of trainees and that scores are normally distributed, find a 0.95 confidence interval estimate of the mean score for the new training course.

8-15 *σ unknown: nonnormal population.* The distribution of lifetimes for a type of transistor has moderate positive skewness. A random sample of 100 from a large shipment of these transistors has a mean lifetime of 1183 hours and a standard deviation of 97 hours. What is the 0.99 confidence interval estimate of the mean lifetime for the shipment?

8-16 *σ unknown: normal population.* The net weight printed on boxes of one brand of butter is 1 lb. A random sample of these boxes has been selected from several supermarkets. The sample mean net weight is 15.94 oz and the sample standard deviation is 0.12 oz.

a. What is the 0.95 confidence interval estimate of the mean net weight if the sample size is 25? 64? 100?

b. What is the expected effect of an increase in sample size on the length of the confidence interval for a specified level of confidence?

8-17 *Real estate appraisal.* A firm specializing in appraisals of single-family residential properties is considering hiring a person who claims to be an expert appraiser of such properties in the local area. The firm has randomly selected seven homes sold at middle range prices during the past two months and has secured the owners' permissions to make appraisals. The differences between the potential employee's appraisals and the sales prices (in dollars) are −443, −102, −1103, −1211, −557, −1044, and −823.

a. Find the 0.95 confidence interval estimate of the mean difference between the appraisal and the selling price. (Assume a normal population.)

b. The errors are less than 3% of the selling price in all seven cases. Is a bias that would produce such errors less important than the relatively small variability in the errors?

8-18 *Emergency vehicle response.* A public safety official is studying the response times of emergency vehicles to calls for help in connection with automobile accidents near a shopping center. A random sample of 18 reports is selected from those submitted during the past two years. The sample mean response time is 9.3 min and the standard deviation is 3.7 min. The official wants a 0.90 confidence interval estimate of the mean response time from this sample. Find this estimate. (Assume a normal population.)

8.3 Confidence Interval Estimation of a Population Proportion

People in business, industry, and government often want an estimate of a population proportion. Firms routinely estimate the proportion of consumers preferring their brands to competing brands. Other firms, which manufacture products such as electrical appliances, must estimate the proportion of defective items in order to control the production process. Candidates for public office hire public opinion pollsters to estimate the proportion of registered voters favoring them. In such cases, proportions from simple random samples can be used to make the estimates of population proportions.

To find a point estimate (p) of a population proportion (π), we said earlier in this chapter that we count the number of "successes" in the sample and divide by the number of observations. If 40 people in a random sample of 100 eligible voters favor a school bond proposal, then p is 0.40 and is a point estimate of π, the proportion of all eligible voters favoring the proposal.

In Section 7.4, we saw that the standard error, or standard deviation, of the sampling distribution of p is

$$\sigma_P = \sqrt{\frac{\pi(1-\pi)}{n}} \tag{8-5}$$

Interval Estimation of π from p with Large Samples

We also saw in Section 7.4 that when the sample contains at least 100 observations and when both $n\pi$ and $n(1 - \pi)$ are at least 10, the sampling distribution of the sample proportion p is essentially normal with mean π and standard error as given in Equation 8-5. When the sample is large, the procedure summarized in connection with Equation 8-4 can be adapted for making a confidence interval estimate of a population proportion. However, there is a complication. We do not know the value of π or we would not be estimating it. In the absence of this knowledge, what is an appropriate guide to see whether the sampling distribution of p is essentially normal, and how can we find the variance under the radical in Equation 8-5?

The answers to these questions are based on the values of r and $n - r$: *For a given sample, both r and $n - r$ must be 50 or greater.* If the sample satisfies this requirement, the sampling distribution of p is essentially normal. Furthermore, $p(1 - p)$ will very nearly equal $\pi(1 - \pi)$, the variance required for the confidence interval estimation procedure. The procedure then becomes:

Given a sample proportion p from a sample large enough to have both r and $n - r$ at least 50, a $1 - \alpha$ confidence interval estimate for the population proportion π is

$$p \pm z_{1-\alpha/2}\sqrt{\frac{p(1-p)}{n}} \qquad \textbf{(8-6)}$$

where z is the standard normal deviate.

Example

In a metropolitan area, a random sample of 500 drivers of cars using premium unleaded gasoline was selected. Of these, 110 reported that they had not been able to buy fuel every time they wanted to during the past month. For all premium unleaded fuel users in the area, what is the 0.95 confidence interval estimate of the proportion who had difficulty getting fuel last month?

Here the sample proportion is $p = r/n = 110/500$, or 0.22. Hence, np, or r, is 110, the observed number of those not always able to buy fuel. Since $n(1 - p) = n - r = 390$, the other criterion is also greater than 50. The normal approximation applies. A z value of 1.96 is correct for a 95% confidence interval. The result is

$$0.22 \pm 1.96\sqrt{\frac{0.22(0.78)}{500}} = 0.22 \pm 0.04, \text{ or } 0.18 \text{ to } 0.26.$$

Interval Estimation of π with Small Samples (Optional)

There is a procedure for making confidence interval estimates of the population proportion which is exact for any sample size, large or small. The procedure is somewhat more complex than the one described in the previous section. For that reason the approach we are going to discuss is usually limited to small samples.

First, we will introduce the exact approach with an example. A random sample of 27 electric toasters from a production line has been tested and 3 have been found defective. A 90% confidence interval estimate of the process proportion defective is required. Here, the sample size is $n = 27$, the number of occurrences is $x = 3$, and the confidence level is $1 - \alpha = 0.90$.

We find the lower (L) and upper (U) limits of the confidence interval from Equations 8-7 and 8-8.

$$L = \frac{x}{x + (n - x + 1)F_1} = \frac{3}{3 + (27 - 3 + 1)3.75} = 0.03$$

where

$$F_1 = F[1 - \alpha/2;\ 2(n - x + 1),\ 2x]$$
$$= F[0.95;\ 50,\ 6] = 3.75$$

This is our first encounter with F distributions. This family of distributions is more fully described in Chapter 13. For the present purpose, we only need to be able to find the required values of F from Appendix Table A-6. The first value in the square brackets above, 0.95, indicates we are to use the lightface numbers in the body of the table rather than the boldface numbers. (The latter would be used if the first number in the brackets were 0.99.) The second number in the brackets is 50, the value of m_1 which is the argument at the top of Table A-6. The final number, 6, is the value of m_2, the argument at the side of the table. The lightface number at the intersection of 50 and 6 degrees of freedom on page A-17 is 3.75, the required value of F_1.

Similarly, the upper end of the 0.90 confidence interval estimate is:

$$U = \frac{(x + 1)F_2}{(x + 1)F_2 + (n - x)} = \frac{4(2.14)}{4(2.14) + 24} = 0.26$$

where

$$F_2 = [1 - \alpha/2;\ 2(x + 1),\ 2(n - x)]$$
$$= [0.95;\ 8,\ 48] = 2.14$$

The 0.90 confidence interval estimate is

$$0.03 \leq \pi \leq 0.26$$

and the sample proportion, p, is $3/27$ or 0.11. Note that the sample proportion is closer to the left end of the interval than to the right end. This reflects the marked positive skewness of the binomial sampling distribution for small values of π and for small samples. The interval is rather long because the sample is quite small.

For the $1 - \alpha$ confidence interval estimate of the population proportion π, the lower (L) and upper (U) limits are:

$$L = \frac{x}{x + (n - x + 1)F_1} \tag{8-7}$$

and

$$U = \frac{(x + 1)F_2}{(x + 1)F_2 + (n - x)} \tag{8-8}$$

where

$$F_1 = F[1 - \alpha/2;\ 2(n - x + 1),\ 2x]$$

and

$$F_2 = F[1 - \alpha/2;\ 2(x + 1),\ 2(n - x)]$$

and where n is the sample size and x is the number of occurrences in the sample.

For the example in the previous section, we can compare the confidence interval produced by this exact approach with that produced by the approximation. There, $x = 110$, $n = 500$ and $1 - \alpha = 0.95$. For F_1 we need the value of $F[0.975;\ 391,\ 220]$, which we approximate as 1.30. For F_2 we want $F[0.975;\ 222{,}780]$, which we estimate to be 1.20. These estimates of F were made by visual inspection of the four tabled values surrounding the desired one. When there are so many degrees of freedom, F-values are nearly constant. Substituting in Equations 8-7 and 8-8, we find $L = 0.18$ and $U = 0.25$ as compared with 0.18 and 0.26 by the previous method. Obviously, the previous method is satisfactory.

Exercises

8-19 *Large sample estimation.* In a random sample of 175 registered voters, 54% were not in favor of raising property tax rates. What is the 0.90 confidence interval estimate of the proportion of all registered voters not in favor of raising these rates?

8-20 *Large sample estimation.* A market survey in a major urban market produced 594 in a random sample of 1350 consumers who stated a preference for a certain brand of dairy products as compared with two other brands. Find the 95% confidence interval estimate of the proportion of all consumers in the urban market who prefer the given brand.

8-21 *Small sample estimation. (Optional)* A random sample of 36 employees is chosen from several hundred in a manufacturing plant. Twelve people in the sample prefer to reduce the payroll deduction for the company retirement program. What is the 0.90 confidence interval estimate of the proportion of all employees preferring such a deduction?

8-22 *Product preference.* In a test city, a random sample of 150 families was selected to estimate what proportion of the families preferred a new syrup to the one being sold by a firm at present. Each family was given two plain cans marked A and B and asked to try both in several ways during the test week. The follow-up found that 72 preferred the new syrup.

a. Estimate the population proportion preferring the new syrup by means of a 0.99 confidence interval.

b. Does the evidence suggest that the new syrup should replace the old type? Should both be produced?

8-23 *Activity analysis. (Optional)* Two hundred fifty people employed as industrial chemists were selected as a random sample of those so employed. Observers dropped in at random times during work days to determine whether each chemist is performing work in the field of chemistry. One such chemist was engaged in professional activity during 85 of the 100 observations made. Find the 0.90 confidence interval estimate for the proportion of time this chemist is working in chemistry.

8-24 *MBA employment field. (Optional)* A random sample of 72 recent graduates of a university's MBA program were surveyed by telephone. Of those sampled, 23.6% said they were currently employed in their field of specialization while in the degree program. What is the 0.90 confidence interval estimate of the percentage of all recent graduates who are working in their field of specialization?

8.4 Determining Sample Size in Advance

Rationale

We pointed out earlier that the objective in confidence interval estimation is a short interval accompanied by a high level of confidence achieved at acceptable cost. The following expression will serve as an illustration of the issues involved in pursuit of this objective. Recall that the confidence interval estimate of the mean of a large population with known standard deviation is

$$\bar{x} \pm z\frac{\sigma}{\sqrt{n}}$$

where $\bar{x}$ is the sample mean and the second term is half the length of the interval. In this second term, the population standard deviation (σ) is a given fixed value when an interval estimate is being made for a specified population. The population standard deviation cannot be used to influence the confidence level or interval length. The standard normal deviate (z) controls the confidence level. Both z and the interval length increase as confidence is raised to a high level. Sample size (n) varies inversely as the interval length. The larger the sample, the shorter the interval for a given confidence level.

Up to now, the confidence level and sample size were given or set arbitrarily and we accepted the resulting interval length. In practice, however, the desired interval length (or half-length) and confidence level often are specified and we want to find what size sample to collect to meet these specifications.

Sample Size to Estimate a Mean

For an example, consider a machine that is filling cans with cream-style corn. From experience with the process it is known that the population of fill weights is normally distributed with a standard deviation of 0.31 oz ($\sigma = 0.31$). The production supervisor wants to collect a sample just large enough to provide a sample mean within 0.25 oz of the true process mean at the 0.99 confidence level. In other words, the supervisor wants the probability to be 0.99 that $\overline{X}$ will be within 0.25 oz of μ. We can say the half-length is to be 0.25 with 99% confidence, that is, in

$$\overline{x} \pm z \frac{\sigma}{\sqrt{n}}$$

$$0.25 \text{ oz} = z \frac{\sigma}{\sqrt{n}}$$

or

$$0.25 = \frac{2.58(0.31)}{\sqrt{n}}$$

This gives

$$\sqrt{n} = \frac{2.58(0.31)}{0.25} = 3.2$$

By squaring both sides,

$$n = 10.2$$

which we increase to 11 cans because we are restricted to integral numbers of cans and we want to maintain at least the specified level of confidence.

The supervisor can now figure the cost of taking periodic samples of 11 cans each to control the true mean fill. If the cost is too high, the sample size must be reduced. Then, as one choice, the supervisor can maintain the same confidence level by keeping z at 2.58 and letting the interval length increase. Alternatively, the short interval can be maintained by reducing z and the confidence level, or both the confidence level and interval length can be changed in order to avoid too much change in either.

We can find a general formula for sample size by letting h be the half-length of a confidence interval estimate. In symbols,

$$h = \frac{z\sigma}{\sqrt{n}}$$

When this is solved for sample size, the result is

$$n = \frac{z^2\sigma^2}{h^2}$$

In our example, the supervisor wanted h to be 0.25 oz and z to be 2.58. The process standard deviation, σ, was 0.31 oz. Hence

$$n = \frac{2.58^2(0.31)^2}{(0.25)^2} = 10.2$$

or 11 observations.

Given a population with known standard deviation σ, the *sample size* (n) required to produce a specified confidence level and interval half-length h for a confidence interval estimate of the population mean is

$$n = \frac{z^2\sigma^2}{h^2} \qquad \textbf{(8-9)}$$

Equation 8-9 assumes that the population standard deviation is known. When the population standard deviation is not known, an estimate must be substituted. Sometimes a satisfactory estimate can be determined from earlier studies. The census, for example, provides distributions of such variables as family size and income for various geographical subdivisions from which a standard deviation can be found. In other instances, a pilot study will not only serve to try data collection procedures but will also provide an estimate of the standard deviation.

Sample Size to Estimate a Proportion

Relying on the normal approximation, the confidence interval estimate of a population proportion (π) is

$$p \pm z\sqrt{\frac{\pi(1-\pi)}{n}}$$

Here the second term is, again, the half-length of the interval estimate. If we let h be a specified half-length and z be the standard normal deviate for the desired confidence level, then

$$h = z\sqrt{\frac{\pi(1-\pi)}{n}}$$

When we solve this for sample size, we get

$$n = \frac{z^2\pi(1-\pi)}{h^2} \qquad \textbf{(8-10)}$$

Equation 8-10 makes use of the unknown population proportion π in determining the necessary sample size. Although π is unknown, we do know that it must be somewhere in the range from 0 to 1. In Table 8-1, values of the variance, $\pi(1-\pi)$, which are required in Equation 8-10, are shown for various values of π. The maximum possible variance is 0.25 and substituting this in the equation will yield

Table 8-1 Variance of a Bernoulli Process for Various Values of the Population Proportion

π	$\pi(1 - \pi)$
0.50	0.2500
0.45 or 0.55	0.2475
0.40 or 0.60	0.2400
0.35 or 0.65	0.2275
0.30 or 0.70	0.2100
0.25 or 0.75	0.1875
0.20 or 0.80	0.1600
0.15 or 0.85	0.1275
0.10 or 0.90	0.0900
0.05 or 0.95	0.0475

the maximum sample size required for any value of the population proportion. Note further that the variance does not differ substantially from 0.25 until π falls from 0.50 to about 0.30. Therefore—though π is unknown—if it can be assumed that π lies between 0.30 and 0.70, we can let the variance be 0.25 without making sample size a great deal larger than necessary.

When we have reason to believe that π is less than 0.30 or more than 0.70, letting the variance be 0.25 will result in a sample that is much too large. Sometimes in such cases a person with experience in the field can supply an appropriate estimate of π. At other times, information from previous studies or from small pilot studies can be used to establish a reasonable value for π.

Example

Student organizations at a university want to build a recreational sports complex to be financed by an increase in student fees and by charging fees to the faculty and staff. As a first step, a questionnaire has been prepared for mailing to determine what proportion of the faculty and staff will say they are willing to pay $150 per year for family-use privileges. A brochure and drawings of the complex are to be included. Those not responding will be combined with those saying they are not willing to pay and the total will be counted as not being in favor of the proposal. There are about 2800 eligible people in the group to be surveyed. What sample size will give a probability of 0.90 that the sample proportion in favor is within 0.05 of the true proportion?

In determining the variance to use in Equation 8-10, we note that experience with mail questionnaires has shown that the nonresponse rate seldom falls below 0.30, with the planned mail (two) and telephone (one) follow-ups. Thus the maximum possible proportion of favorable returns is about 0.70. If those responding split evenly on the issue, the favorable proportion will be 0.35. From these considerations, use of 0.25 for $\pi(1 - \pi)$ seems justified. For a 0.90 confidence level, z must be 1.65. A half-length of 0.05 has been specified. Hence

$$n = \frac{z^2\pi(1 - \pi)}{h^2} = \frac{1.65^2(0.25)}{0.05^2} = 272.25$$

which we round up to 273. After we have selected the sample and found p, both np and $n(1 - p)$ must be checked to see that they are at least 50 to justify using Equation 8-10. Finally, because 273 is less than 10% of the population, we need not use the methods of the next section.

Exercises

8-25 *Sample size to estimate a mean.* At a given setting, the lengths of steel pipe cut by a cutting machine are normally distributed with a standard deviation of 1.1 in. How large a sample is required to be 99% certain that the sample mean cut length is within 0.6 in. of the true mean length?

8-26 *Sample size to estimate a mean.* A filling machine in a food packing plant fills cans of cream-style corn with a standard deviation of 0.37 oz and fill weights are normally distributed. The cans in the next production run are to have a mean fill weight of 20 oz. How many observations must there be in a random sample to make the probability 0.95 that the sample mean will be within 0.1 oz of the true mean fill?

8-27 *Sample size to estimate a proportion.* The chief executive officer of a large corporation wants to use a random sample to estimate the proportion of stockholders who favor a major long-term expansion of the manufacturing plant. The preliminary judgment is that the proportion could well be close to 0.50. How large should the sample be to make the probability 0.90 of being within 0.04 of the correct proportion?

8-28 *Sample size to estimate a proportion.* The poultry and egg purchasing agent for a corporation which owns a large number of supermarkets wants to estimate the proportion of small, grade AA eggs in a shipment from a poultry farm. The agent expects the proportion to be near 0.20. The probability of being within 0.05 of the actual proportion is to be 0.95. How many eggs should there be in the random sample?

8-29 *Household income.* A study of gross income of rural nonfarm households in a state is being initiated. A preliminary sample of 20 such households yielded a mean income of \$7250 and a standard deviation of \$2920. With these preliminary results as a guide, how large a random sample must be taken to be 90% confident that the resulting sample mean is within \$125 of the true value?

8-30 *Public opinion polling.* A polling organization wants to estimate the proportion of registered voters in a large state who currently favor the Republican candidate for governor. The organization wants a sample size that will provide a probability of 0.95 of being within 0.04 of the true proportion. What is this sample size?

8.5 Estimation Using the Finite Population Multiplier (Optional)

Early in this chapter, near the beginning of the discussion of interval estimation, we assumed that any sample from a finite population was a small fraction (at most 10%) of that population. This assumption allowed us to ignore the finite population multiplier in formulas for the standard errors of means and proportions. Now that we have presented this case, we can go on to cover the case in which the sample is an appreciable part of the population from which it came.

All we need to do to adapt our earlier interval estimation procedures to the situation wherein the sample constitutes more than 10% of the population is to multiply the standard errors in Equations 8-4 and 8-6 by the finite population multiplier:

When a random sample constitutes more than 10% of the finite population from which it came and a $1 - \alpha$ confidence interval estimate of the population mean is required, the estimate is

$$\bar{x} \pm t_{1-\alpha/2} \frac{s}{\sqrt{n}} \sqrt{\frac{N-n}{N-1}} \qquad (8\text{-}11)$$

where N is the number of elements in the population and n is the number of elements in the sample.

Similarly, when a $1 - \alpha$ confidence interval estimate of a population proportion π is required, the estimate is

$$p \pm z_{1-\alpha/2} \sqrt{\frac{p(1-p)}{n}} \sqrt{\frac{N-n}{N-1}} \qquad (8\text{-}12)$$

Exercises

8-31 *Interval estimation of a mean.* For one of its products, a firm buys a liquid ingredient in containers. A shipment of 60 containers has arrived. The receiving quality control specialist has weighed a random sample of 10 containers to estimate the mean weight of the containers in the shipment before approving the invoice for payment. The sample mean weight is 24.62 lb and the sample standard deviation is 0.38 lb. What is the 95 percent confidence interval estimate of the mean weight of containers in the shipment?

8-32 *Interval estimation of a mean.* An accountant is auditing 329 sales tickets for a given week in a certain retail shop. The accountant wants a 99% confidence interval estimate of the mean dollar amount sold per ticket. A random sample of 64 tickets has a mean of \$17.31 and a standard deviation of \$5.05. Find the interval estimate.

8-33 *Interval estimation of a proportion.* An organization with 900 eligible employees is considering an increase in group health insurance coverage, to be accompanied by an increased payroll deduction and employer contribution. The proposed change has been widely publicized. A survey of a random sample of 200 eligible employees shows that 135 favor the proposal. What is the 90 percent confidence interval estimate of the proportion of all employees who favor the change?

8-34 *Interval estimation of a proportion.* School officials have just finished interviewing 150 heads of households with children in a certain elementary school. There are 624 such households. Of those interviewed, 68 favor donating $10.00 apiece toward the purchase of microcomputers for the students. Find the 0.95 interval estimate of the proportion of all households whose heads favor the donation. Assume the 150 constitute a random sample.

8.6 Robust Estimation of Means (Optional)

The procedures described above for making confidence interval estimates of population means are based on the assumption that the populations sampled do not depart appreciably from normal distributions, especially with regard to excessive frequencies in one or both tails. When exploratory procedures indicate that this assumption is suspect, alternatives must be used. If the problem is not with the main central portion of the sample but, rather, with one or both tails, the trimmed-mean procedure often is applicable.

The exploratory techniques we will use to decide whether the trimmed-mean procedure is applicable are the stem and leaf display and the boxplot. We discussed these in Chapters 2 and 3, respectively. Recall that a stem and leaf display for a sample from a normal population is expected to have a single mode and to be approximately symmetrical. Similarly, a boxplot for such a sample should show that the median lies about in the middle of the central box, that is, about halfway between the first and third quartiles. In addition, the whiskers and the box should be almost equal in length. Finally, no outliers should be present.

Our example to illustrate the trimmed-mean procedure is based on the situation described in Exercise 2-11. In that exercise, distributions of the ages of minority and nonminority entrepreneurs were presented. We will use the ages of individual nonminority entrepreneurs from that study. The stem and leaf plot for the original observations is shown in Table 8-2(a). The boxplot, without outliers, is shown to the right. Outliers are bold-faced in the stem and leaf plot, Table 8-2(a).

Table 8-2(a) makes it apparent that, except for the presence of one outlier on the low end of the sample and four on the upper end, the sample has the characteristics of one from a normal population. The trimmed-mean procedure was designed for just this type of situation.

The first step is to trim an equal percentage of the observations from each end of the sample. Any percentage can be used, but 10% seems to work well in most instances. Ten percent of 82 is 8.2 and the convention is to round to the next larger number, 9. Nine observations are deleted from each end of the original sample to

Table 8-2 Stem and Leaf Plots of Sample, Trimmed Sample, and Winsorized Sample for Age Data

(a) Original Sample

3L	1
3U	567899
4L	0112233444
4U	5566667777788899999
5L	000000111111222233344444
5U	555566778889
6L	0023
6U	56799
7L	3
7U	
8L	4

(b) Ten Percent Trimmed Sample

3L	
3U	
4L	12233444
4U	5566667777788899999
5L	000000111111222233344444
5U	555566778889
6L	00
6U	
7L	
7U	
8L	

(c) Winsorized Sample

3L	
3U	
4L	111111111112233444
4U	5566667777788899999
5L	000000111111222233344444
5U	555566778889
6L	00000000000
6U	
7L	
7U	
8L	

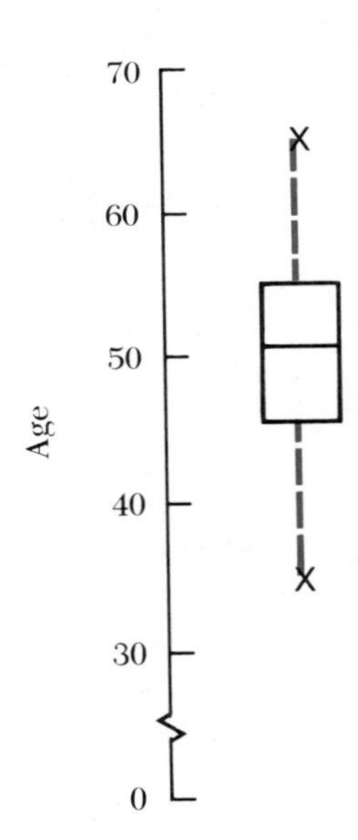

produce the trimmed sample. The stem and leaf plot for the trimmed sample appears in Table 8-2(b).

All we need from the trimmed sample are the *trimmed-sample mean*, $\bar{x}_T$, and the number of observations. For these 64 observations, $\Sigma X = 3236$ so $\bar{x}_T = 50.56$.

The second step is to create the Winsorized sample so that the Winsorized sample standard deviation and standard error of the mean can be found. *The Winsorized sample is found by replacing the trimmed observations in a tail with the most extreme observation remaining in the trimmed sample.* In the lower tail of our trimmed sample, the extreme observation is 41. We will, therefore, add 9 more observations of 41 to this tail. Similarly, we add 9 more observations of 60 to the upper tail. The result is the Winsorized sample shown in Table 8-2(c).

The Winsorized sample variance is the sum of squared deviations from the mean of that sample divided by the degrees of freedom in the ***trimmed*** *sample.* The Winsorized standard deviation is, of course, the square root of this variance. For our sample in Table 8-2(c), $\Sigma X = 4145$ and $\Sigma X^2 = 212583$. From these, the sum of squared deviations is $212583 - (4145^2/82) = 3058.3049$. The trimmed sample has $64 - 1 = 63$ degrees of freedom. We will give this the symbol g. Then the Winsorized sample variance is

$$s_W^2 = \frac{\Sigma X^2 - [(\Sigma X)^2/n]}{g - 1} = \frac{3058.3049}{63} = 48.54$$

the Winsorized standard deviation is

$$s_W = \sqrt{48.54} = 6.97$$

and the standard error of the trimmed mean is

$$SE_T = \frac{s_W}{\sqrt{g}} = \frac{6.97}{\sqrt{64}} = 0.8709.$$

The final step is to find the $1 - \alpha$ confidence interval estimate of the population mean from

$$\bar{x}_T \pm t_{1-\alpha/2} SE_T$$

where t has $g - 1$ degrees of freedom.

The 95% confidence interval estimate of the mean for our example is

$$50.56 \pm 1.9977(0.8709) = 50.56 \pm 1.74$$

or 48.82 to 52.30 years old.

Note that the Winsorized standard deviation, 6.97, is much smaller than the original sample standard deviation, 8.86. As a result, the Winsorized interval estimate is much shorter than the one obtainable from the original sample. The Winsorized interval estimate is not only robust but is resistant to the effect of outliers in the sample. That is, its expected value remains very close to the population mean and its standard error remains relatively small.

Summary

The objective of point estimation is to find the best one-number estimator of a parameter from a random sample. Criteria for point estimators are lack of bias, efficiency, and consistency. The sample mean and sample proportion satisfy all three criteria as estimates of the population mean and proportion, respectively. An unbiased estimator of variance requires a divisor, $n - 1$, rather than n.

Confidence interval estimation of a population parameter consists of a point estimate of the parameter plus or minus a quantity determined by sampling variability and the level of confidence to be placed in the resulting interval. Interval estimation of population means and proportions from either large or small samples is discussed. A procedure using the finite population multiplier is needed if the sample constitutes more than 10% of the population.

The number of observations to be made in a given sampling study can be determined before the sample is collected. Specifying the desired length and level of confidence for a confidence interval in advance permits one to find the sample size to realize these objectives.

Supplementary Exercises

8-35 *Point estimators.* Answer the following questions about point estimators.
  a. How does a biased, efficient estimator differ from an unbiased, less efficient estimator?
  b. How does a biased, consistent estimator differ from an inconsistent estimator?

8-36 *Consumer protection.* To verify the claim that the content weight is 16 oz for a box of cereal, 40 boxes are selected at random from several markets. The mean of this sample is 16.02 oz. A statistics student is asked to interpret this estimate and says, "The sample mean is unbiased, consistent, and absolutely efficient for normal populations such as this. Therefore 16.02 oz is the best possible estimate of the true mean content weight for this package." Comment.

8-37 *Tensile strength.* Steel wire is being manufactured to comply with a certain standard diameter. As a service to potential customers, the manufacturer wants an estimate of the mean tensile strength. A random sample of 120 pieces of this wire has a mean tensile strength of 89.3 lb and a sample standard deviation of 1.8 lb. What is the 0.99 confidence interval estimate of the population mean tensile strength?

8-38 *Television viewing increase.* The manager of a TV station in a large city wants an estimate of the increase in audience viewing time following the introduction of several new programs. Arrangements have been made with a random sample of 50 families in the viewing area to keep records of viewing times before and after the introduction. The sample mean increase in viewing time is 1.22 h per week and the sample standard deviation is 0.29 h per week. Find the 0.99 confidence interval estimate of the increased viewing time per family.

8-39 *Market research reporting.* In a report to the vice president for marketing, a staff analyst states that, "The 95% confidence interval estimate is that from 3.2 to 3.6 packages of the subject product are purchased weekly by the typical household. In other words, the probability is 0.95 that the mean weekly purchase per household lies between 3.2 and 3.6 packages for all households in the market area sampled." Criticize this statement from the viewpoint of the classical relative-frequency definition of probability.

8-40 *Subscriber recreational expenditures.* The 0.99 confidence interval estimate for the mean annual recreational expenditure is $1309.14 to $1327.80 from a random sample of subscribers to a magazine. Can a probability statement be made for this particular interval? If so, what is the statement? Refer to the classical relative-frequency definition of probability for your answer.

8-41 *Vacant housing.* A city planning commission took a random sample of 200 of the 5000 blocks in the city. The numbers of vacant single-family residences were as follows.

Number of Vacancies	*Number of Blocks*
0	21
1	39
2	73
3	44
4	15
5	8
	200

a. What is the 0.90 confidence interval estimate of the mean number of vacancies per block?

b. Multiply your estimate in (a) by the proper factor to get a 0.90 confidence interval estimate of the total number of such vacancies in the city.

8-42 *Absenteeism report.* A report on a study of employee absenteeism stated that a 95% confidence interval estimate of the mean days absent per employee per year is 2.7 to 9.1 days. In attempting to interpret this result, the report made two additional statements given below. Criticize each statement.

a. Ninety-five percent of the company's employees are absent from 2.7 to 9.1 days per year.

b. If a great many random samples of the same size were selected, 95% of them would have sample means between 2.7 and 9.1.

8-43 *Measuring potential market.* A city transportation department is considering adding a new route and discontinuing a current route. The effect on present customers is an important consideration. How large a sample of current bus riders should be selected to estimate the proportion who say they will stop riding the bus if the changes are made? The city wants the probability to be 0.90 that the sample proportion will be within 0.05 of the true proportion. Previous customer studies indicate that about 10% of bus riders would be affected by the changes.

8-44 *Pedestrian safety.** Pedestrian fatalities from motor vehicles in Manhattan were studied in an effort to lessen the number. For a sample of 50 fatalities, 32 occurred within 2500 ft of the pedestrian's residence. Based on these preliminary results, find the necessary sample size (number of fatalities) to estimate the proportion of fatalities in this class within 0.05 for a confidence level of 0.90.

Computer Exercises

1 Source: Real estate data file. For all parts of this exercise, treat the real estate data as if they are a random sample from a large universe of sales in the community.
   a. Find the 95% confidence interval estimates for mean purchase price, mean age of buyer, and mean income of buyer.
   b. Find the 90% confidence interval estimates for the proportion of sales that are new homes and the proportion of sales that are to new residents of the community.

The source for all of the following exercises is the personnel data file. Treat the file as if it is a population.

2 Determine the population mean and the standard deviation for the salary data.

3 Select a random sample of $n = 25$ salaries and find the 0.95 confidence interval estimate of the population mean using (a) the population standard deviation, (b) the sample standard deviation. Which is more realistic?

4 Do your confidence intervals in Exercise 3 include the population mean? If you were to repeat Exercise 3 500 times, how many of the 500 intervals would you expect *not* to include the population mean?

5 Based on your random sample of $n = 50$, find the 95% confidence interval estimate of the proportion of persons who participate in the tax shelter program. Does the interval include the true proportion? Does the sample size justify the normal assumption for the sampling distribution?

6 Based on the results of Exercise 5, find the sample sizes required to produce confidence interval lengths of 0.01 and 0.10. Which sample size is more realistic? Select a random sample of this size and compute the 0.95 interval estimate. Is the confidence interval length specification met?

*Adapted from W. Haddon, Jr. et al., "A Controlled Investigation of the Characteristics of Adult Pedestrians Fatally Injured by Motor Vehicles in Manhattan," *Journal of Chronic Diseases*, 14 (December 1961), pp. 655–678.

9

HYPOTHESIS TESTS FOR MEANS AND PROPORTIONS

In the last chapter, we said that statistical inference involves obtaining information about population characteristics from sample information. We said that the two major branches of statistical inference are estimation and hypothesis testing. Statistical estimation was defined as the use of a statistic to estimate the value of a parameter. For example, we discussed how to use the sample mean $\bar{x}$ as the basis for an estimate of the population mean μ.

This chapter describes hypothesis testing. Like estimation, hypothesis testing is concerned with the value of some parameter. Unlike estimation, hypothesis testing begins with a statement about a value of the parameter. Although we are, in fact, uncertain that the value is correct, we presume it is. We then use this hypothesized value to locate the sampling distribution of a suitable test statistic. Next we calculate the observed value of the test statistic from an actual sample. Finally, we determine whether or not the observed value lies well within the hypothesized sampling distribution. If it does, then the hypothesized value of the parameter cannot be rejected. Its location is consistent with the actual sample evidence. If the observed value of the test statistic is far removed from the hypothesized sampling distribution, on the other hand, we reject the hypothesized value of the parameter.

In hypothesis testing, as in estimation, drawing a conclusion about the value of a population parameter from a sample that constitutes only part of the population always entails a risk. An interval estimate includes the risk that the parameter will actually fall outside the confidence interval. In hypothesis testing, the risk is that we will reach the wrong conclusion about the value of the parameter.

9.1 Testing a Hypothesis about a Population Mean: Two-Sided

Many times and perhaps most often a hypothesis must be tested without knowing the population standard deviation. Nonetheless, we begin with the case in which the population standard deviation is known because of its simplicity. In addition, for all but the last section of the chapter, we assume that the number of observations in any sample is but a small fraction (less than 10%) of the population from which the sample was selected.

We shall illustrate the hypothesis-testing procedure with the help of an example. A filling machine in a cement plant is set to fill bags with a mean of 80 lb of dry cement. Past experience has established that the distribution of fill weights at a given setting is normally distributed with a standard deviation of 0.50 lb. The supervisor wants to take a random sample of 16 fill weights periodically to detect any drift of the mean fill weight from 80 lb. A quality control specialist is directed to set up the appropriate procedure. The specialist does so by following a sequence of steps. These are:

1. State the hypothesis to be tested.
2. Select the test statistic and its sampling distribution.
3. Establish the acceptance and rejection regions.
4. Select a random sample and calculate the sample value of the test statistic.

5. Decide about the hypothesis by comparing the sample value with the acceptance and rejection regions.

Each of these steps will be described in the context of the filling machine example.

The Hypothesis Statement

A question suitable for treatment by hypothesis-testing techniques is a question about a parameter of a relevant statistical population. This type of question can be changed to a statement about the true value of the parameter. The statement constitutes the hypothesis to be tested by comparing what is expected if the hypothesis is true with an actual sample result.

The hypothesis statement is composed of two parts—the *null hypothesis* and the *alternative*. The two parts are mutually exclusive statements about the value of the parameter of interest. We must state the null hypothesis in a manner that permits determining the sampling distribution of the statistic to be used in the test.

A hypothesis statement about a population mean can take one of three forms. These are presented in Table 9-1. The names of the three forms come from the *alternative* hypotheses, rather than from the nulls. The null hypothesis always contains an equal sign. As we just said, this condition must be met to establish a known sampling distribution under the null hypothesis. A two-tailed test sets up a single value for the population mean as the null hypothesis, H_0. The alternative hypothesis, H_A, states that the mean is not equal to the value given in H_0; that is, μ is either greater than or less than the stated value. In a lower-tailed test, the alternative, H_A, states that the population mean is less than a stipulated value. It follows that the null hypothesis, H_0, states that the population mean is equal to or greater than the stipulated value. In an upper-tailed test, the opposite applies.

The first step in the hypothesis-testing procedure is to choose the form of the hypothesis statement. Then the remaining steps lead to a decision about whether to reject the null hypothesis H_0, and thereby accept the alternative H_A, or not to reject H_0.

For the filling machine example, the supervisor selects a two-tailed test for the hypothesis statement. The machine can be presumed to be performing in accor-

Table 9-1 The Three Possible Forms of a Hypothesis Statement about a Population Mean μ

	Two-Tailed	*Lower-Tailed*	*Upper-Tailed*
	(a) General		
Null hypothesis	$H_0: \mu = \mu_0$	$H_0: \mu \geq \mu_0$	$H_0: \mu \leq \mu_0$
Alternative hypothesis	$H_A: \mu \neq \mu_0$	$H_A: \mu < \mu_0$	$H_A: \mu > \mu_0$
	(b) Cement sack filling example		
Null hypothesis	$H_0: \mu = 80$ lb	$H_0: \mu \geq 80$ lb	$H_0: \mu \leq 80$ lb
Alternative hypothesis	$H_A: \mu \neq 80$ lb	$H_A: \mu < 80$ lb	$H_A: \mu > 80$ lb

dance with the setting of 80 pounds most of the time. If a shift in the mean fill does occur, however, it is important to detect a shift in either direction so that remedial action can be taken to return the process mean to 80 lb. A lower-tailed test would be designed to detect a shift to a mean less than 80 lb, which would permit the machine to overfill and give away cement. An upper-tailed test, on the other hand, would be designed to detect a shift of the process mean to a value greater than 80 lb but would allow a downward shift to go undetected. Such a test would not detect short weights.

The Test Statistic and Its Sampling Distribution

Having chosen the two-tailed hypothesis statement

$$H_0: \mu = 80 \text{ lb} \qquad H_A: \mu \neq 80 \text{ lb}$$

the specialist must select a statistic related to the population mean by sampling distribution theory. The sample mean $\overline{X}$ is such a statistic and will be used throughout this chapter in tests concerned with the population mean. The null hypothesis and the assumptions that can be made about the process are needed to establish the sampling distribution of $\overline{X}$. Recall that the population of fill weights in our example is normally distributed around the actual process mean with a standard deviation of 0.50 lb. The null hypothesis is that 80 lb is the process mean. The supervisor has specified that a random sample of 16 fills is to be used for each periodic test.

Given that the null hypothesis is true and that simple random sampling will be used, we learned in Chapter 7 how to specify completely the sampling distribution of $\overline{X}$, the sample mean. Means from random samples of size 16 will be normally distributed with a mean of 80 lb ($\mu_{\overline{X}} = \mu = 80$) and a standard error of 0.125 lb ($\sigma_{\overline{X}} = \sigma/\sqrt{n} = 0.50/\sqrt{16} = 0.125$ lb).

We have now finished the first two steps in the hypothesis-testing procedure: (1) making the hypothesis statement, and (2) selecting the test statistic and defining its sampling distribution under the null hypothesis and the assumptions about the statistical population. In the next step, we establish a range within which the statistic, to be obtained from a sample, can be expected to fall if the sampling distribution is centered on the population mean specified by the null hypothesis, 80 lb.

Acceptance and Rejection Regions

Our progress to this point is summarized by the distribution shown in Figure 9-1. If the null hypothesis is true, the sampling distribution of means from simple random samples of 16 observations each will be normally distributed around a population mean of 80 lb with a standard error of 0.125 lb. When we select our sample, its mean will be one of the values of $\overline{X}$ in this sampling distribution, provided that the population mean fill is 80 lb. Given this hypothesized population mean, we see that 95% of the values of $\overline{X}$ lie between 79.755 and 80.245 lb. These are the values −1.96 and +1.96 standard errors from the hypothesized population mean. It follows that the probability is 0.95 that our forthcoming sample mean will lie in the region from

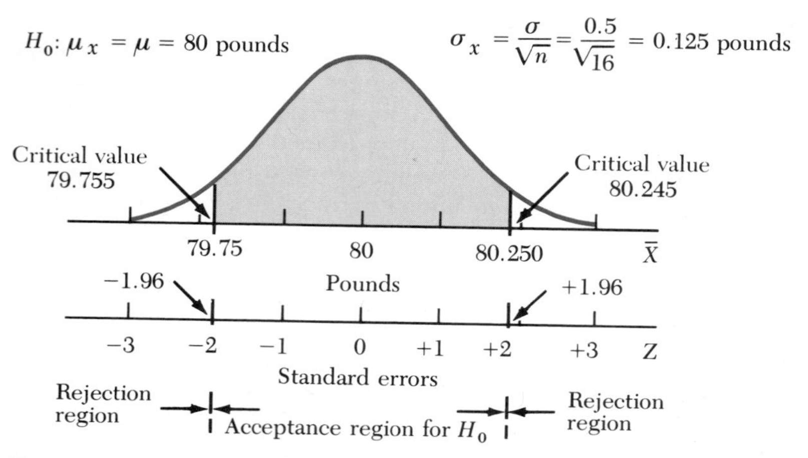

Figure 9-1 Hypothesized Sampling Distribution for $\bar{X}$ and Z for Samples of $n = 16$ Observations (Two-Tailed Test)

79.755 to 80.245 lb when $\mu = 80$ lb. By making this region the acceptance region for the null hypothesis, $H_0{:}\mu = 80$, we make the probability 0.95 for accepting that hypothesis when it is true. We choose a region for which the probability of acceptance of H_0 is high because we want the filling process to continue uninterrupted when the mean fill is correct. We do not want a routine sampling fluctuation in $\bar{X}$ to cause an unwarranted stoppage of production. Some commonly used probabilities of acceptance are 0.90, 0.95, and 0.99. In summary, we have the following.

> An *acceptance region* is a range of values within which the test statistic has a high probability of falling when the null hypothesis is true.

If a high probability of acceptance of H_0 is the only objective in selecting an acceptance region, why stop at 0.95 or 0.99? The longer the acceptance region, the greater is the probability for acceptance. In the extreme, lengthening the acceptance region without limit will make the probability of accepting H_0 approach 1, or certainty, when H_0 is true.

Selecting a high probability of acceptance for H_0 when it is true is not, however, the sole objective in choosing an acceptance region for H_0. We must also provide for the possibility that the null hypothesis is false. We must consider what will happen if, without our knowledge, the sampling distribution in Figure 9-1 is centered on some value other than 80 lb. If, for instance, the process mean fill for some reason has shifted to $\mu = 80.3$ lb, the sampling distribution of $\bar{X}$ will be centered on that value rather than on 80 lb. Furthermore, a sizable proportion of the left tail of the distribution will be to the left of 80.245 lb, the upper limit of the acceptance region for $H_0{:}\mu = 80$ lb. If the null hypothesis is false by reason of the fact that μ is 80.3 lb, our forthcoming random sample of 16 observations very well may have a mean in the lower tail of the distribution, below 80.245 lb. Suppose this were to occur.

Because we would not know that μ has shifted from 80 to 80.3 lb, and because the sample mean would lie in the acceptance region for $H_0:\mu = 80$ lb, we would accept H_0 when it is false. We thus would reach a wrong conclusion. *We want an acceptance region long enough to provide a high probability of accepting H_0 if it is true. At the same time, we must not make this region so long that it includes large portions of sampling distributions with population means substantially different from what H_0 specifies.*

Table 9-2 summarizes what can happen in hypothesis testing.

Table 9-2 The Four Possible Outcomes of a Hypothesis Test

True State \ *Decision*	*Accept H_0*	*Reject H_0*
H_0 true	Correct decision; probability: $1 - \alpha$	Type I error; probability: α
H_0 false	Type II error; probability: β	Correct decision; probability: $1 - \beta$

The null hypothesis is either true or false. We shall make a decision about this on the basis of sample information. We cannot know for sure whether the null hypothesis is true or false on the basis of any sample short of a census. If H_0 is true and we set any realistic critical values for the acceptance region, there is a nonzero probability that a sample mean will fall outside the critical values and lead to erroneous rejection of H_0. This is a type I error, and the area beyond the critical values in the tails of the sampling distribution when H_0 is true is α, the probability of making such an error. The remainder of this sampling distribution lies in the acceptance region, and the probability of getting a sample mean within the acceptance region when H_0 is true is $1 - \alpha$. For our example, when $\mu = 80$ lb and the critical values are 79.755 and 80.245 lb, α is 0.05 and $1 - \alpha$ is 0.95.

When the population mean has any other value than that specified in the null hypothesis, H_0 is false. In this case, the sampling distribution for the test statistic is not centered on the value specified in H_0. Typically, some part of the sampling distribution lies within the acceptance region for the null hypothesis. If the observed sample mean comes from this part of the sampling distribution, a type II error will be made. Because the sample mean, $\bar{x}$, lies in the acceptance region for H_0 and μ is not equal to the value specified in H_0, we erroneously accept the null hypothesis. We shall discuss type II errors again later in this chapter.

We have seen that making a hypothesis statement and selecting a test statistic and acceptance region leads to some risks. If the null hypothesis is true, a sample mean can fall outside the acceptance region, causing a type I error. On the other hand, if the null hypothesis is false, a sample mean can fall inside the acceptance region, causing a type II error.

One approach to balancing the risks stemming from the two possible types of error is to select a significance level and incorporate it into a decision rule. *The significance level of a hypothesis test is the value of α chosen for the probability of a type I error.* Commonly selected values are 0.10, 0.05, or 0.01. Table 9-3 shows commonly selected values of α together with their standard normal deviates that leave a total area of α in the two tails of the standard normal distribution. Half of the area is in each tail. These values of α establish acceptance regions long enough to make the probability of accepting H_0 high when it is true yet not so long as to include large portions of sampling distributions with means greatly different from that specified by H_0.

After the significance level has been selected, the critical values are calculated and used to define the acceptance region. For the example illustrated in Figure 9-1, a significance level of 0.05 was selected. The critical values and acceptance region can be defined in terms of the sample mean, $\bar{x}$. In this case, the critical values are $80 \pm 1.96(0.50)/\sqrt{16}$, or 79.755 and 80.245 lb, and the acceptance region is $79.755 \leq \bar{x} \leq 80.245$ lb.

The foregoing can be summarized in a decision rule.

DECISION RULE (Normal Population; σ Known): For a two-tailed hypothesis test, specify the significance level α and use it to specify the acceptance region lying between critical values

$$\bar{x}_{\text{criterion}} = \mu_0 \pm z_{1-\alpha/2}\frac{\sigma}{\sqrt{n}} \tag{9-1}$$

Accept the null hypothesis if the sample mean falls in the acceptance region. Otherwise, reject the null hypothesis.

The decision rule assumes that the sample size is determined in advance by such factors as availability or cost. By also requiring advance specification of α, the rules explicitly control only the probability of rejecting H_0 when H_0 is true (type I error). Nothing explicit is done to control the probabilities of accepting H_0 when H_0 is false (type II error). A subsequent section of this chapter will consider this topic further.

Table 9-3 Commonly Selected Significance Levels for Hypothesis Tests and Standard Normal Deviations for Two-Tailed Tests

Significance Level Probability of Type I Error: α	*Standard Normal Deviate (z)*
0.10	± 1.65
0.05	± 1.96
0.01	± 2.58

Sample Selection and Decision

Under the approach we are using, the sample size, the significance level, and the acceptance region have all been determined before the sample is selected. The supervisor has specified that a random sample will consist of 16 observations. A significance level of 0.05 also has been selected. These choices, along with the selection of a two-tailed test, result in critical values of

$$\bar{x}_{\text{criterion}} = \mu_0 \pm z_{0.975}\frac{\sigma}{\sqrt{n}}$$

$$= 80 \pm 1.96\left(\frac{0.50}{\sqrt{16}}\right)$$

$$= 80 \pm 0.125$$

or

$$79.755 \quad \text{and} \quad 80.245 \text{ lb}$$

The next step is to select a random sample of 16 observations and find the sample mean. Suppose that the sample selected yields a mean of 79.6 lb. In other words,

$$\bar{x}_{\text{observed}} = 79.6 \text{ lb}$$

Since the sample result falls below the lower critical value, the null hypothesis—that the mean fill is 80 lb—is rejected at the 0.05 level of significance. The sample evidence supports the conclusion that the mean fill weight has drifted to something less than 80 lb. Presumably the machine setting will be checked and reset to the proper position.

Population Standard Deviation Unknown

Because of its simplicity, we began with the situation in which the population standard deviation is known. In practice, the population standard deviation is seldom known.

Population Normally Distributed In Chapter 8, when we found ourselves in the situation where the population could be assumed to be normally distributed with unknown standard deviation σ, we substituted the sample estimate s for σ. When we did so, it was pointed out that the resulting statistic

$$t = \frac{\bar{X} - \mu}{s/\sqrt{n}}$$

is distributed in accordance with a t distribution. Recall that

$$s = \sqrt{\frac{\Sigma(X - \bar{X})^2}{n - 1}}$$

and the appropriate t distribution has $n - 1$ degrees of freedom.

For testing a hypothesis about the mean of a normally distributed population with unknown standard deviation, we can use the 5-step procedure described earlier with only minor modifications. All we need do is change the method of calculating the critical values that define the acceptance region.

DECISION RULE (Normal Population; σ Unknown): For a two-tailed hypothesis test, specify the significance level α and use it to specify the acceptance region lying between critical values

$$\bar{x}_{\text{criterion}} = \mu_0 \pm t_{1-\alpha/2} \frac{s}{\sqrt{n}} \qquad \text{(9-2)}$$

where t has $n - 1$ degrees of freedom and s is the sample standard deviation. Accept the null hypothesis if the sample mean falls in the acceptance region. Otherwise, reject the null hypothesis.

The only change in this decision rule as compared with the one presented in connection with Equation 9-1 is the substitution of t and s for z and σ, respectively. In this book the resulting procedure is applied to all sample sizes because of the extensive t-table in Appendix A-4. Some other books replace t with the standard normal deviate z when the sample size exceeds 30. As was explained in Chapter 8, this practice arose from having only sketchy t-tables in the past and from the fact that the t distribution approaches the z distribution as sample size grows large.

Example

A new training program based on programmed learning materials has been employed for a class of 16 telephone installers. Under the former training program, the mean score of installers on a performance test given after training was 445. The scores of those in the class trained under the new program appear in Table 9-4. The training director wants to know whether this new set of scores on the performance test indicates any change in mean score for the new program.

The 5-step procedure is as follows:

1. The hypothesis statement is

$$H_0: \mu = 445 \qquad H_A: \mu \neq 445$$

2. The test statistic is t with $16 - 1 = 15$ degrees of freedom.
3. Let α be 0.10. Then the critical values of t from Appendix Table A-4 are -1.7531 and $+1.7531$. The sample variance is

$$s^2 = \frac{\Sigma(x - \bar{x})^2}{n - 1} = \frac{5704}{15} = 380.27$$

Hence, $s = 19.5$. From Equation 9-2, the critical values are

Table 9-4 Performance Scores for 16 Telephone Installers

Scores (x)	*Squared Deviations* $(x - \bar{x})^2$
468	81
455	16
457	4
493	1156
491	1024
456	9
464	25
447	144
427	1024
448	121
468	81
492	1089
439	400
444	225
452	49
443	256
7344	5704

$$\bar{x} = \frac{7344}{16} = 459$$

$$\bar{x}_{\text{criterion}} = \mu_0 \pm t_{0.95}\frac{s}{\sqrt{n}}$$

$$= 445 \pm 1.7531\,\frac{19.5}{4}$$

or 436.45 and 453.55.

The acceptance region runs from 436.45 to 453.55 and the rejection region is above and below the acceptance region.

4. The sample mean is $\bar{x}_{\text{observed}} = \dfrac{7344}{16} = 459$

5. Since the observed sample mean falls outside the acceptance region, we reject the hypothesis that the population mean is 445. It appears that the mean is greater than 445.

Population Not Normally Distributed As in Chapter 8, even when the population is not normally distributed, the sampling distribution of t will be essentially as given in Table A-4 of Appendix A, provided the population is not seriously skewed, or the sample size is greater than 50. Because of this, the procedure given in conjunction with Equation 9-2 can be applied in this case too.

DECISION RULE (Nonnormal Population; σ Unknown): If the population is seriously skewed, select a random sample that contains 50 or more observations. Then apply the decision rule given in conjunction with Equation 9-2.

Exercises

9-1 *σ known: normal population.* The net weight printed on a certain brand and size of bottled catsup is 24 oz. The production supervisor on the line has just taken a random sample of 36 bottles and has found the mean net weight to be 24.68 oz. The process is known to have a standard deviation of 0.73 oz. At the 0.05 level of significance, test the hypothesis that the process mean fill weight is as advertised on the label.

9-2 *σ known: normal population.* The distribution of scores on a nationally administered test for admission to graduate study in a certain field has remained normally distributed with a standard deviation of 100 points for many years. The test is designed to have a mean score of 500. This year, a preliminary random sample of 144 tests has been found to have a mean score of 482. Test the hypothesis that this year's mean score nationwide will turn out to be 500 also. Use a 0.01 level of significance.

9-3 *σ unknown: normal population.* The length of a machined part is being checked by the purchasing firm to see if it conforms to a specification of 4 in. The lengths of 9 such parts selected at random from the shipment are 4.067, 4.012, 4.024, 4.081, 3.989, 4.080, 4.053, 4.061, and 4.100. Test the hypothesis that the shipment conforms to the specification at the 0.01 level of significance.

9-4 *σ unknown: normal population.* A coin-operated machine that pours milk by the cup is set to deliver 8 oz per cup. A random sample of the content weights of 12 cups is as follows: 8.40, 8.25, 8.05, 7.84, 7.36, 8.54, 7.56, 7.56, 8.02, 7.39, 8.34, and 8.26. Test the hypothesis that the machine is delivering as set. Use 0.05 as the significance level.

9-5 *Microchip mean life.* The distribution of lifetimes for a type of microchip is positively skewed. A random sample of 100 of these chips has a mean lifetime of 4170 hr and a standard deviation of 90 hr. Test the hypothesis that the mean lifetime for all such chips is 4150 hr. Use a 0.05 level of significance.

9-6 *MBA job offers.* Over the last five years, students who received MBA degrees from a certain university also received an average of 7.2 job offers each. The standard deviation of this job offer distribution is 1.8. The distribution exhibits considerable positive skewness. This year a hypothesis test is to be used to determine whether the mean number of job offers is still 7.2. A staff assistant suggests that t be used with a sample of 15 observations from the 135 students in this year's class. Comment on the proposal.

9.2 Testing a Hypothesis about a Population Mean: One-Sided

In connection with Table 9-1, we pointed out that either a lower-tailed or upper-tailed hypothesis test can be chosen as an alternative to a two-tailed test. To illustrate the one-tailed testing procedure, we shall use another example.

The sales personnel of a retail store record each transaction on a sales ticket. Prior to last month, periodic checks of random batches of 100 tickets produced a mean of 12 errors per batch with a standard deviation of 2 errors per batch. The error distribution is essentially normal. Last month a campaign was carried out to reduce the error rate. A hypothesis test is to be conducted to see if that objective has been realized. Time considerations dictate that a random sample of 9 batches must be used for the test.

The Hypothesis Statement

The manager's only interest is in testing whether the mean error rate has declined. Since this is a movement in only one direction, a one-tailed test is appropriate. We must choose between a lower-tail and an upper-tail test. The hypothesis statements for these are as follows.

Lower-Tail Test	*Upper-Tail Test*
$H_0: \mu \geq \mu_0$	$H_0: \mu \leq \mu_0$
$H_A: \mu < \mu_0$	$H_A: \mu > \mu_0$

Replacing μ_0 with 12 errors in this table, the manager quickly determines that his interest is in H_A in the lower left corner of the table, that is, $H_A: \mu < 12$. This choice leads to selection of a lower-tailed test for which the full hypothesis statement becomes:

$$H_0: \mu \geq 12 \qquad H_A: \mu < 12$$

The Test Statistic and Its Sampling Distribution

Past experience supports the conclusion that errors per batch is a normally distributed variable with an essentially stable standard deviation of 2 errors per batch. Again we begin with the case in which σ is known because of its simplicity, although many, if not most, times the sample standard deviation must be used.

The test statistic will be the sample mean, $\bar{x}$. Under the null hypothesis, we select as the population mean the value when the equal sign (rather than the inequality) applies. Hence we are concerned with the sampling distribution of $\bar{x}$, which has a mean of 12 errors per batch and a standard deviation of $\sigma/\sqrt{n}$, or 2/3.

Acceptance and Rejection Regions

The sampling distribution for our current example is illustrated in Figure 9-2. The sampling distribution has a mean of $\mu = 12$. This value of μ is part of the null hypothesis. Consequently, most values of $\bar{x}$ associated with a μ value of 12 should be in the acceptance region for H_0. This reasoning dictates that the critical value be placed in the lower tail of the sampling distribution—somewhere considerably less than 12. The exact location stems from the choice of α, the level of significance. Recall that the significance level for a hypothesis test is the probability of a type I error, α. A type I error can occur only when the null hypothesis is true. Within this constraint, the relevant sampling distribution is the one for which the equality sign applies. For the lower-tailed test illustrated in Figure 9-2 α is the area in the single tail to the left of the critical value. Indeed, for all one-tailed tests, the critical value is always chosen such that the probability area in just one tail is equal to α. This contrasts with two-tailed tests where α is the area in the two tails combined. Common significance levels for one-tailed tests, together with their standard normal deviates, are shown in Table 9-5.

Our discussion can be summarized in the following decision rule.

DECISION RULE (Normal population; σ known): For a lower-tailed hypothesis test, specify the significance level α and use it to specify the acceptance region lying above the critical value

$$\bar{x}_{\text{criterion}} = \mu_0 - z_{1-\alpha}\frac{\sigma}{\sqrt{n}} \tag{9-3}$$

Accept the null hypothesis if the sample mean falls in the acceptance region. Otherwise reject it.

For an upper-tailed test, change the minus sign in Equation 9-3 to a plus sign and specify the acceptance region as lying *below* the critical value.

Note the difference in the subscripts on z in Equations 9-2 and 9-3. This is the key to the difference between one- and two-tailed hypothesis tests.

Suppose the 0.05 level of significance is selected for our example of the sales tickets. Then the critical value is

$$\bar{x}_{\text{criterion}} = 12 - 1.65\frac{2}{\sqrt{9}} = 10.9 \text{ errors}$$

and the acceptance region is $\bar{x} \geq 10.9$ errors. If the *sample* mean turns out to be 10.9 or greater, we will conclude that the *population* mean is at least 12 errors per batch;

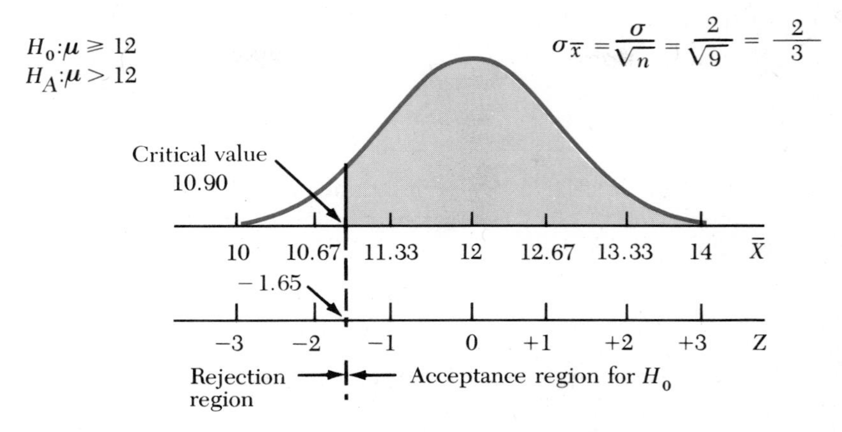

Figure 9-2 Hypothesized Sampling Distribution of $\bar{X}$ for Samples of $n = 9$ Observations

such a sample result will not support the conjecture that the population mean is less than 12.

Sample Selection and Decision

A random sample of 9 batches of 100 is selected from the sales tickets of the most recent month. The sample mean is 8.56 errors per batch. In the decision rule for the lower-tailed test, the critical value is 10.90 errors per batch, and the acceptance region for the null hypothesis is $\bar{x}_{\text{observed}} \geq 10.90$. Since the observed value is less than the critical value, the null hypothesis is rejected at the 0.05 level of significance. The conclusion is that the error rate has fallen to something less than 12 following the campaign.

In this example, batches of 100 sales tickets, rather than single sales tickets, were the units of observation. The distribution of errors on individual tickets would exhibit rather extreme positive skewness. Batching is a useful operational device for utilizing the central limit theorem so that the variable being observed is essentially normally distributed.

Table 9-5 One-Tailed Tests: Commonly Selected Significance Levels and Standard Normal Deviations

Significance Level: α	*Standard Normal Deviate: z*
0.10	+1.28 (upper tail) −1.28 (lower tail)
0.05	+1.65 (upper tail) −1.65 (lower tail)
0.01	+2.33 (upper tail) −2.33 (lower tail)

Population Standard Deviation Unknown

When the population standard deviation is not known, the procedure for a one-tailed hypothesis test can be summarized as follows.

DECISION RULE (Normal population: σ unknown): For a lower-tailed hypothesis test, specify the significance level α and use it to specify the acceptance region lying above the critical value

$$\bar{x}_{\text{criterion}} = \mu_0 - t_{1-\alpha}\frac{s}{\sqrt{n}} \tag{9-4}$$

Accept the null hypothesis if the sample mean falls in the acceptance region. Otherwise reject it.

For an upper-tailed test, change the minus sign in Equation 9-4 to a plus sign and specify the acceptance region as lying *below* the critical value.

Example

A coin-operated coffee machine is designed to deliver 6 oz of coffee. The owner believes the machine is overfilling and decides to run a hypothesis test. She selects $H_A: \mu > 6$ as the statement of interest. She fills 25 cups and finds the mean fill to be 6.5 oz with a sample standard deviation of 0.7 oz.

1. The hypothesis statement is

$$H_0: \mu \leq 6 \text{ oz} \qquad H_A: \mu > 6 \text{ oz}$$

2. The population of fill weights is essentially normal. Since the population standard deviation is not known, the t distribution with $n - 1 = 24$ degrees of freedom applies.

3. Let α be 0.05. Then from Appendix Table A-4, inside the back cover, the critical value of t is 1.7109. For an upper-tailed test, the critical value is

$$\bar{x}_{\text{criterion}} = \mu_0 + t_{0.95}\frac{s}{\sqrt{n}}$$

$$= 6 + 1.7109\frac{0.7}{5} = 6.24 \text{ oz}$$

The acceptance region for the null hypothesis is $\bar{x} \leq 6.24$ oz.

4. The observed sample mean is 6.5 oz.

5. Since the observed sample mean falls in the rejection region, the owner concludes that the machine is overfilling.

Population Not Normally Distributed When the population is not normally distributed, the decision rule cited in conjunction with Equation 9-4 applies, provided that the population is not seriously skewed. If it is, a sample size of 50 will usually be large enough to make the rule apply because of the central limit theorem.

Exercises

9-7 *Upper-tailed test.* The mean contribution to a public television station's financial support campaign was $23.41 before a matching donations pledge was announced. Following inclusion of announcements of the pledge in all appeals for funds, a random sample of 64 donations has a mean of $25.79. Over this and several previous campaigns, the standard deviation of the distribution of donations has remained essentially constant at $9.60. Determine whether the evidence supports the conclusion that the mean amount donated has increased. Let the significance level be 0.05.

9-8 *Lower-tailed test.* An engineer has developed an inexpensive automobile muffler that filters out a large portion of hydrocarbon pollutants. The firm which employs the engineer will go ahead with product development and promotion provided mean hydrocarbon emission is less than 2 parts per million in a specified laboratory procedure. In a random sample of 25 observations, the mean is 1.68 ppm and the standard deviation is 0.80 ppm. Use the 0.01 significance level and test for a decrease in the mean with the new muffler.

9-9 *Cosmetic sales test.* A cosmetic product is worth producing on a large scale only if the mean number of units delivered per retail outlet is no less than 80 per month. A nationwide marketing trial has just been completed in which 208 sample outlets had mean deliveries of 78.3 per month during the trial period. The standard deviation was 7.21. At the 0.05 level of significance, perform the hypothesis test required for this situation.

9-10 *Cost of appliance repair.* The advertising department of a firm that makes a household electrical appliance wants to call attention to the small amount purchasers can expect to spend for repairs over the first three years of ownership. The department claims that the average owner spends less than $20 over that period. To support such a claim, the firm selects a random sample of 225 qualified owners. The sample mean expense for the first three years is $19.49 and the sample standard deviation is $2.18. At the 0.01 level of significance, test the advertising department's claim.

9-11 *Types of error.* For Exercise 9-9, describe the type I and type II errors that are possible. Assume no sample results are available.

9-12 *Types of error.* For Exercise 9-10, describe the type I and type II errors that are possible. Assume no sample results are available.

9.3 Testing a Hypothesis about a Population Proportion

Recall that the sample proportion is p, where $p = r/n$ and r is the number of successes in a sample size of n. The sampling distribution of p has as its mean and standard deviation

$$\mu_P = \pi \quad \text{and} \quad \sigma_P = \sqrt{\frac{\pi(1-\pi)}{n}}$$

where π is the probability of a success on a single trial. The statistic

$$z = \frac{P - \pi}{\sigma_P}$$

is distributed as a standard normal deviate if the sample size n is at least 100 and if both $n\pi$ and $n(1-\pi)$ are greater than 10.

With this as background we can state the decision rule for testing a hypothesis about a population proportion.

> DECISION RULE (Normal Approximation for Binomial Distribution): Select a random sample of at least 100 observations and also large enough to make both $n\pi$ and $n(1-\pi)$ at least 10. Choose the significance level α and, for a two-tailed test, specify the acceptance region lying between critical values
>
> $$p_{\text{criterion}} = \pi_0 \pm z_{1-\alpha/2}\sqrt{\frac{\pi_0(1-\pi_0)}{n}} \tag{9-5}$$
>
> Accept the null hypothesis if the sample proportion, p, falls in the acceptance region. Otherwise reject it.

Example

The manufacturer of a brand of premium ice cream runs periodic market surveys to check the proportion of consumers stating a preference for this brand. The firm wants a test of the hypothesis that the proportion is 0.40, the typical figure for the past year. A random sample of 500 observations is contemplated.

1. The hypothesis statement is

$$H_0{:}\pi = 0.40 \qquad H_A{:}\pi \neq 0.40$$

2. The sample size is 500; $n\pi = 500(0.40) = 200$; since $\pi < 0.5$, $n(1-\pi)$ will be > 200. Hence the normal approximation can be used for the binomial.
3. Let α be 0.01. From Equation 9-5, the critical values are

$$p = 0.40 \pm z_{0.995}\sqrt{\frac{0.40(0.60)}{500}}$$

$$= 0.40 \pm 2.58(0.022) = 0.40 \pm 0.06$$

or 0.34 and 0.46. The acceptance region for the sample proportion p runs from 0.34 to 0.46.

4. In the survey, suppose 238 consumers in the sample state a preference for the brand in question. Then the sample proportion is 238/500, or 0.476.
5. Since the observed sample proportion is greater than the upper critical value, the sample proportion falls outside the acceptance region and the null hypothesis is rejected. It appears that the preference is greater than 0.40.

The decision rules for one-tailed tests for population proportions are apparent by comparison with those for population means described in conjunction with Equation 9-3.

Normal Approximation Not Applicable Two conditions were stated in the above decision rule for application of the normal approximation. Should one or both conditions not be met, we can revert to the relevant binomial distribution. Extensive tables of binomial distributions often are available. Alternatively, many types of pocket calculators have procedures for calculating terms of binomial distributions. Assuming the availability of one or the other, we can illustrate the procedure in the context of the following example.

Example

A random sample of 60 parts is to be selected from a large shipment to see whether the specification that no more than 5% defective is met. The appropriate hypothesis test for a 0.01 level of significance is wanted.

1. The hypothesis statement is

$$H_0: \pi \leq 0.05 \qquad H_A: \pi > 0.05$$

2. Since $n < 100$ and $n\pi = 3$, both conditions for the normal approximation are violated. The binomial distribution with r as the number of defectives in the sample of $n = 60$ and with $\pi = 0.05$ is applicable. Instead of r we can also use $p = r/n = r/60$. Some upper-tail probabilities for the r distribution and the equivalent p distribution are

r:	6	7	8
$P(R \geq r)$:	0.0787	0.0297	0.0098
p:	0.10	0.117	0.133
$P(P \geq p)$:	0.0787	0.0297	0.0098

3. With α specified as 0.01, the acceptance region for H_0 is any value of r less than 8. Alternatively, the acceptance region for the sample proportion of defectives in the sample is $p < 0.133$.

Exercises

9-13 *Normal approximation applicable.* The national headquarters of a large labor union wants to test the claim made by a candidate for public office that 60% of the union's members favor taxing social security payments to those with substantial additional income. In a random sample of 900 members interviewed by telephone on this and other issues, 519 said they favored such a tax. Make a two-tailed test of the claim with a significance level of 0.01.

9-14 *Normal approximation applicable.* The general manager of a minor league baseball team thinks the percentage of people under 25 years old who are attending games has declined from the 20% experienced earlier in the season. Over the past two weeks a random sample of 564 people were asked their age and responded. Of these, 102 or 18.1% were under 25. Test the statement that less than 20% of those attending during the past two weeks were under 25 years old. Use the 0.05 level of significance.

9-15 *Normal approximation not applicable.* From a casual appraisal, it appears that less than 25% of several hundred students who took a certain examination in statistics had the correct concept on the fourth question. Before making a detailed count, the grader takes a random sample of 16 examinations and finds that only 2 people had the correct concept. At the 0.01 level of significance, test the statement that less than 25% of all students had the correct concept.

9-16 *Normal approximation not applicable.* In a hardware store, records indicate that 10% of the shipments from a certain supplier are delayed. Consider a random sample of 9 shipments from this supplier.

a. Discuss the difference between r and p as the designated random variable.
b. From the appropriate appendix table, obtain the sampling distributions of r and of p.
c. What is the probability that r will be greater than 3 for the random sample?
d. Describe the upper-tailed hypothesis test which makes use of the probability found in (c).

9-17 *Dwelling unit survey.* In the last census, 7% of the dwelling units in a major metropolitan area were classified as substandard. Recently, a random sample of 200 dwelling units has been selected from a current list that is essentially complete. A hypothesis test is to be designed to test whether or not there has been a significant increase in the proportion of substandard dwellings since the last census. The 0.05 level of significance is selected for the test.

a. Complete the first three steps of the hypothesis testing procedure for this case.
b. The sample proportion is 0.11. Which of the two possible conclusions in the test does the sample evidence support?

9-18 *Sales expectation.* A retail store owner determined that, in the past, 30% of customers entering the store made a purchase, but the owner thinks the percentage may have increased recently. If 8 or more of the next 15 customers make a purchase, the owner will conclude that the percentage has increased above 30. Assume that the next 15 customers constitute a random sample. Formally state the hypothesis testing procedure that is being employed. What level of significance is applicable?

9.4 Standardized Test Statistics and *P*-Values

So far in this chapter we have been using as test statistics the sample mean $\bar{x}$, the sample proportion p, or the number of defectives in the sample, r. Many of these tests involve either z or t in the specification of the critical value(s) to determine the acceptance region for the null hypothesis. These latter are examples of what are known as *standardized statistics*. In subsequent chapters, it will be far more convenient to state such things as decision rules for hypothesis tests in terms of standardized statistics rather than in terms of the primary random variable.

To convert the 5-step hypothesis testing procedure we have been using to a similar procedure using standardized statistics, we first change the test statistic from $\bar{x}$, p, or r to either z or t and define the acceptance region in terms of the standardized statistic. For example, the decision rule stated in conjunction with Equation 9-2 becomes:

DECISION RULE WITH STANDARDIZED STATISTIC (Normal Population; σ Unknown): For a two-tailed hypothesis test, specify the significance level α and use it to specify the acceptance region

$$t_{\alpha/2} \leq t \leq t_{1-\alpha/2}$$

Then calculate the test statistic

$$t_{\text{observed}} = \frac{\bar{x}_{\text{observed}} - \mu_0}{s/\sqrt{n}} \tag{9-6}$$

where μ_0 is the hypothesized value of the population mean. Accept the null hypothesis if the observed value of the test statistic falls in the acceptance region. Otherwise reject the null hypothesis.

For the example of the training program for telephone installers presented in conjunction with Equation 9-2, the changes are in Steps 2 through 5. Hence, only those steps will be described here.

2. The test statistic is t as defined in Equation 9-6 with 15 degrees of freedom.
3. Let α be 0.10. Then the critical values of t are -1.7531 and $+1.7531$ and the acceptance region is

$$-1.7531 \leq t \leq +1.7531$$

4. From the sample, the mean score is 459. Hence,

$$t_{\text{observed}} = \frac{459 - 445}{19.5/\sqrt{16}} = +2.87$$

5. Since the sample value of the test statistic is greater than the upper critical value, reject the null hypothesis. The population mean appears to be greater than 445.

P-values The 5-step procedure we have been using represents the classical or traditional approach while more recent practice makes use of P-values. *Under the assumption that H_0 is true, a P-value is the probability of getting a value of the standardized test statistic which is more extreme than the one observed.* P-values are prevalent in computer output. Consequently, it is important to become familiar with them.

Using P-values involves some changes in the hypothesis testing procedure just described. Recall that the classical approach establishes an acceptance region and compares the observed value of the standardized test statistic with that region. In the P-value approach, we choose a criterion P-value. Then, after calculating the sample value of the standardized statistic, we find the associated P-value, state it, and reject the null hypothesis if the P-value is smaller than the criterion value.

The new hypothesis testing procedure is:

1. State the hypothesis. (No change.)
2. Select the test statistic and its sampling distribution. (The test statistic is the appropriate standardized statistic.)
3. Choose the criterion P-value. Customary values are 0.10, 0.05, or 0.01.
4. Select a random sample, calculate the observed value of the standardized statistic, find the associated P-value from the appropriate table, and report this P-value along with the observed value of the standardized statistic.
5. Compare the observed P-value with the criterion value and accept the null hypothesis if the observed value is greater than the criterion. Otherwise reject the null hypothesis.

To illustrate this concept we return to the example that dealt with the new training course for telephone installers. Steps 1 and 2 are the same as those used to illustrate Equation 9-6. The remaining steps are:

3. Let $P_{\text{criterion}}$ be 0.10, the same as α was in the previous illustration.
4. As before, the value of the standardized test statistic calculated from the sample is

   $t_{\text{observed}} = +2.87$, as before.

   From Appendix Table A-4, for 15 degrees of freedom, the probability that $t \geq +2.87$ or $t \leq -2.87$ is estimated as $2(1 - .994) = 0.012$. Hence

$P_{\text{observed}} = 0.012.$

5. Since the observed P-value is less than the criterion, reject the null hypothesis. The observed value of t is more extreme than is permitted by the criterion.

In contrast with the two-tailed test just illustrated, for a one-tailed test the observed P-value is that associated with only one tail of the relevant distribution.

Sometimes the P-value for the sample standardized statistic is much smaller than the smallest value attainable from the table. Given this situation, we can report that the observed P-value is less than the one attainable. Had the observed value of t been +4 in the telephone installer example, we would have reported that $P < 0.01$ because $2(1 - 0.995) = 0.01$, the smallest tabled value.

A major reason for using the P-value procedure is that stating the P-value associated with the sample value of the test statistic conveys considerably more information than just stating that the null hypothesis is accepted or rejected "at the α level of significance." In the foregoing example, for instance, the traditional approach would result in a report that the sample t-value is +2.87 while the 0.10 acceptance region lies between t-values of −1.7531 and +1.7531. The null hypothesis is, therefore, rejected at the 0.10 level of significance. With the newer approach we would report that the sample t-value is +2.87, which, for a two-tailed test, has an associated P-value of 0.012. The P-value makes it evident that the null hypothesis could have been rejected at a significance level well beyond 0.10. This report shows that the sample evidence supporting rejection of the null hypothesis is much stronger than can be ascertained from the traditional report.

Exercises

9-19 *P-values.* For the following values of observed standardized test statistics, find the P-values.

a. $z = 2.95$ from a one-tailed test
b. $z = -1.25$ from a two-tailed test
c. $t = 0.6870$ with 20 degrees of freedom; one-tailed test
d. $t = -2.7045$ with 40 degrees of freedom; two-tailed test
e. $t = 4.3333$ with 12 degrees of freedom; one-tailed test

9-20 *P-values.* Find the P-values for the following observed standardized test statistics.

a. $z = -2.78$ in a two-tailed test
b. $z = 0.82$ in a one-tailed test
c. $t = -1.6955$ with 31 degrees of freedom; two-tailed test
d. $t = 1.2909$ with 91 degrees of freedom; one-tailed test
e. $t = -3.500$ with 10 degrees of freedom; two-tailed test

9-21 Replace the hypothesis testing procedure formerly used in Exercise 9-1 with one which incorporates a standardized test statistic and P-value.

9-22 Revise the hypothesis testing procedure formerly used in Exercise 9-4 to include a standardized test statistic and P-value.

9.5 The Power of a Test

Table 9-2 displayed the four possible outcomes of a hypothesis test. If the null hypothesis is true, we make a correct decision by accepting it and an incorrect decision by rejecting it (a type I error). When the null hypothesis is true, the probability of a correct decision is $1 - \alpha$ and the probability of an incorrect decision is α.

Prior to this point in the chapter, we controlled the probability of a type I error by arbitrarily selecting a value for α and the sample size.

The other two outcomes in Table 9-2 are the possible results when the null hypothesis is false. We make a correct decision by rejecting H_0 and an incorrect decision by accepting it (type II error). The probability of a correct decision when H_0 is false is $1 - \beta$, and the probability of an incorrect decision for a specified value of the parameter not included in the null hypothesis is β.

In this section, our objective is to show that an increase in sample size can be used to reduce the probabilities of accepting H_0 when H_0 is false (type II error) without increasing the probability of rejecting H_0 when H_0 is true (type I error). This can best be illustrated in terms of the power of a test procedure.

Power Curve for a Lower-Tailed Test

In Section 9-2, we tested the null hypothesis that the mean number of errors in random batches of 100 sales tickets was 12 or more. The distribution of errors per batch was normal with a known standard deviation of 2 errors per batch. We conducted a lower-tailed test to see whether corrective action had succeeded in significantly lowering the mean number of errors per batch.

For this test, the sample size was set at 9 batches, α was set at 0.05, and μ_0 was set at 12 errors per batch. As a result, the critical value was a mean of 10.90 errors per batch. A sample mean greater than 10.90 would fall in the acceptance region for the null hypothesis. Such sample evidence would indicate no significant shift to a smaller mean error rate. A value of the sample mean less than 10.90 would fall in the rejection region and would indicate a significant shift to a lower error rate.

The reason for developing the power curve for a test procedure such as the one just described is to summarize the performance of the procedure for possible values of the parameter. *When the null hypothesis is false, the power of the test is the probability of making a correct decision.* In Table 9-2, the power of a test can be seen to be $1 - \beta$, the probability in the lower right corner.

> For a specified value of the parameter under H_A, the *power* of a hypothesis test is the probability $1 - \beta$ that the test procedure will correctly reject H_0. The *power curve* for the test procedure is the graph of probabilities for rejecting H_0 for all possible values of the parameter under H_A.

We shall illustrate the concepts of power and the power curve for a hypothesis test with the test for the error rate on sales tickets. The test procedure is as follows.

1. $H_0: \mu \geq 12$; $H_A: \mu < 12$.
2. Test statistic: the standard normal deviate

$$Z = \frac{\overline{X} - 12}{2/\sqrt{9}}$$

3. $\alpha = 0.05$; $\bar{x}_{\text{criterion}} = 12 - 1.65(2/\sqrt{9}) = 10.90$; $z_{\text{criterion}} = -1.65$.
4. Accept H_0 if $\bar{x}_{\text{observed}} \geq 10.90$ or if $z_{\text{observed}} \geq -1.65$. Otherwise accept H_A.

In Figure 9-3, the power of this test procedure is shown for four different values of the parameter less than 12 ($\mu < \mu_0$). For all four of these values, the null hypothesis is false (the condition that must hold in order to find the power, $1 - \beta$). In the top two panels of the figure, the population mean is less than the critical value. Consequently, the major portion of the sampling distribution of $\overline{X}$ falls below the critical value and the power, $1 - \beta$, is greater than 0.50. As we move from the second panel to the top panel, μ moves to the left and $1 - \beta$ grows larger. If we let μ move to the left indefinitely, $1 - \beta$ will come closer and closer to 1, its upper limit. As μ moves to the left, the probability of a correct decision (reject H_0) approaches certainty and the power of the test approaches its upper limit.

Alternatively, as we move down from the second to the third and fourth panels, μ moves to the right toward $\mu_0 = 12$. Less and less of the distribution of $\overline{X}$ remains to the left of the critical value. Hence, $1 - \beta$, the power of the test, decreases. The lower limit for $1 - \beta$ occurs just as μ reaches μ_0 from below. At this point, slightly over 0.05 of the sampling distribution of $\overline{X}$ falls to the left of the critical value. Hence the lower limit for this example is $1 - \beta = 0.05$, the level of significance.

Using Figure 9-3 as a point of departure, we can fill in values of $1 - \beta$ for other values of $\mu < \mu_0$. Then we can plot these on a graph and connect them with a smooth curve. The result will be the power curve.

In Table 9-6, the power of the test for our example is calculated for every one-half standard error that μ lies to the left and right of the critical value (10.90 errors per batch), up to 11.90 errors per batch. Then at $\mu_0 = 12$, the lowest possible value of $1 - \beta$ for this test is calculated.

The next step is to plot the power of the test for each value of μ listed in Table 9-6 and connect these with a smooth curve. The result appears in Figure 9-4. At the critical value ($\mu = 10.90$), the probability of rejecting $\mu_0 = 12$ is $1 - \beta = 0.50$. The complement, β, which is the probability of a type II error, is also 0.50. As μ decreases, $1 - \beta$ increases and approaches its upper limit of 1.00. The smaller the mean error rate, the greater is the power of the test to detect that fact and make the correct decision. On the other hand, as μ moves from 10.90 to the right toward 12, the power to make the correct decision drops from 0.50 to 0.05. The mean error rate must be less than 10.90 for the test to have better-than-even odds of making the correct decision.*

---

*Some branches of applied statistics use the operating characteristic (O-C) curve rather than the power curve. The O-C curve is formed by plotting the appropriate values of β for relevant values of the parameter and shows the probability of accepting the null value of the parameter on the vertical axis. If Figure 9-4 were an O-C curve, it would begin at 0 in the lower left corner and rise in an S-curve to 0.95 at $\mu = 12$.

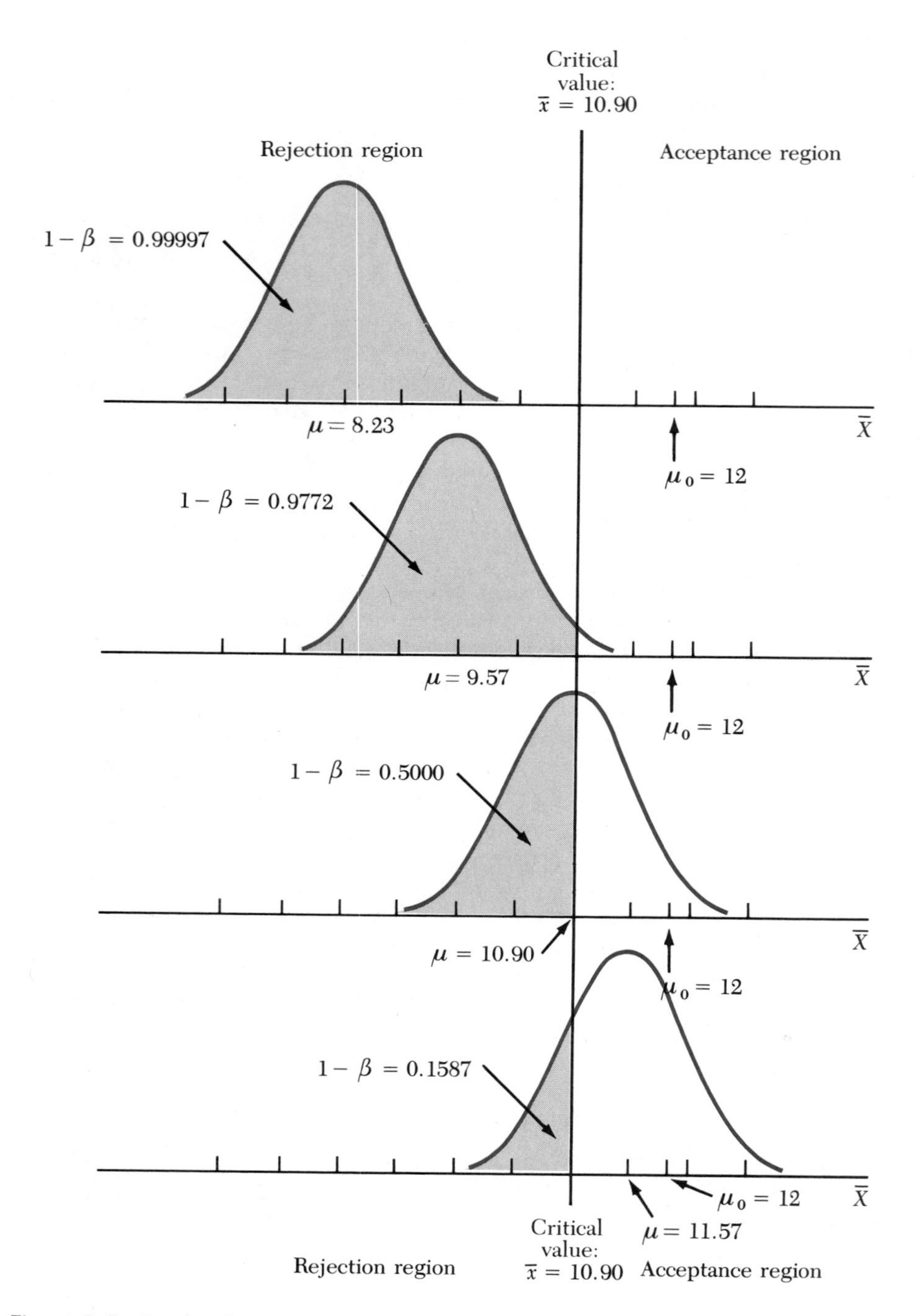

Figure 9-3 Power of a Lower-Tailed Test for Four Values of $\mu < \mu_0$

Table 9-6 Probability of Rejecting $\mu_0 = 12$ for a Lower-Tailed Test

Population Mean μ	*Standard Errors from Critical Value* $z = (10.9 - \mu)/(2/3)$	*Power* $1 - \beta$
8.23	4.0	0.99997
8.57	3.5	0.9998
8.90	3.0	0.9986
9.23	2.5	0.9938
9.57	2.0	0.9772
9.90	1.5	0.9332
10.23	1.0	0.8413
10.57	0.5	0.6915
10.90	0.0	0.5000
11.23	−0.5	0.3085
11.57	−1.0	0.1587
11.90	−1.5	0.0668
12.00	−1.65	0.0500

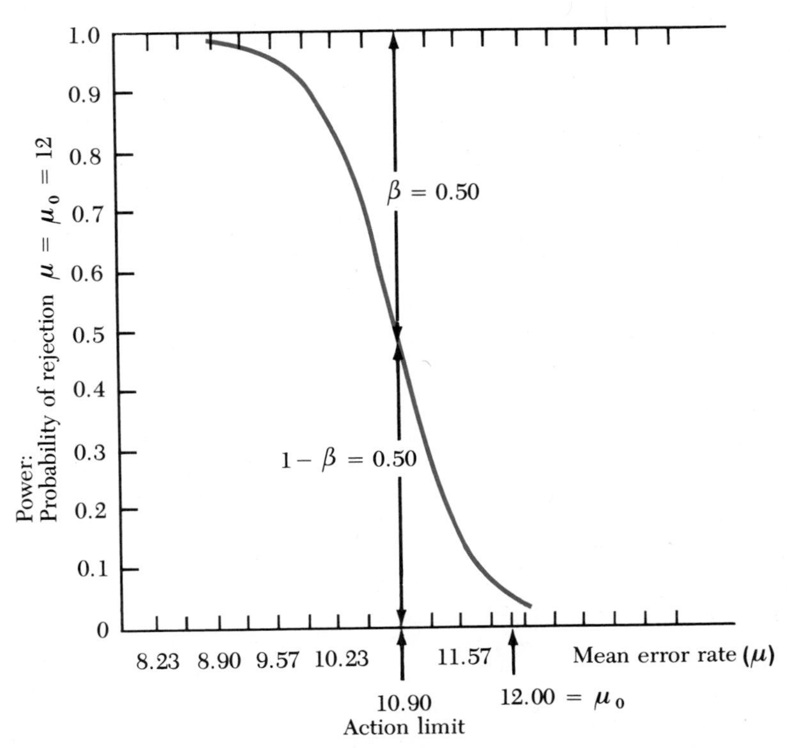

Figure 9-4 Power Curve for Lower-Tailed Hypothesis Test (H_0:$\mu \geq 12$; H_A:$\mu < 12$)

Although we shall not do so, we could develop power curves for upper-tailed and two-tailed tests.

Effect of Changing α Without Changing Sample Size

It is instructive to see what happens to the power curve for our example of the sales tickets when we reduce the probability of a type I error without changing the sample size. This is accomplished by moving the critical value to the left. Suppose we decide to reduce α from 0.05 to 0.005. The critical value must be lowered from 10.90 to 10.29 to leave only 0.005 of the null distribution in the lower tail.

The effect of lowering α to 0.005 on the power of the test is shown in Figure 9-5. The power curve for $\alpha = 0.05$ is also shown for purposes of comparison. It is evident from the figure that reducing α without changing sample size can be accomplished only by reducing the power of the test for all values of the parameter under H_A. This is the trade-off mentioned earlier in the chapter. For a given sample size, the probabilities of type I and type II errors are inversely related. One can be lowered only at the expense of increasing the other.

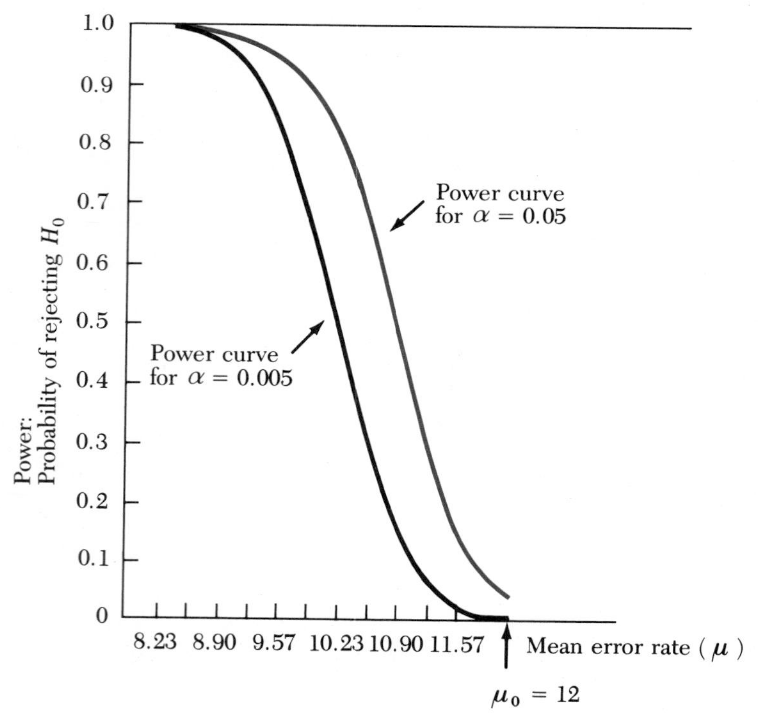

Figure 9-5 Effect of Changing α Without Changing Sample Size

Effect of Changing Sample Size Without Changing α

Suppose in our sales ticket example we increase the sample size to 36 batches of 100 sales tickets each but leave α at 0.05. This will increase the critical value from 10.90 to 11.45 because the standard error of the mean will decrease from 2/3 to 1/3. The critical value is now much closer to $\mu_0 = 12$ than it was for the smaller sample size:

$$\bar{x}_{\text{criterion}} = 12 - 1.65\frac{2}{\sqrt{36}} = 11.45$$

The power curve for this new test is shown in Figure 9-6. The curve for $n = 9$ is also shown. Notice that the probability of a type I error is the same for both tests ($\alpha = 0.05$). But the power $(1 - \beta)$ for the test based on the larger sample size is greater than that for the smaller sample for every relevant value of μ. The probability of a type II error is less for all relevant values of μ with the larger sample; the reduction is rather impressive for values of μ between 9.23 and 11.90.

In Figure 9-6, all of the increased information available from the larger sample is used to reduce the probabilities of type II errors. We also could have used part of the information to reduce the probability of a type II error and the rest of it to reduce

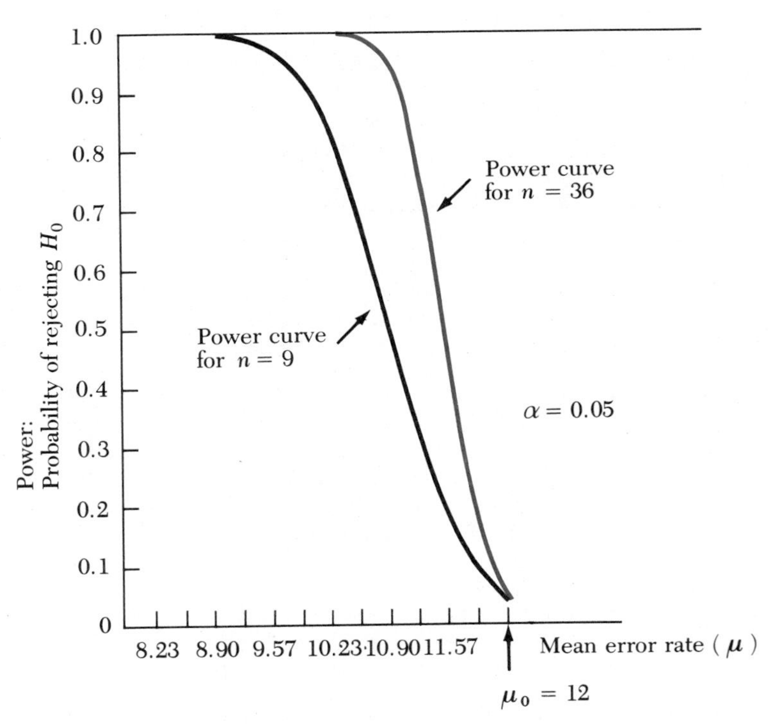

Figure 9-6 Effect of Increasing Sample Size from 9 to 36 for $\alpha = 0.05$

the probability of a type I error. This would be done by not moving the critical value all the way from 10.90 to 11.45.

We have seen that a power curve, in a single graphical display, summarizes the performance of a hypothesis test for all possible values of the parameter covered by the alternative to the null hypothesis, H_A. Power curves for alternative tests provide a convenient way to compare performances and aid in selecting the most appropriate test in a particular application.

Exercises

9-23 *Control of impurities.* The mean weight of all monitored impurities in a cereal product must be controlled at no more than 8 grams per package to satisfy government standards. With the current mix of grains being used as raw material for the product, the mean weight of these impurities is normally distributed and the standard deviation is 0.44 g. A random sample of 16 packages is selected periodically and is used in a one-tailed hypothesis test to control this aspect of product quality.

a. For the 1% level of significance, what is the critical value for the appropriate hypothesis test?
b. The power of the test is to be determined when the mean level of impurities is 8.36 g. As a first step, sketch the sampling distribution of the mean and show its position with respect to the critical value and μ_0.
c. Find the power of the test when μ is 8.36.

9-24 *Pedestrian fatalities and blood alcohol.** A large random sample of pedestrians at sites of pedestrian-automobile accidents fatal to pedestrians was selected. The sample was used as a control group. For the sample, the mean blood alcohol concentration of the pedestrians was 25 parts per 10,000 by weight and the standard deviation was 53. The blood alcohol concentration of a sample of 64 pedestrians who survived less than 6 h after accidents at these sites is to be compared with the control group. An upper-tailed hypothesis test is to be used with $\mu_0 = 25$, $\sigma = 53$, and $\alpha = 0.05$.

a. What is the critical value for the test?
b. Sketch the sampling distribution of $\overline{X}$ for a population mean blood alcohol concentration of 39.2 for fatalities. Show μ_0 and the critical value in the sketch.
c. What is the power of the test when $\mu = 39.2$?

9-25 *Life of transistor radio batteries.* A firm that tests products and is supported by subscriptions from consumers uses the following decision rule to check an advertising claim about the mean life of a certain brand of transistor radio battery: Find the mean life of a random sample of 25 batteries. If the mean of this sample is less than 96.05 h, conclude that the population mean life is less than 100 h. Otherwise conclude that the mean life is 100 h or more. The level of significance is 0.05.

a. What is the standard error of the mean ($\sigma_{\overline{X}}$) for this test? What is σ?
b. What type of error is possible for each of the following values of μ, the true mean life: 102, 99, 96, 94, 92, 90?

*Adapted from W. Haddon, Jr., et al., "A Controlled Investigation of the Characteristics of Adult Pedestrians Fatally Injured by Motor Vehicles in Manhattan," *Journal of Chronic Diseases*, 14 (December 1961), pp. 655-678.

c. Find the power of the test for those values of μ in (b) for which the null hypothesis is not true and sketch the power curve.

9-26 *Charge card balances.* The charge card department of a bank must perform a monthly test concerning the mean unpaid balance on charge card accounts for each branch bank. The mean must be greater than $200 per account to profit from charge card business. The decision rule for the test is: Select a random sample of 100 active charge card accounts from a given branch and record the unpaid balance immediately after the last payment. Find the sample mean. If this mean is greater than $227.96, accept the hypothesis that the population mean unpaid balance is over $200. Otherwise conclude that the population mean is $200 or less. The level of significance is 0.01.

a. What value of the population standard deviation is assumed in this test?
b. What type of error is possible if the true mean balance is equal to each of the following: $190, $200, $205, $215, $227.96, $240, $250, $260?
c. Find the power of the test for those values of μ in (b) for which a type II error is possible. Then sketch the power curve for the test.

9.6 The Finite Population Multiplier (Optional)

Prior to this point in the chapter we have assumed that, for finite populations, the sample being used was only a small portion of the population. Although the cutoff is somewhat arbitrary, we stated that any sample which contains fewer than 10% of the elements in the population is suitable for treatment by the procedures described up to this point in the chapter.

When there are too many observations in the sample for us to ignore the finite population multiplier, the procedures previously described require only that we multiply the relevant standard error by the finite population multiplier. In the cases covered by Equations 9-2 and 9-5, for example, the modifications are:

When the number of observations in a random sample is more than 10% of the number of observations in the population from which it came, and when the level of significance for a two-tailed hypothesis test is α, the critical values for the mean are

$$\bar{x}_{\text{criterion}} = \mu_0 \pm t_{1-\alpha/2}\frac{s}{\sqrt{n}}\sqrt{\frac{N-n}{N-1}} \tag{9-7}$$

and for the proportion the critical values are

$$p_{\text{criterion}} = \pi_0 \pm z_{1-\alpha/2}\sqrt{\frac{\pi_0(1-\pi_0)}{n}}\sqrt{\frac{N-n}{N-1}} \tag{9-8}$$

In either case, accept the null hypothesis if the observed value falls within the acceptance region lying between the critical values. Otherwise reject it.

Example

There are 980 households in a certain village. A random sample of 200 households has been surveyed by telephone to test the statement made by a councilman in the adjacent major city that 60% favor being annexed by the city. Of the 200 households, 102 are in favor of annexation.

The hypothesis test is as follows:

1. The hypothesis statement is:

$$H_0: \pi = 0.60 \qquad H_A: \pi \neq 0.60$$

2. The number of sample observations is over 100. Furthermore, both $n\pi$ and $n(1 - \pi)$ are greater than 10. Hence, the normal approximation and Equation 9-8 apply.

3. Let α be 0.05. From Equation 9-8 the critical value is

$$p_{\text{criterion}} = 0.60 \pm 1.96\sqrt{\frac{0.60(0.40)(980 - 200)}{200(980 - 1)}}$$

$$= 0.60 \pm 0.06$$

The acceptance region runs from 0.54 to 0.66.

4. The sample proportion is $102/200 = 0.51$.

5. Since the observed sample proportion falls below the lower limit of the acceptance region, the 60% statement is rejected. Of course, 60% could have favored annexation earlier.

9.7 Determining Sample Size (Optional)

In the final part of Section 9.5, the sample size for the sales ticket example was arbitrarily increased from 9 batches of 100 sales tickets each to 36 batches of 100 tickets. The probability of rejecting the null hypothesis when it is false increased for all values of μ less than 12 errors per batch. In other words, the probabilities for making the correct decision increased for all values of μ for which the null hypothesis is false. Furthermore, the maximum probability of rejecting H_0 when it is true was left at 0.05, as before. The probability of a type I error was left at a maximum of 0.05 when $\mu = \mu_0 = 12$, and the probabilities of type II errors were reduced by raising the sample size from 9 batches to 36.

Instead of preselecting the significance level and the sample size, as was done in the instance just cited, we begin at a different point in this section. We preselect the significance level and we preselect β, the probability of a type II error, for a designated value of μ not included in the null hypothesis. Then we develop a procedure that will allow us to find the sample size and critical value to meet the preselected specifications. The objective is to protect ourselves against a type I error and a designated type II error at specified probability levels. Alternatively, we want to designate the power, $1 - \beta$, for detecting a designated shift in the parameter.

In the example, we shall keep the batch size constant at 100 tickets per batch, and we shall let the sample size be the number of such batches to be selected at random. Recall that, prior to corrective action, the error rate was 12 errors per batch and the population standard deviation is known to be 2 errors per batch. Furthermore, the population can be assumed to be normally distributed. A lower-tailed hypothesis test is to be conducted to see whether corrective action has reduced the mean error rate to a value significantly less than 12.

The supervisor who will make the hypothesis test judges that lowering the mean error rate by 1.1 errors per batch will justify the time and cost of the corrective action. Consequently the supervisor wants a test with a high probability of detecting a shift from 12 to 10.90 errors per batch. Meanwhile, the probability of detecting no downward shift, should that be the case, is also to be kept high.

The situation is illustrated in Figure 9-7. Here μ_0 is 12, μ_A is 10.90, and the shift, δ, is -1.1. The probability of detecting a shift to 10.90 is the power of the test, $1 - \beta$. Suppose the supervisor chooses 0.99 for $1 - \beta$; then $|z_\beta| = 2.33$. The probability of detecting no downward shift from $\mu_0 = 12$ is $1 - \alpha$ on the figure. Suppose the supervisor sets this probability at 0.95; then $|z_\alpha| = 1.65$. To satisfy these conditions, the figure shows that the sample size must be

$$n = \left[\frac{\sigma}{\delta}(|z_\alpha| + |z_\beta|)\right]^2 \tag{9-9}$$

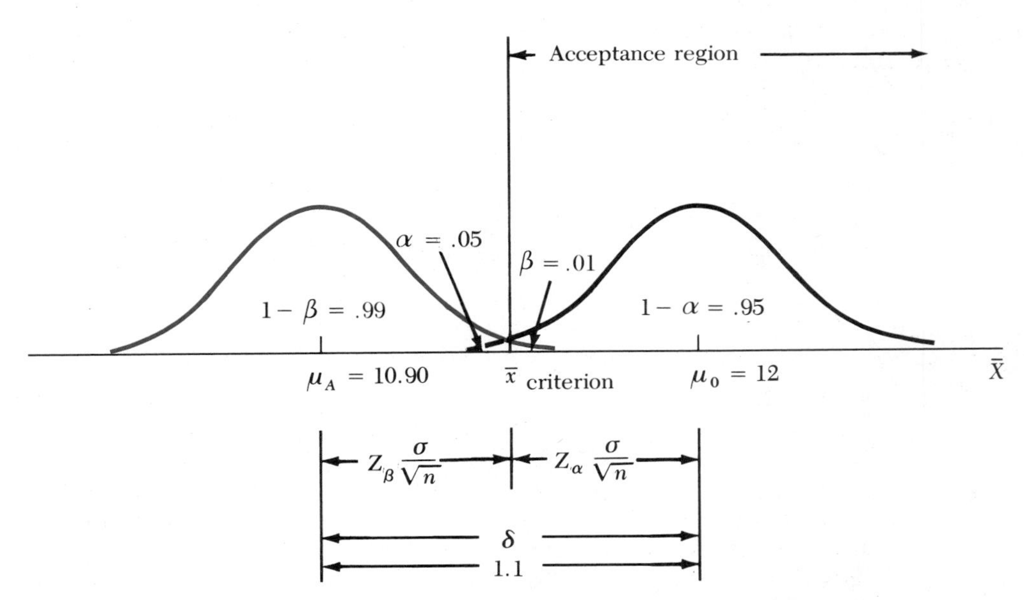

Figure 9-7 Sample Size for Detecting a Shift of δ with Probability $1 - \beta$ (One-Tailed Test)

which for our example is

$$n = \left[\frac{2(1.65 + 2.33)}{1.1}\right]^2$$

$$= 52.4$$

or 53 observations. Given the sample size, the supervisor can now find the critical value:

$$\bar{x}_{\text{criterion}} = \mu_0 - z_\alpha \frac{\sigma}{\sqrt{n}}$$

$$= 12 - 1.65\frac{2}{\sqrt{53}}, \text{ or } 11.55$$

Following corrective action, the test would be completed by finding the mean number of errors in 53 batches of 100 tickets each. If this value, $\bar{x}_{\text{observed}}$, is less than 11.55, the supervisor should conclude that a downward shift has occurred in the mean error rate. Otherwise the conclusion is that no significant downward shift has occurred.

For an upper-tailed test, as well as for a lower-tailed test, Equation 9-9 is used without modification. The absolute values of z_α, z_β, and δ are substituted just as in the lower-tailed case. For a two-tailed test, the probability of a type I error, α, is divided equally between the two tails of the null sampling distribution. Consequently, in Figure 9-7 and in Equation 9-9, z_α must be replaced with absolute $z_{\alpha/2}$. Then the specified shift can be treated as if it were only in the downward direction because the resulting sample size will be exactly the same for an upward shift, or difference, of the same amount.

Summary

Statistical hypothesis testing can be described as a 5-step procedure:

1. State the hypothesis to be tested.
2. Select the test statistic and its sampling distribution.
3. Establish acceptance and rejection regions.
4. Select a random sample and calculate the sample value of the test statistic.
5. Decide about the hypothesis by comparing the observed sample value of the test statistic with the acceptance and rejection regions.

For population means, hypothesis tests use the standard normal deviate, z, and the population standard deviation when these are known. Otherwise t and the sample standard deviation are used. The procedures described in connection with Equations 9-1 through 9-4 apply for all sample sizes from normally distributed populations and for sample sizes of 50 or more from virtually any population.

Equations 9-1 and 9-2 are for two-sided tests about population means, and Equations 9-3 and 9-4 are for one-sided tests. In a two-sided test the probability area equal to the level of significance is divided equally between the two tails in the rejection regions. In a one-sided test, all of the probability area equal to the significance level is put in one tail.

When the normal approximation applies, a test about a population proportion uses the standard normal deviate, z, and the population variance, $\pi(1 - \pi)$. Here the value of π is taken from the null hypothesis. When the normal approximation does not apply, the appropriate binomial distribution probabilities can be used to find the acceptance and rejection regions.

In place of means and proportions, acceptance and rejection regions can be defined in terms of standardized test statistics such as z or t. Furthermore, rather than stop at reporting just the level of significance at which a decision is made, the P-value can also be given. The P-value is the probability of getting a sample result more extreme than the one observed.

The power of a hypothesis test is the probability that the test will reject the null hypothesis when it is, in fact, false. In other words, power is the probability of making a correct decision under the alternative hypothesis, H_A. A power curve can be constructed to display the subject probabilities for all values of the parameter described in H_A. An increased sample size can be used to increase the power without changing the probability of a type I error.

Supplementary Exercises

9-27 *Pharmaceutical manufacturing.* A prescription product is made in the form of a standard pill. The specified weight of the active ingredient is 10 grains per pill. Weights delivered by the production process are normally distributed with a standard deviation of 0.2 grains. The process is checked every quarter hour by weighing the active ingredient in each of a random sample of 16 pills. The sample mean is used in a two-tailed test with critical values of 9.902 and 10.098 grains. If a sample mean falls outside these limits, the process is stopped and checked. Otherwise it continues uninterrupted.

a. If the process is performing according to specification, what is the probability it will not be interrupted after the next check? What is the probability it will be interrupted?

b. Suppose the true mean fill being delivered is 9.95 grains. In this case, the probability of $\overline{X}$ falling in the acceptance region is 0.83. Sketch the sampling distribution of $\overline{X}$ and show the critical values and acceptance region. What is the probability the process (1) will not be interrupted after the next check; (2) will be interrupted after the next check?

c. Prepare a table similar to Table 9-2 for the situations in (a) and (b). Describe each true state with a numerical value in addition to the label. Place a numerical value with each probability label in the table.

9-28 *State revenue department.* A state revenue department is responsible for receiving state income tax returns and checking them for correctness. Control of the checking procedure is maintained by periodically collecting a random sample of four batches of processed returns, finding the mean number of incorrect returns per batch, and comparing the result with a critical value of 1.5 incorrect returns per batch. If the sample mean is greater than this limit, detailed action is taken to reduce the mean number of incorrect returns per batch after processing. Otherwise no action is taken.

For batches of the size being used, the number of incorrect returns after processing is normally distributed with a standard deviation of 0.3 incorrect returns per batch. The critical value was found with an assumed population mean of 1.1505 incorrect returns.

a. What is the significance level of the test?
b. If the checking process is actually producing a mean of 1.6 incorrect returns per batch, what is the probability of taking action to reduce the mean?
c. Prepare a table similar to Table 9-2 for the conditions in (b) and for the condition that $\mu_0 = 1.1505$. Describe each true state with a numerical value in addition to the label. Place a numerical value with each probability label in the table.

9-29 *Decision rule.* State the general decision rule for a lower-tailed test of a hypothesis about the population mean in terms of the standard normal deviate.

9-30 *Accounts payable audit.* A firm's accounting records show that the current total for accounts payable is \$280,516 on 3300 invoices, which is an average of \$85.00 per invoice. A random sample of 50 invoices is selected by an auditor for a two-tailed test that the accounts payable total is correct at the 0.05 level of significance. The sample mean is \$88.11 payable per invoice and the sample standard deviation is \$8.20.

a. Find the acceptance region for the test in terms of the standard test statistic, the mean amount payable per invoice, and the accounts payable total.
b. What conclusion is justified by the sample result?
c. Is the population likely to be normal here?

9-31 *An ideal test.* In Figure 9-6, consider the effect on the power curve of letting n increase indefinitely. Sketch the resulting power curve and state why the test for this curve is an ideal lower-tailed hypothesis test.

9-32 *Quality control for copper wire.* A cutting machine is set to measure 50-foot lengths of copper wire and cut them. The standard deviation of lengths cut at any setting is 0.15 ft. The lengths of a sample consisting of 9 rolls of wire cut by the machine are determined. If the mean length of the sample lies in the interval 49.902 to 50.098 ft, the process is allowed to continue. Otherwise the machine is stopped and checked. Assume the distribution of cuts is normal.

a. What is the probability of a type I error for this test procedure?
b. What is the power of the test for a process mean of 50.01 ft? 49.99 ft? 50.098 ft? 49.902 ft? (Be sure to find the area in both tails of the sampling distribution.)
c. Sketch the power curve for the test.

9-33 *Acceptance sampling construction rods.* A steel supply company received a shipment of 12 dozen 16-foot construction rods. It is important that the rods be neither too long nor too short. The cutting machine at the manufacturing plant is known to cut with a standard deviation of 0.12 ft at any setting. The receiving firm is testing the length specification with a random sample of 25 rods. The 0.05 significance level is used for all acceptance sampling the firm does. The mean length of rods in the sample is 16.06 ft. Does the sample support the conclusion that the shipment conforms to specifications?

9-34 *Wage earners' work week.* A union has complained to the state department of economic security that the mean hours worked per week by wage earners in a large city has declined appreciably from the mean of 42.8 h established during the past year. From previous surveys, the standard deviation is known to remain nearly constant at 3.1 h per week.

a. How large a random sample of wage earners should be chosen to detect a drop of 1.5 in the mean hours worked per week if the probability of a type I error is to be controlled at 0.01 and the probability of a type II error is to be controlled at 0.05?
b. Find the critical value for the resulting hypothesis test.
c. State the decision rule.
d. Is the normal assumption for the sampling distribution of the test statistic justified? Why?

9-35 *Production rate increase.* Workers who assemble a small electrical appliance have made several suggestions for improving working conditions so that production can be increased without greater effort. The management has examined the suggestions and instituted those considered feasible. Before the changes, the distribution for number of assemblies per hour was normally distributed with a mean of 420 and a standard deviation of 27. Management wants a hypothesis test that will detect a shift to a mean of 435 assemblies per hour with probability 0.99. At the same time, the probability of a type I error is to be controlled at 0.05. Assume that the shape and standard deviation of the distribution of assemblies per hour are not affected by the changes.

a. Find the sample size that will satisfy the stated conditions.
b. Find the critical value for the hypothesis test.
c. State the decision rule.

9-36 *Inventory verification.* The accounting records of a firm show that the average cost of the several thousand items in inventory is $182.43 per item. An auditor wants to check this value with a hypothesis test. For the test, the probability is to be 0.99 of detecting a difference of $2.00 in either direction and the probability of a type I error is to be controlled at 0.05. Experience with similar firms leads the auditor to estimate that the standard deviation of the distribution of item costs is $12.00.

a. What sample size is needed to meet the stated specifications for the test?
b. Find the critical values and state the decision rule.
c. Does the fact that the population is severely skewed in the positive direction invalidate your answers to (a) and (b)?

Computer Exercises

1 Source: Real estate data file

For all parts of this exercise, treat the real estate data as if they are a random sample from a large universe of sales in the community.

a. A real estate salesperson maintained that the mean purchase price of single-family homes in the community was no more than $70,000. Test this hypothesis at a significance level of 0.05.
b. Using a 0.01 level of significance, test the hypothesis that the mean age of single-family home buyers in the community is 30 years.
c. Using a 0.05 level of significance, test the hypothesis that the proportion of sales to new residents is 0.40.

The source for all remaining exercises is the personnel data file. Treat the file as if it is a population.

2 According to an industry trade journal, the mean pay for managers is $45,000. Select a random sample of 50 and test the hypothesis that the mean salary is $45,000. Use 0.05 as the level of significance.

3 What is the population mean for the personnel file? (Do you usually know this?) Is the hypothesis in the previous exercise true or false?

4 According to the same trade journal mentioned in Exercise 2, the proportion of persons who participate in tax shelter programs is 0.80. Select a random sample of 100 and test the hypothesis that the proportion in this company is 0.80. Use 0.01 as the level of significance.

Interpreting Computer Output

Computer output may be helpful in testing hypotheses about means and proportions. Suppose a trade association for the industry to which the firm described in Figure 3-6, page 77, belongs, reports that last year the mean salary in the industry was $53,672. In addition, 78% of the industry's employees participated in some form of tax deferral program.

We can treat the output in Figure 3-6 as statistics from a random sample of 672 employees in the industry and use the statistics to compare this firm with the industry on selected variables. For example, to test the hypothesis that the firm has the same mean salary as the industry, we compute the test statistic

$$t_{\text{observed}} = \frac{\bar{x} - \mu_0}{s/\sqrt{n}}$$

$$= \frac{54893.51 - 53672}{382.90} = 3.19$$

where the numerical values came directly from Figure 3-6. At any reasonable level of significance, we would reject the hypothesis that the company has the same mean salary as the industry.

In Figure 3-6, the variable DEFER has the value 1 if a given employee participates in the tax deferral program and 0 if not. To test the hypothesis that the company has the same proportion of participants as the industry, we calculate the test statistic

$$z_{\text{observed}} = \frac{p - \pi}{\sqrt{\pi(1 - \pi)/n}}$$

$$= \frac{0.797619 - 0.78}{\sqrt{0.78(1 - .78)/672}} = 1.10$$

The firm's level of participation does not differ significantly from the industry's at any reasonable level of significance.

The sample size in this example is so large that the normality assumptions for the sampling distribution of salary means and the sampling distribution for the proportion are easily supported. It is important to check such assumptions when using computer output.

10

SIMPLE LINEAR REGRESSION

The two previous chapters demonstrated that answers to many managerial and administrative questions involve estimating a population mean. In those chapters, the estimates were based on observations of the variable in question. For instance, to estimate the mean amount of life insurance sold per policy, an analyst for an insurance company could take a random sample of policies sold recently and observe the amounts of insurance on them. The sample mean and its standard error would then constitute the basis for a confidence interval estimate of the population mean. We call this a *univariate estimate*.

Instead of using just the variable itself, there are many cases where it is more effective to estimate or predict a value of the mean by using one or more variables related to the initial variable. The amount of life insurance a family buys on a policy is related to the family income, age of the head of the family, number and ages of children, and total value of other policies owned, as well as other variables. By basing our estimate of the amount that will be bought on one or more related variables, we may be able to reduce the error of estimation substantially below the corresponding error in the univariate estimate.

Another benefit from using related variables is that we may increase understanding in a given situation. Knowing that policy dollar size is more strongly influenced by income than by family size, for example, can contribute to a decision about the relative emphasis to place on the two factors in marketing plans to find new sales prospects.

Typically, each variable in regression analysis is measured on either an interval or ratio scale. There are three objectives in the analysis. The first is to find a suitable relationship between the variables. The second is to measure the strength and adequacy of the relationship. The third is to make use of the relationship to obtain estimates and predictions for one of the variables. This chapter is concerned with two-variable regression. Chapter 14 extends the techniques to more than two variables.

10.1 Relationships Between Two Variables

Functional and Statistical Relationships

Two variables can be related either functionally or statistically. In algebra, equations express functional relationships among variables. These relationships are exact or deterministic. For example, if the base of a rectangle is 10 in. and the height is designated by the variable h, then the area, A, can be expressed in the functional relationship

$$A = 10h$$

In Figure 10-1, the dots represent areas of rectangles with bases of 10 in. and different heights. One of the dots, for example, shows that such a rectangle with a height of 6 in. has an area of 60 in.^{2} The straight line through these points represents the functional relationship. Because the relationship is exact, the area and height of any rectangle with a 10-inch base can be represented by a point on this line and only by a point on this line.

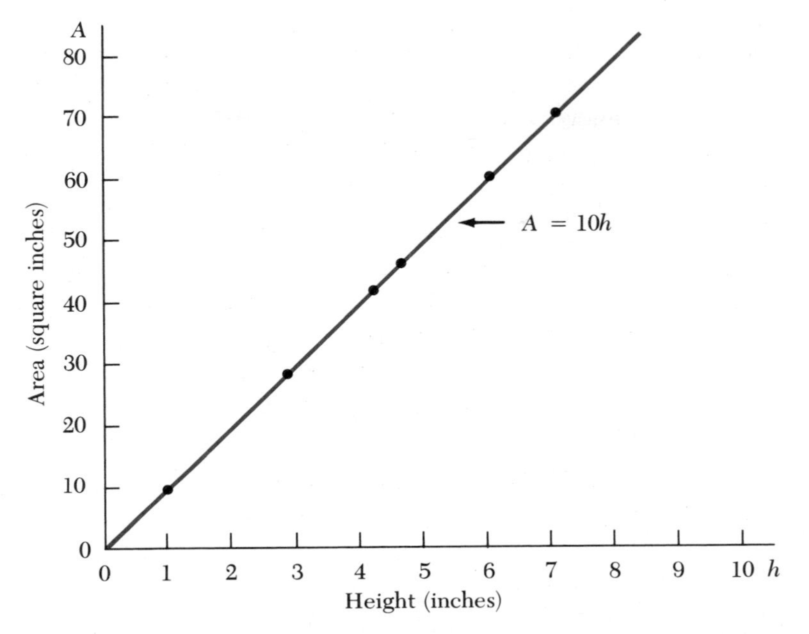

Figure 10-1 Functional Relationship for Rectangles with 10-Inch Bases

In functional relationships, the known variable (height, in our example) is called the *independent variable* and the unknown variable (area) is called the *dependent variable.* (This use of the word *independent* is completely distinct from independence in probability, as discussed in Chapter 4.)

Exact functional relationships among variables are seldom found in business, industry, and government. Instead there often is an underlying or average relationship among the variables, and individual observations are scattered around this relationship. The term *statistical relationship* refers to such cases. For example, scores on a job skills test given to applicants and subsequent performance ratings for those hired might look like those in Figure 10-2. Here the average or underlying relationship moves upward to the right: the greater the test score, the greater the subsequent performance score. But the relationship is not exact. For a given test score, there is not just one possible performance score. Instead, there is a distribution of performance scores around an average. We would not expect any test to perfectly predict a subsequent performance score. Many factors related to performance combine to cause variability around the average performance score for any given test score.

In statistical relationships, the terms *dependent* and *independent variable* carry over from functional relationships. In our example, performance score is the dependent variable and test score is the independent variable.

Direction and Shape of the Relationship

The statistical relationships shown in Figure 10-3 are representative of those encountered in applied work. The one at the upper left is the one shown in Figure 10-2. The relationship at the upper right shows error rates and practice time for people performing a clerical task: the more practice, the fewer errors. The lower left panel shows dollar sales that might result from various advertising expenditures. At first, sales increase rapidly with a given increase in advertising; for higher levels of advertising, the increase in sales is much less per unit increase in advertising. The lower right panel shows a typical relationship between the unit cost of a product and the production rate. In the beginning, the cost drops as resources are added. After the optimal mix is passed, however, the unit cost increases more and more rapidly as resources are added.

The two upper relationships of Figure 10-3 are called *linear* relationships because straight lines are appropriate models for the underlying average relationships between the two variables. Because curved lines are appropriate for the bottom two relationships, they are called *curvilinear* relationships.

The relationship between two variables is said to be *direct* (or *positive*) if the dependent variable increases when the independent variable increases. This is the case in the two left-hand panels in Figure 10-3. On the average, as test scores increase so do performance scores. The same is true for sales and advertising. On the other hand, the relationship between error rates and practice times is said to be *inverse* (or *negative*). The greater the practice time, the less the error rate. The panel in the lower right shows an inverse relationship between unit cost and production rate to the left of the minimum point on the curve and a direct relationship to the right of this point.

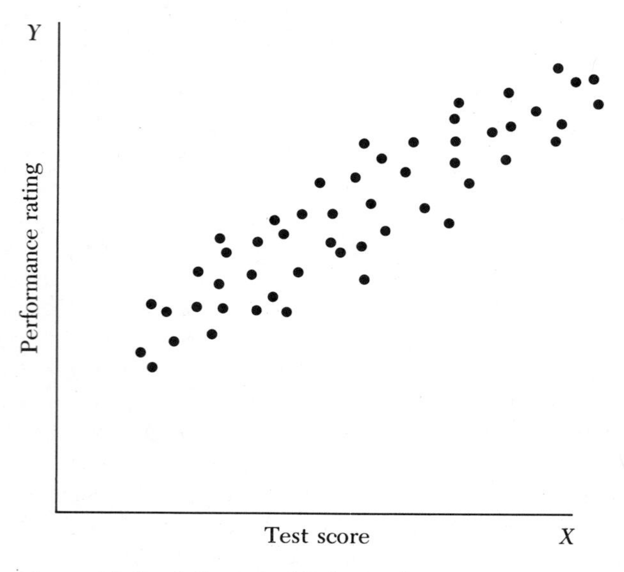

Figure 10-2 A Statistical Relationship

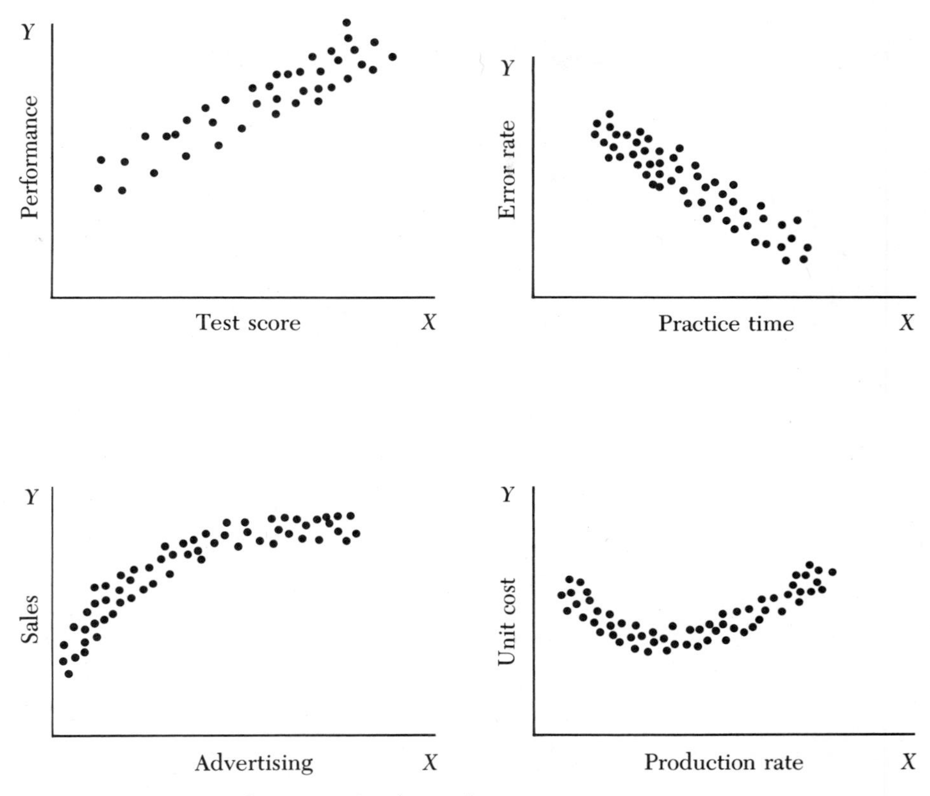

Figure 10-3 Typical Statistical Relationships

We are now in a position to explain *simple linear regression*. *Simple* indicates two variables, an independent and a dependent variable. If more than two independent variables are used, the term *multiple regression* applies. The term *linear* indicates straight line relationships, and *regression* indicates restriction to statistical relationships.

10.2 Fitting a Straight Line

To show how to represent the average relationship between two variables with a straight line, we shall return to the life insurance industry. A company analyst is seeking a relationship between the number of customer calls, or visits, an agent makes in a week and the number of customers to whom the agent makes a sale. The analyst reasons that the number of sales should depend to some degree on the number of calls made. This leads to the designation of sales as the dependent variable and calls as the independent variable. Data for a random sample of 20 agents are given in Table 10-1. It is customary to designate the dependent variable with Y and the independent variable with X, as is done here. The agents are the units

Table 10-1 Observations of Number of Calls and Sales for a Random Sample of Twenty Life Insurance Agents

Agent	*Number of Calls* (X)	*Number of Sales* (Y)
1	26	11
2	13	7
3	21	8
4	37	20
5	17	9
6	20	12
7	17	4
8	28	16
9	28	11
10	6	2
11	23	11
12	25	7
13	38	18
14	33	14
15	12	2
16	30	15
17	30	10
18	10	4
19	18	7
20	21	10

of observation. For each agent, the number of calls and the number of sales are observed. The objective is to find the average relationship between number of sales, the dependent variable, and number of calls, the independent variable.

The Scattergram

It is almost always worthwhile to begin a regression analysis by plotting a *scattergram*. This gives an indication of the shape and direction of the underlying relationship, if any, between the variables. The plot can also call attention to any observations that fall so far away from the main body of the data that they do not seem to belong with that group. Such observations are *outliers* and can influence the subsequent analysis far out of proportion to their number.

The common practice is to plot values of the dependent variable (Y) in the vertical direction and values of the independent variable (X) in the horizontal

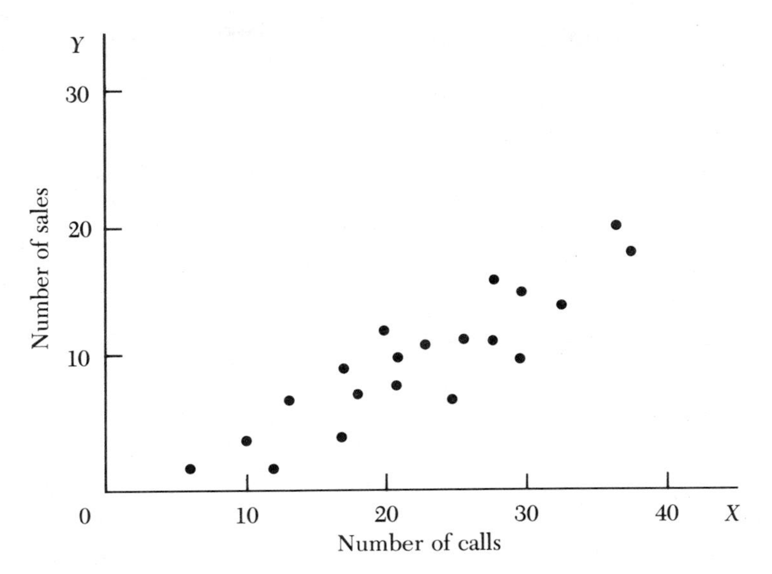

Figure 10-4 Scattergram for Insurance Data

direction. A point is placed at the intersection of these two values for each observation. The scattergram for our example is in Figure 10-4. On the figure the point (10, 4) — where X is 10 and Y is 4 — shows the number of calls and sales for the 18th agent on the list. An agent is the unit of observation that connects a value of Y to the appropriate value of X. If there were an observation where X is 35 and Y is 30, for example, it would clearly be an *outlier*, and steps would be taken to find out whether a recording error or something else had produced an irrelevant observation.

The points in Figure 10-4 appear to fall reasonably close to an imaginary straight line running from lower left to upper right. The relationship appears to be linear and positive. No outliers are apparent. Hence, we can proceed to find a straight line which fits these observations well. It is apparent, however, that there will still be some scatter of the data about such a line. The net effect of many factors other than the number of calls is responsible for their failure to fall even closer to this line.

Linear Equations

One way to describe the underlying relationship in Figure 10-4 is to draw a straight line that we judge to be a good fit by its visual appearance on the scattergram. This technique does not constitute part of what we shall come to know as *least-squares line fitting* but is a useful point of departure. One such line is shown in Figure 10-5. This line intersects the extended Y-axis at −5 and the X-axis at 8. The coordinates of these points are (0, −5) and (8, 0), respectively.

Although drawing a line on a scattergram may be visually satisfactory, we must be able to write an equation for the straight line for later convenience. The *slope-intercept* form of the equation of a line is

$$Y = b_0 + b_1 X$$

where b_0 is the Y-intercept — the value of Y at which the line cuts the Y-axis — and b_1 is the slope. In Figure 10-5, Y is -5 when X is 0. Hence $b_0 = -5$ for the specific line we have drawn.

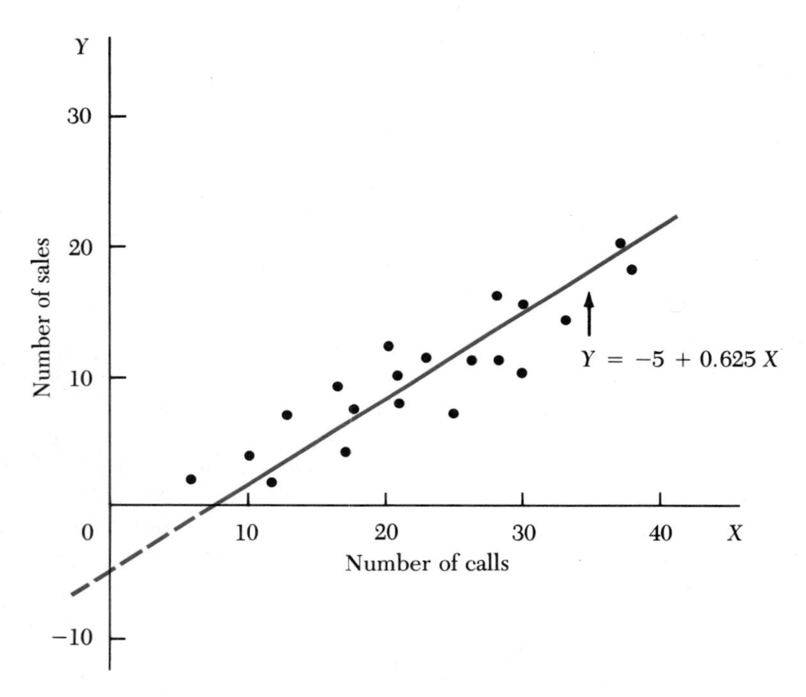

Figure 10-5 Line Fitted by Eye

The slope is the number of units of change in Y relative to a unit increase in X. For example, the line goes through the points (40, 20) and (8, 0). (Two widespread points are picked to improve accuracy.) The value of Y increases by 20 points ($20 - 0 = 20$) while X increases by 32 points ($40 - 8 = 32$). The ratio of the change in Y to the *increase* in X yields both the value and algebraic sign of the slope; that is,

$$b_1 = \frac{20 - 0}{40 - 8} = +0.625$$

We now know that the line in Figure 10-5 has an intercept of -5 and a slope of 0.625. Therefore the equation of the line is

$$Y = -5 + 0.625X$$

This equation summarizes the underlying relationship for the data in Figure 10-5 and can be used to approximate the number of sales resulting from any number of calls within the range of the data. If an agent were planning to make 30 calls in a week, the prediction for the number of sales to expect would be

$$Y = -5 + 0.625(30) = 13.75$$

which rounds to 14 sales.

The line seems to indicate that if a salesperson made no calls during the week, the number of sales to be expected is *negative* 5, a meaningless figure. Similarly, the line shows negative sales for all calls under 8. In reality, the relationship could never drop below zero. This illogical result shows the danger of using a statistical relationship outside the range of the data on which it is based. That is, staying within the observed range of the independent variable is particularly important.

The Least-Squares Line

There are major drawbacks to fitting a line by eye as we did above. No two people can be expected to produce exactly the same line under such circumstances. Furthermore, there is no accepted standard to determine which of several such lines is the best one. We need a procedure that will produce only one line for a given set of data. We also want that line to be, in some sense, the ***best fitting*** line for the given set of points. The method of *least squares* produces such a line and constitutes an important part of regression procedures.

In Figure 10-6, we repeat the scattergram for the insurance example and show the least-squares regression line with equation

$$Y_c = -1.69 + 0.51X \tag{10-1}$$

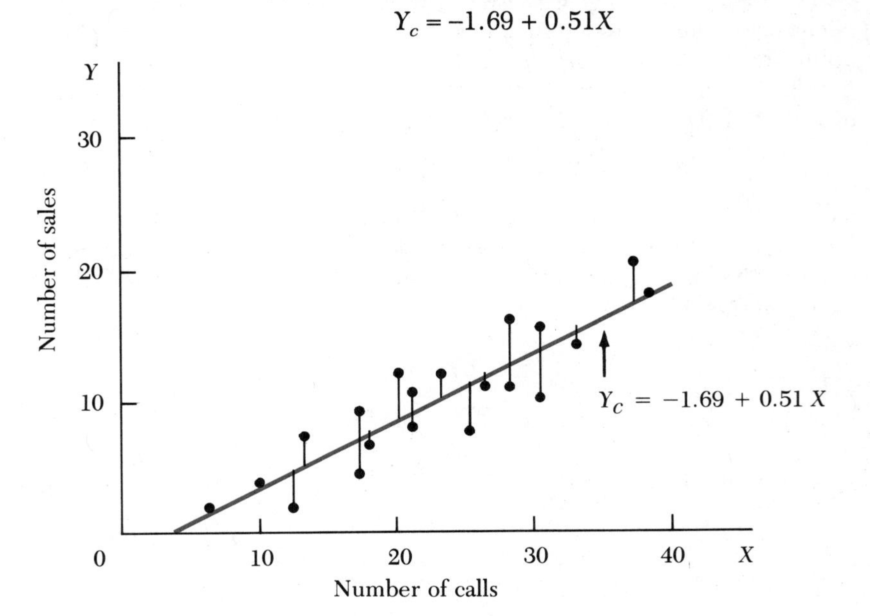

Figure 10-6 Least-Squares Line and Deviations

How was this equation determined? Observe that Figure 10-6 also shows the vertical deviations of the observations from the line. These deviations are the errors that result when we use the least-squares line to approximate the number of sales that will be produced by a given number of calls. For example, the agent who made 20 calls actually made 12 sales. If we substitute 20 for X in Equation 10-1, we find that the approximation from the least-squares equation is $Y_c = -1.69 + 0.51(20) = 8.51$. Hence the deviation is

$$Y - Y_c = 12 - 8.51 = 3.49$$

Actual sales are 3.49 greater than the prediction from the equation.

Together the deviations for the 20 observations shown in Figure 10-6 constitute the dispersion of the data around the line shown. Any other line would have a similar set of deviations. The objective is to find the line for which the dispersion is the least. The dispersion of the observations in the Y-direction is composed of the prediction errors for the set of observations. Reducing dispersion to a minimum also minimizes these errors.

Of all lines that might be selected for the underlying relationship between two variables, *the least-squares line is the one for which the sum of squared deviations of the data from the line is a minimum*. In this sense, the least-squares line is the line of best fit. In symbols, let the equation of the least-squares line be

$$Y_c = b_0 + b_1X \tag{10-2}$$

where b_0 is the intercept and b_1 is the slope. Then for a given set of observations,

$$\Sigma(Y - Y_c)^2$$

is the sum of the squared deviations (predictive errors). In order for Equation 10-2 to be the least-squares line, the sum of squared deviations must be less than that for any other line.

To find the *least-squares* line in the general case, we first substitute the right side of Equation 10-2 into the expression for the sum of squared deviations:

$$\text{SSE} = \Sigma(Y - Y_c)^2 = \Sigma(Y - b_0 - b_1X)^2$$

The set of values of X and Y are fixed for any given set of observations and it is b_0 and b_1 that are the variables in the expression. Calculus can be used to show that the values of b_0 and b_1 that provide the least-squares fit are those satisfying conditions in two equations. These are

$$\Sigma Y = nb_0 + b_1\Sigma X \tag{10-3a}$$

$$\Sigma XY = b_0\Sigma X + b_1\Sigma X^2 \tag{10-3b}$$

The general algebraic solution of these equations yields

$$b_1 = \frac{n\Sigma XY - \Sigma X\Sigma Y}{n\Sigma X^2 - (\Sigma X)^2} \tag{10-4}$$

and

$$b_0 = \overline{Y} - b_1\overline{X} \tag{10-5}$$

To apply these two equations, we obtain the necessary sums from the data and use them to find the slope, b_1. Then we combine b_1 with the means of the two variables to find the intercept, b_0.

Table 10-2 gives the pairs of numbers that constitute the 20 observations in our insurance example. The table also shows how to get the sums needed in Equations 10-4 and 10-5, as well as ΣY^2, which will be needed later. Given these sums, we substitute them in Equation 10-4 to find the slope:

$$b_1 = \frac{20(5238) - 453(198)}{20(11{,}733) - (453)^2} = 0.51156$$

Then we can find the intercept from Equation 10-5:

$$b_0 = 9.9 - 0.51156(22.65) = -1.68687$$

It is best to use five or six significant figures in these calculations to prevent large rounding errors. Once b_0 and b_1 have been found, they can be rounded.

Table 10-2 Calculations for Finding the Insurance Agents' Least-Squares Regression Line

Observation	X	Y	X^2	XY	Y^2
1	26	11	676	286	121
2	13	7	169	91	49
3	21	8	441	168	64
4	37	20	1369	740	400
5	17	9	289	153	81
6	20	12	400	240	144
7	17	4	289	68	16
8	28	16	784	448	256
9	28	11	784	308	121
10	6	2	36	12	4
11	23	11	529	253	121
12	25	7	625	175	49
13	38	18	1444	684	324
14	33	14	1089	462	196
15	12	2	144	24	4
16	30	15	900	450	225
17	30	10	900	300	100
18	10	4	100	40	16
19	18	7	324	126	49
20	21	10	441	210	100
Total	453	198	11,733	5238	2440

$$\overline{X} = \frac{\Sigma X}{n} = \frac{453}{20} = 22.65 \quad \text{and} \quad \overline{Y} = \frac{198}{20} = 9.9$$

To two decimal places, the least-squares regression line for the insurance example is

$$Y_c = -1.69 + 0.51X$$

Because −1.69 and 0.51 satisfy Equations 10-4 and 10-5 for our data, no other straight line can have as small a sum of squared deviations from those data. Hence the line is called the *least-squares regression line.*

Exercises

10-1 *Consumers' automobile preference.* In a large national consumer survey, respondents were given two lists of popular automobiles—a list of lighter automobiles and a list of heavier automobiles. The price, weight, and estimated miles per gallon were stated for each automobile on these lists. Respondents were asked to examine both lists and check the one automobile they preferred. Results were sorted into groups of 100 by age to furnish data for the scattergram and regression line shown below.

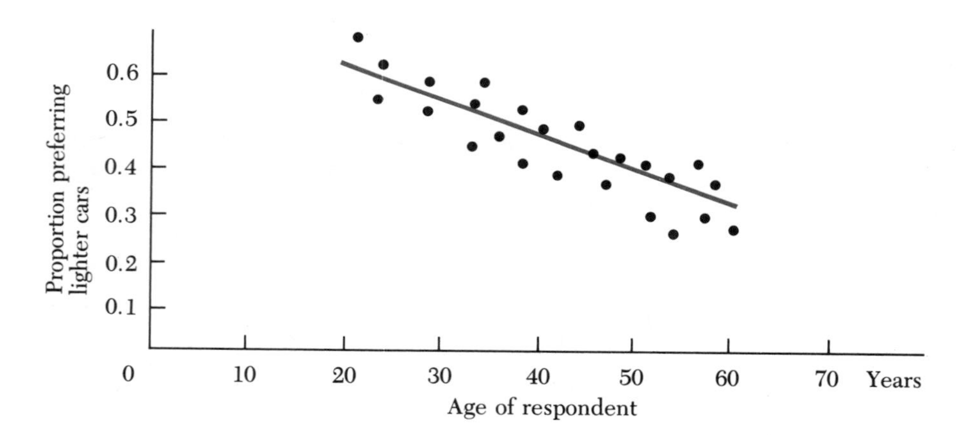

a. From the graph, find the numerical values of the slope and intercept for the regression line.
b. What is the meaning of b_1 in this case?
c. Would it be advisable to use the regression line to estimate the proportion of people 70 years old who prefer lighter automobiles? Why?

10-2 *Construction costs.* A study was made of construction costs for detached single-family dwellings on individual lots in a city. A random sample of such dwellings built during the last six months was tabulated, along with the construction cost and floor area of each. A least-squares regression line fitted to these data showed a direct relationship between floor area and cost. Two points on the regression line were (1000, 21) and (4000, 141), where the first number in each pair is the floor area in square feet and the second is the construction cost in thousands of dollars. Floor area is the independent variable.

a. Find the equation of the regression line by finding numerical values for b_0 and b_1.
b. Discuss the meaning of b_1 by interpreting its physical dimensions.
c. Use the regression line to estimate construction costs for a house with a floor area of 2700 ft^2.

10-3 *Crime control.** An analysis of crime-in-progress calls in a large city found that the probability of an arrest at the scene was $Y_c = 0.266 - 0.111X$ (where X is the response time in minutes) for one-officer units and $Y_c = 0.309 - 0.013X$ for two-officer units. No response time exceeded 10 min.
a. Estimate the probability of arrest for a one-officer unit that arrives in 5 min.
b. Estimate the probability of an arrest for a two-officer unit that arrives in 4 min.
c. For the observed range of response times, which type of unit shows a greater probability of arrest?

10-4 Which of the following is not a suitable observation for a regression analysis? Why?
a. The age and circumference of a type of tree.
b. The age of a wife and that of her husband.
c. An individual's years of schooling and annual income.
d. The weights of two automobiles.

10-5 *Medical expense.* Part of a study to determine factors influencing family medical expenses involves finding a regression relationship between the number of people in a family (X) and average monthly medical expense (Y). The data for a pilot sample of eight families are as follows.

X:	3	2	3	5	4	5	6	8
Y:	8.5	7.5	8.0	12.0	9.5	10.0	13.0	16.5

a. Plot a scattergram and comment on the suitability of a straight line relationship.
b. Assume that a straight line is applicable and find the least squares regression line.
c. Use the regression line to estimate monthly medical expenses for a family of four.

10-6 *Employee performance prediction.* The personnel director for a firm wants a study made of the relationship between scores on a pre-employment test given to potential component assemblers and their performance ratings three months after they were hired. Both test scores and ratings are on 9-point scales. Data for a pilot sample are as follows.

Subject:	A	B	C	D	E	F	G	H	I
Test Score:	3	3	4	4	5	7	7	8	9
Rating:	4	2	4	6	6	5	9	7	8

a. Does a scattergram indicate that a straight line relationship seems to apply?
b. Assume that a straight line is appropriate and find the least-squares regression line.
c. Estimate the performance rating for a person who had a test score of 6.

10-7 *Moving household goods.* A firm specializing in moving local household goods has an accountant who wants to find the relationship between the estimated weight of an

*D. P. Tarr, *Journal of Police Science and Administration*, December 1978.

upcoming job and the time it will take to load the goods on the truck. Data kept on ten recent jobs produced the following sums (Y is the dependent variable):

$$\Sigma X = 21 \qquad \Sigma Y = 28 \qquad \Sigma XY = 66$$
$$\Sigma X^2 = 52 \qquad \Sigma Y^2 = 87$$

a. Which variable should be treated as the dependent variable? Why?
b. Find the equation of the least-squares regression line.
c. If weight is in tons and time is in hours, discuss the meaning of the slope b_1.

10-8 *Overhead expense.* An analyst for a state board of education has collected data on the administrative overhead expense per student (Y) and the number of students (X, in hundreds) for 11 school districts. From these observations, we have

$$\Sigma X = 178 \qquad \Sigma Y = 56.67 \qquad \Sigma XY = 1067.01$$
$$\Sigma X^2 = 3996 \qquad \Sigma Y^2 = 314.5633$$

a. What is the equation for the least-squares regression line?
b. Does the intercept have a useful meaning in this case? Does the slope?
c. Estimate the per-student overhead for a district with 1800 students.

10.3 Sums of Squares Description

In the introduction to this chapter we said that there are three objectives in regression analysis: finding a suitable relationship between the variables, measuring the strength of the relationship, using the relationship to obtain estimates and predictions. The least-squares technique just concluded constitutes a way to find a relationship. Now we must go on to develop measures for the strength of a regression relationship.

One of the most useful methods for measuring the strength of a regression relationship is based on the sum of the squared deviations of the Y values from their mean. To develop this concept, suppose we want to make the best guess of the number of sales, a Y value, for an observation to be randomly selected from the 20 in our insurance example.

In the first case, we presume that we know only that mean sales for the sample is $\overline{Y} = 9.9$. In this case, our best guess is $\overline{Y}$ because the sum of squared errors around the mean is a minimum. In other words, the quantity

$$\text{SSTO} = \Sigma(Y - \overline{Y})^2$$

is less than the sum of squared deviations about any Y-value other than the mean. We will call SSTO the *total sum of squares*.

Partitioning the Total Sums of Squares

In the second case, we suppose that we want to make the best guess of the Y value for a randomly selected observation given that we also know (a) the regression relationship between calls and sales, and (b) the number of calls for the randomly selected observation. For example, suppose someone has randomly chosen Agent 4 as listed in Table 10-1. The person who has made the selection tells us that the number of calls the agent made is 37 ($X = 37$) and wants our best guess as to the number of sales (Y) for this observation. Our guess is

$$Y_c = -1.69 + 0.51(37) = 17.18 \text{ sales}$$

This is our best guess because from our earlier discussion we know that the sum of squared errors around the least-squares regression line is less than for any other line. In other words,

$$\text{SSE} = \Sigma(Y - Y_c)^2$$

is a minimum when Y_c is determined by the least-squares regression line.

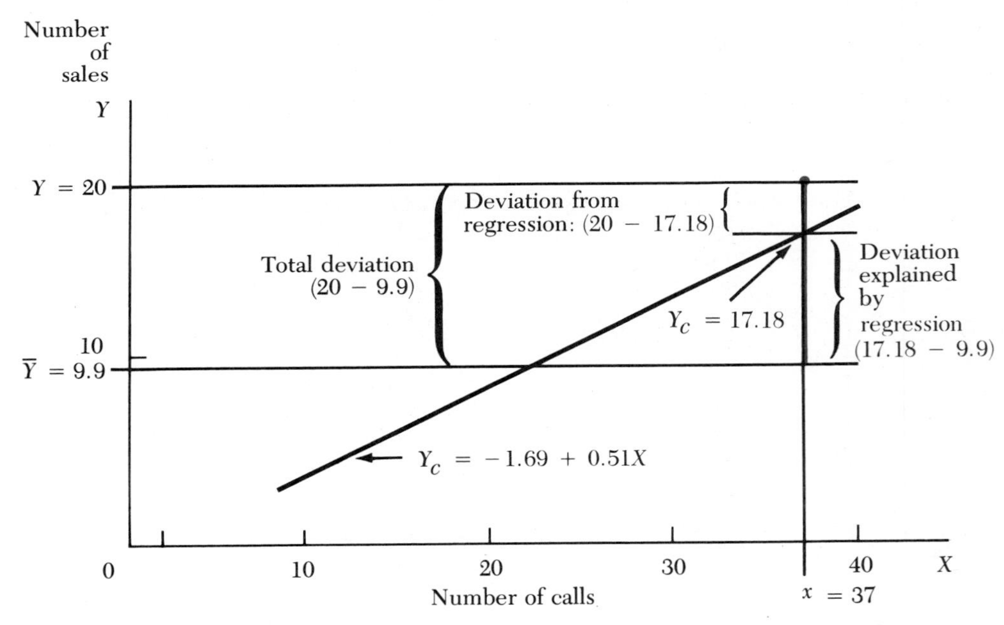

Figure 10-7 Total, Unexplained, and Explained Deviations from the Mean

The situation is illustrated in Figure 10-7. The figure shows the randomly selected observation (37, 20) from our insurance example. Agent 4 made 37 calls from which 20 sales followed. The mean number of sales in the sample is 9.9, shown by a horizontal line. If we wanted to predict the number of sales for an agent in our sample but didn't know the number of calls the agent made, $Y = 9.9$ would be the best guess.

On the other hand, if we did know that the agent had made 37 calls ($X = 37$), then our best guess would be 17.18 sales as found from the regression line.

$$Y_c = -1.69 + 0.51(37) = 17.18$$

This would leave a residual error of 20 − 17.18, or 2.82.

As is apparent in Figure 10-7, the total deviation of the observed Y-value, 20, from the mean of the Y-values can be expressed as follows.

$$(20 - 9.9) = (17.18 - 9.9) + (20 - 17.18)$$

The first component on the right side is the portion attributable to the regression relationship. The second component is the residual error. In general,

$$(Y - \overline{Y}) = (Y_c - \overline{Y}) + (Y - Y_c)$$

The total deviation from the mean is the sum of a deviation attributable to regression and a residual error. Furthermore, it can be shown that

$$\Sigma(Y - \overline{Y})^2 = \Sigma(Y_c - \overline{Y})^2 + \Sigma(Y - Y_c)^2 \qquad \textbf{(10-6)}$$

The sum of squared deviations from the mean equals the sum of squared deviations attributable to regression plus the sum of squared residual errors. *Letting SSR be the sum of squared deviations attributable to regression*, the equation just stated can be written

$$\text{SSTO} = \text{SSR} + \text{SSE}$$

The total sum of squares is the regression sum of squares plus the sum of squared residual errors around regression. Recall that the residual errors we are discussing are those illustrated in Figure 10-6.

While Equation 10-6 is satisfactory for conceptual purposes, it usually proves awkward for computational purposes. With the aid of some algebra it can be shown that

$$\text{SSTO} = \Sigma Y^2 - \frac{(\Sigma Y)^2}{n} \qquad \textbf{(10-7a)}$$

$$\text{SSR} = b_0\Sigma(Y) + b_1\Sigma(XY) - \frac{(\Sigma Y)^2}{n} \qquad \textbf{(10-7b)}$$

and

$$\text{SSE} = \text{SSTO} - \text{SSR} \qquad \textbf{(10-7c)}$$

These three equations are the ones we shall use for computation purposes.

For our example, the sums we need for substitution in the calculation equations are given in Table 10-2. Substituting, we find

$$\text{SSTO} = 2440 - \frac{198^2}{20} = 479.80$$

$$\text{SSR} = -1.69(198) + 0.51(5238) - \frac{198^2}{20} = 376.56$$

$$\text{SSE} = 479.80 - 376.56 = 103.24$$

Coefficients of Determination and Correlation

We learned that SSTO measures the variability of the Y's around their mean, SSR measures the variability attributable to regression, and SSE measures the residual variability of the Y's around the regression equation. Furthermore, we learned that

$$\text{SSTO} = \text{SSR} + \text{SSE}$$

The sum total of squared deviations of the Y values from their mean is the sum of squared deviations attributable to regression and the sum of squared residual errors.

Next, we will divide the equation through by SSTO. The result is

$$1 = \frac{\text{SSR}}{\text{SSTO}} + \frac{\text{SSE}}{\text{SSTO}}$$

In relative terms, total variability is composed of a portion attributable to regression (SSR/SSTO) and a portion attributable to residual variation around the regression line (SSE/SSTO). The ratio SSR/SSTO measures the relative strength of the regression relationship. It is called the *coefficient of determination.*

In symbols, *the coefficient of determination*, r^2, is

$$r^2 = \frac{\text{SSR}}{\text{SSTO}} \tag{10-8}$$

where SSR is the sum of squares attributable to regression and SSTO is the sum of squared deviations of the dependent variable from its mean.

The coefficient of determination tells us how much good regression is doing in either ratio or percentage terms. Because it is expressed in relative terms it is useful for comparing the strengths of two or more regression relationships even though the units of measurement differ.

For our example,

$$r^2 = \frac{376.56}{479.8} = 0.78$$

This tells us that 78% of the total variability in the Y values can be attributed to regression with X. In the context of our example, 78% of the total variability in the number of sales can be attributed to the variation in the number of sales calls.

The coefficient of determination, r^2, can range from 0, when there is no relationship between the variables, to 1, when the relationship explains all of the variation. In the latter case, all of the observations fall on the regression line; there are no residual errors and SSE is 0.

Another measure of the relative strength of a regression relationship is the coefficient of correlation, which is often referred to as the Pearson coefficient of correlation, after the person who developed it. The coefficient of correlation is given the symbol r. As the symbol implies, the coefficient of correlation is the square root of the coefficient of determination. In symbols, *the coefficient of correlation* is

$$r = \pm\sqrt{r^2}$$

where the algebraic sign is chosen to be the same as that of the slope, b_1. When the slope is positive, both the relationship and the correlation are also positive, and vice versa.

For the insurance example,

$$r = +\sqrt{(0.78)} = +0.88$$

which indicates that the relationship between number of sales and number of calls is positive, so that as calls increase so do sales.

Interpretation of the coefficient of correlation isn't as straightforward as was the case for the coefficient of determination. The coefficient of correlation ranges from −1 through 0 to +1. Both +1 and −1 indicate that the regression relationship is perfect in the sense that all the observations fall on the regression line. When the coefficient is 0, there is no relationship between the variables. These interpretations correspond to their counterparts for the coefficient of determination. But for all other values, the coefficient of correlation overstates the degree of relationship expressed by the coefficient of determination. For instance, in the insurance example, we saw that only 78% of the variability is attributable to regression. Yet the coefficient of correlation is +0.88. Nonetheless, the coefficient of correlation is very popular in reporting results of regression analyses.

Exercises

10-9 *Sums of squares description.* For the family medical expense data in Exercise 10-5:

a. Using Equation 10-7 find the total sums of squares (SSTO), the sums of squares attributable to regression (SSR), and the sums of squares attributable to residual errors.

b. What is the coefficient of determination? Describe this coefficient in the context of the data.

c. What is the coefficient of correlation? What sign does it have? Why? What can be inferred from the sign?

10-10 *Sales volume and price.* At a seaside auction market, the pounds of shrimp purchased in a given lot and the price per pound were observed for 14 randomly selected sales during market days over the past 2 weeks. A regression analysis is to be performed to find the effect of lot size on price, if any.

Pounds Bought	*Price per Pound*	*Pounds Bought*	*Price per Pound*
611	$1.50	499	1.96
595	1.63	465	2.02
613	1.58	435	2.14
587	1.69	315	2.22
529	1.74	347	2.37
550	1.81	360	2.43
456	1.90	290	2.51

a. Plot a scattergram for these data. Find the equation of the least-squares regression line and plot it on the scattergram. Comment on the adequacy of the fit.
b. Use Equation 10-7 to find SSTO, SSR, and SSE. Describe each of these sums. Use 5 decimal places with b_0 and b_1.
c. What is the coefficient of determination for this sample? Interpret the result.
d. Find the coefficient of correlation. What sign does it have? Why?

10-11 *Sums of squares description.* For the employee performance data in Exercise 10-6:
a. Find and interpret the coefficients of determination and correlation.
b. For a similar sample with another pre-employment test, the coefficient of determination was 0.85. Compare the performances of the two tests.

10-12 *Direct cost per premium dollar.* A life insurance company is analyzing its direct costs for a certain class of policy with $5000 face value. The net cost per premium dollar per month the policy is in force can be obtained from the firm's computer at considerable expense. Before doing so, a pilot sample has been selected by company analysts for manual calculation. If results are promising, there is a chance for great savings by using age to estimate these costs. The sample data give the net cost per premium dollar (Y) and the age of the policy to the nearest month (X).

X	Y	X	Y	X	Y	X	Y
8	1.26	57	0.61	70	0.58	68	0.62
29	1.15	45	0.88	40	0.74	36	0.91
47	0.81	39	0.99	66	0.67	21	1.10
24	1.14	14	1.11	55	0.70	38	0.78

a. Find the sample regression relationship, plot the scattergram and regression line, and comment on the appropriateness of the regression line.
b. Find the coefficient of determination and interpret it.
c. For another type of policy the coefficient of determination for direct costs and policy age is 0.74. Compare the two regression relationships with regard to their coefficients of determination and explain the difference.

10.4 A Model for Statistical Inference

Typically, we find a regression relationship between a dependent variable and one or more independent variables for a sample rather than for a population. The sample is meant to provide a basis for making estimates and testing hypotheses about the relationship in the population from which the sample came. For example, the sample intercept and slope will be used to estimate the population intercept and slope. It is to be expected that the sample relationship will differ somewhat from the relationship in the population.

Statistical inferences about a population regression relationship based on a sample depend upon assumptions made about the statistical characteristics of the population. The customary model for simple linear regression assumes that each value of the dependent variable, Y, consists of:

1. A linear relationship with the independent variable.
2. A random component which creates the scatter of the observations around the regression line.

In symbols,

$$\underset{\text{relationship}}{Y = B_0 + B_1X} + \underset{\substack{\text{random}\\ \text{component}}}{\epsilon}$$

The random component is often referred to as the *error term*.

Four additional assumptions are made about the error terms associated with the values of the dependent variable (Y) in the population:

1. The error terms are normally distributed.
2. The mean or expected value of the error terms is 0.
3. The variance of the error terms is the same for all values of the independent variable, X.
4. For different observations, the error terms are statistically independent.

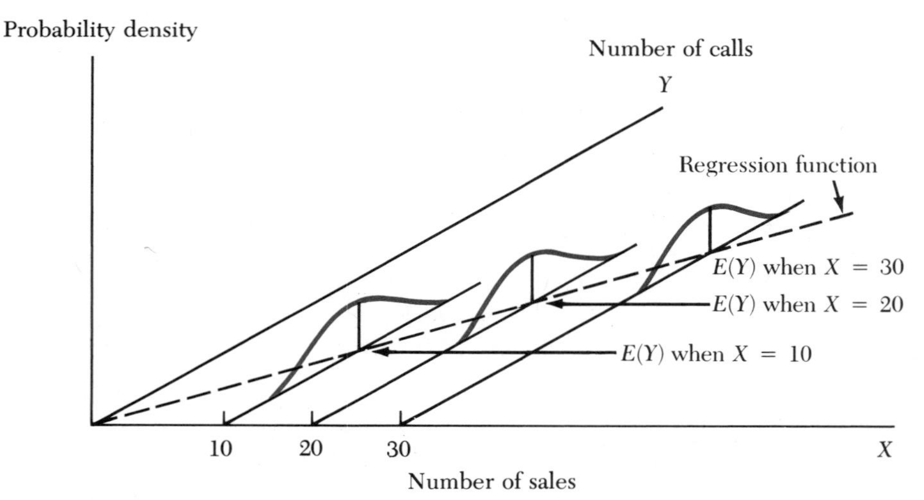

Figure 10-8 Population Model for Insurance Sales versus Calls

This entire set of assumptions is known as the *simple linear regression model*. This model is illustrated in Figure 10-8. In the figure, the regression function is the line which expresses the functional relation between the number of calls made by insurance agents, X, and the resulting number of sales, Y. In general, the line we find for a sample of observations from this population will not be the same as the line for the population.

For any specified value of X in Figure 10-8, the probability distribution of Y values is normal with its mean, or expected, value on the population regression line. The variance of such distributions is indicated with the symbol $\sigma^2_{Y \cdot X}$. This variance is constant for all values of X. We will refer to this as the *variance around the regression function in the population*. The square root of this variance is called the *standard error of estimate* for the population.

Applied research has shown that the simple regression model often is a sound basis for making inferences from samples. One reason this is so is that there often are more independent variables than we wish to include and the net effect of many missing independent variables can be represented by a normally distributed error term. Furthermore, even if the error terms are not normally distributed, the least-squares estimators of the slope and intercept are unbiased and efficient. Least squares provides estimates that are robust in this sense. But these estimates can be distorted beyond acceptable limits by extraneous outliers among the observations. A way to cope with these outliers is given in the final section of this chapter.

10.5 Test for the Presence of a Regression Relationship (Optional)

We want to make use of a sample regression line to make estimates of the dependent variable based on associated values of the independent variable. But before we can do so, there must be a high probability that a relationship between the variables exists in the population from which the sample came. Such assurance can be gained by a hypothesis test concerning the value of the population coefficient of determination. We establish the null hypothesis that this coefficient is 0, that is, that there is no linear relationship between X and Y in the population from which our sample was selected. The alternative hypothesis is that the population coefficient of determination is not 0. We will use the symbol ρ^2 for the population coefficient of determination. The test is:

1. Hypothesis statement $H_0{:}\rho^2 = 0$; $H_A{:}\rho^2 \neq 0$
2. The test statistic is F with 1 degree of freedom in the numerator and $n - 2$ in the denominator.
3. Critical values of F for 0.05 and 0.01 significance levels are read from Appendix Table A-6.
4. The sample value of the test statistic is found from

$$F = \frac{r^2(n-2)}{1-r^2} \qquad \textbf{(10-9)}$$

where r^2 is the sample coefficient of determination and n is the number of sample observations.

5. If the observed value of F is greater than the tabled value, reject the null hypothesis. Otherwise, the sample evidence does not differ from the null hypothesis enough to reject it.

For our insurance example which was based upon 20 sample observations, we found the sample coefficient of determination to be 0.78. We want to test for the presence of a linear relationship in the population using a 0.01 significance level.

1. Hypothesis statement $H_0{:}\rho^2 = 0;\ H_A{:}\rho^2 \neq 0$
2. Use F as defined in Equation 10-9 with 1 and 18 degrees of freedom.
3. From Appendix Table A-6 using 1 for m_1 and 18 for m_2, we read the boldface value as 8.28. This is the F value for the 0.01 significance level while the lightface number is for the 0.05 level. Hence, the critical value for our test is 8.28.
4. The observed value of F is

$$F = \frac{0.78(18)}{1 - 0.78} = 63.8$$

5. Because the observed value is much greater than the critical value, we reject the null hypothesis and conclude that there is a linear relationship in the population for which our sample regression line is an estimate.

10.6 Point Estimates in Regression

Intercept and Slope

We learned in Section 10.2 that Equations 10-4 and 10-5 provide the slope and intercept of the least-squares regression line. For our example, this line is

$$Y_c = -1.69 + 0.51X$$

In that section, we sought the best-fitting straight line, using the criterion of least squares.

In addition to providing the best-fitting regression line, Equations 10-4 and 10-5 are also effective point estimators for the slope and intercept of the population regression line. When the population has the four characteristics listed in Section 10.4, these equations are unbiased, efficient, and consistent estimators. Even when the normality assumption does not hold, these equations are unbiased estimators.

Standard Error of Estimate

In Section 10.4, the standard deviation of a conditional distribution of Y-values around the population regression line was defined to be the standard error of estimate, $\sigma_{Y \cdot X}$. Given the assumptions listed in Section 10.4 and illustrated in Figure 10-8, the standard error of estimate is comparable to standard deviations of other normal distributions. For example, 68.26% of the Y-values in the population will be within ± 1 standard error of the population regression line, 95.44% will be within ± 2 standard errors, and so on. As in earlier cases, the standard error of estimate measures dispersion in the same units as the Y-variable. If Y is measured in feet, the standard error of estimate is also expressed in feet.

Calculating the standard error of estimate is conceptually very similar to calculating the standard deviation for a set of observations of a single random variable. We find the variance of the observations around some reference and then take the square root of the variance. For observations of a single variable, the reference is the sample mean. For observations in a regression analysis, the reference is the least-squares regression line. The deviations that we shall use to find the variance around the regression line for our insurance example are those in Figure 10-6. Symbolically, each of these is a value of $Y - Y_c$. They are squared and summed to obtain the sum of squared deviations of actual values of Y about the regression line values, Y_c. The equation for the *sample standard error of estimate* is

$$s_{Y \cdot X} = \sqrt{\frac{\Sigma(Y - Y_c)^2}{n - 2}} \qquad \textbf{(10-10)}$$

where n is the number of observations. Dividing by $n - 2$ makes $s^2_{Y \cdot X}$ an unbiased estimate of $\sigma^2_{Y \cdot X}$, the population variance around its regression line.

The approach just described for finding the sum of the squared deviations stems directly from the definition of the standard error of estimate. While it is easily understood, it is usually awkward computationally. A more convenient form for computational purposes is the equivalent

$$s_{Y \cdot X} = \sqrt{\frac{\Sigma Y^2 - b_0 \Sigma Y - b_1 \Sigma XY}{n - 2}} \qquad \textbf{(10-11)}$$

From Table 10-2, $\Sigma Y^2 = 2440$, $\Sigma Y = 198$, and $\Sigma XY = 5238$. We found earlier that, for the sample regression line, $b_0 = -1.69$ and $b_1 = 0.51$. We carry several decimal places in the calculations in order to minimize rounding errors. Substituting these in Equation 10-8, we have

$$s_{Y \cdot X} = \sqrt{\frac{2440 - (-1.69)(198) - 0.51(5238)}{20 - 2}}$$

$$= \sqrt{\frac{103.24}{18}}$$

or 2.39 sales.

The standard error of estimate is a special standard deviation. It measures the dispersion of a set of observations of a dependent variable around a least-squares regression line. The larger the standard error of estimate, the more widely dispersed are these observations, and vice versa. If the observations fall exactly on the regression line, there is no scatter around it, and the standard error of estimate is zero.

In addition to its descriptive function, the sample standard error of estimate is one of the sources of sampling error for other measures applicable to regression. We shall discuss this use in the next two sections.

Exercises (Optional)

10-13 *Regression test and standard error of estimate.* For the medical expense data of Exercises 10-5 and 10-9:

a. Assuming the data are from a random sample, use Equation 10-9 to test for the presence of regression in the population. Use the 0.01 level of significance.

b. If the test in (a) is satisfactory, make a point estimate of the standard error of estimate by using Equation 10-11.

10-14 *Sales volume and price.* For the data in Exercise 10-10 on pounds of shrimp offered and price at which sold, perform an F test for the presence of regression at the 0.05 level of significance. Are the results of the test satisfactory for making inferences? Why? If the test is satisfactory, find the sample standard error of estimate. Use 6 places for b_1 to avoid a severe rounding error.

10-15 For the employee performance data in Exercises 10-6 and 10-11: (Assume the data constitute a random sample.)

a. At the 0.05 level of significance test for the presence of a linear relationship in the population from which the sample came.

b. Does the test justify using the sample result to make inferences about the population relationship? Why?

c. If the test result justifies it, find the sample standard error of estimate.

10-16 Assume that the life insurance cost data in Exercise 10-12 came from a random sample of such policies.

a. Use Equation 10-9 to test the null hypothesis that the population coefficient of determination is 0. Let α be 0.01.

b. Does the test justify using the sample regression to make statistical inferences about the linear relationship in the population?

c. If the test result justifies it, find the sample standard error of estimate.

10.7 Inferences about the Slope

Testing the Hypothesis That $B_1 = 0$

A linear relationship between two variables exists only if the slope of the regression line is not zero. If the slope is positive, we know that values of Y associated with large values of X are greater than those associated with small values of X. If the slope is negative, the reverse is true. In either case, knowledge of X contributes to a reliable estimate of Y. This is illustrated in Figure 10-9. When we use the number of calls and the regression relationship to estimate the expected number of sales, the sample standard error of estimate (2.39 sales) is a measure of the dispersion of the actual sales around our estimates.

If we make no use of the regression relationship with X, then all we know is the distribution of Y-values. Hence if we must make a point estimate of the next value of Y to be selected from the population, we choose the mean, $\bar{Y} = 9.9$ sales. The standard deviation of the Y-values ($s_Y = 5.03$ sales) is a measure of the dispersion of the data around our estimates and is much greater than when we use the regression relationship.

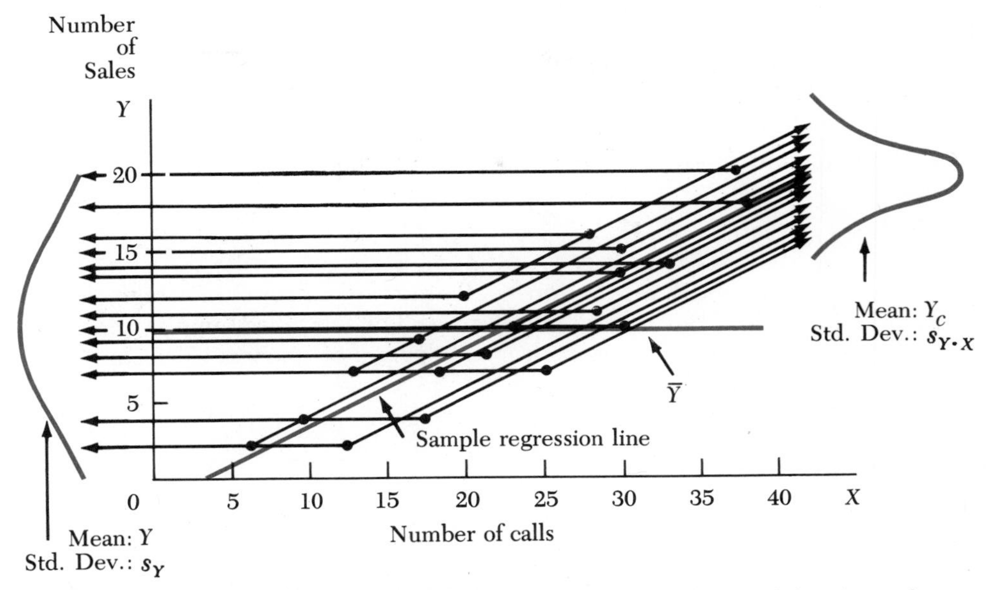

Figure 10-9 Variability of Dependent Variable Compared with Variability Around Regression Line

Another way to look at the situation is shown in Figure 10-10, in which the slope is zero. In this case, knowing the value of the X-variable associated with a particular point is of no help in predicting the value of the Y-variable. In the scattergram, there is no relationship between X and Y. The distribution of the Y-values appears to be the same for all values of X. Consequently, the mean is the best point estimate of Y for any value of X.

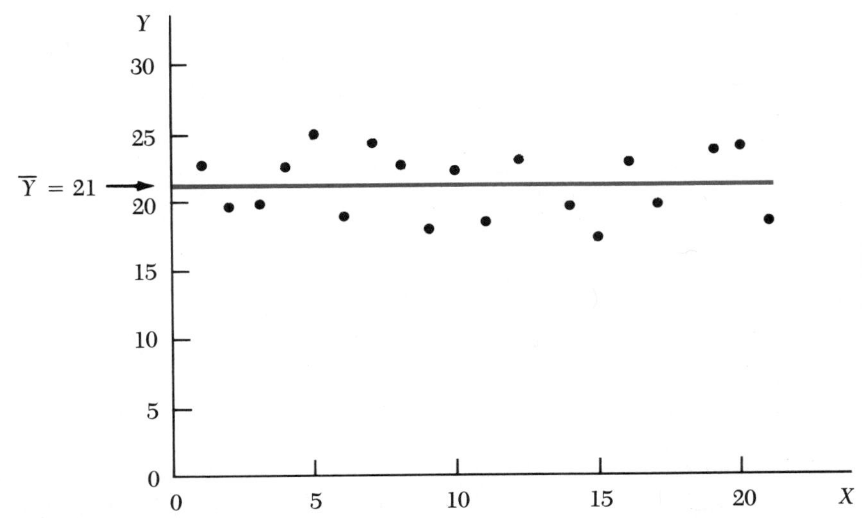

Figure 10-10 Scattergram and Regression Line When Slope Is Zero (No Relationship of Y with X)

To determine whether or not sample evidence supports the existence of a linear relationship in the population, we test the null hypothesis that the slope of the population regression line, β_1, is zero. If we can reject this hypothesis, there is evidence of a relationship between X and Y in the population. The test assumes that the four conditions listed in Section 10-4 hold. The statistic for the test is

$$t = \frac{b_1 - B_1}{s_{b_1}} \tag{10-12}$$

where b_1 and B_1 are the slopes of the sample and population regression lines and s_{b_1} is the standard error of the sample slope. For a specified sample size, n, the test statistic forms a t distribution with $n - 2$ degrees of freedom. The standard error of the sample slope is

$$s_{b_1} = \frac{s_{Y \cdot X}}{\sqrt{\Sigma(X - \overline{X})^2}} \tag{10-13}$$

or, for calculation purposes,

$$s_{b_1} = \frac{s_{Y \cdot X}}{\sqrt{\Sigma X^2 - [(\Sigma X)^2/n]}} \tag{10-14}$$

In Equation 10-13 we see that the standard error of b_1 is directly proportional to the standard error of estimate. As dispersion around the regression line increases, so does the standard error of the sample slope. We also see that the standard error of b_1 is inversely proportional to the dispersion of the X-values in the sample. The more widespread are the sample observations of the independent variable, the more reliable is the sample slope as an estimate of the population slope.

Let us conduct a two-tailed test of the null hypothesis that $B_1 = 0$ for our insurance sales example with a level of significance of 0.01. From Table A-4, the critical values of t are $\pm$ 2.8784 for 18 degrees of freedom. For the sample, we have already found $s_{Y \cdot X} = 2.39$ and, from Table 10-2, $\Sigma X^2 = 11{,}733$ and $\Sigma X = 453$. Hence, from Equation 10-14,

$$s_{b_1} = \frac{2.39}{\sqrt{11{,}733 - (453^2/20)}} = 0.0623$$

Then, from Equation 10-12,

$$t = \frac{0.51156 - 0}{0.0623} = 8.21$$

with $n - 2 = 18$ degrees of freedom. This sample value of t is greater than the upper critical value of t, so we can reject the null hypothesis that $B_1 = 0$. The sample evidence strongly supports the existence of a linear relationship between the variables for the company as a whole.

For the reasons described, we chose to test the null hypothesis that $B_1 = 0$. However, Equation 10-12 can be used to test a null hypothesis for any other value of B_1 as well.

Confidence Interval Estimation for B_1

In addition to testing for the existence of a linear relationship in the population, we can form a confidence interval estimate of the population slope. The general expression is

$$b_1 - ts_{b_1} \leq B_1 \leq b_1 + ts_{b_1} \tag{10-15}$$

For the insurance example, the 0.99 confidence interval estimate of the population slope is

$$0.51156 \pm 2.8784(0.0623)$$

or

$$0.33 \leq B_1 \leq 0.69$$

The confidence interval estimate for B_1 runs from +0.33 to +0.69. This gives us reason to believe there is a positive relationship in the population with a slope in this range. Given this confidence interval, we would not need to test the null hypothesis that $B_1 = 0$. The confidence interval does not contain zero. Therefore, zero is not an acceptable hypothesis for the value of B_1 at the 0.01 level of significance.

Exercises

10-17 *TV spot advertising.* The owner of a retail household appliance store has been collecting data to estimate the effect of local TV spot ads in generating potential customers. For 13 selected weeks during the past year, the number of spot ads per week has been varied from zero to 10. No ads were used for 3 weeks following a week in which an ad was used. A careful count was made of adults who reported coming to the store because of the ad during a 2-week period following the first ad in a series. The observations are given below.

Number of Ads	*Number of Adults*
7	52
3	30
10	60
8	59
0	12
6	49
2	33
1	22
4	37
9	48
2	23
5	53
8	57

For these observations,

$$\Sigma X = 65 \qquad \Sigma Y = 535 \qquad \Sigma XY = 3255$$
$$\Sigma X^2 = 453 \qquad \Sigma Y^2 = 25{,}063 \qquad n = 13$$

a. Find the equation of the least squares regression line.
b. Using Equation 10-11, find the sample standard error of estimate. What percentage of the Y-values lie within one standard error of the regression line? Two? How does this result compare with the expected result?
c. Find the standard deviation of the Y-values and compare the standard error of estimate with it. Does linear regression seem to be effective? Why?

10-18 *Residential construction activity.* Homebuilders' associations in six metropolitan areas conducted a joint study of the influence on residential building permits of the interest rates on mortgage loans. For 12 quarters randomly selected from the past 10 years, average interest rates for the six areas have been collected. Average building permits issued in the quarter following each selected quarter were also collected. (The lag was to allow time for the interest rate to affect building plans.) Quarters selected were at least six months apart in an effort to make each observation independent of the others. The data appear in the following table.

Interest Rates	*Building Permits*
6.00	443
6.75	448
7.00	367
7.50	492
8.25	384
8.75	437
9.00	356
9.50	339
9.75	365
10.25	321
10.75	338
11.25	230

The sums necessary for a regression analysis are:

$$\Sigma X = 104.75 \qquad \Sigma Y = 4520 \qquad \Sigma XY = 38427.25$$
$$\Sigma X^2 = 944.9375 \qquad \Sigma Y^2 = 1{,}757{,}198 \qquad n = 12$$

a. Is a direct or inverse relationship to be expected? Find the sample least-squares regression line and compare it with this expectation.
b. Calculate the sample standard error of estimate. Compare the percentages of the sample data within one and two standard errors of the regression line to the expected percentages for normally distributed errors around a regression line.

c. How does the variance of the sample Y-values around the regression line compare with the variance of the Y-values around their mean? What does this comparison indicate with respect to the effectiveness of the regression relationship?

10-19 *Hypothesis test for B_1*. For the study in Exercise 10-17, test the hypothesis that there is no linear relationship between the dependent and independent variables in the statistical population from which the sample came. Use a two-tailed test and a 0.01 level of significance. What is the statistical population in this case?

10-20 *Hypothesis test for B_1*. Does the sample evidence in Exercise 10-18 support the hypothesis that there is no relationship between interest rates and building permits in the population from which the data came? (Conduct a two-tailed test at the 0.05 level of significance.) Under what conditions can such a test be performed correctly?

10-21 *Confidence interval estimate of B_1*. Find the 0.95 confidence interval estimate for the slope of the population regression line in Exercise 10-17. Discuss the meaning of the result. Is it necessary to conduct a test of the hypothesis that $B_1 = 0$ after the confidence interval estimate has been found?

10-22 *Confidence interval estimate of B_1*. Use a confidence interval estimate as an alternative to the hypothesis test requested in Exercise 10-20. What conclusion is justified when the interval is used that is not justified when only the hypothesis test is conducted?

10.8 Inferences Based on the Regression Line

A typical purpose for finding a sample regression line is to make some inference about the dependent variable from the regression relationship. The insurance company analyst may want to use the sample relationship between number of sales and number of calls to estimate the number of sales for a given number of calls. Inferences of this type fall into two major classes:

1. *Estimating the mean*, or expected value, of the dependent variable for a specified value of the independent variable.
2. *Predicting* the single value of the dependent variable that will be associated with a value of the independent variable.

In the first case, the insurance analyst could be seeking the mean number of sales to be expected if a large number of agents each make 30 sales calls to prospects. This number of calls per week could be the objective set in a new marketing campaign.

In the second case, the analyst could be making a prediction of the number of sales that a single agent will realize as a result of the 30 calls planned for next week.

Estimation of the Conditional Mean

The analyst wants to estimate the mean number of sales to be expected when a large group of agents each makes 30 calls. If the statistical population illustrated in Figure 10-8 were available, the goal would be to find the mean of the conditional distribution of Y when X is 30. If this model is a correct representation of the statistical population, the analyst would only need to find the value of Y on the population regression line when X is 30. In the figure, 15 sales is the mean number of sales per agent when each agent makes 30 calls. Given the population, the conditional mean could be found without error.

In practice, of course, the population relationship is *not* available. The sample provides our only information. As was the case with a single variable, a confidence interval estimate is appropriate. The general form of this estimate is

$$\overline{Y}_c - t(s_{\overline{Y}_c}) \leq E(Y \mid X) \leq \overline{Y}_c + t(s_{\overline{Y}_c})$$

where t has $n - 2$ degrees of freedom and $\overline{Y}_c = b_0 + b_1 X_s$. Here the value on the left of the sample regression equation, $\overline{Y}_c$, is treated as a sample mean. It is the point estimate of the conditional population mean for a specified value of X designated by X_s.

The equation for the standard error of the conditional mean $\overline{Y}_c$ can be expressed as

$$s_{\overline{Y}_c} = \sqrt{\frac{s^2_{Y \cdot X}}{n} + s^2_{b_1}(X_s - \overline{X})^2} \tag{10-16}$$

Although this form is not the most convenient for calculation, it is easy to understand. Another aid to understanding is the sample regression equation expressed as

$$\overline{Y}_c = \overline{Y} + b_1(X_s - \overline{X}) \tag{10-17}$$

This form comes from substituting $\overline{Y} - b_1\overline{X}$ for b_0.

Equation 10-17 states that the sample regression line consists of the mean of the dependent variable corrected by the effect of the slope. This effect is the product of b_1, the rate at which Y changes per unit of X, and $(X_s - \overline{X})$, the number of units that the specified value of X, X_s, departs from the mean of the X-values. In Equation 10-16, the quantity under the radical is the sum of the sampling variances of the two terms on the right in Equation 10-17.

Equation 10-16 can be put in a form more convenient for calculation. The result is

$$s_{\overline{Y}_c} = s_{Y \cdot X}\sqrt{\frac{1}{n} + \frac{(X_s - \overline{X})^2}{\Sigma X^2 - [(\Sigma X)^2/n]}}$$

Then *the confidence interval estimate* of the conditional mean, for a specified value of X, X_s, is

$$\bar{Y}_c \pm t s_{Y \cdot X} \sqrt{\frac{1}{n} + \frac{(X_s - \bar{X})^2}{\Sigma X^2 - [(\Sigma X)^2/n]}} \qquad \textbf{(10-18)}$$

where $\bar{Y}_c$ is the sample regression point estimate, $s_{Y \cdot X}$ is the sample standard error of estimate given in Equation 10-11, and t has $n - 2$ degrees of freedom.

Example

The insurance analyst wants a confidence interval estimate of the conditional mean for 30 calls. Hence $X_s = 30$. We have already found that

$$s_{Y \cdot X} = 2.2906 \qquad \Sigma X = 453 \qquad \Sigma X^2 = 11{,}733 \qquad \bar{X} = 22.65 \qquad n = 20$$

For a 0.95 confidence interval and 18 degrees of freedom, $t = 2.1009$ (Table A-4). Finally,

$$\bar{Y}_c = -1.6869 + 0.5116(30) = 13.6600$$

Substitution in Equation 10-18 gives

$$13.6600 \pm 2.1009(2.2906)\sqrt{\frac{1}{20} + \frac{(30 - 22.65)^2}{11{,}733 - (453^2/20)}} = 13.6600 \pm 4.8125\sqrt{0.0867}$$

or

$$13.6600 \pm 1.4169$$

The 0.95 confidence interval estimate of the expected number of sales that will result from 30 sales calls is

$$12.24 \leq E(Y \mid X) \leq 15.08$$

If many agents each made 30 calls, we would expect the mean number of sales per call to be between 12.24 and 15.08 with 0.95 confidence.

In the instance just described, the analyst specified that X_s was to be 30. Some other value could have been selected for X_s. Equation 10-18 applies to any value of X_s within the range of the sample observations of X. Figure 10-11 shows 0.95 confidence intervals for several values of the independent variable; the end points have been connected with smooth curves. These are the curves above and below the sample regression line. As indicated by the last term under the radical in Equation 10-18, the farther X_s is from $\bar{X}$, the wider is the confidence interval. Thus provision for possible error in the sample slope widens the confidence interval as X_s and $\bar{X}$ become farther apart.

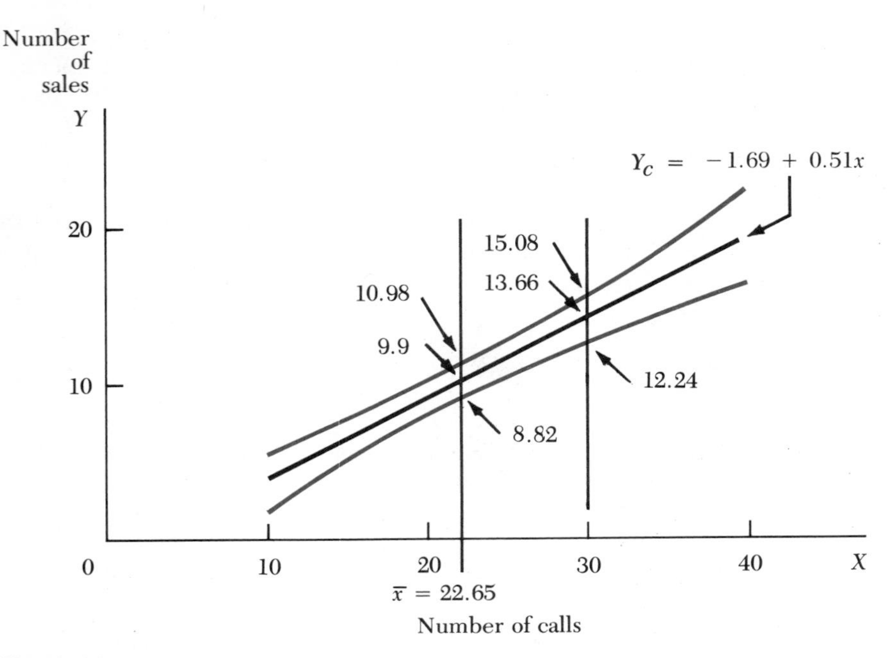

Figure 10-11 0.95 Confidence Intervals for Conditional Mean Sales

Prediction of a Single *Y*-Value

Suppose an agent who plans to make 30 calls next week has asked the insurance company analyst to predict the number of sales which will result. To understand this problem, let us return to the model in Figure 10-8. Once again the focus of interest is the conditional distribution of Y-values for $X_s = 30$ calls. This time, however, the analyst is *not* interested in the mean of the distribution but rather in an unknown value of Y to be selected randomly from this distribution. Even if the population were available, this individual value of Y is subject to prediction error. When only a sample is available, we must add the prediction error from this new source to the sampling error for the mean of the conditional distribution. The sample estimate of this additional source of prediction error is $s^2_{Y \cdot X}$, the variance around the sample regression line. Adding this quantity to those in Equation 10-16, we have

$$s_{Y_c} = \sqrt{s^2_{Y \cdot X} + \frac{s^2_{Y \cdot X}}{n} + s^2_{b_1}(X_s - \bar{X})^2}$$

Given this expression for the standard error of an individual prediction, we can use the process outlined in the previous section to get *the prediction interval* for a single value of Y for a specified value of X, X_s:

$$Y_c \pm t s_{Y \cdot X} \sqrt{1 + \frac{1}{n} + \frac{(X_s - \bar{X})^2}{\Sigma X^2 - [(\Sigma X)^2/n]}} \qquad \textbf{(10-19)}$$

where Y_c is the sample regression point estimate of Y, $s_{Y \cdot X}$ is the sample standard error of estimate, and t has $n - 2$ degrees of freedom.

Example

To find the prediction interval for the number of sales that will result from the 30 calls a single agent will make, the analyst need only add 1 to the quantity under the radical for the confidence interval procedure for $E(Y \mid X)$ in the previous section. Hence the prediction interval for a single Y is

$$13.6600 \pm 4.8125\sqrt{0.0867 + 1} = 13.6600 \pm 5.0168$$

or $8.64 \leq Y \leq 18.68$ sales. The analyst predicts that the agent will make from 8 to 19 sales (to the nearest integer outside the interval) and the confidence coefficient is 0.95.

Given 30 calls, 5.02 sales is the half-width of the 0.95 prediction interval for Y, while 1.42 is the half-width of the 0.95 confidence interval for $E(Y \mid X)$. The greater width of the prediction interval is a result of the dispersion of the individual Y-values around the population regression line. This dispersion is not relevant when we want an interval estimate of $E(Y \mid X)$. If we were to draw a figure similar to Figure 10-11 for prediction intervals, we would get curves similar to those shown, but considerably farther from the sample regression line.

Precautions

We must observe two precautions when making regression analyses. The first is not to infer that statistical association between two variables establishes a causal relationship. Regression simply provides a model that sometimes proves useful for reducing errors of prediction or estimation. As is true for any other model in science, regression models do not claim to be a correct representation of nature as it really exists. Such models are just useful representations. Significant statistical associations have been found in cases where it would be foolish to assume a causal relationship. It would be no surprise to find that, over the years, annual congressional spending on stationery and annual sales of gasoline in the United States have a statistically significant positive relationship. Both are features of underlying growth in the economy. Yet no causal relationship is likely to exist between the two variables. If it does exist, the causal linkage must be supported by logic from some field other than statistics.

The second precaution is that extrapolation of a regression relationship beyond the observed range of the independent variable can lead to highly erroneous predictions and estimates of values of the dependent variable. In a study of the relationship between the rate at which an assembly line operates and output per day, a linear relationship may produce good results over the observed range of assembly line rates. The relationship in this range could well be linear. But it is dangerous to use assembly line rate as an independent variable to estimate mean outputs for rates well above those observed. As the line rate increases, there will come a point beyond which workers cannot keep up as well with the pace. Sheer inability of people to move as rapidly as is required, together with fatigue and stress, prevent the earlier rate of increase in output as line speed increases. The slope of the regression line will no longer be as great. If we happen to select a line speed well beyond the point of diminishing returns, the straight-line estimate of mean output will be greater—possibly much greater—than actual mean output.

Exercises

10-23 *Machine maintenance costs.* A cost analyst and a statistician in a factory are studying factors that affect the costs of maintaining a type of machine. Large numbers of these machines are used in the factory. They are bought and installed in batches. Consequently several machines are available for observation in each of many age groups. Random samples of three machines have been selected for each of five age groups and cost data have been collected in order to find out whether age is a good indicator for these costs.

Age (months)	*Cost (dollars/month)*		
6	5.48	2.24	7.47
15	15.52	20.26	11.64
24	17.80	22.38	16.86
33	14.97	24.19	25.87
42	23.83	24.87	32.12

Note that there are 15 observations. For these observations:

$\Sigma X = 360$ $\Sigma Y = 265.5$ $\Sigma XY = 7711.83$

$\Sigma X^2 = 11{,}070$ $\Sigma Y^2 = 5676.2346$ $n = 15$

a. Plot a scattergram. Then find the regression line and plot it on the same graph. Comment on the adequacy of the fit.
b. For each of the age classes, find the mean cost and plot each mean on the scattergram in (a). Does the least-squares line still appear to be a good fit?
c. Find the point and 0.95 interval estimates for the expected maintenance cost of machines that are 36 months old.

10-24 *Single machine performance.* A single machine that had 11 months of service when it was put in storage is to be put back into service. Use the results of the analysis in Exercise 10-23 to find the 0.95 prediction interval for the maintenance costs this machine will require during its twelfth month of service.

10-25 *Adequacy of regression relationship.* For the situation in Exercise 10-12, the company executives will want to know about the variability to be expected in using the regression relationship to estimate direct costs.
a. As one indication of the variability around the regression line, find the standard error of estimate for the sample and interpret it for executive evaluation.
b. As a second indication of variability, find the 0.95 interval estimate of the conditional mean cost for policies with ages 10, 20, and 36 months and interpret these for executive evaluation.

10-26 *Predicting a single price.* A shrimp fisherman at the seaside auction market described in Exercise 10-10 has a 500-pound catch to sell. Based on the sample regression from the earlier exercise, find a 0.99 prediction interval for the price of this 500-pound catch.

10.9 Resistant Regression (Optional)

A function fitted to data by the least-squares technique is susceptible to great influence from any outliers that are present. For example, when a straight line is the appropriate function, outliers can seriously affect the slope and intercept of the fitted least-squares regression line. Resistant regression techniques are so named because they are not seriously affected by outliers when they are used to fit a function. The resistant technique we will discuss is appropriate after it has been decided on the basis of a scattergram or as a result of other resistant techniques that a straight line is the desired function.

The resistant technique to be described has two advantages. First, it enables us to identify outliers based on the sound theoretical underpinnings of the boxplot. Second, by successive adjustment, it produces a resistant regression line. Often, investigation makes it possible to eliminate outliers from the data. When this can be done, the adjustments tend to bring successive resistant regression lines somewhat closer to the least-squares regression line for the sample purged of outliers. This is highly desirable because, once accomplished, the analysis can shift to the least-squares function. This permits use of statistical inference that is not available without it.

Our example consists of observations of monthly maintenance costs for several machines together with the number of months each machine has been in service. All machines are the same type. The data are shown in Table 10-3. The third data point from the end (34, 48) is not an observation. It has been included only to illustrate the influence of an obvious outlier on the least-squares line.

Table 10-3 Machine Maintenance Costs Ordered by Months of Service

Months in Service	*Maintenance Cost ($ per month)*
22	54
23	59
24	64
25	61
26	71
27	93
29	83
30	91
31	102
32	104
34	48
34	112
35	108

The objective of the analysis is to find a relationship that will permit estimation of maintenance cost by use of months of service. The first step is to check the scattergram for the data to determine whether or not a straight regression line appears suitable. The scattergram is presented in Figure 10-12 along with three regression lines which will be described subsequently.

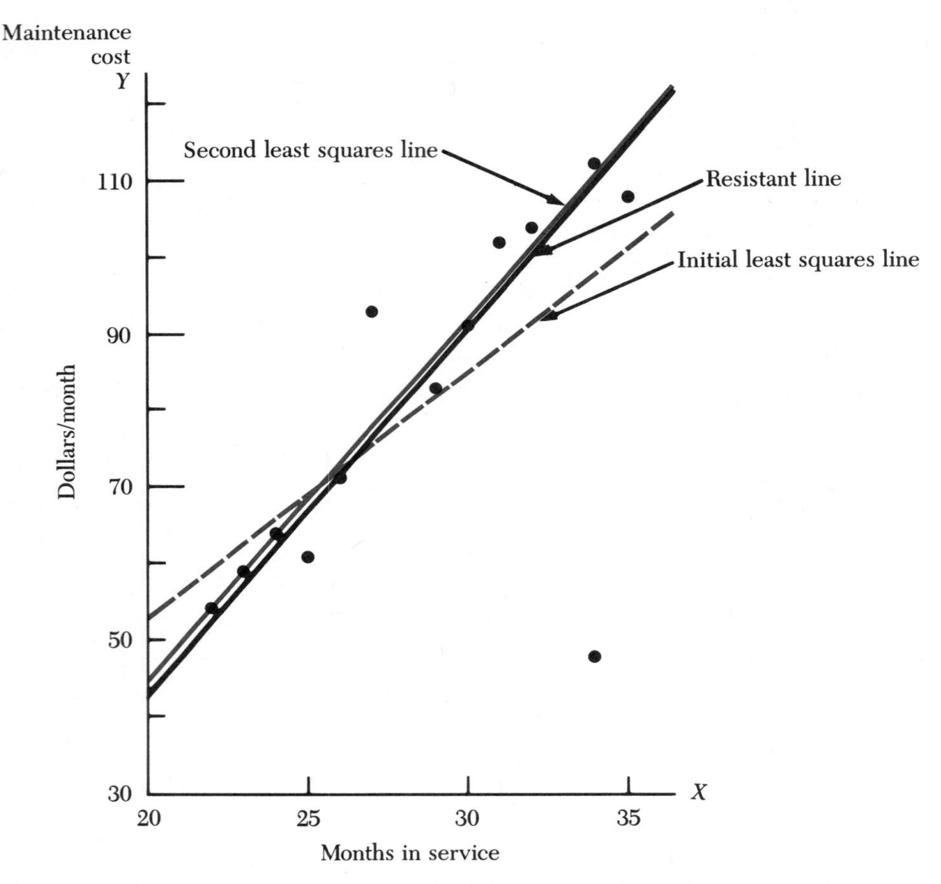

Figure 10-12 Scattergram and Regression Lines for Machine Maintenance Cost Data

First, note that, except for the obvious outlier at (34, 48) and a possible outlier at (27, 93), a straight line appears to be suitable for this range of the service time variable.

We can illustrate the marked influence of an outlier on least-squares regression lines at this point. The least-squares line for the entire set of data is

$$Y_c = -11.43 + 3.22X$$

and is identified as the initial least-squares line in Figure 10-12. After eliminating the point (34, 48), the least-squares line for the remaining 12 observations is

$$Y_c = -47.29 + 4.64X$$

and is the second least-squares line in Figure 10-12. Note the changes in the slope and intercept.

Having eliminated the obvious outlier (34, 48), we can begin the resistant analysis. The method is relatively quick and easy to do by hand so that changes from eliminating outliers identified by the method are easily incorporated, as are subsequent refinements.

We begin by dividing the reduced sample of 12 observations into a left, middle, and right subsample as these would appear on the scattergram. The subsamples should be as nearly equal in number as we can make them. For our example, in place of the scattergram, we can use Table 10-3, in which the values of the independent variable are arranged in ascending order. We can use this sequence to count off the subsamples. We place the top 4 observations in the first subsample, which is to the left in the scattergram, the second 4 in the middle subsample, and the bottom 4 in the right subsample.

We next find the summary points for each group. *The coordinates of a summary point are the respective medians of the X values and of the Y values in a subsample.* Again referring to Table 10-3, we see that, for the first group, the median X value is 23.5. The Y values are 54, 59, 61, and 64, and the median is 60. Thus, the summary point for the first, or left, group is (23.5, 60). Summary points for the other two groups are found with the same procedure and the result is as follows.

	Summary Points	
	X	Y
Left	23.5	60
Middle	28	87
Right	33	106

Now we use the left and right summary points to find the slope of the initial resistant regression line. The equation is

$$b_1 = \frac{Y_R - Y_L}{X_R - X_L} \tag{10-20}$$

where the subscripts refer to the right and left summary points, respectively. For our example,

$$b_1 = \frac{106 - 60}{33 - 23.5} = 4.84$$

For the initial resistant regression line the slope is 4.84.

To get the intercept for the initial resistant regression line we make three estimates based on each of the summary points and average them. The equations are:

$$b_0(\text{left}) = Y_L - b_1 X_L \tag{10-21a}$$

$$b_0(\text{middle}) = Y_M - b_1 X_M \tag{10-21b}$$

$$b_0(\text{right}) = Y_R - b_1 X_R \tag{10-21c}$$

$$b_0 = \frac{b_0(\text{L}) + b_0(\text{M}) + b_0(\text{R})}{3} \tag{10-21d}$$

For our example,

$$b_0(\text{left}) = 60 - 4.84(23.5) = -53.74$$

$$b_0(\text{middle}) = 87 - 4.84(28) = -48.52$$

$$b_0(\text{right}) = 106 - 4.84(33) = -53.72$$

$$b_0 = \frac{-53.74 + -48.52 + -53.72}{3} = -51.99$$

The intercept for our initial resistant line is –51.99. Hence, the equation for the initial resistant regression line is

$$Y_c = -51.99 + 4.84X$$

Our next task is to find the residuals from the initial resistant regression line and make a boxplot of them to identify any outliers which may be present. In Table 10-4, substituting the values of X in the above equation gives us the estimated Y values. These, subtracted from the observed Y values, give us the residuals shown in the final column.

Table 10-4 Residuals from the Resistant Regression Line

Months in Service X	*Maintenance Cost* Y	*Y Estimate*	*Residual*
22	54	54.49	–.49
23	59	59.33	–.33
24	64	64.17	–.17
25	61	69.01	–8.01
26	71	73.85	–2.85
27	93	78.69	14.31
29	83	88.37	–5.37
30	91	93.21	–2.21
31	102	98.05	3.95
32	104	102.89	1.11
34	112	112.57	–.57
35	108	117.41	–9.41

Arranging the residuals in ascending order assists in making the boxplot. Table 10-5 presents the residuals in this order. From the table we find that the median is −0.53, the first quartile is −4.11, the third quartile is +0.47, and the adjacent values are −9.41 and +3.95. The upper outer fence is +14.21. Hence, the residual +14.31 is an extreme outlier. The boxplot is shown to the right of Table 10-5.

Table 10-5 Residuals in Ascending Order

Residual
−9.41
−8.01
−5.32
−2.85
−2.21
−.57
−.49
−.33
−.17
1.11
3.95
14.31

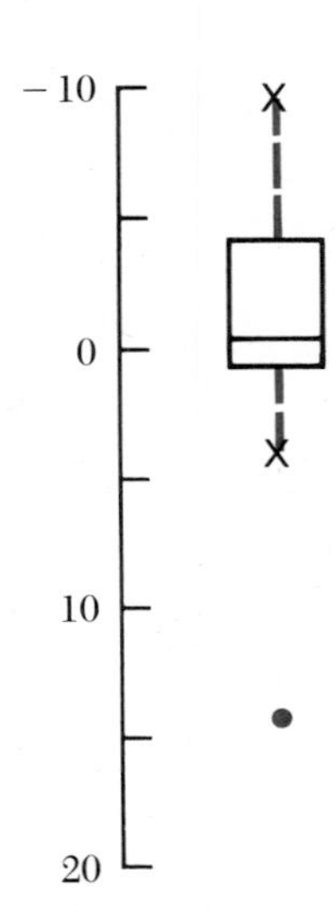

The boxplot establishes the point (27, 93) associated with the residual 14.31 as an extreme outlier. This was hardly apparent in the initial scattergram. Given the information that this point is extremely deviant from the initial resistant regression line, we are motivated to investigate why this is so. The investigation discloses that several months ago a substandard part was used to replace one which had failed in this machine. Subsequently, extra maintenance effort and expense were required to keep the machine running. Because of this anomaly, the decision is made to drop this observation from the analysis.

Having discarded the observation (27, 93), we must now find a resistant regression line for the smaller data set by repeating the process employed to find the initial regression line. We won't go through the steps since they are a direct repetition. The resulting second resistant regression line for the remaining 11 observations is

$$Y_c = -44.23 + 4.45X$$

The residuals from this line are shown in the right column in Table 10-6. The boxplot of these residuals discloses no additional outliers. Consequently, we can begin the adjustments to the slope and intercept which will produce the final resistant regression line.

Table 10-6 Residuals from the Second Resistant Regression Line

Months in Service X	*Maintenance Cost* Y	*Y Estimate*	*Residual*
22	54	53.67	.33
23	59	58.12	.88
24	64	62.57	1.43
25	61	67.02	−6.02
26	71	71.47	−.47
29	83	84.82	−1.82
30	91	89.27	1.73
31	102	93.72	8.28
32	104	98.17	5.83
34	112	107.07	4.93
35	108	111.52	−3.52

Two requirements of a good regression line are (a) that there be no trend in residuals as the independent variable increases, and (b) that the mean of the residuals be 0. If there is no trend, a regression line fitted to the residuals as dependent variable and service months as independent variable will have a slope of 0. If the residuals have a mean of 0, the intercept of this regression line will also be 0. To see whether this is true for our second resistant regression line, we find *the least-squares line* for the *residuals* to be

$$R_c = -7.05 + 0.29X$$

Neither the slope nor the intercept is 0. We can, however, adjust the slope and intercept of the resistant regression line which produced these residuals to make it meet the two requirements.

We use the slope for residuals, 0.29, to adjust the slope of the second resistant regression line, which is 4.45. The latter is adjusted by adding the slope for the residuals to it. Hence, the slope of the third resistant regression line is

$$b_1(\text{adjusted}) = b_1 + b_1(\text{residuals})$$

For our example, $b_1(\text{adj}) = 4.45 + 0.29 = 4.74$.

Similarly, we adjust the intercept of the resistant regression line by adding the intercept of the regression line for residuals to it:

$$b_0(\text{adjusted}) = b_0 + b_0(\text{residuals})$$

For our example, $b_0(\text{adj}) = -44.23 + (-7.05) = -51.28$.

A check is now made to see whether the residuals from the adjusted resistant regression line, $Y_c = -51.28 + 4.74X$, fulfill the two requirements. Since this is a repetition of the process just described, we won't give the details. The result shows the slope and intercept for the new set of residuals to be 0 to 2 decimal places. The requirements have been met.

All adjustments to the slope and intercept have been completed. The equation of the final resistant regression line is:

$$Y_c = -51.28 + 4.74X$$

This line is the third line plotted on the scattergram in Figure 10-12. It is identified as the resistant line.

Although we have completed the process of finding the resistant regression line, we should do one more thing if we are interested in making statistical inferences from our sample. We should see if the least-squares line for this censored sample is close enough to the resistant line to permit shifting to least-squares procedures. Because we found the least-squares fit for the residuals ($R_c = -7.05 + 0.29X$), the final resistant regression line ($Y_c = -51.28 + 4.74X$) is also the least-squares line for the 11 observations in Table 10-6. Consequently, the inferential techniques associated with least squares can be brought into play.

Summary

If justified by a scattergram or other means, a straight-line relationship can be fitted to bivariate data by least-squares procedures. Furthermore, if the assumptions of the normal model with equal variances for conditional distributions of the dependent variable apply, we can test for the existence of such a relationship in the population. When the test supports the existence of a relationship, we can make inferences about its slope and find interval estimates for expected values and predictions.

Supplementary Exercises

10-27 *Goodness-of-fit by direct calculation.* For the data given in Exercise 10-23, find the coefficient of correlation. Does this show the relationship between the variables to be direct or inverse? How does this result compare with that found in Exercise 10-23? Find the coefficient of determination from the coefficient of correlation. Can the direction of the relationship between the variables be inferred from the coefficient of determination alone? Why?

10-28 *Clerical aptitude and spelling.* A personnel director in a large organization wants a correlation analysis of scores on a test of clerical aptitude (Y) and scores on a spelling test (X). A random sample of 10 persons who have taken both tests is selected and the partial results are:

$$\Sigma X = 70 \qquad \Sigma Y = 65 \qquad \Sigma XY = 472$$
$$\Sigma X^2 = 624 \qquad \Sigma Y^2 = 533 \qquad n = 10$$

Find the sample correlation coefficient (r) and interpret its algebraic sign. Would it have made any difference in the value of r if X and Y had been interchanged as symbols for the two variables?

10-29 *Breaking strength of steel cable.* Lengths of steel cable have been subjected to enough force to break them. A scattergram with diameter of cable as the independent variable and breaking strength as the dependent variable suggests that a linear relationship is appropriate. The regression relationship on a larger sample will be used to specify cable sizes for different uses in a large commercial construction project. The results of the pilot study are:

$$\Sigma X = 14.3 \qquad \Sigma Y = 45{,}355 \qquad \Sigma XY = 68{,}098.4$$
$$\Sigma X^2 = 21.99 \qquad \Sigma Y^2 = 211{,}461{,}083 \qquad n = 12$$

Here X is in inches and Y is in pounds.

a. Find the equation of the sample regression line and interpret b_0 and b_1.
b. Find the sample standard error of estimate and interpret it.
c. Find the 0.99 confidence interval estimate of the slope for the population regression line. What does this result imply about the null hypothesis that there is no linear regression between these variables in the population?

10-30 *Suspending a large chandelier.* Part of the construction project described in Exercise 10-29 involves hanging a 2100-pound chandelier in the main lobby. A cable 0.8 in. in diameter is said to be adequate for this purpose. Where does this weight fall relative to the 0.99 prediction interval for the breaking strength for such a cable? Would you recommend a cable of this diameter for the purpose described? Is a larger sample likely to change this conclusion?

10-31 *Operating rates versus unit production costs.* The independent variable in a study of variable unit production costs in a factory is operating rate. Average hourly input rates of raw material corrected by scrap losses are used as a proxy measure of this variable. The factory uses the same raw material in a variety of products produced in essentially constant ratios. The dependent variable is variable production cost per unit of output. Careful cost allocation has isolated costs associated with operating rate. Observations of four cost rates were made for each operating rate at random points in time for which a given operating rate was in effect.

Raw Material (pounds/minute) X	*Operating Cost (cents/minute)* Y			
120	93	95	82	95
150	80	72	77	79
200	71	69	76	75
300	138	139	142	139
400	267	281	272	264

a. Find the slope and intercept of the linear least-squares regression line for these data.
b. Find the coefficient of determination and interpret it. Does it appear that linear regression may be effective?

10-32 For the operating cost study described in Exercise 10-31,

a. Test the hypothesis that there is no linear relationship between the variables.

b. Estimate the conditional mean operating cost when the input rate of raw material is 350 lb/min. Make a 0.99 interval estimate.

c. What is the 0.95 prediction interval for a single production run for which the input rate is 400 lb/min?

10-33 *The significance of linear regression.* For the situation described in Exercise 10-31, find the 0.99 confidence interval estimate of the slope for the population regression line. At the 0.99 level, test the hypothesis that the slope of the population regression line is zero. What additional information is provided by the confidence interval estimate that is not provided by the hypothesis test?

10-34 *The importance of a scattergram.* Plot a scattergram for the data in Exercise 10-31. Does a straight line appear to be a suitable model for these data? At what point in the analysis would it have been wise to plot this scattergram? Would the use of a straight-line relationship be better than assuming no relationship? (A method for testing the hypothesis that a straight line is an adequate fit will be presented in Chapter 14.)

Computer Exercises

1 Source: Real estate data file

For all parts of this exercise, treat the real estate data as if they are a random sample from a large universe of sales in the community.

a. Obtain scattergrams of purchase price versus income of buyer and purchase price versus age of buyer. Comment on the results.

b. Obtain the least-squares regression of purchase price on income of buyer, and execute a plot that contains the original scatter and the regression estimates.

c. Is the regression in (b) significant at the 0.01 one-tailed level? If it is, construct the 95% confidence interval for (1) the price paid in a randomly selected sale by a buyer whose annual income is $40,000, and (2) the mean price paid by buyers whose annual income is $40,000.

d. Obtain a frequency distribution of residuals from the price versus income regression. Also plot the residuals from regression versus income. Do the results appear to support the validity of the assumptions of the linear regression model in this case? Explain.

2 Source: Utilities data file

a. Perform a regression analysis using income as the dependent variable and assets as the independent variable.

b. Repeat (a) after deleting AT&T and GTE from the data.

c. Repeat (a) doing a separate analysis for the communications, electric, and gas utilities.

d. What are your conclusions from the analyses in the previous three parts?

3 Source: Baseball data file

a. Perform a regression analysis using runs as the dependent variable and triples as the independent variable.

b. Repeat (a) using singles as the independent variable.

c. Repeat (a) using "grounded into double play" as the independent variable.

d. Interpret the meaning of the slope for (a) through (c).

Interpreting Computer Output

Computer packages are especially helpful when performing regression analyses. The following table shows the results of a regression analysis based on a random sample of 153 observations from the personnel data file. The independent variable is number of years worked and the dependent variable is salary.

DEP VARIABLE: SALARY

SOURCE	DF	SUM OF SQUARES	MEAN SQUARE	F-VALUE	PROB>F
MODEL	1	135868282	135868282	7.435	0.0072
ERROR	151	2759463201	18274591		
C TOTAL	152	2895331483			

ROOT MSE	4274.879	R-SQUARE	0.0469
DEP MEAN	45163.288	ADJ R-SQ	0.0406
C.V.	9.465385		

VARIABLE	DF	PARAMETER ESTIMATE	STANDARD ERROR	T FOR HO: PARAMETER=0	PROB>|T|
INTERCEP	1	44172.287	501.532	88.075	0.0001
WORKED	1	171.908	63.046586	2.727	0.0072

The top part of the table under SUM OF SQUARES lists SSR, SSE, and SSTO as 135868282, 2759463201, and 2895331483. We will discuss the remaining information in the top part of the table in Chapter 14.

In the middle of the table, ROOT MSE is the standard error of estimate and R-SQUARE is the coefficient of determination. DEP MEAN is the mean of the dependent variable, salary, C.V. is the coefficient of variation (ratio of the standard error of estimate to the mean salary expressed as a percentage), and ADJ R-SQ is a deflated coefficient of determination, a refinement not discussed in this chapter.

The bottom part of the table gives the estimates of the intercept (44172.287) and slope (171.908) under PARAMETER ESTIMATE, and their respective standard errors in the next column. Under T FOR HO: are the t values for testing the hypothesis that each parameter is 0. For H_O: $\beta_1 = 0$ we see that t is 2.727 with a P-value of 0.0072, a significant result for any reasonable level of significance.

Although we would almost certainly reject the hypothesis that β_1 is 0, the coefficient of determination is less than 0.05. The data are very loosely clustered around the regression line. Figure 10-13 is a computer-generated graph of the scattergram, the sample regression line, and the 95% confidence interval estimates for the prediction of individual salaries. The loose clustering around the regression line is apparent along with the wide confidence interval estimates. The figure makes it clear that this model is not very reliable for predicting individual salaries.

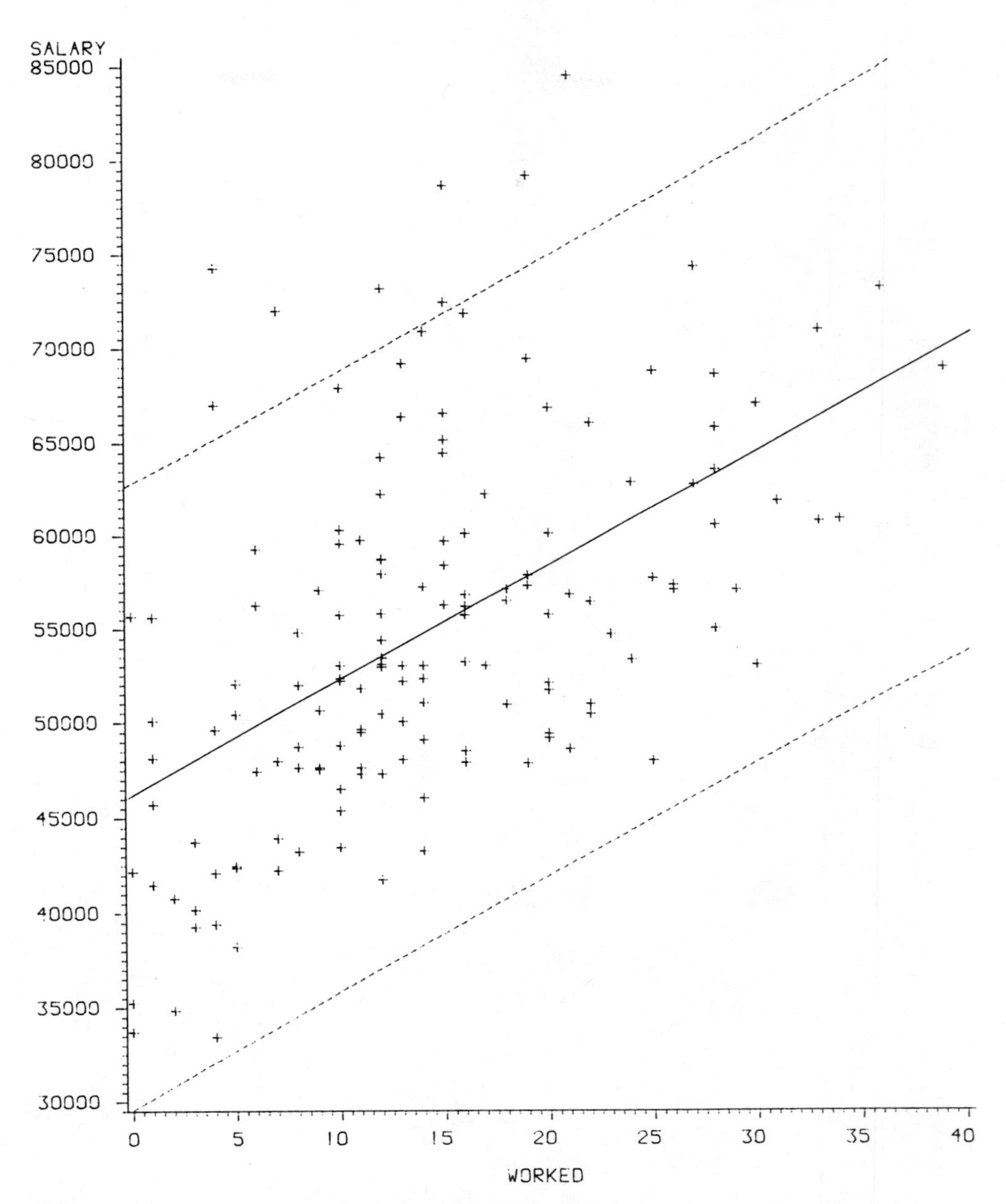

Figure 10-13 Computer-Generated Scattergram Graph

PART FOUR

Further Topics in Statistical Inference

11

INFERENCES FROM TWO SAMPLES

Many occasions arise in which it is important to compare two populations by using a random sample from each population. For example, a training director for a large company wants to know by how much mean performance after training differs from mean performance before training.

Another common type of comparison involves proportions. For instance, advertising campaigns often are accompanied by checks of the proportions of the relevant audiences favorably disposed toward the object advertised before and after advertising.

The two examples just described consist of before-after comparisons. In other cases, the desired comparison is cross sectional (from nearly simultaneous observations). As an example, a construction firm may have narrowed the choice of replacement tires for its fleet of pickup trucks to two brands. A comparison test of the mean miles per dollar under actual performance conditions could be used to settle the issue between the brands. As another example, deciding which of two suppliers should furnish a part may be done by sampling to determine which provides the smaller proportion of defectives.

In this chapter, we shall discuss making statistical inferences about two population means and about two population proportions. Point and interval estimation procedures for the difference between two means and for the difference between two proportions will be described, along with procedures for testing statistical hypotheses about these two types of differences. These topics represent a direct extension of the material covered in Chapters 8 and 9, which were concerned with estimating or testing a hypothesis about the mean or the proportion of a single population. The current concern is with the means or proportions of two populations.

Noting the progression from Chapters 8 and 9 to this chapter leads us to ask about further extensions. What about estimation and hypothesis testing for means and proportions when more than two populations are involved? What if the construction firm mentioned must choose from three or more brands of tires by a comparison experiment? What if there are three or more parts suppliers with products to be compared on proportion defective per shipment? Because such questions are important and are encountered rather frequently in business, industry, and government, a subsequent chapter is devoted to each topic. Chapter 12 discusses statistical inferences about three or more population proportions, and Chapter 13 discusses inferences about three or more population means.

11.1 Estimation of the Difference Between Two Population Means: Independent Samples

Before discussing estimation of the difference between two population means, we shall use an example to describe the sampling distribution of the difference between two sample means.

Example

A firm makes wooden window frames. After it is manufactured, a frame is soaked for some time in a solution that makes it resistant to insect damage and dry rot. The frame is then put in a dryer to remove the moisture in the chemical solution and leave only the dry chemicals. Weights of frames before being put into the dryer and after being removed are used to check on the efficiency of the drying process.

Immediately after soaking, frames are supposed to weigh an average of 42 lb and have a standard deviation of 0.5 lb. These weights are normally distributed. After drying, the frame weights are normally distributed and are supposed to have a mean weight of 35 lb with a standard deviation of 0.4 lb. The expected weight loss of $42 - 35 = 7$ lb is the standard for judging the effectiveness of the dryer.

The Sampling Distribution of $\overline{X}_1 - \overline{X}_2$

Periodic checking on the effectiveness of the dryer begins with a random sample of frame weights as the frames enter the dryer. Frames are hung on hooks, move on a continuous chain through the dryer, and come out to an inspection station. Here a random sample of dried frames is selected periodically. These samples are selected completely independently of those selected for weighing just prior to entering the dryer, except for the requirement that one sample be selected sometime during each hour at both ends of the dryer.

The mean weight of the frames in the dried sample collected during a given hour is subtracted from the mean weight of the wet sample collected during the same hour. This difference in means forms the basis for both a confidence interval estimate of the true mean loss of weight and a hypothesis test that the true mean weight loss is 7 lb.

The statistical situation is illustrated in Figure 11-1. When everything is working as it should, population 1 at the top has a mean of 42 and a standard deviation of 0.5 lb. Population 2 has a mean of 35 and a standard deviation of 0.4 lb. The row below these populations shows three of the samples selected periodically from each population. The mean of each such sample is calculated and recorded. Each sample mean belongs to a sampling distribution of means from similar samples. The two sampling distributions for the sample means are shown in the third row from the top. Note that the sampling distribution of means of samples from population 1 has a mean, $\mu_{\overline{X}_1}$, equal to the mean of population 1 and a standard deviation, or standard error, which is the standard deviation of population 1 divided by the square root of the sample size: $\sigma_{\overline{X}_1} = \sigma_1/\sqrt{n_1}$. The same relationship holds for the sampling distribution of $\overline{X}_2$ and population 2. These are simply new illustrations of what we learned about the sampling distribution of the sample mean in Chapter 7.

The distribution at the bottom of Figure 11-1 illustrates a new concept for our present purpose. During each hour, an independent random sample is selected from each population, the mean of each sample is found, and the mean weight after drying is subtracted from the mean weight before drying.* The *differences* between such means also form a sampling distribution. As shown beneath the lowest distribution, a given difference between a sample mean weight before drying and a sample mean weight after drying belongs to a distribution of such differences with the following three important characteristics:

1. The expected value, or mean, of the distribution of differences in sample means is equal to the difference between the means of the source populations. That is,

$$E(\overline{X}_1 - \overline{X}_2) = \mu_{\overline{X}_1 - \overline{X}_2} = \mu_1 - \mu_2 \tag{11-1}$$

2. The variance of the distribution of differences in sample means is equal to the *sum* of the variances of the individual means. In symbols,

$$\sigma^2_{\overline{X}_1 - \overline{X}_2} = \frac{\sigma_1^2}{n_1} + \frac{\sigma_2^2}{n_2} \tag{11-2}$$

*The subtraction could be done in the reverse order if desired. It is only necessary that the order always be the same.

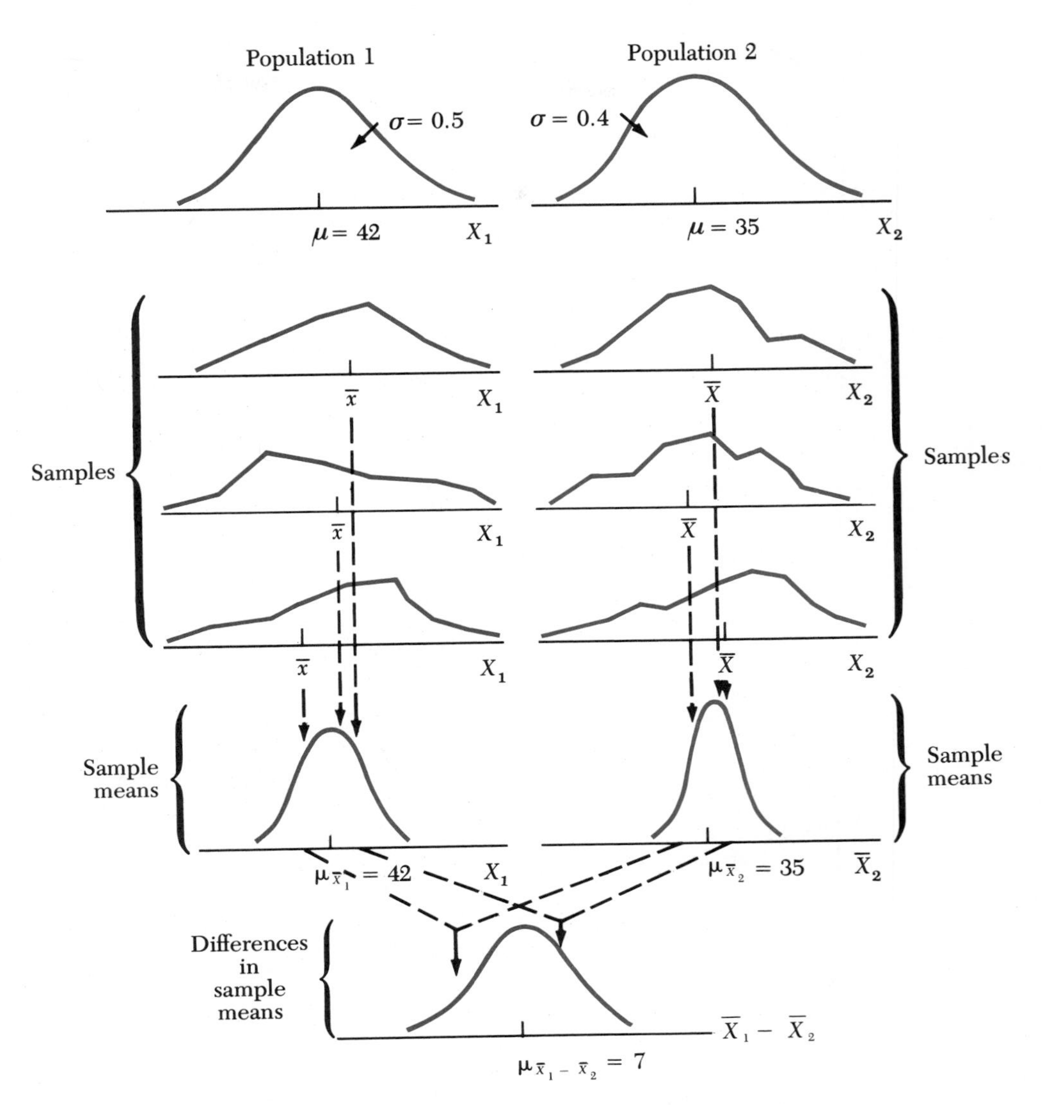

Figure 11-1 Sampling Distribution for Differences in Sample Means*

* *Not to scale.*

The square root of $\sigma^2_{\bar{X}_1 - \bar{X}_2}$ is called *the standard error of the difference between two means*, $\sigma_{\bar{X}_1 - \bar{X}_2}$. It is the standard deviation of the sampling distribution of the differences. It follows from Equation 11-2 that *the standard error of the difference between two means** is

$$\sigma_{\bar{X}_1 - \bar{X}_2} = \sqrt{\frac{\sigma_1^2}{n_1} + \frac{\sigma_2^2}{n_2}} \qquad \textbf{(11-3)}$$

* It will be assumed throughout this chapter that n_1 and n_2 constitute less than 10% of the populations from which they come. Consequently, the finite population multiplier will not be required.

3. If the source populations are normally distributed, then the distribution of differences in sample means is normally distributed. If the source populations are not normally distributed, then the central limit theorem assures us that the distribution of differences in sample means is essentially normal provided that the two sample sizes are large enough. (If each sample is at least 30, this condition will be met for virtually all cases encountered in practice.)

For the example of the wooden window frames, the population of weights before drying has a hypothesized mean of 42 lb and a standard deviation of 0.5 lb. It follows that the distribution of means from samples of 25 will have an expected value, or mean, of 42 lb and a variance of 0.01.

$$\sigma^2_{\bar{X}_1} = \frac{\sigma^2_1}{n_1} = \frac{(0.5)^2}{25} = 0.01$$

Population 2, consisting of weights after drying, has a hypothesized mean of 35 lb and a standard deviation of 0.4 lb. The distribution of means of samples of 20 each from this source population can be expected to have a mean of 35 lb and a variance of 0.008.

$$\sigma^2_{\bar{X}_2} = \frac{\sigma^2_2}{n_2} = \frac{(0.4)^2}{20} = 0.008$$

With the hypotheses and assumptions mentioned, the sampling distribution of differences formed by subtracting the sample mean weight after drying from the sample mean weight before drying has an expected value, or mean, of 7 lb.

$$E(\bar{X}_1 - \bar{X}_2) = \mu_1 - \mu_2 = 42 - 35 = 7 = \mu_{\bar{X}_1 - \bar{X}_2}$$

and a standard error, or standard deviation, of 0.1342 lb.

$$\sigma_{\bar{X}_1 - \bar{X}_2} = \sqrt{0.01 + 0.008} = 0.1342$$

Finally, because the two source populations are normally distributed, the sampling distribution of $\bar{X}_1 - \bar{X}_2$ is also normally distributed.

Interval Estimation Given σ_1, σ_2, and Normal Populations

We learned in Chapter 8 how to form a confidence interval estimate for the mean of a single population. The general form of such an estimate when the population is known to be normal with a standard deviation of σ is

$$\bar{x} - z\left(\frac{\sigma}{\sqrt{n}}\right) \leq \mu \leq \bar{x} + z\left(\frac{\sigma}{\sqrt{n}}\right)$$

The procedure was to find the proper standard normal deviate, z, for the desired confidence level, multiply that value of z by the standard error of the sample mean, $\sigma/\sqrt{n}$, and subtract this product from and add it to the sample mean.

We now have a situation that is very similar. The difference in mean weights is normally distributed with a known standard error. All we need do is find the difference in sample means as a point estimate. Then we add and subtract the appropriate quantity to form the confidence interval estimate. The appropriate quantity is the product of the standard normal deviate for the chosen confidence level and the standard error of the difference in means.

Example

For the example of the window frames, a sample of 25 wet frames has a mean weight of 42.03 lb. A sample of 20 frames dried during the same hour has a mean weight of 35.12 lb. We want a 0.95 confidence interval estimate of the difference in mean weights attributable to the drying process.

On the assumption that the population standard deviations given earlier are correct for the process as it is now operating, we have already found that the standard error of $\overline{X}_1 - \overline{X}_2$ is 0.1342 lb. A 0.95 confidence interval estimate of the difference in mean weights before and after drying is

$$(\overline{x}_1 - \overline{x}_2) - z\sigma_{\overline{X}_1 - \overline{X}_2} \leq \mu_1 - \mu_2 \leq (\overline{x}_1 - \overline{x}_2) + z\sigma_{\overline{X}_1 - \overline{X}_2}$$

or

$$(42.03 - 35.12) - 1.96(0.1342) \leq \mu_1 - \mu_2 \leq (42.03 - 35.12) + 1.96(0.1342)$$

This gives $6.65 \leq \mu_1 - \mu_2 \leq 7.17$. The 0.95 confidence interval estimate extends from 6.65 lb to 7.17 lb.

In summary, if the two source populations are normal with known standard deviations, the confidence interval estimate of the difference between the two population means is

$$(\overline{x}_1 - \overline{x}_2) \pm z\sigma_{\overline{X}_1 - \overline{X}_2} \qquad \textbf{(11-4)}$$

where $\overline{x}_1$ and $\overline{x}_2$ are sample means from the two populations, $\sigma_{\overline{X}_1 - \overline{X}_2}$ is the standard error of the difference as defined in Equation 11-3, and z is the appropriate standard normal deviate for the desired level of confidence.

Normal Populations with Known and Equal Variances As a special case of the above, there are occasions when it can be assumed that the variances of the two populations are equal. When this is the case, the expression in Equation 11-4 for the confidence interval estimate of the difference between the two population means is not changed, but the standard error of the difference is a simplified version of Equation 11-3.

For normally distributed source populations with known and equal standard deviations, the confidence interval estimate for the difference in means is Equation 11-4 but with a standard error

$$\sigma_{\bar{X}_1 - \bar{X}_2} = \sigma\sqrt{\frac{1}{n_1} + \frac{1}{n_2}} \tag{11-5}$$

Example

The mean fills of two machines of the same model that fill soft drink bottles are being compared. A 0.99 confidence interval estimate of the difference in mean fills is desired. A sample of 16 fills from machine A has a mean of 8.02 oz, and a sample of 25 fills from machine B has a mean of 7.97 oz. Both machines are known to have a normal distribution of fill weights with a standard deviation of 0.07 oz. What is the 0.99 confidence interval estimate of the difference in mean fills?

From Equation 11-5, the standard error of the difference in means is

$$\sigma_{\bar{X}_1 - \bar{X}_2} = \sigma\sqrt{\frac{1}{n_1} + \frac{1}{n_2}} = 0.07\sqrt{\frac{1}{16} + \frac{1}{25}} = 0.0224$$

The point estimate of the difference in means is

$$\bar{x}_A - \bar{x}_B = 8.02 - 7.97 = 0.05$$

From Equation 11-4, the 0.99 confidence interval estimate is

$$0.05 \pm 2.58(0.0224)$$

or −0.008 to 0.108 oz. This interval includes zero, so it is consistent with the conclusion that the two machines are delivering the same mean fill.

It is, of course, only under rather special circumstances that we can assume the two source populations are normally distributed, and, at the same time, know the values of their standard deviations. It is even less often that we know the two standard deviations are equal and know their values. We shall discuss situations that do not meet these special requirements next.

Interval Estimation When Variances Are Unknown

In the first case we shall discuss, the difference in means will be estimated for two populations for which the variances are not known. But the populations are assumed to be normally distributed, or nearly so, with equal variances.

In Chapter 8, given a normal population with unknown variance, the interval estimate of the mean had the general form

$$\bar{x} - t\left(\frac{s}{\sqrt{n}}\right) \leq \mu \leq \bar{x} + t\left(\frac{s}{\sqrt{n}}\right)$$

Here $\bar{x}$ is the sample mean and s is the standard deviation for the same sample. The sample standard deviation is a point estimate of the unknown population standard deviation. It is necessary to use t in this expression to allow for the sampling variability in s. The degrees of freedom for t are $n - 1$, one less than the sample size.

Normal Populations with Unknown but Equal Variances The present situation is similar to the one just described. Because the two populations are normally distributed, the sampling distribution of $\bar{X}_1 - \bar{X}_2$ is normal. Because we do not know the common population variance, however, we cannot use Equation 11-3 for the standard error of the difference and Equation 11-4 for the confidence interval.

In place of the standard error of the difference in Equation 11-3, we substitute the estimate

$$s_{\bar{X}_1 - \bar{X}_2} = \sqrt{\left[\frac{(n_1 - 1)s_1^2 + (n_2 - 1)s_2^2}{n_1 + n_2 - 2}\right]\left[\frac{1}{n_1} + \frac{1}{n_2}\right]} \quad \textbf{(11-6a)}$$

where

$$s_1^2 = \frac{\Sigma x_1^2 - n\bar{x}_1^2}{n_2 - 1} \quad \textbf{(11-6b)}$$

and

$$s_2^2 = \frac{\Sigma x_2^2 - n\bar{x}_2^2}{n_2 - 1} \quad \textbf{(11-6c)}$$

The first quantity under the radical in Equation 11-6a is a pooled estimate of the common variance of the two populations. Furthermore, we use t in place of z in Equation 11-4.

> When two populations are normal or nearly so with equal but unknown variances, the confidence interval estimate of the difference in their means is
>
> $$(\bar{x}_1 - \bar{x}_2) \pm ts_{\bar{X}_1 - \bar{X}_2} \quad \textbf{(11-7)}$$
>
> where the standard error of the difference is defined in Equation 11-6, and t has $n_1 + n_2 - 2$ degrees of freedom.

Example

An industrial plant discharges liquid waste into a large lake at a constant rate. Periodically an inspector compares the quantity of a pollutant within 600 yd of the plant with the quantity at the point in the lake least affected by human habitation. The comparison consists of estimating the difference in means based on two samples collected at the two locations at randomly selected times during the same day. The most recent results are as follows.

	Location	
	Near Plant	*Far from Plant*
Mean pollutant concentration (ppm)	$\bar{x}_1 = 17.2$	$\bar{x}_2 = 8.3$
Sample size	$n_1 = 12$	$n_2 = 10$
Standard deviation (ppm)	$s_1 = 2.4$	$s_2 = 2.1$

The inspector must find the 0.95 confidence interval estimate of the mean difference. Assume the populations are normally distributed with equal variances.

From Equation 11-6a, the standard error of the difference is estimated to be

$$s_{\bar{X}_1 - \bar{X}_2} = \sqrt{\left[\frac{11(2.4^2) + 9(2.1^2)}{12 + 10 - 2}\right]\left[\frac{1}{12} + \frac{1}{10}\right]}$$

$$= 0.97 \text{ ppm}$$

There are $12 + 10 - 2 = 20$ degrees of freedom. From Appendix Table A-4, $t = 2.086$ for 20 degrees of freedom and a 0.95 confidence interval. From Equation 11-7, the 0.95 confidence interval estimate of the difference in means is

$$(17.2 - 8.3) \pm 2.086(0.97) = 8.9 \pm 2.02$$

or 6.9 to 10.9 ppm. There were an estimated 6.9 to 10.9 more parts per million of the pollutant in the lake near the plant on the specified day.

Nonnormal Populations with Unequal, Unknown Variances

The population variances were unknown but equal in the case we just discussed, and the populations were essentially normal. In the case we shall consider next, the population variances not only are unknown but also may not be equal, and the populations cannot be assumed to be normally distributed. Indeed, the two populations may have distributions that are different from each other.

For this case of two nonnormal populations, the interval estimation procedure imposes the requirement that the samples from the two populations be large. In accordance with the central limit theorem, a minimum of 30 observations in each sample should suffice to make the sampling distribution of the difference in sample means essentially normal.

Given large samples, the standard error of the difference is estimated as follows.

$$s_{\bar{X}_1 - \bar{X}_2} = \sqrt{\frac{s_1^2}{n_1} + \frac{s_2^2}{n_2}} \tag{11-8}$$

where s_1^2 and s_2^2 are the variances of the two samples. These variances were defined in Equations 11-6b and 11-6c.

Given two populations not restricted to having equal variances or normal distributions, the confidence interval estimate of the difference in their means is

$$(\bar{x}_1 - \bar{x}_2) \pm z s_{\bar{X}_1 - \bar{X}_2} \qquad \textbf{(11-9)}$$

where each sample has at least 30 observations, where the standard error is defined by Equation 11-8, and where z is the appropriate standard normal deviate for the specified confidence level.

Example

A member of the planning department in a city wants to compare the present mean family income in a certain residential area with the mean three years ago, when a similar study was made. It can be assumed that the two income distributions are positively skewed and the degrees of skewness may be different. The sample results are as follows.

	Present Survey	*Earlier Survey*
Mean family income	$\bar{x}_1 = \$16{,}410$	$\bar{x}_2 = \$14{,}750$
Sample size	$n_1 = 150$	$n_2 = 100$
Standard deviation	$s_1 = \$4{,}960$	$s_2 = \$3{,}440$

The analyst wants a 0.99 confidence interval estimate of the difference in mean family incomes.

From Equation 11-8, the estimated standard error of the difference is

$$s_{\bar{X}_1 - \bar{X}_2} = \sqrt{\frac{4960^2}{150} + \frac{3440^2}{100}} = \$531$$

For a 0.99 confidence level, the value of z from Appendix Table A-2 is 2.58. From Equation 11-9, the 0.99 confidence interval estimate of the mean difference is

$$(\$16{,}410 - \$14{,}750) \pm 2.58(\$531) = \$1660 \pm \$1370$$

or \$290 to \$3030. At the 0.99 level of confidence, there has been an increase of from \$290 to \$3030 in mean family income.

Exercises

11-1 *Computer operator training.* As a customer service, a large computer manufacturing firm has developed a training course for operators that includes two tests of competence. The tests have been checked extensively and are essentially equivalent. One test is to be administered to operators before the training program and the other is to be administered a month after the program. The tests are known to have equal standard deviations of 5 points. One customer firm has had the following results from the program and tests:

Test Scores	
Before Training	*After Training*
85	88
77	84
86	86
69	86
81	90
80	97
83	
86	
73	

Assume that test scores are normally distributed. Find the 0.95 confidence interval estimate of the increase in mean score as a result of the training program.

11-2 *Metal working operations.* A metal plate manufactured as a support in automobiles has a hole drilled in it for a brace to pass through. The quality control department has gathered data on the diameters of the hole and of the brace to make an interval estimate of the mean difference in these diameters. The standard deviation of hole diameters is 0.05 cm. The standard deviation of brace diameters is 0.03 cm. The data from two recent samples are as follows.

Hole Diameters (X_1)	*Brace Diameters* (X_2)
2.57 cm	2.28 cm
2.49	2.26
2.55	2.20
2.44	2.26
2.58	2.27
2.48	2.26
2.50	2.24
2.51	2.27
	2.25
	2.30

Assuming that both sets of diameters are normally distributed, find the 0.99 confidence interval estimate of the difference in mean diameters of the hole and the brace, $(\mu_1 - \mu_2)$. Does it appear that the mean diameter of the holes should be increased?

11-3 *Health club reducing plans.* The physical director of a health club, in collaboration with a physician, developed two exercise and diet programs to help people lose excess weight. Twenty-seven club members expressed a desire to participate and were cleared by a physician to do so. The director randomly assigned participants to the two programs. For those who completed a program, results after one month were as follows.

	Program A	*Program B*
Sample size	9	11
Mean weight loss	4.20 lb	5.64 lb
Sample standard deviation	0.73 lb	0.91 lb

At the 0.99 level, what is the confidence interval estimate of the difference in mean weight loss for the two programs? Assume normally distributed populations with equal standard deviations.

11-4 *Product spoilage rates.* A component for stereo receivers is produced on an assembly line by two shifts of workers. The mean spoilage rates for the two shifts are to be compared by an interval estimate of their difference. The daily number of defective assemblies per thousand produced by each shift is calculated for two random samples of daily production records.

Shift A (X_1)	*Shift B* (X_2)
30.10	42.25
31.01	35.76
21.06	32.33
33.41	34.20
35.61	41.81
27.68	39.51
31.41	23.67
29.20	35.21
27.41	35.40
30.82	29.19
25.65	36.07
	40.31

Assume that the populations are normally distributed with equal variances. What is the 0.95 confidence interval estimate of the difference between mean spoilage rates?

11-5 *Weekly earnings in construction.* A large residential construction firm is considering opening operations in another large urban center. A major consideration is any difference that may exist in mean weekly earnings by construction workers between the proposed area and the firm's current base. Random samples of construction workers in the two centers produced the following information.

	Current Center	*Proposed Center*
$\bar{x}$	\$410	\$380
n	156	169
s	\$27	\$31

Both populations can be assumed to have considerable positive skewness and their standard deviations may not be equal. At the 0.99 level of confidence, estimate the difference in mean earnings for the two urban centers.

11-6 *Mean lives of energy cells.* A manufacturer of pocket calculators is comparing the mean lives of two rechargeable energy cells to help in deciding which cell to use. The populations are assumed to have positive skewness, and little is known about their standard deviations. Consequently, the lives of two large random samples, one from each type of cell, are tested with the following results.

	Cell A	*Cell B*
Sample mean	1872 h	1848 h
Sample standard deviation	55 h	37 h
Sample size	46	49

What is the estimated difference in mean lives of the cells? Use the 0.95 confidence level.

11.2 Testing the Difference Between Two Population Means: Independent Samples

The following general procedure for testing statistical hypotheses was described in Chapter 9.

1. State the null and the alternative hypotheses.
2. Specify the test statistic and its sampling distribution under the null hypothesis.
3. Choose the level of significance and the acceptance region for the test statistic under the null hypothesis.
4. Calculate the observed value of the test statistic.
5. Accept the null hypothesis if the observed value falls within the acceptance region; accept the alternative hypothesis if it does not.

The test can be either two-tailed, upper-tailed, or lower-tailed. An upper-tailed test is designed to detect a value of the parameter substantially greater than the value specified in the null hypothesis. A lower-tailed test is designed to detect a substantially smaller value than is specified. A two-tailed test is designed to detect substantial difference in either direction from a specified value.

Normal Populations with Known Variances

In Section 11.1, we discussed using the difference in two sample means to make an interval estimate of the difference in the means of two normal populations from which the samples came. Each sample was composed of randomly and independently selected observations. We saw that the expected value of the sampling distribution of this difference is the difference in the means of the two populations,

$$E(\overline{X}_1 - \overline{X}_2) = \mu_1 - \mu_2$$

and the standard error of difference is

$$\sigma_{\overline{X}_1 - \overline{X}_2} = \sqrt{\frac{\sigma_1^2}{n_1} + \frac{\sigma_2^2}{n_2}}$$

Hence we have the following general statement.

> Given normally distributed differences in the means of independent samples from two populations with known standard deviations, the test statistic
>
> $$z = \frac{(\overline{X}_1 - \overline{X}_2) - (\mu_1 - \mu_2)}{\sigma_{\overline{X}_1 - \overline{X}_2}} \qquad \textbf{(11-10)}$$
>
> is used to test a hypothesis about the difference in the two population means. Here z is the standard normal variate.

Example

A stamping machine makes metal trays that are placed inside tool boxes made by another process. The first machine is set to make trays with an average outside length of 19.80 in. The machine has a known standard deviation of 0.05 in. The second process is designed to make tool boxes with a mean inside length of 20.00 in. The standard deviation is known to be 0.04 in. At the 0.05 significance level, a two-tailed test is to be made for the hypothesis that the difference in mean lengths is

$$20.00 - 19.80 = 0.20 \text{ in.}$$

A sample of 25 trays has a mean outside length of 19.81 in. and a sample of 20 tool boxes has a mean inside length of 19.98 in.

The hypothesis statement is

$$H_0\colon \mu_1 - \mu_2 = 0.20 \quad \text{and} \quad H_A\colon \mu_1 - \mu_2 \neq 0.20$$

The test statistic is z, as defined in Equation 11-10. The acceptance region for H_0 is

$$-1.96 \leq z \leq 1.96$$

From Equation 11-3,

$$\sigma_{\overline{X}_1 - \overline{X}_2} = \sqrt{\frac{0.05^2}{25} + \frac{0.04^2}{20}} = 0.0134 \text{ in.}$$

Then, from Equation 11-10,

$$z_{\text{observed}} = \frac{(19.98 - 19.81) - (0.20)}{0.0134} = \frac{-0.03}{0.0134} = -2.23$$

The observed value of z is less than the lower limit of the acceptance region for H_0, so the null hypothesis is rejected in favor of H_A. It appears that mean clearance has dropped below the desired 0.20 in.

Normal Populations with Equal, Unknown Variances

Another case discussed in Section 11.1 assumed that the populations from which the two independent samples come are essentially normally distributed. Furthermore, although the population standard deviations are not known, they are assumed to be equal.

When two populations are normally distributed with equal but unknown variances, the statistic to test for a significant difference in their means is

$$t = \frac{(\overline{X}_1 - \overline{X}_2) - (\mu_1 - \mu_2)}{s_{\overline{X}_1 - \overline{X}_2}} \qquad \textbf{(11-11)}$$

in which the standard error of the difference, $s_{\overline{X}_1 - \overline{X}_2}$, is defined by Equation 11-6. Here, t has $n_1 + n_2 - 2$ degrees of freedom, where n_1 and n_2 are the sizes of the two samples.

Example

A marketing research firm has been hired to determine whether consumers' attitudes toward a newly designed toaster are more favorable than toward an earlier model. Only 30 of the newly designed toasters are available for the study. Consumers selected for the two samples use one of the two models for two weeks. Then they score the model on specified characteristics. High total scores are favorable. The scoring instrument is known to produce scores that are essentially normally distributed with equal standard deviations for all products being scored. The sample results are as follows.

	Old Model	*New Model*
Mean score	147 ($\overline{X}_1$)	158 ($\overline{X}_2$)
Sample size	40	30
Sample standard deviation	13.2	12.6

Make a one-tailed test of the null hypothesis that the new model is no better than the old. Assume population variances are equal. Use the 0.01 level of significance.

The hypothesis statement is

$$H_0: \mu_1 - \mu_2 \geq 0 \quad \text{and} \quad H_A: \mu_1 - \mu_2 < 0$$

The test statistic is t, as defined in Equation 11-11, with 40 + 30 − 2 = 68 degrees of freedom. At the 0.01 level of significance, the acceptance region for H_0 is

$$t \geq -2.3824$$

because this is a lower-tailed test.

From Equation 11-6,

$$s_{\bar{X}_1 - \bar{X}_2} = \sqrt{\frac{39(13.2^2) + 29(12.6^2)}{40 + 30 - 2}\left(\frac{1}{40} + \frac{1}{30}\right)} = 3.13$$

From Equation 11-11,

$$t_{\text{observed}} = \frac{(147 - 158) - (0)}{3.13} = \frac{-11}{3.13} = -3.51$$

The observed value of t is smaller than the minimum value in the acceptance region. The null hypothesis is rejected in favor of the alternative. At the 0.01 level, the new model had a mean score significantly greater than the mean score of the old model.

Nonnormal Populations with Unequal, Unknown Variances

The last case discussed in Section 11.1 involved two nonnormal populations that have unknown, unequal variances. Under such conditions, the independent samples from the two populations must be large so that the distribution of the differences in sample means is essentially normal.

> For two populations not normally distributed and with unknown, unequal variances, the statistic to test for a significant difference in their means is
>
> $$z = \frac{(\bar{X}_1 - \bar{X}_2) - (\mu_1 - \mu_2)}{s_{\bar{X}_1 - \bar{X}_2}} \qquad \textbf{(11-12)}$$
>
> in which z is a standard normal variate, the standard error of the difference is defined in Equation 11-8, and each sample contains at least 30 observations.

Example

In Connecticut, a major study* was conducted to determine whether increasing the

* Connecticut Public Utilities Control Authority, and others, *Connecticut Peak Load Pricing Test—Final Report.* Hartford, May 1977, pp. 148–152.

price of electricity during peak-load periods of the day and reducing the price during off-peak periods would reduce nonfarm residential consumption peaks. Two customer samples were used. The differential pricing described was applied in the experimental sample and traditional pricing was continued in the control sample. Each sample was subdivided into five strata based on the amount of electricity consumed.

There is a question of interest that is unrelated to simply shifting the timing of electrical consumption. For customers in the highest consumption stratum, did differential pricing have any effect on total consumption? We would like to see the daily pattern level off, provided there is no appreciable increase in overall consumption. It would also be worthwhile to know whether total consumption declined in conjunction with differential pricing. Although the question can be asked for any stratum, it is particularly important to ask it of the heavy users. For this stratum the results were as follows.

	Experimental Sample	*Control Sample*
Sample size	35	30
Sample mean*	30,066 ($\bar{x}_1$)	30,563 ($\bar{x}_2$)
Standard deviation(s)*	7,872	6,011

* June 1976 consumption in annual terms (KWH per year).

At the 0.05 level of significance, the hypothesis that mean consumption is equal in the two populations from which the samples came will be tested with a two-tailed test.

The hypothesis statement is

$$H_0\colon \mu_1 = \mu_2 \quad \text{and} \quad H_A\colon \mu_1 \neq \mu_2$$

The test statistic is given in Equation 11-12, where z is the standard normal variable. The acceptance region for H_0 at the 0.05 level of significance is

$$-1.96 \leq z \leq 1.96$$

From Equation 11-8,

$$s_{\bar{X}_1 - \bar{X}_2} = \sqrt{\frac{7872^2}{35} + \frac{6011^2}{30}} = 1724.8$$

The sample value of the test statistic is

$$z_{\text{observed}} = \frac{(30{,}066 - 30{,}563) - (0)}{1724.8} = -0.29$$

The observed value of z falls in the acceptance region for H_0. As measured during the peak summer month, total consumption for the average top-stratum customer in the experimental group did not differ significantly from that of the average top-stratum customer in the control group. The study did show, however, that there was a marked shift to off-peak period consumption for the experimental group that did not occur in the control group.

Exercises

11-7 *Computer operator training.* Exercise 11-1 presented before-and-after examination results for a computer operator training program. At the 0.01 level of significance, make an upper-tailed test of the null hypothesis that there was no increase in mean examination score.

11-8 *Metal working operations.* In Exercise 11-2, holes drilled in a metal plate are to be large enough for a brace to pass through. Conduct a test for which the hypothesis statement is

$$H_0: \mu_2 - \mu_1 \leq 0.15 \quad \text{and} \quad H_A: \mu_2 - \mu_1 > 0.15 \text{ cm.}$$

Here, μ_1 is the mean diameter of braces and μ_2 is the mean diameter of holes. Use the 0.005 level of significance.

11-9 *Advertising claim.* A consumer research organization wants to test the claim made in a television advertisement by a manufacturer of a leading headache remedy. The claim is that the advertised brand contains more of the active ingredient than does another brand named in the ad. The consumer group assumes the claim refers to aspirin. The sample information for the test is as follows.

	Advertised Brand	*Competing Brand*
Sample size	24 tablets	21 tablets
Mean (grains)	5.53 ($\bar{x}_1$)	5.04 ($\bar{x}_2$)
Standard deviation	0.22 (s_1)	0.13 (s_2)

Do the sample results support the claim? Conduct a one-tailed test at the 0.01 level of significance. Assume normally distributed populations with equal variances.

11-10 *Collective bargaining.** A study was made of criteria when collective bargaining over firefighters' wages reached an impasse and was submitted to either compulsory arbitration or a fact-finding procedure, depending on the collective-bargaining law. The change in the wage offer from the initial bargaining to the point of impasse was measured as a percent of the existing wage level, with the following results.

Type of Law	*Management*		
	n	$\bar{x}$	s
Compulsory arbitration	25	+1.14	2.49
Fact-finding	21	+2.69	2.96

Is this evidence that managements behave differently under a prospect of compulsory arbitration from when the law calls for a fact-finding procedure? Assume normal populations with equal variances. Make a two-tailed test at the 0.05 significance level.

* H. N. Wheeler, *Industrial Relations*, February 1978.

11-11 *Housing prices.* A large electronics firm is considering establishment of a branch plant in an urban center where it presently has no facilities. Higher mean housing prices will cause the firm to make higher transfer compensation to over 100 key employees. Random samples of appropriate housing for sale in the new location and in the one where the employees now live are selected from lists compiled from newspapers, multiple-listing publications, and similar sources. Sample results are as follows.

	n	$\bar{x}$	s
Present location	150	$82,500	$5000
New location	180	$84,750	$6000

Assume that the two populations are positively skewed with unequal variances. Conduct a one-tailed test at the 0.05 level of significance to determine whether the mean price of housing in the new location is significantly greater than in the present location.

11-12 *Population growth versus stabilization.*† County commissioners in Georgia and Florida responded on a scale from 1 to 13 to indicate where they stood on stabilization (1) versus growth (13) as policy principles in their counties. The following results were recorded.

	n	$\bar{x}$	s
Georgia	135	8.7	2.0
Florida	114	5.8	3.0

Assume these are random sampling results from populations that have different variances and are not normally distributed. At the 0.01 level of significance, make a two-tailed test of the hypothesis that there is no difference in mean response from the two states.

11.3 Inferences from Differences in Matched Pairs

For the two-sample inferences in Sections 11.1 and 11.2, the observations were independently selected. The values in one sample were selected randomly from one population. Then, independently, sample values were selected randomly from the second population. In this section, observations within the two samples are not independent. Instead, the two samples consist of related *pairs* of observations. In many such cases, one sample is then subjected to a treatment different from that used for the other. The variable of interest is the combined difference in response for all the pairs in the two samples.

† V. L. Marando and R. O. Thomas, *Social Sciences Quarterly*, June 1977.

The first technique for obtaining matched, or related, pairs is to make repeated measurements on the same unit of observation. For example, a new error-checking procedure is being considered for those who write sales tickets in a restaurant chain. Twelve people who write such tickets are selected to test the new procedure. Three weeks' tickets prepared by each person before the new procedure was installed are collected and three weeks' tickets are collected after installing it. Then, for each person in the study, the error rate after installing the new procedure is subtracted from the error rate before installing it. The two error rates and the differences are shown in the first three columns of Table 11-1 for each of the 12 persons in the study.

Table 11-1 Error Rates for Matched Pairs Before and After Installing the New Procedure

		Errors per 100 Tickets			
Employee	*Before* (X_1)	*After* (X_2)	$d = X_1 - X_2$	$(d - \bar{d})^2$	d^2
1	14.5	7.3	7.2	12.25	51.84
2	6.1	6.5	−0.4	16.81	0.16
3	8.9	3.5	5.4	2.89	29.16
4	5.8	4.3	1.5	4.84	2.25
5	13.0	6.8	6.2	6.25	38.44
6	6.4	5.7	0.7	9.00	0.49
7	12.0	10.2	1.8	3.61	3.24
8	13.2	6.9	6.3	6.76	39.69
9	11.3	7.0	4.3	0.36	18.49
10	2.8	3.2	−0.4	16.81	0.16
11	10.2	4.9	5.3	2.56	28.09
12	11.0	4.5	6.5	7.84	42.25
			44.4	89.98	254.26

$n = 12$ pairs $\quad \bar{d} = \frac{44.4}{12} = 3.7$

$$s_d = \sqrt{\frac{\Sigma(d - \bar{d})^2}{n - 1}} = \sqrt{\frac{89.98}{11}} = 2.86$$

or

$$s_d = \sqrt{\frac{\Sigma d^2 - (\Sigma d)^2/n}{n - 1}} = \sqrt{\frac{254.26 - (44.4^2/12)}{11}} = 2.86$$

Notice that—in general—employees above the median error rate before the new procedure was installed are also above the median after, and vice versa. Basic personal differences tend to keep people at about the same relative positions in the before-and-after groups. As a result, the two samples are not independent, and the techniques of Sections 11.1 and 11.2 do not apply. Instead, it can be assumed that subtracting the "after" observation from the "before" has removed practically all effects of factors other than that of the new procedure from the differences, d.

We can then find the mean difference

$$\bar{d} = \frac{\Sigma d}{n} \tag{11-13}$$

which is shown in Table 11.1 to be 3.7 for our example. On the average, there are 3.7 less errors per 100 tickets with the new procedure. The central limit theorem indicates that the sampling distribution of $\bar{d}$ can be considered to be normal for reasonably large samples.

Typically, the population standard deviation of the differences is not known. Consequently, the sample estimate, s_d, must be used. In turn, this will be used to estimate the standard error of the mean of the differences, $\sigma_{\bar{d}}$. The estimated standard deviation of the distribution of differences, d, is either

$$s_d = \sqrt{\frac{\Sigma(d - \bar{d})^2}{n - 1}} \tag{11-14a}$$

or

$$s_d = \sqrt{\frac{\Sigma d^2 - (\Sigma d)^2/n}{n - 1}} \tag{11-14b}$$

where the latter form usually results in less rounding error. Table 11-1 shows that $s_d = 2.86$ for our example. It follows that the standard error for the mean difference is

$$s_{\bar{d}} = \frac{s_d}{\sqrt{n}} \tag{11-15}$$

When the differences can be assumed to be normally distributed, *the confidence interval estimate of the population mean difference, μ_d, is*

$$\bar{d} - t\left(\frac{s_d}{\sqrt{n}}\right) \leq \mu_d \leq \bar{d} + t\left(\frac{s_d}{\sqrt{n}}\right) \tag{11-16}$$

where s_d is defined in Equation 11-14. Here t has $n - 1$ degrees of freedom, where n is the number of differences.

For the program to reduce the error rate in sales tickets, there are 12 observations and 11 degrees of freedom for t. For a 0.95 confidence interval estimate of the mean difference in error rates, Appendix Table A-4 gives t as 2.201. Then the interval estimate is

$$\bar{d} \pm t\left(\frac{s_d}{\sqrt{n}}\right) = 3.7 \pm 2.201\left(\frac{2.86}{\sqrt{12}}\right)$$

$$= 3.7 \pm 1.82$$

or 1.88 to 5.52. The new procedure results in an estimated mean-error reduction of from 1.88 to 5.52 errors per 100 sales tickets. Since the interval lies entirely above zero, it can be concluded that the new procedure has made a reduction in the mean-error rate.

The second technique for obtaining pairs for a study of differences does not make use of repeated measurements on the same individual or unit of observation. Instead, pairs of such units are matched on one or more extraneous factors that affect the variable being observed. Then one member of each pair is randomly assigned to each of the two samples.

As an example of the second technique, an engineer for an oil company is comparing a newly developed unleaded gasoline with the company's current unleaded gasoline. Instead of a confidence interval estimate, the engineer wants the comparison to consist of testing the null hypothesis that the mean miles per gallon (mpg) with the new gasoline is no greater than with the present gasoline. The significance level is 0.05. The alternative hypothesis is that mileage with the new gasoline is greater. Rather than making the test with two independent samples of automobiles, the engineer selects a pair of identical automobiles in each of five weight classes. One member of each pair is assigned the new gasoline at random and the other uses the present gasoline. Note that each pair consists of two units of observation; in the sales ticket case, the same unit of observation was used twice.

Table 11-2 gives the data for the gasoline comparison study. The sample mean difference is 1.8 mpg. The new gasoline produced 1.8 more miles per gallon, on the average, than the present type.

As before, sample differences will be assumed to be normally distributed.

> With normally distributed differences from matched pairs, the statistic
>
> $$t = \frac{\bar{d} - \mu_d}{s_d/\sqrt{n}} \qquad \textbf{(11-17)}$$
>
> belongs to a t-distribution with $n - 1$ degrees of freedom. Here, μ_d is the expected value of the mean difference, $\bar{d}$, and s_d is the standard error of the differences as defined in Equation 11-14, for n differences.

For the example involving gasoline mileage, the hypothesis statement is

$$H_0\colon \mu_d \leq 0 \quad \text{and} \quad H_A\colon \mu_d > 0$$

The test statistic is t with $5 - 1 = 4$ degrees of freedom where t is defined in Equation 11-17. For the 0.05 level of significance, the critical value of t is +2.132, and H_0 will be accepted if the observed value of t is 2.132 or less. From Equation 11-14, s_d = 0.65, as shown in Table 11-2. From Equation 11-17,

$$t_{\text{observed}} = \frac{1.8 - (0)}{0.65/\sqrt{5}} = 6.19$$

The observed value of t is much greater than the upper limit of the acceptance region for the null hypothesis. At the 0.05 level of significance, the new gasoline produces a greater mean number of miles per gallon than does the company's present gasoline.

Table 11-2 Mileages from Two Types of Gasoline for Matched Pairs

Automobile Weight Class	*Gasoline* New X_1	Present X_2	$d = X_1 - X_2$	$(d - \bar{d})^2$	d^2
Very Heavy	11.3	9.5	1.8	0	3.24
Heavy	13.4	12.5	0.9	0.81	0.81
Medium	15.7	13.7	2.0	0.04	4.00
Light	17.6	14.9	2.7	0.81	7.29
Very Light	19.0	17.4	1.6	0.04	2.56
			9.0	1.70	17.90

$$n = 5 \qquad \bar{d} = \frac{9.0}{5} = 1.8$$

$$s_d = \sqrt{\frac{1.70}{4}} = 0.65 \quad \text{or} \quad s_d = \sqrt{\frac{17.90 - (81/5)}{4}} = 0.65$$

When it can be applied, the design using matched pairs offers a distinct advantage over the design involving independent samples. The advantage can be illustrated by referring to the example of the gasoline mileage. Suppose two independent random samples of the same size had been selected from the entire population of automobiles and that no attempt had been made to match the pairs of automobiles before finding a column of differences as in Table 11-2. This procedure is the equivalent, for two samples of equal size, of the procedures described in Sections 11.1 and 11.2 for two independent samples. With unmatched pairs, a very heavy automobile using the new gasoline could have been paired with a very light automobile using the present gasoline. Typically, the difference in mileage would favor the present gasoline—not because the present gasoline is better, but because the car with the present gasoline weighs less. The extraneous factor, weight, has overshadowed the superiority of the new gasoline. Of course, such an unfavorable match would not always occur, but weight fluctuations present in the differences would make it more difficult to detect the real effect of the new gasoline.

By matching cars by weight, the effect of this factor is removed as a source of variability in the differences. The standard error for such differences does not include variability caused by weight differences.

It is not always possible to identify important extraneous factors that are related to the variable being studied and match pairs on such a factor. For instance, there are no obvious factors for matching pairs of flour sacks filled by two different filling machines that are set to the same mean fill weight. Testing the hypothesis that the mean fill weights are equal would make use of two independent random samples. Erroneously using matched pairs here puts one in a worse position because the degrees of freedom for the test are reduced from $2(n - 1)$ to $n - 1$.

11.4 Inferences from Differences in Two Proportions

Statistical inferences about the difference in two population proportions are very similar in procedure to those for the difference in two population means, provided that the sample proportions come from samples large enough for the normal approximation to apply.

In this section, all procedures are based on the assumption that the normal distribution is a satisfactory approximation to the sampling distribution under discussion.

Estimation of the Difference in Two Population Proportions

The procedure for making a confidence interval estimate of the difference in two population proportions begins with independent random samples from the two populations. It is assumed that the sampling distribution of the differences in sample proportions is essentially normal. To justify the normality assumption both n_1 and n_2 should be at least 100, both n_1p_1 and n_2p_2 should be at least 50, and both $n_1(1 - p_1)$ and $n_2(1 - p_2)$ should be at least 50.

Given two sample proportions, we must estimate the standard error of the difference in two population proportions. Because these two population proportions need not be equal, the estimate is

$$s_{P_1 - P_2} = \sqrt{s_{p_1}^2 + s_{p_2}^2} \qquad \textbf{(11-18a)}$$

where

$$s_{P_1}^2 = \frac{p_1(1 - p_1)}{n_1 - 1} \qquad \textbf{(11-18b)}$$

and

$$s_{P_2}^2 = \frac{p_2(1 - p_2)}{n_2 - 1} \qquad \textbf{(11-18c)}$$

In these equations, p_1 and p_2 are the two sample proportions based on samples of n_1 and n_2 observations, respectively. It is assumed that each of the two samples constitutes less than 10% of its population so that the finite population multiplier can be ignored.

> When the differences in sample proportions, $p_1 - p_2$, can be assumed to be normally distributed, the confidence interval estimate of the difference in the population proportions, $\pi_1 - \pi_2$, is
>
> $$(p_1 - p_2) - zs_{P_1 - P_2} \leq \pi_1 - \pi_2 \leq (p_1 - p_2) + zs_{P_1 - P_2} \qquad \textbf{(11-19)}$$
>
> where the standard error of the difference, $s_{P_1 - P_2}$, is defined in Equation 11-18 and z is the standard normal variate for the appropriate confidence level.

Example

An industrial designer has created two package designs for a new cold breakfast cereal. A marketing executive believes the proportions of consumers favoring one of the designs over the other may differ in the eastern and western regions of the nation. A package preference study has been carried out in the two regions. In the East, 420 people in a random sample of 500 preferred design A to design B. In the West, 219 people in an independent random sample of 300 preferred A to B. What is the 0.99 confidence interval estimate of the difference in population proportions for the two regions?

Let $p_1 = 420/500$, or 0.84, and let $p_2 = 219/300$, or 0.73. Then, from Equation 11-18,

$$s_{P_1}^2 = \frac{0.84(0.16)}{499} \qquad s_{P_2}^2 = \frac{0.73(0.27)}{299}$$

and

$$s_{P_1 - P_2} = \sqrt{\frac{0.84(0.16)}{499} + \frac{0.73(0.27)}{299}} = 0.0305$$

For a 0.99 confidence level, $z = 2.58$. Using Equation 11-19, the 0.99 confidence interval estimate of the difference in population proportions is

$$(0.84 - 0.73) - 2.58(0.0305) \leq \pi_1 - \pi_2 \leq (0.84 - 0.73) + 2.58(0.0305)$$

which is 0.11 ± 0.079, or 0.031 to 0.189. The percentage of people preferring design A in the East exceeds the percentage preferring that design in the West by from 3.1 to 18.9.

Hypothesis Test for the Equality of Two Population Proportions

Testing a hypothesis about the difference in two population proportions, rather than making an interval estimate, requires only slight modification of the estimation procedures. To test the null hypothesis

$$H_0\colon \pi_1 = \pi_2$$

the standard error of the difference in proportions is estimated by

$$s_{P_c} = \sqrt{p_c(1 - p_c)\left(\frac{1}{n_1} + \frac{1}{n_2}\right)} \tag{11-20a}$$

where

$$p_c = \frac{n_1 p_1 + n_2 p_2}{n_1 + n_2} \quad \text{or} \quad p_c = \frac{r_1 + r_2}{n_1 + n_2} \tag{11-20b}$$

where r_1 and r_2 are the number of successes in the two samples. Then the test statistic is the standard normal variate

$$z = \frac{(p_1 - p_2) - (0)}{s_{P_c}} \tag{11-21}$$

Once again, the sample sizes should be large. Both n_1p_c and n_2p_c should be at least 5, and both $n_1(1 - p_c)$ and $n_2(1 - p_c)$ should be at least 5. In Equation 11-20(b), p_c is the pooled estimate of the common population proportion posited by the null hypothesis. Given that hypothesis, the two populations are assumed to have the same proportion, and the test is conducted to see whether that assumption is justified.

Example

On a section of interstate highway far removed from any urban center, the proportions of eastbound and westbound automobile traffic exceeding 55 mph are to be compared by testing the null hypothesis that the two population proportions are equal. Observers are placed out of sight and collect observations during short randomly selected periods during daylight hours. The process is repeated in other nearby locations. The sample observations are as follows.

Number of Observations	*Eastbound Traffic*	*Westbound Traffic*
Exceeding 55 mph	81	90
Total	125	150

Using a two-tailed test, test the null hypothesis of equal proportions at the 0.05 level of significance.

Let $p_1 = {}^{81}/_{125}$, or 0.648, and let $p_2 = {}^{90}/_{150}$, or 0.600. The hypothesis statement is

$$H_0\colon \pi_1 = \pi_2 \quad \text{and} \quad H_A\colon \pi_1 \neq \pi_2$$

or, alternatively,

$$H_0\colon \pi_1 - \pi_2 = 0 \quad \text{and} \quad H_A\colon \pi_1 - \pi_2 \neq 0$$

The test statistic is z, which has the standard normal distribution. For a two-tailed test at the 0.05 level of significance, the acceptance region for the null hypothesis is $-1.96 \leq z \leq +1.96$. From Equation 11-20,

$$p_c = \frac{0.648(125) + 0.600(150)}{125 + 150} = \frac{81 + 90}{275} = 0.6218$$

Then, using fractions for p_c and $1 - p_c$ to reduce rounding error,

$$s_{p_c} = \sqrt{{}^{171}/_{275}({}^{104}/_{275})({}^{1}/_{125} + {}^{1}/_{150})} = 0.0587$$

From Equation 11-21,

$$z = \frac{(0.648 - 0.600) - (0)}{0.0587} = 0.818$$

that is,

$$z_{\text{observed}} = 0.818$$

The observed value of z falls within the acceptance region for H_0. At the 0.05 level, there is no significant difference in the proportions of eastbound and westbound traffic exceeding 55 mph. If efforts are to be made to reduce the percentages exceeding the speed limit, it would be reasonable to use equal effort in both directions of traffic flow.

Exercises

11-13 *Assembly production.* A number of changes designed to improve productivity have been made in assembly procedures for a component of an AM-FM radio. The mean numbers of assemblies completed per hour during a five-day period before the changes and a similar period after the changes were recorded for each of 8 assemblers.

	Assemblies per Hour	
Assembler	*Before*	*After*
1	37	40
2	41	40
3	28	32
4	32	35
5	38	41
6	35	37
7	30	36
8	34	38

What is the 0.95 confidence interval estimate of the change in mean number of assemblies per hour?

11-14 *Telephone sales appointments.* The employees of a marketing firm telephone people at their homes for the purpose of making an appointment for a subsequent product demonstration and possible purchase. Two different telephone messages are being tried. The two messages have been used alternately by employees for the past week. Random samples of reports from 100 calls have been selected for each employee with each message. The numbers of appointments made per 100 calls are given below.

Employee	*Message A*	*Message B*	*Employee*	*Message A*	*Message B*
1	8	10	10	8	15
2	35	18	11	14	12
3	7	19	12	28	14
4	15	1	13	36	11
5	17	7	14	27	3
6	39	25	15	41	29
7	21	10	16	24	16
8	36	19	17	5	20
9	19	13	18	24	6

a. Should these be treated as independent samples or as matched pairs? Why?
b. At the 0.90 level, what is the confidence interval estimate of the difference in mean number of appointments per 100 calls?

11-15 *Real estate training.* A firm prepares prospective real estate agents for the licensing examination given by a state agency. The lecture method has been used in the course and is in use at present. Recently, a programmed-learning version of the course has been prepared. The firm wants a comparison of the two approaches. Twelve matched pairs of beginning students were selected on the basis of verbal and arithmetic reasoning tests given before the students began the course. Upon completion of the course, students were given a practice licensing examination. The scores for the twelve matched pairs were as follows.

Pair	*Examination Scores*	
	Programmed Course	*Lecture Course*
1	33	38
2	60	59
3	59	63
4	94	92
5	89	87
6	68	62
7	81	74
8	82	78
9	74	72
10	79	71
11	88	83
12	55	49

Test the null hypothesis that the mean score with the programmed learning approach is no greater than with the lecture approach. The alternative hypothesis is that the mean score with the programmed learning approach is greater than with the lecture approach. The level of significance is to be 0.05.

11-16 *Brand preference survey.* The manufacturer of a leading brand of toothpaste hires a market research firm to make an annual survey of brand preference. A random sample of consumers is chosen. Each person is shown pictures of the five leading brands and is asked to choose the preferred brand. Pictures are presented to subjects in different orders to prevent bias from that source. In last year's survey, 760 of 1900 persons in the sample preferred the manufacturer's brand. In this year's survey, 990 of 2200 persons preferred it. What is the 0.95 confidence interval estimate of the change in the proportion preferring the brand in question?

11-17 *Automobile seat belt usage.** Seat belt usage was observed in a large study of 1973 and later models of automobiles. For Volvos, 233 of 523 drivers were observed to be using their seat belt systems. The next highest seat belt use was among drivers of

* Contract No. DOT-HS 6-01340, January 1978.

Dodge Colts, where 90 of 266 drivers were wearing belts. Find the 0.90 confidence interval estimate for the difference in percentage use between Volvo and Colt drivers.

11-18 *Health plan preference.* The employee benefits section of a company that has several thousand employees is developing a new group health benefits plan for all wage- and salaried workers. A comparison sheet has been printed to show the potential costs and benefits under the proposed plan and under the present plan. In a random sample of 400 employees on wages, 264 prefer the proposal to the present plan. In a random sample of 320 salaried employees, 192 prefer the proposal. At the 0.01 level of significance, make a two-tailed test of the hypothesis that the proportions preferring the proposal are equal for the two groups.

Summary

$\mu_1 - \mu_2$; Independent Samples

Normal populations with known variances

Confidence interval:

$$(\bar{x}_1 - \bar{x}_2) \pm z\sigma_{\bar{X}_1 - \bar{X}_2}$$

Test statistic:

$$z = \frac{(\bar{X}_1 - \bar{X}_2) - (\mu_1 - \mu_2)}{\sigma_{\bar{X}_1 - \bar{X}_2}}$$

In both of the above,

$$\sigma_{\bar{X}_1 - \bar{X}_2} = \sqrt{\frac{\sigma_1^2}{n_1} + \frac{\sigma_2^2}{n_2}}$$

Normal populations with equal, unknown variances

Confidence interval:

$$(\bar{x}_1 - \bar{x}_2) \pm ts_{\bar{X}_1 - \bar{X}_2}$$

Test statistic:

$$t = \frac{(\bar{X}_1 - \bar{X}_2) - (\mu_1 - \mu_2)}{s_{\bar{X}_1 - \bar{X}_2}}$$

In both of the above, t has $n_1 + n_2 - 2$ degrees of freedom and

$$s_{\bar{X}_1 - \bar{X}_2} = \sqrt{\frac{(n_1 - 1)s_1^2 + (n_2 - 1)s_2^2}{n_1 + n_2 - 2}\left(\frac{1}{n_1} + \frac{1}{n_2}\right)}$$

Nonnormal populations with unknown variances (n_1 and $n_2 \geq 30$)

Confidence interval:

$$(\bar{x}_1 - \bar{x}_2) \pm zs_{\bar{X}_1 - \bar{X}_2}$$

Test statistic:

$$z = \frac{(\bar{X}_1 - \bar{X}_2) - (\mu_1 - \mu_2)}{s_{\bar{X}_1 - \bar{X}_2}}$$

In both of the above,

$$s_{\bar{X}_1 - \bar{X}_2} = \sqrt{\frac{s_1^2}{n_1} + \frac{s_2^2}{n_2}}$$

$\mu_1 - \mu_2$; Matched Pairs with Normally Distributed Differences (d)

Confidence interval:

$$\bar{d} \pm t\frac{s_d}{\sqrt{n}}$$

Test statistic:

$$t = \frac{\bar{d} - \mu_d}{s_d/\sqrt{n}}$$

In both of the above, n is the number of pairs, t has $n - 1$ degrees of freedom, and

$$s_d = \sqrt{\frac{\Sigma(d - \bar{d})^2}{n - 1}}$$

$\pi_1 - \pi_2$; Proportions (n_1 and n_2 must be large; see text)

Confidence interval:

$$(p_1 - p_2) \pm z s_{P_1 - P_2}$$

where

$$s_{P_1 - P_2} = \sqrt{s_{p_1}^2 + s_{p_2}^2}$$

and

$$s_{P_i}^2 = \frac{p_i(1 - p_i)}{n_i - 1}$$

Test statistic:

$$z = \frac{(p_1 - p_2) - (\pi_1 - \pi_2)}{s_{P_c}}$$

where

$$s_{P_c} = \sqrt{p_c(1 - p_c)\left(\frac{1}{n_1} + \frac{1}{n_2}\right)}$$

and

$$p_c = \frac{r_1 + r_2}{n_1 + n_2}$$

with r_1 and r_2 the number of successes in each sample.

Supplementary Exercises

11-19 *Commercial truck weight.* A state highway department engineer is estimating the difference in the mean gross weight of trucks using a northern and a southern interstate highway in the state. The mean gross weight for a random sample of 60 trucks using the northern interstate was 13.5 t with a standard deviation of 3.6 t. The mean gross weight for a random sample of 40 trucks using the southern interstate was 16.3 t with a standard deviation of 2.9 t. Construct the 0.95 confidence interval estimate of the difference in mean weight between the southern and northern interstates.

11-20 *Children's television programs.* Surveys of households during two children's television programs were conducted to determine the viewing behavior of children under 12 years old. The results of the two surveys were as follows.

	Program A	*Program B*
Number of households in which children under 12 were watching	108	96
Number of households in survey	600	480

At the 0.05 significance level, test the hypothesis that there is no difference in viewing behavior for the two programs. Use a two-tailed test.

11-21 *Weight gain in pigs.* A commercial pig farm is comparing the mean weight gains for pigs on two different feeding programs. A matched-pairs test is carried out by using littermates for each pair. The gains, in pounds, for 45 days are as follows.

Feeding Program	*Pair*							
	A	*B*	*C*	*D*	*E*	*F*	*G*	*H*
I	93	79	74	90	58	87	57	63
II	75	84	62	80	44	78	58	70

At the 0.10 level of significance, conduct a two-tailed test of the null hypothesis that there is no difference in mean weight gain for the two programs.

11-22 *Electronic component life.* Two types of machine can be purchased to manufacture an electronic component. Type A costs more than type B but will be the best buy if the mean life of the component it makes is at least 60 h greater than that made from the type B machine. Type A and B machines are known to have standard deviations of 5.0 and 6.2 h, respectively, for their component lifetime distributions. Fifteen components produced by type A machines had a mean lifetime of 1556 h, while 25 components produced by type B machines had a mean lifetime of 1490 h. Conduct the correct one-tailed test to decide which type of machine to buy. Use the 0.01 significance level and assume that the two lifetime distributions are normal.

Computer Exercises

1 Source: Real estate data file.
For all parts of this exercise, treat the real estate data as if they are a random sample from a large universe of sales in the community.
   a. At the 0.05 significance level, test the null hypothesis that the mean home price paid by new residents is the same as the mean price paid by old residents of the community.
   b. Find the 95% confidence interval estimate of the difference between mean prices paid for old and new homes.
   c. An economist maintains that purchase prices of homes average 2.5 times annual income regardless of income level. Construct the variable "2.5 times income," and conduct a matched-pairs test for mean difference between purchase price and 2.5 times income.

2 Source: Personnel data file. Assume it is the population of interest.
   a. Use a random sample of 30 from each group to find the 0.95 confidence interval estimate of the difference in mean salaries for those who do and those who do not participate in the tax shelter plan.
   b. Use a random sample of 25 from each group to test the null hypothesis that persons with graduate degrees receive the same mean salary as do persons with bachelor's degrees at the 0.05 level of significance.
   c. Repeat (b) using a matched pairs t-test. Use number of years since highest degree as the basis for matching. (*Hint*: Sort the data by degree type and within these groups sort by number of years since receiving highest degree.)

Interpreting Computer Output

Computer packages usually contain procedures for performing many of the analyses discussed in this chapter. The following shows the results of a t-test for a difference between the means of two groups. In the personnel data file, the two groups are those who participate in a tax deferral plan (Y) and those who do not (N). The null hypothesis is that the mean salary difference for the two groups is 0, that is, that the two groups have the same mean salary.

PART I

VARIABLE: SALARY

DEFER	N	MEAN	STD DEV	STD ERROR
N	53	51243.9622642	3953.16534303	543.00902090
Y	180	52324.6055556	5023.17882115	374.40564357

PART II

VARIANCES	T	DF	PROB >|T|
UNEQUAL	-1.6384	106.2	0.1043
EQUAL	-1.4396	231.0	0.1513

PART III

```
FOR HO: VARIANCES ARE EQUAL, F'=       1.61 WITH 179 AND 52 DF
        PROB > F'=0.0448
```

The three parts of the table show relevant parts of the output. Part 1 gives the sample statistics for the two groups: the mean salaries, the standard deviations, and the standard errors of the means. Part 2 shows the values of t, the degrees of freedom, and the P-values for the hypothesis test under two different assumptions. If the population variances are unequal, the t statistic is -1.6348 with 106.2 degrees of freedom and the P-value is 0.1043. We would reject the null hypothesis of equal mean salaries for levels of significance of 0.1043 or less. For equal population variances, the P-value is 0.1513 so we would reject the null hypothesis for levels of significance of 0.1513 or less.

Part 3 presents the results of an F test of the null hypothesis that the two populations have equal variances. Although we will not cover the F distribution until Chapter 13, the test is easy to interpret. Small P-values indicate that the hypothesis of equal variances should be rejected. Here, the P-value 0.0448 indicates that we would reject the hypothesis of equal variances for any significance level equal to or less than 0.0448. If, at the outset, we had selected the 0.05 level of significance we would reject the hypothesis of equal variances. If we had selected the 0.01 level, we would not reject it. In either case, the associated t statistic in Part 2 is not significant at the 0.10 level. There is no evidence here to support a significant difference in mean salary.

12

ANALYSIS OF FREQUENCIES

In Chapter 9 we learned how to test a hypothesis about a population proportion based on a random sample of observations, and in Chapter 11 we saw how inferences could be made about the difference between two population proportions. In this chapter we shall take up further inference procedures often required for nominal data. The situations giving rise to these procedures are briefly described below.

1. Multinomial Distribution Tests for proportions thus far have been based on the binomial distribution. Nominal data for which binomial tests apply are restricted to two classes. For example, suppose we wish to test that the proportion of beer drinkers who have ever tried a light beer is 0.50 or more. The underlying binary classification is triers and nontriers. But suppose a taste preference test involves three brands of light beer and the hypothesis that the proportion of light-beer drinkers preferring each brand is $1/3$. Because there are three classes rather than two, we need an inference procedure that extends to three or more classes.

2. Association Cross-classification tables in Chapter 4 introduced the idea of association in nominal data. Tests for association in interval data were discussed in Chapter 10, but we need such a test for nominal data. Such a test would allow us to decide whether preference for light over regular beer was associated with the region of residence of beer drinkers in the United States given an appropriate national sample.

3. Multiple Samples So far we have applied tests for proportions to only one or two samples at a time. For example, we might test that 30% of beer drinkers in Chicago prefer light to regular beer. Suppose, however, that random samples of beer preference in Chicago, Cleveland, and Minneapolis are available. The question has been raised whether the 30% preference for light beer prevails in all three cities. This involves *multiple samples*.

4. Goodness-of-Fit We encounter analysis of frequencies in interval data as well as nominal data. The particular case in point is the frequency pattern of an interval (or ratio) variable. The question is whether, given observed sample data, we can conclude that the frequency distribution of the population follows a specified theoretical probability distribution. For example, we might wonder whether monthly beer consumption of beer consumers in Chicago can be considered to be normally distributed, or whether the number of male pups in five-pup litters of thoroughbred Great Danes can be considered to be binomially distributed. If sample data establish the *goodness-of-fit* of a probability distribution model to the observed data, then the model can be used in further analysis.

12.1 Test for Multinomial Distribution

Suppose 150 randomly selected beer consumers are asked to taste three brands of light beer and then indicate their preferences for one of the brands. To avoid positional bias, the various possible orders in which the beers can be tasted are randomly assigned to equal numbers of the subjects. The data are as follows.

Brand	*Number Preferring*
A	68
B	43
C	39
Total	150

Hypothesis

As with tests involving binary distributions, we can test any hypothesis about the relative frequencies of these classes in the population. A possible hypothesis would be

$$\pi_A = 0.50 \qquad \pi_B = 0.30 \qquad \pi_C = 0.20$$

where π, as before, stands for a proportion. Because the three proportions in our example must have a sum of 1.0, the hypothesis can be completely stated by specifying any two of the proportions. In a binomial test, we did not state the proportion of failures because it is always known when the proportion of successes is specified for the hypothesis. In the binomial test, there was only one *independent* parameter—the proportion of successes. In a multinomial test, the number of independent parameters is one less than the number of classes involved.

Suppose the sponsor of the taste preference test is interested in whether consumer preference for the several brands is equal. Further, the sponsor is willing to run a 0.05 risk of rejecting this hypothesis when it is true. This means the sponsor is willing to accept a one-in-twenty chance of concluding the brands are not equally preferred when they really are. We then have, for the null hypothesis and significance level,

$$H_0\colon \pi_A = \pi_B = \pi_C = 1/3 \qquad \alpha = 0.05$$

Expected Frequencies

Once a null hypothesis has been specified, we can derive a set of *expected* frequencies under the hypothesis. We reason that if H_0 were true, repeated samples of 150 consumers would produce $150(1/3) = 50$ preferences for each brand *on the average*. We thus have now a table of *observed* frequencies and a table of frequencies *expected* under the null hypothesis.

Brand	*Preference* Observed f_o	Expected f_e
A	68	50
B	43	50
C	39	50
Total	150	150

We can generalize how to determine the expected frequencies in a multinomial test by saying that the *i*th expected frequency in a multinomial test with *r* classes is

$$(f_e)_i = n\pi_i \qquad \text{where} \qquad i = 1, 2, \ldots, r \tag{12-1}$$

For our example,

$$(f_e)_2 = (f_e)_B = n\pi_B = 150(1/3) = 50$$

Chi-Square Test Statistic

At this point we have a set of observed frequencies and a set of frequencies that would be expected if the null hypothesis about the set of relative frequencies (proportions) were true. If the two sets agree, our null hypothesis is supported by the data; if the two sets disagree, the null hypothesis lacks support. For example, if the sample frequencies had been 49 for A, 51 for B, and 50 for C, the actual frequencies would be considered to be close to the expected frequencies. The sample evidence would be taken as supporting the null hypothesis of equal preference in the population.

But how different from the expected frequencies can the observed frequencies be and still support the null hypothesis? This question can be answered only in terms of a test statistic in relation to the α-level specified along with the null hypothesis. A significant difference between observed and expected frequencies is a difference that has a small (less than α) probability of being exceeded. If the probability associated with the sample outcome is less than the predesignated α-level, we reject the null hypothesis. Otherwise the null hypothesis is accepted. Remember that the larger the difference, the smaller the probability of the difference being exceeded when H_0 is true.

The following measure of difference between observed and expected frequencies for a multinomial distribution follows a probability distribution form called a *chi-square* (χ^2) distribution.

$$\chi^2_{df} = \sum\left[\frac{(f_o - f_e)^2}{f_e}\right] \tag{12-2}$$

where $df = r - 1$.

From the definition of the measure, we see that chi-square will be zero if the observed frequencies correspond to the expected frequencies. The chi-square statistic takes on larger values as the observed frequencies differ more from the expected frequencies. If the null hypothesis were true, different repeated random samples would lead to different values for χ^2. The probability distribution of these values is known for any specified value of *df*, the *degrees of freedom*. The number of degrees of freedom is one less than the number of classes in a multinomial distribution table.

Chi-Square Table The degrees of freedom, *df*, can be thought of as the number of independent opportunities for differences between observed and expected frequencies. There is a different chi-square distribution for each value of *df*. The chi-square distributions for various degrees of freedom are tabulated in Table A-5 of Appendix A. The column headings are lower-tail cumulative probabilities; the body of the table contains values of χ^2 that cut off these cumulative probabilities for different values of *df*. For example, the graph at the top of Table A-5 shows the chi-square distribution for 5 degrees of freedom. The probability is 0.90 that a value of χ^2 selected at random from this distribution will be equal to or less than 9.24. The value 9.24 appears in the table under 0.90 in the row for 5 degrees of freedom.

Figure 12-1 shows the density functions for the chi-square distributions for 5 and 10 degrees of freedom. The mean of a chi-square distribution is equal to the degrees of freedom. Chi-square distributions are skewed to the right but, as the figure shows, the skewness diminishes as the degrees of freedom increase.

Chi-Square Multinomial Test

We are now ready to carry out the hypothesis test for our example. Table 12-1 shows the original sample data, together with the expected frequencies and the calculations to obtain the observed value of the chi-square statistic. Recall that the null hypothesis to be tested is

$$H_0\colon \pi_A = \pi_B = \pi_C = 1/3$$

with an acceptable risk level of $\alpha = 0.05$.

The criterion, or critical, chi-square value stems from the α-level and the degrees of freedom. In our example with three classes, $r = 3$ and $df = r - 1 = 2$. The chi-square test for agreement between observed and expected frequencies is an upper one-tailed test. That is, we will reject the null hypothesis that gave rise to the expected frequencies only when the measure of difference, χ^2 (observed), is so large

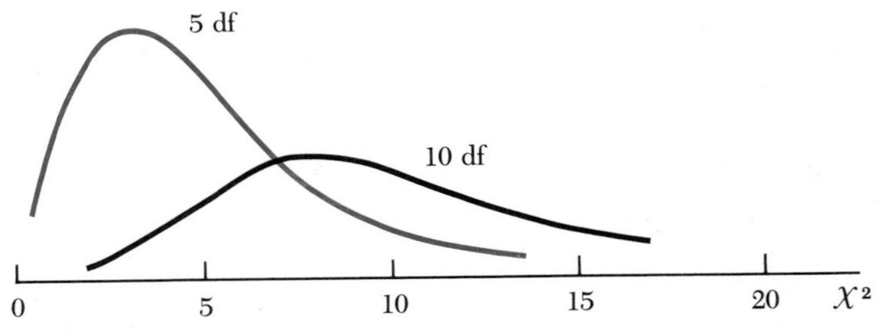

Figure 12-1 Two Chi-Square Distributions

Table 12-1 Calculation of Chi-Square for Multinomial Test

Brand	f_o	f_e	$(f_o - f_e)$	$(f_o - f_e)^2/f_e$
A	68	50	18	6.48
B	43	50	− 7	0.98
C	39	50	−11	2.42
Total	150	150		$9.88 = \chi^2_{\text{observed}}$

that the probability of a larger discrepancy *when the null hypothesis is true* is less than α, the allowable type I error risk.

With $\alpha = 0.05$, the level specified earlier, the critical value of χ^2 with $df = 2$ is $\chi^2(0.95) = 5.99$. The observed value of χ^2 is 9.88, which exceeds the critical level. So we reject the hypothesis. The sample data are sufficient to conclude that taste preferences for the three brands are not the same in the underlying population.

The *P*-value associated with our observed value of chi-square is less than 0.01. The 0.99 level of chi-square for 2 degrees of freedom is 9.21.

Cautions

The chi-square multinomial test is essentially a large sample test. The minimum sample size is governed by the smallest expected frequency in any single class. A conservative rule is that the minimum expected frequency in any class should be 5. When there are a large number of classes in the frequency table, however, it is permissible for several of them to have expected frequencies as low as 2. For small samples, there are exact hypothesis testing procedures based on combinatorial probability calculations. One such procedure is known as *Fisher's exact test*.

Exercises

12-1 *Chi-square distribution.* A company converted the annual pay of its wage and salary personnel to hourly rates and discovered that the hourly rates formed a chi-square distribution with $df = 14$.
a. What is the fifth percentile of the hourly rates?
b. What is the mean hourly rate?
c. What is the 99th percentile of hourly rates?

12-2 *Chi-square distribution.* Answer each question for a chi-square distribution with 6 degrees of freedom.
a. Find $\chi^2(0.99)$.
b. What value of chi-square is exceeded with probability 0.05?

12-3 *Baseball World Series.* In the World Series, the team winning four out of seven games wins the championship. The numbers of games played in 67 World Series are given below. Also given are the probabilities for lengths of a series between evenly matched teams, that is, with a probability of one-half that either team will win any given game.

Number of Games	4	5	6	7	*Total*
Number of Series	12	16	15	24	67
Probability	0.1250	0.2500	0.3125	0.3125	1.0000

a. Find the expected number of series of each length for 67 series between evenly matched teams.
b. At the 0.05 significance level, can the hypothesis be rejected that the 67 World Series described were played with evenly matched teams?

12-4 *Labor force survey.* A survey research firm was employed to conduct a study of attitudes of the labor force in a community toward compulsory retirement. The age distribution of the respondents and the age distribution of the labor force from a recent census are given below.

Age	*Labor Force* (%)	*Respondents* (*number*)
15–24	16.4	34
25–34	16.4	27
35–44	27.2	42
45–54	26.2	30
55–64	13.8	15
Total	100.0	148

A critic of the survey claimed that the survey underrepresented persons aged 55 through 64 and overrepresented those aged 15 through 24, and that a proper random sample would more closely reproduce the age distribution of the population.

a. If the sample were random, what would be the expected numbers of respondents in each of the age groups?
b. What is the approximate probability of a random sample producing larger differences from the expected numbers of respondents by age group?

12.2 Association in Cross-Classification Tables

An example in Chapter 11 involved the question of whether there was a significant difference in sample proportions of eastbound and westbound traffic exceeding the 55 mph speed limit at an interstate highway observation point. The following data were given in that example.

Speed	*Direction*		*Total*
	Eastbound	*Westbound*	
Exceeding limit	81	90	171
Not exceeding limit	44	60	104
Total	125	150	275

This table emphasizes that we have 446 sample observations cross-classified by categories of speed and direction. In Chapter 11, we conducted a test for the difference in the proportions of eastbound and westbound traffic exceeding the speed limit. That test can be thought of as a test for association between speed and direction of traffic. If we conclude that there is a difference between the population proportions of eastbound and westbound traffic exceeding the speed limit, then we conclude there is an association between speed and direction of traffic.

In the case of a two-by-two classification table, the test for difference between two proportions is equivalent to a test for association between the two sets of categorical variables. When the cross-classification table exceeds this dimension, a more general chi-square method must be employed, and we now turn to this more general situation.

Table 12-2 is a three- (rows) by-four (columns) cross-classification table. It is the result of a random sample of families in the circulation area of a metropolitan daily newspaper. The advertising department of the newspaper wants to demonstrate to potential advertisers that frequent readers of the newspaper have higher educational levels than infrequent readers, so that advertisers interested in appealing to an educated audience should consider newspaper advertising.

Table 12-2 Educational Levels of Householders by Frequency of Newspaper Purchase

Educational Level	*Frequency of Purchase*				*Total*
	Daily	*Frequently*	*Occasionally*	*Never*	
Primary	42	24	17	59	142
High School	47	32	11	21	111
Post High School	96	35	12	15	158
Total	185	91	40	95	411

For the sample data, we can easily calculate the percentage distribution of educational levels for the householders in each frequency of purchase class, as shown at the top of page 379. For the sample data, the educational distributions differ markedly for the several classes of readership frequency. The advertising department should satisfy themselves that the differences represent real differences in the population. That is, they must be satisfied that the sample differences could not have arisen as a result of chance from a population in which educational level is independent of readership frequency.

Educational Level	*Frequency of Purchase*				*Total*
	Daily	*Frequently*	*Occasionally*	*Never*	
Primary	22.7	26.4	42.5	62.1	34.6
High School	22.4	35.2	27.5	22.1	27.0
Post High School	51.9	38.4	30.0	15.8	38.4
Total	100.0	100.0	100.0	100.0	100.0

Hypothesis

Recall the concept of independence from Chapter 4. Independence of educational level and readership prevails when the conditional probabilities of educational levels given readership are the same as the marginal probabilities of educational levels. In our example, educational level is the variable of interest and we want to know if educational level is independent of readership class. If we always arrange the variable of interest in the rows of the cross-classification table, then the null hypothesis of independence, or no association, can be stated as

$$H_0: \pi_i \mid j = \pi_i$$

where i indexes the rows of the table from 1 to R and j indexes the columns of the cross-classification table from 1 to C. R is the number of rows and C is the number of columns in the table.

Expected Frequencies

If we knew the probabilities π_i, the expected frequencies under the null hypothesis of independence would be

$$(f_e)_{ij} = \pi_i \, n_j$$

where n_j is the total number of observations in the jth of the C columns of the cross-classification table. Typically, we do not know these probabilities. Because the data of Table 12-2 are from a random sample, we can use the marginal relative frequencies for each educational level to estimate them. For our example,

$$p_{(\text{primary})} = p_1 = 0.346$$

$$p_{(\text{high school})} = p_2 = 0.270$$

$$p_{(\text{post high school})} = p_3 = 0.384$$

With these estimates for the marginal probabilities, we are in a position to estimate the frequencies that would be expected in the cross-classification table if educational level were independent of readership frequency. We simply multiply the marginal relative frequencies just given by the total numbers of observations in each readership class. The results are in Table 12-3. A sample calculation of an expected frequency is

$$(f_e)_{11} = p_1 n_1 = 0.346(185) = 64$$

Table 12-3 Expected Frequencies Based on Marginal Relative Frequencies for Educational Levels

Educational Level	*Frequency of Purchase*				*Marginal Probability*
	Daily	*Frequently*	*Occasionally*	*Never*	
Primary	64	31	14	33	0.346
High School	50	25	11	26	0.270
Post High School	71	35	15	36	0.384
Total	185	91	40	95	

Chi-Square Test

The observed and expected frequencies for the chi-square test of association between educational level and readership are brought together in Table 12-4.

The squared difference between each observed and expected frequency is divided by the expected frequency, and these quantities are added to obtain the observed χ^2.

$$\chi^2_{\text{observed}} = \sum \left[\frac{(f_o - f_e)^2}{f_e} \right]$$

$$= \frac{(42-64)^2}{64} + \frac{(24-31)^2}{31} + \cdots + \frac{(15-36)^2}{36}$$

$$= 55.0$$

The degrees of freedom for χ^2 in a test of association in a two-way table are $(r - 1) * (c - 1)$, where r is the number of rows in the table and c is the number of

Table 12-4 Observed and Expected Frequencies for the Chi-Square Test of Association

Educational Level	*Frequency of Purchase*							
	Daily		*Frequently*		*Occasionally*		*Never*	
	f_o	f_e	f_o	f_e	f_o	f_e	f_o	f_e
Primary	42	64	24	31	17	14	59	33
High School	47	50	32	25	11	11	21	26
Post High School	96	71	35	35	12	15	15	36

columns. The loss of "rows" degrees of freedom came when we estimated the marginal probabilities with an effective restriction that they have a sum of 1.0. Thus there are only two independently estimated marginal probabilities. The loss in the "columns" degrees of freedom came when we estimated expected frequencies in each column subject to a restriction that the total expected frequency for each column add up to the total observed frequency for the column.

In the education-by-readership table, there are $(3 - 1)(4 - 1) = 6$ degrees of freedom. The (99.5)th percentile of χ^2 with these degrees of freedom is 18.55. An observed χ^2 of 55.0 or more could hardly ever occur by chance in sampling from a population in which there was no association between education and readership. The newspaper's advertising department can be confident that there is some association between these variables in the population.

Having rejected the possibility that random sampling errors are the cause of the observed association, the advertising department should now present the sample data as the best estimate of the pattern of association in the population. The table that gave the percentage distribution of educational levels for each readership frequency class brings out this pattern. A graphic presentation of this table is even more effective—Figure 12-2 shows that the educational level of daily readers is much higher than those only occasionally or never reading the newspaper.

Alternative Calculation

In presenting the test for association, we have emphasized the logic of the determination of expected frequencies. This lies in the common set of conditional probability distributions that would be expected under independence, or lack of association. In calculating expected frequencies, this set of marginal probabilities does not have to be written down. Using the original table of observed frequencies, the expected frequency for a given cell in the table can be found from

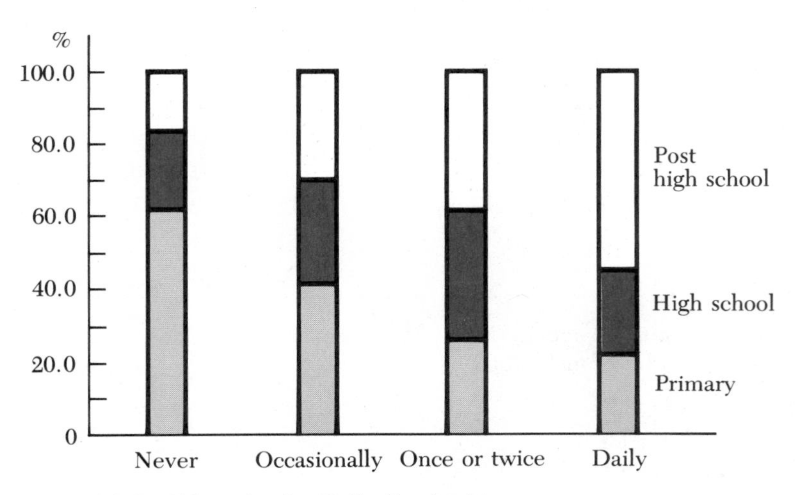

Figure 12-2 Education by Daily Readership

$$(f_e)_{ij} = \frac{n_i n_j}{n} \tag{12-3}$$

For example, the first entry in Table 12-2, $f_{11} = 42$, refers to householders who have a primary educational level and are daily readers of the newspaper. The total number of householders with a primary level education, n_i, is 142 and the total number of householders who are daily readers, n_j, is 185. The grand total for the table, n, is 411. Therefore,

$$(f_e)_{11} = \frac{142(185)}{411} = 64$$

This value is the estimated marginal probability of a primary educational level, $(142/411) = 0.346$, multiplied by the total number of daily readers, 185, as we emphasized earlier. With a pocket calculator, however, we do not have to write down this intermediate step.

Cautions

The test for association between two sets of categorical variables based on a random sample of observations from the population is a test of the null hypothesis that the variables of a bivariate multinomial population are statistically independent. Generally speaking, the test requires large samples. A conservative rule is that individual expected frequencies equal or exceed 5. When there is a large number of cells in the table, it is permissible for several not to exceed 5.

Exercises

12-5 *Teen-age shopping companions.** Random samples of teen-agers in Sacramento, California, were conducted in three different years. The respondents were asked with whom they preferred to shop. A tabulation of the results is as follows.

Preferred Companion	*1964*	*1970*	*1974*
Self	130	132	101
Friends and other	213	272	232
Relatives	182	193	232
Total	525	597	565

a. Combine the results for all three years to estimate the marginal probabilities for preferred companions.

*D. H. Tootelian and H. N. Windeshausen, *Journal of Retailing*, 52, 2 (Summer 1976).

b. Find the expected frequencies for preferred companions in the given years under the hypothesis of no association between preferred companion and time.
c. Calculate χ^2 and indicate your conclusion, using a significance level of 0.05.

12-6 *Manpower program outcomes.* The success of a federally funded, locally administered manpower program was measured by the proportion of clients who moved from subsidized employment into unsubsidized (private sector) employment and remained there for a certain length of time. A random sample of 376 clients of the program produced the following results.

Education	*Success*	*Failure*
8 years or less	13	19
9 to 11 years	76	45
12 years	107	65
13 years or more	32	19

a. Estimate the marginal probabilities of success and failure.
b. Test the hypothesis that program outcome is independent of educational level.

12-7 *Warranty card analysis.* A firm marketed a new model of portable typewriter through office supply stores, typewriter shops, and discount department stores. Warranty cards sent back by buyers included a question on the reason for the purchase. A cross tabulation of the first 1091 warranty cards returned showed the following.

Type of Purchase	*Office Supply Store*	*Typewriter Shop*	*Discount Department Store*	*Total*
Gift	66	106	135	307
Personal use	124	241	183	548
Business use	63	117	56	236
Total	253	464	374	1091

a. Conduct a chi-square test for association of the type of purchase with the kind of outlet. Use $\alpha = 0.01$.
b. If association is inferred, estimate the conditional probability distributions of the type of purchase for each kind of outlet on the assumption that the first 1091 warranty cards are a random sample from the long-run population.

12-8 *Income of patrons.** In a random sample of families in a midwestern city, respondents indicated the fast-food restaurant they most frequently patronized. The results were tabulated by income group for the two most popular restaurants, as follows.

*R. T. Justice and B. Jackson, *Journal of Small Business Management*, October 1978.

Income Group	*A*	*B*
Under $7000	61	13
$7000–$14,999	49	26
$15,000–$24,999	45	28
$25,000 and over	34	41
Total	189	108

a. Find the expected income distribution for each set of patrons on the assumption that they are random samples from the same income population.
b. What is the approximate probability of greater differences in the distributions if they are from the same population?
c. Would a significance test based on (b) above answer the same question as a significance test of difference between mean incomes? Discuss.

12.3 Multiple Independent Samples

We introduce this topic with the example mentioned earlier.

Example

Among random samples of beer drinkers in Chicago, Cleveland, and Minneapolis, preference for light and regular beer was distributed as follows.

	Chicago	*Cleveland*	*Minneapolis*
Light	202	316	226
Regular	298	624	524
Total	500	940	750

An East Coast brewer was planning to extend distribution to these Midwest cities. Thirty percent of the brewer's East Coast sales were light beer, and this figure was being considered in planning shipments into the three midwest cities. Do the data support the assumption that preference for light beer is uniformly 30% in the three cities?

The following are the sample percentages for the data.

	Chicago	*Cleveland*	*Minneapolis*
Light	40.4	33.6	30.1
Regular	59.6	66.4	69.9

The sample percentages suggest that preference for light beer may not be 30% in all the cities. In order to confirm this suspicion statistically, we must be able to reject the possibility that the sample differences are nothing more than random fluctuations from populations where the preference level *is* 30%. Such a procedure requires an appropriate test statistic.

Hypothesis

The null hypothesis in this example is

$$H_0: \pi_{\text{Chicago}} = \pi_{\text{Cleveland}} = \pi_{\text{Minneapolis}} = 0.30$$

and the alternative hypothesis is that at least one of the cities' preference levels for light beer is not equal to 0.30. The null hypothesis is a multiple hypothesis, which states that in all three cities the preference proportion is 0.30.

Expected Frequencies

As before, the expected frequencies are derived from the null hypothesis. If the null hypothesis were true, we would expect to find 0.30(500) = 150 preferences for light beer in a random sample of 500 Chicago beer consumers, 0.30(940) = 282 preferences for light beer in the Cleveland sample, and 0.30(750) = 225 preferences for light beer in the Minneapolis sample. The table of observed and expected frequencies is as follows.

	Chicago		*Cleveland*		*Minneapolis*	
	f_o	f_e	f_o	f_e	f_o	f_e
Light	202	150	316	282	226	225
Regular	298	350	624	658	524	525
Total	500	500	940	940	750	750

Chi-Square Test

With a set of observed and expected frequencies, we are ready to calculate an observed value of χ^2. To compare the observed to a critical value of χ^2, we need the degrees of freedom and a specified α-level for the test. For the multiple-samples test, consider the samples to be designated by the columns in the original frequency table and the classes of the nominal variable to be represented by the rows of the original table. The degrees of freedom are then

$$df = c(r - 1)$$

For our test, suppose an α-level of 0.05. The degrees of freedom are $3(2 - 1) = 3$, and the critical value of χ^2 is

$$\chi^2(0.95) = 7.81$$

Our observed value of χ^2 is

$$\chi^2_{\text{observed}} = \sum\left[\frac{(f_o - f_e)^2}{f_e}\right]$$

$$= \frac{(202 - 150)^2}{150} + \frac{(298 - 350)^2}{350} + \frac{(316 - 282)^2}{282} + \frac{(624 - 658)^2}{658}$$

$$+ \frac{(226 - 225)^2}{225} + \frac{(524 - 525)^2}{525}$$

$$= 18.03 + 7.73 + 4.10 + 1.76 + 0.004 + 0.002$$

$$= 31.63$$

At the 0.05 significance level, we can reject the null hypothesis stated earlier. We conclude that at least one of the cities' preference levels for light beer is not equal to 30%. Examination of the contribution of each city's data to the total observed χ^2 suggests the city, although this is not an integral part of the composite test. The sample proportions and the contributions to the observed χ^2 are given in the following table.

Chicago	(0.404)	18.03 + 7.73 =	25.76
Cleveland	(0.336)	4.10 + 1.76 =	5.86
Minneapolis	(0.301)	0.004 + 0.002 =	0.006
		Total	31.626

Each of these separate sums is a chi-square statistic with one degree of freedom for testing the hypothesis that the population proportion *in that city* is 0.30. Notice, however, that our original question and the resulting null hypothesis did not treat the cities separately, but rather in a composite manner.

It is worth noting an important difference between the multiple-samples test and the test for association in the preceding section. A hypothesis that could be tested with the three-city beer preference data is that the proportion of consumers preferring light beer is the same in all three cities. This is the null hypothesis that preference is independent of (not associated with) city. The procedure of Section 12-2 would be followed and the degrees of freedom would be $(r - 1)(c - 1) = (2 - 1)(3 - 1) = 2$. The degrees of freedom were 3 for the multiple-samples test that the proportion preferring light beer was 0.30 in all three cities. The difference is that in the test for association, a common proportion is estimated from the data under the null hypothesis of no association, while in the multiple-samples test, a common proportion was specified directly in the hypothesis, that is, external to the sample data.

Finally, a multiple-samples test does not require a uniform proportion or set of proportions. For example, the hypothesis that the preference level in Chicago is 0.40 *and* the preference level in Cleveland is 0.40 *and* the preference level in Minneapolis is 0.30 would be tested in the same manner as the hypothesis that all three cities had preference levels of 0.30.

The guidelines for minimal expected frequencies in the cells for the multiple-sample test are the same as for the previous two chi-square tests.

Exercises

12-9 *Retirement employment.* The developers of a large retirement community were told by a professional planner that two-thirds of the male residents would desire suitable part-time employment regardless of their age. A random sample of male residents produced the following results.

Age	*Desiring Employment* *Yes*	*No*
50–59	37	15
60–69	135	82
70–79	36	36
80–89	10	14

a. Calculate χ^2 for a test of the planner's assertion.
b. At an α-level of 0.05, should the planner's assertion be accepted?

12-10 *Test for randomness.* In the first five rows of the random-number table (Table A-8, Appendix A), the number of odd integers out of 100 is as follows.

Row 1	*Row 2*	*Row 3*	*Row 4*	*Row 5*
48	39	42	55	51

a. Calculate the chi-square statistic for the differences from the expected number of odd (and even) integers by rows. Assume a Bernoulli process with $\pi = 1/2$.
b. At $\alpha = 0.05$, can the hypothesis of a Bernoulli process with $\pi = 1/2$ be rejected?
c. Why is a test like this one by rows better than testing the number of odd and even integers for the whole table?

12-11 *Independent market studies.* The marketing manager in a food manufacturing firm obtained the following data from four independent consumer preference studies on a cereal product.

	Sample			
	1	2	3	4
Prefer brand A	248	260	252	258
Not prefer brand A	152	140	148	142

Test the hypothesis that all four samples come from populations in which the proportion preferring brand A is 0.60. Use a significance level of 0.05.

12-12 *Interviewer performance test.* A survey of dwelling units was made in an area where it was known that there were equal numbers of dwelling units on the four sides of each city block. Interviewers were assigned blocks at random and were given instructions for selecting dwelling units at random from the assigned blocks. For the five interviewers in the survey, the number of dwelling units surveyed by block face is as follows.

Block Face	*Interviewer*				
	A	*B*	*C*	*D*	*E*
North	9	14	12	6	14
South	8	10	16	10	6
East	7	6	8	11	14
West	16	10	4	13	6
Total	40	40	40	40	40

Test the hypothesis, using $\alpha = 0.10$, that the instructions were uniformly carried out, that is, that the interviewer samples are random samples from a population in which the probability of a dwelling unit from each block face is equal.

12.4 Chi-Square Goodness-of-Fit Tests

In earlier chapters, we encountered several distribution models for variable characteristics. The binomial distribution and the Poisson distribution are two models for discrete variables and we have encountered the normal model for a continuous variable. When a variable is normally distributed, the normal table of areas can easily be used to find probabilities for desired ranges of the continuous variable. This is because the normal distribution is a fixed form. Knowing only the mean and standard deviation for the population, we can find required standard normal deviates (z); their corresponding probabilities are found from the cumulative standard normal probability table. When we do this, we are using a distribution *model* (normal in this case) in place of the actual distribution for the variable. In earlier examples, we have been content to say that a variable has been found to be normally distributed or that it appeared reasonable to assume the variable was normally distributed. The chi-square goodness-of-fit test is a procedure for using a sample distribution of a variable to test the hypothesis that the sample came from a population distributed according to a specified distribution model. If the hypothesis is accepted, the case for using the model to obtain detailed probabilities is strengthened. If the hypothesized model is rejected, detailed probabilities will have to be found in some other way. Either a more acceptable model will have to be found, or enough data will have to be collected to provide direct empirical estimates of sufficient accuracy for all the desired probabilities.

Binomial Goodness-of-Fit Test

Example

County health inspectors check food-service establishments for compliance with five conditions. In the most recent 200 inspections, the number of noncomplying conditions out of five were as follows.

Number of Conditions	0	1	2	3	4	5	Total
Number of Inspections	20	58	64	28	20	10	200

Using an α-level of 0.01, test the hypothesis that the number of noncomplying conditions is a binomially distributed random variable, that is, that the 200 inspections are a random sample from a population in which the number of noncomplying conditions out of five is binomially distributed.

Hypothesis The null hypothesis must be a statement which, if true, leads to a test statistic. In a chi-square goodness-of-fit test, the null hypothesis is that the underlying distribution *does* conform to the model of interest. Under this hypothesis, we can estimate the parameters of the model and use these parameters to calculate expected frequencies for repeated samples of the same size as the observed sample.

The key parameter for a binomial distribution is π, the probability of occurrence of the event in question on a single Bernoulli trial. Recall that the mean of a binomial random variable is $n\pi$. We can estimate the binomial mean from the observed distribution and then divide by n to obtain an estimate of π.

The mean number of noncomplying conditions for the sample distribution is

$$\frac{0(20) + 1(58) + 2(64) + 3(28) + 4(20) + 5(10)}{200} = 2.0$$

Our estimate of π is, then, $2.0/5 = 0.40$.

Expected frequencies To get expected frequencies under the null hypothesis that the number of noncomplying conditions out of 5 in an inspection is a binomially distributed random variable, we need only multiply the binomial probabilities for $n = 5$ and $\pi = 0.40$ by 200, the total number of sample inspections. This is done in Table 12-5. The observed chi-square value is also calculated in the table. It will be discussed in the next section.

Chi-Square Test The observed value of χ^2 is calculated just as for other chi-square tests from

$$\chi^2 = \sum\left[\frac{(f_o - f_e)^2}{f_e}\right]$$

In Table 12-5, the last 2 classes have been combined in the calculation of $(f_o - f_e)^2/f_e$. This is in accordance with a rule to combine adjacent classes when the expected

Table 12-5 Calculation of Chi-Square for Binomial Goodness-of-Fit Test

r	f_o	*Binomial Probability*	*	$\Sigma f_o =$	f_e	$f_o - f_e$	$(f_o - f_e)^2/f_e$
0	20	0.0778	*	200 =	15.6	4.4	1.24
1	58	0.2592	*	200 =	51.8	6.2	0.74
2	64	0.3456	*	200 =	69.1	− 5.1	0.38
3	28	0.2304	*	200 =	46.1	−18.1	7.11
4	20	0.0768	*	200 =	15.4 }	12.6	9.12
5	10	0.0102	*	200 =	2.0 }		
	200	1.0000			200.0		18.59 = χ^2_{observed}

frequency is less than 5. In goodness-of-fit tests, the number of degrees of freedom is the number of classes (after any are combined) less 1 degree of freedom for each independent way in which the expected frequencies are made to agree with the observed frequencies. Generally, 1 degree of freedom will be lost for each of the parameters of the model distribution that has to be estimated and 1 for agreement between the total expected and total observed frequencies.

We estimated one parameter of the binomial distribution, namely π. Thus 2 degrees of freedom are lost. There are 5 classes, and the number of degrees of freedom for the chi-square statistic is $5 - 2 = 3$ for our example. The significance level for the test was specified as 0.01. The criterion level for χ^2 is, then, $\chi^2(3,0.99) = 11.34$. Therefore we reject the hypothesis of an underlying binomial distribution. We might speculate that the probabilities of noncompliance differ for different conditions or vary among groups of food-service establishments. Alternatively, the 5 trials (conditions) at each establishment may not constitute independent events. For example, not complying with one condition may increase the probability of not complying with another condition.

Normal Goodness-of-Fit Test

Table 12-6 gives data on energy comsumption for heating 359 industrial buildings during one season in a large city. The data are a random sample and the question is whether they can be regarded as a random sample from a normal distribution of consumption (BTU/ft^2) measures. The table carries through to the calculation of χ^2 for the test, and will be explained in detail below.

Hypothesis The null hypothesis is that the population from which the sample came is normally distributed. We will use the 0.05 level of significance for the test.

To estimate probabilities for a normal distribution, we need the mean and standard deviation. Using the methods of Chapter 3 for frequency distributions, we find the estimated mean and standard deviation from Table 12-6 to be 43.0 and 5.0, respectively.

Table 12-6 Fitting a Normal Distribution

Interval	f_o	*z(upper)*	*Normal Probability*		f_e	$(f_o - f_e)^2/f_e$
			Cumulative	*Interval*		
Under 29.0	2	−2.80	0.0026	0.0026	0.9	
29.0–31.0	4	−2.40	0.0082	0.0056	2.0	1.88
31.0–33.0	6	−2.00	0.0228	0.0146	5.2	
32.0–35.0	9	−1.60	0.0548	0.0320	11.5	0.54
35.0–37.0	14	−1.20	0.1151	0.0603	21.7	2.73
37.0–39.0	34	−0.80	0.2119	0.0968	34.8	0.02
39.0–41.0	51	−0.40	0.3446	0.1327	47.6	0.24
41.0–43.0	55	0.00	0.5000	0.1554	55.8	0.01
43.0–45.0	58	0.40	0.6554	0.1554	55.8	0.09
45.0–47.0	52	0.80	0.7881	0.1326	47.6	0.41
47.0–49.0	38	1.20	0.8849	0.0968	34.8	0.29
49.0–51.0	22	1.60	0.9452	0.0603	21.7	0.00
51.0–53.0	8	2.00	0.9772	0.0320	11.5	1.07
53.0–55.0	3	2.40	0.9918	0.0146	5.2	
55.0–57.0	1	2.80	0.9974	0.0056	2.0	0.54
Over 57.0	2		1.0000	0.0026	0.9	
	359			1.0000	359.0	$7.82 = \chi^2_{\text{observed}}$

Expected Frequencies Having estimated the mean and standard deviation of the underlying population, we can now estimate the frequencies that would be expected in each of the class intervals if the population were normally distributed. In the column labeled z (upper), we indicate the standard score for the upper limit of each frequency class. This is calculated from

$$z(\text{upper}) = \frac{x(\text{upper}) - \bar{x}}{s} = \frac{x(\text{upper}) - 43.0}{5.0}$$

The next column gives the cumulative normal probability for each z (upper) from Appendix A. By subtracting successive cumulative normal probabilities, the normal probabilities in each interval can be obtained. For example,

$$P(29.0 \leq X \leq 31.0) = 0.0082 - 0.0026 = 0.0056$$

Finally, the estimated normal probabilities are multiplied by n, in this case 359, to determine the expected frequencies under the normal model. For example,

$$0.0056(359) = 2.0$$

Chi-Square Test In the final column of Table 12-6, the observed value of χ^2 is calculated. The first three and the last three classes in the distribution are combined to secure expected frequencies of at least 5 for each class in the chi-square summation.

The degrees of freedom for χ^2 are the total number of classes in the chi-square summation less 1 degree of freedom for each independent way in which the expected frequencies have been made to conform with the observed frequencies. In our fitted normal example, we forced such agreement with respect to (1) the mean, (2) the standard deviation, and (3) the total of the frequencies. In fitting a normal distribution, 3 degrees of freedom are thus lost. In our case, $df = 12 - 3 = 9$.

The observed χ^2 of 7.82 is somewhat below the 0.95 level (16.92) of χ^2 for 9 degrees of freedom. It is a value of χ^2 consistent with the hypothesis that the underlying energy consumption population is normally distributed.

Having accepted the hypothesis of normality, an agency could now use the normal distribution model in place of the observed data. For example, suppose an energy conservation agency had decided to target buildings exceeding the 90th percentile of consumption per square foot for special attention in a voluntary energy conservation program. From the normal model, the agency would determine this level to be

$$43.0 + 1.28(5.0) = 49.4 \text{ BTU/ft}^2\text{/mo}$$

Poisson Goodness-of-Fit Test

Our final goodness-of-fit test is for a Poisson distribution. In a California driver-record study, 6938 accidents occurred among 86,726 drivers. The observed distribution is shown in Table 12-7.

Hypothesis The hypothesis is that the population distribution is Poisson. The Poisson distribution has but one parameter, the mean number of occurrences, μ. From the sample data, we estimate this to be 6,938/86,726 = 0.08 accidents per driver per year.

Expected Frequencies Poisson probabilities for each value of r when $\mu = 0.08$ can be found in Appendix A. Then these probabilities are multiplied by the total frequency (86,726 drivers) to obtain the expected numbers of drivers with 0, 1, 2, and 3 or more accidents in a year if the underlying population were Poisson distributed.

Chi-Square Test The observed value of χ^2 is calculated in the usual way in the final column of Table 12-7. The degrees of freedom for a Poisson test will be the number of classes after any grouping is done, minus two. The degrees of freedom lost are for the one estimated parameter and for agreement between the totals for estimated and actual frequencies. Our chi-square statistic has 1 degree of freedom. The observed chi-square of 176.00 exceeds χ^2 (0.995) for 1 degree of freedom, so we reject the hypothesis that accidents occurred to California drivers according to a Poisson distribution. Note that more drivers than expected had no accidents and more

Table 12-7 Chi-Square Goodness-of-Fit for a Poisson Distribution

Accidents per Year r	*Number of Drivers** f_o	*Poisson Probability*	*Expected Frequency* f_e	$f_o - f_e$	$(f_o - f_e)^2/f_e$
0	80,369	0.923	80,048	321	1.29
1	5,910	0.074	6,418	–508	40.21
2	415	0.003	260 }	187	134.50
≥ 3	32	0.000	0 }		
	86,726	1.000	86,726		$176.00 = \chi^2_{\text{observed}}$

*Donald C. Weber, *Journal of the American Statistical Association*, Vol. 66, No. 334 (June 1971).

drivers than expected had 2 or more accidents. Insurance companies have learned that when drivers are divided into classes by age, sex, marital status, and so on, the Poisson distribution *is* an appropriate model. Then they can conveniently estimate the probabilities of 0, 1, 2, and 3 or more accidents to a policyholder of a certain class over a specified time period. All that is needed is an estimate of μ, the number of accidents per policyholder for the class during the period.

Specification of Parameters

In all of the foregoing goodness-of-fit tests, we have estimated the parameters of the hypothesized distribution from the sample data. These parameters were π for the binomial, the mean and standard deviation for the normal distribution, and the mean for the Poisson.

In some cases the hypothesis of interest may include parameters as well as the distribution form. For example, a breeder may want to know if the number of males in five-pup litters of Great Danes is binomially distributed *with* π *equal to 0.50*, or an inspector of weights and measures may want to test the hypothesis that a certain class of scale is calibrated so that errors of measurement are normally distributed *with a mean of zero and a specified standard deviation*.

When parameters are specified in the hypothesis, those values are used in deriving expected frequencies. However, the degrees of freedom are different from when the parameters are estimated from the sample data. Degrees of freedom are lost only for parameters estimated from the sample data. When a parameter is specified in the hypothesis, no degrees of freedom are lost for that parameter.

Exercises

12-13 *Predicting sports outcomes.* The sportscaster for a local TV station predicted the outcomes of four contests each week for a period of 36 weeks. The number of correct calls for each week was as follows.

Number correct	0	1	2	3	4
Number of weeks	4	3	7	11	11

a. Estimate the probability of a correct call.
b. Find the expected frequencies under the hypothesis that the number of correct calls per week is binomially distributed.
c. Conduct a goodness-of-fit test for the binomial model, using $\alpha = 0.05$.

12-14 *Temperatures in May.* Exercise 3-26 deals with a distribution of daily low temperatures in May for a particular growing region. The sample mean and standard deviation are 39.17° and 6.313°, respectively.
a. Test the hypothesis that the observed data are a random sample from a normally distributed population. Use a significance level of 0.10.
b. If the hypothesis is acceptable, estimate the probability of a daily low temperature in excess of 50°.

12-15 *Ages of chief executives.* Ages of 754 chief executives are given in Exercise 3-31. The mean and standard deviation of the 754 ages are 49.00 and 7.545 years, respectively.
a. After grouping the data by 5-year age groups beginning with 20 and under 25, fit a normal distribution to the class intervals.
b. Conduct a chi-square test of goodness-of-fit, using an α-level of 0.05. State your conclusion.
c. If your conclusion is that the data are not a random sample from a normally distributed population, describe how the distribution form differs from the normal form.

12-16 *Outpatient arrivals.** The number of patients arriving during the morning at an outpatient clinic was observed by half-hour intervals for a period of 13 days. The average arrival rate was 5.4 patients per interval and the distribution was as follows.

Number arriving	≤ 2	3	4	5	6	7	≥ 8
Frequency	6	9	14	9	3	7	12

a. Find the expected frequencies under the hypothesis that arrivals constitute a Poisson process.
b. At $\alpha = 0.05$, test the hypothesis that the underlying process is Poisson.

Summary

In this chapter, we departed from statistical inference for interval-scaled data to consider inference for frequencies. The analysis of frequencies is a principal means of studying data that are nominally scaled, or categorical.

All of the tests used a statistic that measures the differences between a set of observed frequencies and a set of frequencies that would be expected to occur if the

*T. F. Keller and D. J. Laughunn, *Decision Sciences*, 4, 3 (July 1973), pp. 379–394.

null hypothesis under consideration were true. The statistic follows a family of distributions called *chi-square distributions*. The relevant chi-square distribution in any particular situation depends on the degrees of freedom, which is related to the number of pairs of observed and expected frequencies that are included in the chi-square statistic.

Three chi-square tests involving categorical data were examined. They are (1) a test for whether an observed multinomial distribution came from a specified parent population, (2) a test for whether the observed data in a cross-classification table came from a population in which association between the characteristics prevails, and (3) a test for whether several observed multinomial distributions came from a common specified population.

The last section of the chapter considered chi-square goodness-of-fit tests. This procedure permits a test of the hypothesis that a set of sample data came from a population distributed according to a specified probability model.

Supplementary Exercises

12-17 *Taste preferences.* A manufacturer of a meat seasoning is satisfied with the spiciness of a branded product when consumers are equally divided among those desiring more spice, less spice, and the same amount of spice. Recently, taste tests were conducted in three cities with the following results.

Reaction	*City A*	*City B*	*City C*
Want more spice	17	27	25
About right	21	17	22
Want less spice	22	16	13

a. Test the data for each city against the distribution desired by the manufacturer.
b. Test the hypothesis that reactions of consumers are independent of city of residence.

12-18 *New residents.* Random samples of families in four neighborhoods produced the following data on new arrivals.

Lived in	*Neighborhood*			
Neighborhood	*1*	*2*	*3*	*4*
Less than 1 year	24	2	19	15
1 year or more	126	78	71	65

a. Test the hypothesis that the four samples are from populations with 1/6 new residents.
b. Test the hypothesis that the proportion of new residents is the same in the four neighborhoods.

12-19 *Pretrial releases.** In an experimental pretrial release program for defendants accused of nonviolent crimes in a large city, 328 defendants were released pending trial who would not have been released by the Superior Court under its usual program. Subsequent appearance history for this experimental group was compared with a control sample of those released under the Superior Court criteria, as follows.

Outcome	*Experimental*	*Control*
No failure to appear	159	151
Voluntary return (late)	81	21
Warrant served	70	23
At large	18	6
Total	328	201

a. Test the hypothesis that there are no outcome differences associated with the groups.
b. Construct the 95% confidence interval for the proportion failing to appear for each group.

12-20 *In-home food shopping.*† A comparison was made between the characteristics of a sample of 797 patrons of an in-home phone order and delivery service and a random sample of nonusers of the service. Auto ownership and age distributions of the groups are compared below.

	Number of Automobiles in Family					
	0	*1*	*2*	*3*	*4*	*Total*
Patrons	64	215	390	96	32	797
Nonpatrons	13	80	136	29	8	266

	Age of Respondent					
	20–25	*26–30*	*31–40*	*41–50*	*Over 50*	*Total*
Patrons	64	183	255	120	175	797
Nonpatrons	8	45	56	59	98	266

a. Use χ^2 to test whether patrons (1) differ from nonpatrons in regard to auto ownership and (2) in regard to age. Indicate your conclusions.
b. Construct a graph to emphasize any differences inferred in (a) above.

*M. R. Gottfredson, *Journal of Criminal Justice*, 2 (1974), pp. 287–304.

†E. N. Berkowitz, J. R. Walton, and O. C. Walker, Jr., *Journal of Retailing*, 55, 2 (Summer 1979).

12-21 *Content weights.* Table 2-5 presents a frequency distribution of content weights of bags of flour. The mean and standard deviation of the tabled data are 100.0275 and 0.3374 lb, respectively.

a. Determine expected frequencies for the class intervals under the hypothesis that the 80 observations are a random sample from a normally distributed population of content weights.
b. At the 0.05 significance level, can the hypothesis of normality be rejected?

12-22 *Political kidnappings.** During a 56-month period from September 1969 to April 1974, there were 39 recorded political kidnappings involving ransom demands. The frequency distribution of the number of kidnappings by months was as follows.

Number of kidnappings	0	1	2	3	4	5
Number of months	34	9	9	4	0	0

a. Estimate the mean number of kidnappings per month and the expected frequencies under the hypothesis that the observations are a sample from a Poisson-distributed population.
b. Conduct a goodness-of-fit test for the Poisson distribution, using an α-level of 0.01.

12-23 *Farm survey.*† A 1979 statewide survey of farms in Kentucky was conducted with a sampling frame of registered voters. The size distribution of farms in the survey and the percentages from the 1978 Census of Agriculture are given below.

Farm Acreage	*Sample Number*	*Census (Percent)*
Less than 50	110	35
50–179	183	43
180–499	85	18
500 & over	28	4
Total	406	100

Test the hypothesis, using $\alpha = 0.05$, that the method of drawing the sample is unbiased in regard to the size distribution of farms.

12-24 *Jazz listening.*‡ The following data on attending jazz performances was obtained by a random sample of households in 1982.

*R. E. Peterson, *Interfaces*, 8, 2 (February 1978).
†C. M. Coughenour and L. Swanson, *Rural Sociology* 48, 1 (1983), pp. 23–33.
‡J. P. Robinson, *American Behavioral Scientist*, 26, 4 (1983), pp. 543–552.

	Age					
	18–25	*26–35*	*36–45*	*46–55*	*56–65*	*Total*
Total surveyed	158	190	163	147	129	787
Attended jazz	30	23	8	19	3	83

Is it safe to conclude that the sample data come from a population in which the percentage attending jazz performances differs by age group? Explain.

12-25 *Measuring income.** Two different question sequences were designed to find the income levels of respondents to personal interviews. A split-panel test was conducted in which respondents were randomly assigned to receive the two test question forms. Results appear below.

Income ($ thousands)	*Under 5*	*5– 7.49*	*7.5– 9.99*	*10– 14.99*	*15– 19.99*	*20– 24.99*	*25 and over*	*Total*
Form 1	22	8	31	50	33	10	23	177
Form 2	18	12	17	48	43	23	28	189

At $\alpha = .05$, test the hypothesis that the forms elicit the same income information.

12-26 *International comparison.*† Random samples of consumers in the United States and England were asked to indicate agreement or disagreement with the statement, "The games and contests that manufacturers sponsor to encourage people to buy their products are usually dishonest." The results were:

	Strongly Agree	*Agree*	*Uncertain*	*Disagree*	*Strongly Disagree*	*Total*
England	12	49	131	104	9	305
United States	38	88	276	207	19	628

Do these data justify the conclusion that attitudes toward the question posed are different in the two countries? Explain.

Computer Exercises

1 Source: Real estate data file. Regard the 100 observations as a random sample from a large population.

*R. A. Herriot, *Journal of Marketing Research*, 14 (August 1977), pp. 322–329.

†W. A. French, H. C. Barksdale, and W. D. Perreault, Jr., *European Journal of Marketing*, 16, 6 (November 1982), pp. 20–30.

a. Using six classes for selling price, test the hypothesis that the distribution of prices is the same for old and new homes. What P value is associated with your conclusion?
b. Using six classes for income, test the hypothesis that the distribution of incomes is the same for previous residents and nonresidents. What P value is associated with your conclusion?

2 Source: Personnel data file.
a. Conduct a chi-square test for association of management level and degree type.
b. Perform a chi-square test for association of degree type and department. Is the result reliable?

Interpreting Computer Output

Figure 12-3 shows the computer output for the analysis of a cross-classified table. The variables are management level versus participation in a tax shelter plan (Y or

LEVEL DEFER

FREQUENCY EXPECTED CELL CHI2 PERCENT ROW PCT COL PCT	N	Y	TOTAL
LEVEL1	8 2.2 15.0 1.19 72.73 5.88	3 8.8 3.8 0.45 27.27 0.56	11 1.64
LEVEL2	61 31.2 28.6 9.08 39.61 44.85	93 122.8 7.2 13.84 60.39 17.35	154 22.92
LEVEL3	53 47.2 0.7 7.89 22.75 38.97	180 185.8 0.2 26.79 77.25 33.58	233 34.67
LEVEL4	14 55.5 31.0 2.08 5.11 10.29	260 218.5 7.9 38.69 94.89 48.51	274 40.77
TOTAL	136 20.24	536 79.76	672 100.00

STATISTICS FOR 2-WAY TABLES

CHI-SQUARE 94.335 DF= 3 PROB=0.0001

Figure 12-3 Computer Output for Analysis of a Cross-Classified Table

N) for 671 employees in a firm. Each cell shows (from top to bottom) the frequency in the cell, the expected frequency, the contribution of that cell to the chi-square statistic, and the cell, row, and column percentages. The chi-square statistic is 93.993 with 3 degrees of freedom. The associated P-value is 0.0001, which means we would reject the null hypothesis of independence at levels of significance of 0.0001 and smaller.

The conditional percentages are shown in Figure 12-4, and these give some insight into why we would reject the hypothesis of independence. For those who participate in the tax shelter (Y), the percentages *increase* from level 2 through level 4. For those who don't participate (N), the percentages *decrease* for the same levels.

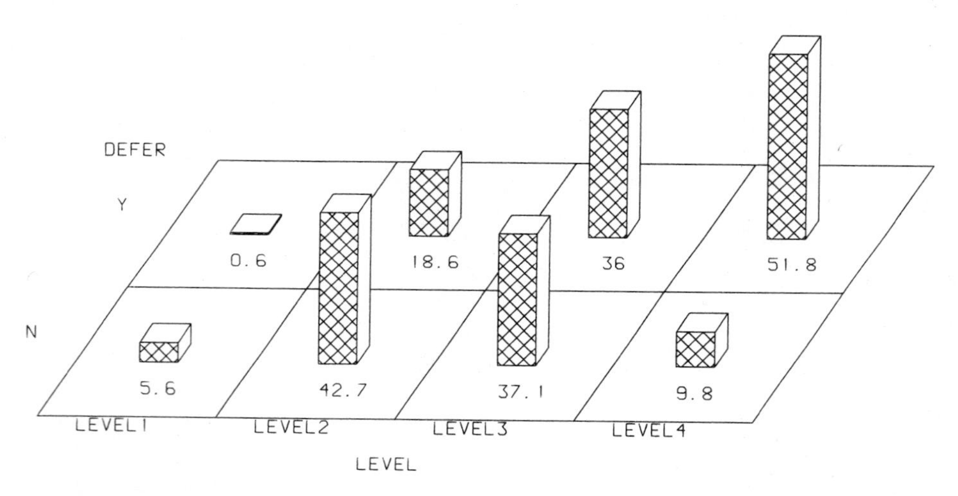

Figure 12-4 Conditional Percentages

13

ANALYSIS OF VARIANCE

In Chapter 11, we discussed how to test for a significant difference between two means. We also discussed how to test for a significant difference between two proportions. Chapter 12 was an extension of Chapter 11 in that it presented a set of procedures, called chi-square tests, for testing more than two proportions for a significant difference. Now we shall discuss an extension of the other topic in Chapter 11: We shall present a set of procedures for testing more than two means for a significant difference. These procedures are called *analysis of variance* tests.

13.1 One-Way Analysis of Variance

Our example to explain analysis of variance concerns a marketing executive for a chain of small convenience markets who has had four different posters prepared to advertise a brand of canned salted nuts. Once the most effective poster has been selected, it will be printed in a large quantity and distributed to all markets to be displayed over the shelf where the product appears. For the present, however, only three copies of each type of poster are available for testing. Twelve markets have been selected at random, and one of the posters has been left on display for a week in each market.

The sales results for this experiment are given in Table 13-1. Poster D had mean sales of 50 packages per week associated with it. Poster A was nearly as effective, with a mean of 47 packages sold per week. Posters B and C, at 38 and 37, are very close to each other and well below the other two. The purpose of analysis of variance is to test whether such differences in sample means reflect differences in the means of the statistical populations from which they came or are merely sampling fluctuations.

Notation

To discuss analysis of variance effectively, we must introduce some notation. Table 13-2 is a general representation of the experimental results reported in Table 13-1. In general, let r represent the number of rows in such a table, and let c represent the number of columns. To begin with, we shall consider only tables in which there are an equal number of observations in each column. Hence the total number of observations in the table is $r * c$. For our example, $r * c = 3 * 4 = 12$ observations. Let X_{ij} be the observation in the ith row and the jth column. This is the system used to determine the subscripts for each of the X-values in Table 13-2. One of the observations in Table 13-1 is represented by each X in the body of Table 13-2. For instance, $X_{32} = 40$ and $X_{23} = 35$ in Table 13-1.

Table 13-1 Number of Packages Sold

	Poster			
	A	*B*	*C*	*D*
	55	32	38	47
	43	42	35	48
	43	40	38	55
Total	141	114	111	150
Mean	47	38	37	50

$$\text{Grand mean} = \frac{47 + 38 + 37 + 50}{4} = 43$$

Table 13-2 General Notation for Table 13-1

$i \backslash j$	1	2	3	4
1	X_{11}	X_{12}	X_{13}	X_{14}
2	X_{21}	X_{22}	X_{23}	X_{24}
3	X_{31}	X_{32}	X_{33}	X_{34}
Total	C_1	C_2	C_3	C_4
Mean	$\bar{X}_1$	$\bar{X}_2$	$\bar{X}_3$	$\bar{X}_4$
	Grand mean		$\bar{\bar{X}}$	

Because each of the four columns represents sales results from a different poster, our interest focuses on the four column means, the $\overline{X}_j$. In Table 13-1, $\overline{X}_1$ is 47 and $\overline{X}_4$ is 50; these means are found by dividing the column totals by the number of observations in the column, where a column total is represented by C_j. The grand mean is denoted by $\overline{\overline{X}}$, "X double-bar."

In this chapter, the columns of a data table will represent different treatments. A treatment is a value or class of the independent factor whose influence on the dependent variable is being analyzed. Here the independent factor is poster advertising, and the treatments are the different posters used in the experiment. The rows within a column are random observations made with a given treatment.

The Hypothesis Statement

We said earlier that analysis of variance is designed to test for a significant difference among several means. In Table 13-1, the mean sales for the four types of posters range from 37 to 50, and analysis of variance will indicate whether these are real differences or just sampling fluctuations. Statistically speaking, are these samples from four populations with the same mean or do at least two of the populations have different means? Formally, the null hypothesis is

$$H_0\colon \mu_1 = \mu_2 = \mu_3 = \mu_4$$

and the alternative hypothesis is

$$H_A\colon \text{The population means are not all equal.}$$

Two assumptions are made about these populations. They are assumed to have equal variances, and they are assumed to be normally distributed. With the addition of these assumptions, the test is equivalent to asking if the four samples come from four normally distributed populations with equal means and variances.

Rationale

If the null hypothesis and the assumptions are true, the 12 observations in Table 13-2 can be thought of as coming from the same population. We shall refer to the variance of this common population as σ^2.

To test the null hypothesis, analysis of variance compares two estimates of σ^2. The first estimate is based on the variability of the column means around the grand mean, and the second is based on the variability of the observations in each column around the mean of that column. Because both are estimates of σ^2, they should be essentially equal if the null hypothesis is true. On the other hand, if the populations have different means, then the variability of the column means around the grand mean will be much greater than the estimate from within the columns. Hence the null hypothesis will be accepted when the two estimates of σ^2 are nearly equal, and the null hypothesis will be rejected when the estimate based on the variability of the column means around the grand mean is much greater than the second estimate.

Let us first consider the estimate of σ^2 based on the variability of the four column means around the grand mean. If the assumptions and the null hypothesis are true, then the four sample means (47, 38, 37, and 50) belong to the sampling distribution of the mean for samples of three observations each. We know from Chapter 7 that the variance of the sampling distribution of the mean, $\sigma^2_{\overline{X}}$, is the variance of the population from which the samples came, σ^2, divided by the sample size. In symbols,

$$\sigma^2_{\overline{X}} = \frac{\sigma^2}{n} \tag{13-1}$$

If we multiply both sides of this equation by n, we have

$$n\sigma^2_{\overline{X}} = \sigma^2 \tag{13-2}$$

In other words, multiplying the variance among sample means by the number of observations in each sample gives a result equal to the population variance.

Although we do not know the value of the variance among sample means, we can estimate it from the four means at our disposal. This estimate is

$$s^2_{\overline{X}} = \frac{\Sigma(\overline{X} - \overline{\overline{X}})^2}{c - 1} \tag{13-3}$$

The numerator is the sum of the squared deviations of the column means from the grand mean, and the denominator is the number of degrees of freedom applicable to the four column means. For our poster experiment in Table 13-1,

$$s^2_{\overline{X}} = \frac{(47 - 43)^2 + (38 - 43)^2 + (37 - 43)^2 + (50 - 43)^2}{4 - 1} = \frac{126}{4 - 1}$$

There are $r = 3$ observations in each of the samples with means being tested. Hence if we multiply the variance among sample means by r, we shall have an estimate of the population variance σ^2.

The between-means estimate of σ^2 is

$$r(s^2_{\overline{X}}) = \frac{r\Sigma(\overline{X} - \overline{\overline{X}})^2}{c - 1} \tag{13-4}$$

For our posters,

$$\text{between-means estimate} = \frac{3(126)}{4 - 1} = 126$$

We next need to find an estimate of σ^2 based on variability of the observations within columns around their column means. We shall call this the *within-columns estimate*. Consider just one column. We learned in Chapter 8 that an unbiased estimate of population variance from such a sample is

$$s^2 = \frac{\Sigma(X - \overline{X})^2}{r - 1}$$

where $\overline{X}$ is the column mean and r, the number of rows, is also the number of observations in the column. For poster A in Table 13-1,

$$s^2 = \frac{(55 - 47)^2 + (43 - 47)^2 + (43 - 47)^2}{3 - 1} = 48$$

Similarly, the other three posters will each furnish an unbiased estimate of σ^2. The four estimates are

$$s_1^2 = 48 \qquad s_2^2 = 28 \qquad s_3^2 = 3 \qquad s_4^2 = 19$$

We can form a pooled estimate by averaging these four sample estimates. The pooled estimate is

$$\frac{48 + 28 + 3 + 19}{4} = 24.5$$

The within-columns estimate of σ^2 is

$$\frac{\sum_j s_j^2}{c} = \frac{\sum_j \sum_i (X_{ij} - \overline{X}_j)^2}{c(r-1)} \tag{13-5}$$

To find the numerator in Equation 13-5 we add the squared deviations of the observations about the column mean for each column and then add these sums for all columns. The quantity $r - 1$ is the degrees of freedom in each column, which is the required divisor to get the estimate of σ^2 for each column. We then divide by c, the number of columns, to produce the mean sample variance within the columns.

For the poster experiment, we found that the within-columns estimate of σ^2 is 24.5. Earlier, in conjunction with Equation 13-4, we found that the between-means estimate of σ^2 is 126. It certainly appears that there is far more variability between means than there is within columns. At this point, we might well suspect that the null hypothesis of equal population means for columns will be rejected. But such action should be based on an appropriate statistical test.

The *F* Test

To compare the two estimates of population variance, a ratio is formed. The numerator of this ratio is the between-means estimate of σ^2 and the denominator is the within-columns estimate. The result is called an *F ratio*. That is,

$$F = \frac{\text{estimate of } \sigma^2 \text{ based on variance among sample means}}{\text{estimate of } \sigma^2 \text{ based on variance within columns}}$$

In terms of Equations 13-4 and 13-5,

$$F = \frac{[r\Sigma(\overline{X} - \overline{\overline{X}})^2]/[c-1]}{\left[\sum_j \sum_i (X_{ij} - \overline{X}_j)^2\right]/[c(r-1)]} \tag{13-6}$$

For the poster example,

$$F = \frac{3(126)/(4-1)}{(96 + 56 + 6 + 38)/4(3-1)} = \frac{126}{24.5} = 5.14$$

In the denominator, the values 96, 56, 6, and 38 are the sums of the squared deviations of the observations in each column from the column mean. Because there are 2 degrees of freedom in each column, dividing each of these values by 2 gives the four column variances found earlier. These values of s^2 were 48, 28, 3, and 19 for the four posters.

In Equation 13-6, the numerator consists of the sum of squares between means multiplied by r and divided by the degrees of freedom that apply to the sum of squares, $c - 1$. Similarly, the denominator consists of the sum of squares within columns divided by the degrees of freedom within columns. The degrees of freedom within columns is $r - 1$ per column times c columns, for a total of $c(r - 1)$.

We know that the numerator and the denominator in Equation 13-6 are estimates of the same population variance when the null hypothesis is true. In this case, the F ratio should be near one. When the null hypothesis is not true, the variability among means will be much larger than that within columns, and the F ratio will be considerably greater than one. For the posters, the F ratio is 5.14; we wonder if this is so large that the null hypothesis should be rejected. Should we conclude that the mean sales produced by the four posters are significantly different?

The *F* Distribution

Like z, t, and χ^2, F is a statistic with a sampling distribution that depends upon the characteristics of the sample and the assumptions about the population from which the sample or samples came. As was the case with t and χ^2, there is a whole family of F distributions, one distribution for each pair of degrees of freedom that appear in the numerator and denominator of an F ratio.

Three F distributions are shown in Figure 13-1. For each pair of numbers identifying each distribution, the first number is the degrees of freedom for the numerator of the F ratio and the second number is the degrees of freedom for the denominator. The lower limit of these F distributions is zero and they are usually skewed in the positive direction.

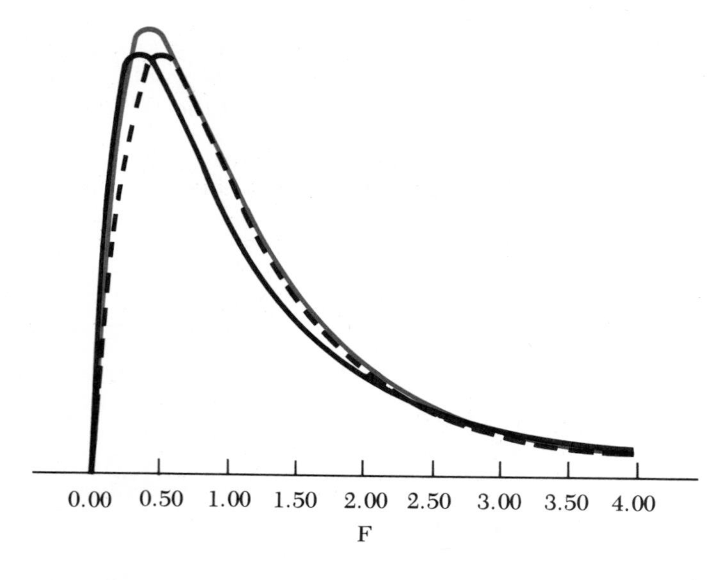

Figure 13-1 Three *F* Distributions

Table A-6 in Appendix A identifies different F distributions by showing the degrees of freedom (df_1) for the numerator of a given F ratio along the top of the table and the degrees of freedom for the denominator (df_2) along the sides. For each pair of values of df_1 and df_2, two numbers appear in the body of the table. The top number is the value of F such that 5% of the values in the distribution are greater. The bottom number is the value of F such that 1% of the values in the distribution are greater. For example, consider the F distribution for which df_1 is 4 and df_2 is 8. Table A-6 shows that 5% of this distribution lies above an F-value of 3.84 and 1% lies above an F-value of 7.01. In an analysis of variance, only 1% of the time will the observations produce an F ratio greater than 7.01 when the column populations all have the same mean. Consequently, if the analysis produces an F ratio greater than 7.01, we can reject the null hypothesis of equal population means at the 0.01 level of significance.

We can now complete the test for a significant difference in effectiveness for our posters. It has already been shown that the data in Table 13-1 produced the F ratio

$$F_{\text{observed, 3, 8}} = 5.14$$

where 3 is the number of columns less 1 and is the degrees of freedom for the numerator. Eight is the degrees of freedom for the denominator. Hence, $df_1 = 3$ and $df_2 = 8$. In Table A-6, we find that an F distribution with these specifications has 5% of its area in the upper tail beyond a value of 4.07. The 0.05 significance level for the test is

$$F_{\text{criterion, 3, 8}} = 4.07$$

Our sample result is so large that we cannot accept the null hypothesis of equal means for the four populations from which the data in the columns came. Subject to a risk of 0.05 of rejecting H_0 when it is true, we conclude that the sales population for at least one poster has a mean that is different from the means of the other poster sales populations.

Exercises

13-1 *Practice with the F table.* Use Appendix Table A-6 to answer the following questions.

a. What percentage of an F distribution with 10 degrees of freedom in the numerator and 20 degrees of freedom in the denominator lies below 3.37?

b. For an F distribution with $df_1 = 10$ and $df_2 = 20$, above what F-value does 1% of the distribution lie? How does this answer compare with the result given in (a)? Is this result to be expected?

c. For an F distribution with $df_1 = 20$ and $df_2 = 10$, what is the value of F such that the probability of exceeding that value is 0.01? Is this the same as the answer for (b)? Why?

13-2 *Testing pharmaceuticals.* A company that manufactures a specific prescription drug has conducted an experiment on three dosage concentrations. With the help of cooperating physicians, four samples of 100 patients taking the drug have been selected for each dosage concentration. The number of patients in each sample who report persistent nausea are shown in the table below.

Dosage Concentration A	B	C
12	14	13
7	6	16
9	13	22
4	11	17

a. Find the mean number of patients who suffered the undesirable side effect for each dosage concentration. Also find the grand mean.
b. Use Equation 13-4 to calculate the between-means estimate of population variance. How many degrees of freedom are associated with the estimate? Use Equation 13-5 to find the within-columns estimate. Also find the within-columns degrees of freedom.
c. Calculate the F ratio for this set of observations. How does it compare with what one would expect if all dosage concentrations produce equivalent side effects?
d. Test the null hypothesis of equal mean effects for the dosage populations at the 0.05 significance level.

13-3 *Automobile traffic planning.* Three traffic observers are being trained to count automobiles making right turns at key intersections in a city. To check on their accuracy, the trainer has them count simultaneously but independently at the same site for four one-hour periods.

	Observer		
Period	*A*	*B*	*C*
1	51	50	52
2	36	46	40
3	35	31	26
4	62	29	34

a. Calculate the observer means and the grand mean.
b. Use Equations 13-4 and 13-5 to estimate population variance with between-means and within-columns sample variance.
c. At the 0.05 level of significance, test the null hypothesis of no differences between observer means. Is this a desirable outcome?
d. Why should observer A's count in Period 4 be investigated?

13-4 *Effect of within-columns variance.* Suppose the four column populations from which observations came for an analysis of variance are normally distributed and have equal variances but different means. Next, suppose that the population variance in the fourth column is doubled. How would the F ratio in the second case be expected to differ from the F ratio in the first case? (*Hint:* Would the result from Equation 13-5 be expected to change much? Would the result from Equation 13-4 change?)

13.2 Calculation Procedures

To reduce the difficulties encountered in calculating the sample F ratio and to achieve reasonable consistency in reporting results of analyses of variance, a more-or-less standard set of procedures has developed over the years. There are two main parts to these procedures. The first part consists of an efficient way to find the sums of squared deviations used in the two estimates of population variance that were described in the previous section of this chapter. The second part consists of showing the several intermediate values needed to calculate a sample F ratio in a standardized format.

The Sums of Squares

In Table 13-2, we can view the deviation of each of the observations from the grand mean as follows.

$$(X_{ij} - \bar{\bar{X}}) = (X_{ij} - \bar{X}_j) + (\bar{X}_j - \bar{\bar{X}}) \tag{13-7}$$

The deviation of an observation from the grand mean is equal to the sum of its deviation from its column mean and its column mean's deviation from the grand mean. Furthermore, with the aid of some algebra, it can be shown that

$$\sum_j \sum_i (X_{ij} - \bar{\bar{X}})^2 = \sum_j \sum_i (X_{ij} - \bar{X}_j)^2 + r \sum_j (\bar{X}_j - \bar{\bar{X}})^2 \tag{13-8}$$

That is, the sum of squared deviations of the observations from the grand mean is equal to the sum of the squared deviations of the observations from their column means added to the sum of the squared deviations of the column means from the grand mean multiplied by the number of rows in the table.

The quantity on the left of Equation 13-8 is called the *total sum of squares*, and we shall denote it by SST. The first quantity to the right of the equal sign in that equation is called the *within sum of squares*, or SSE. The final quantity on the right is called the *between sum of squares*, or SSA. In terms of these symbols,

$$\text{SST} = \text{SSE} + \text{SSA} \tag{13-9}$$

The total sum of squares equals the within sum of squares plus the between sum of squares.

As shown on the left side of Equation 13-8, SST is the sum of the squares of the deviations of all the observations from their mean. It is not difficult to show that

$$\text{SST} = \sum_j \sum_i (X_{ij}^2) - \frac{T^2}{n} \tag{13-10}$$

where T is the grand total of all the observations in the table. The right side of this equation is the sum of the squares of all the observations in the table less a correction term, which is the square of the grand total for the table divided by the number of observations in the table. With Equation 13-10, SST can be calculated relatively easily and with minimal rounding error. Using Table 13-1,

$$\text{SST} = (55^2 + 43^2 + 43^2 + \cdots + 47^2 + 48^2 + 55^2) - \frac{(55 + 43 + 43 + \cdots + 47 + 48 + 55)^2}{12}$$

$$= 22{,}762 - \frac{516^2}{12} = 574$$

In Equation 13-8 the between sum squares (SSA) is the sum of squared deviations of column means from the grand mean, weighted by the number of observations in each column. The between sum of squares can be calculated relatively easily and with small rounding error from the relationship

$$\text{SSA} = \sum_j \left(\frac{C_j^2}{n_j}\right) - \frac{T^2}{n} \qquad \textbf{(13-11)}$$

Here the C_j are the respective column totals, as shown in Table 13-2, and n_j is the number of observations in the column with total C_j. Again, the correction term used in Equation 13-10 is used here. For the poster experiment in Table 13-1,

$$\text{SSA} = \left(\frac{141^2}{3} + \frac{114^2}{3} + \frac{111^2}{3} + \frac{150^2}{3}\right) - \frac{516^2}{12} = 378$$

Finally, Equation 13-9 can be rearranged to obtain the within sum of squares (SSE) by subtraction. That is,

$$\text{SSE} = \text{SST} - \text{SSA} \qquad \textbf{(13-12)}$$

Hence, for the poster experiment,

$$\text{SSE} = 574 - 378 = 196$$

The connection of this discussion about sums of squares with our earlier discussion about estimates of the population variance can now be established. We can compare Equation 13-8 with Equations 13-4 and 13-5. The numerator of Equation 13-4, the between-means estimate, is the same as the last term on the right in Equation 13-8, which we call SSA, the between sum of squares. Similarly, the numerator of Equation 13-5, the within-columns estimate of population variance, is identical to the first term to the right of the equal sign in Equation 13-8, which is SSE, the within sum of squares.

In summary, we have seen that efficient calculation procedures exist to find the sums of squared deviations required for the between-means and the within-columns estimates used to find the sample F ratio.

The Analysis of Variance Table

The steps remaining in this approach to finding the F ratio for the sample can be described by referring to a general format, shown in Table 13-3. The general format applied to the poster experiment is given in Table 13-4.

In both tables, the mean squares are found by dividing the respective sums of squares by the applicable degrees of freedom. Division of SSA by $c - 1$ yields MSA, which is just the between-means estimate of population variance described earlier in connection with Equation 13-4. MSE, the mean squared error, is the within-columns estimate described in connection with Equation 13-5. The value of F for the experiment is found by dividing MSA by MSE. It is helpful to show the criterion values of F just beneath the table, as is done in Table 13-4. This makes it apparent that F_{observed} falls between the criterion values for the 0.05 and 0.01 levels of significance.

Table 13-3 General Format for One-Way Analysis of Variance with the Same Number of Observations per Treatment

Source of Variation	*Sums of Squares*	*Degrees of Freedom*	*Mean Squares*	F
Factor	SSA	$c - 1$	MSA	MSA/MSE
Residual	SSE	$c(r - 1)$	MSE	
Total	SST	$n - 1$		

Table 13-4 Analysis of Variance for Poster Experiment

Source of Variation	*Sums of Squares*	*Degrees of Freedom*	*Mean Squares*	F
Posters	378	3	126	5.14
Residual	196	8	24.5	
Total	574	11		
	$F_{0.95,3,8} = 4.07$		$F_{0.99,3,8} = 7.59$	

Unequal Numbers of Observations in Columns

In all of the procedures described up to this point, we have assumed the same number of observations for each treatment, as was true in the poster example. If possible, we should obtain an equal number of observations for all treatments. Under certain conditions, this arrangement gives maximum protection against violation of the assumptions of normality and equality of variance for the separate treatment populations. This issue is covered under "Violation of Assumptions" in Section 13.4.

Sometimes, however, it is just not possible to obtain the same number of observations for all treatments. At other times one or more observations are spoiled during the course of the data-collection process, and we end up with unequal numbers of observations by accident.

When the numbers of observations in all treatments are not equal, a slight modification of the procedures already described is necessary. The only required change occurs in Table 13-3, the general format. In that table, there are $c(r - 1)$ degrees of freedom associated with residual variation. For unequal observations, this must be changed to $n - c$, where n is the total number of observations in the experiment and c is the number of columns, or treatments, as before.

With regard to sums of squares, Equations 13-10, 13-11, and 13-12 apply to both equal and unequal numbers of observations per treatment.

Exercises

13-5 *Traffic planning revisited.* Show that the methods of Section 13.2 yield the same results for the traffic planning study in Exercise 13-3 as did the methods used there.

a. Find the column sums and the column sums of squares. Then find the grand total and the grand sum of squares for the entire table.
b. Find SST, SSA, and SSE, in that order. Then complete an analysis of variance table.
c. What conclusion is justified by the analysis?
d. Compare these results with those for Exercise 13-3.

13-6 *Brand preference.* A panel of five consumers judged four brands of peanut butter with regard to flavor, texture, packaging, and reputation of the manufacturer. Up to 10 points could be awarded for each characteristic. The composite score is the sum of the points awarded by one of the consumer judges. The composite scores for each judge on each brand are given in the following table.

	Brand			
Judge	*A*	*B*	*C*	*D*
1	60	41	47	58
2	83	34	62	21
3	69	35	53	18
4	69	39	32	50
5	72	50	54	25

a. Apply Equations 13-10, 13-11, and 13-12 to these observations.
b. Prepare an analysis of variance table for this set of observations and interpret the result.

13-7 *Comparing pocket calculators.* An organization is going to purchase several dozen programmable pocket calculators for use by analysts in all major divisions of the organization. A committee has narrowed the choice to three calculators. A representative set of problems has been designed, and three samples of five analysts unfamiliar with the three calculators have been selected. Each analyst was given the problem set, one of the three types of calculators, and the manufacturer's instructions. All were directed to do the problems without any additional aid. Effectiveness was measured by an equally weighted combination of points for percentage of the problem set done correctly and for time spent. One subject dropped out because of illness. Another could not finish because of a serious family problem. A third lost the completion time slip. The remaining results were as follows.

Calculator		
A	*B*	*C*
83	96	86
70	90	90
72	80	74
	95	84
	88	

a. Find n_j, C_j, and ΣX^2 for the columns. Then find n, T, and $\Sigma\Sigma X^2$.
b. Calculate SST, SSA, and SSE. Then prepare an analysis of variance table.
c. What can be concluded from this analysis?

13-8 *Career guidance.* A person in the final semester of an MBA program has chosen real estate as a career field. Rapid advancement to the executive level is a key consideration in the decision about which firm to join. Ages at which people in real estate firms reached the executive level were gathered for as many firms in the state as was reasonably convenient. These data were then classified by the size of firm in terms of total numbers of employees, agents, and associates.

Size of Firm		
Small	*Medium*	*Large*
52	34	48
39	42	52
47	45	50
53	36	53
50	46	
53		
46		

a. Find the column sums and sums of squares. Then find the table sum and sum of squares.
b. Use the procedures of Section 13.2 to construct an analysis of variance table.
c. What conclusion is justified by the F test?

13.3 Estimates and Comparisons

We can begin by summarizing the result of the poster experiment up to this point. The four sample mean sales were as follows.

Poster:	A	B	C	D
Mean sales:	47	38	37	50

In Table 13-4, the analysis of variance table for the experiment, we rejected the hypothesis that the four population means are equal at the 0.05 level of significance.

Concluding that all population means are not the same is hardly enough, however. One additional type of knowledge that is often useful consists of point and interval estimates for a column, or treatment, mean. Should the decision be made to use poster D, it would be helpful to have a confidence interval estimate of the mean sales to be expected. Such knowledge would also be useful in planning amounts of stock to order and similar tasks.

A second type of knowledge also is required. Once it has been established that there is a significant difference among the treatment means, it becomes important to identify the sources of that significance. We have already said that it looks as if there is no significant difference between the means for posters A and D, nor between the means for posters B and C. But it does look as if the mean for the first of these two sets of posters is significantly different from the mean of the second set. Such judgments need to be justified by confidence interval estimates for the various possible differences.

Estimates of a Treatment Mean

In analysis of variance, as elsewhere in earlier chapters, the mean of a random sample is an unbiased, efficient, and consistent point estimate of the mean of the population from which the sample came. Because each of the columns in our one-way analysis of variance experiment is such a sample, the column mean is the stated type of point estimate for the mean of the treatment population. In symbols,

$$\text{point estimate of } \mu_j = \overline{X}_j \tag{13-13}$$

Up to this point in our intuitive interpretations of sample treatment means, we have acted as if this were true. Now the soundness of our intuition has been established. The four sample means given for the poster experiment at the beginning of this section are appropriate point estimates of their respective population means.

It was stated earlier that all of the treatment populations in an analysis of variance are assumed to have the same variance. This assumption was used as the basis for the pooled within-columns estimate of the common population variance that constitutes the denominator of an F ratio. In Table 13-4, we saw that this denominator is

$$\text{MSE} = \frac{\text{SSE}}{df} \tag{13-14}$$

In this equation, SSE consists of the pooled sum of the squared deviations of the observations in each treatment from the treatment mean. When SSE is divided by the applicable degrees of freedom, the result is MSE, called the *mean squared residual error*, which is an unbiased, efficient, and consistent point estimate of σ^2, the common variance of the treatment populations. That is,

$$\text{point estimate of } \sigma^2 = \text{MSE} \tag{13-15}$$

Again, we have made the assumption that this was the case in our earlier discussions. We shall shortly use this estimate as one of the ingredients in the confidence interval estimate of a treatment mean.

The sampling distribution of $\overline{X}$, the sample mean, was discussed in several contexts in previous chapters. The standard deviation of such a distribution has been designated as the *standard error of the mean*, $\sigma_{\overline{X}}$. In every case, the standard error of the mean when the sample was less than 10% of the population was

$$\sigma_{\overline{X}} = \sqrt{\frac{\sigma^2}{n}}$$

The standard error of the mean is the square root of the population variance divided by the sample size. When the population variance is unknown, the sample variance is used as a point estimate and the estimated standard error of the mean becomes

$$s_{\overline{X}} = \sqrt{\frac{s^2}{n}}$$

The situation is no different in the case of the estimated standard error of a treatment mean in an analysis of variance. The best estimate available for the common variance of the treatment populations is divided by the number of observations in the specified treatment, and the square root of this result is the desired standard error. Symbolically,

$$s(\overline{X}_j) = \sqrt{\frac{\text{MSE}}{n_j}} \tag{13-16}$$

For the poster experiment, it was determined in conjunction with Table 13-4 that MSE = 24.5. Furthermore, for any one of the treatment means in Table 13-1, there are three observations in the sample. Hence $n_j = 3$. It follows that the estimated standard error for any one of the four poster means is

$$s(\overline{X}_j) = \sqrt{\frac{24.5}{3}} = 2.8577$$

Given a point estimate of a treatment mean and its standard error, we can readily obtain a confidence interval estimate for that mean. In Chapter 8, we formed a confidence interval estimate of a population mean by using the general expression

$$\overline{X} - ts_{\overline{X}} \leq \mu \leq \overline{X} + ts_{\overline{X}}$$

Here, $\overline{X}$ is the sample mean, $s_{\overline{X}}$ is the estimated standard error of the mean, and t is the appropriate table value for the confidence level. The degrees of freedom for t were $n - 1$, one less than the number of sample observations used to find the sample mean and standard deviation.

Forming a confidence interval estimate of a treatment mean in one-way analysis of variance requires only slight modification of the procedure just discussed.

The general expression for the *confidence interval estimate of a treatment mean* is

$$\bar{X}_j - t\sqrt{\frac{\text{MSE}}{n_j}} \leq \mu_j \leq \bar{X}_j + t\sqrt{\frac{\text{MSE}}{n_j}} \qquad \text{(13-17)}$$

where t has $n - c$ degrees of freedom, the degrees of freedom applicable to MSE.

We want a 0.95 confidence interval estimate of the mean sales for poster D in our example. We have already found that $\bar{X}_4 = 50$ and that $\sqrt{\text{MSE}/3} = 2.8577$. From Table A-4 in Appendix A, we find that $t_{0.975,8} = 2.3060$. Substituting these in Equation 13-17, we have

$$50 - 2.3060(2.8577) \leq \mu_D \leq 50 + 2.3060(2.8577)$$

or

$$43.41 \leq \mu_D \leq 56.59$$

At the 0.95 level of confidence, we estimate that the mean sales for poster D are between 43.41 and 56.59. This interval is rather wide. But it is based on only a small number of observations. Equation 13-16 makes it apparent that the way to reduce the width of an interval estimate for a given level of confidence is to increase the number of observations.

Contrasts

Finding the confidence interval for a given treatment mean is helpful. But such an interval contributes little to identifying the sources of the significant differences in those cases where the null hypothesis has been rejected. The usual question of interest is: What is the pattern of significant differences among the treatment means that gave rise to overall significance? For example, the following three questions would occur to most people with regard to the poster experiment treatment means.

1. Is the difference between the means for posters A and D significant, where the sample estimate of the difference is

$$\bar{X}_D - \bar{X}_A = 50 - 47 = 3$$

2. Is the difference between the means for posters B and C significant, where the sample estimate of the difference is

$$\bar{X}_B - \bar{X}_C = 38 - 37 = 1$$

3. Suppose, as we suspect, the answer to both of the previous questions is *No*. Then it may be that we have just two poster populations with respect to sales effects: posters A and D and posters B and C. Then a third question arises. Is the

difference between the means for the two groups of posters significant, where the sample estimate of the difference* is

$$\frac{\overline{X}_D + \overline{X}_A}{2} - \frac{\overline{X}_B + \overline{X}_C}{2} = \frac{50 + 47}{2} - \frac{38 + 37}{2} = 48.5 - 37.5 = 11$$

If the answer to the last question is *Yes*, then—in combination with the negative answers to the first two questions—we have an explanation for the overall significance of the F test. Posters A and D constitute one family with regard to effect on sales, and posters B and C constitute another. Furthermore, the confidence interval for the difference in the means of these two families must show that the difference is positive. This is a necessary consequence of finding the difference of +11 to be significant.

To answer questions such as the three above, we must know how to find the applicable confidence interval estimates in order to determine whether or not these estimates include zero. If one does, then the population difference it estimates is not significant.

The first step in developing the necessary confidence intervals is to observe that the differences shown in the three questions posed above are special types of differences known as *contrasts*.

In general, the *sample estimate of the ith contrast* is

$$K_i = \sum_j b_j \overline{X}_j \qquad \textbf{(13-18a)}$$

where the b_j are arbitrarily chosen coefficients subject to the constraint that

$$\sum_j b_j = 0 \qquad \textbf{(13-18b)}$$

We can restate the differences shown in our original three questions as contrasts.

K_1: $(+1)\overline{X}_D + (-1)\overline{X}_A = 50 - 47 = 3$

K_2: $(+1)\overline{X}_B + (-1)\overline{X}_C = 38 - 37 = 1$

K_3: $(+1/2)\overline{X}_D + (+1/2)\overline{X}_A + (-1/2)\overline{X}_B + (-1/2)\overline{X}_C = 50/2 + 47/2 - 38/2 - 37/2 = 11$

In each case, a term in the sum is the product of a coefficient and a treatment mean; the sum of the coefficients is zero. The coefficients are chosen to make the contrasts meaningful weighted averages of the treatment means.

* It can be shown that differences between sample treatment means of the type shown are minimum variance, unbiased point estimates of the corresponding differences between population means.

The Sampling Error of a Contrast

Given a sample estimate of a contrast, we must next find the estimated sampling error of that contrast. We shall first find the sampling variance of a contrast and then take its square root to get the standard error.

The variance of the sampling distribution for a contrast bears a strong resemblance to the variance of the sampling distribution of two means, which was discussed in Chapter 11. In that chapter, we saw that the sampling variance of the difference between two means is the sum of the sampling variances of each mean. The sample estimate is

$$s^2_{\bar{X}_1 - \bar{X}_2} = \frac{s_1^2}{n_1} + \frac{s_2^2}{n_2}$$

To get the sampling variance of a contrast, we square the coefficients and use these squares as weights for the sampling variances of the treatment means in the contrast. Then we sum the weighted sampling variances of the treatment means. In symbols, the sampling variance of a contrast as estimated from the treatment samples is

$$s^2_{K_i} = b_1^2\frac{\text{MSE}}{n_1} + b_2^2\frac{\text{MSE}}{n_2} + \cdots + b_c^2\frac{\text{MSE}}{n_c} \qquad \text{(13-19)}$$

where K_i identifies a particular contrast, and there are c treatment means in the contrast. For the posters, we saw in conjunction with our discussion of Equation 13-16 that

$$\frac{\text{MSE}}{n_j} = \frac{24.5}{3} = 8.167$$

for each of the treatment means. The sampling variance of the contrast between posters A and D is, therefore,

$$\begin{aligned} s^2(K_1) &= (+1)^2(8.167) + (-1)^2(8.167) \\ &= (2)(8.167) = 16.33 \end{aligned}$$

The sampling variance for the contrast between the means of posters B and C is found in an identical manner. Hence, it is also 16.33. The estimated sampling variance for the third contrast is

$$\begin{aligned} s^2(K_3) &= (+1/2)^2(8.167) + (+1/2)^2(8.167) + (-1/2)^2(8.167) + (-1/2)^2(8.167) \\ &= 4(1/4)(8.167) = 8.167 \end{aligned}$$

Stating Equation 13-19 more succinctly, we have the following.

The sampling variance of a contrast is

$$s^2(K_i) = \text{MSE}\sum_j \frac{b_j^2}{n_j} \tag{13-20}$$

from which it follows that the *standard error* of a contrast is

$$s(K_i) = \sqrt{\text{MSE}\sum_j \frac{b_j^2}{n_j}} \tag{13-21}$$

where the standard error has the same number of degrees of freedom as does MSE, that is, $df_2 = n - c$. As before, n is the total number of observations in the analysis of variance table and c is the number of columns, or treatments.

For the posters, the standard errors of the three contrasts are

$$s(K_1) = s(K_2) = \sqrt{16.33} = 4.04145$$

and

$$s(K_3) = \sqrt{8.167} = 2.8577$$

In each case, there are $12 - 4 = 8$ degrees of freedom associated with the standard error.

We can now state the interval estimate of a contrast. The confidence interval estimate of a contrast is

$$K - t * s(K) \leq \mathbf{K} \leq K + t * s(K) \tag{13-22}$$

where K and $s(K)$ are defined in Equations 13-18 and 13-21 and t has $n - c$ degrees of freedom. Here, **K** is the population value of the contrast.

This procedure can be used to make just one such interval estimate for a given set of observations because the t distribution assumes that each estimate is based on a random sample that is independent of any other random sample. Since all of the observations in one set are used to estimate MSE, it would take a new set to get an independent estimate.

A 0.95 confidence interval estimate will be found for the contrast that includes posters D and A in our example. We have already found that $K_1 = 50 - 47 = 3$, $s(K_1) = 4.04145$, and $t_{0.975,8} = 2.3060$. The 0.95 confidence interval estimate of the difference in mean sales for posters D and A is

$$3 - 2.3060(4.04145) \leq \mu_D - \mu_A \leq 3 + 2.3060(4.04145)$$

or

$$3 - 9.32 \leq \mu_D - \mu_A \leq 3 + 9.32$$

which is

$$-6.32 \leq \mu_D - \mu_A \leq 12.32$$

Since this interval includes zero, we cannot conclude that these two population means are different.

Multiple Comparisons

We could form a confidence interval estimate for any one of the contrasts we have been considering by selecting a suitable value of t, multiplying that value by the standard error, and adding the product to and subtracting it from the point estimate of the difference. But we seldom want a confidence interval for just one contrast. Instead, as in the poster experiment, we want to look at *an entire family of contrasts*. Consequently, because we are dealing with the probability of a joint event, we must replace t with a value that gives us a specified level of confidence for the entire set of contrasts. In such a situation, we are making multiple comparisons, and we need a procedure that affords the necessary protection for the combined set of confidence intervals, where there is one such interval for each contrast.

There are several alternative replacements for t in the situatin described. The one we shall use is attributable to Scheffé, and we shall use T to indicate that it is a substitute for t. In multiple comparisons, the multiplier for the standard error of each contrast is

$$T = \sqrt{(c-1)F_{df_1,df_2}} \tag{13-23}$$

where $df_1 = c - 1$ and $df_2 = n - c$, as in the analysis of variance table. The value of F is selected from Appendix Table A-6 to establish an overall confidence level of either 0.95 or 0.99. For the family of contrasts in the poster experiment, suppose we select 0.95 as the overall confidence level. Then

$$\begin{aligned} T &= \sqrt{(4-1)F_{0.95,3,8}} \\ &= \sqrt{(3)(4.07)} = \sqrt{12.21} = 3.4943 \end{aligned}$$

Alternative confidence levels can be obtained from more extensive F tables.

Finally, for multiple comparisons, the confidence interval for each contrast in the set is

$$K_i - T * s(K_i) \leq \mathbf{K}_i \leq K_i + T * s(K_i) \tag{13-24}$$

where $\mathbf{K}_i$ is the population value for the contrast, K_i is the point estimate of $\mathbf{K}_i$, and T is the multiplier defined in Equation 13-23. In the poster example, we have already found that

$$K_1 = 50 - 47 = 3 \qquad K_2 = 1 \qquad K_3 = 11$$

$$s(K_1) = s(K_2) = 4.04145 \qquad s(K_3) = 2.8577$$

and that $T = 3.4943$ for all three contrasts. The family of confidence intervals is, therefore,

$$3 - 3.4943(4.04145) \leq \mathbf{K}_1 \leq 3 + 3.4943(4.04145)$$

$$1 - 3.4943(4.04145) \leq \mathbf{K}_2 \leq 1 + 3.4943(4.04145)$$

$$11 - 3.4943(2.8577) \leq \mathbf{K}_3 \leq 11 + 3.4943(2.8577)$$

When simplified, these intervals are

$$-11.12 \leq \mathbf{K}_1 \leq 17.12$$

$$-13.12 \leq \mathbf{K}_2 \leq 15.12$$

$$1.01 \leq \mathbf{K}_3 \leq 20.99$$

The first two intervals contain zero. Hence they indicate that the first two contrasts in our set of three are not significant. The last contrast does not contain zero. Hence it indicates a significant result. We can, therefore, make the composite statement that posters A and D constitute one group with respect to sales effect, posters B and C constitute another, and the first group produces more sales than the second. The level of confidence for this entire statement is 0.95. The outcome confirms our intuitive judgment made at the beginning of this section.

Earlier, when we found the 0.95 interval estimate for the single contrast for posters D and A, the interval ran from −6.32 to +12.32. In the family of three contrasts, for which the combined confidence level is 0.95, the contrast for posters D and A runs from −11.12 to +17.12, a much greater range. This is a consequence of providing 0.95 confidence for an entire family of contrasts in the latter case and for only a single contrast in the former case.

The Scheffé T-multiplier defined in Equation 13-23 permits us to place as many contrasts as we wish into the composite statement without affecting the overall level of confidence. Furthermore, any additional contrasts may be stated at the outset or included subsequently, as may happen when the original contrasts are interpreted. For example, suppose that results had been different in our composite statement for the poster experiment. Specifically, suppose that, in the first contrast, poster D with a mean of 50 had turned out to be significantly different from poster A with a mean of 47. In the second contrast, suppose there were no change in the conclusion: posters B and C, with means of 38 and 37, are not significantly different. It follows that we are not interested in the third contrast we stated previously, which was the group A and D versus the group B and C. We cannot assume that posters A and D constitute a group, and this nullifies our interest. Instead, we now want to see whether or not poster A is significantly better than the group containing posters B and C. In the composite statement, we would change the third contrast to

$$K_3 = 47 - \frac{37 + 38}{2}$$

and construct the appropriate confidence interval. Suppose that the intervals for K_3 and K_1 do not now include zero, but the interval for K_2 (for posters B and C) does. Then we conclude that the overall significance comes from poster D being better than A and poster A being better than the group composed of posters B and C. A confidence level of 0.95 would still apply to this new composite statement.

Throughout this section, we have used an example in which there is an equal number of observations per treatment. We could just as well have used an example with different numbers of observations in the treatments; all procedures described would still apply.

13.4 Some Final Comments

Replication, Randomization, and Cross Classification

There is an ever-present problem in designing an experiment to be analyzed by analysis of variance. It is similar to the problem of establishing a cause-and-effect relationship between the independent and dependent variables in regression. The problem is whether the overall significance in an experiment is actually the result of the explanation furnished by the multiple comparisons, or is, in fact, the result of some related factor not included in the experiment.

In the poster experiment, we have attributed the result to differences in such things as poster design and message. But it could happen that the markets selected for posters D and A were in areas higher on the socioeconomic scale than the areas around the markets selected for the other two posters. The product, a canned mixture of salted nuts, may sell better to persons from one income group than to those from another income group. Consequently, the reason for the differences may not be differences in poster characteristics at all. Instead, the explanation may lie in local market characteristics.

There are three important ways of ruling out other factors as being responsible for observed results. None of the three is absolutely guaranteed, but all are highly effective when used properly.

The first is by randomly and independently selecting more than one store for each type of poster. We did this when we selected three stores for each type of poster. This is called *replication*. By selecting an independent random sample of several stores for each poster, we are providing an opportunity for extraneous factors to average out when we calculate the mean effect for a poster. In addition, the variability attributable to all factors other than type of poster is present in the within sum of squares (SSE) that constitutes the denominator of the F ratio we discussed. The numerator contains not only the effects of poster type but also those of all unobserved factors. Hence a significant F ratio indicates a poster effect that is large enough to be detected over and above the variability or noise produced by all other factors in the operating environment. This gives us reason to expect the same effect when we put the findings into practice.

The second way to rule out possible alternative factors as an explanation for observed differences in results is through *randomization*. To understand randomization, suppose the 12 stores chosen were selected at random from a roster of stores listed in order of geographic location. Each store on the roster has a number assigned to it. Stores on the eastern edge of the city are listed first and have lower numbers. Those successively farther west are farther down the list and have higher numbers. Suppose the stores are assigned to poster types as shown in Table 13-5.

With such an arrangement, there is a high probability that stores on the east side will be assigned posters A and B, while stores on the west side will be assigned C and D. Suppose the east side is far less affluent than the west side and the product sells primarily to people with higher incomes. If the pattern in the list of stores was not brought to light, the researcher could erroneously conclude that poster D is

Table 13-5
Posters Assigned as Listed

Poster	*Store*
A	3
A	5
A	9
B	10
B	11
B	15
C	18
C	21
C	24
D	30
D	32
D	33

Table 13-6
Randomization of Experimental Units

Poster	*Store*
A	24
A	32
A	9
B	30
B	33
B	11
C	3
C	15
C	5
D	21
D	10
D	18

responsible for better sales results when, in fact, the higher average income on the west side is responsible.

Randomization means using a table of random numbers or the equivalent to assign experimental units to treatments. In our experiment, the 12 stores are the units and the poster types are the treatments. In Table 13-6, the first store number encountered in a table of random numbers was 24, the second was 32, and so forth.

We can expect incomes and any other systematic factors from the list to be equalized in the four poster-type subsamples. There is, of course, no guarantee that randomization will produce subsamples that are all representative of the entire urban area, but the larger the subsamples, the more likely this is to happen.

It is important to note that random sampling and randomization are two entirely different concepts. On the one hand, we used random sampling to select 12 stores for the experiment from the population of stores in the urban area. On the other hand, we used randomization to assign stores in the sample to poster types. This difference is true in general. Random sampling is a technique for selecting experimental units. Randomization is a technique for assigning units in the sample to factor levels.

A third way to rule out a possible alternative factor as an explanation for observed differences in results is to include that factor in the design. Including the factor will make possible a statistical test of its effect, if any. Suppose the income level of the neighborhood in which a store operates is considered too important for us to rely on replication and randomization alone for control. Suppose instead that we use the latest census to classify the neighborhood for each store as having income either above or below the median. We will call the first *high-income neighborhoods* and the second *low-income neighborhoods*. Next we select a random sample of four stores from the high-income portion of the population and another random sample of four stores from the low-income portion.

First, let us assume that the eight stores in our sample are assigned by a randomization process to treatments as shown in Table 13-7. Here we have an example of the *confounding* of factor effects. Suppose that mean daily sales for posters A and B are much larger than for posters C and D. Since only high incomes are associated with A and B and only low with C and D, we cannot know whether to attribute the response differences to poster type or income. This arrangement is not acceptable.

Table 13-7 Incorrect Inclusion of Second Factor

Income	*Poster*	*Store*
High	A	36
High	A	41
High	B	39
High	B	28
Low	C	14
Low	C	6
Low	D	19
Low	D	11

Table 13-8 Cross Classification of Factor Levels

Income	*Poster*	*Store*
High	A	36
High	B	41
High	C	39
High	D	28
Low	A	14
Low	B	6
Low	C	19
Low	D	11

Now consider the arrangement in Table 13-8. Here every possible combination of income and poster appears. We describe such arrangements by using the term *cross classification*. The reason for using this term becomes apparent when we use an alternative display for the arrangement in Table 13-8. This alternative is shown in Table 13-9, where each cell represents a unique combination of income level and poster.

When treatments are cross-classified, analysis of variance makes it possible to test for the effect of one factor *independent* of the effects of all other factors included in the design. In such a test for our example, if the poster effect is statistically significant this conclusion holds regardless of any possible effect of income on sales.

Besides replication, randomization, and cross classification, there is one more principle of sound research design. The experimental units should come from a *homogeneous* statistical population of units. For our example, it would be an obvious violation to include supermarkets and drugstores along with convenience stores on the list from which the sample of experimental units will come. Again, there is no way to guarantee homogeneity, but important violations can frequently be brought out by exploratory analysis.

Two-Factor Analysis of Variance

As Table 13-9 and the accompanying discussion imply, it is possible to include row effects as well as column effects in a single analysis of variance. There are several types of such designs. They are known collectively as *two-factor analysis of variance*. Chapter 15 is devoted to two-way analysis of variance. Designs also exist for including more than two factors in the experiment. These are commonly referred to as *multifactor designs*.

Table 13-9 Alternative Display of Cross Classification

		Poster Type			
		A	B	C	D
Income	High				
	Low				

Violation of Assumptions

As we pointed out earlier in this chapter, there are three assumptions underlying the sampling theory upon which analysis of variance is based. The observations are assumed to be randomly and independently selected from a treatment population. Treatment populations are assumed to be normally distributed. Treatment populations are assumed to have equal variances. But what happens to the F tests, the estimates, and the Scheffé multiple comparisons if one or more of these assumptions do not hold?

In order to answer the question, we must distinguish between two models used for analysis of variance, regardless of how many factors are involved. Model I is illustrated by the examples and exercises in this chapter. In this model, interest lies only in the treatments included in the experiment. For example, the four posters we have discussed represent the extent of our interest in posters. They are not considered to be random selections representing four general classes of posters, for instance. Model II, on the other hand, considers treatments to be a random sample from a population of such treatments. Selecting one machine from each of four age groups of machines in a factory and collecting ten productivity observations per machine selected in order to study the effect of age on productivity could be such an example. If the intent is to extend the conclusions to the four age groups and not confine the conclusions just to the four particular machines that constitute the treatments, then model II applies.

Regardless of which model applies, random and independent selection of observations must be observed. This is not usually difficult to accomplish. Use of a table of random numbers and similar procedures can be expected to provide such observations.

For model I, called the *fixed effects model*, departure of the treatment populations from normality in their distributions is of no major importance. It takes extreme departures seldom met in practice to cause noticeable changes in the sampling distributions underlying the described inference procedures.

Also for model I, there is an essentially negligible effect if the variances depart from equality *so long as there is the same number of observations in each treatment.* If it cannot be established that there is good reason to assume equality of population variances, then it is very important to make the number of observations per treatment equal. Of course, if some 20 observations per treatment were planned and an observation happened to be destroyed, one can still assume there is ample protection against departure from equal variances. The greater the (equal) number of planned observations per treatment, the smaller the effect of a missing observation.

Generally speaking, model II is very sensitive to departures from the normality and equal variance assumptions. There are many excellent references on this subject that can be consulted to select the most appropriate of the several techniques for handling such departures.

Exercises

13-9 *Comparisons of brand preferences.* In Exercise 13-6, a panel of five judges scored each of four brands of peanut butter. The sample value of F was significant at the 0.01 level. Multiple comparisons are now to be made among the four sample means to identify the sources of the overall significance.

a. Find point estimates for the following contrasts.

$$K_1 = \overline{X}_A - \overline{X}_C$$

$$K_2 = \overline{X}_C - \overline{X}_B$$

$$K_3 = \overline{X}_B - \overline{X}_D$$

$$K_4 = \overline{X}_C - \overline{X}_D$$

b. Apply Equations 13-16 and 13-21 to find the standard errors of the contrasts in (a).

c. Use Equations 13-23 and 13-24 to calculate interval estimates for the four contrasts. Assume a 0.90 confidence level for the set of contrasts ($F_{0.90,3,16} = 2.46$).

d. Why is the following contrast suggested by the above analysis?

$$K_5 = \overline{X}_A - \frac{(\overline{X}_B + \overline{X}_C + \overline{X}_D)}{3}$$

Add the interval estimate for this contrast to the set.

e. What conclusion is justified by the set of five interval estimates?

13-10 *Starting ages of executives.* Continue the analysis started in Exercise 13-8 by making multiple comparisons of the mean ages at which real estate executives began their first executive job with the firm.

a. Make point estimates for the contrasts

$$K_1 = \overline{X}_S - \overline{X}_M$$

$$K_2 = \overline{X}_L - \overline{X}_S$$

where S, M, and L represent small, medium, and large firms.

b. Use Equations 13-16, 13-21, 13-23, and 13-24 to form a set of confidence intervals for this set of contrasts. Use the 0.95 level of confidence.

c. Interpret the results in (b). Why are no additional contrasts required?

13-11 *Incentive systems and productivity.* Three different incentive systems are being investigated with regard to their effect on productivity of assembly workers in a plant. Workers are to be assigned to two categories: those with young children and those without.

a. List all combinations of these two factors that are needed to accomplish complete cross classification.
b. What is the total number of workers required to use complete cross classification and include four replications in an experiment? How many workers will be subject to each combination of factor levels?

13-12 *Random sampling and randomization.* Distinguish between the roles of random selection and randomization with regard to the workers needed in Exercise 13-11(b).

Summary

Analysis of variance consists of procedures for making statistical inferences concerning a set of more than two sample means. Typically the different means constitute responses of a dependent variable to different treatments or levels of one or more independent variables or factors. An *F statistic* is used to test the null hypothesis that the population means for the different treatments are equal. The statistic compares variability among the treatment means with variability of observations within treatments.

When the hypothesis of equal treatment means has been rejected, it is usual to make estimates of these means. The objective is to find the pattern of differences among the means. *Contrasts* and *multiple comparisons* are used to determine the pattern. They make it possible to control the confidence level of estimates for the entire pattern.

Supplementary Exercises

13-13 *Estimated product development costs.** Ratios of actual to estimated cost of developing new products in an ethical pharmaceutical firm for three types of products were reported as follows.

Product Type	$\overline{X}$	s	n
Chemical entities	2.25	0.72	13
Compounded products	1.70	0.72	23
Alternative dosage forms	1.51	0.71	13

Given these results, we can reconstruct the quantities needed for an analysis of variance.

* J. E. Schnee, *Journal of Business*, July 1972.

	Product		
	A Chemicals	*B* Compounds	*C* Dosages
n_j:	13	23	13
C_j:	29.25	39.10	19.63
ΣX^2	72.0333	77.8748	35.6905

a. Calculate the total sum of squares, sum of squares between means, and within sum of squares.
b. Prepare an analysis of variance table and compare the sample F ratio with 0.05 and 0.01 criterion values.
c. Discuss the possible import of these results for company operations.

13-14 *Multiple comparisons for traffic observers.* For the situation described in Exercise 13-3, suppose that the period 4 observation for observer A is found to be in error. Instead of 62, the observation should be 26.
a. Complete an analysis of variance table for the revised data and test the null hypothesis at the 0.05 level.
b. What does the analysis in (a) suggest with regard to multiple comparisons?

13-15 *Comparing product development costs.* For the analysis of variance reported in Exercise 13-13, make multiple comparisons of the treatment means.
a. Find point estimates for the following contrasts.

$$K_1 = \overline{X}_A - \overline{X}_B$$

$$K_2 = \overline{X}_B - \overline{X}_C$$

$$K_3 = \overline{X}_A - \frac{\overline{X}_B + \overline{X}_C}{2}$$

b. Calculate interval estimates for the contrasts in (a). Use a 0.90 confidence level for the multiple comparisons ($F_{0.90,2,46} = 2.42$).
c. Does any single product type seem to be worse than the other two with respect to cost estimating procedures?

13-16 *Household laundry cleaner.* A consumer testing service in a large city is testing the claim made by the manufacturer of a cleansing product for home laundry use. The claim is that the product removes stains better than three other brands named in the claim. The testing service director is concerned that the very hard water in the city may nullify the product's performance. A panel of six judges scores four carefully matched loads of stained material. High scores indicate low residual staining. One load was washed by each product. The product advertised as being better is C in the following table.

	Product			
Judge	*A*	*B*	*C*	*D*
1	31	23	36	16
2	29	31	33	21
3	32	20	39	24
4	34	25	37	14
5	29	34	43	21
6	31	29	40	18

a. Calculate SST, SSA, and SSE.
b. Prepare an analysis of variance table and state the conclusion with respect to the null hypothesis.
c. Can the manufacturer's claim be evaluated just from the results in (b)?

13-17 *Selecting the best pocket calculator.* For the experiment described in Exercise 13-7, multiple comparisons are needed to determine which calculator appears to be best.

a. Find sample point estimates for the following contrasts.

$$K_1 = \overline{X}_B - \overline{X}_A$$

$$K_2 = \overline{X}_B - \overline{X}_C$$

$$K_3 = \overline{X}_C - \overline{X}_A$$

$$K_4 = \frac{\overline{X}_B + \overline{X}_C}{2} - \overline{X}_A$$

b. Calculate interval estimates for the contrasts in (a) for an overall confidence level of 0.90 ($F_{0.90,2,9} = 3.00$).
c. What composite statement is justified by the analysis in (b)? Are any additional contrasts suggested?

13-18 *Evaluating manufacturer's claim.* For the laundry cleaner claim data in Exercise 13-16, multiple comparisons are to be used to evaluate the manufacturer's claim for product C.

a. Find point estimates of treatment mean differences for the following pairs: C and A, B and D, A and B.
b. For a 0.95 composite confidence level, make interval estimates for the contrasts described in (a) and interpret the result.
c. Are any additional contrasts necessary? Why?

13-19 Why are all the F tests for this chapter upper one-tailed tests? That is, why test only for excessively large values of F?

13-20 Is the assumption of equal variances likely to be an underlying problem that could invalidate the conclusions reached in Exercise 13-13?

Computer Exercises

Source: Personnel data file. Treat this file as the population of interest.

1 Test the null hypothesis that all management levels in the firm have equal mean salaries with a random sample of five observations from each management level.

2 Construct a bar chart showing the four sample means from Exercise 1. Does this chart help explain the result in Exercise 1?

3 Find the population mean salary level for each management level. Is the null hypothesis in Exercise 1 true? Are you usually able to know this?

Interpreting Computer Output: A Mini Case

Use the following material to form your own conclusion with regard to the issue presented.

In 1981 Maine, Connecticut, and Vermont had bottle laws that required a deposit on beverage containers. The Massachusetts Legislature overrode the governor's veto of a bottle bill, but the law would not take effect until 1983. The Rhode Island Legislature was considering a bottle bill. Opponents of the bill claimed that bottle laws would drive up costs to the consumer. In an attempt to determine whether bottle laws influence consumer prices, the Providence *Evening Bulletin* conducted a price survey in four states on November 19, 1981. The methods and results are shown in Figure 13-2, and the relevant ANOVA tables are shown in Figures 13-3 and 13-4.

What conclusions do you draw about the survey?

True cost of bottle bill: Unknown

Six-pack of Coca-Cola in 12-ounce cans

STATE	Store 1	Store 2	Store 3	Store 4	Store 5	Avg.
Mass.	1.69	1.69	1.89	1.79	1.69	1.75
R.I.	1.99	1.69	1.99	1.88	1.69	1.85
Conn.	2.29	1.99	2.29	1.99	2.29	2.17
Vt.	2.19	2.15	2.25	2.25	2.19	2.21

Six-pack of Budweiser in 12-ounce bottles

STATE	Store 1	Store 2	Store 3	Store 4	Store 5	Avg.
Mass.	2.80	2.70	2.49	2.65	2.90	2.71
R.I.	2.68	2.73	2.57	2.68	2.68	2.67
Conn.	2.72	2.56	2.68	2.56	2.56	2.62
Vt.	2.70	3.00	2.89	2.59	3.00	2.84

How the survey was conducted

Five stores were selected at random in the Providence, Boston, Hartford and Burlington, Vt. areas and asked the price of the cans. Five other stores were asked the price of the bottles. Taxes and deposits were subtracted from the price in states where applicable. Prices are as of November 19.

Beer cheaper in Conn.; soda cheaper in Vermont

By PAUL CARRIER
Journal-Bulletin Staff Writer

Soft drink prices are higher in two New England states that have returnable bottle laws than in Rhode Island and Massachusetts, which allow throwaway containers, according to a telephone survey of 40 retailers.

The survey also showed that beer costs less in Connecticut, a state with a returnable container law, than in any of the other states. However, beer prices are highest in Vermont, which has the oldest returnable bottle law in the region.

Those findings are subject to conflicting interpretations, as the Rhode Island battle lines are drawn anew for a perennial legislative debate full of opinions but devoid of clear conclusions.

Maine, Connecticut and Vermont are the only New England states with mandatory bottle deposit laws on the books. Rhode Island and New Hampshire do not have similar statutes. The Massachusetts Legislature recently overrode Governor Edward King's veto of a bottle bill, and a law is scheduled to take effect in 1983.

Opponents of laws that require mandatory deposits on beverage containers — metal, glass or plastic — contend they lead to higher prices for consumers.

To test this assertion, the *Journal-Bulletin* compared prices among the four states, at sales outlets selected at random from the Providence, Boston, Hartford and Burlington, Vt., telephone directories.

In each state, five stores were asked how much they charged for a six-pack of Coca-Cola in 12-ounce cans. Five other stores in each state were contacted to determine the retail price for a six-pack of Budweiser in 12-ounce bottles. Sales taxes and deposits were subtracted from the selling prices.

All of the prices included in the survey were in effect Nov. 19.

In the Coca-Cola survey, prices in Connecticut and Vermont were consistently the same as, or higher than, the top price in Massachusetts and Rhode Island. On average, prices ranged from $1.75 in Massachusetts and $1.85 in Rhode Island to $2.17 in Connecticut and $2.21 in Vermont.

The beer survey, however, showed an average price in Vermont higher than in Connecticut, Rhode Island and Massachusetts. But some stores in Vermont undersold their counterparts in both throwaway bottle states.

Connecticut registered the lowest average price for beer among the four: $2.62 per sixpack, compared to $2.67 in Rhode Island, $2.71 in Massachusetts and $2.84 in Vermont.

* * *

Figure 13-2 Bottle bill survey from the Providence *Evening Bulletin*, November 23, 1981

```
DEPENDENT VARIABLE: PRICE

SOURCE                        DF     SUM OF SQUARES                MEAN SQUARE

MODEL                          3         0.78385500                 0.26128500

ERROR                         16         0.23880000                 0.01492500

CORRECTED TOTAL               19         1.02265500

MODEL F =                  17.51                               PR > F = 0.0001

R-SQUARE                    C.V.           ROOT MSE                 PRICE MEAN

0.766490                  6.1283         0.12216792                 1.99350000

SOURCE                        DF           ANOVA SS      F VALUE      PR > F

STATE                          3         0.78385500        17.51      0.0001


SCHEFFE'S TEST FOR VARIABLE: PRICE
NOTE: THIS TEST CONTROLS THE TYPE I EXPERIMENTWISE ERROR RATE
      BUT GENERALLY HAS A HIGHER TYPE II ERROR RATE THAN REGWF
      FOR ALL PAIRWISE COMPARISONS.

ALPHA=0.05  DF=16  MSE=0.014925
CRITICAL VALUE OF T=1.79969
MINIMUM SIGNIFICANT DIFFERENCE=0.240849

MEANS WITH THE SAME LETTER ARE NOT SIGNIFICANTLY DIFFERENT

SCHEFFE  GROUPING            MEAN       N   STATE

                A          2.2060       5   VT
                A
                A          2.1700       5   CONN

                B          1.8480       5   RI
                B
                B          1.7500       5   MASS
```

Figure 13-3 ANOVA Procedure: Prices for a Six-Pack of Coca-Cola (12-ounce cans)

```
DEPENDENT VARIABLE: PRICE

SOURCE                   DF      SUM OF SQUARES               MEAN SQUARE

MODEL                     3          0.13222000                0.04407333

ERROR                    16          0.27020000                0.01688750

CORRECTED TOTAL          19          1.40242000

MODEL F =              2.61                              PR > F = 0.0873

R-SQUARE               C.V.            ROOT MSE                PRICE MEAN

0.328562             4.8006          0.12995191                2.70700000

SOURCE                   DF            ANOVA SS     F VALUE       PR > F

STATE                     3          0.13222000        2.61       0.0873


SCHEFFE'S TEST FOR VARIABLE: PRICE
NOTE: THIS TEST CONTROLS THE TYPE I EXPERIMENTWISE ERROR RATE
      BUT GENERALLY HAS A HIGHER TYPE II ERROR RATE THAN REGWF
      FOR ALL PAIRWISE COMPARISONS.

ALPHA=0.05  DF=16  MSE=0.0168875
CRITICAL VALUE OF T=1.79969
MINIMUM SIGNIFICANT DIFFERENCE=0.256195

MEANS WITH THE SAME LETTER ARE NOT SIGNIFICANTLY DIFFERENT.

SCHEFFE  GROUPING             MEAN      N  STATE

                A           2.8360      5  VT
                A
                A           2.7080      5  MASS
                A
                A           2.6680      5  RI
                A
                A           2.6160      5  CONN
```

Figure 13-4 ANOVA Procedure: Prices for a Six-Pack of Budweiser (12-ounce bottles)

14

MULTIPLE REGRESSION

Chapter 10 discussed regression in which there is a dependent and a single independent variable. This chapter extends the analysis to include more than one independent variable. In Chapter 10, we considered techniques that could be used, for example, to find a relationship between automobile mileage as the dependent variable and weight as the independent variable. In this chapter, the analysis is extended so that additional independent variables such as automobile age, engine size, and speed can be included. There are good reasons for making this extension. One of the most important is that including several independent variables can sometimes substantially reduce errors in predictions, as compared with those made with only one independent variable.

Applications of multiple regression are widespread in business, industry, and government. In marketing, independent variables such as price, amount of advertising, and sales of competing products are used to predict sales for various consumer products. Labor, material, and machinery inputs are used in production to predict manufacturing costs. Applications in accounting, finance, organization behavior, personnel, administration, and other functional areas are also common. The illustration we shall use comes from the field of real estate.

Real estate appraisers, brokers, and agents are interested in ways to estimate current selling prices for different types and pieces of property. Property tax collection and real estate sales are just two activities within this industry that depend heavily upon good price estimates. We shall be concerned with housing prices in a certain urban area. A random sample of 25 homes sold during 1980 is available. The homes are single-family residences with one house per lot. We shall restrict ourselves to just two of the many potentially useful independent variables that could be chosen for predicting selling price: floor area and age of the house. The observations appear in Table 14-1.

14.1 Descriptive Procedures for the Sample

We shall select the simplest of several models that could be used for this first application. We shall assume that the dependent variable, Y, is a random variable that satisfies the relationship

$$Y = B_0 + B_1X_1 + B_2X_2 + \epsilon \tag{14-1}$$

or

$$Y = \mu_{Y\cdot12} + \epsilon$$

In the first equation, Y is the sum of a constant (B_0), the products of the values of two independent variables and their coefficients (B_1X_1 and B_2X_2), and a normally distributed random error term with a mean of zero and a standard deviation ($\sigma_{Y\cdot12}$), which is called the *standard error of estimate*. The second equation shows that all but the final random term in the first equation can be considered a mean, or expected, relationship between Y and the independent variables X_1 and X_2. The model indicates that, for a given set of values for X_1 and X_2, the values of Y are normally distributed around the mean, $\mu_{Y\cdot12}$, with a standard deviation of $\sigma_{Y\cdot12}$. It is also assumed that values of Y are independent of each other with regard to their probability of selection. The entire model is a straightforward extension of the model in Chapter 10.

Initial Scattergrams

When there are only two or three independent variables in the regression relationship, scattergrams for each pair of variables provide a practical check on the applicability of the model selected. For instance, the model in Equation 14-1 specifies linear relationships.

Table 14-1 Selling Price and Related Variables for a Sample of 25 Single-Family Houses

Observation	*Price (thousands of dollars)*	*Floor Area (hundreds of ft²)*	*Age (years)*
1	32	11	21
2	33	10	12
3	35	10	18
4	37	11	10
5	38	8	17
6	40	9	13
7	44	15	10
8	44	18	21
9	45	12	26
10	46	15	21
11	47	17	14
12	48	27	22
13	49	23	20
14	50	22	15
15	50	18	14
16	52	22	12
17	52	18	18
18	56	20	17
19	57	23	9
20	58	26	17
21	60	24	14
22	61	16	17
23	64	19	16
24	65	23	17
25	70	19	2

Figure 14-1 presents the scattergrams for all three pairs of variables in the real estate illustration. The relationship between selling price and floor area appears to be linear, positive, and stronger than the other two relationships. There may be a weak negative linear relationship between age and selling price. The relationship between floor area and age is so weak that its direction cannot be detected. In no case is there evidence of a nonlinear relationship between the variables, and the directions of the relationships are consistent with what experience leads us to expect.

Another visual check on the suitability of the model will be made after the sample regression relationship has been found. The check consists of plotting scattergrams of residual errors against each of the variables. Residual errors are the differences between actual selling prices in the sample and prices as estimated by the regression relationship.

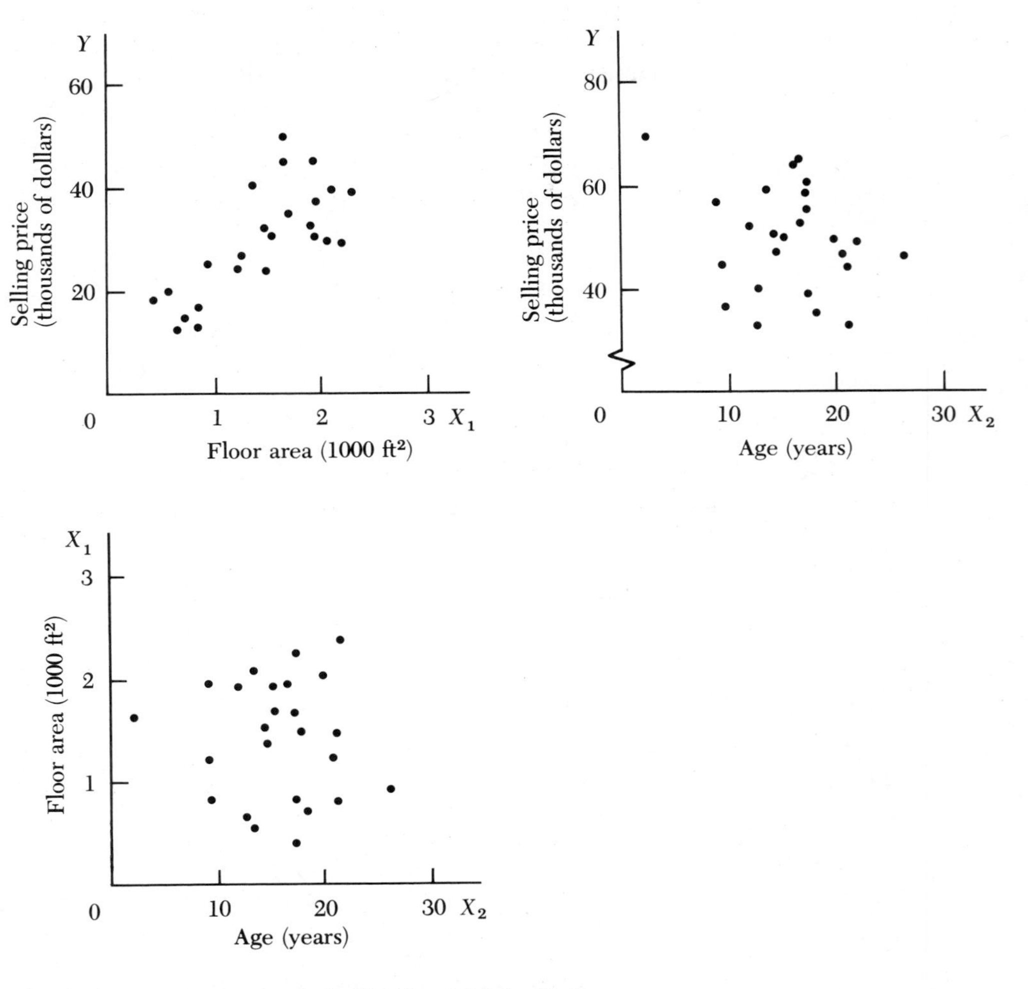

Figure 14-1 Scattergrams for Residential Sales Data

The initial scattergrams have given us no reason to doubt the suitability of the model selected. Therefore the next step is to find the sample multiple regression relationship.

The Sample Least-Squares Relationship

As in Chapter 10, the objective is to find a sample regression relationship for which the sum of squared deviations of the dependent variable from the regression estimates is a minimum. As implied by the model selected, we want a sample relationship that has the following form.

$$Y_c = b_0 + b_1X_1 + b_2X_2 \tag{14-2}$$

From the sample, we want to find numerical values for b_0, b_1, and b_2 that will minimize the sum of squared deviations, that is, minimize $\Sigma(Y - Y_c)^2$.

For two-variable regression, the values of b_0 and b_1 that minimized the sum of squared deviations were found by solving a set of normal equations, Equation 10-3. For multiple regression, also, the values of b_0 and the b coefficients are found from sets of normal equations. The set of normal equations for the relationship in Equation 14-2 is

$$\Sigma Y = nb_0 + b_1\Sigma X_1 + b_2\Sigma X_2 \tag{14-3a}$$

$$\Sigma X_1 Y = b_0\Sigma X_1 + b_1\Sigma X_1^2 + b_2\Sigma X_1 X_2 \tag{14-3b}$$

$$\Sigma X_2 Y = b_0\Sigma X_2 + b_1\Sigma X_1 X_2 + b_2\Sigma X_2^2 \tag{14-3c}$$

Comparison of these equations with Equation 10-3 shows them once again to be extensions of the simple linear regression case.

In Equation 14-3, b_0, b_1, and b_2 are the unknowns. All other quantities are sums obtained from the sample observations. Table 14-2 gives these sums as column totals and also shows how to calculate them. Although not required at present, the sum of squares for the dependent variable, ΣY^2, is found in the table.

When the sums from Table 14-2 are substituted in Equation 14-3, the result is:

$$\begin{aligned} \text{I:}\quad 1233 &= 25b_0 + 436b_1 + 393b_2 \\ \text{II:}\quad 22{,}483 &= 436b_0 + 8356b_1 + 6841b_2 \\ \text{III:}\quad 19{,}035 &= 393b_0 + 6841b_1 + 6787b_2 \end{aligned}$$

In general, the least-squares values of b_0 and the b coefficients come from solving a set of k equations in k unknowns. To solve such a set of equations, computer library programs provide a practical alternative for more than three variables or more than 20 to 25 observations.

Two common techniques for solving a set of normal equations are the elimination technique and matrix algebra. We will illustrate the elimination process. To do so with paper and pencil requires the use of a pocket calculator. Any divisions that do not give exact answers must be carried to 6 or 7 places so that cumulative rounding errors will not seriously affect the solution. In the illustration, all calculations were carried out to several more places than are shown.

To solve the normal equations for the real estate example by elimination, we can begin by eliminating b_2 from Equations I and II. We multiply Equation II by $393/6841$ to get

$$\text{IV: } 1291.598 = 25.047b_0 + 480.033b_1 + 393b_2$$

This now has the same coefficient for b_2 as Equation I. We subtract I from IV to eliminate b_2:

$$58.598 = 0.047b_0 + 44.033b_1$$

We then solve this for b_1 in preparation for eliminating this variable:

$$\text{V: } b_1 = 1.331 - 0.0011b_0$$

Table 14-2 Calculations Required for Multiple Regression

Y	X_1	X_2	X_1Y	X_2Y	X_1X_2	Y^2	X_1^2	X_2^2
32	11	21	352	672	231	1024	121	441
33	10	12	330	396	120	1089	100	144
35	10	18	350	630	180	1225	100	324
37	11	10	407	370	110	1369	121	100
38	8	17	304	646	136	1444	64	289
40	9	13	360	520	117	1600	81	169
44	15	10	660	440	150	1936	225	100
44	18	21	792	924	378	1936	324	441
45	12	26	540	1170	312	2025	144	676
46	15	21	690	966	315	2116	225	441
47	17	14	799	658	238	2209	289	196
48	27	22	1296	1056	594	2304	729	484
49	23	20	1127	980	460	2401	529	400
50	22	15	1100	750	330	2500	484	225
50	18	14	900	700	252	2500	324	196
52	22	12	1144	624	264	2704	484	144
52	18	18	936	936	324	2704	324	324
56	20	17	1120	952	340	3136	400	289
57	23	9	1311	513	207	3249	529	81
58	26	17	1508	986	442	3364	676	289
60	24	14	1440	840	336	3600	576	196
61	16	17	976	1037	272	3721	256	289
64	19	16	1216	1024	304	4096	361	256
65	23	17	1495	1105	391	4225	529	289
70	19	2	1330	140	38	4900	361	4
1233	436	393	22483	19035	6841	63377	8356	6787

In the same manner, we can eliminate b_2 from Equations II and III. We first multiply II by $6787/6841$, which gives

$$22305.529 = 432.558b_0 + 8290.041b_1 + 6787b_2$$

We then subtract Equation III from this result to get

$$3270.529 = 39.558b_0 + 1449.041b_1$$

or

$$\text{VI: } b_1 = 2.257 - 0.0273b_0$$

Because they are both equal to b_1, we can equate the right sides of Equations V and VI. This eliminates b_1 and leaves

$$1.331 - 0.0011b_0 = 2.257 - 0.0273b_0$$

which simplifies to

$$0.0262b_0 = 0.926$$

or

$$b_0 = 35.3$$

Now that we know the value of b_0 that satisfies the normal equations, we can return to Equation VI to find b_1:

$$b_1 = 2.257 - 0.0273(35.3) = 1.293$$

Finally, we can use these two values in Equation I to find b_2:

$$1233 - 25(35.3) - 436(1.293) = 393b_2$$

or

$$b_2 = -0.543$$

Having found b_0, b_1, and b_2, we can now write the least-squares regression equation:

$$Y_C = 35.3 + 1.293X_1 - 0.543X_2 \tag{14-4}$$

Here, b_0 is 35.3 thousands of dollars ($35,300), b_1 is 1.293 thousands of dollars per 100 ft^2 of floor area, and b_2 is -0.543 thousands of dollars per year of age. For a 5-year-old house with a floor area of 2200 ft^2, the regression estimate of the selling price is

$$Y_C = 35.3 + 1.293(22) - 0.543(5) = 61.031$$

or $61,031.

In multiple regression equations, as in two-variable regression equations, b_0 is the Y-intercept. It is the value of Y_C when all of the independent variables are equal to zero. The b-values are called *net regression coefficients*. For instance, b_1 is the change in Y_C per unit increase in X_1 with all other independent variables held fixed. In Equation 14-4, consider a set of houses all of the same age, such as 6 years old. Here X_2 is held fixed at 6. For this set, the equation states that average selling price increases by $1293 for each increase in floor area of 100 ft^2. Similarly, for all houses with a given floor area, expected selling price declines $543 for each additional year of age.

The Standard Error of Estimate

As was the case with two-variable regression, the standard error of estimate measures dispersion from the multiple regression relationship. The standard error of estimate is the standard deviation of the observed values of the dependent variable (Y) from the regression estimates (Y_C).

When there are three variables in the regression equation (one dependent and two independent variables), the equation for the *standard error of estimate* is

$$s_{Y \cdot 12} = \sqrt{\frac{\Sigma(Y - Y_C)^2}{n - 3}} \tag{14-5}$$

The subscript on the left indicates that Y is the dependent variable and X_1 and X_2 are the independent variables. The numerator on the right, $\Sigma(Y - Y_C)^2$, is the sum of squared deviations of the sample observations from the estimates provided by the regression equation. This quantity has the same definition as it did in two-variable regression. The denominator on the right is the degrees of freedom, $n - 3$. Each unknown we must find in the normal equations costs us a degree of freedom.

If we square the sample standard error of estimate, the result is the variance around the sample regression relationship, $s^2_{Y \cdot 12}$. Because the divisor is the degrees of freedom, this is an unbiased estimate of the variance around the population regression relationship, $\sigma^2_{Y \cdot 12}$.

In general, when there are k unknowns in the set of normal equations (b_0, $b_1, \ldots, b_{k-1}$), there are $n - k$ degrees of freedom in the sample standard error of estimate. Hence $n - k$ replaces $n - 3$ in Equation 14-5.

The regression equation for the real estate example is given in Equation 14-4. The variance around this relationship is

$$s^2_{Y \cdot 12} = \frac{\Sigma(Y - Y_C)^2}{n - 3}$$

An equivalent expression for the *variance around the regression plane is*

$$s^2_{Y \cdot 12} = \frac{\Sigma Y^2 - b_0 \Sigma Y - b_1 \Sigma X_1 Y - b_2 \Sigma X_2 Y}{n - 3} \tag{14-6}$$

where b_0, b_1, and b_2 are the coefficients in the least-squares regression equation and the remaining sums in the numerator are shown as column totals in Table 14-2. In general, for the k unknowns in the set of normal equations, the denominator of Equation 14-6 is $n - k$. In the numerator, terms of the form $b_i \Sigma X_i Y$, where i runs from 1 through $k - 1$, must be subtracted from $\Sigma Y^2 - b_0 \Sigma Y$. Hence there is one such term for each independent variable in the regression equation.

For the residential housing price study, the variance around the regression relationship is

$$s^2_{Y \cdot 12} = \frac{1}{22}[63{,}377 - 35.317(1233) - 1.2929(22{,}483) - (-0.54357)(19{,}035)]$$

$$= \frac{1109.72}{22} = 50.44$$

Then the standard error of estimate is

$$s_{Y \cdot 12} = \sqrt{50.44} = 7.10 \text{ (thousands of dollars)}$$

The standard deviation of the observations in the sample around the regression relationship is \$7100.

The Coefficient of Multiple Determination

As a descriptive measure of dispersion and goodness-of-fit for the regression relationship, the standard error of estimate has the same physical dimensions as the dependent variable. Because housing prices are in thousands of dollars, so is the standard error of estimate.

By contrast, the coefficient of multiple determination is a dimensionless measure of the strength of the multiple regression relationship, or goodness-of-fit. Consequently, it can be used to compare the strengths of two or more relationships where the dependent variables have different units of measure.

Recall from two-variable regression that the coefficient of determination is the proportion of Y-variability in the sample that is accounted for by regression. This statement is also true for the coefficient of multiple determination. In fact, with one slight change, Equation 10-8 becomes a symbolic definition for the coefficient of multiple determination:

$$r^2_{Y \cdot 12} = \frac{\text{SSR}}{\text{SSTO}} \tag{14-7}$$

The right side of this equation is identical to Equation 10-8, but SSR is calculated differently. The subscript on the left side of Equation 14-7 indicates that Y is the dependent variable and X_1 and X_2 are the independent variables.

Where there are two independent variables in the multiple regression equation, a convenient form for the *coefficient of multiple determination* is

$$r^2_{Y \cdot 12} = \frac{b_0 \Sigma Y + b_1 \Sigma X_1 Y + b_2 \Sigma X_2 Y - [(\Sigma Y)^2/n]}{\Sigma Y^2 - [(\Sigma Y)^2/n]} \tag{14-8}$$

where b_0, b_1, and b_2 are the constants in the multiple regression equation and the sums are the same as those used in the normal equations. In general, when there are $k - 1$ b-coefficients in the regression relationship, one term of the form $b_i \Sigma X_i Y$ must be added for each such coefficient to the numerator on the right side of Equation 14-8.

In the real estate example, the coefficient of multiple determination is

$$r^2_{Y \cdot 12} = \frac{35.317(1233) + 1.2929(22{,}483) + (-0.54357)(19{,}035) - (1233^2/25)}{63{,}377 - (1233^2/25)}$$

$$= \frac{1455.717}{2565.440} = 0.567$$

The sample regression relationship in Equation 14-4 accounts for 56.7% of the total Y-variation in the sample.

The sample coefficient of multiple determination is a descriptive measure that ranges from zero to 1. The lower limit indicates there is no regression relationship at all. The values of all the b-coefficients (but not b_0 necessarily) are zero in this case. The upper limit indicates that the relationship is perfect in the sense that all the sample observations fall exactly on the regression relationship.

Typical Computer Output

This is a convenient place to illustrate how our analysis to this point appears in a typical computer output from a software program. Although the entire computer solution is shown in Figure 14-2, our current discussion will include only those portions based on what we have done so far. The remaining portions of the output will be covered in the next section, Statistical Inference.

```
MULTIPLE REGRESSION

INDIVIDUAL VARIABLES

VARIABLE     COEFFICIENT   STANDARD ERROR   T STAT
 CONSTANT     35.316991
 FLR AREA      1.292886        0.25905        4.99
 AGE          -0.543570        0.28788       -1.89

ANALYSIS OF VARIANCE

                  SS        DF       MS      F STAT
REGRESSION   1455.3878     2  727.69391  14.422
RESIDUAL     1110.0522    22   50.45692
  TOTAL      2565.4400    24

STD ERROR OF ESTIMATE    =7.103303
MULT R SQ (COEFF DETERM)=0.567305
MULT R (COEFF CORREL)    =0.753197

SIMPLE CORRELATION MATRIX
        Y       X1       X2
 Y   1.000   0.705   -0.278
 X1          1.000   -0.019
 X2                   1.000

ESTIMATE RESPONSE VARIABLE FOR
          FLR AREA=23
          AGE     =10

POINT ESTIMATE                      =59.6177
STD ERROR OF MEAN RESPONSE          = 2.5911
STD ERROR OF INDIVIDUAL RESPONSE= 7.5611

RESIDUALS
 OBSERVED   ESTIMATED   RESIDUAL
    32       38.1238     -6.1238
    33       41.7230     -8.7230
    35       38.4616     -3.4616
    37       44.1030     -7.1030
    38       36.4194      1.5806
    40       39.8866      0.1134
    44       49.2746     -5.2746
    44       47.1740     -3.1740
    45       36.6988      8.3012
    46       43.2953      2.7047
    47       49.6861     -2.6861
    48       58.2664    -10.2664
    49       54.1820     -5.1820
    50       55.6069     -5.6069
    50       50.9790     -0.9790
    52       57.2376     -5.2376
    52       48.8047      3.1953
    56       51.9340      4.0660
    57       60.1612     -3.1612
    58       59.6913     -1.6913
    60       58.7363      1.2637
    61       46.7625     14.2375
    64       51.1847     12.8153
    65       55.8127      9.1873
    70       58.7947     11.2053
```

Figure 14-2 Computer Output for Real Estate Data

The section titled *individual variables* at the top of Figure 14-2 presents, in order, the values of b_0, b_1, and b_2 for the least-squares regression relationship. The constant and b-coefficients shown here are the same as those in Equation 14-4, except that those in the equation were rounded to fewer decimal places. The columns of standard errors and t statistics that complete the output for individual variables will be discussed in the section on statistical inference. This is also what will be done in the case of the *analysis of variance* output.

Just beneath the analysis of variance material in Figure 14-2, the standard error of estimate and the coefficient of multiple determination are given. Once again, the values as found by the computer agree with those found earlier, except for differences in the degree of rounding. The coefficient of multiple correlation follows the coefficient of multiple determination. The *coefficient of multiple correlation*, $r_{Y \cdot 12}$ in this case, is the square root of the coefficient of multiple determination. Unlike the coefficient of correlation for two variables, the coefficient of multiple correlation cannot be negative. Like the coefficient of multiple determination, it is restricted to a range from zero to 1. The coefficient of multiple determination can be interpreted as the proportion of Y-variation explained by regression. But there is no such direct interpretation for the coefficient of multiple correlation.

For the time being, we shall omit discussion of the next two sections of Figure 14-2, *simple correlation matrix* and *estimate response variable for*. This brings us to the final section, *residuals*.

Checking Residuals

In the section concerned with residuals, the observed values of the dependent variable, Y, appear in rank order. If necessary, we accomplish this by ranking the data and entering the observations for that variable into the computer in that order. Doing so makes possible further checks on the adequacy of the model initially selected for the analysis.

One check comes from noting the sequence of plus and minus signs in the column of residuals. The first four values of Y calculated from the regression equation are larger than the observed values. Hence the residuals are preceded by minus signs. There is another collection of six successive minus signs in the middle, and a final set of five plus signs at the end. It appears that there is greater bunching of signs than is consistent with a sequence of random, independent residuals for which the mean value is zero. In the chapter on nonparametric statistics, we shall introduce a runs test. With this, we can test the hypothesis that the observed pattern of signs represents random, independent deviations from the model we selected and defined in Equation 14-1. The test will show that the number of runs is significantly smaller than one would expect for random, independent observations. We shall return to this result in a moment.

The residuals are the basis for still another check on the adequacy of the model. The check consists of plotting scattergrams of the residuals and each of the variables in the regression relationship. These scattergrams for our residential price data are presented in Figure 14-3. From top to bottom, residuals are plotted against selling prices, floor areas, and ages of the houses. The dashed horizontal lines on the

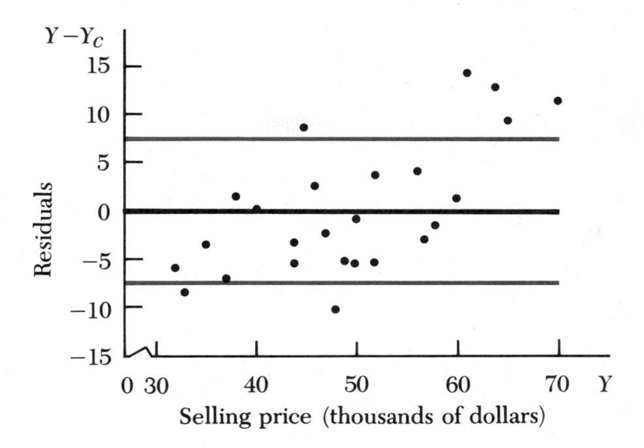

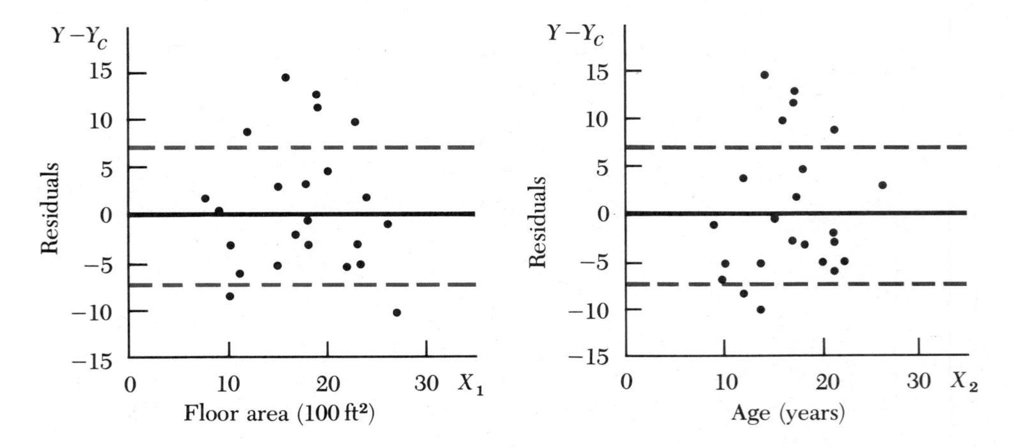

Figure 14-3 Residuals Plotted with Each Variable in the Regression Relationship

scattergrams are plotted a distance of one standard error of estimate, $s_{Y \cdot 12}$, above and below the zero residual line. Seventeen, or 68%, of the 25 observations fall between the dashed lines. For normally distributed residuals, one would expect 68% to be within one standard error of the regression relationship. Hence the residuals do not suggest that anything is wrong in this respect.

In the top scattergram, however, there appears to be a positive linear relationship between the residuals and selling prices. In fact, the slope of this relationship is significantly different from zero at the 0.01 level of significance. If the original model, Equation 14-1, were correct, one would not expect any relationship between residuals and the variables. A frequent cause of such a relationship is the omission of an important independent variable from the model.

Both the runs test and the plot of residuals have cast doubt on the adequacy of Equation 14-1 as a model for a regression relationship to predict selling prices for single-family residences. We shall return to this set of observations and attempt to improve the relationship after we have discussed statistical inference.

Exercises

14-1 *Warehouse space control.* A supervisor, charged with using the space in a warehouse effectively, has recorded three characteristics for a sample of 12 incoming shipments. They are

X_1: number of crates in shipment,

X_2: weight of shipment (hundreds of pounds),

Y: warehouse space required (cubic feet).

The regression equation for the observations is

$$Y_C = 4.817 + 3.500X_1 + 5.297X_2$$

In addition,

$$\Sigma X_1 Y = 17{,}111 \qquad \Sigma X_2 Y = 15{,}924$$

$$\Sigma Y = 1187 \qquad \Sigma Y^2 = 150{,}101$$

a. Interpret the quantities 4.817, 3.500, and 5.297 in the regression equation.
b. Find the coefficient of multiple determination. Does it indicate that the regression relationship may be useful for estimating space required for future incoming shipments? Why?
c. Estimate the space needed for a shipment containing 19 crates and weighing 16 hundred pounds.

14-2 *Toothpaste preference.* Eight different mixtures of toothpaste were prepared for a consumer preference study. The mixtures differed in the quantities of flavoring and astringent they contained. Tubes of each mixture were given to random samples of potential purchasers. After several uses, these people scored the particular mixture being used. The regression relationship is

$$Y_C = 47.25 + 0.4475X_1 + 0.5375X_2$$

where Y is the mean preference score for a mixture, X_1 is milligrams of flavoring, and X_2 is milligrams of astringent. Preference scores can range from 0 to 100.

a. Explain the meaning of the net regression coefficients in this application.
b. For the sample, $n = 8$, $\Sigma Y = 643$, $\Sigma Y^2 = 52{,}769$, $\Sigma X_1 Y = 33{,}940$, and $\Sigma X_2 Y = 13{,}290$. Calculate the standard error of estimate and the coefficient of multiple determination. Do these measures suggest that the regression equation will provide practical estimates of preference scores?
c. What is the estimated preference score for a mixture with 65 mg of flavoring and 20 mg of astringent?

14-3 *Electrical energy consumption.* An electric utility company collected observations on total consumption in all-electric homes during December, January, and February. The average settings of water heater and space heating thermostats were also observed. The data from a small pilot sample are as follows.

Consumption (kWh/100)	Thermostat Setting in Degrees Fahrenheit	
	Space Heating	Water Heater
16	66	130
18	69	125
24	63	145
26	71	140
29	75	135
35	67	155
39	76	140
44	74	150

a. Is there any evidence of a nonlinear relationship in the scattergrams for all pairs of variables?
b. Using Equation 14-1 as a model, find the normal equations and solve them to determine the sample least-squares regression relationship.
c. What percentage of variation in electricity consumption is attributable to regression with the two thermostat settings?

14-4 *Performance rating.* The personnel department of a firm that assembles a household appliance wants to find the regression relationship between assembly workers' job performance as rated by their supervisors, scores on a manual dexterity test, and years of schooling. A random sample of nine observations is selected for preliminary study.

Performance Rating Y	*Manual Dexterity* X_1	*Years of Schooling* X_2
30	15	13
20	10	14
70	20	8
60	22	11
40	12	14
80	21	10
50	19	12
30	12	16
80	24	9

a. Do scattergrams for the three pairs of variables give evidence of a nonlinear relationship?
b. Assume that the model $\mu_{Y \cdot 12} = B_0 + B_1X_1 + B_2X_2$ is appropriate, and find the normal equations for the sample data.
c. Determine the numerical values for b_0, b_1, and b_2, and state the multiple regression equation for the sample.

14-5 *Electrical energy consumption residuals.* Consumption estimates from the regression relationship found in Exercise 14-3 are given as values of Y_C in the table below. Actual consumption, Y, is also shown. The differences are the residuals shown in the final column.

Consumption (KWH/100) Y	*Estimated Consumption (KWH/100)* Y_C	*Residual* $Y - Y_C$
16	15.877	0.123
18	16.154	1.846
24	23.078	0.922
26	30.046	–4.046
29	31.661	–2.661
35	35.908	–0.908
39	36.738	2.262
44	41.538	2.462

a. What do visual examinations of the residuals and scattergrams of the residuals against the variables suggest about the suitability of the model?
b. Use the sum of the squared residuals to show that the standard error of estimate is 2.82 hundred KWH.
c. Is the percentage of observations within one standard error of the regression relationship about what one would expect?

14-6 *Job performance rating residuals.* Residuals are to be examined for the study relating job performance rating to manual dexterity and years of schooling in Exercise 14-4. The following table lists the observed ratings, the ratings as estimated by the regression equation, and the differences.

Job Performance Ratings		
Observed	*Estimated*	*Residuals*
20	25.26	–5.26
30	24.09	5.91
30	41.67	–11.67
40	30.54	9.46
50	55.44	–5.44
60	66.57	–6.57
70	70.97	–0.97
80	78.29	1.71
80	67.16	12.84

a. Plot a scattergram with residuals as the dependent variable and performance rating as the independent variable. Then plot similar scattergrams with manual dexterity and years of schooling as independent variables. Is there visual evidence suggesting a relationship in any of the scattergrams?

b. From the sum of squared residuals, determine the variance around the regression relationship, $s^2_{Y \cdot 12}$.

c. The standard error of estimate is 9.40. Determine what percentage of the residuals lie within one standard error of the regression relationship and comment on how this compares with what one would expect.

d. Given the sum of squared residuals from (b), find the sum of squared deviations of the observed performance ratings from their mean. Then apply Equation 14-7 to calculate the coefficient of multiple determination. Interpret the result.

14-7 *Explained variability in income.* A class project was undertaken to develop a regression relationship to estimate family income from four family characteristics. One progress report on the project stated, "The coefficient of multiple correlation for the sample is –0.73. This means that the four family characteristics together explained 73% of the total variation in family income." Do you agree with this statement? Why? How should it be stated if it is incorrect?

14-8 *Signs of coefficients.* For the family income project described in the preceding exercise, another statement in the report was, "The coefficients for the four independent variables in the regression relationship should all be negative. Yet they are not. The project team is checking its work." What could have led the reporter to expect negative coefficients? Is that expectation justified?

14.2 Statistical Inference

The descriptive approach taken in Section 14.1 furnishes a general impression about the adequacy of a regression relationship and may suggest some remedial measures if the relationship is inadequate. But such an approach should be accompanied by information from statistical tests and estimates whenever possible.

As applied to multiple regression, the statistical inference procedures we shall discuss fall into three categories, which were illustrated in Figure 14-2. A comprehensive test for the utility of the regression relationship constitutes the first. The second deals with making confidence interval estimates and predictions for the dependent variable based on the sample regression relationship. The last considers the contribution of independent variables, both individually and in combination with each other, to the regression relationship.

Comprehensive Analysis of Variance Test for Regression

In Section 10.7, we discussed a test of the hypothesis that the slope, B_1, of the population regression line for two-variable regression is equal to zero. We showed this to be logically equivalent to testing for the absence of a linear regression relationship between the variables in the underlying statistical population.

The analysis of variance test to be introduced now is the counterpart for multiple regression of the one described in Section 10.5. Here the test is a single, comprehensive test of the null hypothesis that all B-coefficients except B_0 in the population model are simultaneously equal to zero, that is, that

$$H_0: B_1 = B_2 = B_3 = \cdots = B_{k-1} = 0$$

is true. The alternative is that at least one of the coefficients for the independent variables is not zero.

If this null hypothesis cannot be rejected, no relationship of the form contained in the model can be presumed to exist in the statistical population. Hence it would be a contradiction to use the sample regression relationship as a basis for making estimates and predictions of values of the dependent variable. Instead, the multiple regression analysis is dropped or a search is made for a more adequate model.

Table 14-3 repeats the *analysis of variance* portion of Figure 14-2; the numerical values in the top of the table are represented symbolically in the lower part of the table.

In general, the sum of the squares column contains the sum of squares for regression, SSR, the sum of squared residuals, SSE, and the total sum of squares, SST. In Figure 10-7 and the accompanying discussion (page 305), we saw that a deviation of an observed value of the dependent variable, Y, from the sample mean, $\overline{Y}$, can be stated as the sum of two parts:

$$(Y - \overline{Y}) = (Y_C - \overline{Y}) + (Y - Y_C)$$

The total deviation is the sum of a deviation explained by regression and a residual error not so explained. Furthermore, we said that the following can be shown with some algebra:

$$\Sigma(Y - \overline{Y})^2 = \Sigma(Y_C - \overline{Y})^2 + \Sigma(Y - Y_C)^2 \quad \textbf{(14-9)}$$

The sum of squared deviations of Y-observations from their mean is equal to the sum of squares attributable to the regression relationship plus the sum of squared deviations from the regression relationship. For convenience, typical symbols for Equation 14-9 are

$$\text{SSTO} = \text{SSR} + \text{SSE} \quad \textbf{(14-10)}$$

The total sum of squares, SSTO, is equal to $\Sigma(Y - \overline{Y})^2$. Similarly, SSR $= \Sigma(Y_C - \overline{Y})^2$ and SSE $= \Sigma(Y - Y_c)^2$. Although Equation 14-10 was developed in the context of two-variable regression, it also applies to multiple regression. Hence these are the quantities shown in Table 14-3.

The DF column in Table 14-3 defines the degrees of freedom applicable to each sum of squares. In general, n is the number of observations in the sample and k is the number of unknowns in the normal equations. These are b_0, and a b-coefficient for each of the $k - 1$ independent variables. For our example, $n = 25$ and $k = 3$, the constant b_0 and the coefficients b_1 and b_2.

In Table 14-3, the mean squares for regression (MSR) and for residual error (MSE) are the respective sums of squares divided by their degrees of freedom. When the null hypothesis mentioned earlier is true, all B-coefficients in the regression model are zero and there is no underlying regression relationship. In this case, both MSR and MSE are independent estimates of the variance of the dependent variable,

Table 14-3 Analysis of Variance in Multiple Regression

	Analysis of Variance Portion of Figure 14-2			
	SS	*DF*	*MS*	*F Stat*
Regression	1455.3878	2	727.69391	14.422
Residual	1110.0522	22	50.45692	
Total	2565.4400	24		
	General Contents of Analysis of Variance Portion			
Regression	SSR	$k-1$	MSR = SSR/$(k-1)$	MSR/MSE
Residual	SSE	$n-k$	MSE = SSE/$(n-k)$	
Total	SST	$n-1$		

$\sigma^2_{Y \cdot}$. Under this condition, the ratio of MSR to MSE has an F distribution as its sampling distribution, with an expected value not far from 1. This distribution was discussed in Chapter 13 (page 406). On the other hand, when regression is present, MSR can be expected to be much larger than MSE and the F ratio will fall far in the upper tail of the relevant F distribution.

For the example at the top of Table 14-3, the F distribution with which the sample results are to be compared has the 2 degrees of freedom associated with MSR in the numerator and the 22 degrees of freedom associated with MSE in the denominator. Table A-6 shows that an F distribution with 2 degrees of freedom in the numerator and 22 in the denominator has a critical value of 5.72 at the 0.01 level of significance. Only 1% of the area in such a distribution is found above an F value of 5.72. Because our sample F value is 14.422, far greater than the 0.01 critical value, we reject the null hypothesis of no regression at the 0.01 level.

The comprehensive F test supports the use of the sample regression relationship in Equation 14-4 as a basis for predictions and estimates of the selling prices of single-family residences. We must qualify this finding, however, with the adverse result from our analysis of residuals in Figure 14-3. There we found positive correlation, instead of a random pattern in the scattergram of residuals versus the dependent variable. We shall continue with the description of the computer output in Figure 14-2 and then return to this problem.

Estimates of Values of the Dependent Variable

The real estate sample has passed the analysis of variance test for the existence of an underlying regression relationship. Had we also not found anything wrong with the residuals, we would now be justified in using the sample regression relationship to estimate selling prices for similar houses on the basis of their floor areas and ages. As it is, because of the positive relationship of residuals to the dependent variable, we can expect estimates in the lower \$30,000 to \$40,000 range to be somewhat high and those in the upper \$60,000 to \$70,000 range to be somewhat low. Nevertheless, our example will still serve to complete the explanation of the computer output.

The section of Figure 14-2 concerned with making estimates and predictions from the sample regression relationship is repeated here.

```
        ESTIMATE RESPONSE VARIABLE FOR
                 FLR AREA=23
                 AGE     =10

       POINT ESTIMATE                     =59.6177
       STD ERROR OF MEAN RESPONSE         = 2.5911
       STD ERROR OF INDIVIDUAL RESPONSE= 7.5611
```

This section begins by specifying that estimates and predictions are to be made for the selling price of one or more single family residences with floor areas of 2300 ft^2 and ages of 10 years. These are substituted in Equation 14-4 for X_1 and X_2, using the full number of decimal places found by the computer. The result is the point estimate of 59.6177, or \$59,618, for the selling price, as shown.

Beneath the point estimate, the standard error of the mean response, 2.5911, is used to construct a confidence interval estimate of the conditional mean selling price of all 10-year-old houses with 2300 ft^2 of floor area. The 0.95 confidence interval estimate is

$$59.6177 \pm 2.074(2.5911)$$

or

$$\$54{,}244 \quad \text{to} \quad \$64{,}992$$

Here, 2.074 is from a t distribution with 22 degrees of freedom. In general, we use a t distribution with the $n - k$ degrees of freedom associated with MSE to form such confidence interval estimates.

The estimate for the conditional mean parallels the process for two-variable regression (page 320). With two variables, we were able to state the formula for the standard error of the conditional mean. With the addition of more independent variables, the formula is most easily expressed in matrix notation, which is beyond the scope of this presentation. Even so, we can easily understand how to apply the result, and at least a few of the available computer software programs provide the output.

The second standard error under the point estimate of 59.6177 in the response variable estimation section is 7.5611. This is the standard error associated with a prediction interval of the selling price for a single house not in the original sample for which the floor area is 2300 ft^2 and the age is 10 years.

We will also find a 0.95 prediction interval for this selling price. Once again, t has $n - k$, or 22 degrees of freedom, and a t-value of 2.074 is correct for an 0.95 interval. The prediction interval is

$$59.6177 \pm 2.074(7.5611)$$

or

$$\$43{,}936 \quad \text{to} \quad \$75{,}299$$

A much wider interval, of course, is associated with this prediction of a single

Y-value than was associated with the estimate of the mean selling price for all houses with this same floor area and age. Again, the situation is analogous to the earlier discussion for two variables.

For several independent variables, the formula for the standard error of a single prediction is again best expressed in matrix notation, and it will not be given here. As in the case of the standard error of a mean response, there are software programs that do the calculations and present the results.

Figure 14-2 presents estimated response results for only one set of values for the independent variables. For programs that produce such output, it is possible to obtain response estimates for several sets of values of the independent variables. Care must be exercised to make certain the point represented by each set of values lies within the observed range of all independent variables. Otherwise, extrapolation can lead to erroneous results. This point was discussed in some detail in conjunction with two-variable regression (page 323).

Inferences about the *B*-Coefficients

The *individual variables* section of Figure 14-2 provides the information needed for confidence interval estimates and hypothesis tests for the B-coefficients for independent variables in the model for the residential sales example. That section is repeated here.

VARIABLE	COEFFICIENT	STANDARD ERROR	T STAT
CONSTANT	35.316991		
FLR AREA	1.292886	0.25905	4.99
AGE	-0.543570	0.28788	-1.89

The estimated standard errors of the coefficients for floor area and age are $s_{b_1} = 0.25905$ and $s_{b_2} = 0.28788$, respectively. These are analogous to the standard error of the slope for two-variable regression.

To test the null hypothesis that a given B-coefficient in a multiple regression relationship is zero, we use the statistic

$$t = \frac{b_i - 0}{s_{b_i}}$$

When the assumptions stated in connection with the model, Equation 14-1, apply, this statistic has a t distribution.

To find the acceptance region for a hypothesis test, we must know the degrees of freedom for the standard error of a B-coefficient. As was true for the residuals in analysis of variance for regression and for the standard errors of a mean and individual response, the degrees of freedom for this case is the number of observations, n, less the number of unknowns to be found from the normal equations, k. Alternatively, the degrees of freedom for a t test of a B-coefficient is $n - k$, where n is the number of observations and k is one more than the number of independent variables.

The t statistics for floor area and age are shown above as 4.99 and −1.89. The calculations are

$$t_1 = \frac{1.292886}{0.25905} = 4.99 \quad \text{and} \quad t_2 = \frac{-0.543570}{0.28788} = -1.89$$

Suppose we make a two-tailed test of the hypothesis that the coefficient for floor area, B_1, is zero, with a 0.05 level of significance. The acceptance region comes from a t distribution with $n - k = 25 - 3 = 22$ degrees of freedom. Table A-4 shows the acceptance region to be $-2.074 \leq t \leq 2.074$. Since the sample value is 4.99, the B_1-coefficient is significantly different from zero at the 0.05 level.

If we make the same test on the age coefficient, B_2, the t statistic from the sample is −1.89, well within the acceptance region for the null hypothesis. Thus B_2 is not significantly different from zero at the 0.05 level.

We concluded from the analysis of variance test conducted earlier that not all B-coefficients in the model are zero. Now a t test on B_2 indicates that age does not make a significant additional contribution in explaining Y-variation beyond that made by floor area alone. Alternatively, the t test on B_1 indicates that floor area makes a significant marginal contribution beyond that made by age. Even so, the t statistic for age is close to being significant here, and it probably would be significant in a somewhat larger sample. Hence it would not be advisable to drop age from any further consideration as a useful independent variable. This opinion is reinforced by our earlier discovery that the residuals appear to violate the assumptions in the model.

A confidence interval estimate for one of the B-coefficients is found in the usual way. The 0.95 confidence interval estimate for the age coefficient, B_2, is

$$b_2 \pm ts_{b_2}$$

or

$$-0.543570 \pm 2.074(0.28788)$$

which is

$$-1.141 \leq B_2 \leq 0.053$$

The interval includes zero, so we reach the same conclusion about age here as we did with the hypothesis test.

We shall not give the calculation formula for the standard error of a B-coefficient. The formula is quite complex unless it is expressed in matrix notation.

14.3 The Problem of Correlated Independent Variables

Section 14.2 implied that the effect of the independent variables floor area and age can be assessed separately in regard to the regression relationship with selling price. If there is little or no correlation between the two independent variables, the implication is correct. When the correlation is zero ($r_{12} = 0$), it can be shown that

$$r^2_{Y\cdot12} = r^2_{Y1} + r^2_{Y2}$$

In this case, the coefficient of multiple determination is the sum of the coefficients of determination for the independent variables. The contribution of each does not depend on the other. If $r_{Y1} = 0.6$ and $r_{Y2} = 0.7$, then X_1 explains 36% of Y-variation, X_2 explains 49%, and together they explain 85% in a multiple regression relationship based on the model in Equation 14-1.

From Figure 14-2, the *simple correlation matrix* section for our example is

	Y	X1	X2
Y	1.000	0.705	-0.278
X1		1.000	-0.019
X2			1.000

Here we see that $r_{12} = -0.019$. There is practically no correlation between floor area and age. This justifies assuming virtually independent contributions of each variable to the multiple regression relationship.

When there is appreciable correlation between the independent variables, the condition is called *multicollinearity*. As we shall see, multicollinearity causes problems in assessing the role of individual independent variables. For the most part, such variables vary jointly and exert a combined effect on the dependent variable. Typically, it is not possible to isolate the contribution of any one variable alone. Furthermore, the b-coefficients for two correlated independent variables can differ markedly from one random sample to the next. The value of the coefficient for any such variable can vary over a wide range because of mere random sampling fluctuation. In the presence of multicollinearity, great care must be taken to avoid misinterpreting the t statistics associated with the b-coefficients, as the following example illustrates.

Example

Several stores in a national cooperative for buying hardware are participating in a study of advertising effectiveness on sales of a certain product. Three types of advertising were used: local direct mail, local newspaper, and local television spot ads. A multiple regression relationship of the form

$$Y_C = B_0 + B_1X_1 + B_2X_2 + B_3X_3$$

was calculated, in which the variables are in the order listed above. For a sample of 12 stores, the coefficient of multiple determination was 0.88. In the comprehensive analysis of variance test, F was significant at the 0.001 level.

The following are the relevant sections from the computer printout.

INDIVIDUAL VARIABLES

VARIABLE	COEFFICIENT	STD ERROR	T STAT
CONSTANT	0.746514		
MAIL	0.734536	5.69265	0.13
PAPER	3.858099	5.86676	0.66
TV	3.353757	1.46623	2.29

SIMPLE CORRELATION MATRIX

	Y	X1	X2	X3
Y	1.000	0.891	0.897	0.525
X1		1.000	0.993	0.281
X2			1.000	0.290
X3				1.000

The correlation matrix shows that both X_1 and X_2 are highly correlated with sales and with each other ($r_{Y1} = 0.891$, $r_{Y2} = 0.897$, and $r_{12} = 0.993$). Direct mail advertising and newspaper advertising varied jointly in the 12 stores. By contrast, television advertising is less highly correlated with sales than the other two types of advertising, and television advertising has little multicollinearity with them ($r_{Y3} = 0.525$, $r_{13} = 0.281$, and $r_{23} = 0.290$).

The t statistics for the coefficients are somewhat surprising. The coefficients for mail and newspaper advertising are not significant, even though each correlates highly with sales. Television advertising is almost significant at the 0.05 level, even though it is not as highly correlated with sales as are the other two forms of advertising.

This example indicates what can happen when multicollinearity is present. It is incorrect to conclude that mail and newspaper advertising are not important in the relationship. Instead, they are so highly correlated that including just one of them would contribute almost as much to the relationship as does including both. Yet we know from general experience that it is reasonable to assume each of the two types of advertising has an important effect on sales. Had the 12 stores in the sample not varied mail and newspaper advertising jointly, the importance of each could have been more clearly assessed.

It should be emphasized again that the t statistics in the table above have a special interpretation. Each t statistic tests only the *marginal* contribution of the independent variable to which it applies. That is, it tests whether the independent variable with which it is associated makes a significant incremental contribution to the regression relationship over and above that made by the other independent variables present.

We might ask what to do to reduce multicollinearity in a case such as the advertising example. If several additional stores can be found in which there is little correlation between the independent variables, a new analysis can be conducted. It may also be possible to conduct a designed experiment in which levels of the three types of advertising are deliberately chosen to keep multicollinearity low.

Exercises

14-9 *F test for electricity consumption.* The regression equation for the electrical consumption study in Exercise 14-3 has two independent variables: thermostat settings for space heating (X_1) and for water heaters (X_2). There are eight observations. The partially completed analysis of variance results are as follows:

```
ANALYSIS OF VARIANCE
                SS        DF   MS   F STAT
REGRESSION   645.1467     2
RESIDUAL
 TOTAL       684.8750
```

a. Complete the table.
b. State the null hypothesis to be tested.
c. At the 0.01 level of confidence, test the null hypothesis and state the conclusion.

14-10 *F test for toothpaste preference regression.* The partially completed *analysis of variance* section of the computer printout for the toothpaste preference study described in Exercise 14-2 is as follows:

```
ANALYSIS OF VARIANCE
                SS        DF   MS   F STAT
REGRESSION   1032.150
RESIDUAL
 TOTAL       1087.875     7
```

The regression equation for the sample is

$$Y_C = 47.25 + 0.4475X_1 + 0.5375X_2$$

where X_1 and X_2 are the amounts of flavoring and astringent in the toothpaste mixture.

a. Complete the table.
b. To what null hypothesis does the F statistic apply? Interpret the meaning of the null hypothesis in this case.
c. Using the 0.01 level of significance, complete the hypothesis test and state your conclusion.

14-11 *Estimated warehouse space.* In Exercise 14-1, the number of crates in a shipment (X_1) and the weight of the shipment (X_2) are the independent variables in a regression relationship with warehouse space requirements. Weight is measured in hundreds of pounds and space is measured in cubic feet. There are 12 observations in the sample. The estimation section of the computer printout for this regression is as follows:

```
 ESTIMATE RESPONSE VARIABLE FOR
             CRATES=20
             WEIGHT=18
POINT ESTIMATE                    =170.1573
STD ERROR OF MEAN RESPONSE        =  2.1187
STD ERROR OF INDIVIDUAL RESPONSE=   4.6035
```

a. Refer to Exercise 14-1 and verify the point estimate.
b. Find and interpret a 0.99 confidence interval estimate for the mean response.
c. How does an individual response differ from the mean response in this case? Why is the standard error larger for the individual response?

14-12 *Estimating performance rating.* In Exercise 14-4, job performance rating was the dependent variable in a regression relationship with score on a manual dexterity test (X_1) and years of schooling completed (X_2). There were nine observations in the sample. Part of the computer output was as follows.

```
ESTIMATE RESPONSE VARIABLE FOR
          DEXTERITY=21
          SCHOOLING=10
POINT ESTIMATE                        =67.1596
STD ERROR OF MEAN RESPONSE            = 4.0254
STD ERROR OF INDIVIDUAL RESPONSE=10.2235
```

a. Interpret the meaning of the point estimate and each of the standard errors.
b. An employee not in the original sample scored 21 on manual dexterity and had 10 years of schooling. Make a 0.95 interval estimate of the job performance rating this employee will receive from the supervisor.

14-13 *B-coefficients for electricity consumption.* There are eight observations in the electric utility study of Exercises 14-3 and 14-9. Two sections of the computer analysis are given below.

```
INDIVIDUAL VARIABLES
 VARIABLE   COEFFICIENT   STD ERROR   T STAT
 CONSTANT   -169.654273
 SPACE         1.338361    0.228087     5.87
 WATER         0.747691    0.106591     7.01

SIMPLE CORRELATION MATRIX
          Y        X1        X2
 Y        1      0.609     0.737
 X1                 1     -0.031
 X2                          1
```

a. Verify the calculations for the t statistics applicable to the coefficients of the independent variables.
b. Test the hypothesis that $B_1 = 0$. Use the 0.01 level of significance. Perform the same test for B_2.
c. Taking into account the degree of multicollinearity between the independent variables, discuss the role of each in the regression relationship. The value of F for this relationship is 40.60 with 2 and 5 degrees of freedom.

14-14 *Contributions of independent variables to performance.* Supervisor's job performance rating as a function of employee's manual dexterity and years of schooling is the subject of Exercise 14-4. The sample regression relationship is based on nine observations. The following are two sections of the software program output for this study.

```
INDIVIDUAL VARIABLES
 VARIABLE    COEFFICIENT   STD ERROR   T STAT
 CONSTANT     44.014610
 DEXTERITY     2.636773     1.30200     2.03
 SCHOOLING    -3.222723     2.51962    -1.28

SIMPLE CORRELATION MATRIX
          Y        X1        X2
 Y        1      0.914    -0.884
 X1                 1     -0.864
 X2                          1
```

a. At the 0.05 level of significance, test the hypothesis that $B_1 = 0$. Repeat the test for B_2.
b. Interpret the simple correlation matrix.

c. Discuss the marginal contributions of dexterity and schooling to the multiple regression relationship. (For this relationship, $F = 20.149$ with 2 and 6 degrees of freedom.)

14.4 Indicator Variables

Our previous discussion of regression has used only quantitative variables such as price, floor area, and age. Many times, however, there is a need to include one or more qualitative variables, such as the sex of an employee or the make of an automobile. This section describes how to include a qualitative variable in a regression relationship.

As we mentioned earlier, there is a problem with the residential selling price regression analysis at the beginning of this chapter, as shown by the linear relationship between selling price and the residuals in Figure 14-3. Because omitting an important variable is one thing that can cause such a result, we now want to try a new independent variable to see if it will improve the analysis.

Many people experienced in real estate sales think that the socioeconomic division of the urban area exerts an important influence on residential prices. Virtually the same house will sell for a much higher price in a section where managerial or professional workers live than in a section where wage and salaried workers live.

A second regression analysis was thus carried out involving floor area and socioeconomic section as the independent variables. Information from the city planning office was used to determine the socioeconomic section within the urban area for each of the 25 houses in the original sample. It turned out that these homes are located in three socioeconomic areas.

Socioeconomic section is a qualitative variable which, in this instance, has three classes. When the qualitative variable consists of three classes, two indicator variables must be included in the regression relationship.

We might think that three "dummy" variables coded 0, 1 for each of the three sections of the city could be used instead of the two indicator variables. Generally this cannot be done because such dummy variables make it impossible to find a least-squares regression relationship. There are also other considerations that make it advisable to use indicator rather than such dummy variables.

> In general, $c - 1$ indicator variables are needed in a regression model to introduce a qualitative variable with c classes.

Indicator variables X_3 and X_4 will be used to identify the three socioeconomic areas. The coding scheme for accomplishing this is shown in Table 14-4.

With floor area represented by X_1 and with the two new indicator variables, the residential selling price model we will use to replace Equation 14-1 is

$$Y = B_0 + B_1X_1 + B_3X_3 + B_4X_4 + \epsilon \qquad \textbf{(14-11)}$$

For any given set of values for X_1, X_3, and X_4, the dependent variable, Y, is assumed to be normally and independently distributed around the regression function with a standard deviation of $\sigma_{Y \cdot 134}$.

The sample observations are shown in Table 14-5, and the sums required for the normal equations are given at the bottom of that table. Solving the normal equations, we have the sample regression equation

$$Y_C = 20.316 + 1.230X_1 + 16.375X_3 + 6.759X_4 \tag{14-12}$$

From the regression equation and the codes in Table 14-4, we can gain useful interpretations of the indicator variables and their coefficients. For residents in the southwestern section of the city, $X_3 = 0$, $X_4 = 0$, and the sample regression equation, Equation 14-12, reduces to

$$Y_C = 20.316 + 1.230X_1$$

Table 14-4 Indicator Variable Codes for Socioeconomic Sections

	Variable	
Section	X_3	X_4
Southwest	0	0
North	1	0
East	0	1

This is now a two-variable regression between price and floor area. For residences in the northern section, $X_3 = 1$ and $X_4 = 0$. In this case, the regression equation becomes

$$Y_C = 20.316 + 1.230X_1 + 16.375$$

which simplifies to

$$Y_C = 36.691 + 1.230X_1$$

This is a two-variable relationship with a greater intercept but the same slope as the equation for the southwestern section. Finally, for the eastern section, $X_3 = 0$ and $X_4 = 1$. The equation is

$$Y_C = 20.316 + 1.230X_1 + 6.759$$

or

$$Y_C = 27.075 + 1.230X_1$$

This has the same slope as the equations for the other two sections of the city, but it has a different intercept from either.

Table 14-5 Selling Price, Floor Area, and Socioeconomic Indicator Variables for 25 Single-Family Residences

Price (thousands of dollars) Y	*Floor Area (hundreds of ft²)* X_1	*Indicator Variables* X_3	X_4
32	11	0	0
33	10	0	0
35	10	0	0
37	11	0	1
38	8	0	1
40	9	0	1
44	15	0	1
44	18	0	0
45	12	1	0
46	15	0	1
47	17	0	1
48	27	0	0
49	23	0	0
50	22	0	0
50	18	0	1
52	22	0	1
52	18	1	0
56	20	1	0
57	23	0	1
58	26	0	1
60	24	0	1
61	16	1	0
64	19	1	0
65	23	1	0
70	19	1	0
1233	436	7	11

$n = 25$ $\quad \Sigma Y^2 = 63{,}377 \quad \Sigma X_1^2 = 8356 \quad \Sigma X_3^2 = 7 \quad \Sigma X_4^2 = 11$

$\Sigma X_1 Y = 22{,}483 \quad \Sigma X_3 Y = 413 \quad \Sigma X_4 Y = 529$

$\Sigma X_1 X_3 = 127 \quad \Sigma X_1 X_4 = 188 \quad \Sigma X_3 X_4 = 0$

The three regression lines just described are plotted along with the sample observations in Figure 14-4. Selling price is a linear function of floor area for each of the three sections of the city. All three functions have the same slope but different intercepts. Typically, houses in the eastern section sold for $6759 more than houses

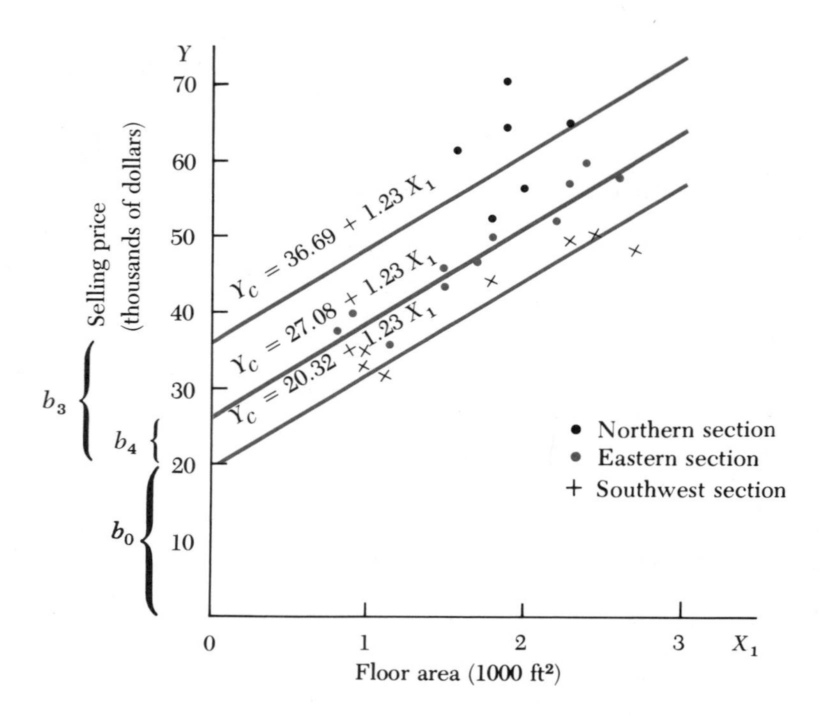

Figure 14-4 Interpretation of Indicator Variables for Section Location

in the southwestern section ($27,075 – $20,316 = $6759). Similarly, houses in the northern section sold for $16,375 more than those in the southwestern section ($36,691 – $20,316 = $16,375). These differences are the coefficients b_4 and b_3, respectively, in Equation 14-12. They are shown on the vertical axis in the figure.

The indicator variables have served to establish a separate regression line for each class of the qualitative variable by giving each line a different intercept. Indicator variables can also be used to establish separate regression functions that differ in both slope and intercept. Many additional variations are also possible.

The coefficients of multiple determination for our first and second residential selling price regression relationships are $r^2_{Y\cdot 12} = 0.567$ and $r^2_{Y\cdot 134} = 0.866$. The second analysis explains considerably more Y-variation. The F statistic and the t statistics for all B-coefficients are also larger and more significant statistically than for the earlier analysis. The number of runs in the residuals is 14, well within the acceptance region for random fluctuations.

The only feature of the second analysis that causes some doubt appears in Figure 14-4. The regression lines for the southwestern and eastern sections fit their observations well, so that relationships based on floor area may provide adequate estimates of selling prices within these sections. But the observations for the more expensive homes in the northern section show considerably greater dispersion around their regression line. Thus a separate analysis based on additional independent variables must be considered for residences in this section. Special features such as saunas, covered patios, or large lots may be important. The type and number of rooms may also have to be included.

In this application, the use of indicator variables proved very helpful. The original problem was subdivided into three parts, and two of these parts turned out to have rather simple solutions. Indicator variables served to isolate the source of the problem in the third part, the northern section. In this northern section, it looks as if a completely separate regression study is called for to find important variables in addition to floor area.

14.5 Nonlinear Regression

Sometimes it is apparent that the relationship between an independent and a dependent variable is not linear. Consider, for instance, people learning to assemble a component for television sets. The number of correctly assembled components produced per hour is likely to be relatively low, then increase rapidly as the necessary skills are acquired. After a few days, the growth in output should level off at a steady, long-term production rate.

Table 14-6 and Figure 14-5 show observations for seven assemblers who are new on the job, as well as the regression relationship. On their first day, assemblers produce an estimated 59 assemblies per hour. Productivity then climbs to a plateau of nearly 84 assemblies per hour, which is reached after 4 days on the job. Prior to that time, the function exhibits definite curvature in order to fit the observations.

Table 14-6 Productivity and Time on the Job

Time on Job (days) X	*Productivity (assemblies/hour)* Y
0	58
0	60
1	70
2	74
3	83
4	82
5	84

A model that should fit the productivity observations reasonably well is the parabola

$$Y = B_0 + B_1X + B_2X^2 + \epsilon \tag{14-13}$$

This is perhaps the simplest nonlinear model and is often used in this type of regression.

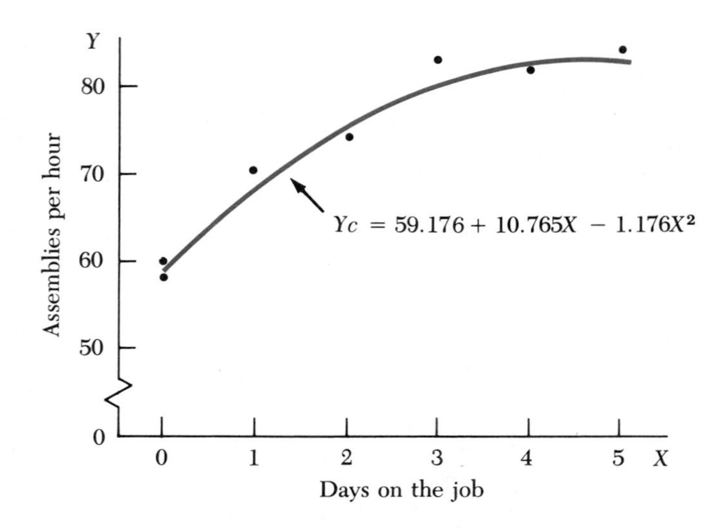

Figure 14-5 Productivity and Experience

By making two substitutions, we can adapt the procedures for multiple linear regression given in Section 14.1 for use in our present analysis. Suppose we let $X_1 = X$, the number of days on the job, and let $X_2 = X^2$. Then Equation 14-13 becomes

$$Y = B_0 + B_1X_1 + B_2X_2 + \epsilon$$

which is the same as the model of Equation 14-1 that we used at the outset of this chapter.

The observations and the necessary sums for the normal equations, Equation 14-3, are shown in Table 14-7; the calculations needed to get the sums are also shown. Note that $X_2 = X_1^2$ for each observation.

Table 14-7 Calculations for Nonlinear Regression Analysis

Y	X_1	X_2	X_1^2	X_2^2	YX_1	YX_2	X_1X_2
58	0	0	0	0	0	0	0
60	0	0	0	0	0	0	0
70	1	1	1	1	70	70	1
74	2	4	4	16	148	296	8
83	3	9	9	81	249	747	27
82	4	16	16	256	328	1312	64
84	5	25	25	625	420	2100	125
511	15	55	55	979	1215	4525	225

The solution of the normal equations is $b_0 = 59.176$, $b_1 = 10.765$, and $b_2 = -1.176$. Hence the sample regression relationship is

$$Y_C = 59.176 + 10.765X_1 - 1.176X_2$$

or

$$Y_C = 59.176 + 10.765X - 1.176X^2$$

where X is the original independent variable, number of days on the job.

The model fits the data well. The coefficient of multiple determination is 0.9798, the F statistic is 97.2 with 2 and 4 degrees of freedom, and the t statistics for B_1 and B_2 are 7.79 and −4.17, respectively. All three statistics are significant at the 0.02 level. The fact that both of the b-coefficients are significant means that both the linear and squared terms in the regression equation make significant marginal contributions to explaining variation in productivity during the first few days on the job.

A single chapter can only begin to suggest the many ways in which regression techniques can be applied in the analysis of observations of a dependent variable and several independent variables. One important class of techniques we have omitted, for example, is used to screen a large number of independent variables being considered for inclusion in a regression relationship. Sometimes one only wants a few of the more important variables from such a list. Because of multicollinearity, it is not obvious in most instances how to select those variables.

Two approaches appear in most computer software programs. The first computes a regression relationship for every possible subset of the listed independent variables, along with several measures of the quality of each relationship. This approach is preferable when the list of variables is short enough to prevent exceeding expense limits.

The second approach is called *stepwise regression*. In forward stepwise regression, independent variables are added one at a time, and a regression relationship is computed at each step. The variable selected for a given step is the one that accounts for the largest proportion of Y-variation not already explained by regression with variables already included. The process continues until the first variable that does not make a statistically significant contribution is found. Other versions of stepwise regression also exist. All work on the principle of examining one additional independent variable at a time.

Exercises

14-15 *Indicator variables for employees.* Employees in a company are classified as apprentice, journeyman, craftsman, foreman, or management. Prepare a coding table similar to Table 14-4 for the indicator variables required to include employee classifications in a regression analysis.

14-16 *Place of manufacture and accident rate.* In addition to other independent variables, the part of the world in which an automobile was manufactured is to be used as a qualitative independent variable in a regression study of accident rates. The classes for this variable are United States, England, Western Europe, Japan, and other. How many indicator variables are needed to put this qualitative variable in the regression relationship? Prepare a suitable set of 0, 1 codes for the indicator values.

14-17 *Variable cost of a manufactured part.* A plant manufactures a small automobile part. The cost accountant wants the regression relationship between the variable unit production cost of the part (Y) and the plant's operating rate (X). A small random sample is collected for a pilot study:

X	Y	X	Y	X	Y
1	12	6	7	17	13
1	9	8	10	20	17
2	11	9	6	23	18
3	8	11	8	24	20
4	9	14	10	28	25

Cost is stated as dollars per part, and the operating rate is stated as hundreds of parts per day.

a. Plot the data.

b. The regression line for the data is

$$Y_C = 6.5295 + 0.4974X$$

and the least-squares parabola is

$$Y_C = 10.68455 - 0.63091X + 0.04204X^2$$

Plot these on the graph prepared in (a).

c. From the graph prepared in (a) and (b), which equation appears to fit the data better?

14-18 *Comparison of goodness-of-fit.* For the situation described in Exercise 14-17, the plant manager wants a specific measure of fit for the straight line and the parabola. Find the coefficients of determination for each and explain the results in non-statistical language for the plant manager.

Summary

Multiple regression techniques are used to find least-squares relationships between a dependent variable and more than one independent variable. The objective is to find a relationship that results in significantly better predictions of values of the dependent variable than would be the case without such a relationship. Scattergrams, checks on residuals, and a comprehensive analysis of variance test all contribute to evaluating the suitability and utility of a given relationship.

When an independent variable is qualitative rather than quantitative, multiple regression can still be used by employing indicator variables to represent the categories of the qualitative variable. Multiple regression can also be adapted to handle curvilinear relationships.

Supplementary Exercises

14-19 *Selection of independent variables.* For each of its 120 sales districts throughout the United States, a company has gathered last year's data on sales, population, and average age. A linear least-squares regression relationship has been found for the data. The sales director says the relationship would be much better if advertising expenditures were included as an additional independent variable. All advertising is done at the district level. The home office allocates funds for advertising to the districts in proportion to district populations. What do you think of the director's suggestion?

14-20 *Toothpaste preference study design.* The observations for the toothpaste preference study of Exercises 14-2 and 14-10 are given below. Here X_1 is milligrams of flavoring, X_2 is milligrams of astringent in a particular mixture, and Y is preference score.

X_1	X_2	Y
20	10	59
20	30	71
40	10	72
40	30	87
60	10	79
60	30	88
80	10	90
80	30	97

Explain why the eight combinations of values for X_1 and X_2 were chosen in the pattern given. (*Hint:* Find the coefficient of correlation for the independent variables.)

14-21 *Residuals in the computer printout.* In Figure 14-2, the mean square for residuals (MSE) is 50.45692, which is equal to the square of the standard error of estimate, 7.103303, shown beneath the analysis of variance table. Explain why this relationship exists by referring to Equations 14-5, 14-9, and 14-10.

14-22 Why is the coefficient of multiple correlation defined to have no negative sign?

14-23 *Contribution of independent variables.* For the toothpaste preference analysis of Exercise 14-20, $r^2_{Y\cdot 12} = 0.94878$, $r_{Y1} = 0.85809$, $r_{Y2} = 0.46093$, and $r_{12} = 0$. In Section 14.3, the equation

$$r^2_{Y\cdot 12} = r^2_{Y1} + r^2_{Y2}$$

held under certain conditions. Make the necessary calculations to determine if the equation holds in this case. Explain why it does or does not hold. Then discuss the contributions of the independent variables to the relationship.

14-24 *Independent variables and electricity consumption.* The dependent variable (Y) in the analysis of Exercise 14-3 is electricity consumption. The independent variables are thermostat settings on space heating (X_1), and water heaters (X_2). In this analysis, $r^2_{Y \cdot 2} = 0.94199$, $r_{Y1} = 0.60921$, $r_{Y2} = 0.73657$, and $r_{12} = -0.03057$. In view of the text discussion in Section 14.3, can the equation

$$r^2_{Y \cdot 12} = r^2_{Y1} + r^2_{Y2}$$

be expected to hold in this case? Does it? Can the contributions of the independent variables to the regression relationship be separated?

Computer Exercises

1 Source: Real estate data file

For all parts of this exercise treat the real estate data as if they are a random sample from a large universe of sales in the community.

a. Obtain the multiple regression of purchase price of homes on income and age of buyer. Give an interpretation of each of the regression coefficients.

b. Plot the residuals from the regression against purchase price, income, and age (three plots). Mark off one standard error of estimate above and below the zero residual line. Do the plots suggest any failures to meet the regression model assumptions? Explain.

c. Can the null hypothesis that all of the regression coefficients are equal to zero be rejected at the 0.05 significance level? Can the null hypothesis that the individual regression coefficients are equal to zero be rejected at the 0.05 significance level? Explain.

d. Estimate the purchase price given an income of $25,000 and an age of 35 years. Distinguish between the standard error of this estimate as an individual prediction and as a mean response, illustrating with specific figures if your computer program provides them.

e. Obtain the regression of purchase price (Y) on income (X_1), residence (X_2), and old-new homes (X_3). What do the coefficients for X_2 and X_3 represent? Are the coefficients statistically significant?

f. Plot residuals from the regression in (e) against Y, X_1, X_2, and X_3 (four plots). Are there any suggestions of failure of regression assumptions? Explain.

2 Source: Utilities data set

a. Delete AT&T and GTE. Perform a multiple regression analysis using income as the dependent variable and assets and number of employees as the independent variables.

b. Make scattergrams for each pair of variables in (a). Do the independent variables appear to be related?

c. If your program does stepwise regression, use it to analyze the data for the model in (a). How do the scattergrams in (b) help explain the results of the stepwise regression?

d. Restore AT&T and GTE to the data file. Define an indicator variable to describe industry type (electrical, gas, communications). Perform a regression analysis using income as the dependent variable and assets and the industry type as independent variables. What do you conclude about the relationship?

e. Perform a simple regression analysis between income and assets for each industry type. Make scattergrams. Does the use of the indicator variable in (d) seem justified?

3 Source: Baseball data file. Use runs as the dependent variable throughout.

a. Perform a multiple regression analysis incorporating all 12 independent variables. Interpret the meaning of the regression coefficient for each variable.

b. If your program has stepwise regression use it to find the "best" set of independent variables.

c. If your program has the "all possible regressions" procedure, use it to determine the best model with one variable, with two variables, and so forth. Find the adjusted coefficient of determination and/or Mallow's Cp statistic for each model.

d. For the best models in (c), find the one with the highest adjusted coefficient of determination (the lowest Cp). Which appears to be the best of the best models?

e. Suppose you want to use a regression model to evaluate the performance of individual players. What is the practical utility of using a model with 5 or 6 variables versus a model with twice as many variables? Which model would you use?

Interpreting Computer Output: A Mini Case

Use the following material to form your own conclusion with regard to the issue presented.

Figure 14-6 shows the ratings of macaroni and cheese dishes as determined by Consumers Union (*Consumer Reports* Volume 48, Number 6, p. 309, June 1983). Use any statistical technique you have learned to analyze these data. Your analysis should attempt to determine what factors influence cost and quality.

Ratings

Macaroni & cheese

Listed in order of estimated overall quality based on comparisons of commercial products with CU's recipe on page 308. This was judged excellent, having a pronounced cheese flavor, a generous amount of sauce, and appropriately firm pasta. Nutritional values are based on lab analysis and/or nutritional labeling. Cost per serving is calculated using average package price paid by CU plus cost of any added ingredients.

Excellent	Very good	Good	Fair	Poor
● (with white center)	◓ (top half dark)	○	◒ (bottom half dark)	●

Products	Sensory judgments: Cheese flavor	Amount of sauce	Firmness of pasta	Type [1]	Labeled weight	Package cost	Cooked weight	Estimated servings per package [2]	Per cooked 8-oz. serving: Cost	Protein	Fat	Carbohydrate	Calories	Sodium
STOUFFER'S	Excellent [3]	Excellent	Very good	F	12 oz.	$1.27	12.2 oz.	1½	85¢	15 g.	18 g.	30 g.	340	1080 mg.
KRAFT DELUXE	Excellent [3]	Good	Excellent	M	14	1.14	26.4	3¼	35	17	12	52	380	920
HOWARD JOHNSON'S	Very good	Very good	Good	F	10	1.00	9.8	1¼	80	12	17	32	330	770
LIPTON	Good [3]	Good	Excellent	M	5¼	.75	17.1	2¼	36	9	19	47	390	1020
GREEN GIANT BOIL 'N BAG	Good	Good	Good	F	9	.93	9.3	1¼	74	12	11	34	280	950
PRINCE SHELLS & CHEDDAR	Fair	Good	Excellent	M	7¼	.32	19.9	2½	18	14	23	56	480	990
TOWN HOUSE (Safeway)	Fair	Good	Excellent	M	7¼	.35	22.6	2¾	18	11	15	54	390	910
MORTON	Very good	Fair	Good	F	8	.54	7.6	1	54	12	7	44	290	1210
CREAMETTES	Fair	Good	Very good	M	7¼	.40	22.1	2¾	20	11	20	53	410	970
KROGER	Fair	Good	Very good	M	7¼	.29	22.5	2¾	16	11	18	56	430	870
ANN PAGE (A&P)	Fair	Good	Good	F	8	.45	7.1	1	45	10	5	40	250	1260
KRAFT	Fair	Fair	Excellent	M	7¼	.36	21.7	2¾	18	12	21	53	440	910
PRINCE	Fair	Fair	Excellent	M	7¼	.35	20.1	2½	20	14	22	56	480	960
BANQUET	Fair	Very good	Fair	F	8	.49	7.3	1	49	10	13	30	270	810
BETTY CROCKER MUG-O-LUNCH	Poor	Excellent	Excellent	M	4	.73	15.3	2	37	7	5	43	250	790
FOOD CLUB	Poor	Good	Excellent	M	7¼	.28	22.6	2¾	15	10	20	48	410	960
GOLDEN GRAIN	Poor	Good	Excellent	M	7¼	.39	19.4	2½	22	13	21	64	510	940
GOLDEN GRAIN STIR-N-SERV	Poor	Good	Good	M	3⅞	.30	9.7	1¼	28	10	15	51	380	910
SPRINGFIELD	Fair	Poor	Excellent	M	7¼	.29	21.7	2¾	16	12	19	53	430	850
HEINZ MINUTE MEALS	Fair	Excellent	Poor	C	7½	.46	7.1	1	46	9	9	27	220	950
ANN PAGE (A&P)	Poor	Fair	Excellent	M	7¼	.25	22.2	2¾	15	10	17	53	420	1000
COST CUTTER (Kroger)	Poor	Fair	Excellent	M	7¼	.23	20.7	2½	15	12	15	58	410	1010
GRAND UNION	Poor	Fair	Excellent	M	7¼	.28	22.1	2¾	15	11	20	46	410	860
SCOTCH BUY (Safeway)	Poor	Fair	Excellent	M	7¼	.25	21.9	2¾	14	10	15	56	410	860
SWANSON	Poor	Very good	Fair	F	7	.50	6.9	¾	67	10	6	34	230	1100
BASICS (Grand Union)	Poor	Fair	Very good	M	7¼	.22	21.0	2¾	13	11	15	57	400	910
ECONO BUY (Alpha Beta)	Poor	Fair	Very good	M	7¼	.25	22.2	2¾	14	11	14	54	380	1030
PATHMARK	Poor	Fair	Very good	M	7¼	.23	19.9	2½	15	13	16	61	430	810
GOLDEN WHEAT	Poor	Fair	Good	M	7¼	.28	22.2	2¾	16	10	19	47	380	950
KROGER	Poor	Fair	Fair	F	8	.42	7.7	1	42	11	8	38	270	860
P & Q (A&P)	Poor [4]	Poor	Excellent	M	7¼	.23	23.0	3	12	10	17	53	400	770
FRANCO-AMERICAN	Poor	Poor [5]	Poor	C	14¾	.54	13.7	1¾	31	7	4	26	170	990
BEL-AIR (Safeway)	Poor [4]	Poor	Fair	F	8	.43	7.7	1	43	11	7	41	270	780
BASICS (Grand Union)	Poor [4]	Poor	Poor	F	8	.34	7.0	1	34	17	25	44	470	710

[1] F—frozen; M—dry mix; C—canned.

[2] Servings based on 8-oz. portion; figures rounded to nearest quarter-serving.

[3] Product had tart flavor typical of cheddar cheese

[4] Product was noticeably lacking in salt flavor.

[5] Product had far too much sauce.

Figure 14-6 Macaroni and Cheese Ratings. Copyright 1983 by Consumers Union of United States, Inc., Mount Vernon, N.Y. 10553. Reprinted by permission of *Consumer Reports*, June 1983.

15

TWO-WAY ANALYSIS OF VARIANCE

Chapter 13 introduced and explained one-way analysis of variance. Main topics were the calculation of sums of squared deviations attributable to treatments and error, the use of the F test for significance of treatment effects, and estimates and comparisons of treatment means. The final comments in that chapter touched on the issue of the confounding of treatment effects with a second factor not included in the experimental design. The principle of two-factor and multi-factor design was introduced as a way of improving designs to forestall such a confounding effect. We now want to move directly to consider two-factor analysis of variance designs. These are introduced first using conventional analysis of variance computations. In the final section of the chapter we present alternative calculation methods for both one- and two-way analysis of variance that use multiple regression.

15.1 Two-Factor Analysis without Replication

Two major types of design apply when two or more factors (independent variables) are included in an experiment suitable for analysis of variance. We can distinguish between these types with an example which involves two factors and a response variable. Suppose the product safety group in an automobile manufacturing firm is investigating the effect of differences in automobile dashboard light intensity on reaction times of drivers. A series of night-driving emergencies is programmed in driving simulators. Drivers are to be exposed to the emergencies under different levels of dashlight intensity. The response variable will be the sum of a driver's reaction times to the series of emergencies under a given level of intensity.

A consulting physiologist points out that characteristic reaction times vary among individuals and that age is a primary cause of such variation. In fact, this effect on reaction times could obscure any effect from differences in dashlight intensity. To prevent such a result, the experimenters decide to include age as a second factor.

The two major types of design mentioned earlier arise from the way an additional factor such as age is incorporated. One design is called a **completely randomized design**. In this type experimental units are randomly selected within predetermined levels of the additional factor. The other design is called a **randomized block design**. Here the additional factor is controlled by arranging the experimental units in blocks according to this factor. Our example can be used to illustrate the difference.

A completely randomized design for the reaction time experiment with dashlight intensity and driver age as factors could be accomplished with the following steps. First, a master list of drivers is obtained from the state. Second, the list is divided into four lists by age group (29 or less, 30 through 44, 45 through 60, over 60). Third, three drivers are selected from each age group by using a table of random numbers. Fourth, levels of the first factor (low, medium, high dashlight intensity) are assigned to the three drivers in each age group by a randomization process. Note that the second step could incur considerable cost to make up four long lists from each of which only three drivers are selected. On the other hand, a legitimate random sample from the population lists by age makes it proper to estimate such age effects in the population from sample results.

A randomized block design for the reaction time experiment with light intensity and age as factors could be accomplished with the following steps. First, the original master list of drivers is obtained as in the previous design. Second, 12 drivers are randomly selected from the list without regard to their ages. Third, the 12 drivers are arranged in four age groups. The three youngest drivers are put in group A, the three next youngest in group B, the three next youngest in group C, and the three oldest in group D. This step is called **blocking** (on age in this case). The final step is the same as the final one in the completely randomized design: Dashlight intensity levels are assigned to the three drivers in each age group by a randomization process. Because of their practicality, randomized block designs are used frequently in applied research to control nuisance factors that may obscure differences attributable to a factor of interest. Blocking is cheap and effective. But it violates the random sampling assumption on which statistical inferences about the blocked factor can be made.

Note that the experimenter preselected the age categories in the completely randomized design. Also note that drivers were randomly selected from the entire population in each age group. On the other hand, in the randomized block design the experimenter did not sample age groups randomly; he had to arrange whatever ages occurred in the sample for the randomized block design. He did not sample from the entire population in a preselected age group.

A Randomized Complete Block Design

We will illustrate two-factor analysis of variance without replication by using a randomized block design rather than a completely randomized design. After presenting the data we can more easily point out what qualifies the example as a randomized *complete* block design.

Twelve drivers were selected at random from the state's list of registered passenger car operators. The 12 drivers were then arranged in four age groups as described above. Next the three levels of light intensity were assigned to drivers randomly with results as shown in Table 15-1. The table shows that, for instance, in the youngest age group (block A) the first driver will be subjected to medium dashlight intensity, the second to low, and the third to high. The orders for the other blocks are also shown. This is a randomized *complete* block design, because the number of experimental units in each block is equal to the number of levels of light intensity. There are three levels of intensity and three drivers in each age block. Here the words *randomized complete* indicate full cross-classification. In a completely randomized design, on the other hand, the words *completely randomized* indicate that all independent factors included in the design were randomly sampled.

Table 15-1 Random Arrangement of Dashlight Intensities

Age Block	*Intensity Sequence*		
A	Medium	Low	High
B	Medium	High	Low
C	High	Low	Medium
D	Low	High	Medium

The observations of reaction time appear in Table 15-2. The means for the three intensities appear to be considerably different, and the means for blocks show an apparent increase in reaction time with age. The analysis of variance to which we turn next tests the column means to see whether the differences attributable to light levels are statistically significant.

Table 15-2 Reaction Times* Classified by Light Intensity and Age Block

Age Block	*Intensity*				
	Low	*Medium*	*High*	*Totals*	*Means*
A	22	11	25	58	19.33
B	29	18	26	73	24.33
C	33	18	35	86	28.67
D	36	22	39	97	32.33
Totals	120	69	125	314	
Means	30.00	17.25	31.25		26.17

*In tenths of a second.

What is needed is an estimate of the variability of reaction times under essentially constant conditions. Recall that in the one-way analysis of variance such an estimate came from the within treatment sum of squared deviations. In the current example we have only one observation for each treatment and level of the blocking factor. How can an estimate of experimental error be derived?

If it is assumed that the different light intensities produce the same *differences* in reaction times for all the age groups, and that the *differences* associated with the age groups are the same for each level of light intensity, an estimate of experimental error can be made. The condition just stated amounts to assuming that the column and row effects are additive in a model which states that an observation in any cell of the two-way table is comprised of a column (light intensity) effect plus a row (age block) effect plus experimental error.

$$X_{ij} = \mu + \Delta_i + \Delta_j + e_{ij} \tag{15-1}$$

where Δ_i = a constant effect of the treatment in the ith row and $\Sigma\Delta_i = 0$,
Δ_j = a constant effect of the treatment in the jth column and $\Sigma\Delta_j = 0$, and
e_{ij} = independent residual variation.

The model assumes, further, that the e_{ij} are normally distributed with zero mean and variance independent of the treatments. This assumption of equal residual variance within treatment cells is analogous to the assumption in one-way ANOVA of equal variances within populations represented by the column classifications.

The estimates of Δ_i and Δ_j are implicit in Table 15.2. They are the difference between each column mean and the grand mean for light intensity levels and the difference between each row mean and the grand mean for age groups. For example, observation D(low) is estimated to be comprised as follows.

$$X_{41} = \overline{\overline{X}} + d_{i=4} + d_{j=1} + e_{41}$$

$$36 = 26.17 + (32.33 - 26.17) + (30.00 - 26.17) + (-.17)$$

$$= 26.17 + \quad 6.16 \quad + \quad 3.83 \quad - \ .17$$

Here, 6.16 is the sample effect of age block D (32.33 – 26.17), 3.83 is the sample effect of low light intensity (30.00 – 26.17), and –.17 is the deviation of the actual value from the value produced by the grand mean plus the sum of the row and column effects (36.00 – 36.17).

The residuals, e_{ij}, sum to zero and the variance estimated from them is a variance not affected by differences in row means or column means. The estimated residual variance is attributable to uncontrolled factors in the experiment and chance variation. The degrees of freedom for the estimated residual variance are the product of the number of rows less 1 and the number of columns less 1, or $(r - 1)(c - 1)$. In estimating the residuals by cells, we have used the estimate of the grand mean, $(r - 1)$ independent row means, and $(c - 1)$ independent column means. Thus, $1 + (r - 1) + (c - 1)$ degrees of freedom are lost and $(r - 1)(c - 1)$ remain,

$$rc - 1 - (r - 1) - (c - 1) = (r - 1)(c - 1).$$

The equations for finding the total sum of squares (SST) and the sum of squares among column means (SSA) in this analysis are identical to Equations 13-10 and 13-11 as used in one-factor analysis of variance. They are repeated below as Equations 15-2(a) and 15-2(b). Equation 15-2(c) is used to find the sum of squares among block means (SSB). The final equation is used to find the residual sum of squares (SSE).

$$\text{SST} = \sum_j \sum_i (X_{ij}^2) - \frac{T^2}{n} \qquad \textbf{(15-2a)}$$

$$\text{SSA} = \frac{\sum_j (C_j^2)}{n_j} - \frac{T^2}{n} \qquad \textbf{(15-2b)}$$

$$\text{SSB} = \frac{\sum_i (R_i^2)}{n_i} - \frac{T^2}{n} \qquad \textbf{(15-2c)}$$

$$\text{SSE} = \text{SST} - \text{SSA} - \text{SSB} \qquad \textbf{(15-2d)}$$

where

$$T = \sum_j \sum_i (X_{ij})$$

In Equation 15-2(a) recall that the summation is found by squaring all the observations in the table and then summing these squares. The second term (T^2/n) is found by summing all the observations in the table to get T, squaring to get T^2, and dividing by the number of observations in the entire table (n). In Equation 15-2(b) the first term is found by getting column totals, C_j, squaring the column totals, adding the squares, and dividing by the number of observations in each column, which is the same as the number of rows (r). The first term in Equation 15-2(c) is found by getting the row totals, R_i, squaring them, adding the squares, and dividing by the number of observations in each row, which is the same as the number of columns (c).

By rearranging Equation 15-2(d), we find that

$$\text{SST} = \text{SSA} + \text{SSB} + \text{SSE}.$$

The total sum of squares can be decomposed into a sum attributable to column effects, a sum attributable to row effects, and the residual. The degrees of freedom can be separated in a similar manner. The result can be shown in a general format for two-factor ANOVA without replication. The general format is shown in Table 15-3.

Table 15-3 Format for Two-Factor ANOVA without Replication

Source of Variation	*Sums of Squares*	*Degrees of Freedom*	*Mean Squares*
Columns factor	SSA	$c - 1$	$\text{SSA}/(c - 1)$
Rows factor	SSB	$r - 1$	$\text{SSB}/(r - 1)$
Residual	SSE	$(r - 1)(c - 1)$	$\text{SSE}/(r - 1)(c - 1)$
Total	SST	$n - 1$	

We can now find the numerical values of the sums of squares for our dashlight intensity experiment by referring to Table 15-2 and Equation 15-2. We can begin by finding the correction term:

$$\frac{T^2}{n} = \frac{314^2}{12} = 8216.33.$$

Then

$$\begin{aligned}
\text{SST} &= \sum_j \sum_i (X_{ij}^2) - \frac{T^2}{n} \\
&= (22^2 + 11^2 + 25^2 + \ldots + 36^2 + 22^2 + 39^2) - 8216.33 \\
&= 9010 - 8216.33 = 793.67;
\end{aligned}$$

$$\begin{aligned}
\text{SSA} &= \frac{\sum_j (C_j^2)}{n_j} - \frac{T^2}{n} \\
&= \frac{120^2 + 69^2 + 125^2}{4} - 8216.33 \\
&= 8696.50 - 8216.33 = 480.17;
\end{aligned}$$

$$\text{SSB} = \frac{\sum_i (R_i^2)}{n_i} - \frac{T^2}{n}$$

$$= \frac{58^2 + 73^2 + 86^2 + 97^2}{3} - 8216.33$$

$$= 8499.33 - 8216.33 = 283.00;$$

and

$$\text{SSE} = \text{SST} - \text{SSA} - \text{SSB}$$
$$= 793.67 - 480.17 - 283.00 = 30.50.$$

The general format in Table 15-3 applied to our light intensity example is shown in Table 15-4. The sums of squares just found appear in the first column. The degrees of freedom in the second column are found by subtracting one from the number of columns ($c - 1 = 2$), one from the number of rows ($r - 1 = 3$), and multiplying these two quantities for the residual degrees of freedom.

As was the case in one-factor ANOVA, we can test the null hypothesis of no difference in column means for the population by using an F ratio. As in that earlier case, the F ratio was formed by putting the mean square for column means in the numerator and the mean square for residual variation in the denominator. Hence, for our light intensity experiment,

$$F = \frac{240.08}{5.08} = 47.3.$$

In this F ratio 2 degrees of freedom apply to 240.08 and 6 apply to 5.08. The F table in Appendix A shows that $F_{2,6,.99}$ is 10.92. We conclude that differences in light intensity have produced differences in mean reaction times significant at well beyond the 0.01 level. The mean reaction time for medium intensity is 17.25 seconds in Table 15-4. Apparently, this is significantly smaller than the other two means, which are nearly equal (30.00 and 31.25, respectively).

It may occur to you that the ratio of the mean square for age blocks to the residual mean square could be found by dividing 94.33 in Table 15-4 by 5.08. In a randomized complete block design the result cannot be interpreted as an F ratio, however. Recall that experimental units (drivers) were not selected for each age

Table 15-4 Two-Factor ANOVA for Light Intensity

Source of Variation	*Sums of Squares*	*df*	*Mean Squares*	*F Ratio*
Light intensity	480.17	2	240.08	47.3
Age block	283.00	3	94.33	
Residual	30.50	6	5.08	
	793.67	11		

level by random sampling from the population of all drivers in that age level. Had this been done, we would have a completely randomized design and it would be correct to interpret the ratio of 94.33 to 5.08 as an F statistic.

The advantage of blocking can be brought out by comparing the result of a one-factor analysis on light intensity with what we have just done. Suppose we had elected not to put our drivers into age blocks, but had left them in the same columns as in Table 15-2. We have already seen that SST and SSA are exactly the same for one-factor ANOVA as they are for the two-factor analysis just completed. Hence, for the one-factor analysis,

$$\text{SST} = 793.67,$$
$$\text{SSA} = 480.17,$$

and, from Equation 13-12,

$$\begin{aligned}\text{SSE} &= \text{SST} - \text{SSA} \\ &= 793.67 - 480.17 = 313.50.\end{aligned}$$

Table 15-5 shows the sums of squares and degrees of freedom for the one-factor ANOVA. Before we test light intensity for significance, note that the residual sum of squares (SSE) for the one-factor analysis was partitioned into a sum of squares for age blocks and a smaller sum of squares for residual error in the two-factor analysis shown in Table 15-4. The same relationship exists among degrees of freedom in the two analyses.

Table 15-5 One-Factor ANOVA for Light Intensity

Source of Variation	*Sums of Squares*	*df*	*Mean Squares*	*F Ratio*
Light intensity	480.17	2	240.08	6.89
Residual	313.50	9	34.83	
	793.67	11		

The test statistic for the one-factor experiment is

$$F = \frac{240.08}{34.83} = 6.89,$$

with 2 and 9 degrees of freedom. Appendix A shows that $F_{2,9,.95} = 4.26$ and $F_{2,9,.99} = 8.02$. The level of significance has dropped sharply compared with the result of the randomized complete block analysis. Ninety percent of the one-factor residual sum of squares (313.50) was partitioned to the two-factor age blocks (283.00). Only one-third of the degrees of freedom associated with the residual in the one-factor analysis (9) was partitioned to the age blocks in the two-factor analysis (3). Hence the residual mean square in the two-factor analysis (5.08) is much smaller than the corresponding one-factor mean square (34.83). Since the numerators of the two F

ratios are the same (240.08), the two-factor analysis is much more sensitive to differences in light intensity than is the one-factor analysis.

The foregoing comparison is an example of the general objective in blocking. We want to find a blocking factor that is associated with the response variable as we suspected might be the case with age and reaction time. Such a factor will account for more variability than it costs in degrees of freedom. On the other hand, a completely unrelated factor can be expected to account for the same proportions of variability and degrees of freedom. As a result, the F ratios with and without such a useless blocking factor can be expected to be almost the same.

At the beginning of this section we pointed out that there are two major types of two-factor analysis without replication—a completely randomized design and a randomized block design. We illustrated the randomized block design. The analysis of variance calculations for a completely randomized design are the same. Instead of a blocking factor, a second treatment factor is present, and the experimental units would have been selected randomly within each treatment combination.

15.2 Two-Factor Analysis with Replication

For each combination of factor levels in Table 15-2, there was only one observation. There were four age blocks and three levels of dashlight intensity. These conditions resulted in a cross-classified table in which there were 12 cells. Each cell contained a single observation.

To accomplish *balanced replication* in the light intensity experiment, there would have to be more than one observation per cell and the same number of observations in each cell. Replication not only requires more than one observation per cell but also complete independence among the observations in a cell. The same driver could not be used a second time, for example. Balancing requires the same number of replications in all cells. We will point out the advantages of balancing later. When balanced replication is carried out for every possible combination of factor levels, many statisticians call the result a **factorial experiment**.

A major advantage of a factorial experiment over a two-factor experiment without replication is that the former type makes it possible to test for interaction between the two factors, whereas the latter type is based on the assumption that there is no interaction. To illustrate the difference we can consider two different experiments concerned with automobile gasoline consumption.

The data for the first experiment appear in Table 15-6. The body of the table shows the miles per gallon for four sets of conditions. A heavy car with a high-compression engine gets 14 miles per gallon, for instance. Note that there is an increase of 3 miles per gallon from high compression to regular engines, regardless of weight. Note also that there is a difference of 6 miles per gallon attributable to weight differences, regardless of engine type. This is an example of an experiment in which the effects are *additive*. In such an experiment there is no interaction between the factors.

Table 15-6 Miles per Gallon Classified by Engine and Weight

Curb Weight	*Engine*	
	High Compression	*Regular*
Heavy	14	17
Light	20	23

Now consider an experiment on gasoline consumption where one factor is engine type, as before, and the second factor is type of fuel. The data appear in Table 15-7. For fuel without the new chemical, gasoline performance changes from 15 to 16 miles per gallon with a change from a high compression to a regular engine, an increase of 1 mile per gallon. By contrast, with the new chemical the comparable increase is from 18 to 22, a gain of 4 miles per gallon. Similarly, there is a difference of 3 in the first column and a difference of 6 in the second. The new chemical provides twice as much mileage improvement for regular engines as it does for high-compression engines. The effects are not additive. Fuel and engine type *interact* to the special benefit of the regular engine. To test whether such an interaction is statistically significant, we must have replications in each of the four cells of such a table.

Table 15-7 Miles per Gallon Classified by Engine and Fuel

Fuel	*Engine*	
	High Compression	*Regular*
Without new chemical	15	16
With new chemical	18	22

For an illustration of how to carry out an analysis of variance for a factorial experiment we return to the poster example discussed in Chapter 13. In Chapter 13, the one-factor analysis produced significantly different sales results for the four types of posters. The marketing executive decides to discard posters B and C from further consideration because of their low sales.

The marketing executive knows that the canned salted nuts being advertised sell primarily to households with high incomes. He wants to learn more about the combined effects of poster and income on sales. For this experiment, six convenience markets in low-income neighborhoods and six markets in high-income neighborhoods are randomly and independently selected. A randomization process is used to assign posters A and D to three markets in each income class. The results of the sales experiment are shown in Table 15-8(a). The three observations, X_{ijk}, in each of the four cells are totaled. These sums are designated as K_{ij}. Cell totals are summed horizontally to produce row totals, R_i, and vertically to produce column totals, C_j. The grand total (T) of all 12 observations is 600.

We can carry out a simple informal analysis before doing the analysis of variance. This simple analysis will tell us what we might expect from the later one. In Table 15-8(b) we show the mean sales for each of the four cells in Table 15-8(a) together with the row means, column means, and grand mean. Mean sales in high-income neighborhoods are 61, while those in low-income neighborhoods are only 39. There appears to be an income effect on sales.

The two column means are 52 and 48. On this evidence we may be tempted to conclude that if there is a poster effect, it isn't very important. But look at the cell means. Poster A appears to perform much better than D in high-income neighborhoods and somewhat worse in low-income neighborhoods. There appears to be a

Table 15-8(a) Sales Classified by Poster and Neighborhood Income

	Poster		
Income	*A*	*D*	*Total*
	68 (X_{111})	51	
High	65 (X_{112})	53	
	74 (X_{113})	55	
	207 (K_{11})	159 (K_{12})	366(R_1)
	39	44	
Low	32	47	
	34	38	
	105 (K_{21})	129 (K_{22})	234(R_2)
Total	312 (C_1)	288 (C_2)	600(T)

Table 15-8(b) Mean Sales Classified by Poster and Neighborhood Income

	Poster		
Income	*A*	*D*	*Mean*
High	69	53	61
Low	35	43	39
Mean	52	48	50

marked interaction between poster and income level. If this interaction proves statistically significant, we will conclude that poster A should be used in markets in high-income neighborhoods and poster D should be used in low-income neighborhoods.

A way to illustrate interaction is shown in Figure 15-1. Here the two sales response lines for the posters are not parallel as they would be if the effects were additive. Instead, poster A's sales response to increased income is a much more dramatic increase in sales than is D's response. Figure 15-1 brings out clearly that poster D may be slightly better than poster A in low-income neighborhoods but A performs much better than D in high-income neighborhoods. The statistical analysis to which we now turn will tell us whether these differences are likely to be dependable guides to action or are mere sampling fluctuations.

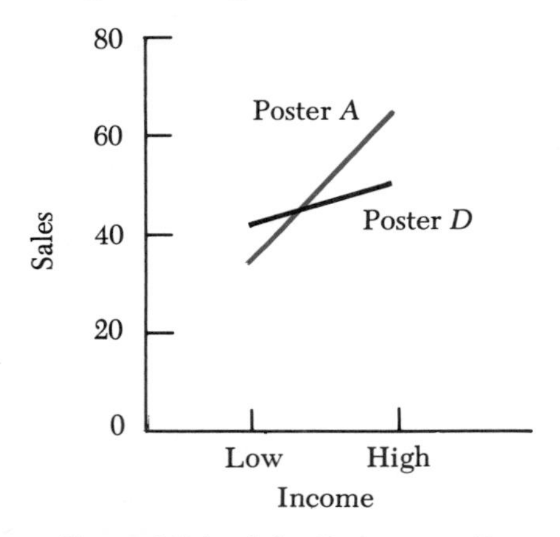

Figure 15-1 Sales Response to Posters and Income Levels

The analysis of variance model for a two-way factorial with allowance for interaction is

$$X_{ijk} = \mu + \Delta_i + \Delta_j + \Delta_{ij} + e_{ijk} \tag{15-3}$$

where Δ_i and Δ_j = constant effects of the ith row and jth column treatments,
$\Sigma\Delta_i$ and $\Sigma\Delta_j = 0$,
Δ_{ij} = interaction effect for the ith row and jth column treatments,
$\Sigma_{ij}\Delta_{ij} = 0$,
and e_{ijk} = deviation of kth observation in a cell from the cell mean, μ_{ij}.

The e_{ijk} are assumed to be independent and normally distributed with zero mean and equal variance for all ij.

Here we have a constant effect of the ith row treatment and a constant effect of the jth column treatment. The next term, Δ_{ij}, is an effect of the treatment combination in cell ij which is *not* the additive result of the effect of the treatment in row i and the treatment in column j. It is the interaction term. In terms of the sample values just discussed, consider $X_{111} = 68$.

This value can be comprised as

$$\begin{aligned} X_{111} &= \bar{\bar{X}} + d_{i=1} + d_{j=1} + (d_{11} - d_{i=1} - d_{j=1}) + e_{111} \\ 68 &= 50 + (61 - 50) + (52 - 50) + (19 - 11 - 2) + (-1) \\ &= 50 + 11 + 2 + 6 - 1 \end{aligned}$$

The deviation of the high income-poster D cell mean of 69 from the grand mean $(69 - 50 = 19)$ is not the sum of the deviation of the high income mean from the grand mean $(61 - 50 = 11)$ and the deviation of the poster D mean from the grand mean $(52 - 50 = 2)$. The effect that is not additive $(19 - 11 - 2 = 6)$ is associated with the particular *combination* of income level and poster. It is the (sample) interaction effect.

The estimate of experimental error comes from the sum of squared deviations of the within cell observations from their respective means. There are $rc(k - 1)$ degrees of freedom for this estimate, where k is the number of observations per cell.

The equations for finding the necessary sums of squares in a factorial experiment are

$$\text{SST} = \sum_{ijk} (X_{ijk}^2) - \frac{T^2}{n} \qquad \textbf{(15-4a)}$$

$$\text{SSA} = \frac{\sum_j (C_j^2)}{n_j} - \frac{T^2}{n} \qquad \textbf{(15-4b)}$$

$$\text{SSB} = \frac{\sum_i (R_i^2)}{n_i} - \frac{T^2}{n} \qquad \textbf{(15-4c)}$$

$$\text{SSK} = \frac{\sum_{ij} (K_{ij}^2)}{n_{ij}} - \frac{T^2}{n} \qquad \textbf{(15-4d)}$$

$$\text{SSI} = \text{SSK} - \text{SSA} - \text{SSB} \qquad \textbf{(15-4e)}$$

$$\text{SSE} = \text{SST} - \text{SSK} \qquad \textbf{(15-4f)}$$

where

$$T = \sum_{ijk} (X_{ijk})$$

$$C_j = \sum_{ik} (X_{ijk})$$

$$R_i = \sum_{jk} (X_{ijk})$$

$$K_{ij} = \sum_{k} (X_{ijk})$$

The total sum of squares is the same in this case as it was earlier. The sum of squares for factor A is the same as before. Now, however, each column total is based on kr observations, where k is the number of observations per cell and r is the number of cells per column. For our example, there are three observations in each of two vertically arranged cells per column. Similarly, the denominator of the first term for SSB is the product of k observations per cell and c cells per row. SSK is the sum of squares for cells. The cell totals are squared, summed, and divided by the $k = n_{ij}$ observations per cell before subtracting the correction factor. Equation 15-4(e) is the sum of squares for **interaction**. It is found by subtracting the sums of squares for rows and columns from the sum of squares for cells. The row and column effects are often called **main effects**. Finally, as shown in Equation 15-4(f), the residual sum of squares for factorial experiments (SSE) is the difference between the total sum of squares (SST) and the total sum of squares for cells (SSK). Notice that SSK is the sum of squares which contains both main effects and the interaction effect. It is sometimes referred to as the row-by-columns, or cell means effect.

From the data in Table 15-8, we find that the numerical values of the sums of squares are

$$\text{SST} = (68^2 + 65^2 + \ldots + 47^2 + 38^2) - \frac{600^2}{12}$$

$$= 32{,}050 - 30{,}000 = 2050$$

$$\text{SSA} = \frac{312^2 + 288^2}{6} - 30{,}000 = 48$$

$$\text{SSB} = \frac{366^2 + 234^2}{6} - 30{,}000 = 1452$$

$$\text{SSK} = \frac{207^2 + 159^2 + 105^2 + 129^2}{3} - 30{,}000 = 1932$$

$$\text{SSI} = 1932 - 48 - 1452 = 432$$

and

$$\text{SSE} = 2050 - 1932 = 118$$

As before, these sums can be arranged in a table along with their degrees of freedom and mean squares. The general format for two factors with replication appears in Table 15-9. The numerical results for our example appear in Table 15-10. Each mean square is divided by the mean square for the residual to obtain the F ratio. From the significance of the interaction F ratio we conclude that the way in which the posters affect sales depends on income (or vice versa). To describe this pattern we must look at the four income-by-poster means just as we did earlier. When the interaction is significant some analysts go no further for the reason just given.

Others will look also at the main effects. When the interaction is significant, we know that there are no main effects in the sense of constant treatments differences for

Table 15-9 General Format for Two-Factor Factorial Experiment

Source of Variation	*Sum of Squares*	*Degrees of Freedom*	*Mean Squares*
Main effect A	SSA	$r - 1$	SSA$/(r - 1)$
Main effect B	SSB	$c - 1$	SSB$/(c - 1)$
Interaction	SSI	$(r - 1)(c - 1)$	SSI$/(r - 1)(c - 1)$
Cell means	SSK	$rc - 1$	
Residual	SSE	$rc(k - 1)$	SSE$/rc(k - 1)$
Total	SST	$n - 1$	

Table 15-10 ANOVA Table for Poster and Income Factorial Experiment on Sales

Source of Variation	*Sums of Squares*	*df*	*Mean Squares*	*F Ratios*
Poster	48	1	48	3.25
Income	1452	1	1452	98.44**
Interaction	432	1	432	29.29**
Cell means	1932	3		
Residual	118	8	14.75	
Total	2050	11		

**Significant at the 0.01 level.

the classes of factor A over all classes of factor B (or vice versa). Therefore, the tests of main effects have meaning only in the sense of average effects. In this sense the marginal income means in our example are significant, but not the marginal poster means.

Complete factorial experiments provide more information than either one-factor designs or two-factor designs without replication. If we had not chosen to conduct a factorial experiment, our analysis probably would have led us to quite different conclusions.

Looking at Table 15-8, we see the means along the margins (61, 39, 52, and 48). These are the only means we could test in a two-factor experiment without replication. In such an experiment, we would expect to conclude that income has a significant effect on sales, but the two posters are equally effective. You should be quite certain that effects are additive before rejecting a factorial approach. Of course, with the highly apparent interaction pattern in Table 15-8, one could hardly miss the correct approach. In real applications, however, the correct conclusions are not always so obvious in a preliminary analysis.

We have illustrated the analysis of variance calculations for a two-factor analysis with replication in the context of a completely randomized factorial design. The analysis of a randomized block design with replication in which the analyst wants to allow for the possibility of interaction between the blocking factor and the treatments being investigated would proceed in the same way. The purpose of including the blocking factor and its possible interaction with the treatments is to permit a more sensitive test of treatment effects rather than to test the significance of the blocking factor and interaction. Indeed, such tests could be questioned because random selection of experimental units from classes formed by the blocking factor is not required in the randomized block design.

Assumptions and Violations

In all ANOVA designs, procedures for making statistical inferences such as hypothesis tests depend on the same set of three assumptions. Two of these assumptions apply to the statistical characteristics of the populations from which the observations come, and the final assumption is concerned with the method of selecting observations. The assumptions are:

1. The statistical population of measurements for any combination of factor levels is normally distributed.
2. The statistical population of measurements for any combination of factor levels has the same variance as does the population for any other combination.
3. Sample observations are selected at random and every selection is independent of every other selection.

We will call these assumptions the normality, homogeneity of variance, and random-independent selection assumptions.

Only the third assumption can be and must be fully satisfied for valid inferences. The first half of this chapter, together with earlier discussions of sampling techniques, described how to do so. There are statistical tests to check the first two assumptions and there are mathematical transformations that can be made to bring some data into compliance with these assumptions. Both procedures have rather important shortcomings, however.

Fortunately, inferential procedures for analysis of variance can be made reliable by taking some simple precautions. Inferential procedures for the fixed effects model are generally more sensitive to departures from the second assumption than the first. Making sure to include an equal number of observations in each cell (for each combination of factor levels) will furnish adequate protection against the effects of violations of the second assumption. Effects of violating the first assumption are negligible even if the populations within cells are moderately skewed.

15.3 Estimates and Comparisons

Estimates and comparisons of treatment means from randomized block and factorial experiments are carried out in essentially the same manner as in one-way analysis of variance. Illustrations are carried out below for the examples used earlier in this chapter, and special problems are pointed out.

Randomized Blocks

In the driver reaction times example we concluded from the analysis of variance that mean reaction times differed by levels of light intensity. The means (in 10ths seconds) were

Low intensity:	30.00
Medium intensity:	17.25
High intensity:	31.25

Let us use multiple contrasts to establish a confidence interval for differences among pairs of these means. Following Equation 13-10, the variance of a contrast involving a difference between treatment means is

$$s^2(K_i) = \frac{\text{MSE}}{n_j} + \frac{\text{MSE}}{n_j}$$

The number of observations per treatment is 4, and from Table 15-4 we find that MSE is 5.08. Then,

$$s^2(K_i) = \frac{5.08}{4} + \frac{5.08}{4} = 2.540$$

The Scheffé T multiple requires the numerator degrees of freedom used in the test for means and the F value for those degrees of freedom in conjunction with the degrees of freedom for the error mean square. In the current example, these lead to

$$T = \sqrt{(3-1)F_{.95,2,6}}$$

for 95% confidence. The multiple is

$$T = \sqrt{2(5.14)} = 3.205$$

and the half-width of the 95% confidence interval for the family of contrasts (differences) is

$$T * s(K_i) = 3.205\sqrt{2.540} = 5.10$$

The differences among pairs of means are

Low–medium:	30.00 − 17.25 = 12.75
High–medium:	31.25 − 17.25 = 14.00
High–low:	31.25 − 30.00 = 1.25

Adding and subtracting the half-width calculated above, we see that we can assert with at least 95% confidence that the average reaction times at low and at high light intensity exceed the average reaction time at medium intensity, but we cannot assert any difference between reaction times at high and at low intensity.

Two-Way Factorials

Our example in this chapter has been the poster experiment involving two posters (A and D) and two neighborhood income levels (high and low). From the ANOVA in Table 15-10 we failed to detect any average (or marginal) poster effect, but did find the income effect and the interaction between posters and income highly significant. When the interaction is significant, one must look to the cell means to estimate effects even if one of the main effects is not found significant. Therefore, we are concerned with the four treatment means (sales volumes)

	Poster	
Income	A	D
High	69	53
Low	35	43

Let us again use Scheffé multiple contrasts for differences in pairs of treatment means. From Table 15-10 we find MSE = 14.75, and the resulting $s(K_i) = 3.136$. Then, the half-width of the 95% confidence interval for multiple contrasts among pairs of means, following Equation 13-3, is

$$3.136\sqrt{3(F_{.95,3,8})} = 3.136\sqrt{12.21} = 10.96$$

Some differences are

$$\begin{aligned} &A(\text{high}) - D(\text{high}): && 69 - 53 = 16 \\ &D(\text{low}) - A(\text{low}): && 43 - 35 = \ \ 8 \\ &A(\text{high}) - A(\text{low}): && 69 - 35 = 34 \\ &D(\text{high}) - D(\text{low}): && 53 - 43 = 10 \end{aligned}$$

Therefore, we can assert with 95% confidence that under conditions of the experiment the long-run average sales using poster A would exceed those using poster D in high income neighborhoods and that the sales volume associated with poster A in high income neighborhoods would exceed that following the use of poster A in low income neighborhoods.

If the interaction alone or the interaction along with one or both of the main effects is found significant in a factorial study, then one must estimate cell means as we did above. Confidence intervals for means, contrasts, or multiple comparisons are based on the mean square from the error component.

When the interaction is not significant but both main effects are significant, the mean for the treatment combination in the ith row and jth column is estimated by

$$\overline{X}_{ij} = \overline{\overline{X}} + (\overline{X}_i - \overline{\overline{X}}) + (\overline{X}_j - \overline{\overline{X}})$$

which accords with the conclusion that row and column treatment effects are present in an additive form.

If only one main effect is significant, then the means for the different treatment levels are estimated from the relevant row or column means.

Establishment of confidence intervals for means, contrasts, or multiple comparisons in the two cases just mentioned is not a settled matter. The reason is that we have used the results of the factorial experiment to suggest a simpler model for the data. To establish proper confidence intervals for estimates from this simpler model requires, strictly speaking, a new experiment in which the simpler model is tested and confirmed.

Exercises

15-1 *Basic concepts*

a. In an experiment, learning time for memorized material is being studied with human subjects. The units of observation are regular, part-time, and off-campus students in a large university. Enrollment data are processed to produce six lists of students: females under 20 years of age; females 20 to 30; females 30 and over; males under 20; males 20 to 30; males 30 and over. One student is selected at random from each list for the experiment. Age and sex effects on learning time will be analyzed. Is this a completely randomized or a randomized block experiment? Why?

b. Describe the change that would be required in selection of human subjects to change the experiment from (randomized block, completely randomized) to (completely randomized, randomized block).

c. State the general feature in selection of units of observation which distinguishes a completely randomized from a randomized block experiment. Assume that both experiments are for two factors without replication.

d. What is the major advantage of a completely randomized two-factor experiment without replication over a randomized block experiment with two factors and no replication?

e. In (d), what is the chief advantage of the latter type of experiment over the former?

15-2 *Training methods.* In a large organization, people hired to be personnel counselors must take a screening test. Then they are placed in one of four training courses. At the end of their training courses, the new counselors all take the same standardized examination, the scores of which are correlated with subsequent job performance. The scores of a sample of 12 new counselors selected for an experiment are:

Grade on Screening Test	*Training Method* A	B	C	D
Superior	20	33	21	29
Excellent	17	28	19	21
Average	15	22	19	20

a. Find the column means and state a null hypothesis concerning these means that can be tested by two-way analysis of variance.

b. For the experiment described, what must you assume to test the null hypothesis that row means are equal in the population of trainees?

c. Find the sums of squares, degrees of freedom, and other necessary numerical values to complete the statistical test of the null hypothesis. What conclusion is justified by the test?

d. Find the half-width of the 95% confidence interval for multiple comparisons of differences between pairs of training method means. What can be asserted about the difference?

15-3 *Machine tool modifications.* Five modifications of a machine tool are being tested for their effect on the outputs of operators. Three operators who happen to have time have each used all five tools in randomized order. The output results for a five-minute test period on each machine are given in the following table:

	Modification				
Operator	*A*	*B*	*C*	*D*	*E*
I	9	11	12	10	12
II	6	7	10	9	9
III	2	5	8	6	6

Treating this as a randomized block experiment, conduct an analysis of variance to test whether mean outputs from all modifications are equal.

15-4 *Tomato seed plants.* A firm that grows tomato plants for seed is testing two types for yield. The type constitutes one factor and the greenhouses in which the test plants were grown constitute a second factor. The yield data from the experiment are shown in the following table:

	Plant Type	
Greenhouse	*A*	*B*
I	10	8
	8	10
II	6	10
	8	12
III	4	14
	2	16

a. Find the mean yield for each of the six combinations of experimental conditions.
b. Use the means to draw a graph in which plant type is on the horizontal axis, mean yield is on the vertical axis, and each greenhouse is represented by a line on the graph. Interpret the result in detail.
c. Perform a complete analysis of variance for the experiment and state your conclusions.
d. Find the differences between all pairs of greenhouse by plant type means. What can be asserted with 95% confidence about these multiple comparisons?

15-5 *One-way analysis.* Assume in Exercise 15-2 that the observations are classified only by training method. Conduct a one-way analysis of variance, interpret the results, and compare the results with those for Exercise 15-2.

15-6 *Marketing experiment.* A firm that does all its advertising and selling by direct mail is investigating alternative marketing programs for a new product. The product is a box of crystals which, when dissolved in water, forms a household metal-cleaning solution. An experimental marketing program consisted of three different advertising appeals and three different product sizes. The product treatments were 2-inch, 4-inch, and 9-inch squares of cloth moistened with the solution and sealed in plastic bags. Each combination of type of appeal and product were sent to three randomly selected mailing lists of 100 each. The sales results are given below.

Advertising Appeal	*2-inch*	*Product Size* *4-inch*	*9-inch*
	4	9	4
A	7	15	4
	8	11	3
	2	9	3
B	4	8	3
	10	7	6
	10	20	13
C	8	12	12
	10	16	12

a. Analyze these data by means of a two-factor ANOVA with interaction.
b. What conclusions do you draw from your ANOVA table?
c. Make point estimates of the nine treatment means.

15-7 *Data processing.* Three different data entry programs were developed for the billing function of a health agency. These programs could be used with two different data entry terminals. Eighteen data entry clerks were randomly assigned three to each of six treatment groups. The number of errors made in a test assignment was recorded for each. The results were:

Terminal	*1*	*Program* 2	3
	6	4	8
I	9	6	10
	3	2	6
	13	7	15
II	11	6	11
	12	5	10

a. Conduct an analysis of variance to test for main program effects, main terminal effects, and interaction between program and terminal effects. What are your conclusions?
b. Construct point estimates of the six treatment means.

15.4 Case Studies

In this section we present two cases of applications of analysis of variance. The first is a production control example and the second is from the field of advertising research.

Control of Multiple Head Machines*

Multiple head machines are common in repetitive operations in industry such as in drilling holes to a desired dimension or in filling containers.

Filling operations in pharmaceutical, beverage, and food-processing industries often use filling machines with multiple heads. Operation of a 24-head machine, for example, is often so fast that a quality control inspector can identify and remove at one time only the containers filled by six heads. Efforts to determine if all 24 heads are filling containers to a uniform standard are thus complicated by time difference in sampling. In Table 15-11 we see one day's results from a series of samples in which six heads at a time are sampled at 15-minute intervals.

Each group of heads can be analyzed separately with the different sampling times forming a second factor whose influence we want to eliminate—much in the manner of a randomized block experiment. Subject to an assumption of no interaction between time and heads, that is, that there are no machine changes over time that affect different heads differently, the mean square calculated from the interaction term gives an estimate of the variance of fills under constant conditions. Against this measure of experimental error, the differences in mean fills for the different heads (and time periods) can be tested for significance.

The results for the heads in Group III are shown in the ANOVA table below.

Source	*SS*	*df*	*MS*	*F*
Heads	1582.5	5	316.5	9.28
Time	1277.2	4	319.3	9.36
Residual	682.0	20	34.1	
Total	3541.7	29		

*Adapted, with permission, from Ellis R. Ott and Ronald D. Snee, "Identifying Useful Differences in a Multiple Head Machine." Copyright American Society of Quality Control, Inc., reprinted by permission from *Journal of Quality Technology*, Vol. 5, No. 2 (April 1973), pp. 48–57.

Table 15-11 Filling Weights (grams) from a 24-Head Machine

	Time of Samples	*Sampling Time Order*	*Head* 1	2	3	4	5	6	$n = 6$ $\overline{X}$
Group I	9:00 AM	1	1211	1222	1225	1207	1230	1229	1220.7
	12:40 PM	5	1229	1233	1222	1211	1242	1236	1228.8
	1:40 PM	9	1210	1227	1228	1235	1234	1228	1227.0
	2:40 PM	13	1241	1246	1247	1239	1254	1227	1242.3
	3:40 PM	17	1238	1239	1236	1238	1222	1245	1236.3
	$\overline{X}$	=	1225.8	1233.4	1231.6	1226.0	1236.4	1233.0	$\overline{\overline{X}} = 1231.02$
			7	8	9	10	11	12	
Group II	9:15 AM	2	1268	1265	1275	1257	1232	1270	1261.2
	12:55 PM	6	1256	1252	1255	1248	1265	1247	1253.8
	1:55 PM	10	1240	1251	1252	1236	1249	1245	1245.5
	2:55 PM	14	1284	1287	1288	1273	1234	1270	1272.7
	3:55 PM	18	1250	1252	1252	1250	1245	1261	1251.7
	$\overline{X}$	=	1259.6	1261.4	1264.4	1252.8	1245.0	1258.6	$\overline{\overline{X}} = 1256.98$
			13	14	15	16	17	18	
Group III	9:30 AM	3	1259	1232	1232	1239	1229	1229	1236.7
	1:10 PM	7	1252	1234	1235	1230	1233	1239	1237.2
	2:10 PM	11	1256	1240	1244	1246	1232	1235	1242.2
	3:10 PM	15	1261	1249	1255	1260	1260	1240	1254.2
	4:10 PM	19	1253	1224	1243	1246	1232	1232	1238.3
	$\overline{X}$	=	1256.2	1235.8	1241.8	1244.2	1237.2	1235.0	$\overline{\overline{X}} = 1241.72$
			19	20	21	22	23	24	
Group IV	9:45 AM	4	1221	1226	1241	1220	1218	1229	1225.8
	1:25 PM	8	1235	1216	1239	1230	1219	1220	1226.5
	2:25 PM	12	1259	1250	1268	1255	1247	1240	1253.2
	3:25 PM	16	1253	1247	1263	1249	1240	1254	1251.0
	4:25 PM	20	1241	1228	1247	1238	1227	1226	1234.5
	$\overline{X}$	=	1241.8	1233.4	1251.6	1238.4	1230.2	1233.8	$\overline{\overline{X}} = 1238.20$
									$\overline{G} = 1242.1$

Source: Ellis R. Ott and Ronald D. Snee, "Identifying Useful Differences in a Multiple Head Machine." Copyright American Society for Quality Control, Inc., reprinted by permission from *Journal of Quality Technology*, Vol. 5., No. 2 (April 1973) pp. 48–57.

To compare the mean fills for the different heads, the 95% confidence interval for multiple contrasts among pairs of the four group means is appropriate. The half-width of the interval, following Equation 13-24, is

$$T * s(K_i) = \sqrt{(6-1) * 2.71} * \sqrt{\frac{2 * 34.1}{5}} = 13.58$$

Using this criteria, the population mean for head 13 is judged to be different from heads 14, 15, 17 and 18. Similar calculations show no differences among heads in Groups I and II and that head 21 is significantly different from four of the five other heads in Group IV. In each case the statement about multiple comparisons is made subject to a 5% risk of the combination of differences occurring as a result of chance.

The estimates of residual variance in Groups I and II were notably larger than in Groups III and IV, and this was cause for concern about the variability of the fills from particular heads in those groups — a subject that we will not go into further, except to note that these large variances affect the power of the ANOVA tests to detect differences among head means. Heads 4, 5, 6, and 11 were the heads suspected of unusually high variability, and possibly in need of a change of diaphragm or other maintenance.

The ANOVA table for all 24 heads and groups of five time periods is:

Source	*SS*	*df*	*MS*	*F*
Heads	25624.4	23	679.3	6.66
Times	5791.8	4	1448.0	14.20
Residual	9383.8	92	102.0	
Total	30800.0	119		

Both heads and times are significant at the 0.01 level. The 95% confidence interval for multiple contrasts among times is

$$T * s(K_i) = \sqrt{(5-1) * 2.47} * \sqrt{\frac{2(102.0)}{24}} = 9.16$$

and for comparing means from groups of heads the 95% multiple contrast interval is

$$T * s(K_i) = \sqrt{(4-1) * 2.71} * \sqrt{\frac{2(102.0)}{30}} = 7.44$$

In the accompanying graphs, means for groups of times and groups of heads are shown, and the half-width of the intervals just found are measured off above and below the plotted means. The set of pairs of means with non-overlapping intervals are judged to differ. Perhaps the pumping mechanism fails to deliver equal pressure to the groups of heads, and needs attention. Time period IV differs from all the other time periods. It would seem that some adjustment was made by the operator after the 2:40-3:30 period to bring the machine back to its earlier settings.

The treatment means for heads and times shown in Figure 15-2 are displayed in the table below.

	Treatment Means				
	Group I	*Group II*	*Group III*	*Group IV*	*Group V*
Heads	1231.02	1256.98	1241.72	1238.20	
Times	1236.10	1236.58	1241.98	1255.05	1240.20

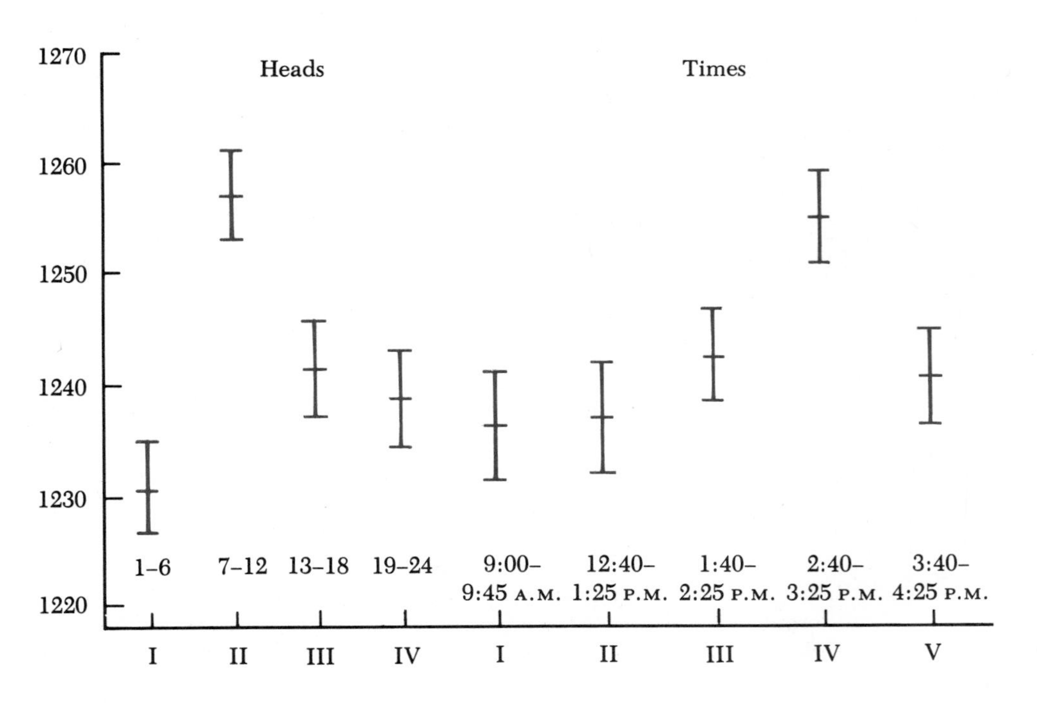

Figure 15-2 Multiple Contrasts for Mean Fills by Groups of Heads and Times

Effectiveness of Endorsements*

Advertisements featuring endorsements are a popular promotional tool in merchandising many products. A key to the success of endorsements is the credibility of the endorser. But the credibility of an endorser is likely to depend on the kind of endorser and the kind of product endorsed.

Drawing on previous studies and on psychological theory in communications, two investigators hypothesized that celebrity endorsers would be most effective with products that present the consumer with psychological and social risk of damage to his or her self perception or perception by others. Expert endorsers will be most effective with products high in the financial, performance, and physical risks of money loss, failure to work properly, and chance of physical harm. Typical consumer endorsers, on the other hand, were thought to be more effective with products carrying little risk of any of the kinds mentioned.

To test these ideas, 12 experimental black and white printed advertisements were professionally prepared featuring 12 treatment combinations of 4 endorsers and 3 products. The products were costume jewelry, a vacuum cleaner, and a box of cookies, selected to be appropriate to the celebrity endorser, the expert endorser, and the consumer endorser, respectively. The fourth endorser treatment was a no endorser, or control advertisement. The advertisements were similar in layout and illustration, featuring a photograph of the endorser (except for the control ad), a photograph of the product, and copy. The celebrity chosen was Mary Tyler Moore, and the expert and the consumer endorsers were presented as a woman named Joan Greene. Table 15-12 shows the captions used under the photograph for the several endorser and product types.

The sample was 360 housewives from a white, middle-class area of Brooklyn, N.Y. The housewives were approached by a female interviewer to give their opinions about a proposed advertisement for a new product that would be on the market soon. In addition to rating the advertisement on scales such as honesty, clarity, informativeness, originality, and so forth, an overall attitude toward the product was obtained, using a six-point scale from "not at all favorable" to "extremely favorable."

With random assignment of subjects to the 12 treatment groups, there were 30 observations in each treatment cell. The ANOVA table is given at the bottom of Table 15-12. The interaction between endorser and product was significant, and the kinds of differences observed supported the theories of the investigators.

The means for the several endorsers are plotted for each product in Figure 15-3. The results are in the directions predicted. In each case the attitude toward the product associated with the most effective endorser is significantly more favorable (0.05 level) than that created by the least effective endorser, but not significantly different from the next best endorser. These statements are based on the 95% confidence intervals for individual treatment differences. The 95% confidence interval is

*Adapted, with permission, from Hershey H. Friedman and Linda Friedman, "Endorser Effectiveness by Product Type," *Journal of Advertising Research*, Vol. 19, No. 5 (October 1979), pp. 63–71.

$$t_{.975,348}\sqrt{\frac{2(2.26)}{30}} = 0.76$$

Table 15-12 Endorser Effectiveness by Product Type

Endorser	*Caption to the Photograph*
Celebrity	Mary Tyler Moore star of the CBS hit series "The Mary Tyler Moore Show"
Professional/Expert (Vacuum Cleaner	Joan Greene, well-known appliance expert, author of the best-selling fix-it book, *A Woman's Guide to Home Appliances*
Professional/Expert (Cookies)	Joan Greene, director, Metropolitan Cooking School, author of the best-selling cookbook, *The Joy of Creative Cooking*
Professional/Expert (Costume Jewelry)	Joan Greene, well-known jewelry expert, author of the best-selling book, *Make Your Own Costume Jewelry*
Typical Consumer	Joan Greene, housewife, Clifton, New Jersey

Dependent Variable: Overall Attitude

A. Cell Means

	Endorser			
Product	*Celebrity*	*Expert*	*Typical Consumer*	*Control*
Vacuum Cleaner	2.83	3.77	3.10	3.50
Cookies	3.87	2.97	4.30	4.00
Costume Jewelry	4.10	3.23	2.50	3.17

Note: 1 = "not at all favorable"; 6 = "extremely favorable."

B. ANOVA Table

Source	*df*	*Mean Square*	*F Value*	*Significance Level*
Product (P)	2	10.41	4.61	.0106
Endorser (E)	3	2.17	.96	.5870
$P \times E$	6	12.89	5.70	.0001
Error	348	2.26		
Total	359			

Source: Hershey H. Friedman and Linda Friedman, "Endorser Effectiveness by Product Type," *Journal of Advertising Research*, Vol. 19, No. 5 (October 1969), pp. 63–71. Reprinted from the *Journal of Advertising Research* ©Copyright 1979, by the Advertising Research Foundation.

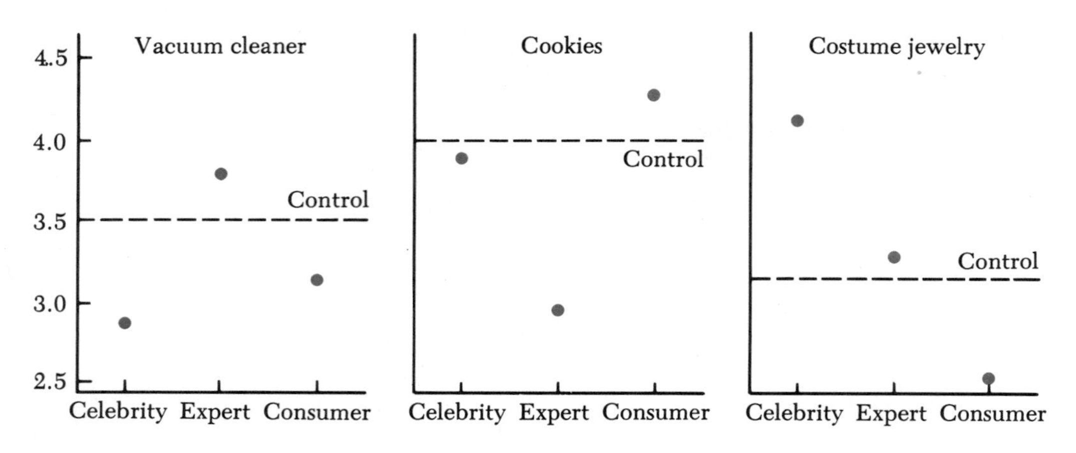

Figure 15-3 Mean Attitude Scores by Product and Endorser

15.5 Regression Approach To Analysis Of Variance

In Chapter 14 on multiple regression and correlation we introduced the topic of indicator variables. It was brought out in Section 14.4 that indicator (0, 1) variables are appropriate for introducing qualitative variables into a regression relationship.

One-Way Analysis

One-way analysis of variance, covered in Chapter 13, examines the relation between an interval variable and a nominal variable that classifies the observations into meaningful groups. The first example of one-way analysis of variance was the sales experiment involving four point-of-sale posters. The calculations for that analysis of variance can be carried out by defining three (0, 1) indicator variables. Letting the indicator variables stand for use of posters B, C, and D, the data are repeated in Table 15-13.

The multiple regression equation for the poster data as organized above is

$$Yc = 47.0 - 9.0\,X_1 - 10.0\,X_2 + 3.0X_3$$

When X_1, X_2, and X_3 are all equal to zero, we are talking about poster A and $Yc =$ 47.0. The set of treatment means are:

Poster A

$X_1 = 0, \quad X_2 = 0, \quad X_3 = 0; \qquad Yc = 47.0$

Poster B

$X_1 = 1, \quad X_2 = 0, \quad X_3 = 0; \qquad Yc = 47.0 - 9.0 = 38.0$

Poster C

$X_1 = 0, \quad X_2 = 1, \quad X_3 = 0; \qquad Yc = 47.0 - 10.0 = 37.0$

Poster D

$$X_1 = 0, \quad X_2 = 0, \quad X_3 = 1; \qquad Yc = 47.0 + \ 3.0 = 50.0$$

The sum of squared deviations explained by the three indicator variables and the residual sum of squared deviations will appear in the analysis of variance output of standard multiple regression programs as shown in Table 15-14. This is the same as Table 13.4 shown earlier. We find, then, that calculations for analysis of variance problems can be carried out as multiple regressions.

Table 15-13 ANOVA Example with Indicator Variables

Poster B	C	D	*Packages Sold*
X_1	X_2	X_3	Y
0	0	0	55
0	0	0	43
0	0	0	43
1	0	0	32
1	0	0	42
1	0	0	40
0	1	0	38
0	1	0	35
0	1	0	38
0	0	1	47
0	0	1	48
0	0	1	55

Table 15-14 ANOVA Output for Poster Data

Source	*SS*	*df*	*MS*	*F*
Regression	378	3	126.0	5.14
Residual	196	8	24.5	
Total	574	11		

Randomized Block Design

In Section 15.1, data on driver reaction times by light intensity and age blocks were analyzed. Interest centered on the effects of light intensity on reaction times, and age of driver was a disturbing factor whose influence was controlled by blocking. Using the indicator variables X_1, X_2, and X_3 for age blocks, B, C, and D, and X_4 and X_5 for medium and high light intensity, the data appear in Table 15-15 for multiple regression.

Table 15-15 Reaction Times Data Organized for Regression

Age Block			*Light Intensity*		*Reaction Time*
B	*C*	*D*	*Medium*	*High*	
X_1	X_2	X_3	X_4	X_5	Y
0	0	0	0	0	22
0	0	0	1	0	11
0	0	0	0	1	25
1	0	0	0	0	29
1	0	0	1	0	18
1	0	0	0	1	26
0	1	0	0	0	33
0	1	0	1	0	18
0	1	0	0	1	35
0	0	1	0	0	36
0	0	1	1	0	22
0	0	1	0	1	39

It is instructive to examine the correlation matrix of the independent variables, remembering that the first three are the indicator variables for age blocks and the last two are indicator variables for light intensity levels. The matrix is

	X_2	X_3	X_4	X_5
X_1	−.333	−.333	0.0	0.0
X_2		−.333	0.0	0.0
X_3			0.0	0.0
X_4				−0.5

Here we see that there is zero correlation between each of the age block variables and the light intensity variables. This occurs because the design included low, medium, and high light intensity treatment within each age block. The feature of zero correlation between the age block and light intensity variables means that the effects of age can be separated unambiguously from the effects of light intensity.

The multiple regression equation for the reaction times data is

$$Yc = 23.1667 + 5.0000\, X_1 + 9.3333\, X_2 + 13.0000\, X_3 - 12.7500\, X_4 + 1.2500\, X_5$$

The sum of squared deviations explained by the age block variables is

$$b_1\,\Sigma(X_1-\bar{X}_1)(Y-\bar{Y})+b_2\,\Sigma(X_2-\bar{X}_2)(Y-\bar{Y})+b_3\,\Sigma(X_3-\bar{X}_3)(Y-\bar{Y})$$
$$=5.0000(-5.5)+9.3333(7.5)+13.0000(18.5)$$
$$=283.0$$

The sum of squared deviations explained by the light intensity variables is

$$b_4\,\Sigma(X_4-\bar{X}_4)(Y-\bar{Y})+b_5\,\Sigma(X_5-\bar{X}_5)(Y-\bar{Y})$$
$$=-12.7500(-35.6667)+1.2500(20.3333)$$
$$=480.17$$

Given the total sum of squared deviations of 793.67, this leads to the analysis of variance in Table 15-16. This same table was produced by ANOVA formulas as Table 15-4 earlier. Again we find that the analysis of variance problem can be approached through multiple regression.

Table 15-16 Regression ANOVA Table for Reaction Times Data

Source	*SS*	*df*	*MS*	*F*
Regression				
Age Blocks	283.00	3		
Light Intensity	480.17	2	240.08	47.3
Residual	30.50	6	5.08	
Total	793.67	11		

Two-Way Factorial Design

In Section 15.2 the example of two-factor analysis with replication was the point-of-sale experiment involving two posters and two neighborhood income levels. The posters were installed in three convenience markets in each of the resulting four treatment categories.

The test for interaction in the two-way factorial design can be interpreted in the regression context as follows. It is a test of whether the regression

$$Yc=a+b_1X_1+b_2X_2+c(X_1X_2)$$

explains significantly more variability in sales levels than the regression

$$Yc=a+b_1X_1+b_2X_2$$

The second regression represents the situation in which the effects of neighborhood income and poster type are independently additive, while the first regression includes an interaction effect. For the reaction times data the two regressions are

$$Yc = 35.0 + 34.0X_1 + 8.0X_2 - 24.0X_1X_2$$

and

$$Yc = 41.0 + 22.0X_1 - 4.0X_2$$

Here, $X_1 = 1$ represents high-income neighborhood and $X_2 = 1$ stands for poster D. Thus, the means for the treatment combinations from the alternative regressions are:

Income Level	*Additive Model*		*Interaction Model*	
	Poster		*Poster*	
	A	*D*	*A*	*D*
High	63.0	59.0	69.0	53.0
Low	41.0	37.0	35.0	43.0

As we saw earlier, there is a strong interaction in this example. The poster effect depends on income level of the neighborhood. One way to emphasize this is to restate the regression equation for the interaction model as

$$Yc = 35.0 + 34.0X_1 + (8.0 - 24.0X_1)X_2$$

Here, the effect of poster D over poster A is seen to be 8.0 when $X_1 = 0$, that is, for low-income areas, and $8.0 - 24.0 = -16.0$ when $X_1 = 1$, that is, for high-income areas.

A regression for the Treatment A main effect as well as the two regressions above must be run to produce the explained sums of squared deviations needed for the complete ANOVA table. In Table 15-17 we indicate the regression sources.

Table 15-17 Source of Sums for Poster Experiment Factorial ANOVA

Source	*Regression Source*	*SS*
A	$Yc = 39.0 + 22.0\ X_1$	1452
A and B	$Yc = 41.0 + 22.0\ X_1 - 4.0\ X_2$	1500
A × B	$Yc = 35.0 + 34.0\ X_1 + 8.0\ X_2 - 24.0\ X_1X_2$	1932
Total		2050

The main effects are unambiguously assignable to A and B. Because of the factorial design, the correlation between the independent variables representing the A treatments with those representing the B treatments will be zero. The sum of squared deviations attributable to the cell totals is the explained sum of squares from the regression that includes interaction. In Table 15-18 we make the subtractions to establish the traditional ANOVA table, which we saw earlier in Table 15-10.

Table 15-18 ANOVA for Poster Experiment Factorial from Regressions

Source		*SS*	*df*	*MS*	*F*
Treatments A		1452	1	1452	98.44
Treatments B	1500 − 1452 =	48	1	48	3.25
Interaction	1932 − 1500 =	432	1	432	29.29
Residual	2050 − 1932 =	118	8	14.75	
Total		2050	11		

A Three-by-Three Example

In the poster example there were only two categories of each treatment type. That is, there were two posters (*A* and *D*) and two neighborhood income levels (low and high). When there are more classes of each treatment variable, more indicator variables and interaction terms are required for the multiple regression solution. In general, with r classes of Treatment A and c classes of Treatment B there will be required:

$r - 1$ indicator variables for Treatment A
$c - 1$ indicator variables for Treatment B
$(r - 1)(c - 1)$ interaction variables

These numbers of variables will be seen to correspond with the number of degrees of freedom for main effects and interaction.

A three-by-three example deals with three types of display for packaged jellies utilizing small, medium, and large supermarkets as the B treatments. The data on sales movement are in Table 15-19.

Table 15-19 Sales Movement of Packaged Jellies

	Store Size		
Display	*Small*	*Medium*	*Large*
End-aisle	9	17	18
	11	20	21
Gondola	12	21	35
	8	23	42
Check-out	10	17	19
	11	21	22

To handle the three display modes we need two indicator variables. Let us designate X_1 for gondola and X_2 for check-out displays. In like fashion designate X_3 for medium and X_4 for large stores. Then, four product variables are needed for the interactions. We designate $X_5 = X_1X_3$, $X_6 = X_1X_4$, $X_7 = X_2X_3$, and $X_8 = X_2X_4$. The data are set up for regression in Table 15-20. The coefficients for the three regressions needed are given in Table 15-21.

Table 15-20 Sales Movement for Packaged Jellies

	Small	*Medium* $X_3 = 1$	*Large* $X_4 = 1$
End-aisle			
Gondola $X_1 = 1$		$X_5 = 1$	$X_6 = 1$
Check-out $X_2 = 1$		$X_7 = 1$	$X_8 = 1$

X_1	X_2	X_3	X_4	X_5	X_6	X_7	X_8	Y
0	0	0	0	0	0	0	0	9
0	0	0	0	0	0	0	0	11
0	0	1	0	0	0	0	0	17
0	0	1	0	0	0	0	0	20
0	0	0	1	0	0	0	0	18
0	0	0	1	0	0	0	0	21
1	0	0	0	0	0	0	0	12
1	0	0	0	0	0	0	0	8
1	0	1	0	1	0	0	0	21
1	0	1	0	1	0	0	0	23
1	0	0	1	0	1	0	0	35
1	0	0	1	0	1	0	0	42
0	1	0	0	0	0	0	0	10
0	1	0	0	0	0	0	0	11
0	1	1	0	0	0	1	0	17
0	1	1	0	0	0	1	0	21
0	1	0	1	0	0	0	1	19
0	1	0	1	0	0	0	1	22

Since our purpose was the regression setup for a three-by-three factorial, we do not carry this example any further here.

We have seen in this section how regression with indicator variables can be used to find the explained sums of squares that then yield ANOVA tables for one-way, two-way randomized block, and two-way factorial analysis of variance. A principal practical reason for showing these relationships is that packaged computer pro-

Table 15-21 Coefficients and Explained Sums of Squares for Sales Movements of Packaged Jellies

Source	*Constant*	X_1	X_2	X_3	X_4	X_5	X_6	X_7	X_8	*SS*
A	16.0	7.5	.67							206.8
A and B	7.44	7.5	.67	9.67	16.0					985.9
A × B	10.0	0.0	.50	8.50	9.5	3.5	19.0	0.0	.50	1251.1
Total										1309.7

grams for multiple regression are more common and more standardized than analysis of variance programs.

Estimation of Treatment Means from Regression

In the one-way analysis of alternative posters, we illustrated the use of the indicator variable regression to obtain the treatment means. The mean for the treatment level not coded will be equal to the intercept term in the regression. The means for the remaining treatments are found by adding the coefficients for the relevant variable to the constant term.

In a randomized block experiment only the experimental treatment means (and not the means for the separate blocks) are of direct interest. These treatment means can be found from the regression which includes only the indicator variables for the experimental treatments. This is the same as the regression for a one-way analysis of variance. Therefore, a convenient way to do randomized block analysis by regression is to do the regression for the treatment variables and then the regression for all variables (treatment plus blocks). The relevant ANOVA sums are the explained (treatment) sum of squares from the first regression, the additional sum explained by the blocks and the error sum of squares from the second regression.

The use of the regression equations to calculate treatment means in a factorial experiment was illustrated by the example of sales movement associated with posters *A* and *D* used in stores in high- and low-income neighborhoods. In general, if the interaction is significant, we use the regression which includes interaction (product) terms. If the interaction is not significant but both main effects are significant, we use the corresponding additive effects regression. If only one main effect is significant, we use the corresponding one-way indicator variables regression. As mentioned earlier, it is advisable not to go beyond point estimates in the latter two cases because results of testing one model have been used to shift to another model.

Summary

Analysis of variance in which two independent variables or factors are being examined for their effect on a response variable can be conducted with or without replication. Without replication, there is only one observation for each combination of factor levels. The design for an experiment without replication can be a completely randomized experiment or a randomized block experiment. Completely randomized designs require random sampling from population strata that represent every combination of factor levels. In randomized block designs, one factor of secondary interest is controlled by placing observations in homogeneous blocks with respect to that factor. Both designs assume that factor effects are additive rather than interactive.

When two factors are to be tested and the investigator wants to test for the presence of interaction between the factors, replication, or multiple observations for each treatment combination, are required. The essence of interaction is that the way in which the response variable is affected by variable A depends upon variable B.

Estimates and comparisons among treatment means in two factor analysis of variance are accomplished by methods which are extensions of those used in one-way analysis of variance.

Calculations for both one- and two-way analysis of variance can be carried out by regression methods. The technique involves coding the treatment classes by indicator variables. Generally, there are as many indicator variables as the number of degrees of freedom for treatment effects, including interaction.

Exercises

15-8 *Regression for one-way ANOVA.* Set up indicator variables and run a regression analysis on a suitable statistical computer program to produce the ANOVA in Exercise 13-6.

a. Identify in your output the values that would result from Equations 13-10, 13-11, and 13-12.
b. Show by using the regression equation how the means for the several brands could be obtained.

15-9 *Unequal observations per treatment.* Use indicator variables and regression to find the relation between the calculator ratings and the brands of calculators in Exercise 13-7.

a. What are the values for the sums of squares explained by regression, the unexplained sum of squared deviations, and the total sum of squared deviations? Are these the same as the ANOVA sums of squares?
b. What values for the treatment means result from the regression? Are these correct?

15-10 *Regression for randomized block design.* Set up the required indicator variables to obtain ANOVA totals for Exercise 15-2.

a. Do your results agree with your earlier answers to Exercise 15-2?
b. If your regression program outputs a correlation matrix, what is the correlation between the two indicator variables for screening level? Between your training method indicator variables? Between a training method and a screening test variable?

15-11 *Machine tool modifications.* Verify the answers to Exercise 15-3 using a packaged computer regression program. Use the regression equation to calculate the operator and modification treatment means.

15-12 *Planning and layout.* Teams of three persons each work in a large job shop where the tasks assigned are not strictly repetitive but require some group planning. Two layouts have been in use in the work areas, and two planning systems. Under Plan A the team takes 15 minutes at the beginning of each day to survey and plan the day's assignments. Under Plan B no special time is allotted. Twelve teams are randomly assigned to four treatment combinations with the following results for output.

	Layout I	*Layout II*
	89.9	82.5
Planning A	96.2	84.8
	92.5	90.5
	89.6	76.4
Planning B	90.2	75.4
	89.2	73.9

a. Use regression analyses with indicator variables to find the sums of squared deviations attributable to layout, planning, and layout-planning interaction from the data above.

b. Complete the ANOVA table with necessary F tests. What are your conclusions about significance?

15-13 *Hydroponic gardening.* A hydroponic gardener had a choice of two seed stocks and two feeds for growing a certain plant. In a preliminary experiment the gardener assigned seeds and feed materials randomly to the proper cells in the two-by-two design below with the resulting figures for the number of days required for growth to maturity.

	Feed A	*Feed B*
Seed A	20	22
	22	25
	17	26
	17	19
Seed B	23	30
	22	28
	22	28
	20	28

a. Use regression analysis to find the totals for an ANOVA table for the results above that includes main effects and interaction. Use 0.05 as the significance level.

b. Construct 95% confidence intervals for means of each significant treatment group.

15-14 *Tomato seed plants.* Do Exercise 15-4 using regression analysis with indicator variables to find totals for the ANOVA table and to find appropriate treatment means.

15-15 *Regression computation.* Use a packaged computer regression program to duplicate the ANOVA totals in Exercise 15-7.

Supplementary Exercises

15-16 *Crop yields.* Four varieties of strawberries are each assigned randomly to five out of twenty experimental plots at an agricultural experiment station. Yields in bushels from the experiment are given below.

Type A 7, 9, 9, 9, 13
Type B 10, 11, 13, 16, 17
Type C 10, 11, 12, 14, 17
Type D 13, 14, 16, 17, 17

a. Use indicator variables and regression analysis to find the sums of squares for an analysis of variance test of difference among mean yields.
b. At alpha = 0.05, what conclusion should be reached?
c. Determine the 95% confidence interval half-width for multiple contrasts among pairs of means of the varieties.

15-17 *Display locations.* A chain of convenience food outlets wished to feature a special display promoting sales of boxed apples over a fall season. To decide on a location for the display, four stores from each of four volume classes were randomly selected, and four display locations were assigned randomly to the four stores. The numbers of boxes of apples sold during the trial period are given below.

Display Location	*Store Sales Volume Class*			
	Low	*Medium*	*Average*	*High*
Entrance	3	7	15	13
Exit	10	10	21	22
Cash register	20	10	22	20
Interior	24	16	17	31

a. Perform the appropriate analysis to test for difference among mean sales by location of display.
b. If you had failed to take account of the blocking factor, would your conclusion about effects of the display locations have been any different? Explain.

15-18 *Advertising copy.* Three different appeals were devised for TV advertisements for a shampoo. Ads with these appeals were produced with the appeal presented by an attractive model and a TV personality. Each advertisement was shown to three different test audiences with the following percentage approval ratings.

a. Set up the indicator variables required for a regression solution of an ANOVA that will include a test for interaction between appeal and presentor.
b. Process the regressions and present the ANOVA table with appropriate F statistics.

c. Establish the confidence interval for percentage approval of the best sample combination.

	Model	*Personality*
Appeal A	48	77
	69	52
	68	57
Appeal B	62	53
	49	37
	55	31
Appeal C	35	76
	36	66
	55	69

15-19 *Production methods.* A new order received by a job shop required a rather complex assembly. Three methods were suggested for the assembly operation. Only two teams were available for testing the methods. Each of the teams completed three assemblies in the times (minutes) given below.

	Method I	*Method II*	*Method III*
Team A	4, 11, 5	2, 7, 19	4, 10, 7
Team B	14, 18, 19	21, 23, 21	4, 8, 10

a. Conduct an ANOVA for the data that will include a test for interaction between method and team. What is your conclusion?
b. Regroup the sums of squares in your ANOVA table to represent a one-way ANOVA for difference among methods. What conclusion would have been reached had a one-way test (ignoring the teams factor) been made?

15-20 *Sales territories.* A company introduced a new model of personal computer with advertising in all regions where the company operated. During the first five business days following the introduction, sales were as follows in the six regions in which the company operated.

Region	*Daily sales (number)*				
	Mon	*Tues*	*Wed*	*Thurs*	*Fri*
A	46	26	13	20	23
B	19	41	30	31	18
C	35	33	40	29	28
D	73	41	26	42	33
E	20	63	65	47	31
F	52	39	52	68	40

Assume the first five days to be a random sample of daily experience.

a. Use indicator variables and multiple regression to find the sums of squares for a one-way ANOVA on daily sales rates by territories. What conclusion would be drawn?
b. Use the regression equation to derive estimates of mean daily sales for each territory.
c. Establish the half-width of the 95% confidence interval for any territorial mean daily sales rate.

Computer Exercises

1 Source: Baseball data file
  a. Use leagues and divisions as factors and runs as the dependent variable. Perform two-factor analyses.
  b. Use indicator variables and a regression package to answer part (a).
  c. Use a regression routine to perform a one-way analysis on the number of runs scored by the teams in the Eastern Division of the National League. How many indicator variables are needed?

2 Source: Real estate data file
Perform a two-factor analysis using type of home and residence status as factors and purchase price as the dependent variable.

3 Source: Salaries data file
Use management level and tax deferral status to explain any relationship between these factors and salaries. Construct a chart of means to assist in explaining any relationship.

16

NONPARAMETRIC STATISTICS

In earlier chapters, many procedures for statistical inference relied on the assumption that the statistical population of concern has a known distribution, usually a normal distribution. Sometimes this assumption applied to the observations themselves. In other cases, when little could be assumed about the distribution of the observations, large sample sizes and the central limit theorem allowed us to assume that a normal distribution applied to the statistic used for inferences.

On many occasions, the assumptions required for earlier procedures cannot be justified. When this is true, we want a procedure that makes only minimal assumptions about the population distribution. Such techniques, which typically apply to both large and small samples, belong to a class known as *nonparametric statistics*.

One situation in which nonparametric procedures can replace those previously covered is when only weak assumptions about the distribution are justified. For example, family income distributions typically have marked positive skewness. Suppose family incomes in two cities are to be tested for a difference in central location and the only samples available are both small. Because the samples are small, differences in sample means may not be normally distributed, and the t-tests for small samples discussed in Chapter 11 would not apply. But the first procedure to be described below, the rank sum test, does apply.

Another situation in which nonparametric procedures can be used is when the data available for making inferences consist only of ranks. For instance, some studies of product preference require respondents only to rank the specified products. Procedures discussed earlier are not suitable for such data, although most nonparametric procedures can be used with them.

It is only fair to point out that nonparametric techniques often have disadvantages when compared with the procedures discussed earlier. Many nonparametric tests are less efficient than their parametric counterparts in that the sampling variability of the test statistic is greater, with a given sample size, for the nonparametric test. Another weakness is that nonparametric techniques consist primarily of hypothesis tests. The number of available nonparametric estimation procedures is rather limited.

16.1 Comparison of Two Populations: Independent Samples (the Rank Sum Test)

A test that compares two populations for a difference in location is the *Mann-Whitney-Wilcoxon rank sum test*, which we shall call simply the *rank sum test*. This test is based on independent random samples from the two populations. It is similar in purpose to the two-sample t-test in Section 11.2.

We shall illustrate this test by applying it to two training programs for salespeople. A performance rating is available for each of three salespeople trained under the new program. Ratings are also available for five people trained under the old program.

Assumptions about the Population Distributions

The rank sum test assumes that the two populations have distributions that are the same shape, but it is not necessary to know what shape they have. For our illustration, the two distributions of performance ratings will be assumed to be positively skewed to the same degree. If the two training programs are equally effective, the two population distributions will be identical in every respect. This is illustrated in Figure 16-1(b). If the old training program typically produces higher performance ratings than the new one, the distribution for the old program is offset to the right of the new one. This is illustrated in (a). Finally, if the new program is more effective than the old one, the rating distribution for the new program is offset to the right of the old one, as shown in (c).

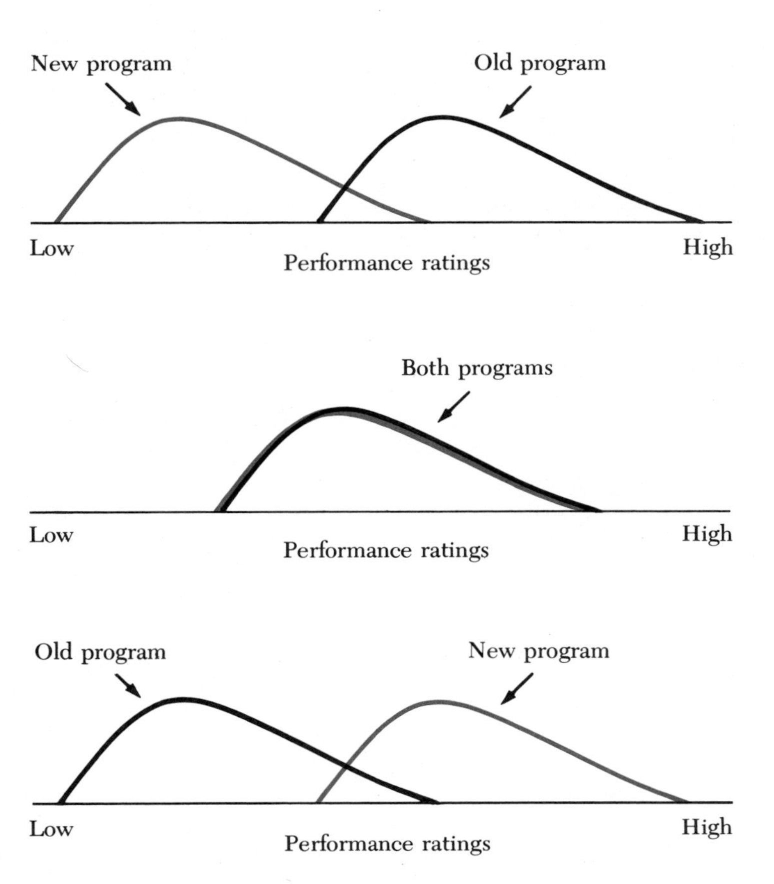

Figure 16-1 Illustrative Distributions for Two Performance Rating Populations

The Hypothesis Statement

The training director for the company in which the two training programs are used elects to make a statistical test that will detect a difference in effectiveness in either direction between the two programs, if there is such a difference. The null hypothesis is that the two programs are equally effective, and the alternative hypothesis is that they are not equally effective. This is a two-tailed test; Figure 16-1(b) illustrates the situation when the null hypothesis is true and the other two parts illustrate situations typical under the alternative hypothesis.

The Test Statistic

The first step in finding the test statistic is to determine the sample sizes for the two populations. We have already pointed out that a random sample of $n_1 = 3$ ratings is available from trainees under the new program and a similar sample of $n_2 = 5$ ratings is available from trainees under the old program. The total number of observations is, of course, $n = 8$. We shall define n_1 to be the number of observations in the smaller sample, n_2 to be the number of observations in the larger sample, and n to be the total observations in both samples. In cases where $n_1 = n_2$, either symbol may be assigned to either sample.

The next step in finding the test statistic is to choose a ranking rule and apply it to all n observations in the two samples as a single set. In the case of the training programs, suppose we elect to rank from high ratings to low; we shall then assign rank 1 to the highest rating in the set and rank 8 to the lowest rating.

Now, for the moment, let us assume the rating distribution for the new program lies completely to the right of that for the old program. This means that the two distributions at the bottom of Figure 16-1 do not overlap. All ratings under the new program would be higher than any under the old one. Under this assumption, ratings with ranks 1, 2, and 3 must come from the new program and ratings ranked 4 through 8 must come from the old one.

In the third step, the ranks just assigned are listed separately according to the distribution from which they came. For the two training programs under the situation assumed above, the table of ranks would be as follows.

Program	*Ranking of Ratings (High Rating = Rank 1)*				
New	1	2	3		
Old	4	5	6	7	8

The final step is to determine the value of the test statistic, W, by summing the ranks in the smaller sample. Here, $W = 1 + 2 + 3 = 6$. As we have shown, this is the smallest possible value of W for samples of $n_1 = 3$ and $n_2 = 5$. For the same sample sizes, the largest possible value for W is 21 ($W = 8 + 7 + 6 = 21$). Under the ranking rule we chose, this situation is certain to occur when the distribution for the new program is offset to the left of that for the old program with no overlap. In practice, it would seem that the typical situation would lie between these two extremes.

The Distribution of *W* under the Null Hypothesis

For the null hypothesis pictured in Figure 16-1(b), the two rating distributions are identical in all respects. Because of this identity, we can assume a common population of ratings for the two training programs. Now suppose we select a long series of random samples, each of which contains 8 observations, from this common population. Furthermore, suppose we randomly and independently subdivide each sample of 8 into subsamples of $n_1 = 3$ and $n_2 = 5$.

Given the sampling conditions described, any three of the rank numbers 1 through 8 are as likely to be assigned to the smaller subsample as are any other three. As we saw, W can range from 6 to 21. We can list the ${}_8C_3 = 56$ equally likely combinations of three rank numbers and group combinations with the same value of W to produce the sampling distribution for W. The result for the case with $n_1 = 3$ and $n_2 = 5$ is shown in Table 16-1.

Table 16-1 Probability Distribution of *W* under the Null Hypothesis for $n_1 = 3$ and $n_2 = 5$

W	*Probability*	W	*Probability*
6	0.018	14	0.107
7	0.018	15	0.107
8	0.036	16	0.089
9	0.054	17	0.071
10	0.071	18	0.054
11	0.089	19	0.036
12	0.107	20	0.018
13	0.107	21	0.018

The Acceptance Region and Level of Significance

There are several possible choices for the acceptance region in a two-tailed test based on the sampling distribution of W in Table 16-1. For instance, we can accept the null hypothesis for values of W from 7 through 20 and reject the null hypothesis if W is either 6 or 21. The level of significance is the sum of the probabilities for these latter two values of W. Hence $\alpha = 0.018 + 0.018 = 0.036$.

A second choice for the acceptance region is

$$8 \leq W \leq 19$$

The level of significance for this case is

$$\alpha = 0.018 + 0.018 + 0.018 + 0.018 = 0.072$$

In our example, the training director elects to use the acceptance region specified in the second choice.

The Observed Value of the Test Statistic

The observed performance ratings for the eight salespeople in the available sample, arranged by training program, are shown in Table 16-2. The new program appears to have higher ratings in general than does the old program. The purpose of the rank sum test is to determine whether this apparent difference in location for the two distributions is statistically significant.

Table 16-2 Performance Ratings under Two Training Programs

Program			*Ratings*			*Sample Sizes*
New	82	79	71			$n_1 = 3$
Old	75	68	62	59	42	$n_2 = 5$

Table 16-3 Rank Assignments and Calculation of W

Program			*Rank*			
New	1	2	4			$W = 1 + 2 + 4 = 7$
Old	3	5	6	7	8	

To find the observed value of W, we rank all the observations as a single set and sum the ranks assigned to the smaller sample. Recall that with our chosen ranking rule, we must rank from largest to smallest. Thus the 82 rating is assigned rank 1, and so on. The result, given in Table 16-3, shows that

$$W_{\text{observed}} = 7$$

Conclusion of the Test

The null hypothesis for the two-tailed test is that the population distributions of performance ratings for the two training programs are identical. The test statistic is W, the sum of ranks for the three ratings from the new program. The acceptance region for the null hypothesis is $8 \le W \le 19$, and the level of significance is 0.072. For the particular random sample selected, W_{observed} is 7. Because this value is not in the acceptance region, the null hypothesis is rejected. The two-tailed test has established that performance rating distributions for the two programs have different locations at the 0.072 level of significance.

Applications to Large Samples

Some important general properties have been developed for the rank sum test. Under the null hypothesis, the two populations from which the independent random samples come are assumed to be identical in shape and location. With this assumption, it can be shown that the sampling distribution of W under the null hypothesis has a mean

$$\mu_W = \frac{n_1(n+1)}{2} \tag{16-1}$$

and a standard deviation

$$\sigma_W = \sqrt{\frac{n_1 n_2(n+1)}{12}} \tag{16-2}$$

Here, n_1 and n_2 are the number of observations in the two samples with $n_1 \leq n_2$, and n is the total number of observations in the two samples ($n = n_1 + n_2$).

A further development has established that, when n_1 and n_2 are large and the null hypothesis is true, the statistic

$$Z = \frac{W - \mu_W}{\sigma_W} \tag{16-3}$$

is essentially normally distributed with mean zero and standard deviation 1. The standard normal distribution is sufficiently accurate if n_1 and n_2 are both at least 10. Given such large samples, Z can replace W as the test statistic in the rank sum test. For smaller samples, tables of the exact sampling distribution of W are available in texts devoted to nonparametric statistics.

Example

A mail-order company prepares specialty packs of fruit for sale as holiday gifts. The company's contract with Ace Delivery Service is coming up for renewal, and Nationwide Delivery Service is being considered as an alternative. Nationwide will be selected only if its typical delivery times are less than those for Ace. Otherwise, the contract with Ace will be renewed. A random sample of shipments is selected for delivery by each service. Transit times, in hours, for the two samples are as follows.

Nationwide	19	22	26	31	42	44	45	50	57	60		
Ace	12	27	34	39	41	43	48	49	54	55	61	68

A one-tailed test is called for. The null hypothesis is that delivery time distributions for the two services are identical, and the alternative hypothesis is that Nationwide's delivery times are generally less than Ace's.

Low rank numbers will be assigned to faster delivery times (12 h is assigned rank 1, and so on). If the alternative hypothesis is true, the ranking rule leads us to expect that the ranks assigned to Nationwide will be smaller than would occur under the null hypothesis. In this eventuality, we can expect W to be smaller under the alternative hypothesis than under the null hypothesis.

Because the sample sizes are $n_1 = 10$ for Nationwide and $n_2 = 12$ for Ace, the test statistic is z, as defined in Equation 16-3. If the alternative hypothesis is true, the observed value of W can be expected to be less than μ_W, the expected value under the null hypothesis, and z will be negative. The test is lower-tailed. It is decided that the level of significance will be 0.05. Consequently, the acceptance region for the null hypothesis is $z \geq -1.65$.

From Equations 16-1 and 16-2, the mean and standard deviation for W under the null hypotheses are

$$\mu_W = \frac{n_1(n+1)}{2} = \frac{10(22+1)}{2} = 115$$

and

$$\sigma_W = \sqrt{\frac{n_1 n_2(n+1)}{12}} = \sqrt{\frac{10(12)(23)}{12}} = 15.166$$

The following are the ranks for the observed delivery times.

Nationwide	2	3	4	6	10	12	13	16	19	20		
Ace	1	5	7	8	9	11	14	15	17	18	21	22

The observed value of W is

$$W = 2 + 3 + 4 + 6 + 10 + 12 + 13 + 16 + 19 + 20 = 105$$

and the observed value of the test statistic is

$$z_{\text{observed}} = \frac{W - \mu_W}{\sigma_W} = \frac{105 - 115}{15.166} = -0.66$$

The observed value of the test statistic falls within the acceptance region for the null hypothesis. Observed delivery times for Nationwide are not significantly less than those for Ace at the 0.05 level of significance.

Tied Observations

The rank sum test assumes that tied observations are not possible. In practice, such observations do occur, however. When only a small number of ties are present, these observations can each be given the average of the ranks that would be assigned if there were no identical values. In the delivery time example just discussed, suppose the shortest delivery time for Ace had been 19 hours instead of 12. Then both delivery services would have the shortest delivery time. The ranks in question are 1 and 2. The mean is 1.5, and this is the rank that would be assigned for both observations. Large numbers of ties call for special methods, which are discussed in texts on nonparametric statistics.

Exercises

16-1 *Abrasive material in toothpaste.* The number of milligrams of abrasive material per tube for two brands of toothpaste is as follows.

Brand A	28	26	25	23	21
Brand B	24	20	16		

A consumer group wants to test for significantly less abrasive material per tube in Brand B. In a rank sum test, rank 1 is to be assigned to 28 mg, and so on, with rank 8 assigned to 16 mg. Conduct the test at the 0.072 level of significance; refer to Table 16-1 in the text to establish the acceptance region.

16-2 *Strength of materials.* The impact strengths of two materials being considered for milk cartons are given below. At the 0.01 level of significance, conduct a two-tailed rank sum test of the hypothesis that the materials are equally strong.

Material A	95	98	99	96	90	89	92	94	97	91
Material B	93	88	91	83	87	86	82	79	85	83

16-3 *Investment.* Random samples of 12 and 16 firms from two highly competitive industries are available. Annual reports of the sample firms show the following capital expenditures for last year, in millions of dollars.

Industry A	0.6	1.7	1.8	2.0	2.1	2.2	2.3	2.4	2.5	2.6	2.9	3.9	4.3	5.0	5.1	5.5
Industry B	2.4	3.3	3.4	3.5	3.6	3.7	4.3	4.4	4.7	5.3	6.2	6.8				

Assign smaller ranks to smaller expenditures and larger ranks to larger expenditures. Then perform a one-tailed rank sum test of the null hypothesis that the two expenditure distributions are identical against the alternative that expenditures for firms in Industry B are greater than those for A. Use the 0.025 level of significance.

16-4 *Reversed ranking rule (two-tailed test).* For the two sales training programs in Table 16-2, rank the observations from lowest rating to highest by assigning rank 1 to 42 and continuing up to rank 8 for 82.

a. Carry out a two-tailed rank sum test for identical distributions where the level of significance is 0.072.
b. Does the reversed ranking rule change the conclusion? How does it change the test?

16-5 *Reversed ranking rule (one-tailed test).* Reverse the ranking rule used in the delivery service example of this section by assigning rank 1 to 68 h and continuing up to rank 22 for 12 h. Then perform a one-tailed test at the 0.05 level of significance to detect shorter delivery times for Nationwide Delivery Service if the difference is significant. What changes must be made to reach the correct conclusion when ranking is reversed in a one-tailed test?

16.2 Comparison of More than Two Populations: Independent Samples (the Kruskal-Wallis Test)

The *Kruskal-Wallis test* is a direct extension of the rank sum test described in the previous section of this chapter. It applies when the data consist of independent random samples from each of three or more statistical populations. To illustrate the Kruskal-Wallis test, we shall consider a case with three populations.

A firm that grows rosebushes for wholesale to nurseries wants to determine which of three climates produces the best bushes. Nineteen bushes of comparable ages and varieties are used in the experiment. Six bushes were raised in region A, five in region B, and eight in region C. A panel of qualified judges scores each bush on a composite of characteristics. Then, without regard to the region from which they came, the 19 bushes are ranked on the basis of the judges' scores. The smallest rank, 1, is assigned to the poorest bush and rank 19 is given to the best bush.

Initially, suppose the population distribution of scores for region A lies entirely below or to the left of the distribution for region B. Similarly, suppose the distribution for region B is entirely to the left of that for region C. Then any eight bushes from region C would be judged better than any five from B, and any five from B would be judged better than any six from A. The rank table is shown below. All the poor bushes with low ranks are from A, all the best bushes with high ranks are from C, and the middle ranks are in B.

Growing Region		
A	*B*	*C*
1	7	12
2	8	13
3	9	14
4	10	15
5	11	16
6		17
		18
		19

Under the null hypothesis, all three distributions of scores are presumed to be identical in shape and location. In essence, they can be considered a common population. Suppose 19 scores are randomly selected from this common population. Then six of the scores selected are randomly and independently assigned to column A, five to column B, and eight to column C. Any of the rank numbers 1 through 19 is as likely to occupy any of the 19 positions in the three columns as is any other rank number. All possible equally likely combinations of the 19 rank numbers can be listed and used to determine the exact sampling distribution of the test statistic, H. Tables of the exact sampling distributions of H for varying numbers of populations and varying small numbers of observations per population are given in non-parametric texts. We shall shortly describe how to calculate H from the sample results given above.

First, we shall mention an important generalization. When sampling for the Kruskal-Wallis test satisfies three conditions, the sampling distribution of the test statistic H can be adequately approximated by a chi-square distribution with $k - 1$ degrees of freedom, where k is the number of populations sampled. The three conditions are as follows.

1. The null hypothesis that all populations are identical must be true.
2. At least three populations must be sampled ($k \geq 3$).
3. At least five observations must be randomly and independently selected from each of the k populations.

Now we return to the table of ranks to calculate H and test the null hypothesis of identical distributions of population scores for the three growing regions. The table of ranks given earlier is repeated and the calculations are shown in Table 16-4.

Before we discuss Table 16-4 in detail, let us mention what to expect. We can expect the null hypothesis to be rejected at a very low level of significance, such as $\alpha = 0.005$. Under the null hypothesis, the expected value for each of the 19 positions in the table is the mean rank, $(n + 1)/2 = 10$. Hence the expected mean rank for any one of the three columns is also 10. The ranks in Table 16-4 do not fit this expected pattern at all. Consequently, the null hypothesis should be rejected at a very low level of significance.

To check our expectation, we begin in Table 16-4 by adding the ranks in each column to get the values of T_j for the three regions. Each of these column totals is squared to get T_j^2 for each region. Then the squared total for each region is divided by the number of observations for that region, n_j.

Continuing with the calculations in Table 16-4, we find the total number of observations in the entire table, n. Next, the column values of T_j^2/n_j are added to find S. Then the number of populations sampled, k, is counted. This is the same as the number of columns in Table 16-4.

With these inputs, the test statistic is H, where

$$H = \frac{12S}{n(n + 1)} - 3(n + 1) \tag{16-4}$$

As given in Table 16-4, n is the total number of observations for all populations sampled and $S = \Sigma(T_j^2/n_j)$. The acceptance region for H under the null hypothesis of identical populations is $H \leq 10.60$. For the extreme sample, $H_{\text{observed}} = 15.81$, well out of the acceptance region. As anticipated, the three score distributions have significantly different locations at the 0.005 level.

We now turn to the actual data for the rosebush experiment. We decide to test the null hypothesis of identical score distribution shapes and locations at the 0.05 level of significance. Since there are three regions ($k = 3$) and at least five observations per region, the chi-square approximation can be used for the sampling distribution of the test statistic H under the null hypothesis. For $k - 1 = 2$ degrees of

Table 16-4 Kruskal-Wallis Test for a Significant Difference in Location for Three Nonoverlapping Populations

	Ranks by Growing Region (Rank 1 = Poorest Rosebush) A	B	C
	1	7	12
	2	8	13
	3	9	14
	4	10	15
	5	11	16
	6		17
			18
			19
T_j:	21	45	124
T_j^2:	441	2025	15,376
n_j:	6	5	8
T_j^2/n_j:	73.5	405	1922

$$n = \sum n_j = 6 + 5 + 8 = 19$$

$$S = \sum \left(\frac{T_j^2}{n_j}\right) = 73.5 + 405 + 1922 = 2400.5$$

$$k = 3$$

$$H = \frac{12S}{n(n+1)} - 3(n+1)$$

$$= \frac{12(2400.5)}{19(20)} - 3(20) = 15.81$$

$$H_{\text{observed}} = 15.81$$

$$H_{\text{criterion}} = \chi^2_{0.995,(k-1)} = \chi^2_{0.995,2} = 10.60$$

Acceptance region for null hypothesis: $H_{\text{observed}} \leq 10.60$

Conclusion: Reject null hypothesis at 0.005 level.

freedom and for the 0.05 level of significance, the acceptance region for the null hypothesis is

$$H \leq \chi^2_{0.95,2} \qquad H \leq 5.99$$

The data for the rosebush experiment appear in Table 16-5, along with the necessary calculations to find the observed value of H for the sample results. In this table we see that H_{observed} is 6.44, which falls outside the acceptance region for the null hypothesis. At the 0.05 level of significance, we conclude that the score distributions for the three growing regions have different locations.

A search for an explanation of the significant difference just found begins with a comparison of the following mean ranks for the three regions.

$$\bar{r}_A = 34/6 = 5.667 \qquad \bar{r}_B = 71/5 = 14.2 \qquad \bar{r}_C = 85/8 = 10.625$$

It appears that region A has the poorest bushes, B has the best, and C is in the middle. But multiple comparison procedures similar to those discussed in Section 13.3 for analysis of variance must be applied to the three differences in mean ranks to determine which differ significantly. These procedures are discussed in standard nonparametric texts.

Table 16-5 Composite-Score Ranks of 19 Bushes

	Growing Region		
	A	B	C
	2	10	1
	3	11	5
	4	15	8
	6	16	9
	7	19	13
	12		14
			17
			18
T_j:	34	71	85
T_j^2:	1156	5041	7225
n_j:	6	5	8
T_j^2/n_j:	192.7	1008.2	903.1

$$n = 6 + 5 + 8 = 19$$

$$k = 3$$

$$S = 192.7 + 1008.2 + 903.1 = 2104$$

$$H_{\text{observed}} = \frac{12(2104)}{19(20)} - 3(20) = 6.44$$

It was pointed out in the previous section of this chapter that the rank sum test is a replacement for the t-test when the assumptions for the t-test cannot be met, but those for the rank sum test can be. A parallel situation exists between the Kruskal-Wallis test just described and one-way analysis of variance as described in Chapter 13.

Both of the nonparametric tests discussed so far are highly effective. Even when conditions for the t and F tests just mentioned are met and those two tests are the most effective of all possible tests, the effectiveness of the respective nonparametric replacements is close to 95%.

When there are only a few ties in the sample data for a Kruskal-Wallis test, these data are handled as in the rank sum test. That is, each observation in a given set of ties is assigned the mean of the rank numbers that would apply if the observations were not tied.

Exercises

16-6 *Packaging effects on sales.* Transparent sacks, paper bags, and bulk display were the three methods used to market oranges in 20 supermarkets, all belonging to the same chain. The same grade of oranges was sold at the same price in all markets during the test period. Sales in pounds per day for each type of display were as follows.

Transparent Sacks	*Paper Bags*	*Bulk Display*
36	29	16
40	30	26
47	45	17
49	39	22
23	13	28
41	32	
38	18	
	31	

Use the Kruskal-Wallis procedure to test the null hypothesis that all three displays result in the same sales per day. Use 0.05 as the level of significance. What display would appear to be the best? Can you be sure it is best? Discuss.

16-7 *Comparison of sources for professional athletes.* Scouts from the headquarters of a national professional baseball league have evaluated selected college baseball players at four universities. One analysis to be performed on composite scores computed from scouting reports for the sampled athletes is a comparison of the four universities with respect to the class of athlete being produced. The scores for each athlete, arranged by university, are as follows.

School A	*School B*	*School C*	*School D*
41	17	87	84
54	36	84	24
56	52	28	74
62	58	19	63
60	71	46	31
20		78	18
			76

High scores indicate better athletes. At the 0.05 level of significance, use a Kruskal-Wallis test for the hypothesis that all four schools are equally effective in producing baseball players. Assign rank 1 to the score 87, and so on, from the largest to the smallest score.

16-8 *Adjustment for tied observations.* Suppose the first observation under Transparent Sacks (Column 1) in Exercise 16-6 is 39 instead of 36. Repeat the test called for in that exercise with this change in data. Discuss the effects of this change on the test and its earlier results.

16-9 *Effect of reversed ranking rule.* In Exercise 16-6, reverse the ranking rule; then repeat the test described. Discuss the effect of this change on the test results in this case. Can this same effect be expected in the general case?

16-10 *Interview time.* Four people are hired and trained to interview heads of households with regard to credit card usage. For a six-day period, mean times per interview in minutes were recorded each day for each interviewer. The data for interviewer C for one day were found to be wrong. Hence that observation was dropped. The other observations are given below.

Interviewer			
A	*B*	*C*	*D*
14.2	19.1	40.3	5.0
18.3	26.2	26.0	23.5
4.9	8.3	29.1	14.8
8.9	6.7	4.1	15.1
12.2	9.2	10.4	53.5
20.1	16.8		32.0

Are the observations consistent with the hypothesis that interview time distributions are the same for all four interviewers? Use the 0.05 level of significance for the test.

16.3 Comparison of Two Populations: Matched Pairs (the Signed Rank Test)

Both tests described earlier in this chapter are designed to detect significant differences in the locations of statistical populations. Both tests are based on *independent* random samples from each population. The rank sum test applies to only two populations, while the Kruskal-Wallis test applies to three or more populations.

The Wilcoxon signed rank test, which will be discussed in this section, is designed to detect a significant difference in the locations of two statistical populations. Unlike either of the preceding tests, the signed rank test is not based on independent observations from the populations. Instead it makes use of matched pairs. It is similar in purpose to the matched-pairs test discussed in Section 11.3.

Matched Pairs

In the rank sum test, a random sample of observations was selected from one population, and this was followed by selecting a random sample from the other population. The two samples and all of the observations are independent of one another.

The signed rank test uses two samples composed of matched pairs. One member of each matched pair belongs to one population and the other member of the pair belongs to the other population. To be a matched pair, the two individuals in the pair must have some identical characteristics. Hence the two samples are not independent. Furthermore, to be of any use in a statistical study, at least one of the characteristics must be related to the variable being studied.

An example should clarify the concept of matched pairs. Suppose two new types of white house paint are to be compared for wearing quality. Adjacent patches of the two types will be painted on a number of houses in different parts of the country. On any one such house, the adjacent patches constitute a matched pair. Each pair will be matched with respect to climate, chemical pollutants in the environment, exposure to sun, and many other conditions that could affect wear.

Assignment of paint type to the left or right position within each pair will be done by a random process. This should eliminate any effect that position might have on wear.

The months of satisfactory service are to be observed for both paint types on all patches. The distribution of service times for one paint type constitutes one statistical population, and the distribution for the second type is the other population. The objective of the study is to determine whether or not the two distributions have the same location. The research question is: Do the two paint types wear equally well?

Rather than using the two distributions of service times, the signed rank test uses the differences in service times obtained from the matched pairs. The population of such differences is the basis for the test. Of course, the service time for one paint type must be subtracted from the service time for the other consistently for all pairs.

Each matched pair is subjected to essentially the same treatment with respect to many factors affecting wear. Matching these factors cuts down variation in factors other than any average difference attributable to the paint types. Both members of a pair will tend to have very similar service lives unless the two paint types in general have different service lives.

Sample Data and the Test Statistic

Table 16-6 presents the sample service lives in months for the two paint types in the study. There are ten matched pairs and each pair contains an observation from each population.

The sample value of the test statistic, V, is calculated in four steps. First, the absolute difference in service lives is found for each pair by subtracting Y from X and ignoring the algebraic sign of the difference. Second, all ten absolute differences are ranked as a single set. Third, the algebraic sign of the original difference in service lives is attached to the rank found in the previous step. Fourth, the sample value of the test statistic is found by adding the signed ranks. As shown in Table 16-6, $v = +43$ for the ten matched pairs.

Table 16-6 Matched-Pairs Signed Rank Test for a Significant Difference in Location

	Paint Type A	*Paint Type* B			
	Months of Service		*Absolute Difference*	*Rank of*	*Signed*
Pair	X	Y	$\|X - Y\|$	$\|X - Y\|$	*Rank*
1	22.3	19.1	3.2	7	+7
2	26.7	27.6	0.9	1	−1
3	41.0	38.1	2.9	6	+6
4	13.4	8.3	5.1	10	+10
5	30.2	28.5	1.7	4	+4
6	18.4	19.8	1.4	3	−3
7	16.5	12.3	4.2	8	+8
8	24.6	19.8	4.8	9	+9
9	13.0	14.2	1.2	2	−2
10	19.7	17.9	1.8	5	+5

$$v = +7 - 1 + 6 + 10 + 4 - 3 + 8 + 9 - 2 + 5 = +43$$

$$n = 10 \text{ pairs}$$

Given the null hypothesis:

$$\mu_V = 0$$

$$\sigma_V = \sqrt{\frac{n(n+1)(2n+1)}{6}} = \sqrt{\frac{10(11)(21)}{6}} = 19.62$$

$$z_{\text{observed}} = \frac{v - \mu_V}{\sigma_v} = \frac{43 - 0}{19.62} = +2.19$$

Two-tailed test:

Significance level: 0.05

Acceptance region for H_0: $-1.96 \le z \le 1.96$

Conclusion: Reject H_0.

The Hypothesis Statement and Sampling Distribution

The null hypothesis for the matched-pairs experiment is that the medians of the two populations are equal. In this case, the null hypothesis is that the median service lives of the two paint types are equal. Given this hypothesis, the only assumption required is that the underlying population of signed differences is randomly and symmetrically distributed around zero. Random assignment of paint types to left and right positions in each pair assures us that this assumption is satisfied.

For a sample of n pairs, the ranks of the absolute differences run from 1 through n. There are 2^n possible ways to assign signs, and all assignments are equally likely under the null hypothesis. All assignments can be listed and those with the same value of the test statistic, V, can be combined to produce the sampling distribution of V for any specified value of n. Tables of these sampling distributions of V for small values of n are available in nonparametric texts.

Given a sample of n from a distribution of differences that is symmetrical around zero, the distribution of the sum of signed ranks, V, has a mean

$$\mu_V = 0 \tag{16-5}$$

and a standard deviation

$$\sigma_V = \sqrt{\frac{n(n+1)(2n+1)}{6}} \tag{16-6}$$

Furthermore, when n is sufficiently large, the statistic

$$Z = \frac{V - \mu_V}{\sigma_V} \tag{16-7}$$

is approximately normally distributed with mean zero and standard deviation 1. Here, μ_V and σ_V are as defined in Equations 16-5 and 16-6. When n is at least 10, the normal approximation can be applied.

Both types of paint in the example are new. There is no reason to anticipate that one type will be better than the other. Consequently, a two-tailed test is performed. The null hypothesis is that the service life distributions for the two types of paint do not differ in location. The alternative hypothesis is that they do differ in location. The selected level of significance is 0.05.

Completion of the Test

Because there are ten matched pairs in Table 16-6, the normal approximation defined in Equation 16-7 applies. The test statistic is the standard normal variate, Z, where

$$Z = \frac{V - \mu_V}{\sigma_V}$$

is calculated from $n = 10$ differences. From Equation 16-5,

$$\mu_V = 0$$

and, from Equation 16-6,

$$\sigma_V = \sqrt{\frac{10(11)(21)}{6}} = 19.62$$

The sum of signed ranks for the sample observations is shown to be +43 in the table. That is,

$$v_{\text{observed}} = +43$$

Hence

$$z_{\text{observed}} = \frac{43 - 0}{19.62} = +2.19$$

For a two-tailed test with 0.05 as the significance level, the acceptance region for the null hypothesis is $-1.96 \leq z_{\text{observed}} \leq 1.96$. The observed value, 2.19, is well outside the acceptance region, and the null hypothesis is rejected. At the 0.05 level, we conclude that the two types of paint have service life distributions that differ in general location. In Table 16-6, it appears that type A lasts longer.

Tied Observations

As in the previous two tests in this chapter, the signed rank test assumes that tied observations are impossible. Two types of ties do occur in practice, however. The first type produces a difference of zero for the matched pair being considered. For this situation, the observation is ignored and n, the number of differences, is decreased by 1.

The second case occurs when two or more matched pairs have equal, nonzero absolute differences. As in the previous two tests, the tied observations can each be given the mean of the ranks that would be assigned if there were no such tie. Then the test can be carried out as described earlier in this section.

Handling the two situations in the manner just given presents no special problems so long as only a small number of ties is present. Special methods are given in nonparametric texts to handle large numbers of ties.

Exercises

16-11 *Automobile fuel comparison.* The transportation director for a city government has selected 12 of the city's automobiles for a comparison of mileages using two different fuels, gasoline and a mixture of gasoline and alcohol. Each automobile uses one fuel for a week and the other fuel for the following week. Miles per gallon are calculated for each type of fuel. The order in which fuels are used is decided by flipping a coin. Each pair of observations is matched because the same automobile is used for the pair. The data are as follows.

	Mileage	
Automobile	*Gasoline*	*Mixture*
1	15.8	14.6
2	18.3	16.2
3	5.6	3.9
4	14.2	15.5
5	7.8	7.2
6	18.9	17.1
7	22.4	22.0
8	15.3	16.9
9	5.6	5.8
10	21.9	21.8
11	12.3	12.0
12	7.5	7.0

At the 0.05 level of significance, conduct a two-tailed test of the hypothesis that mileages are the same for the two fuels.

16-12 *Product shelf life.* A snack food manufacturer is testing the effect of a new packaging material on ten products. The new material will replace that currently being used if it can be established that the shelf lives of products are significantly longer with the new material. For each of the 10 products, 15 packages are prepared using each of the packaging materials. Then the mean shelf life in days is found. The observations are given below.

	Mean Shelf Life (Days)	
Product	*New Package*	*Present Package*
1	29.2	30.5
2	12.4	9.7
3	18.3	13.0
4	38.8	36.2
5	22.1	20.2
6	16.3	14.0
7	31.9	28.2
8	27.2	29.2
9	35.3	29.0
10	17.1	16.6

Test the null hypothesis that shelf lives are the same for the two packages against the alternative that the new packaging material is better. Use a signed rank test and a 0.05 level of significance.

16-13 *Nutrition and productivity.* A personnel analyst for a large branch banking firm is studying productivity in the firm's central computer center. Fourteen employees are participating in the study. Each employee uses a high protein diet (diet A) for two weeks and a high carbohydrate diet (diet B) for another two weeks. The order in which the diets are used is randomly determined. Mean productivities for the two-week periods are observed and given below.

	Productivity			*Productivity*	
Employee	*Diet A*	*Diet B*	*Employee*	*Diet A*	*Diet B*
1	35.8	33.4	8	42.2	42.9
2	35.1	40.6	9	14.9	14.7
3	50.4	48.8	10	34.8	31.8
4	19.9	19.8	11	10.9	16.8
5	27.4	32.5	12	13.0	15.2
6	21.4	26.4	13	24.2	23.5
7	19.0	18.5	14	29.5	29.8

Use the two-tailed signed rank procedure to test the null hypothesis that there is no difference in productivity for the two diets. Use a 0.10 level of significance.

16-14 *Rate of inflation.* A national association of business leaders conducts a year-end opinion poll of 15 influential economists. One item in the poll is a forecast of the percentage increase in the general price level during the coming year. Forecasts from both this year and last are available for only 11 economists.

	Forecast Inflation Rate	
Economist	*This Year*	*Next Year*
1	8.9	8.9
2	18.8	21.9
3	7.6	11.9
4	7.9	12.0
5	14.3	14.7
6	10.3	9.9
7	17.1	18.4
8	16.3	14.5
9	10.7	13.3
10	16.6	17.8
11	12.6	16.2

Use the signed rank test to determine whether the hypothesis of no difference in forecast inflation rate should be accepted or rejected in favor of the forecast that the inflation rate will be greater next year. The significance level is 0.025.

16.4 A Test for Randomness

All procedures for making statistical inferences about a population characteristic from a sample assume that the sample is randomly selected from the population. This assumption applies to both parametric and nonparametric techniques. In many cases, it is possible to test the assumption of randomness.

Several tests for randomness have been developed. Among these are tests based on runs, and it is one such test with which we shall be concerned. The test is known as the *Wald-Wolfowitz number of runs test*, which we shall call the *number of runs test*.

The Number of Runs

Our illustration for this test is based on the current membership list of a national professional association. Names, addresses, and several personal characteristics of members are carried on a computer file. One characteristic accompanying each member's name on the file is that member's date of birth. Using the date of birth, the computer will calculate the age of each member in days and also will calculate the median age. The median age will be used as the basis for the test of randomness.

The test of randomness will be applied to a computer program devised to select a random sample of members from the master file. A pilot sample of 20 members will be chosen and checked for randomness before the full sample is selected. The ages of members in the pilot sample will be listed in the order of their selection. If a member is older than the median age, a plus sign will be recorded, and if a member is younger than the median, a minus sign will be recorded. Should a member's age equal the median, the member will not be eligible for selection for the randomness test.

Suppose we were to get a pilot sample in which ten members are older than the median and ten are younger. Suppose also that the sequence of signs listed in order of selection is

How likely is such a sequence if ten of each sign happened to be selected by a random sampling process? Intuitively, it appears that there is too much clustering of signs. At the other extreme, the sequence would be

+ − + − + − + − + −

+ − + − + − + − + −

Here the perfect alternation of signs suggests that more mixing of signs has occurred than might be expected from a random selection process.

The test to be described is based on the number of runs in the sequence of observations. Only two symbols are permitted in a given sequence. *A run is defined as a succession of identical symbols.*

For the first example just above, the sequence begins with a run of ten plus signs and is followed by a run of ten minus signs, for a total of two runs. For the second example of 20 alternating plus and minus signs, there are 20 runs. The number of runs observed in a given sequence is the statistic in the number of runs test.

The Hypothesis and Sampling Distribution

The null hypothesis for the number of runs test is that the number of runs in a given sequence of observations is consistent with the number to be expected from a random process. The alternative hypothesis for our membership example is two-tailed. Either too much clustering or too much mixing will lead to rejection of randomness.

For a sequence of n observations selected by a random process, the probability distribution for the number of runs, R, has an expected value (mean)

$$\mu_R = \frac{n + 2n_1 n_2}{n} \tag{16-8}$$

and a standard deviation

$$\sigma_R = \sqrt{\frac{2n_1n_2(2n_1n_2 - n)}{n^2(n-1)}} \tag{16-9}$$

where n_1 is the number of occurrences of one of the two symbols in the sequence and n_2 is the number of occurrences of the other symbol. In addition, when n is sufficiently large, the statistic

$$Z = \frac{R - \mu_R}{\sigma_R} \tag{16-10}$$

is approximately normally distributed with mean zero and standard deviation 1. If n_1 and n_2 are equal or nearly so, the normal approximation is adequate for a total sample of 20 or more observations. Tables of the exact sampling distribution of R for small samples are available in texts devoted to nonparametric statistics.

In general, an upper-tailed, a lower-tailed, or a two-tailed test for randomness can be conducted. For all three, the null hypothesis is that the sequence is random. The alternative for an upper-tailed test for too many runs is that too much mixing or alternation of signs is present. The alternative for a lower-tailed test for too few runs is that too much clustering is present.

Completion of the Test

As indicated previously, a two-tailed test of the randomness hypothesis will be carried out for the membership sample. The level of significance is 0.05.

In order of selection, the sequence of ages for the pilot sample of 20 members selected from the membership file is

$$\underbrace{+\ +\ +}_{1}\ \ \underbrace{-}_{2}\ \ \underbrace{+}_{3}\ \ \underbrace{-\ -\ -\ -}_{4}\ \ \underbrace{+\ +}_{5}\ \ \underbrace{-\ -}_{6}\ \ \underbrace{+}_{7}\ \ \underbrace{-\ -}_{8}\ \ \underbrace{+\ +}_{9}\ \ \underbrace{-\ -}_{10}$$

Each of the ten runs is underlined. Let n_1 be the number of plus signs and let n_2 be the number of minus signs. Then $n_1 = 9$, $n_2 = 11$, and $n = n_1 + n_2 = 20$. The number of runs, r, is 10. From Equations 16-8 and 16-9,

$$\mu_R = \frac{n + 2n_1n_2}{n} = \frac{20 + 2(9)(11)}{20} = 10.9$$

and

$$\sigma_R = \sqrt{\frac{2n_1n_2(2n_1n_2 - n)}{n^2(n-1)}} = \sqrt{\frac{2(9)(11)[2(9)(11) - 20]}{20^2(20-1)}} = 2.15$$

Using the normal approximation, we have

$$z_{\text{observed}} = \frac{10 - 10.9}{2.15} = -0.42$$

For the 0.05 level of significance, the acceptance region for the null hypothesis is $-1.96 \leq z \leq 1.96$. The observed value from the sample, -0.42, is well within the acceptance region. The result of the pilot sample supports the conclusion that the computer sample selection program produces a random sequence.

Exercises

16-15 *Residential selling prices.* In Figure 14-2 (page 443), the observed selling prices of 25 residences were compared with the prices estimated by a multiple regression equation. The observed values were first written in sequence in order of magnitude. Then the estimated values were subtracted from the observed values to get a sequence of residuals. The sequence of algebraic signs for these residuals is

− − − − + + − − + + − − − − − −
+ + − − + + + + +

At the 0.05 level of significance, test the number of runs for randomness. Use a two-tailed test.

16-16 *Calls for jury duty.* Jurors in a certain judicial district are supposed to be selected at random from the list of those eligible. Of the last 24 people called, the sequence of males and females selected is as follows.

F F F F M M M M M F F F M M F F F F
M M M M M M

Perform a two-tailed test of the null hypothesis that the sequence is random. Let the level of significance be 0.01.

16-17 *Breaking strength.* Two machines, A and B, are producing a part. A random sample of 12 parts is chosen from the output of A and a similar sample of 10 parts is chosen from the output of B. The breaking strengths of all 22 parts are determined and arranged in order of magnitude. Then the machine that produced each part is identified. The sequence of machine source is as follows.

B B B B B B B A A B B B B A A B A A A A A A

Use the number of runs of the two machine symbols to test the null hypothesis that the breaking strength distributions are identical against the alternative that the strengths for machine B are less than those for A. Use 0.05 as the level of significance.

16-18 *Political opinion.* A local television reporter has just interviewed a sequence of 20 adults in a shopping mall. The people were asked whether or not they thought the city council is doing a better job now than it was doing a year ago. The sequence of *Yes* and *No* responses is as follows.

Y Y Y N N N N Y Y Y Y Y N N Y Y Y Y Y Y

Can this be considered a random sequence? Use a two-tailed test and an 0.05 level of significance.

16.5 Rank Correlation

In Section 10.3 we described the sample correlation coefficient, r, as a measure of association between two variables. To distinguish it from the measure we discuss below, we shall refer to r as the *Pearson correlation coefficient*. The Pearson correlation coefficient can be applied only to variables measured on interval or ratio scales if inferences about the population correlation are to be made from a sample result. In addition, the statistical population of paired observations must satisfy a number of assumptions.

When observations are judgments, attitudes, preferences, or aspects of behavior, for example, the data often consist of ranks. Such data come from ordinal scales and are not suitable for use in Pearson correlation procedures. In other instances, ratings constitute the observations, and there is usually a degree of uncertainty about the scale. Is a 5-point increase from 85 to 90 on an employee performance scale that can run from 0 to 100 equivalent to an increase from 40 to 45? When we are uncertain about the measuring scale, it is preferable to use a measure of association that uses only the ranks of the observations. A measure that satisfies this requirement is the *rank correlation coefficient*.

To illustrate rank correlation procedures, we shall consider an industrial training program in which employees are selected and trained at company expense. In an effort to improve the program, a test designed to measure motivation was given to 12 trainees before they entered the program. At the end of the program, comprehensive course performance scores were collected for the same people. Both motivation and performance scales contain arbitrary rating scales for use by supervisors and instructors.

If there is a strong relationship between motivation and performance scores, the motivation test may be helpful in selecting future trainees. The null hypothesis is that there is no correlation between motivation upon entering the course and performance in the course. Uncertainty about the motivation and performance measuring scales leads to the choice of rank correlation to test for a significant relationship between the two variables.

The sets of motivation and performance scores are each ranked, with rank 1 assigned to the person with the greatest motivation and to the person with the best performance score. The two sets of ranks for the 12 trainees in the study are given in Table 16-7.

Given two sets of ranks, the first step in calculating the rank correlation coefficient is to find the difference between motivation and performance ranks for each trainee. These differences are denoted by d in Table 16-7. Then the differences are squared and summed. For our example, the result is $\Sigma\, d^2 = 82$, as shown in the table. The last quantity needed to calculate the coefficient is the number of paired observations, n. We have a pair of motivation and performance ranks for each of 12 trainees, so $n = 12$.

Given n paired ranks, the rank correlation coefficient is

$$r_s = 1 - \frac{6\, \Sigma\, d^2}{n(n^2 - 1)} \tag{16-11}$$

where $\Sigma\, d^2$ is the sum of the squared difference in ranks for each paired observation.

For our example,

$$r_s = 1 - \frac{6(82)}{12(143)} = +0.713$$

As with the Pearson correlation coefficient, the rank correlation coefficient can range from −1 through zero to +1. In Table 16-7, suppose the motivation ranks run in sequence from 1 at the top to 12 at the bottom. Suppose also that the column of performance ranks has the same sequence. Then the columns of differences and squared differences will contain nothing but zeros. Hence $\Sigma\, d^2 = 0$ and the rank correlation coefficient, r_s, is +1. A rank correlation of +1 indicates perfect agreement between the two sets of ranks. On the other hand, if the sequence of performance ranks is exactly the reverse of the sequence for motivation, it can be shown that $6\, \Sigma\, d^2 = 2n(n^2 - 1)$. Substituting this in Equation 16-11, we have $r_s = 1 - 2 = -1$. A rank correlation of −1 indicates a perfect inverse relationship between the two sets of ranks. Between these two extremes, a rank correlation of zero indicates no association between the two sets of ranks. In our example, $r_s = +0.713$. This suggests that strong motivation at the outset of the training program is followed by high performance in the program, and vice versa.

Table 16-7 Calculation of the Rank Correlation Coefficient

	Rank			
Trainee	*Motivation*	*Performance*	*Difference* (d)	d^2
1	3	7	−4	16
2	7	2	5	25
3	1	4	−3	9
4	11	9	2	4
5	2	1	1	1
6	9	12	−3	9
7	6	5	1	1
8	4	6	−2	4
9	5	3	2	4
10	10	8	2	4
11	12	11	1	1
12	8	10	−2	4
				$\sum d^2 = 82$

$n = 12$ paired observations

$$r_s = 1 - \frac{6\, \Sigma\, d^2}{n(n^2 - 1)} = 1 - \frac{6(82)}{12(144 - 1)} = +0.713$$

The null hypothesis states that there is no association between motivation and performance in the population. Under this hypothesis, for a given sample set of n rank scores for motivation, there are $n!$ equally likely arrangements of n rank scores for performance. Each arrangement will produce a value for the rank correlation coefficient. The number of such arrangements that produce a given value for the coefficient can be divided by $n!$ to obtain the probability for that value under the null hypothesis. When this is done for all possible values of the coefficient, the result is the sampling distribution of the statistic r_s when there is no correlation in the population. The distribution for r_s ranges from −1 to +1 and is symmetrical around zero for any value of n. As sample size increases, values of r_s become more concentrated around zero and probabilities of values in the vicinity of +1 and −1 become very small.

For small numbers of paired observations, tables of the sampling distribution of the rank correlation coefficient are available in nonparametric references.

When there are ten or more pairs, and when the null hypothesis of no correlation in the population is to be tested, the statistic

$$t = r_s\sqrt{\frac{n-2}{1-r_s^2}} \quad \textbf{(16-12)}$$

can be assumed to belong to a t distribution with $n - 2$ degrees of freedom.

For our example, we obtained a sample rank correlation coefficient of +0.713 from 12 pairs of ranks. We want to test the null hypothesis that the population rank correlation coefficient is zero against the two-tailed alternative that it is not zero. The level of significance is 0.02. For $n - 2 = 10$ degrees of freedom, we find from Appendix Table A-4 that the acceptance region for the null hypothesis is $-2.764 \leq t \leq +2.764$. From Equation 16-12,

$$t_{\text{observed}} = r_s\sqrt{\frac{n-2}{1-r_s^2}}$$

$$= 0.713\sqrt{\frac{10}{1-0.713^2}} = +3.22$$

We can reject the null hypothesis and accept this result as evidence of association between motivation and performance in the population.

As with the other ranking techniques in this chapter, the assumptions underlying rank correlation exclude ties in ranks. Should a small percentage of ties be encountered, the mean rank procedure discussed for the previous tests can be employed to assign ranks, and then the procedures described in this section can be used with these ranks. A large percentage of ties calls for special methods that are available in standard nonparametric texts.

Exercises

16-19 *Television program rating.* A panel of housewives and a panel of college students rated 12 television programs. The mean ratings for the programs were as follows.

Program	Housewives' Rating	Students' Rating
1	61	86
2	55	70
3	83	58
4	42	87
5	31	52
6	64	48
7	69	78
8	78	92
9	92	67
10	94	82
11	70	75
12	28	59

a. Calculate the rank correlation coefficient as a measure of consistency for the two groups. Interpret the coefficient obtained from this sample.
b. Make a two-tailed test of the null hypothesis that there is no consistency in the ratings for the two groups. Use the 0.05 level of significance.

16-20 *Hiring criterion.* One criterion being considered by a large corporation for guidance when hiring new electrical engineering graduates is IQ, as measured by a battery of standardized tests. The company has records of 16 such graduates who were working for it between one and two years following graduation. The records provide both IQ scores and competency ratings. The ranked results appear below.

	Rank			Rank	
Employee	IQ	Competency	Employee	IQ	Competency
1	15	5	9	16	7
2	13	4	10	10	9
3	7	2	11	11	8
4	3	1	12	5	15
5	8	3	13	6	13
6	9	11	14	4	16
7	1	12	15	12	10
8	2	14	16	14	6

a. Find the sample value of the rank correlation coefficient for these 16 observations.
b. Test the null hypothesis that there is no association between IQ and competency against the two-tailed alternative that there is association. Set the significance level at 0.02.

16-21 *Product rating and price.* A consumer organization rated 14 electric toasters for quality and also reported the selling price of each toaster. The observations are given in the following table. Prices are rounded to the nearest dollar.

Rating	*Price*	*Rating*	*Price*
1	19	8	42
2	22	9	36
3	31	10	33
4	23	11	27
5	22	12	29
6	40	13	35
7	38	14	22

At the 0.05 level of significance, is there any evidence here that quality rating is associated with price?

16-22 *Sales promotion plans.* Two executives have ranked 10 sales promotion plans prepared for a new product. The results were as follows.

	Rank	
Plan	*Executive A*	*Executive B*
1	3	1
2	4	4
3	1	2
4	7	6
5	2	5
6	9	8
7	6	7
8	10	9
9	5	3
10	8	10

a. Calculate the rank correlation coefficient for these observations.
b. Test the null hypothesis of independence against the one-tailed alternative of positive agreement. Make the level of significance 0.01.

Summary

Nonparametric counterparts exist for many of the statistical procedures discussed in earlier chapters. For the most part, nonparametric techniques require less rigorous assumptions about the underlying statistical populations than do their parametric counterparts. Furthermore, many nonparametric techniques use only nominal or ordinal scale measurements rather than the interval or ratio scale measurements typical of parametric procedures. On the other hand, nonparametric estimation procedures are few in number, and nonparametric techniques often are less efficient than their parametric counterparts.

Supplementary Exercises

16-23 *Useful life of light bulbs.* Two processes for manufacturing light bulb filaments are being compared. Bulbs from each process were selected randomly and submitted to life testing. The number of hours the bulbs burned were as follows:

Process A	*Process B*	*Process A*	*Process B*
124.7	104.6	125.1	106.0
119.9	131.7	129.2	136.3
138.5	119.6	174.6	139.9
185.6	107.7	151.9	122.5
141.6	148.2	135.8	111.8
144.8	113.8	148.9	127.1

a. Are these matched pairs or independent samples? Why?
b. At the 0.05 level of significance, test the null hypothesis that the service life distributions have the same location against the alternative that process B produces bulbs with shorter lifetimes. Assign rank 1 to the shortest lifetime.

16-24 *Safety management.* Two methods for improving the safety record in an industrial plant are being compared at 10 locations within the plant. Each method was used for three months in each location. The order of use was selected at random. The number of lost-time accidents were as follows:

Location	*Method A*	*Method B*	*Location*	*Method A*	*Method B*
1	3	2	6	6	3
2	19	15	7	1	0
3	2	5	8	13	7
4	16	14	9	5	6
5	9	4	10	7	0

a. Should the data be treated as matched pairs? Why?
b. Make a two-tailed test of the null hypothesis that the medians of the two lost-time accident distributions are equal. Use the 0.05 level of significance.

16-25 *Runs test for useful life.* Put the 24 observations in Exercise 16-23 in order from the shortest bulb lifetime to the longest. Then identify each observation as coming from process A or B.

a. Use the runs test to determine if there is sufficient evidence to conclude that lifetimes from process B are generally shorter than those from process A. Use the 0.05 level of significance.
b. Compare this result with that obtained in Exercise 16-23.

16-26 *Justifying the matched pairs approach.* If location is associated with accident expectation in Exercise 16-24, we can expect the observed number of accidents under the two methods of safety management to exhibit positive correlation. At the 0.025 level of significance, perform a one-tailed rank order correlation hypothesis test to check this supposition.

PART FIVE

Additional Topics

17

TIME SERIES AND INDEX NUMBERS

Time-series analysis is the study of how values of a variable change over time. Many business decisions are made in response to changes in critical operating variables such as sales and inventory levels. Planning for a business operation over future months and years involves anticipating, or *forecasting*, the future levels of these and other variables. Interpreting the changes in variables that influence a firm's economic health — such as total personal income in the nation or total sales in the firm's industry — can be important in this planning process. The following are some examples of business problems involving time-series data.

> The number of orders last month for a tractor replacement shaft doubled over the preceding month. Should the manufacturer anticipate that this increased level of orders will continue?
>
> What level of packaged beer sales should a manufacturer of beer containers anticipate in the four quarters of the coming year?
>
> Is a recent increase in domestic sales of television sets by U.S. producers a resumption of long-term market growth or a reflection of a general cyclical upturn in domestic business conditions?
>
> How can a manufacturer explain a sudden decline in the sales of corrugated paper shipping containers?

As these examples suggest, changes in a time series are the outcome of a variety of forces that make themselves felt in different ways over time. First, there may be long-term growth associated with basic factors such as population increase and market expansion. Other changes in a series might reflect specific causes, such as gaining or losing major customers to competing firms, industries, or nations. Still other changes may reflect general business recessions or recoveries. Cyclical changes of varying intensity and duration are common in business and economic series. Other changes in time series occur within a year or shorter periods; for example, the demand for fuel oil for home heating rises in the winter and slackens during warmer months. Beer sales rise during the warmer months.

In analyzing and forecasting time series, the forces that produce change are categorized according to the lengths of time over which their effects are felt. The different classes of forces produce different *patterns of variation* in a time series. Segregating these patterns is the key to interpreting past changes in the series and forecasting future changes. There are four kinds of variations.

1. *Trend* in a time series is a long-term, smooth, underlying pattern of change from period to period in the series.
2. *Cyclical variation* in a time series is a wavelike or oscillating pattern of variation in the series. The duration and amplitude need not be repetitive but can vary from cycle to cycle. Variation associated with short repetitive calendar periods is excluded (see seasonal variation).
3. *Seasonal variation* is the repetitive pattern of variation occurring within a year (or shorter repetitive calendar period) in a time series.
4. *Irregular variation* in a time series is composed of changes that cannot be described as trend, cyclical variation, or seasonal variation.

Some Time-Series Examples

Figure 17-1 shows some time series in which the sources of variation are quite different. Part (a), which shows beer consumption by quarters from 1967 to 1976, shows a series with a consistent moderate upward trend and a very marked and

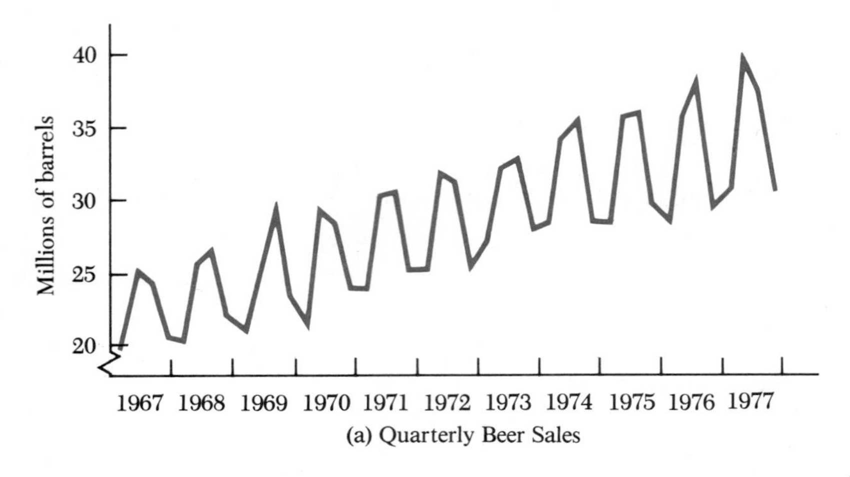

(a) Quarterly Beer Sales

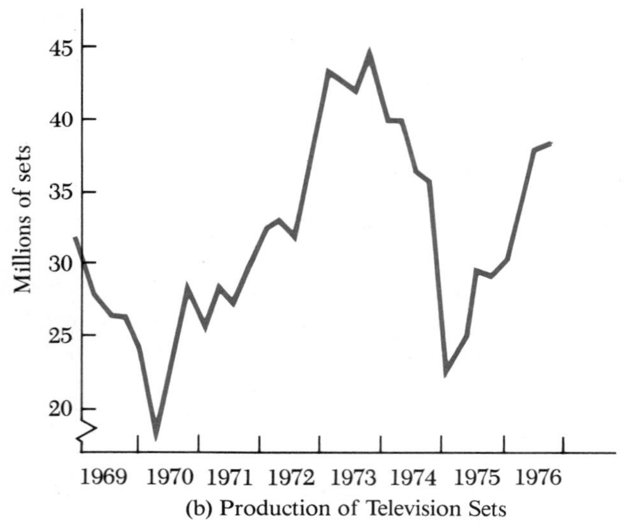

(b) Production of Television Sets

Figure 17-1 Selected Time Series

regular seasonal pattern. In the second and third quarters of the year, beer consumption is higher than in the first and last quarters. Market production of television sets, shown in (b), is an example of a series dominated by cyclical swings. Through 1969, quarterly production declines, and then falls to a cyclical low in 1970. Recovery occurs until 1973, when production again peaks and then falls to a low point in 1975. By the end of 1976, substantial recovery seems to have taken place. In (c), replacement orders for a tractor shaft, while having some upward trend, seem to be dominated by large, irregular variation, particularly in the last half of the period shown. In (d), corrugated box shipments have a strong upward trend until 1974, when a marked change occurs. This change cannot really be labeled as a cyclical variation because no cyclical change this large occurred during the earlier growth period. We would look for some fundamental change in market conditions to explain the interruption of growth in this industry.

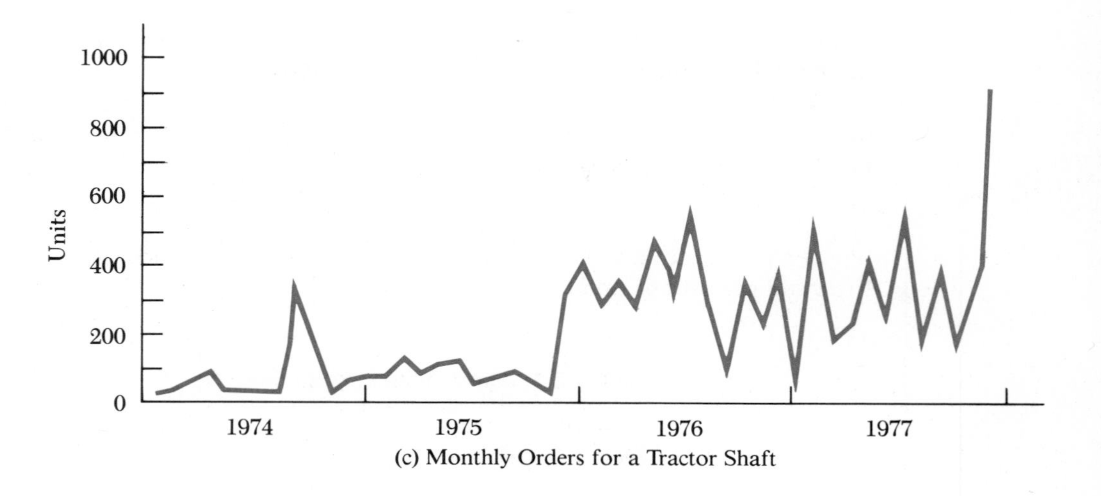

(c) Monthly Orders for a Tractor Shaft

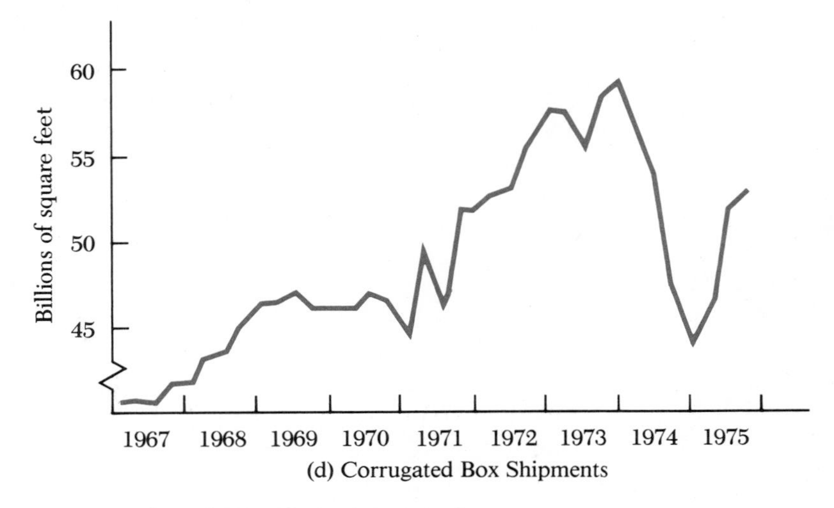

(d) Corrugated Box Shipments

Figure 17-1 Selected Time Series (continued)

17.1 Trend Analysis

The methods for analyzing time series presented in this chapter are descriptive techniques designed to bring out the different kinds of variation already mentioned. The approach is sometimes called *time-series decomposition*. We seek a separate description of trend, of cycles, and of seasonal variation. In the next chapter, we shall discuss the problem of *forecasting* a future value of a time series. We begin with trend analysis.

Arithmetic Trend

Fitting a simple equation to a time series is the most common way of describing long-term trend. A linear equation with a constant absolute change in level per time period is the simplest trend model. Table 17-1 and Figure 17-2 show annual data for packaged beer sales from 1976 to 1982. Obtaining an arithmetic linear trend for the data amounts to finding the linear regression between the dependent variable *sales* and the independent variable *time*. The regression will be

$$\hat{y} = a + bx$$

where a is the estimated sales (or trend of sales) at time zero and b is the increase in the trend per unit of time.

Table 17-1 Calculation of Arithmetic Trend for Beer Sales*

Year	y	x	xy	T
1976	132.2	3	−396.6	135.1
1977	138.7	−2	−277.4	138.9
1978	144.1	−1	−144.1	142.6
1979	148.7	0	0.0	146.4
1980	152.4	1	152.4	150.1
1981	154.5	2	309.0	153.9
1982	153.9	3	461.7	157.6

$$a = \frac{\Sigma\, y}{n} = \frac{1024.5}{7} = 146.4$$

$$b = \frac{\Sigma\, xy}{\Sigma\, x^2} = \frac{105.0}{28} = 3.75$$

*Millions of barrels.

It is convenient to place the time zero within the span of years covered by the data. We can do this by writing the variable time as a deviation from the mean time value; this has the added advantage of simplifying the equations for a and b. In our example, the mean time value is 1979, and the x-values in Table 17-1 are the deviations from that time. The equations for the constants in the least-squares straight line then become

$$a = \frac{\Sigma\, y}{n} \tag{17-1}$$

$$b = \frac{\Sigma\, xy}{\Sigma\, x^2} \tag{17-2}$$

where n is the number of data points. The calculations needed for these equations are carried out in Table 17-1, and we obtain the trend

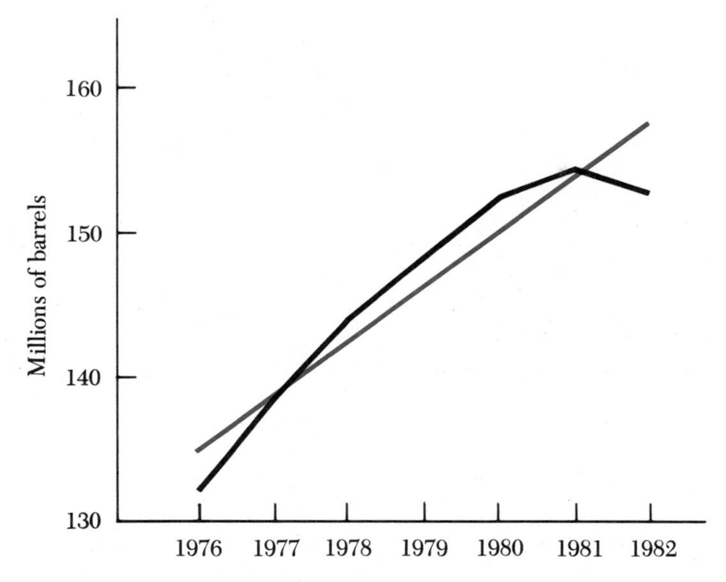

Figure 17-2 Annual Packaged Beer Sales and Arithmetic Trend

$$T = a + bx$$

$$T = 146.4 + 3.75x$$

where $x = 0$ in 1979 and x is in 1-year units. Thus the trend level for 1979 is 146.4 million barrels and the trend of beer sales increases by 3.75 million barrels a year.

The trend values for each year are shown in the final column of Table 17-1. They are obtained by substituting the appropriate x-values in the trend equation. For example, the trend for 1982 is obtained by

$$T(1982) = 146.4 + 3.75(3)$$

$$= 157.6$$

In Figure 17-2, the trend line is plotted along with the actual yearly beer sales. A plot of the trend and the original data is always useful for judging whether the fitted trend is a satisfactory description of the underlying growth pattern of the series. For the beer sales data, the linear arithmetic equation appears to be a good description of the underlying trend of the series.

Table 17-2 presents a situation in which the number of data points is even. There are six annual averages for factory sales of gas warm-air furnaces.

Here, the mean time value is $1966\frac{1}{2}$ and the coded x-deviations, selected so that the smallest deviations are plus and minus 1, represent deviations from the mean time point in $2\frac{1}{2}$-year units. Calculation of the trend equation yields

$$a = \frac{\Sigma y}{n} = \frac{7361}{6} = 1226.8$$

$$b = \frac{\Sigma\, xy}{\Sigma\, x^2} = \frac{6993}{70} = 99.9$$

Thus the trend is

$$T = 1226.8 + 99.9x$$

where x is zero in $1966\frac{1}{2}$ and x is in $2\frac{1}{2}$-year units. The increase in the trend values is 99.9 (thousand furnaces) per $2\frac{1}{2}$-year period or 199.8 per 5-year period.

To obtain trend values from packaged computer programs, it is not necessary to code time values in the way we have done. We could code the first time point as zero and the successive points as 1, 2, 3, ... For the gas furnace series, this coding would yield the trend equation

$$T = 727.3 + 199.8x$$

In this equation, x is zero in 1954 and is in 5-year units. Application of this equation produces the same trend values as in Table 17-2.

Table 17-2 Factory Sales of Gas Warm-Air Furnaces*

Year	y	x	xy	T
1954	667	−5	−3335	727
1959	888	−3	−2664	927
1964	1155	−1	−1155	1127
1969	1477	1	1477	1327
1974	1600	3	4800	1527
1979	1574	5	7870	1726

*Five-year averages centered at middle year in thousands.

Exponential Trend

In both examples—beer sales and gas furnace sales—a trend increasing by a constant absolute amount per time period reasonably expresses the long-term growth of the series. However, it is not uncommon to encounter series where a linear arithmetic trend would be inappropriate. The dishwasher sales series of Figure 17-3 is such an example. The series appears to be growing by *increasing* amounts in each 5-year period. Frequently in such cases, an exponential trend—one that increases by a constant *percentage rate* of change per time period—will provide a suitable growth model.

The trend equation that describes constant percentage rates of change is the exponential expression

$$T = a(b)^x$$

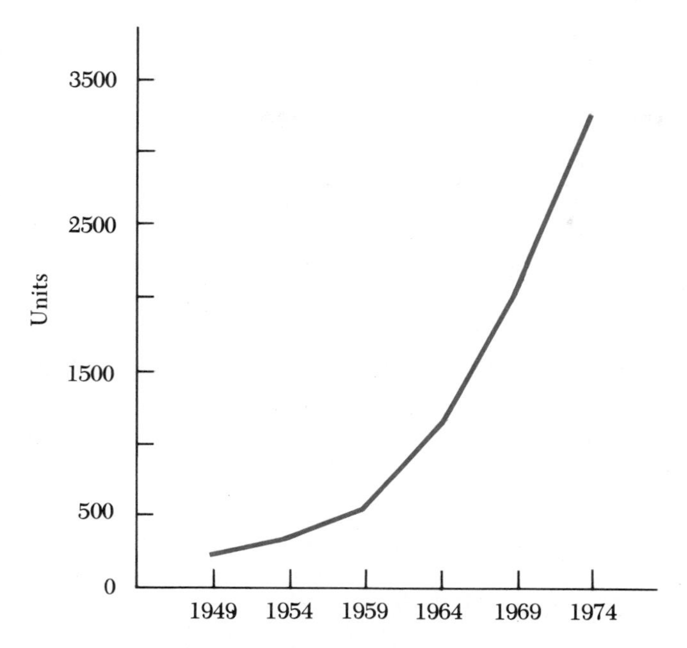

Figure 17-3 Annual Factory Sales of Dishwashers

where a is the trend value at a time origin (where $x = 0$) and b is a constant multiplier per time period. For example, if \$2000 were invested at time $x = 0$ to yield 8% compound interest per time period, then $a = \$2000$, $b = 1.08$, and T is the value of the investment after x time periods.

The exponential equation is linear in terms of logarithms.

$$\log T = \log a + (\log b)x$$

This means that the equations for least-squares linear regression can be applied to the logarithms of the time-series values to find $\log a$ and $\log b$. When we let x be zero at the mean time point, the equations are

$$\log a = \frac{\Sigma\ (\log y)}{n} \tag{17-3}$$

$$\log b = \frac{\Sigma\ (x * \log y)}{\Sigma\ x^2} \tag{17-4}$$

Table 17-3 shows the calculations needed to obtain these values for the dishwasher sales series.

The constants are

$$\log a = \frac{\Sigma\ (\log y)}{n} = \frac{17.265}{6} = 2.878$$

$$\log b = \frac{\Sigma\,(x * \log y)}{\Sigma\, x^2} = \frac{9.101}{70} = 0.130$$

Table 17-3 Factory Sales of Dishwashers*

Year	y	$\log y$	x	$x * \log y$	$\log T$	T
1949	199	2.299	−5	−11.495	2.228	169
1954	253	2.403	−3	−7.209	2.488	308
1959	507	2.705	−1	−2.705	2.748	560
1964	1094	3.039	1	3.039	3.008	1019
1969	2051	3.312	3	9.936	3.268	1854
1974	3213	3.507	5	17.535	3.528	3373
		17.265		9.101		

*Centered five-year annual averages in thousands.

The logarithmic trend is, then,

$$\begin{aligned}\log T &= \log a + (\log b)x \\ &= 2.878 + 0.130x\end{aligned}$$

If we substitute the x-values in the logarithmic trend equation, we have the logarithms of trend given in the next to the last column of the table. The final step is to take the antilogarithms of these values to obtain the trend values.

All the calculations above can be done on a pocket calculator with a logarithm key and an exponent, or y^x, key. The latter is used for antilogarithms. For example, $10^{3.528} = 3373$, the trend value for 1974. Packaged computer programs typically have a logarithmic transformation feature that can be employed for exponential trends.

To express the trend equation in the form $T = a(b)^x$, we take the antilogarithms of the logarithmic constants found from Equations 17-3 and 17-4. Using a pocket calculator we have

$$10^{2.878} = 755 \quad \text{and} \quad 10^{0.130} = 1.349$$

Thus the arithmetic form of the exponential trend is

$$T = 755(1.349)^x$$

where 755 is the trend in 1961½ (the mean time point) and x is the deviation of any time point from 1961½ in 2½-year time units. The constant 1.349 is the trend "multiplier" per 2½-year time unit. The constant percentage increase in trend is 34.9% every 2½ years. To find the annual multiplier, we divide 0.130 by 2½, obtaining 0.052. The antilogarithm of 0.052, that is, $10^{0.052}$, is 1.127. The annual percentage rate of increase in the trend of dishwasher sales is, then, 12.7%.

In Figure 17-4, the dishwasher sales series is plotted in two ways. The first is the usual plot of sales against time, and the second is a plot of the logarithms of sales

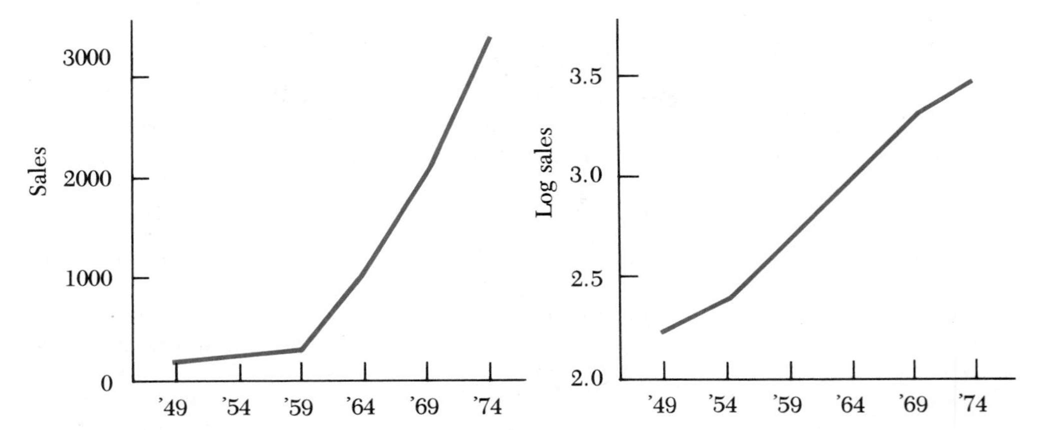

Figure 17-4 Arithmetic and Logarithmic Plots of Dishwasher Sales

against time. Such pairs of plots are useful in deciding the type of trend that is most suitable for a series. If the graph of the logarithms of the series form a straight line, the exponential trend will be a better fit. This is the case with the dishwasher series.

It can, of course, happen that a series will not be linear either in terms of the original numbers or the logarithms of these numbers. In these cases, curvilinear equations can be fit. They will not be discussed further here, except to mention that packaged computer programs have the capability of providing such fit under program names such as *polynomial regression* or *polynomial trend*.

Exercises

17-1 *Time-series changes.* Identify the kind of time series changes (trend, cycle, seasonal, irregular) you would associate with each of the following business situations.

a. Producers of woolen goods are concerned about the continued erosion of their markets stemming from the substitution of synthetic for natural products.
b. The world price for coffee rises following an unusual winter frost in Brazil.
c. A manufacturer of skiing equipment decided to develop a line of tennis rackets and accessories in order to utilize the firm's manufacturing and marketing capacities more effectively.
d. A recent lowering of interest rates following measures by the Federal Reserve Board to stimulate the economy is expected to stimulate the demand for new residential construction.

17-2 *Time-series changes.* Identify the force (trend, cycle, seasonal, irregular) or combination of forces that could give rise to the following changes in sales for an industry.

a. In the 1970's, average annual sales exceeded the annual sales in the 1960's by 25%.
b. For three years in a row sales have increased.
c. Despite the fact that the long-term trend of sales is downward, sales in the second half of this year showed an increase over the first half of the year.

d. In an industry with stable long-term sales, a normal seasonal increase in the summer does not occur.

17-3 *Cable TV subscribers.* The following data refer to cable TV subscribers (millions of households) in the United States.

1973	7.3	1975	9.8	1977	11.9	1979	14.1
1974	8.7	1976	10.8	1978	13.0	1980	15.5

a. Determine the least-squares arithmetic trend for the series.
b. Graph the series and draw in the trend line by connecting trend values for 1973 and 1980 with a straight line.
c. Does the arithmetic trend appear to be a suitable trend for the series?

17-4 *Divorces to 1950.* The following are annual numbers of divorces, in thousands, in the United States at five-year intervals.

1925	175	1935	218	1945	485
1930	196	1940	264	1950	385

a. Graph the series on arithmetic graph paper.
b. Find the least-squares arithmetic trend and plot it on the same graph. Comment on the suitability of the trend.

17-5 *Divorces since 1950.* The following are annual numbers of divorces, in thousands, in the United States at five-year intervals.

1955	377	1965	479	1975	1036
1960	393	1970	708		

a. Fit an exponential trend to the series and find the trend values for each time point.
b. Draw a graph of the original data. Locate the trend values on the same graph and draw in a curve connecting them.

17-6 *Gross national product.* The following are three-year averages for gross national product (GNP) in the United States in billions of dollars.

1965	692	1971	1072	1977	1911
1968	867	1974	1416	1980	2663

a. Find the logarithmic constants for the exponential trend of the series.
b. Find the antilogarithm of the value for log b. Find the antilogarithm of ⅔ of log b. What do these multipliers represent?

17-7 *Growth patterns of synthetics.* The following table gives polystyrene and polyethylene production (millions of pounds) in the United States (three-year averages) as well as the logarithms of the production figures.

	1956	*1959*	*1962*	*1965*	*1968*	*1971*
Polystyrene	660	934	1305	2049	2877	4143
(logarithm)	2.820	2.970	3.116	3.312	3.459	3.617
Polyethylene	559	1132	1964	3073	4739	6623
(logarithm)	2.747	3.054	3.293	3.488	3.676	3.821

a. Plot the original values and the logarithms of each series on a pair of graphs.
b. Can the growth pattern of each series be well described by either a linear arithmetic or an exponential trend? Explain.

17.2 Trend-Cycle Analysis

One reason for fitting a trend to a time series is to study the deviations of the actual series from the long-term trend. Under the proper circumstances, the deviations from the trend are indicative of the influence of cyclical forces on the series. One of these circumstances is that the trend used be a good representation of the long-term growth pattern for the series.

A trend-cycle analysis for market sales of television sets in the United States from 1965 to 1981 is illustrated in Table 17-4 (p. 557) and Figure 17-5(a) and (b). The trend is a least-squares straight line fitted by the method described earlier. The trend (millions of television sets) is

$$\begin{aligned} T &= \quad a \quad + \quad bx \\ &= 13.92 + 0.456x \end{aligned}$$

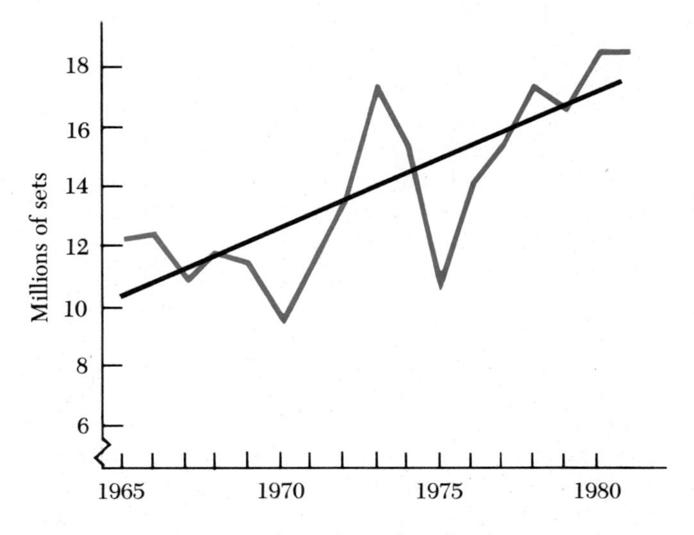

Figure 17-5(a) Annual Market Sales of Television Sets

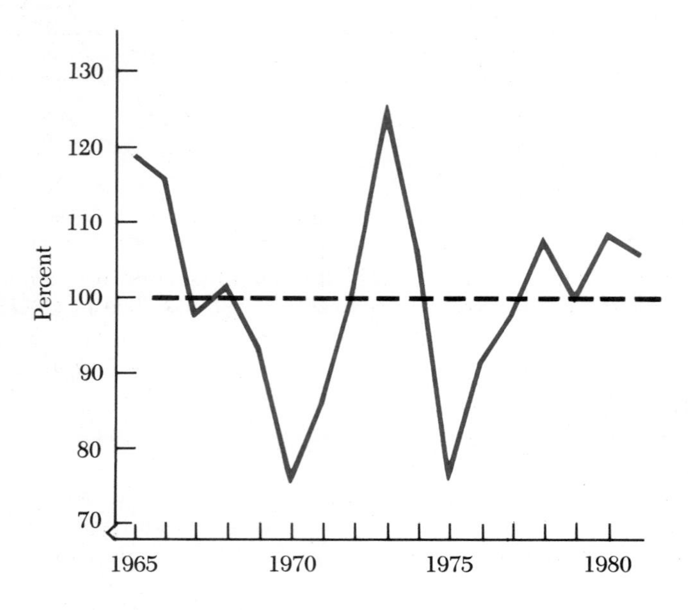

Figure 17-5(b) Annual Market Sales of Television Sets

where x is zero in 1973 and x is in yearly units. Substitution of x-values in the trend equation yields the trend values given in the table.

Percentages of Trend

The most common form for expression of deviations from the trend is *percentage relatives*. For example, actual production in 1975 was 10.6 million sets, while the trend for 1975 was 14.83 million sets. Actual production was 71.5% of the trend level. The percentage of trend for each of the annual production figures is given in the final column of Table 17-4.

In Figure 17-5, the actual data and the trend are plotted. Part (b) shows the percentages of trend. In both cases, the wave-like cyclical variation is evident. In the percentage of trend form, the cyclical variation is separated. We see that cyclical peaks are reached in 1973 and 1980, and low points in 1970 and 1975. The most recent years suggest recovery from the cyclical low of 1975.

Moving Averages

Another approach to trend-cycle analysis is use of *moving averages*. This method does not separate trend from cyclical variation, but rather separates the combined trend and cycle from the shorter-term variations in a series. The shorter-term variations are seasonal and irregular movements, so moving averages are more often used with quarterly and monthly data than with annual data.

Figure 17-1(c) presented an order series for tractor shafts for 48 months. It was given as an example of a series with a large element of irregular variation. We shall

Table 17-4 Market Sales of Television Sets, Linear Trend, and Percentages of Trend, 1965–1981

Year	*Millions of Sets*	x	*Linear Trend*	*Percentage of Trend*
1965	12.2	−8	10.27	118.8
1966	12.4	−7	10.72	115.6
1967	10.9	−6	11.18	97.5
1968	11.8	−5	11.64	101.4
1969	11.3	−4	12.09	93.4
1970	9.5	−3	12.55	75.7
1971	11.2	−2	13.01	86.1
1972	13.5	−1	13.46	100.3
1973	17.4	0	13.92	125.0
1974	15.3	1	14.37	106.4
1975	10.6	2	14.83	71.5
1976	14.1	3	15.29	92.2
1977	15.4	4	15.74	97.8
1978	17.4	5	16.20	107.4
1979	16.6	6	16.65	99.7
1980	18.5	7	17.11	108.1
1981	18.5	8	17.57	105.3

use the moving average method to smooth out the irregular variations and achieve an expression of the longer-term levels of the series. We call this *smoothing*. An analyst who is more interested in the changing *level* of the series than in irregular monthly variation would seek to smooth out the irregular component.

Table 17-5 shows the first 12 months of the order series and shows the calculations necessary for a 3-month and a 5-month moving average. Since irregular variations are thought to be of short-term duration, moving averages of corresponding duration should provide the desired smoothing. The 3-month moving average for any given month is simply the average of the number of orders for the preceding month, the given month, and the following month. The 5-month moving average uses the given month, the two preceding months, and the two following months.

For month 5, the 3-month moving average is

$$MA = \frac{80 + 33 + 31}{3} = \frac{144}{3} = 48$$

and the 5-month moving average is

$$MA = \frac{54 + 80 + 33 + 31 + 31}{5} = \frac{229}{5} = 46$$

Table 17-5 Calculation of 3- and 5-Month Moving Averages

Month	Number of Orders	3-Month		5-Month	
		Moving Total	*Moving Average*	*Moving Total*	*Moving Average*
1	27				
2	35	116	39		
3	54	169	56	229	46
4	80	167	56	233	47
5	33	144	48	229	46
6	31	95	32	211	42
7	31	98	33	476	95
8	36	412	137	617	123
9	345	555	185	605	121
10	174	538	179	638	128
11	19	257	86		
12	64				

The 5-month moving total for month 6 is

$$80 + 33 + 31 + 31 + 36 = 211$$

The term *moving* average comes from the idea that this total, and hence the resulting average of 42 for month 6, differs from the average for month 5 because 36 has been added to the moving total and 54 has been dropped. Four of the five terms are the same. In the 3-month moving average, only two of the three terms in two successive averages will be the same. For this reason, the 5-month moving average will change less rapidly as it moves through the series. It is said to have more *smoothing power*. On the other hand, a shorter-term moving average will be more *responsive* to the changing levels of the series.

The results of our two moving averages are shown in Figure 17-6. The original order series is plotted, but the points not connected. The 5-month moving average succeeds better in smoothing out the irregular ups and downs in the series, but responds less quickly to the change in level of orders that takes place near the middle of the series. The 5-month moving average is a better representation of the trend-cycle levels in the series.

In fitting a moving average to the tractor shaft order series, we wished to smooth out irregular variations. There was no seasonal variation evident in the series. To smooth out seasonal variation, we must use a moving average that covers a period of one year because seasonal variation takes place over that time span. Table 17-6 and Figure 17-7 present the results of such an analysis. The series is quarterly consumption of gasoline in the United States from 1971 to 1976.

The moving average in Table 17-6 is constructed differently from those calculated earlier. This is because the moving average must cover four quarters, but we want the averages to center on the quarters rather than between them. A simple four-quarter moving average would center between the second and third of the

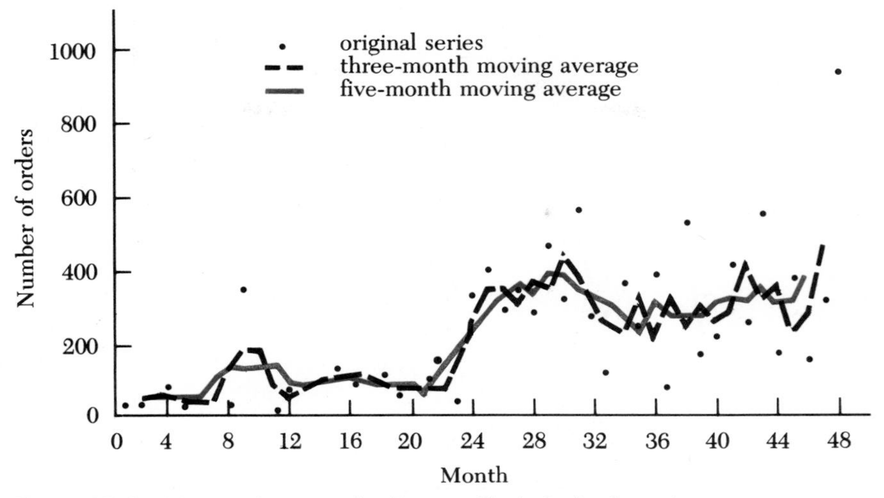

Figure 17-6 Moving Averages for Tractor Shaft Order Series*

*Original data courtesy of William F. Fulkerson, Deere and Company

Table 17-6 Quarterly Gasoline Consumption and Yearly Moving Average

Quarter		Consumption (million barrels)	Four-Quarter Moving Total	Two-Term Moving Total	Moving Average
1971	1	502			
	2	567			
			2213		
	3	582		4465	558
			2252		
	4	562		4534	567
			2282		
1972	1	541		4602	575
			2320		
	2	597		4671	584
			2351		
	3	620		4737	592
			2386		
	4	593		4799	600
			2413		
1973	1	576		4850	606
			2437		
	2	624		4890	611
			2453		
	3	644		4876	610
			2423		
	4	609		4836	604
			2413		
1974	1	546		4813	602
			2400		
	2	614		4803	600
			2403		
	3	631		4822	603
			2419		
	4	612		4854	607
			2435		
1975	1	562		4881	610
			2446		
	2	630		4896	612
			2450		
	3	642		4934	617
			2484		
	4	616		4993	624
			2509		
1976	1	596		5041	630
			2532		
	2	655		5098	637
			2566		
	3	665			
	4	650			

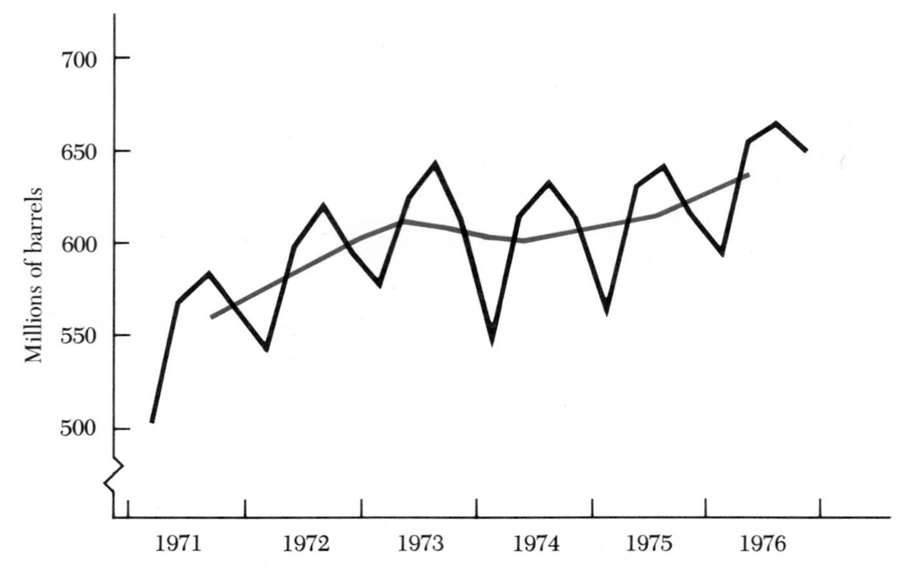

Figure 17-7 Quarterly Gasoline Consumption and Yearly Moving Average

included quarters, which is indicated in the table by the placement of the four-quarter moving totals. But if we run a two-term moving total of the four-quarter moving total, the resulting totals and corresponding moving average will be centered on the quarters. Since there are two four-quarter totals in each two-term moving total, the two-term totals are divided by eight to obtain the final moving average.

The results are shown graphically in Figure 17-7. We see a seasonal pattern in gasoline consumption in which use is high in the spring and summer quarters and low in the winter. The moving average smooths out the seasonal variation and traces a line that is called the *trend-cycle base* for the series. We see a general upward trend in gasoline consumption until 1973, after which the effects of the oil embargo and the subsequent economic slump appear to have both slowed the trend and induced a cyclical downturn in consumption.

Exercises

17-8 *Automotive replacement batteries.* The following are factory shipments of automotive replacement batteries in the United States (millions of units).

1971	39.1	1974	43.7	1977	54.6	1980	50.1
1972	43.2	1975	42.6	1978	56.4	1981	53.6
1973	43.5	1976	49.2	1979	53.7		

a. Find the least-squares arithmetic trend for the data.
b. Plot the trend along with the original data.

c. Determine the percentages of trend for each year.
d. Plot a graph of the percentages of trend and describe it.

17-9 *Weekly moving average.* A merchandising firm open five days a week has the following sales (in hundreds of dollars).

	Week 1	*Week 2*	*Week 3*	*Week 4*
Monday	33	36	38	47
Tuesday	29	34	40	44
Wednesday	27	31	38	45
Thursday	27	31	32	41
Friday	25	28	33	36

a. Fit a 5-day moving average to the sales series.
b. Plot the series and the moving average on the same graph. Comment on the result.
c. Comment on the likely smoothing power of a 3-day moving average fitted to this series.

17-10 *Tractor filter replacement demand.** The following are shipments supplied from a regional depot for 36 months of a replacement filter for a tractor.

Year	*J*	*F*	*M*	*A*	*M*	*J*	*J*	*A*	*S*	*O*	*N*	*D*
1	231	80	270	170	169	230	90	62	158	135	120	210
2	136	140	179	119	154	93	97	102	212	103	145	117
3	149	90	227	376	185	126	85	101	151	90	125	119

a. Fit a 3-month moving average to the series.
b. Does the moving average appear to smooth out the irregular variation in the series? Explain.

17-11 *Quarterly shipments of replacement batteries.* The following are quarterly factory shipments of automotive replacement batteries, in millions of units.

Quarter	*1977*	*1978*	*1979*	*1980*	*1981*
1	13.1	12.9	13.7	10.3	11.7
2	10.0	10.5	10.5	8.8	10.6
3	14.3	15.0	13.8	13.8	15.4
4	17.1	17.9	15.7	17.1	15.9

*Data provided by William F. Fulkerson, Deere and Company.

a. Plot a graph of the series.
b. Fit a centered 4-quarter moving average to the data. Plot the moving average line on your graph.
c. Describe the trend-cycle movement revealed by the moving average.

17-12 *Stock exchange trading volume.* Trading volume on the New York Stock Exchange varies by day of the week and by season of the year, and in addition, it is affected by long-term trend and business cycles. Comment on the kinds of variation that would and would not be present in each of the following.
a. A weekly moving average (five trading days).
b. A 13-week moving average.
c. A 52-week moving average.

17.3 Index Numbers

An *index number* should express the aggregate behavior of a collection of related time series. The series may describe prices, quantities, or dollar values. For reasons to be developed, price indexes are of greatest importance, and we begin with a simple example.

Price Indexes

The ingredients for a pound cake, excluding 1 tsp of vanilla and ½ tsp of salt, are 1 lb of butter, 1 lb of sugar, 9 eggs, and 2 lb of flour. The prices for these ingredients in 1972 and 1983 are shown below.

	Butter (pound)	*Sugar (pound)*	*Eggs (dozen)*	*Flour (10-pound bag)*
1972	\$0.87	\$0.14	\$0.52	\$1.20
1983	\$1.87	\$0.27	\$0.88	\$2.75

We might well ask what the change in the cost of making the pound cake from 1972 to 1983 is. This is easily answered; we simply calculate the cost of the recipe at the two different sets of prices.

$$1972 \quad \$0.87(1) + \$0.14(1) + \$0.52(0.75) + \$1.20(0.2) = \$1.64$$

$$1983 \quad \$1.87(1) + \$0.27(1) + \$0.88(0.75) + \$2.75(0.2) = \$3.35$$

The cost (or price) of making the pound cake has increased from \$1.64 to \$3.35. To express this as an index number, we say that the cost of making the cake in 1983 is $\$3.35(100)/\$1.64 = 204\%$ of the cost of making the same cake in 1972.

This example illustrates in miniature the basic idea behind a comprehensive price index such as the consumer price index (CPI) of the Bureau of Labor Statistics.

This index shows the change in price of a fixed market basket of goods and services. In the cake example, the market basket contains the quantities of the ingredients needed to make the cake. In a comprehensive price index, the market basket is a set of quantities of goods and services consumed by a typical family. These would be priced at two different time periods, and the cost (or price) of filling the market basket could be determined in the same way that we found the cost of filling the cake recipe. The formula for this "fixed-weight" price index is

$$PI_n = \frac{\Sigma\,(p_n q)}{\Sigma\,(p_0 q)} * 100 \tag{17-5}$$

In this equation, the summation is understood to be over the several commodities; q refers to the quantity weights for each item; p refers to prices, with p_n indicating prices for any given period and p_0 referring to prices in the period used as the *base* for the index.

A different representation of the fixed-weight formula will show the actual method used to find the CPI. First, let us ask how important the different ingredients are to the total cost of the cake in 1972, our base year. Using the symbols introduced above, we have the following.

	p_0	$p_0 q$	% of \$1.64 = w
Butter	\$0.87	\$0.87(1) = \$0.87	53.05
Sugar	0.14	0.14(1) = 0.14	8.54
Eggs	0.52	0.52(0.75) = 0.39	23.78
Flour	1.20	1.20(0.2) = 0.24	14.63
		\$1.64	100.00

Next, we express the price for each of the ingredients in 1983 relative to the price for the same ingredient in 1972. These are called *price relatives* and are calculated as p_1/p_0. The index we calculated earlier can also be calculated as the weighted average of the price relatives, where the weights are the relative importance of the ingredients in the base year.

	Relative Importance w	Price Relative p_1/p_0	Product $w(p_1/p_0)$
Butter	53.05	1.87/0.87	114.03
Sugar	8.54	0.27/0.14	16.47
Eggs	23.78	0.88/0.52	40.24
Flour	14.63	2.75/1.20	33.53
			204.27 = PI(1983)

This calculation can be expressed in the following formula.

$$PI_n = \Sigma\left[w\left(\frac{p_n}{p_0}\right)\right] \tag{17-6}$$

where $w = p_0q/\Sigma(p_0q)$.

The reason the weighted average of relatives method is used in the CPI is that the *quantities* of the items in the consumer market basket are not known in detail. What is known is a typical pattern of *dollar expenditures* on fairly narrow classes of goods and services, which is determined from a large survey of households for a particular year. The Bureau of Labor Statistics then samples price changes within each expenditure class to determine price relatives for each class. Additional details on the CPI will be given later in this section.

Fixed weight, or market basket price indexes, do not take account of certain changes related to the *cost of living*. For example, the use of a fixed market basket does not allow for the possibility that consumers will change their buying patterns in response to changing price relationships in the economy. In inflationary times, consumers might alter their consumption patterns in favor of lower priced substitutes, and thus limit the impact of rising prices. Changes in the quality of items in the market basket are another problem. How they are handled (or ignored) affects the use of the index as a measure of the cost to maintain a fixed level of living.

Quantity Indicators

In the cake example, we saw that a price index basically is a measure of the change in price of a fixed set of goods or services. Indeed, in that example, the question of changes in quantities did not arise. In the following example on fossil fuel prices and production (Table 17-7), both prices and production change from 1965 to 1975.

A price index for 1975 on 1965 as a base, using 1965 quantities, is constructed by first pricing the 1965 quantities (production) at 1975 prices.

2.8 billion bbl at \$7.67 per bbl = \$21.476 billion
16.0 trillion ft^3 at \$0.445 per ft^3 = \$7.120 billion
Total 1965 production at 1975 prices = \$28.596 billion

Table 17-7 Prices, Production, and Value of Petroleum and Natural Gas, United States*

	Average Price		*Production*		*Value*	
	1965	*1975*	*1965*	*1975*	*1965*	*1975*
Petroleum	\$2.86	\$7.67	2.8	3.5	\$8.008	\$26.845
Natural Gas	\$0.156	\$0.445	16.0	21.9	2.496	9.746
					\$10.504	\$36.591

*Production for petroleum given in billion barrels; for natural gas, in trillion cubic feet. Value is given in billions of dollars.

The total dollar value of 1965 production at 1975 prices is then related to the actual value of production in 1965 to form the price index.

$$PI_{75} = \frac{\Sigma\ (p_{75}q_{65})}{\Sigma\ (p_{65}q_{65})}(100) = \frac{\$28.596}{\$10.504}(100) = 272.2$$

The effect of price changes of the fuels has been to increase the dollar value of 1965 production by 172.2%.

Suppose we now want a production index for the two fuels. The *production*, or *quantity*, index should give us a single figure indicating the percentage increase in production of the two fuels over the period. Adding barrels of petroleum to cubic feet of natural gas makes no sense. It ignores the fact that a barrel of petroleum has a much greater economic value than a cubic foot of natural gas. Weighting production quantities by unit prices will compensate for the economic value of different production units. In a quantity index, prices become the appropriate weights.

Using 1965 price weights, a production index for 1975 using 1965 as a base is

$$QI_{75} = \frac{\Sigma\ (p_{65}q_{75})}{\Sigma\ (p_{65}q_{65})}(100)$$

We already have the denominator for this index. The numerator calls for the value of 1975 production at 1965 prices.

3.5 billion bbl at \$2.86 per bbl = \$10.010 billion
21.9 trillion ft^3 at \$0.156 per ft^3 = \$3.416 billion
Total 1975 production at 1965 prices = \$13.426 billion

The quantity index is then

$$QI = \frac{\$13.426}{\$10.504}(100) = 127.8$$

We say that production of the two fuels has increased by 27.8%. More specifically, the effect of quantity changes on the value of production (at 1965 prices) has been to increase the dollar value of the quantities produced by 27.8%.

There is a quick way to approximate the quantity index if a price index is already available. This method stems from the fact that price times quantity equals dollar value, which can be restated in index terms as

$$\text{quantity index} = \frac{\text{value index}}{\text{price index}}(100)$$

The value index for 1975 on 1965 as a base is obtained from the value totals in Table 17-7.

$$\text{value index} = \frac{\$36.591}{\$10.504}(100) = 348.4$$

The dollar value of production increased by 248.4% over the period. We then approximate the quantity index by

$$\text{quantity index} = \frac{348.4}{272.2}(100) = 128.0$$

The difference between 128.0 and the 127.8 obtained earlier is more than a rounding error. The two ways of getting a quantity index are not mathematically equivalent. In most practical situations, the approximation obtained by dividing the value index by a base-weighted price index will be very good.

A closely related procedure is known as *adjusting a dollar value aggregate for price changes.*

$$\text{dollar value (adjusted for price)} = \frac{\text{dollar value}}{\text{price index}}(100) \qquad \textbf{(17-7)}$$

Returning to the fuels example, suppose we read in a secondary source that the value of petroleum and natural gas production in 1975 was \$36.591 billion, up from \$10.504 billion in 1965, and that the 1975 price index for these fuels on a 1965 base was 272.2. To find the increase in production quantity, we calculate the following.

1965 \$10.504/1.000 = \$10.504
1975 \$36.591/2.722 = \$13.443

Because the base year of the price index was 1965, the effect of the price adjustment is to restate production values in terms of 1965 prices. The 1975 production value, at prices prevailing in 1965, is \$13.443 billion. This is a 28.0% increase over the 1965 level, a change attributable to changes in production quantities of barrels of petroleum and cubic feet of natural gas.

Real Income and the Purchasing Power of the Dollar

A common application of Equation 17-7 is adjusting dollar income for price changes to obtain a measure of "real" income. Table 17-8 shows personal income in the United States from 1976 to 1980. Also shown is the CPI for the same years, with 1976 as a base.

Table 17-8 Real Per Capita Income and the Purchasing Power of the Dollar

Year	*Per Capita Income*	*CPI (1976 = 100)*	*Real Per Capita Income (1976 Dollars)*
1976	\$6379	100.00	\$6379 = \$6379/1.0000
1977	6982	106.45	6559 = 6982/1.0645
1978	7734	114.55	6752 = 7734/1.1455
1979	8635	127.68	6763 = 8635/1.2768
1980	9489	144.87	6550 = 9489/1.4487

Per capita income in 1980 is \$9489 compared with \$6379 in 1976. But the price of the market basket of goods and services consumed by urban wage earners in 1980 is 144.87% of the price in 1976, that is, up 44.87%. Dividing $9489 by 144.87/100.00, or 1.4487, yields $6550, a measure of 1980 per capita income in terms of dollars of constant (1976) purchasing power. Per capita income in 1980 is capable of purchasing only modestly more goods and services than in 1976 and slightly less, in fact, than in 1978. The purchasing power of the dollar was eroded by inflation over the period.

GNP Implicit Price Deflators

The Office of Business Economics of the U.S. Department of Commerce prepares estimates of gross national product, national income, and related series. The series, known collectively as the *national income accounts*, are published monthly in the *Survey of Current Business*. *Gross national product* (GNP) is the market value of all goods and services produced by the labor and property of the nation's residents. *National income* is the sum of labor and property earnings from current production of goods and services in the nation, and *personal income* is the income currently received by individuals and noncorporate enterprises in the form of wages and salaries, proprietors' and rental income, dividends, interest, and other (transfer) payments from business and government.

Gross national product and other major series in the national income accounts are estimated in terms of current dollars and constant dollars. The constant-dollar estimates are obtained by dividing components of current-dollar GNP by appropriate price indexes, using as fine a product breakdown as possible. Then the constant-dollar figures are totaled to obtain the GNP in constant dollars.

Table 17-9 The GNP Implicit Price Deflator

Year	*Gross National Product (billions of dollars)*		*Implicit Price Deflator (1972 = 100)*
	Current Dollars	*Constant (1972) Dollars*	
1972	1185.9	1185.9	100.0
1973	1326.4	1254.3	105.8
1974	1434.2	1246.3	115.1
1975	1549.2	1231.6	125.8
1976	1718.0	1298.2	132.3
1977	1918.3	1369.7	140.0
1978	2163.9	1438.6	150.4
1979	2417.8	1479.4	163.4
1980	2633.1	1474.0	178.6
1981	2937.7	1502.6	195.5

Constant-dollar GNP is the value of gross national product in dollars of constant purchasing power. If the current-dollar GNP is divided by the constant-dollar

GNP, the result is the *implicit GNP price deflator*. It is a weighted price index of the components of gross national product—ouputs in mining, agriculture, forestry and fisheries, values added in construction, manufacturing, wholesaling and retailing, personal and business services, governmental services, and so on. Table 17-9 shows the GNP in current and in constant (1972) dollars for selected years. In the final column, the implicit price deflator is obtained. It represents a comprehensive price index for the national economy.

From 1972 to 1981, GNP increased from $1186 to $2938 billion in current dollars, but increased to only $1503 billion in constant (1972) dollars. The implicit increase in price levels was thus 95.5% over the period. The percentage changes in the implicit price deflator from 1979 to 1980 and 1980 to 1981 represent inflation rates of 9.3% and 9.5%, respectively.

Shifting the Base of an Index

The base given in an index is not the one we want to use in all comparisons. For example, the implicit GNP price deflator has a 1972 base and the CPI has a 1967 base. To compare the two price indexes, it is necessary to establish a common base.

To change the base of an index series, we divide the series by the index for the desired new base period and state the results as a percentage. In Table 17-10, we show this process for shifting the base of the CPI from its published base to a 1972 base. Each published index is divided by the published index for 1972 and restated in index form.

With the base of the CPI changed to 1972, we can easily compare the CPI as an inflation indicator with the GNP implicit price deflator in Table 17-9. For example, the percentage increase in the CPI from 1979 to 1980 was 13.5, while the GNP implicit price deflator increased by 9.3%. Prices paid by urban wage earners and clerical workers rose somewhat faster than prices in the general economy.

Table 17-10 Shifting the Base of a Published Index

Year	*CPI (1967 = 100)*	*CPI (1972 = 100)*
1972	125.3	125.3(100)/125.3 = 100.0
1973	133.1	133.1(100)/125.3 = 106.2
1974	147.7	147.7(100)/125.3 = 117.9
1975	161.2	161.2(100)/125.3 = 128.7
1976	170.5	170.5(100)/125.3 = 136.1
1977	181.5	181.5(100)/125.3 = 144.9
1978	195.3	195.3(100)/125.3 = 155.9
1979	217.7	217.7(100)/125.3 = 173.7
1980	247.0	247.0(100)/125.3 = 197.1
1981	272.3	272.3(100)/125.3 = 217.3

Important United States Indexes

Producer Prices The Bureau of Labor Statistics has prepared monthly indexes of wholesale prices since 1902. The indexes are designed to measure price movements in primary markets—that is, the first important commercial transaction for each commodity. The index is a weighted average of price relatives where the weights are net sales values in the weight-base reference period. Periodic revisions of the weights are made in accordance with information from the U.S. Censuses of Agriculture, Mineral Production, and Business. Monthly indexes are prepared for major groups, such as farm products, metals and metal products, and chemicals and allied products, for sub-groups, and even for product classes. In addition, special indexes, such as the index of wholesale prices of building materials, are prepared and published.

Consumer Prices The CPI was initiated during World War I to provide a basis for "cost-of-living" adjustments for wage-earners in the shipbuilding industry. The historic focus of the CPI until 1978 was on a market basket of goods and services purchased by urban wage and clerical worker families and single individuals. Escalator clauses in wage agreements and pensions are commonly tied to changes in this CPI. Since 1978, the Bureau of Labor Statistics has published a broader index applicable to all urban families and individuals. It is called CPI-U, while the historic index is called CPI-W. Figures given earlier in this chapter were CPI-W.

The weights employed in the consumer price indexes are based on a large scale consumer expenditure survey conducted in 1972 and 1973. Relative-importance weights were established for 265 product and service classes. The ongoing pricing activity of the Bureau of Labor Statistics is directed toward obtaining prices so that price relatives can be established for each of the categories. Prices are collected in 85 urban areas throughout the country by checking outlets that represent a probability sample of points-of-purchase for families and individuals. Items priced in stores are designed to provide representation of brands and varieties in proportion to sales importance. About 2300 food stores are included in the pricing program.

The most recent change in CPI methodology is in the treatment of housing. Instead of surveying home prices, interest rates, property taxes, and so forth, rental equivalents are now used to measure the price of housing services. The new approach was introduced into CPI-U in January 1983, and will be incorporated into CPI-W in January 1985.

Consumer price indexes are published monthly for the nation and for major cities. Breakdowns are available for major expenditure categories and for cities classified by region and population size. Current CPI data are summarized in the *Survey of Current Business* and the *Federal Reserve Bulletin*, and detailed data appear in a publication of the Bureau of Labor Statistics called *CPI Detailed Report*.

Industrial Production The Board of Governors of the Federal Reserve System compiles a monthly index called the FRB index of industrial production. The index is based on over 200 series collected by government agencies and trade organizations. It covers quantities of output in manufacturing, mining, and utilities. In forming the index, quantity relatives for various outputs are weighted by value added in the product or industry classification, as determined from periodic census data.

Exercises

17-13 *Corn and wheat prices.* In 1970, the average price in dollars per bushel received by farmers in the United States for corn and for wheat was $1.33. In 1975, corn was $2.54 and wheat was $3.56 a bushel. Over the period, average annual production was 5500 million bushels for corn and 1800 million bushels for wheat.

a. Use the quantity weights to develop a price index for 1975 compared with 1970 for the two crops.
b. Calculate the same price index using an average of price relatives with relative importance weights.
c. A farmer sold about 8000 bushels of wheat and 4000 bushels of corn in both years. Does the index calculated above represent a good index of prices received by this farmer? Explain.

17-14 *Gold and silver production.* Production (millions of fine ounces), price (dollars per ounce), and value (millions of dollars) for gold and silver in the United States are given below.

	1975			*1980*		
	Production	*Price*	*Value*	*Production*	*Price*	*Value*
Gold	1.053	161.5	170	0.952	612.6	583
Silver	34.64	4.42	154	31.36	20.63	647

a. Find the index of value for 1980 compared with 1975.
b. Calculate a quantity index for 1980, using 1975 as a base with 1975 price weights.
c. Approximate the quantity index from the value index and a price index.

17-15 *Changing consumption habits.* Per capita consumption of beef and pork products in the United States is given below for selected years.

	1970	*1975*	*1980*
Beef	113.7 lb	120.1 lb	103.4 lb
Pork	72.7 lb	56.1 lb	73.4 lb

Beef and pork products constitute major subgroups in a price index for meat products. If 1975 expenditures on beef and pork products were used to obtain relative importance weights, what would be the implications for a price index for 1980 in each case?

a. If beef prices rose faster than pork prices.
b. If pork prices rose faster than beef prices.

17-16 *Earnings in contract construction.* Average gross weekly earnings of workers in contract construction are given below, along with the CPI for selected years.

	1965	*1970*	*1975*	*1980*
Earnings ($)	138	195	265	367
CPI-W (1967 = 100)	94.5	116.3	161.2	247.0

a. Convert the price index to a 1970 base.
b. Convert the average weekly earnings to dollars of 1970 purchasing power. Interpret the results.

17-17 *Personal consumption expenditures.* The following table gives major components of personal consumption expenditures in the United States in current and constant (1972) dollars (billions).

Year	*Durable Goods*		*Nondurable Goods*	
	Current	*Constant*	*Current*	*Constant*
1972	111	111	301	301
1974	123	112	373	303
1976	157	127	442	323
1978	199	146	530	346
1980	212	136	676	358

a. Find the implicit price deflator for each series.
b. Compare the price changes for the durable and nondurable goods expenditures.

17-18 *Disposable income per capita.* Disposable income per capita in the United States in current and constant dollars is given below.

	1972	*1976*	*1980*
Current Dollars	3837	5477	8012
Constant (1972) Dollars	3837	4158	4472

a. The constant-dollar disposable income per capita in 1980 was 1.166 times the figure for 1972. What does this figure mean?
b. Find the implicit price deflator for 1980 compared to 1972.

Summary

Time-series analysis in business and economics is a collection of procedures for studying the systematic time patterns in a series. The time patterns most commonly studied are *trend*, *cyclical*, and *seasonal variation*.

The trend element in a series can be identified by equations describing the long-term relationship between the series values and time. An *arithmetic trend equation* describes a growth pattern of constant absolute change, and an *exponential trend* describes a growth pattern of a constant percentage rate of change. For more complex trend patterns, more complex equations may be required.

If a satisfactory trend equation has been found for a series of annual data, the deviations from trend will often reveal the cyclical element in the series. The most common form for expressing cyclical variation is as percentages of trend for the various time points.

Moving averages can eliminate short-term variations and leave only the longer-term elements. For monthly data, a several-term moving average will eliminate irregular variation. A moving average of a year's duration will eliminate seasonal variation, leaving trend and cyclical variation combined.

Price, *quantity*, and *value* indexes are used to express the aggregate behavior of a collection of related series. The most common form of price index measures the change over time in the aggregate price of a fixed collection of goods and services. Quantity indicators are designed to remove the effect of price changes on the changing value of an economic aggregate. A widely used price index is the *consumer price index* of the Bureau of Labor Statistics, and a widely used quantity index is the *FRB index of industrial production*.

Supplementary Exercises

17-19 *Growth pattern for kitchen disposals.* The following are centered five-year averages of annual unit factory sales (thousands) of disposals in the United States.

1959	1964	1969	1974	1979
703	1209	1876	2579	3142

a. Fit a linear arithmetic least-squares trend.
b. Fit an exponential least-squares trend.
c. Determine the trend values from each of your trends.
d. Plot the actual data and the trends on one graph. Which trend is the better fit to the underlying growth pattern?

17-20 *New plant and equipment expenditure cycles.* New plant and equipment expenditures in the United States are given annually below in billions of dollars.

1967	65.5	1971	81.2	1974	112.4
1968	67.8	1972	88.4	1975	112.8
1969	75.6	1973	99.7	1976	120.5
1970	79.7				

a. Determine the least-squares arithmetic trend.
b. Plot the original data along with the trend line.
c. Calculate and plot a graph of the percentages of trend. Describe the cyclical variation in the series.

17-21 *Cycles in GNP.* A trend for GNP in billions of 1972 dollars is $T = 1285 + 35.8x$, where $x = 0$ in 1975 and is in yearly units. Annual data for 1968 to 1982 are given below.

1968	1052	1973	1254	1978	1439
1969	1079	1974	1246	1979	1479
1970	1075	1975	1232	1980	1474
1971	1108	1976	1298	1981	1503
1972	1186	1977	1370	1982	1476

a. Find the trend values and the percent of trend for each year.
b. Plot the percentages of trend and briefly describe the cyclical behavior of GNP.

17-22 *Tractor tire chains.** Dealer demand for tractor tire chains from a manufacturer is given below by quarters.

	1975	*1976*	*1977*	*1978*
First Quarter	707	775	1105	1669
Second Quarter	566	888	579	1008
Third Quarter	2091	1867	2200	5520
Fourth Quarter	3170	3051	5685	5897

a. Fit a centered four-quarter moving average to the data.
b. Plot the original data along with your moving average. Describe the changes brought out by the moving average.

17-23 *Cotton picker spindle replacement demand.*† The data below (read across) give 36 months of dealer demand for a replacement part (left picker spindle) for a cotton picker.

9,826	2,200	9,700	1,450	6,500	10,000
16,509	4,526	3,084	999	37,851	19,600
18,400	2,699	10,700	13,903	18,200	22,352
26,154	23,885	4,527	1,305	45,658	26,923
22,400	17,300	31,350	20,160	62,831	80,697
127,589	62,228	43,867	24,902	32,727	71,791

a. Plot a graph of the 36-month demand history.
b. Fit a three-month moving average and plot it on the graph constructed in (a).
c. What changes are brought out better by the moving average than by the original data? Describe them.

*Data courtesy of William F. Fulkerson, Deere and Company.

†Data courtesy of William F. Fulkerson, Deere and Company.

17-24 *Factor prices and quantities.* The operator of an amusement park had the following expenses for major factors of production in 1977.

Land (rental)	5 acres at $400 per acre	$ 2,000
Labor	3000 h at $4.00 per hour	12,000
Capital	$100,000 loan, at 9% interest	9,000

By 1980 the operation had been expanded to utilize 15 acres, 4000 h of labor, and $50,000 additional capital. Rental per acre was up 50%, the average labor rate was $5, and the original loan as well as an additional $50,000 loan carried interest at 12%. As a result, total factor costs in 1980 were $47,000.

a. Calculate the 1980 price index for factor inputs using 1977 relative importance weights.
b. Adjust the 1980 factor expenditures for price changes as revealed by the price index.
c. Calculate costs of the 1980 factor inputs at 1977 prices. Does the total agree more or less with the result in (b)?

17-25 *Manufacturers' shipments.* Manufacturers' shipments (billions of dollars) and a price index (1967 = 100) for industrial commodities are given below for the United States in selected years.

	1965	*1970*	*1975*	*1980*
Shipments	492	634	1039	1846
Price Index	96.4	110.0	171.5	274.8

a. Adjust the value of shipments for price changes. Indicate what the adjusted series represents.
b. If you were to change the base of the price index to 1970 and then deflate the shipments series by that index, what would the deflated series represent?

17-26 *Shellfish catch.* The quantities (millions of pounds) and value (millions of dollars) of the U.S. shellfish catch in 1975 and 1980 are given below.

	Quantity		*Value*	
	1975	*1980*	*1975*	*1980*
Clams	112	95	141	90
Crabs	301	523	84	291
Lobsters	30	37	49	75
Oysters	53	49	43	70
Scallops	13	30	22	114
Shrimp	344	340	226	403

a. Find the quantity relatives for 1980 on 1975 as a base and the relative importance weights for 1975.

b. Calculate the weighted average of the relatives in (a) using the relative importance weights. What index is this?
c. Contrast the influence on the index that doubling the lobster catch would have compared with doubling the shrimp catch.

17-27 *Home values, prices, and costs.* The average price of a new single-family home in a community was $13,500 in 1958, $18,200 in 1968, and $52,000 in 1978. The Department of Commerce construction-cost index (1972 = 100) was 63.9 in 1958, 76.1 in 1968, and 176.0 in 1978.
a. How much of the change in average price between 1958 and 1968 can be attributed to changes in construction costs? To what can you attribute the remainder of the price change?
b. Analyze the change from 1968 to 1978 in a similar fashion.

17-28 *Gasoline consumption update.* Quarterly gasoline consumption in the United States for 1977–1981 is given below.

Quarter	*1977*	*1978*	*1979*	*1980*	*1981*
1	611	627	644	589	571
2	670	696	654	616	616
3	684	705	650	614	620
4	668	690	632	602	607

a. Fit a centered 4-quarter moving average to the data.
b. Plot a graph of the series along with the moving average.
c. Use your graph and Figure 17-7 to describe trend-cycle movement in gasoline consumption from 1971 through 1980.

17-29 *Beer barrels.* Three-year averages for malt beverage production, in millions of barrels, are as follows.

Fiscal Year	*1963*	*1966*	*1969*	*1972*	*1975*	*1978*
Production	99.1	111.4	124.9	139.1	157.2	175.8

a. Find the logarithms of the production figures.
b. Make time plots of the original series and the logarithms.
c. Fit an exponential trend to the production series.
d. Plot the trend on your logarithmic plot. Comment on the fit.

18

TIME SERIES AND FORECASTING

We have seen that a time series can be viewed as a composite of trend, cyclical, seasonal, and irregular forces. In Chapter 17, we dealt mainly with the trend and cyclical components. Therefore the first subject on the agenda in this chapter is seasonal variation.

When we have seen how seasonal variation is measured, we will be able to combine trend, cyclical, and seasonal variation to *explain* the past behavior and *forecast* the future values of a time series. Two questions posed at the beginning of Chapter 17 illustrate these uses.

Explanation Is a recent increase in domestic sales of TV sets by U.S. producers a resumption of long-term market growth or a reflection of a general cyclical upturn in domestic business conditions?

Forecasting What level of packaged beer sales should a manufacturer of beer containers anticipate in the four quarters of the coming year?

We shall find that cyclical variation is often a troublesome problem in forecasting a time series; we study the use of regression of the time series against past values of an "indicator" series to help solve this problem.

The final topic in the chapter is exponential smoothing, which is a useful alternative approach to short-term forecasting.

18.1 Seasonal Indexes

Seasonal variation is the pattern of variation within a calendar year in a time series. Behind the method for isolating the seasonal element in a time series is the idea that seasonal variation takes place around a base level established by the two longer-term elements of variation, trend and cycle. In the previous chapter we saw how, in the case of quarterly data, the centered four-quarter moving average describes the trend-cycle levels of a series.

Quarterly factory sales of room air conditioners from 1970 to 1976 are shown in Table 18-1 and Figure 18-1. There is a marked within-year pattern of variation in which first and second quarter sales are typically high and third and fourth quarter sales are generally low. The high level of dealer orders in the winter and spring quarters anticipates their peak selling seasons of spring and summer. Our problem is to describe the seasonal pattern of factory sales more precisely.

Table 18-1 shows the centered, four-quarter moving average. In Figure 18-1, we can see that the moving average describes a downward trend and a cycle that peaks in 1973.

The Ratio-to-Moving-Average Method

The moving average represents a combination of trend and cyclical variation. These elements are assumed to be related by multiplication.

$$\text{moving average} = \text{trend} * \text{cycle}$$

Table 18-1 Quarterly Factory Sales of Room Air Conditioners, Moving Average, and Ratios of Actual Values to Moving Average

Quarter		*Thousands of Units*		*y/MA* %
		Actual *y*	*Moving Average* *MA*	
1970	1	1798		
	2	2286		
	3	840	1484	56.6
	4	963	1494	64.5
1971	1	1899	1458	130.2
	2	2265	1392	162.7
	3	573	1326	43.2
	4	701	1234	56.8
1972	1	1635	1166	140.2
	2	1793	1141	157.1
	3	493	1137	43.3
	4	587	1198	49.0
1973	1	1718	1259	136.5
	2	2194	1303	168.4
	3	581	1326	43.8
	4	852	1268	67.2
1974	1	1633	1217	134.2
	2	1816	1176	154.4
	3	555	1040	53.4
	4	546	862	63.3
1975	1	856	746	114.7
	2	1169	689	169.7
	3	274	680	40.3
	4	371	702	52.8
1976	1	958	720	133.1
	2	1244	735	169.3
	3	342		
	4	418		

or

$$MA = T * C$$

The central step in establishing the seasonal pattern is the calculation of ratios of the original series values (y) to the moving average values (MA).

$$\frac{y}{MA} = \frac{\text{trend} * \text{cycle} * \text{seasonal} * \text{irregular}}{\text{trend} * \text{cycle}}$$

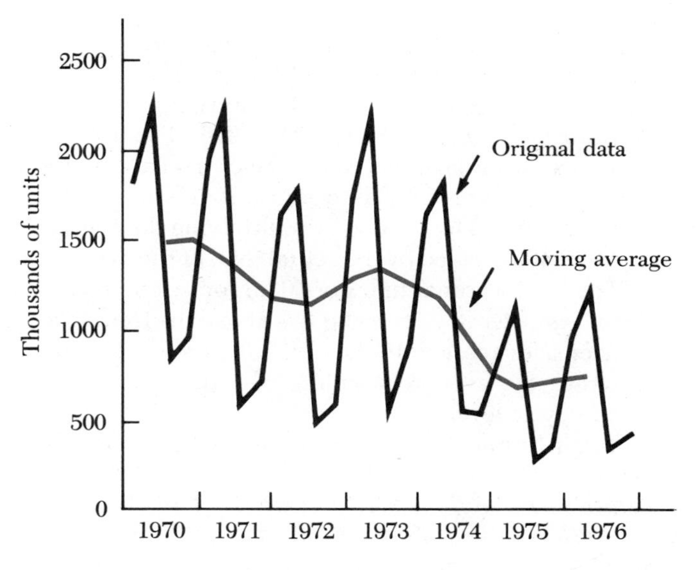

Figure 18-1 Unit Factory Sales of Room Air Conditioners, 1970–1976

Table 18-2 *y/MA* Ratios by Quarters and Calculation of Seasonal Indexes

Year	*Quarter*				*Total*
	1	*2*	*3*	*4*	
1970			56.6	64.5	
1971	130.2	162.7	43.2	56.8	
1972	140.2	157.1	43.3	49.0	
1973	136.5	168.4	43.8	67.2	
1974	134.2	154.4	53.4	63.3	
1975	114.7	169.7	40.3	52.8	
1976	133.1	169.3			
Total	788.9	981.6	280.6	353.6	
Average	131.5	163.6	46.8	58.9	400.8
Seasonal Index*	131.2	163.3	46.7	58.8	400.0

* Average (400/400.8).

$$= \frac{T * C * S * I}{T * C} = S * I$$

The y/MA ratios for the room air conditioner example are shown in Table 18-2. We now have a set of values containing a combination of seasonal and irregular variation.

The next step in isolating the seasonal pattern is to average the values of y/MA for each term (quarter) in the seasonal period. The purpose of this step is to eliminate the irregular elements present in individual y/MA ratios. For example, first quarter y/MA ratios range from 114.7 in 1975 to 140.2 in 1972. The average first-quarter y/MA ratio of 131.5 (percent) is assumed to be free of irregular variation.

The final step (shown in Table 18-2) is to adjust the average y/MA ratio for each seasonal period to equal 100.0. This is done by multiplying the average ratio of each period by the quotient of 400 divided by the actual total of the averages in the case of quarterly indexes. The final seasonal index of 163.3 for second quarter tells us that unit sales in the second quarter have been 63.3% above the trend-cycle base on the average. Third-quarter sales average 46.7% of (53.3% below) the trend-cycle base. A seasonal index is thus the average standing of a quarter's actual series value compared to the trend-cycle level.

Besides providing a format for averaging the y/MA ratios for each quarter, organizing the ratios in the manner of Table 18-2 provides an opportunity to examine the appropriateness of deriving an *average* seasonal pattern. Presence of wide variation in the y/MA ratios for given quarters would suggest that the seasonal indexes are of limited usefulness because irregular variation dominates the within-year changes. If the ratios for particular quarters move steadily upward or downward, this suggests that the seasonal pattern is undergoing change. There are extensions of the ratio-to-moving-average method that are appropriate to changing seasonal patterns.

In the example involving air conditioner sales, the y/MA ratios for each quarter appear neither to have large variability nor to change persistently over time. So we can accept the set of four seasonal indexes as describing a stable seasonal pattern for the period.

Adjusting Data for Seasonal Variation

An important benefit from deriving seasonal indexes for a series is the ability to adjust the data for seasonal variation. This means *removing* the seasonal influence from the levels of the series. Data are adjusted by dividing original values by the seasonal relatives for corresponding periods. The seasonal relatives, S, are the decimal equivalents of the seasonal indexes.

$$y_{\text{adjusted for seasonal}} = \frac{y}{S} \tag{18-1}$$

The data adjusted for seasonal variation reflect all sources of variation *except* seasonal:

$$\frac{y}{S} = \frac{T * C * S * I}{S} = T * C * I$$

Suppose an analyst, after having determined seasonal indexes for sales of air conditioners based on 1970 to 1976 quarterly data, is now looking at the quarterly sales data for 1977. These figures are given in the first column of Table 18-3.

The analyst might be as interested in the sales levels adjusted for seasonal variation as in the actual sales levels. From 1970 to 1976, a consistent pattern of high sales levels in the first two quarters and lower levels in the second two quarters was observed. Thus a drop in sales in the second half of the year is to be expected. When this normal seasonal influence is discounted by dividing actual sales by the seasonal relatives, the resulting figures describe the product of the trend, cycle, and irregular levels for the periods. On this basis, sales for the second half of the year are *up* from sales of the first half of the year. Adjusted for seasonal variation, sales for the second half of the year are 2048 thousand, compared with 1496 thousand for the first half of the year. Since the trend of sales of room air conditioners was downward, this increase appears to represent continued recovery from the cyclical lows established in 1975.

Figure 18-2 shows the entire series, adjusted for seasonal variation, including the 1977 data just discussed. The downward trend and wavelike movement associated with cyclical changes are evident, as are some short-term ups and downs suggestive of irregular variation. Removal of the seasonal element that dominated the original series has helped to bring out these remaining elements of change.

Table 18-3 Adjusting Unit Sales of Air Conditioners for Seasonal Variation

Quarter	*Actual Sales (thousands)*	*Seasonal Relative*	*Adjusted Sales (thousands)*
1977 1	900	1.312	686
1977 2	1322	1.633	810
1977 3	608	0.467	1301
1977 4	439	0.588	747

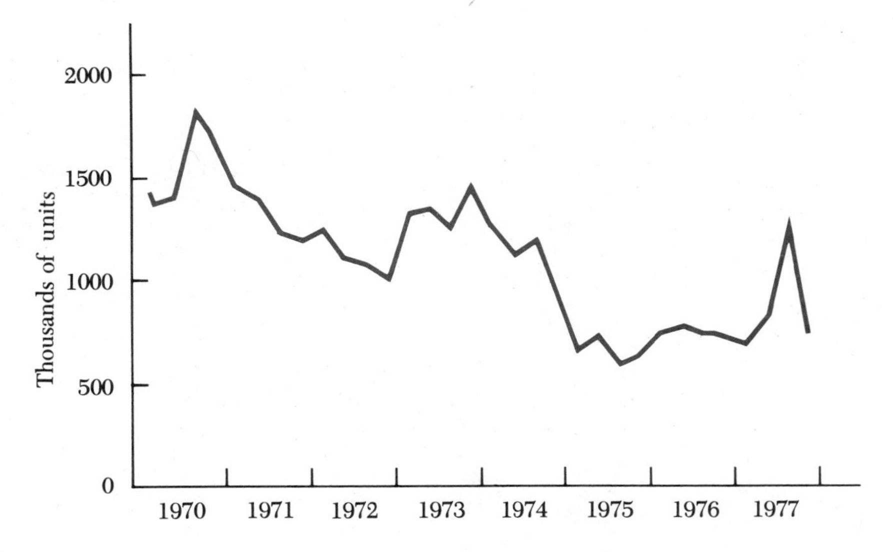

Figure 18-2 Sales of Room Air Conditioners, Adjusted for Seasonal Variation, 1970–1977

Exercises

18-1 *Gasoline consumption.* Table 17-6 gives a centered, four-quarter moving average for quarterly consumption of gasoline.

a. Calculate the ratios of the original data to the moving average and organize them in a table like Table 18-2.

b. Calculate the seasonal indexes for each quarter. Does the seasonal pattern appear to be stable? Explain.

18-2 *Beer consumption.* The data below are quarterly totals for packaged beer consumption (millions of barrels) in the United States.*

	First Quarter	*Second Quarter*	*Third Quarter*	*Fourth Quarter*
1978	31.9	39.6	40.2	32.4
1979	35.2	41.2	38.9	33.4
1980	34.8	41.5	41.8	34.3
1981	34.7	43.7	42.0	34.1
1982	35.5	43.7	40.9	33.8

a. Find seasonal indexes for each quarter by the ratio-to-moving-average method.

b. Give an interpretation of the seasonal indexes for the third and fourth quarters.

18-3 *Woolen goods sales.* A study of factory sales of woolen goods in the United States from 1970 to 1976 showed the following actual data from 1974 to 1976 (millions of yards).

	First Quarter	*Second Quarter*	*Third Quarter*	*Fourth Quarter*
1974	23.6	22.7	17.6	17.1
1975	17.3	19.4	20.4	21.7
1976	26.0	26.4	22.9	21.9

a. The seasonal indexes were 107.4, 110.5, 90.8, and 91.3 for the first through fourth quarters. Adjust the 1974–1976 data for seasonal variation.

b. The long-term trend for the series was downward. In view of this, how would you interpret the movement in the adjusted data for 1974–1976?

18-4 *Quarter-to-quarter changes.* Seasonal indexes for United States consumption of fuel oil and gasoline are given below.

	First Quarter	*Second Quarter*	*Third Quarter*	*Fourth Quarter*
Fuel Oil	128.2	85.7	74.9	111.2
Gasoline	93.1	102.5	104.8	99.6

*Courtesy of Elmer P. Lotshaw, Owens Illinois Corporation.

a. Suppose consumption of fuel oil increased by 10% from fall to winter (fourth to first quarter), while gasoline consumption fell by 5%. Should you conclude that the nation was doing a better job of conserving gasoline? Explain.
b. If the trend-cycle level were constant, what is the normal percentage change in consumption between the first and the second half of the year in fuel oil? Gasoline?

18-5 *Moving average and seasonally adjusted data.* Adjust quarterly gasoline consumption for 1974 through 1976 given in Table 17-6 for the seasonal variation given in Exercise 18-4. Graph the moving average values and the adjusted data for the period.
a. What accounts for the difference between the moving average and the adjusted values?
b. Which of the two series could be used to study trend-cycle changes for the most recent two quarters?

18.2 Projecting Trend and Seasonal Patterns

Our concern to this point has been with methods for analyzing the past variation of a time series. Now we turn to anticipating, or forecasting, future levels of a series. What we have learned about separating out the trend, cyclical, and seasonal components of past variation is helpful in forecasting. Concentrating on separate elements of variability is helpful in two ways: First, persistent patterns of variation are often found when a series is broken down into these components. If these patterns are strong and consistent, as with high beer sales in the summer, we may feel some confidence about projecting them into the future. Second, separation into components forces us to consider the different sets of causes underlying the several components of variation. Even when a pattern has shown consistency in the past, we should ask whether there are any new forces at work that might cause the pattern to change as the future unfolds. If the pattern in question is a long-term trend, the kinds of forces to look for will be different from when the pattern in question is a seasonal one. In the example of beer sales, long-term growth might be modified by changing population growth rates and changes in the age composition of the population, while the seasonal pattern might undergo some change in response to brewers' advertising and pricing strategies directed toward increasing off-season consumption.

Long-Term Projections

Past analyses of time series can be used in forecasting by projecting the patterns of past variation into the future. The components that we choose to project depend on the period for which the forecast is intended. Long-range forecasts (of five years or more into the future) are usually made in annual terms, so no projection of the seasonal pattern would be required. Cyclical patterns vary in their amplitude (amount of change from peaks to troughs) and duration (time to complete a cycle). Therefore a projection more than five years into the future would normally not consider cyclical variation. Long-range forecasts, then, are forecasts of the trend-level of a series.

If a trend has been fitted to the past values of a time series, projecting the trend into the future is a simple algebraic operation. For example, a trend was fitted to energy consumption (million BTU's per capita) in the United States from 1920 to 1970 with the following results.

$$T = 151.68 + 14.064x$$

Here x is zero in 1920 and x is in 5-year units.

To project this trend to 1985 required only that the user determine the appropriate x-value and substitute it in the trend equation. The x-value is $(1985 - 1920)/5 = 13$, and the trend for 1985 is

$$T = 151.68 + 14.064(13) = 334.51$$

Other trend equations gave quite different results, however. Figure 18-3 shows the series and the projections to 1985 for three other trend lines, as well as the arithmetic trend above. One of them (c) is the exponential trend discussed in Chapter 17 and the other two are second-degree equations. We referred to the latter in Chapter 17, but did not give examples. A computer program worked out all the trend equations.

We see from Figure 18-3 that the projections of the different trends diverge rapidly as the trends are projected farther into the future. The example illustrates the hazards of mechanically projecting a past trend. Surely not all of the projections in Figure 18-3 could be "on the mark," but how does one choose among them? Each represents a projection of some feature of the past growth in energy consumption per capita. The highest projected per-capita use comes from an equation that projects

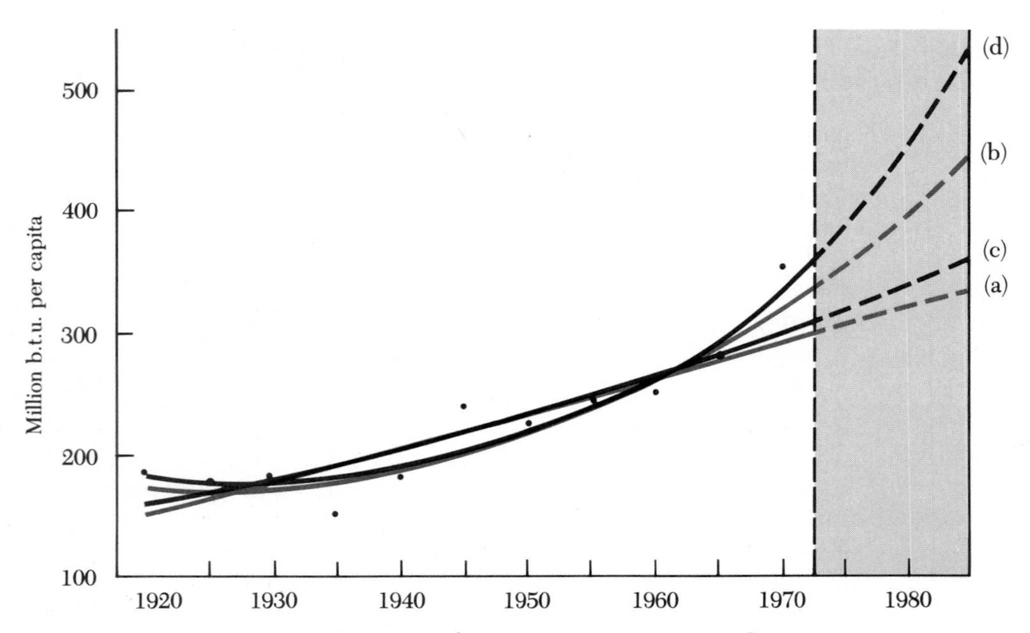

Figure 18-3 Long-Range Projections of Energy Consumption per Capita

the past tendency for the percentage rates of increase in energy usage to become larger and larger over time. Persons knowledgeable about energy in the early 1970's may have known that such rates of increase could not be sustained because of limitations on supplies of traditional fuels and the state of technological development in the utilization of nontraditional alternatives. Thus we see that a past trend should not be projected "blindly" to form a long-range forecast.

Short-Term Projections

We now turn to projecting past patterns for short-term forecasts. By short-term, we generally mean up to a year, with primary interest in forecasts for individual months or quarters. Projecting a trend for such a limited period is not nearly as hazardous as for long-term forecasting. If the past seasonal pattern has been shown to be stable and there are no reasons for anticipating a sudden change in the seasonal pattern, we would then anticipate seasonal variation corresponding to the past average seasonal change. This pattern is the one captured by the seasonal indexes described earlier in the chapter. Short-term forecasts must also consider cyclical variation. The methods for incorporating the cyclical element will be discussed in the next section.

Consider a series in which trend and seasonal variation is dominant. The packaged beer sales series shown in Figure 17-2 is a good example. The historical series there was given through 1977. We shall form a quarterly forecast for 1978 that incorporates past trend and seasonal variation by projecting the trend for each quarter of the forecast period and then applying the seasonal indexes to these trend levels. We found the following arithmetic trend for the beer sales series from 1971 through 1977.

$$T = 27.27 + 0.2905x$$

Here x is zero in the first quarter, 1971, and x is in quarterly units. Seasonal indexes found for the period 1971 to 1977 were

First quarter	89.6
Second quarter	110.4
Third quarter	110.9
Fourth quarter	89.1

Table 18-4 and Figure 18-4 show the development of the forecast.

Table 18-4 Forecast Incorporating Trend and Seasonal Variation

Quarter		*x*	*Trend*	* *Seasonal*	= *Forecast**
1978	1	28	27.27 + 0.2905(28) = 35.40	0.896	31.7
	2	29	27.27 + 0.2905(29) = 35.69	1.104	39.4
	3	30	27.27 + 0.2905(30) = 35.98	1.109	39.9
	4	31	27.27 + 0.2905(31) = 36.28	0.891	32.3

*Actual 1978 sales for the four quarters were 31.9, 39.6, 40.2, and 32.4.

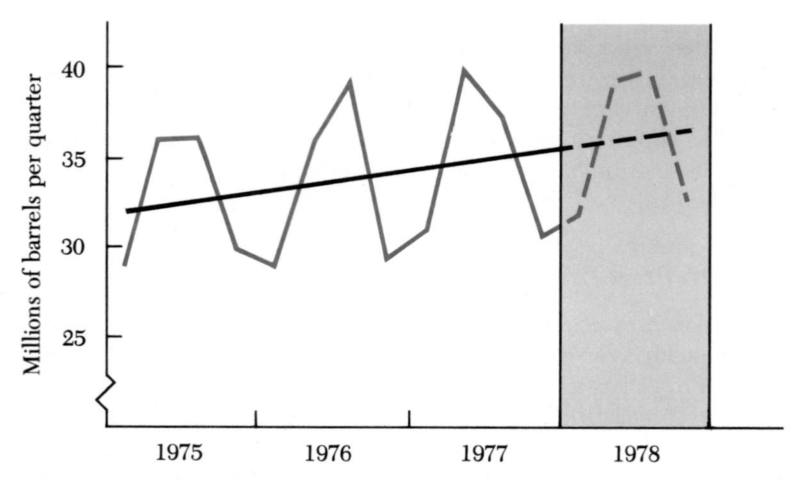

Figure 18-4 Trend-Seasonal Projection of Packaged Beer Consumption for 1978

As the table indicates, the first step is to project the trend for the quarters of the forecast period. The seasonal indexes for the past data tell us that first quarter sales have averaged 89.6% of the trend-cycle base, second quarter sales have averaged 110.4% of the trend-cycle base, and so on. These percentages are applied as decimal multipliers to the trend projections. The effect is to impose the past seasonal pattern on the base levels provided by the trend projection. The process is shown graphically in Figure 18-4, which includes three years of past data to add to the perspective.

The trend and seasonal elements in a time series are the components that lend themselves to straightforward projection of past patterns of variation into the future. While such projections are mechanically easy to make, they should be treated as serious forecasts only when the reasonableness of assuming a continuation of past patterns is supported by the best information and qualitative reasoning that can be brought to bear.

Exercises

18-6 *Projections.* Table 17-2 gives 5-year averages for sales of gas furnaces from 1954 to 1979. The linear trend for the series is $T = 1226.8 + 99.9x$, where x is zero in 1966½ and is in 2½ year units. Sales are in thousands. The linear trend for millions of cable TV subscribers in Exercise 17-3 for annual data from 1973 to 1980 is $T = 11.38 + 0.566x$, where x is zero in 1976½ and is in half-year units.

a. Project the trend of the gas furnace sales series to 1980.

b. Project the trend of the cable TV subscriber series to 1984.

18-7 *Trend-seasonal projections.* Table 18-1 gives quarterly data for unit sales of room air conditioners; seasonal indexes are calculated in Table 18-2. A quarterly trend for the series is $T = 1598.17 - 33.37x$, where x is zero in the final quarter of 1969 and is in quarterly units.

a. Make a trend-seasonal projection of quarterly sales for the year 1977.

b. Does the graph of Figure 18-1 suggest that one can safely ignore cyclical fluctuations in forecasting sales for room air conditioners? Explain.
c. The actual 1977 data are given in Table 18-3. How do the trend-seasonal projections compare with them?

18-8 *Home heating oil consumption.* Consumption of distillate (home heating) fuel oil from 1970 to 1976 in millions of barrels is shown below.

1970	1971	1972	1973	1974	1975	1976
927.2	971.1	1066.1	1128.7	1076.0	1040.6	1125.6

The linear trend for these data is $T = 1047.9 + 26.58x$, where x is zero in 1973 and is in yearly units. Seasonal indexes for quarterly consumption are 128.2, 85.7, 74.9, and 111.2 for first through fourth quarters.
a. Project the trend for consumption of home heating oil for the year 1977.
b. In the statement of the trend, x is zero in mid-1973, or July 1, 1973. The middle of the first quarter of 1977 is 3.625 yearly units removed from the middle of the year 1973. Project the trend of *annual* consumption for each quarter of 1977.
c. Divide your results in (b) by four, and complete a trend-seasonal projection of quarterly consumption for 1977.

18-9 *Projecting gross national product.* The exponential trend for GNP in current dollars from Exercise 17-6 was $\log T = 3.111 + 0.058x$, where x is zero in 1972½ and is in 1½-year units. The data below are three-year averages for GNP in billions of *constant* 1972 dollars.

1965	927	1971	1118	1977	1369
1968	1046	1974	1244	1980	1485

a. Fit a linear arithmetic trend to the constant-dollar GNP series.
b. Project the current-dollar and the constant-dollar trends to 1984.
c. Check a source such as the *Survey of Current Business* to find actual values for 1984. Which of the projections proved to be closer?

18.3 Time-Series Regression

We now need to consider what can be done to forecast the level of a series in which the cyclical component is important. In forecasting over a short to intermediate time period—less than a year to around three years—the cyclical element in many economic and business series can hardly be ignored.

A common approach in forecasting many series is to find a general economic indicator with changes, especially of a cyclical nature, that can be used to forecast changes in the series of interest. Various series in the national income accounts, maintained by the U.S. Department of Commerce and published in the *Survey of Current Business*, components of the index of industrial production from the Federal Reserve Board, and certain employment and earnings series maintained by the Bureau of Labor Statistics can be used in this manner. We hope to find a strong relationship between the series to be forecast and an economic indicator to which the series is logically related.

Various organizations devote considerable effort to forecasting general economic indicators. These forecasts are usually made with the assistance of complex econometric models. If a reliable relationship can be found between a specific series and a general economic indicator, then the regression can be used to forecast the level of the specific series. The value of the independent variable that will be used in the forecast of the specific series is the *forecast* for the indicator series.

To illustrate time-series regression as an aid in forecasting, we have selected unit factory sales of food waste disposers as our specific series to be forecast and durable goods expenditures as the related economic indicator series. Table 18-5 shows annual figures for factory sales of food waste disposers (thousands of units) and durable goods expenditures (billions of constant 1972 dollars). Also shown in Table 18-5 are the year-to-year changes in both series, called *first differences*.

The first differences are simply the year-to-year changes in the original data. For example, $d_y(1966) = y(1966) - y(1965) = 1410 - 1355 = 55$.

In Figure 18-5, the annual data for sales of food waste disposers and expenditures for durable goods are plotted. We see that the behaviors of the two series over the period from 1965 to 1976 exhibit marked, though not complete, similarity. We want to forecast sales of food waste disposers for 1977 using the historical relationship between the two series and a forecast of durable goods expenditures for 1977. We shall use a forecast of expenditures for durable goods in 1977 of 137.8 billion (1972) dollars. This is the actual expenditure for durable goods in 1977, so we have a perfect forecast of the indicator series.

Table 18-5 Annual Data and First Differences: Food Disposer Sales and Expenditures for Durable Goods

Year	*Original Data*		*First Differences*	
	Sales y	*Expenditures* x	*Sales* d_y	*Expenditures* d_x
1965	1355	73.4		
1966	1410	79.0	55	5.6
1967	1356	79.7	−54	0.7
1968	1812	88.2	456	8.5
1969	1943	91.9	131	3.7
1970	1976	88.9	33	−3.0
1971	2292	98.1	316	9.2
1972	2771	111.2	479	13.1
1973	2974	121.8	203	10.6
1974	2553	112.5	−421	−9.3
1975	2080	112.7	−473	0.2
1976	2515	127.5	435	14.8

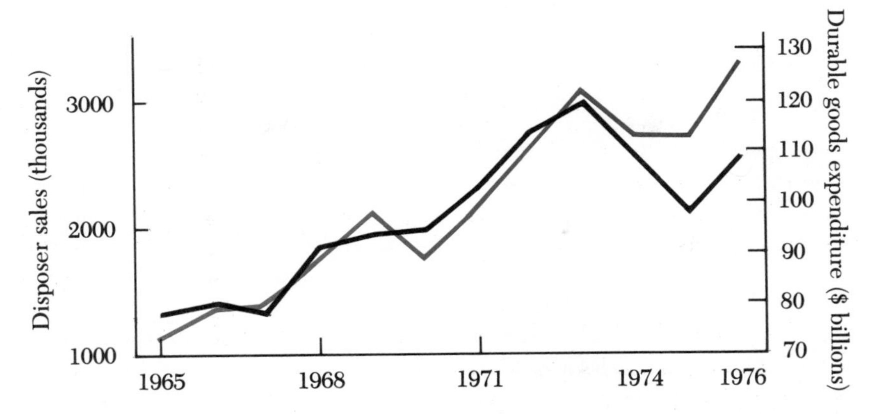

Figure 18-5 Sales of Disposers and Expenditures for Durable Goods

Original-Data and First-Difference Regressions

We wish to consider two forms of the relationship between the two series. The first is the relation between the *levels*, or the original data, of the two series. From the data of Table 18-5, the regression of y (sales of food waste disposers) on x (expenditures for durable goods) is

$$\hat{y} = a + bx$$
$$= -648.80 + 27.70x$$

For this regression,

$$r^2 = 0.816 \quad \text{and} \quad s_{y \cdot x} = 247$$

The regression coefficient (b) tells us that when expenditures for durable goods change by 1 billion (1972) dollars, the expected change in sales of food waste disposers is 27.70 thousand waste disposers. The estimate of 1977 disposer sales from this regression is

$$\hat{y} = -648.80 + 27.70(137.8)$$
$$= 3168$$

The second regression form we shall examine is the *first-difference* regression. The first differences, or year-to-year changes, in sales of food waste disposers and expenditures for durable goods were given in Table 18-5. In the original data regression, we estimated the *level* of sales given the level of expenditures. From the first differences, we shall estimate the *change* in sales given the *change* in expenditures.

The regression for estimating the *change* in sales given the *change* in expenditures for the given data is

$$\hat{d}_y = a + bd_x$$
$$= -82.3 + 38.17d_x$$

Other measures associated with this regression are

$$r^2 = 0.728 \quad \text{and} \quad s_{y \cdot x} = 180$$

The regression tells us that if there is no change in durable goods expenditures, we would anticipate a decrease of 82.3 (thousand) in annual sales of disposers, and for every billion (1972) dollars of change in expenditures for durable goods, we would anticipate a further change in the same direction of 38.2 (thousand) in the number of disposers sold. We shall comment on the coefficient of determination and the standard error of estimate later.

To project 1977 disposer sales, we need a projection of the change in expenditures for durable goods from 1976 to 1977. Expenditures in 1977 were, in fact, 137.8 billion (1972) dollars, a change from the 1976 level of

$$\begin{aligned} d_x(1977) &= x(1977) - x(1976) \\ &= 137.8 - 127.5 \\ &= 10.3 \end{aligned}$$

We would then estimate the change in sales of disposers from 1976 to 1977 to be

$$\begin{aligned} \hat{d}_y &= -82.3 + 38.17(10.3) \\ &= 311 \end{aligned}$$

The projection of 1977 sales is, then,

$$\begin{aligned} \hat{y}(1977) &= y(1976) + \hat{d}_y(1977) \\ &= 2515 + 311 \\ &= 2826 \end{aligned}$$

Actual sales of food waste disposers in 1977 were 2953 thousand. The projection made from the original data regression was 3168 thousand, which is 7.3% higher than actual 1977 sales. The projection of 2826, which utilized the first-difference regression, is 4.3% below actual 1977 sales. The first-difference regression leads to a closer forecast in this instance. We now want to compare the two forms of regression in more general terms.

Table 18-6 presents the actual sales data for food waste disposers from Table 18-5, along with estimated values and residuals from each of the regressions that we have just considered.

The first column contains the original sales data. The second column contains the results of the regression of sales levels with expenditures for durable goods. The original-expenditure series is in Table 18-5. For 1966, the estimated expenditure level ($\hat{y}$) is

$$\begin{aligned} \hat{y} &= -648.80 + 27.70(79.0) \\ &= 1540 \end{aligned}$$

In the third column are the residuals from this regression, $e = y - \hat{y}$. For 1966,

$$\begin{aligned} e &= y - \hat{y} \\ &= 1410 - 1540 \\ &= -130 \end{aligned}$$

Table 18-6 Residuals from Original-Data and First-Difference Regressions

Year	*Original-Data Regression*			*First-Difference Regression*		
	y	$\hat{y}$	e	d_y	$\hat{d}_y$	e
1965	1355	1384	−29			
1966	1410	1540	−130	55	131	−76
1967	1356	1559	−203	−54	−56	2
1968	1812	1794	18	456	242	214
1969	1943	1897	46	131	59	72
1970	1976	1814	162	33	−197	230
1971	2292	2069	223	316	269	47
1972	2771	2431	340	479	418	61
1973	2974	2725	249	203	322	−119
1974	2553	2467	86	−421	−437	16
1975	2080	2473	−393	−473	−75	−398
1976	2515	2883	−368	435	483	−48

The next three columns contain the corresponding data for the first-difference regression. For 1966, the actual change from the preceding year in disposer sales (d_y) is 1410 − 1355 = 55. The change in expenditures for durable goods from 1965 to 1966 can be found in Table 18-5 to be 79.0 − 73.4 = 5.6. Given this change (d_x), the estimated change in sales from the first-difference regression is

$$\begin{aligned}\hat{d}_y &= -82.3 + 38.17(5.6)\\ &= 131\end{aligned}$$

The residual from this regression for 1966 is $d_y - \hat{d}_y = 55 - 131 = -76$. Actual sales for 1966 are 130 below the estimate based on the original-data regression and 76 below the estimate based on the first-difference regression. The estimated sales level from the first-difference regression would be 1355 + 131 = 1486, and the actual sales (1410) are 76 below that.

The appropriate measure to compare the two relationships for the purpose of estimating the *levels* of the dependent variable series (disposer sales) is the standard error of estimate. Using first differences, our errors in estimating disposer sales are the same as our errors in estimating the year-to-year *change* in disposer sales. The standard error of estimate of 180 reflects these differences for the historical series. For the original-data regression, the standard error of estimate is 247, which is substantially larger. On this basis, the difference regression is better than the original-data regression.

Notice that r^2 for the original-data regression is 0.816, while the coefficient of determination for the first-difference regression is 0.728. The coefficients of determination are not comparable. The coefficient for the original-data regression tells us that 81.6% of the variation in *levels* of disposer sales is explained by that regression, while the first-difference coefficient tells us that 72.8% of the variability in year-to-year *changes* in disposer sales is explained by the regression with changes in expenditures for durable goods. The bases of the coefficients are different.

It is important not to be misled by high coefficients of determination that are sometimes produced in time-series regression. In a series with a strong trend, the coefficient of determination with respect to an indicator series that also has a strong trend can be very high when the coefficient of determination for the corresponding first-difference regression is relatively modest. The variation in the *levels* of the two series is highly correlated, but the variation in the period-to-period *changes* in the two series is not as highly correlated. This can happen when the difference regression, as in the present case, produces better estimates for the dependent variable series. The appropriate comparison is between the sizes of the standard errors of estimate, and not between the coefficients of determination.

In Figure 18-6, the original-data and first-difference scatter diagrams, along with the regressions, are shown for the example involving sales of disposers. The scales are the same for the two diagrams. The smaller standard error of estimate for the first-difference regression is reflected in the smaller vertical scatter about the first-difference regression.

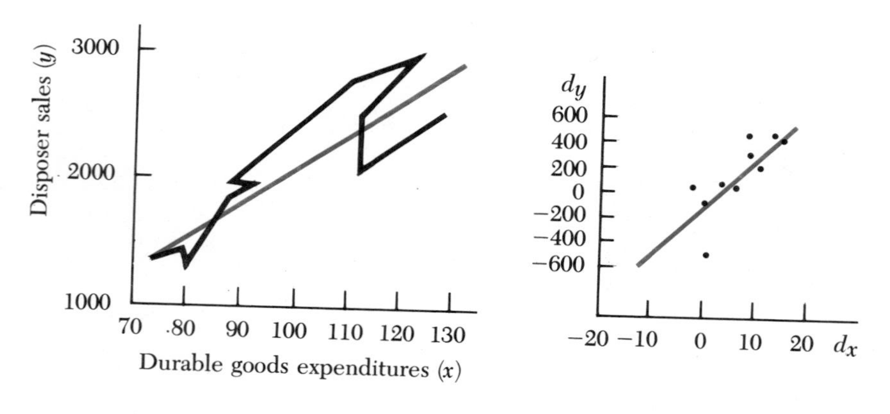

Figure 18-6 Original Series and First-Difference Regressions

Another related factor favors the difference regression. Notice in our example that the independent variable value used to project disposer sales for 1977 from the original data regression lies outside the range of past values. That is, expenditures for durable goods of 137.8 projected for 1977 is higher than any previous level. However, the projected *change* in expenditures of 10.3 from 1976 to 1977 is not outside the range of values used to estimate the first-difference regression. This situation is not an uncommon one in historical time series.

Serial Correlation of Residuals

We saw that the smaller standard error of estimate of the first-difference regression in our example led to a preference for the first-difference regression over the original-data regression. In many series, it happens that the value of the independent variable to be inserted in the regression to produce a forecast is more apt to be within the range of historical values when first differences are used than when the original-series values are used. Another important consideration that favors the use of first-difference regressions is the character of the residuals from the regressions. In Table 18-6, the residuals from the original-data regression and from the first-difference regression for sales of food waste disposers versus expenditures for durable goods were shown. Consider now the original series regression.

A glance at the residuals from the regression of the original series reveals a feature that occurs frequently in time-series regression. The residuals from the regression are not independent of time; they are serially correlated. For the first three years, actual sales are below the regression line. This is followed by seven years when sales are above the regression line. In the last two years, they swing below again. This means that if sales estimated from the regression were below the actual sales in a given year, we would anticipate that our estimate based on the regression would be below actual sales in the following year.

The presence of serially correlated residuals invalidates a major assumption in regression analysis. Recall from Chapter 10 that the validity of the t-test for significance of a regression coefficient and the construction of confidence intervals for a prediction based on the regression depend on residuals from the true regression being normally distributed and *independent* of one another.

First-difference regression offers a solution to the problem of serially correlated residuals in the original-data regression. It can be shown that if we take first differences, we remove what is called a *first-order serial correlation* in the residuals from the original-series regression. At first glance (Table 18-6), it might seem that the residuals from the first-difference regression are also serially correlated. However, r^2 for that correlation is only 0.03, despite the run of six positive residuals in succession. In general, then, we recommend that first-difference regressions be used in time-series analysis in preference to regressions of the original time series.

Incorporating Seasonal Variation

In Section 18.2, we saw how to apply seasonal multipliers to a trend projection to form a forecast for a series in which cyclical variation appeared unimportant. In this section, we have seen how regression of a time series against an economic indicator can assist in forecasting the trend-cycle level of the series. Seasonal variation has not been of concern because the examples have been carried out with annual data. In this section, we shall illustrate one way to incorporate the seasonal element into forecasts of the trend-cycle level of a series. We continue with the example of the food waste disposers.

Our forecast of sales of food waste disposers for 1977 from the regression of first differences of the annual data for disposer sales and expenditures for durable goods was 2826 (thousand). We used quarterly data and the ratio-to-moving-average method to find the following seasonal indexes.

	Quarter			
	1	*2*	*3*	*4*
Index	94.9	99.9	104.5	100.7

Recall that seasonal indexes represent (approximately) the standing of each quarter's sales compared with average quarterly sales for a year. Average quarterly forecast sales for 1977 are 2826/4 = 706.5. The quarterly seasonal pattern can then be imposed by multiplying average quarterly forecast sales by the decimal equivalent of the seasonal indexes. The results are shown below, along with actual quarterly disposer sales for 1977.

Quarter	*Forecast*		*Actual*
1	706.5(0.949) =	670	703
2	706.5(0.999) =	706	711
3	706.5(1.045) =	738	773
4	706.5(1.007) =	711	764
Total		2825	2951

Utilizing Quarterly Data

The time-series regression methods have been illustrated in this section with annual data. Quarterly data could be used as well. The dependent variable would then be the first differences in the seasonally adjusted quarterly series and the independent variable would be the first differences in the seasonally adjusted values for the indicator series. The most commonly used indicator series are published in adjusted (for seasonal) form. The regression would predict the change in the dependent series (adjusted for seasonal) given the forecast of the change in the indicator series (adjusted for seasonal). We would apply the historical seasonal factor to the forecast of the (adjusted for seasonal) level of the dependent series to obtain a forecast of the actual quarterly value in question.

In the above method, seasonally adjusted data are used. An alternative is to fit the difference regression directly to quarterly data, including additional variables to identify the seasons. The result is a multiple regression to predict the dependent variable, given a value for the indicator series and the seasonal period (1st to 2nd quarter, 2nd to 3rd quarter, 3rd to 4th quarter, or 4th to 1st quarter).

Universes Past and Future

We need to consider some final cautions about the use of regression of time series as an aid in forecasting, which can be summarized in the phrase "universes past and future." The data analyzed of necessity relate to the past, and the inference desired is a forecast about the future. Relationships that have held in the past may not persist into the future because the business and economic systems that produced the historical relationships are subject to change. The causal universes in business and economic time series do not stand still. Even if the data meet all the formal assumptions for regression analysis, a calculated standard error of a prediction is subject to the assumption that the causal universe will not change. Thus, although we have emphasized the standard error of estimate as a criterion for comparing alternative regressions, we have not constructed formal confidence intervals for predicted (forecast) values. The major protection against a forecast being upset by a sudden change in historical relationships is close up-to-date knowledge about the area of application.

Other Applications of Time-Series Regression

Applications of time-series regression are not limited to explicit forecasting situations. In this section, we shall give two examples.

The first example is reported by the South Central Bell Telephone Company.* A business activity index for the five-state region served by the company is produced. Personal income would be a good general measure of economic activity, but it is available only annually for the states. So, using annual data, the company searched for other series, available on a monthly basis, that would have annual values closely estimating annual personal income over the years 1956 to 1978. Once such variables were found, their monthly values could be used to estimate monthly personal income in the five-state region (adjusted for seasonal variation).

The regression that was finally developed to produce the monthly index of personal income included two independent variables. The first of these was the product of non-farm employment and hourly earnings in manufacturing in the states, and the second was personal income in the nation. The first variable is a substitute for nonagricultural payrolls in the region, and the second variable picks up the dependence of the region's activity on national economic conditions. The regression equation developed was

$$\hat{y} = -4.55 + 0.5313x_1 + 0.5141x_2$$

where $\hat{y}$ is the index of real personal income in the south central region, x_1 is the index of real nonfarm payrolls, and x_2 is the index of real U.S. personal income. The regression has to be continually reevaluated as new data become available and as the statistical agencies make revisions in the historical series. Figure 18-7 shows the historical behavior of the index, which is used by the company and its customers as a current economic "barometer" for the region.

Our second regression application involves the corrugated boxes, presented in Figure 17-1.† Based in part on a regression analysis to be described, an economist for the company concluded the following.

*Courtesy of George F. Kirsten, South Central Bell Telephone Company.

†The example was provided by Elmer P. Lotshaw of Owens Illinois Corporation.

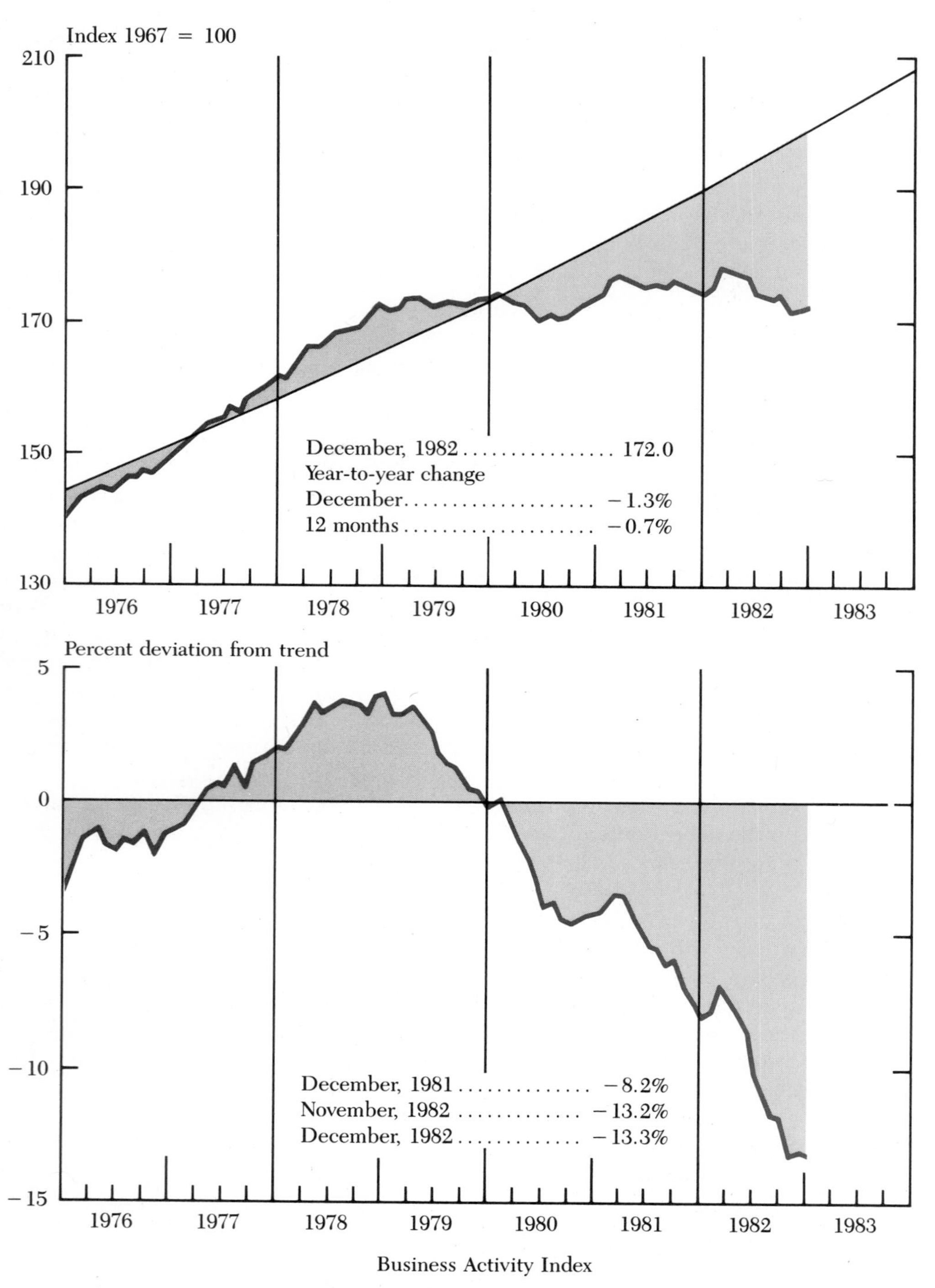

Figure 18-7 South Central Regional Business Index*

*Source: *Business Trends*, South Central Bell Telephone Company, February, 1983

> Since 1973 there has been a marked slowdown in the rate of growth of corrugated box shipments. Only part of this is due to the sluggish and incomplete recovery from the 1974–75 recession. The rest appears to be the result of a downward shift in demand precipitated by the sharp increase in corrugated prices in 1974. Also, despite the fact that corrugated prices have declined since 1974 while most other prices have increased, there is no evidence that corrugated has regained the markets lost since 1974.

The supporting regression analysis used two variables to explain variation in quarterly demand (shipments in billions of square feet) for corrugated boxes. The first independent variable was an index of output for industries using corrugated boxes. The index was made up of components of the Federal Reserve Board's index of industrial production, with each component weighted in accordance with the proportion of total corrugated-box shipments consumed in 1976. The second independent variable was an index of corrugated-box prices relative to the producers' price index of the Bureau of Labor Statistics for industrial products. The first variable captures demand for packaging materials, and the second was designed to measure comparative prices of corrugated packaging materials. The price variable was "lagged" three quarters. That is, the effect of changes in relative prices on corrugated-box shipments was assumed to show up three quarters after the price change occurs.

The regression was fitted to quarterly data from the last quarter of 1965 to the last quarter of 1975. Figure 18-8 shows the time plot of actual and estimated shipments. The coefficient of determination was 0.932, and the standard error of estimate was 1.54 billion square feet per quarter.

Further analysis in the late 1970's convinced the analysts that the apparent downward shift in demand in the mid-1970's was attributable in part to earlier

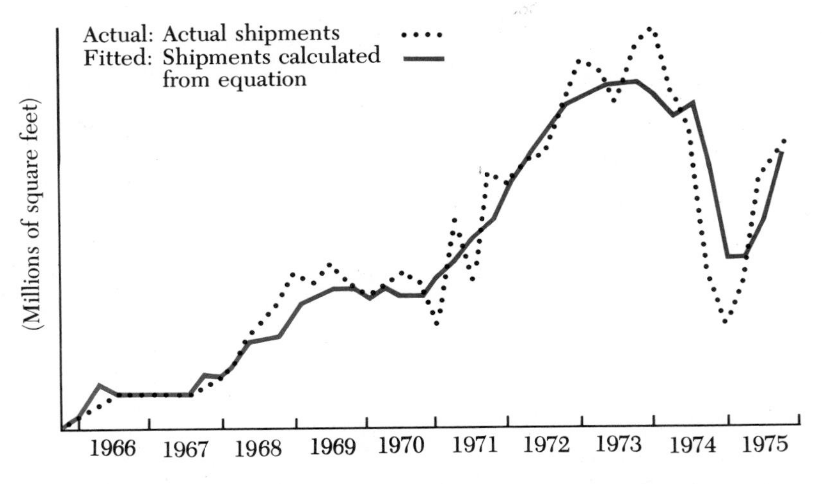

Figure 18-8 Exhibit 3: Actual and Estimated Corrugated-Box Shipments (Millions of Square Feet)*

*Source: Elmer P. Lotshaw, Owens Illinois Corporation.

inventory accumulations by customers. The analysis was reworked to include an indicator variable for the period when inventories were being accumulated and liquidated. The price variable was not lagged and logarithms of the output index were used.

A comparison of the old and new models was made in predicting 1980 shipments, as follows.

	1980 (billions of square feet)
Actual	241.4
Old Equation	283.3
New Equation	246.9

The importance of continuous checking and updating of an analysis is apparent in this example.

Exercises

18-10 *Forecast for room air conditioners.* Factory sales of room air conditioners (thousands) and nonresidential investment (billions of 1972 dollars) from 1965 to 1977 are given below.

Year	*Sales*	*Investment*
1965	2960	95.6
1966	3345	106.1
1967	4129	103.5
1968	4026	108.0
1969	5459	114.3
1970	5886	110.0
1971	5438	108.0
1972	4508	116.8
1973	5346	131.0
1974	4564	130.6
1975	2670	112.7
1976	2962	116.8
1977	3270	129.8

The first difference regression of sales on investment is $\hat{d}_y = -140.77 + 58.46d_x$, and the linear trend for sales is $T = 4423 - 32.31x$, where $x = 0$ in 1964 and is in yearly units.

a. Plot a scatter diagram for the difference relation and draw in the regression line.

b. Given a forecast for nonresidential investment in 1978 of 140.2, prepare forecasts of room air conditioner sales using the difference relation and the trend equation.

18-11 *Forecast for food waste disposer sales.* The following are residential investments (billions of 1972 dollars) from 1965 to 1976.

1965	43.2	1969	43.2	1973	59.7
1966	38.5	1970	40.4	1974	45.0
1967	37.2	1971	52.2	1975	38.8
1968	42.8	1972	62.0	1976	47.7

Sales of food waste disposers for the same period are given in Table 18-5.

a. Determine the first differences for residential investment and plot a scatter diagram of first differences in food waste disposer sales (Table 18-5) versus first differences in residential investment.

b. The regression for estimating change in disposer sales given change in residential investment is $\hat{d}_y = 90.78 + 35.87d_x$, with a standard error of estimate of 167.6. Plot the regression line on your scatter diagram.

c. Use the regression to prepare a forecast of 1977 sales of food waste disposers given a forecast for residential investment of $57.6 billion.

18-12 *Forecast for annual sales of television sets.* Annual U.S. market sales of television sets (millions), disposable personal income, and expenditures for durable goods (billions of 1972 dollars) from 1965 to 1976 are shown below.

Year	*Sales of TVs*	*Disposable Personal Income*	*Expenditures for Durable Goods*
1965	11.0	612.4	73.4
1966	12.4	643.6	79.0
1967	10.9	669.8	79.7
1968	11.8	695.2	88.2
1969	11.3	712.3	91.9
1970	9.5	741.6	88.9
1971	11.2	769.0	98.1
1972	13.5	801.3	111.2
1973	17.4	854.7	121.8
1974	15.3	842.0	112.5
1975	10.6	857.3	112.7
1976	14.1	890.3	127.5

a. A first-difference regression of sales of TVs on disposable personal income for this data is $\hat{d}_y = -2.55 + 0.112d_x$ with $r = 0.68$ and $s_{y \cdot x} = 2.05$. Using expenditures for durable goods, the first-difference equation is $\hat{d}_y = -1.19 + 0.300d_x$ with $r = 0.83$ and $s_{y \cdot x} = 1.56$. Plot a scatter diagram for each of the difference relationships and draw in the regression line on the diagram.

b. Interpret the coefficients in the regression equation for each relationship.

c. Assume a forecast for 1977 of 926.2 for disposable personal income and 137.8 for durable goods expenditures. Estimate the change in sales of TVs in each case and show the estimate on your scatter diagram.
d. Based on each relationship, forecast sales of TVs for 1977.

18-13 *Dishwasher regressions.* Factory sales of dishwashers (thousands), residential investment, and durable goods expenditures (billions of dollars) are given below for selected years.

Year	*Sales of Dishwashers*	*Residential Investment*	*Expenditures for Durable Goods*
1973	3702	59.7	123.3
1974	3320	45.0	121.5
1975	2702	38.8	132.2
1976	3140	51.2	156.8
1977	3356	60.7	178.2
1978	3558	62.4	200.2
1979	3488	59.1	213.4
1980	2738	47.2	214.3
1981	2484	44.9	234.6

a. Find the first-difference regression for sales of dishwashers versus each of the indicator series.
b. Which indicator series appears to have the more promise for use in forecasting sales of dishwashers? Explain.

18.4 Forecasting by Exponential Smoothing

In Chapter 17 we saw that moving averages can be used to smooth out irregular fluctuations in a time series. The example used was a series of monthly orders for a tractor replacement part. Except for an apparent shift upward in the level of orders in the middle of the series, irregular variation seemed to dominate the series. Moving averages of three-month and five-month durations succeeded in smoothing out the irregular variations and giving a better expression of the underlying levels of monthly orders over time.

A somewhat different kind of moving average is often used for very short-term forecasting, that is, forecasting an activity level for the next week, next month, or possibly the next quarter. The new moving average is an exponentially weighted moving average, and is called *exponential smoothing*. Forecasts by exponential smoothing are most useful when irregular variation is substantial. In general, the shorter the time periods considered, the more important irregular variation will be.

To illustrate forecasting by exponential smoothing, we have constructed two time series in Table 18-7. Suppose series A represents 30 weeks of sales figures for an electric motor sold by a manufacturer's representative. The representative wants to forecast each week's sales based on the sales record up to and including the preceding week. The purpose is to forecast, at the close of business on Friday, the coming week's sales so that enough motors can be ordered for delivery by the following Tuesday to meet the anticipated demand for the week.

We shall look at several different ways of making this forecast. In each case, we shall use the series to find how successful a suggested method of forecasting would have been had it been used as the series developed. We shall do this by looking at the forecast errors that would have been made.

Our first forecasting method will be to forecast that next week's sales will be the same as the most recent week's sales. This is called the *naive forecast*. It is a reasonable rule-of-thumb that anyone might try. A more complicated method would be futile if it did not provide a better solution than this simple rule. The rule can be expressed as

$$\hat{y}_t = y_{t-1}$$

which says that the forecast for period t, made at the end of period $t-1$, is the actual sales in period $t-1$.

The second column in Table 18-7 records the errors that would have been made by applying the naive forecasting rule. They are

$$\text{forecast error}_t = \text{forecast}_t - \text{actual}_t$$

or

$$e_t = \hat{y}_t - y_t$$

For the naive forecasting rule, $\hat{y}_t = y_{t-1}$ and $e_t = y_{t-1} - y_t$. For example:

$$\begin{aligned} e_2 &= y_1 - y_2 = 50 - 51 = -1 \\ e_3 &= y_2 - y_3 = 51 - 42 = 9 \\ e_{30} &= y_{29} - y_{30} = 42 - 43 = -1 \end{aligned}$$

A measure of the errors of a forecasting method is the *root mean squared error* (RMSE). This is the square root of the average squared error.

$$\text{RMSE} = \sqrt{\frac{\Sigma(\hat{y}_t - y_t)^2}{n}} \tag{18-2}$$

The sum of squared errors (SSE) is given at the bottom of Table 18-7. Thus the RMSE for naive forecasting of series A would have been

$$\text{RMSE}_{\text{naive}} = \sqrt{\frac{4209}{29}} = 12.0 \text{ motors}$$

Series A was actually constructed by random draws from a normal population with a mean of 50 and a standard deviation of 10. There is, in fact, no systematic variation in the series. Although our forecaster could not know it, a constant forecast of 50 motors each week would have produced a smaller RMSE than the naive forecasts. The RMSE of a constant forecast of 50 motors weekly is actually 8.6. In naive forecasting, the random element is compounded by its presence in both the

Table 18-7 Forecast Errors for Naive and Exponential Smoothing Forecasts

Time	*Series A*	*Naive Error*	*w = 0.5*		*w = 0.1*		*Series B*	*Naive Error*	*w = 0.5*		*w = 0.1*	
			Forecast	*Error*	*Forecast*	*Error*			*Forecast*	*Error*	*Forecast*	*Error*
1	50						50					
2	51	−1	50	−1	50	−1	51	−1	50	−1	50	−1
3	42	9	50	8	50	8	42	9	50	8	50	8
4	52	−10	46	−6	49	−3	52	−10	46	−6	49	−3
5	45	7	49	4	50	5	45	7	49	4	50	5
6	51	−6	47	−4	49	−2	51	−6	47	−4	49	−2
7	50	1	49	−1	49	−1	50	1	49	−1	49	−1
8	47	3	50	3	49	2	47	3	50	3	49	2
9	52	−5	48	−4	49	−3	52	−5	48	−4	49	−3
10	48	4	50	2	49	1	58	−6	50	−8	49	−9
11	56	−8	49	−7	49	−7	66	−8	54	−12	50	−16
12	60	−4	53	−7	50	−10	70	−4	60	−10	52	−18
13	62	−2	56	−6	51	−11	72	−2	65	−7	54	−18
14	59	3	59	0	52	−7	69	3	69	0	55	−14
15	62	−3	59	−3	53	−9	72	−3	69	−3	57	−15
16	43	19	61	18	54	11	53	19	70	17	58	5
17	56	−13	52	−4	53	−3	46	7	62	16	58	12
18	46	10	54	8	53	7	36	10	54	18	57	21
19	43	3	50	7	52	9	33	3	45	12	55	22
20	48	−5	46	−2	51	3	38	−5	39	1	52	14
21	53	5	47	−6	51	−2	43	−5	38	−5	51	8
22	54	−1	50	−4	51	−3	44	−1	41	−3	50	6
23	62	−8	52	−10	51	−11	52	−8	42	−10	50	−2
24	29	33	57	28	53	24	24	28	47	23	50	26
25	67	−38	43	−24	50	−17	62	−38	36	−26	47	−15
26	59	8	55	−4	52	−7	54	8	49	−5	49	−5
27	52	7	57	5	53	1	47	7	51	4	49	2
28	34	18	55	21	52	18	29	18	49	20	49	20
29	42	−8	44	2	51	9	37	−8	39	2	47	10
30	43	−1	43	0	50	7	38	−1	38	0	46	8
SSE		4209		2761		2290		3804		3323		4411
RMSE		12.0		9.7		8.8		11.5		10.7		12.3

forecast and the actual value realized for the forecast period. Thus forecast error, which is the difference between these two values, becomes greater than the inherent variability in the series.

In a forecast by exponential smoothing, the forecast for any period is a weighted average of the forecast for the previous period and the actual value for the previous period.

$$\hat{y}_t = (1 - w)\hat{y}_{t-1} + wy_{t-1} \tag{18-3}$$

where $w \leq 1.0$.

For the first forecast period ($t = 2$), $\hat{y}_{t-1}$ is taken from $y_{t-1} = y_1$. It is as if the forecast for the second period were a naive forecast. Thereafter, the forecasts follow Equation 18-3. For example, using $w = 0.5$,

$$\hat{y}_3 = (1 - 0.5)\hat{y}_2 + 0.5y_2 = 0.5(50) + 0.5(51) = 50.5$$

which is rounded down to 50 motors. Then

$$\hat{y}_4 = (1 - 0.5)\hat{y}_3 + 0.5y_3 = 0.5(50.5) + 0.5(42) = 46.25$$

which is recorded in the forecast column as 46 motors, and so on. The forecast errors are then calculated and recorded, and the RMSE for exponential smoothing forecasts using $w = 0.5$ turns out to be 9.7 motors. This is certainly better than the naive forecast.

The exponential smoothing forecast equation is reasonable because while the value of the most recent period has a bearing on the level of activity to anticipate in the next period, so also do the values prior to the most recent period. These earlier experienced values are embedded in an exponentially declining weighting pattern in the forecast for the period $t - 1$. For example, consider the exponential smoothing forecast for the period $t = 10$. With $w = 0.5$, the actual sales of the most recent period (sales in the period $t = 9$) has a weight of 0.5 in the forecast. All the previous periods have a weight of $1 - 0.5$. Period $t = 8$ has a weight of 0.25, period $t = 7$ has a weight of 0.125, period $t = 6$ has a weight of 0.0625, and so on.

An exponential smoothing forecast rule with a weight of 0.9 on the actual activity of the most recent period would yield results very close to our naive forecasting rule, which in effect puts a weight of 1.0 on the most recent actual value. The weights decline very rapidly as older periods are considered. On the other hand, if a weight of 0.10 is given to activity of the most recent period, the weights decline very slowly as older periods are considered. These patterns are shown in Table 18-8.

Table 18-7 and Figure 18-9 show the results of exponential smoothing forecasts with $w = 0.1$ and $w = 0.5$ for the data of series A. The results with $w = 0.1$ are approaching what would have happened with a fixed forecast of 50 motors each week. Past values weigh more heavily in the current forecast than the value of the single most recent week. The result is that the forecast values change very slowly. They are not sensitive to the random variation of the sales of the most recent week from past average weekly sales.

So far we would conclude that an exponential smoothing forecast with a low value of w is best when random variation predominates in a series. While this is true, there is another problem to consider. This problem is introduced by series B.

Series B is identical to series A with these exceptions: from week 10 through week 16, the values are 10 greater than series A; from week 17 to week 23, they are 10 less than series A; from week 24 through week 30, there are 5 fewer motors than series A. These changes create several shifts in the underlying demand level for the

Table 18-8 Exponential Smoothing Weighting Patterns for Forecast for Period t

w	$t-1$	$t-2$	$t-3$	$t-4$	$t-n$
0.90	0.09	0.009	0.0009	$0.90(0.10)^4$	$0.90(0.10)^n$
0.50	0.25	0.125	0.0625	$0.50(0.50)^4$	$0.50(0.50)^n$
0.10	0.09	0.081	0.0729	$0.10(0.90)^4$	$0.10(0.90)^n$

motors. Instead of random variation about a constant underlying demand level, we now have several changes in the demand level itself. The effect on the forecasts of using different w-coefficients is shown graphically in Figure 18-10. The forecasts with $w = 0.5$ respond fairly readily to the changes in underlying demand level, but the forecasts with $w = 0.1$ respond very slowly. The result is that the RMSE for forecasts with $w = 0.1$ is larger (12.3) than for the forecasts using $w = 0.5$.

From our two examples we can conclude that exponential smoothing forecasts with high w-coefficients will pick up both real changes and false signals (random changes) quicker and more completely than forecasts with low w-coefficients. The choice of what w to use in forecasting a particular series depends on the mix of random and systematic variation in the series. We can experiment with a particular series to find the best value of w in the past and use that value for future forecasts. There is no guarantee that it will continue to be best; we can check that as time goes along.

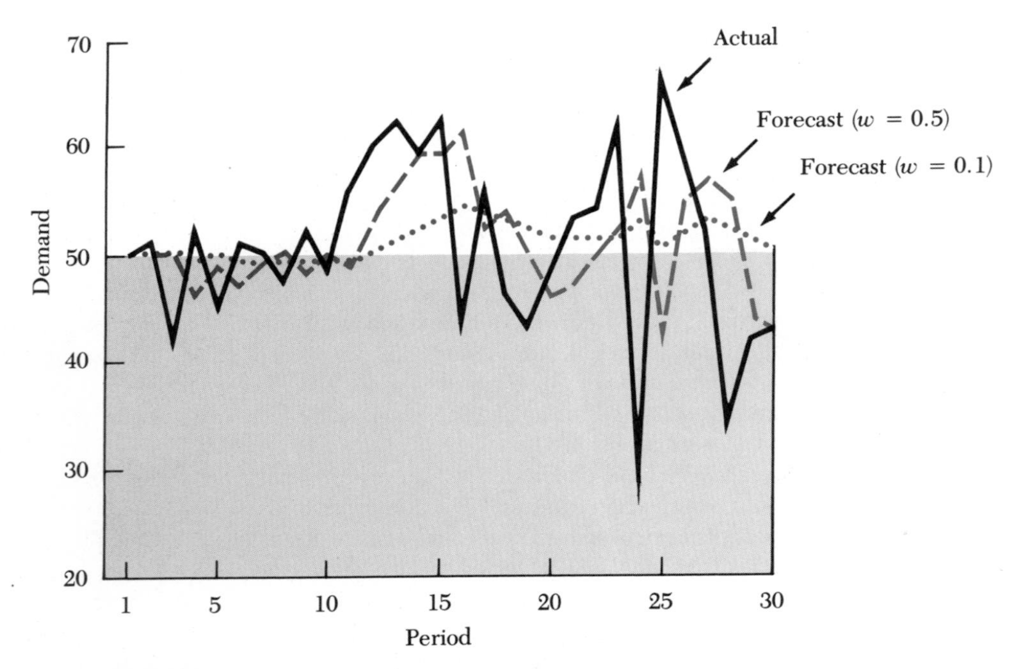

Figure 18-9 Demand Series A with Exponential Smoothing Forecasts

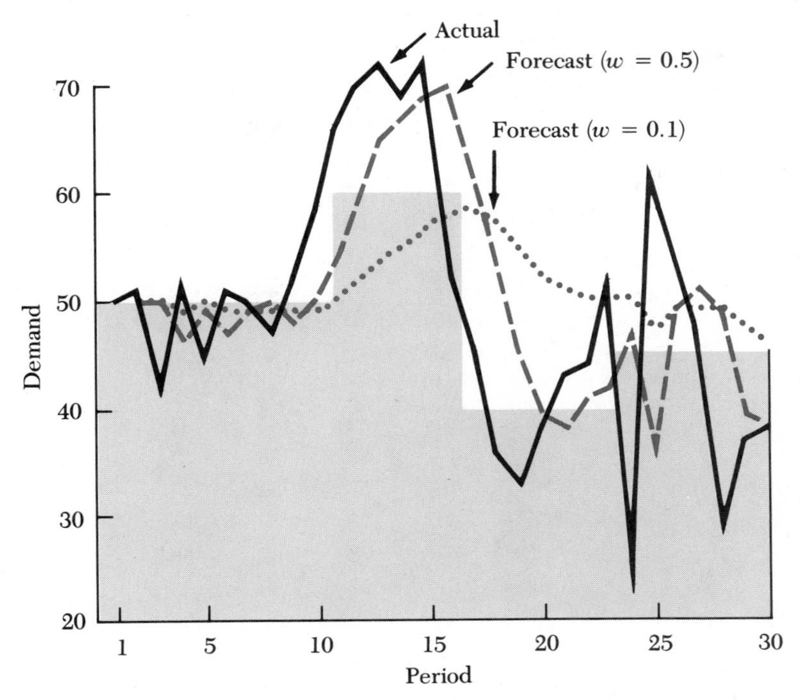

Figure 18-10 Demand Series B with Exponential Smoothing Forecasts

Exponential smoothing forecasts are attractive when many forecasts must be routinely made. In many inventory control and ordering systems, forecasts for hundreds of items are required. Exponential smoothing forecast systems are easily computerized because the current forecast requires only the actual and the forecast value for the previous period. When the next period's actual values are known, the entire data base can be updated.

This section has introduced the basic elements in the use of exponential smoothing as a forecasting tool. Extensions of the basic ideas are required and available to employ exponential smoothing in connection with series with persistent trends or seasonal variation.

Exercises

18-14 *Smoothing and naive forecasts.* Given below are unit sales for the past 20 periods of four different products marketed by a firm.

Period	*Product*			
	A	*B*	*C*	*D*
1	31	46	60	22
2	39	49	81	11
3	37	52	68	19
4	24	49	78	24

Period	*Product*			
	A	B	C	D
5	24	54	55	5
6	25	50	62	32
7	45	65	74	42
8	36	51	64	37
9	36	46	40	43
10	36	51	40	21
11	45	70	41	35
12	20	50	36	30
13	52	77	41	18
14	64	84	45	41
15	47	62	30	37
16	45	55	34	34
17	63	78	39	45
18	53	78	42	58
19	61	91	39	35
20	45	70	46	39

a. For each product as assigned, find the root mean square forecast error that would have resulted if we forecast that sales in the coming period will be the same as sales in the current period.
b. For each product as assigned, find the root mean square forecasting error for exponential smoothing forecasts with $w = 0.5$.

18-15 *Varying the smoothing coefficient.* For the series in Exercise 18-14, as assigned, find whether a smoothing coefficient of $w = 0.25$, 0.50, or 0.75 would have resulted in the smallest past forecast error as measured by root mean squared error.

18-16 *Tractor replacement shaft forecasts.** The following data are the tractor shaft orders from Figure 17-1(c) grouped by three different segments of 16 months each.

First 16 months: 27, 35, 54, 80, 33, 31, 31, 36, 345, 174, 19, 64, 83, 81, 129, 87

Second 16 months: 108, 123, 49, 63, 96, 67, 33, 324, 419, 286, 349, 286, 460, 320, 562, 282

Third 16 months: 82, 365, 224, 385, 41, 532, 169, 214, 411, 249, 557, 161, 370, 154, 314, 932.

a. For each of the segments, find whether an exponential smoothing forecast with $w = 0.2$ or $w = 0.5$ would have resulted in the smaller root mean square forecast error.
b. Explain your results by referring to the behavior of the series.

*Courtesy of William F. Fulkerson, Deere and Company.

18.5 Further Time-Series Methods

A number of time-series methods have been introduced in recent years that extend the alternatives available to the analyst. Some of these are extensions of methods that we have covered earlier, and some involve new elements. They are accessible through computer program packages that often contain the traditional techniques as well. In this section we catalogue some of these further methods.

Extensions of Exponential Smoothing

Our discussion of exponential smoothing did not extend to series with persistent trend-cycle or seasonal components. Extensions called *double* and *triple* exponential smoothing are available to adapt this technique to series with linear trend or curvilinear trend-cycle components. In addition, there are programs for exponential smoothing that include terms that incorporate quarterly seasonal differentials.

Extensions of Time-Series Decomposition

The segregation of variation into trend, cyclical, seasonal, and irregular variation is a traditional approach to time-series analysis and forecasting. The methods have been refined in programs such as Census Method II, developed by the Bureau of the Census. Among the refinements are provision for changing seasonal indexes rather than constant seasonal indexes, the use of a weighted moving average that permits a sharper measurement of the cyclical component, projection of seasonal indexes, and the provision of statistics that describe the cyclical component in some detail.

Box-Jenkins Methods

A unified set of techniques called Box-Jenkins methods attempts to extract the systematic, or nonrandom, elements in a time series through a combination of moving averages and auto-correlation terms. When an appropriate model has been found, it is used for forecasting future values. Commonly the procedures begin with the first differences of a series, which eliminates the first-order auto-correlation, or trend. An important tool in identifying the appropriate model is the auto-correlation pattern. This is a diagram showing the correlations of the series with values 1, 2, 3, . . . , n periods preceding.

Spectral Analysis

Another method for modeling the cyclical elements, including seasonal variation, in a time series is spectral analysis. The series is viewed as a composite outcome of sine waves of varying duration and amplitude, plus random variation. If a model can be found that fits the systematic elements of the past series, it may be useful in forecasting. Again, the auto-correlation pattern is critical to identifying a promising model.

18.6 Company Examples

Statistical analysis and projection of time series do not constitute forecasting, but they are a useful and sometimes indispensable tool for the forecaster. A recent publication by The Conference Board, Inc., on company practices in sales forecasting underscores this viewpoint.* Practices range from judgmental methods and use of estimates by the sales force through traditional time-series analysis (decomposition and projection) and time-series regression, to more elaborate econometric models and economic input-output analysis. The analytical aids presented in this chapter lie in the "mainstream" of company practice as revealed by this publication. In this section we present some examples drawn from this report (unless otherwise noted) with the permission of the authors and The Conference Board, Inc.

Carrier Corporation

Carrier Corporation forecasts industry sales by projecting a historical regression of sales on real disposable income in the United States and a measure of construction activity, such as housing starts. Since housing starts lead changes in industry sales by several months, this variable is entered in the regression with an equivalent lag. Projections of the independent variables are provided by an econometric forecasting service, but company forecasters may develop sales forecasts based on alternative assumptions about these independent variables. As a check on this aggregate forecast, company forecasters also estimate sales generated by individual sectors of the economy.

Radio Corporation of America

To forecast sales of color TV sets, RCA has developed a long-run growth trend for the industry. Deviations of seasonally adjusted quarterly sales from this trend are studied for past relationships with changes in gross national product, the money supply, and past deviations, that is, serial correlation. Annual forecasts are made for four years into the future based on "most likely," "optimistic," and "pessimistic" projections of the key variables. Quarterly forecasts are made two years ahead, based on the most likely set of values.

Joseph E. Seagram and Sons, Inc.

Forecasters at Seagram develop monthly forecasts six months to one year ahead for a large number of brands of distilled spirits by divisions of the company and by major markets. The principal computational tool used is the X-11 version of Census Method II, a computer package developed by the Bureau of the Census for time-

*David L. Hurwood, Elliott S. Grossman, and Earl L. Bailey, *Sales Forecasting* (New York: The Conference Board, Inc., 1978).

series decomposition. The methods employed by this computer program are refinements of the basic decomposition methods that we have described. For each series to be forecast, computer graphs of the past data adjusted for seasonal variation and the moving average (trend-cycle) line are produced. The forecaster then makes a projection of the trend-cycle base either by visual inspection or by projecting a least-squares equation fitted to past values. Projections of seasonal factors provided by the program are then applied to the projected trend-cycle base for the forecast months.

Uniroyal, Inc.

About 100 series representing major products or product lines in Uniroyal's diversified lines of rubber, fiber, plastic, textile, and chemical products are forecast each month for the remaining months of the current year and monthly for the coming year. Forecasts are prepared using three different methods: exponential smoothing, regression with logical indicator series, and least-squares trend projection. Company experience has been that exponential smoothing (with trend and seasonal adjustments) is the most reliable method for periods up to around six months into the future, while the regression method has an advantage for 6 to 18 months out, and the least-squares trends are preferred for longer periods. Accordingly, forecasts that average the results of the three methods are developed, using weights that are different for groups of the forecast months.

Summary

In this chapter, we completed the analysis of time-series components by examining the construction of seasonal indexes and discussed the major methods of forecasting future values of a time series. Seasonal indexes constructed by the *ratio-to-moving-average* methods express the seasonal element as the typical standing of the actual values of each period in comparison to a trend-cycle base. An important use of seasonal indexes is to adjust actual values for the typical seasonal effect.

Long-term projections of future time-series values can be made by extension of past trends. Different trend equations will produce different projections, however, and changing conditions can upset the continuation of a past pattern of growth. In series without much cyclical variation, short-term projections of trend combined with application of past seasonal indexes for quarterly or monthly values can provide useful forecasts.

The *cyclical component* is critical to many short- and intermediate-term forecasts. The past relation of the series to a logically related economic indicator series can often handle this problem. *First-difference regressions* are more useful than regressions of original values for this purpose.

Exponential smoothing is a widely used short-term forecasting method. A forecast by exponential smoothing is a weighted average of the most recent forecast and the actual value for the same period. The best weights depend on the nature of the specific series, and can be estimated by analysis of the historical record for the series.

Methods used in forecasting range from judgmental techniques through conventional time-series methods to more elaborate statistical and econometric methods. Some of the more elaborate methods were briefly catalogued, and some examples of current company practices in forecasting were given.

Supplementary Exercises

18-17 *Projecting new hires.* A time-series analysis was performed on quarterly new hires in manufacturing establishments from 1971 through 1977. Some results of the analysis were as follows.

Quarter		*New Hires per 100 Employees*	*Moving Average*	*Ratio to Moving Average*	*Seasonal Relative*	*New Hires Adjusted for Seasonal*
1971	1	6.1	———	———	0.865	7.1
	2	8.5	———	———	1.128	7.5
	3	9.5	7.850	1.210	1.247	7.6
	4	6.5	8.312	0.782	0.760	8.6
1972	1	7.7	8.900	0.865	0.865	8.9
	2	10.6	9.500	1.116	1.128	9.4
	3	12.1	10.075	1.201	1.247	9.7
	4	8.7	10.675	0.815	0.760	11.5
1973	1	10.1	11.187	0.903	0.865	11.7
	2	13.0	11.475	1.133	1.128	11.5
	3	13.8	11.400	1.211	1.247	11.1
	4	9.3	11.062	0.841	0.760	12.2
1974	1	8.9	10.625	0.838	0.865	10.3
	2	11.5	9.912	1.160	1.128	10.2
	3	11.8	8.812	1.339	1.247	9.5
	4	5.6	7.500	0.747	0.760	7.4
1975	1	3.8	6.437	0.590	0.865	4.4
	2	6.1	6.025	1.012	1.128	5.4
	3	8.7	6.375	1.365	1.247	7.0
	4	5.4	7.125	0.758	0.760	7.1
1976	1	6.8	7.600	0.895	0.865	7.9
	2	9.1	7.737	1.176	1.128	8.1
	3	9.5	8.350	1.138	1.247	7.6
	4	5.7	9.450	0.603	0.760	7.5
1977	1	11.4	10.537	1.082	0.865	13.2
	2	13.3	11.525	1.154	1.128	11.8
	3	14.0	———	———	1.247	11.2
	4	9.1	———	———	0.760	12.0

a. Describe the seasonal pattern of new hires.
b. Plot the seasonally adjusted series and comment on the cyclical behavior of new hires over the period.
c. A quarterly trend for the series is $T = 8.775 + 0.0303x$, where x is zero in the first quarter of 1971. Make a projection for the quarters of 1978 that takes account of trend and seasonal. Criticize this forecast.

18-18 *Enrollment forecasts.** The data below give fall undergraduate enrollment at the University of New Mexico and the unemployment rate in Albuquerque in January of the same year.

Year	*Enrollment*	*Unemployment Rate (%)*
1971	13,746	6.8
1972	13,656	6.4
1973	13,850	6.8
1974	14,145	8.0
1975	14,888	10.1
1976	14,991	9.8
1977	14,836	8.1
1978	14,478	5.6

a. Fit a least-squares trend to the enrollment data and use the trend to project enrollment for the fall of 1979.
b. Calculate the first differences for enrollment and the unemployment rate. Find the regression for predicting change in enrollment from change in unemployment rate. Plot the regression, along with a scatter diagram of the first differences.
c. January 1979 unemployment was 5.3%. Project fall 1979 enrollment using the first-difference regression.

18-19 *Daily variation.* A merchandising firm open five days a week has the following sales (in hundreds of dollars).

	Week 1	*Week 2*	*Week 3*	*Week 4*
Monday	33	36	38	47
Tuesday	29	34	40	44
Wednesday	27	31	38	45
Thursday	27	31	32	41
Friday	25	28	33	36

a. Fit a 5-day moving average to the sales series. Plot the moving average and the original series on the same graph.

*Source: R. Cady, Office of Institutional Research, University of New Mexico.

b. Use ratios of the original data to the moving average to determine a set of "seasonal" indexes for days of the week. Interpret the index for Monday.

18-20 *Exponential smoothing.* From the following order series, calculate exponentially weighted moving averages using $w = 0.2, 0.4, 0.6$, and 0.8. Explain the difference in the responsiveness of the different smoothed series to the apparent change in the level of orders after period 5.

Period	1	2	3	4	5	6	7	8	9	10
Orders	16	14	15	17	13	8	9	11	5	7

18-21 *Room air conditioners update.* The following is the continuation of the series in Table 18-1.

Quarter	*1977*	*1978*	*1979*	*1980*	*1981*
1	900	1072	1216	979	1215
2	1322	1923	1720	1076	1733
3	608	509	337	726	399
4	439	533	468	418	347

a. Find seasonal indexes by quarters using the ratio-to-moving average method.
b. Adjust the original data for seasonal variation.
c. Plot the seasonally adjusted data as a continuation of Figure 18-2 and describe the series from 1971 to 1981.

18-22 *Food waste disposer forecasts.* Data for food waste disposer sales and residential investment (see Exercise 18-11) for 1977 to 1981 follow.

	1977	*1978*	*1979*	*1980*	*1981*
Sales	2941	3312	3317	2962	3178
Investment	60.7	62.4	59.1	47.2	44.9

a. Use the regression in Exercise 18-11 to estimate change in sales over each of the four periods above.
b. Compare estimated with actual changes in sales. Do the results support the validity of the regression?

18-23 *Sea-Tac Originations.** The following are adjusted passenger originations (thousands of persons) at the Seattle-Tacoma International Airport and the regional personal income (millions of 1967 dollars) for the Northwest.

*Data furnished by Susan Doolittle, Chief Economist, Port of Seattle.

Year	*Originations*	*Income*	*Change in* *Originations*	*Income*
1967	1459	6820		
1968	1712	7348	253	528
1969	1896	7428	184	80
1970	1925	7289	29	−139
1971	1890	7283	−35	−6
1972	1963	7528	73	245
1973	2082	7855	119	327
1974	2303	7995	221	140
1975	2359	8286	56	291
1976	2610	8720	251	434
1977	2817	8982	207	262
1978	3372	9585	555	603

a. Find the first-difference regression of originations on income.
b. Give an interpretation of the constant and the slope of the regression.
c. Assume a forecast of $10 billion for income in 1979 and forecast originations.
d. Draw a scatter diagram of the first-difference relation, showing the regression and a range of plus and minus two standard errors of estimate.

19

INTRODUCTION TO DECISION MAKING

In earlier chapters on statistical inference, we learned procedures for forming estimates and testing hypotheses about population means. The use of the sample mean as a point estimator of the population mean was justified on the grounds of the statistical behavior of repeated estimates. It was in this sense that the sample mean was recommended as an unbiased, minimum variance estimator of the population mean.

In testing a hypothesis, we learned how to design a test so that the probability of rejecting the null hypothesis when it is true (type I error) is kept to a desired low level. Where we were concerned about falsely accepting the alternative hypothesis (type II error), we saw how sample size and decision limits could be found so that the power of a test to reject the null hypothesis when it is false met predetermined levels.

We should recognize two features about the above considerations regarding inference. First, the criteria leading to the procedures stem from conditional probability distributions of the sample statistic given specified values of the parameter. Second, consequences of error are considered only in the sense of risk, or relative frequency of error. The seriousness of error, in monetary or other terms, is not explicitly considered.

In contrast to these features of classical statistical inference, it is often useful in business decision making (1) to consider the true value of a parameter as an uncertain quantity that can be described by a probability distribution, and (2) to take the cost consequences of errors of inference about the parameter explicitly into account.

The approach that includes these features is often called *Bayesian decision making* or *Bayesian inference*. In this and the following chapter, we explore the basic ideas in this field.

19.1 Elements of Decision Making

First we shall consider a problem that involves only concepts introduced in Chapter 5. The example will set the stage for introducing the additional elements of decision making that are new to this chapter.

Example

About a year ago, a newsstand operator agreed to carry a weekly magazine called *Recworld* on consignment. The concept of *Recworld* is that each issue is devoted to a particular recreation. For example, an issue on golf features articles on the history of the game, famous players, outstanding courses, golfing resorts and vacations, and so on. Each week the publisher provided the newsstand operator with ten copies of *Recworld*. The copies sold at $5, and the operator paid $2 a copy for the copies that were sold and returned any unsold copies to the publisher when the next week's supply was received. The following numbers of copies were sold by the newsstand per week over a period of 50 weeks.

Number of copies	4	5	6	7
Number of weeks	20	15	10	5

What is the expected weekly return to the newsstand operator?

The problem deals with the expected value of a random variable. Weekly return is an uncertain numerical outcome that depends on the number of copies sold. In the language employed in decision making, the numbers of copies sold will be called *states*. In any selected week, the true *state* of affairs may have been that 4 copies were sold. Or the true state may have been that 5, 6, or 7 copies were sold. If we are looking at a future week, the state that will determine the operator's return for that week is the number of copies sold.

Probability Distribution of States

If the number of copies sold per week has been the result of a stable random process over the 50 weeks, the operator might use the past record as the basis for assigning probabilities to states.

In this case we have the following.

States (number of copies sold)	S	4	5	6	7
Probabilities of States	$P(S)$	0.40	0.30	0.20	0.10

The return, should each state occur, is simply $3 per copy. Bringing together the probabilities of states and the returns *conditional* on each state, we have the following.

States (number of copies sold) S	*Probability* $P(S)$	*Return* $V \mid S$
4	0.40	$12
5	0.30	$15
6	0.20	$18
7	0.10	$21

The expected return is the sum of the products of returns weighted by their probabilities. Since returns depend on the states (number of copies sold), we can say that the expected return, $E(V)$, is

$$E(V) = \Sigma\ [P(S) * V \mid S]$$

It is customary to remove the parentheses in the expectation notation and write

$$EV = \Sigma\ [P(S) * V \mid S] \qquad \textbf{(19-1)}$$

where EV means expected value. The expected value of selling *Recworld* under the consignment arrangement is

$$EV = 0.40(\$12) + 0.30(\$15) + 0.20(\$18) + 0.10(\$21) = \$15 \text{ per week}$$

Alternative Acts

We found that the weekly return to the newsstand operator is a random variable and calculated its expectation. This tells the operator what the long-run average return per week will be under the particular (consignment) arrangement. We now introduce the decision element of alternative acts.

Example

The publisher of *Recworld*, having successfully introduced the magazine, decides that the consignment arrangement is no longer necessary. The newsstand operator is informed that future issues will cost \$2 and no unsold copies may be returned. Instead, they are to be destroyed when the new issues come in. How many copies should the newsstand operator order each week?

Now we have a decision problem because the newsstand operator is faced with a choice among alternative acts. The operator must decide on the number of copies to order. The viable alternative acts are: order 4 copies, order 5 copies, order 6 copies, and order 7 copies. We assume the operator wishes to maximize profit and would not consider ordering 3 copies or less because the return would always be less than the \$12 return from ordering 4 copies. Similarly, the return from ordering 8 copies is never better than the return from ordering 7 copies, as we shall see.

Arranging the states vertically and the alternative acts horizontally, we can develop a *payoff table* for the possible state versus act combinations. For the *Recworld* problem, the payoff table is as follows.

States (*number demanded*)	*Acts: Stock*					
	3	4	5	6	7	8
4	\$9	\$12	\$10	\$ 8	\$ 6	\$4
5	\$9	\$12	\$15	\$13	\$11	\$9
6	\$9	\$12	\$15	\$18	\$16	\$14
7	\$9	\$12	\$15	\$18	\$21	\$19

For example, if the newsstand operator stocks 5 copies and 4 are demanded, the return will be 4(\$5) − 5(\$2) = \$10. If 5 are stocked and 7 are demanded, the return will be 5(\$5) − 5(\$2) = \$15 because there is no stock to meet the demand for the sixth and seventh copies. We have drawn vertical lines to set off the acts that need not be considered further. No matter what the demand, the act *stock 3* is never better than the act *stock 4*, as we mentioned. Also, the act *stock 8*, or any larger number for that matter, is never better than the act *stock 7*. Therefore we limit further consideration to the acts we earlier called the viable alternatives.

Just as we calculated the expected return under the consignment arrangement, we can calculate the expected return, or payoff, under each of the viable acts. Given the probability distribution of the number of copies demanded in a week, each act is a *gamble* made up of the payoffs under that act, should various states occur, and the probabilities of the states leading to the payoffs. For example, the act *stock 7* has payoffs of \$6, \$11, \$16, and \$21 for the states of demand 4, 5, 6, and 7, respectively. Since the probabilities of these states are 0.40, 0.30, 0.20, and 0.10, respectively, the expected value of the act, *stock 7*, is

$$\begin{aligned} EV(\text{stock } 7) &= 0.40(\$6) + 0.30(\$11) + 0.20(\$16) + 0.10(\$21) \\ &= \$11 \end{aligned}$$

Similarly, the expected values of *stock 6* and *stock 5* are

$$EV(\text{stock } 6) = 0.40(\$8) + 0.30(\$13) + 0.30(\$18) = \$12.50$$

$$EV(\text{stock } 5) = 0.40(\$10) + 0.60(\$15) = \$13$$

The alternative of stocking 4 copies has a certain return of $12, since the return will be $12 no matter how many are demanded.

We can now put the elements of Bayesian decision making together. As we have seen, the elements are as follows.

1. A set of possible events, called *states*.
2. Uncertainty concerning the state that will occur.
3. A set of alternative acts that can be adopted in the face of uncertainty about the states.
4. A payoff table giving the value of each state-act combination.
5. A probability distribution of the states.

All these elements are brought together for the *Recworld* weekly stocking problem in Table 19-1. In addition, the expected values of the alternative acts are shown at the bottom of the table.

The act *stock 5* has a greater expected value than any of the other alternatives. In Bayesian terms, it is called the *optimal* act. The optimal act is the act that maximizes expected value. This tells us the best act for the newsstand operator under the new terms established by the publisher of *Recworld*. The operator should stock 5 copies, and the expected value of adopting this best alternative is

$$EV(A_{\text{opt}}) = EV(\text{stock } 5) = \$13$$

Table 19-1 Summary of *Recworld* Stocking Problem

S (*demand*)	*P*(*S*)	*Alternative Acts: Stock*			
		4	5	6	7
4	0.40	$12	$10	$8	$6
5	0.30	$12	$15	$13	$11
6	0.20	$12	$15	$18	$16
7	0.10	$12	$15	$18	$21
Expected values		$12	$13	$12½	$11

The Cost of Uncertainty

We are now in a position to examine an important idea in Bayesian decision making. When the publisher of *Recworld* changed the basis of sales from consignment selling to outright purchase by the buyer, the consequences of uncertainty about the demand at the newsstand were shifted from the publisher to the newsstand operator. Under the consignment arrangement, the newsstand operator had no decision problem about how many copies to stock. The newsstand made the \$3 margin on the number of copies demanded and returned unsold copies at no penalty. The expected return per week under this arrangement is \$15, which we calculated earlier. Now the operator is being asked to assume the risks connected with uncertainty in demand, and the best that can be done is to stock 5 copies each week with an expected weekly return of \$13. The difference of \$2 is the expected cost of assuming these risks since nothing else has been changed.

Let us examine this idea from another vantage point. Look at Table 19-1 and ask how well the newsstand operator could do if a perfect prediction of the number of copies that would be demanded in any week were available prior to that week. If it were predicted that 4 copies would be demanded, that number would be ordered and sold; if it were predicted that 5 copies would be demanded, the best act would be to order 5 and the week's return would be \$15. In short, perfect advance knowledge about the state of demand each week would allow the operator to adjust the number of copies ordered so that the number sold was equal to the number demanded *and the maximum return for that state was achieved*. But the state is still a random variable. Demand will change from week to week in accordance with the probability distribution of demand, but the operator will know what demand is forthcoming each week. The operator's expected return under this condition of perfect advance knowledge is

$$0.40(\$12) + 0.30(\$15) + 0.20(\$18) + 0.10(\$21) = \$15$$

This is the same as the calculation for the consignment arrangement. What we have calculated is the *expected value with certain prediction*. Here certain prediction means certain foreknowledge of the value of the random variable (demand in our example).

In general terms, the expected value with certain prediction, $EV(CP)$, is obtained by finding the maximum payoff given each state, $V_{max} \mid S$, and weighting these maximum payoffs by the probabilities of states.

$$EV(CP) = \Sigma\,[V_{max} \mid S * P(S)] \tag{19-2}$$

The difference between the expected value with certain prediction and the expected value of the optimal act under uncertainty is called the *expected value of perfect information*, $EV(PI)$.

$$EV(PI) = EV(CP) - EV(A_{opt}) \tag{19-3}$$

In the *Recworld* example, $EV(CP) = \$15$ and $EV(A_{opt}) = \$13$, and

$$EV(PI) = \$15 - \$13$$
$$= \$2$$

As we have emphasized, \$2 is the additional amount the operator could expect to gain (per week) if perfect advance knowledge were available about the state of demand each week. It is also the cost of assuming the risk of acting without any predictive information beyond the original probability distribution of states.

The operator should be willing to pay any sum up to the expected value of perfect information to obtain perfect information about each week's demand in advance. If the cost were anything less than $2 for a perfect forecast each week, the newsstand could increase its expected weekly return by buying it.

The expected value with certain prediction, $EV(CP)$, under the purchase arrangement is the same as the expected return under the consignment arrangement because there was no risk to the newsstand operator under the consignment arrangement. The operator could take advantage of whatever demand level occurred during a week by selling that number of copies and returning any unsold copies with no penalty. Perfect advance forecasting under the purchase arrangement would also allow the operator to avoid risk.

Exercises

19-1 *Competitive strategies.* A soft-drink bottler packages cola in standard bottles at present. The bottler is considering several packaging strategies for the coming year. These are A_1, to adopt a lift-top can; A_2, to adopt a screw-top bottle; and A_3, to keep the standard bottle. Profits from each alternative depend on what the bottler's major competitor does. The payoff table (in $10,000 units) and the probabilities assigned for the competitor's actions are as follows.

	Competitor's Action (S)	*P(S)*	*Bottler's Strategy*		
			A_1	A_2	A_3
S_1	Standard bottle	0.5	14	13	16
S_2	Screw-top bottle	0.3	9	8	6
S_3	Lift-top can	0.2	6	8	5

a. Determine the expected payoff from each strategy.
b. Find the expected value with certain prediction.
c. What is the expected value of perfect advance information about the competitor's actions, $EV(PI)$, to the bottler?

19-2 *Selling insurance.* An insurance agent handles two different insurance plans. Customers usually prefer one of the two types, but if the nonpreferred plan is presented first in the sales presentation, the amount of coverage purchased and the agent's commissions tend to be reduced. The agent has prepared the following commission payoff table.

Plan Preferred	*Probability*	*Present First*	
		Plan 1	*Plan 2*
Plan 1	0.6	$40	$20
Plan 2	0.4	$30	$40

a. Determine the expected value of each plan of presentation.

b. What would be the agent's expected commission if perfect information were available about a customer's preference before making the sales presentation?

19-3 *Optimal stock level.* Skipper's Marina stocks a limited number of inboard and outboard motor boats. Skipper's makes a net profit of $1000 on the 16-foot "Sea Breeze" model if a boat is sold the same month it is put in stock. If the boat is held over into the second month, Skipper's incurs insurance and storage charges of $100. The probability distribution of monthly demand for the Sea Breeze is 0.80, 0.15, and 0.05 for 0, 1, and 2 demands, respectively.

a. Construct the payoff table for alternative acts to stock 0, stock 1, and stock 2 boats monthly.

b. What is the expected monthly return for the optimal monthly stock level?

19-4 *Sorting cherries.* The East Slope Cherry Orchard sells a large part of its annual harvest at its own roadside stand. If the cherries are not sorted and graded, they can be sold for $0.60 a pound. If the cherries are sorted as grade A and grade B, the grade A cherries can be sold for $1.20 a pound and the grade B can be sold for $0.40 a pound. The cost of sorting and grading is $0.05 a pound. Find the expected revenue per pound, after sorting costs, for the acts of grading and not grading, as follows.

a. If the probabilities of grade A and grade B cherries are 0.60 and 0.40, respectively.

b. If the probabilities of grade A and grade B cherries are 0.20 and 0.80, respectively.

19-5 *Checking experimental components.* Prior experience with shipments of four experimental components led a research and development department to establish the following probability distribution for number defective in a shipment of four components.

Number defective	0	1	2	3	4
Probability	0.6	0.3	0.1	———	———

A shipment of four components is received. If the department uses the components without any further checking, it will incur a cost of $500 for each defective component used. On the other hand, all the components can be tested at a cost of $50 each. Any defectives found will be put in working order at a cost of $200 each.

a. Find the expected cost of the alternative acts.

b. Find the expected value of perfect information.

19-6 *Bonus and penalty contract.* In negotiations with a city to outfit a fleet of specialized buses, a firm was offered a flat payment of $500,000, or $500,000 and a bonus and penalty schedule depending on the time of delivery, as follows.

Early	$50,000	Late	–$50,000
On time	$25,000	Very late	–$100,000

Company executives were uncertain about the firm's ability to meet the delivery date. They assessed probabilities of 0.2 and 0.4 for "early" and "on time," and 0.3 and 0.1 for "late" and "very late."

a. Which method of payment is optimal given the executives' belief about their ability to meet the delivery date?

b. What is the expected value of perfect information about their ability to meet the delivery date?

19.2 The Utility of Money Gains and Losses

In the previous section, we used expected money value as the criterion for selecting acts. In the *Recworld* example, we did not question whether money value was the appropriate common denominator for evaluating the alternative acts. Where the amounts of money are relatively small, as they were in that case, few would question the use of expected money gain as the basis for decision making. Consider, however, a different example.

Example

Mercury Products, Inc., is a small firm run by two brothers, one recently graduated in engineering and one in business administration. The firm develops and markets active solar energy systems for home heating uses. The firm has recently developed a product that gives it a unique opportunity to exploit a new regional market. The key to the product is a device that may be patentable. A national firm holds a patent on a somewhat similar device, however, and it is not clear whether Mercury's patent application would be granted. If Mercury proceeds and their device proves to be not patentable, the brothers will be liable to a suit for infringement of the national firm's patent. An alternative is for Mercury to purchase rights from the national firm for the use of their device in the new product. The brothers have agreed that the following payoff table properly represents the situation they face.

	State	*Action*	
		A_1 *Apply for Patent*	A_2 *Purchase Rights*
S_1	Device patentable	\$100,000	\$50,000
S_2	Device not patentable	–\$60,000	\$50,000

If Mercury applies for the patent and the device proves to be patentable, a \$100,000 profit can be made in the market venture. Damages from a patent infringement suit, should the device prove not to be patentable, would wipe out the profit and establish a final loss of \$60,000. If Mercury purchases the rights from the national firm, profit from the venture will be \$50,000, regardless of whether Mercury could have secured a patent on their device.

The range of payoffs in this example is such that the brothers might have good reason for not accepting expected *money* value as the criterion for making a decision. For example, a 0.50 probability of gaining \$100,000 and a 0.50 probability of losing \$60,000 has an expected money value of \$20,000, but that might be an unacceptable gamble. The possibility of gain might not be high enough to offset the risk of losing \$60,000, which could put the firm out of business.

When considerations like this are present, money gains and losses are not a satisfactory measure of the desirability, liking, or preference for the gains and losses.

To illustrate this idea, called the *utility curve for money*, the relation between the desirability and amounts of monetary gains and losses is shown in Figure 19-1 for a prospective student at a state university. The student has about $700 saved for expenses and needs approximately $400 more for tuition and books.

The units of utility are arbitrary, like the units of a temperature scale. It is required that equal differences on the utility scale represent equal differences in the desirability of their corresponding money amounts. For example, the utility of zero dollars to the student is zero (an arbitrary origin), and the utility of $300 is 2.0 (utility units). The utility of $500 is 4.0. This means that the $200 difference between $300 and $500 is as important to the student as the first $300 (the difference between zero and $300). An additional $300 will not be adequate, but an additional $500 will meet the student's need. After this, the additional utility for further increases in money amounts tapers off. Additional money would certainly be welcome in providing some flexibility in the budget, but such money does not have as much *marginal utility* as the incremental dollars between $300 and $500.

On the side of money losses, the curve shows the student's feelings that some modest losses up to around $200 are not critical, but beyond that, the *dis*utility of further losses increases rapidly. Losing a large part of the $700 savings could upset the student's plans to the point of dropping out for the semester.

Let us examine the implications of the student's utility curve when it comes to choices between gambles involving various payoff and probability combinations. First, suppose the student is offered a choice between (1) a gamble with a 50%

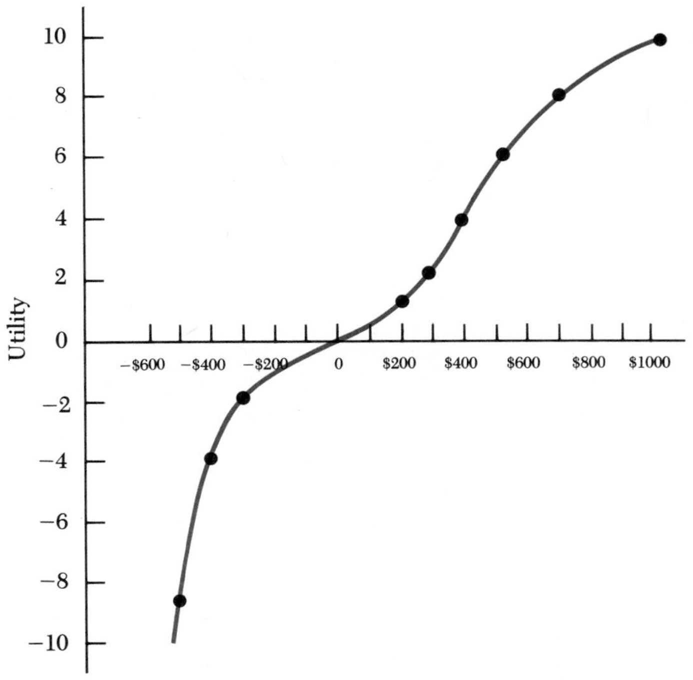

Figure 19-1 Utility Curve for a Student

chance of winning zero and a 50% chance of winning \$400, and (2) a certain payoff of \$200. From the viewpoint of expected money return, the alternatives are equivalent because the expected money value of the gamble is \$200. But, given the utility relation in Figure 19-1, the student would prefer the gamble because the expected utility of the gamble is

$$\begin{aligned} 0.50 * U(\$0) + 0.50 * U(\$400) &= 0.50(0) + 0.50(4) \\ &= 2.0 \text{ (utility units)} \end{aligned}$$

while the utility of the certain payment of \$200 is only 1.0 unit. This makes sense in view of what we said about the student's situation. Consider another gamble versus a certain money alternative: Would the student prefer a gamble with an equal chance of winning either \$400 or \$1000 to a certain gain of \$700? Again, the decision based on expected money value would be indifference. But for the student, the expected utility of the gamble is

$$\begin{aligned} 0.50 * U(\$400) + 0.50 * U(\$1000) &= 0.50(4.0) + 0.50(10) \\ &= 7.0 \text{ (utility units)} \end{aligned}$$

while the utility of a sure gain of \$700 is 8.0 units. Here, the student should take the sure amount rather than the gamble.

What the two examples illustrate is that when the *marginal* utility to the decision maker is increasing between two uncertain payoffs, the decision maker will prefer gambling on the payoffs to a sure return equal to the expected money value of the gamble. The decision maker is said to *prefer risk*. When the uncertain payoffs lie on a curve of decreasing marginal utility, the decision maker will not take the gamble at "fair" money value. The decision maker is said to be *risk averse*.

We now use the utility curve to find out what sure payments would make our student indifferent to the gambles mentioned. In the first gamble, the expected utility of the gamble is 2.0 units. This utility corresponds to \$300, so we know the student would have to be offered \$300 for sure before the gamble would be abandoned. In the second case, the expected utility is 7.0 units, which has a money equivalent of \$600. Thus the student would decline the gamble unless the sure alternative were less than \$600.

We can also use the utility curve to find what probability of the larger payoff in a two-sided gamble would make the decision maker indifferent between the gamble and a specified sure value. Take the first gamble, posed earlier. The uncertain payoffs were \$400 and zero and the sure amount was \$200. The utilities from the curve are $U(\$0) = 0$, $U(\$400) = 4.0$, and $U(\$200) = 1.0$. What probability of \$400 would make the student indifferent between the gamble and the sure payment of \$200? The probability is

$$\begin{aligned} P * U(\$400) + (1 - P) * U(\$0) &= U(\$200) \\ P(4) \quad + \quad (1 - P)(0) &= 1 \\ P &= 0.25 \end{aligned}$$

For any probability for the gain of \$400 exceeding 0.25, the student would prefer the gamble to receiving \$200 for sure. For the second gamble, which was on the decreasing marginal utility portion of the curve, a probability of just over

$$
\begin{aligned}
P * U(\$1000) + (1 - P) * U(\$400) &= U(\$700) \\
P(10) \quad + \quad (1 - P)(4) &= 8 \\
P &= 0.67
\end{aligned}
$$

for receiving the larger amount would be required before the gamble would be preferred to receiving $700 for sure.

In the foregoing examples, we have seen that if we have a utility curve we can answer two kinds of questions:

1. What sure money amount will make the decision maker indifferent between the money amount and a twofold gamble with specified money amounts and probabilities?
2. What probability of the larger gain in a twofold gamble with specified amounts will make the decision maker indifferent between the gamble and a specified sure money gain?

As we shall see, when we do not have the utility curve for a decision maker but can get answers to either of these questions, we can find the utility curve for the decision maker. We are now ready to return to the Mercury Products example.

In the Mercury Products example, there are three payoffs: –$60,000, $50,000, and $100,000. We can get answers leading to utility numbers for these three payoffs by asking questions in which only one of the payoffs is present in any one question. First, establish two arbitrary endpoints. We will let the utility of $0 be zero and the utility of $100,000 be 1000 units. Suppose we then ask what certain money amount will make the brothers indifferent to a 50-50 gamble for $0 and $100,000. Suppose the answer is $30,000. We then know

$$
\begin{aligned}
U(\$30{,}000) &= 0.50 * U(\$0) + 0.50 * U(\$100{,}000) \\
&= \quad 0.50(0) \quad + \quad 0.50(1000) \\
&= 500 \text{ (utility units)}
\end{aligned}
$$

Knowing this, we could ask what probability of $50,000 and complementary probability of $0 would make the brothers indifferent between the gamble and a sure $30,000. Suppose they answer 5/8, or 0.625. We then know

$$
\begin{aligned}
5/8 * U(\$50{,}000) + 3/8 * U(\$0) &= U(\$30{,}000) \\
5/8 * U(\$50{,}000) &= 500 \\
U(\$50{,}000) &= 800
\end{aligned}
$$

Now we need the utility of –$60,000. It would be helpful to have the utility of $60,000 so we could ask the brothers what probability of $60,000 and complementary probability of –$60,000 would make them indifferent between the gamble and a sure $0, because that is an easy gamble to contemplate. To get the utility of $60,000, we could ask what probability of $60,000 would be required in a gamble with $60,000 and $0 to make the brothers indifferent between the gamble and a sure $30,000. Suppose the probability is 5/9. Then we know

$$5/9 * U(\$60{,}000) + 4/9 * U(\$0) = U(\$30{,}000)$$
$$U(\$60{,}000) = 500(9/5)$$
$$= 900$$

Now we ask for the probability of $60,000 required in a gamble for $60,000 and −$60,000 that would make the brothers indifferent between the gamble and $0 for sure; that is, at any greater probability of $60,000, the gamble would be preferred. Suppose the answer is 0.80. Then

$$0.80 * U(\$60{,}000) + 0.20 * U(-\$60{,}000) = 0$$
$$U(-\$60{,}000) = \frac{-(0.80)(900)}{0.20}$$
$$= -3600$$

Now we have utilities for all the dollar payoffs. There are other sequences of questions that might do just as well, and further questions could be asked to check on the consistency of the answers and help in removing any inconsistencies. Let us assume that has been done and the above answers hold. We have

$$U(-\$60{,}000) = -3600 \qquad U(\$50{,}000) = 800$$
$$U(\$0) = 0 \qquad U(\$60{,}000) = 900$$
$$U(\$30{,}000) = 500 \qquad U(\$100{,}000) = 1000$$

The utility curve for the brothers is plotted in Figure 19-2. We see that the brothers' marginal utility for money is diminishing throughout the range of money payoffs in their decision problem. They are risk averse.

Once utilities have been established, the decision problem is solved in terms of expected utility. Suppose the brothers have established a subjective probability of 0.75 that their device is patentable. The decision problem is summarized in Table 19-2.

Table 19-2 Probabilities and Utility Payoffs for Mercury Products Problem

	State	*P(S)*	*Utility Payoffs If*	
			A_1 *Apply for Patent*	A_2 *Purchase Rights*
S_1	Device patentable	0.75	1000	800
S_2	Device not patentable	0.25	−3600	800

The expected utility of act A_1, to apply for the patent, is

$$EU(A_1) = 0.75(1000) + 0.25(-3600) = -150$$

and the expected utility of act A_2, to purchase rights, is

$$EU(A_2) = 0.75(800) + 0.25(800) = 800$$

Thus the optimal act is to purchase rights from the national firm and

$$EU(A_{opt}) = 800$$

The expected utility with certain prediction is

$$\begin{aligned} EU(CP) &= \Sigma\,[U_{max} \mid S * P(S)] \\ &= 0.75(1000) + 0.25(800) = 950 \end{aligned}$$

The amount the brothers would be willing to pay for perfect information about the patentability of their device is the difference in the amount of money that has a utility of 950 to them and the amount of money that has a utility of 800 to them. From the graph of their utility function, this is about \$30,000. This means that, while the optimal act if a decision must be made now is for the brothers to purchase the rights, they should be willing to pay up to \$30,000 for the opportunity to obtain perfect information about patentability prior to making a decision.

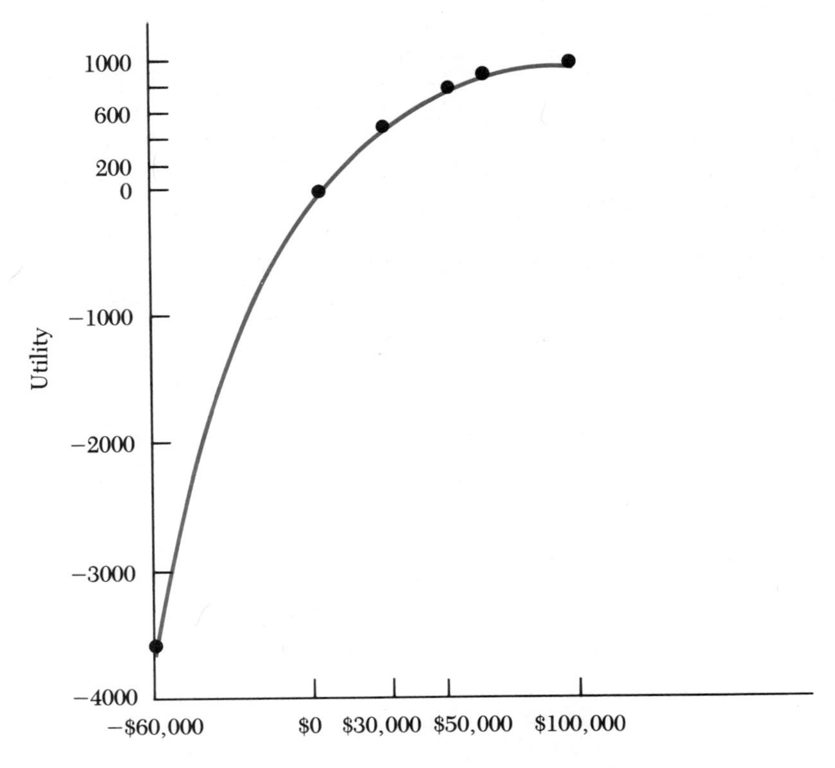

Figure 19-2 Utility Curve for Mercury Products Problem

These results are very different from the results that would prevail if the decision maker's utility were proportional to money return. In the latter case, money return itself could be used as the measure of value. The expected money value solution in the Mercury Products example would be as follows.

State S	$P(S)$	*Action*	
		A_1	A_2
S_1	0.75	\$100,000	\$50,000
S_2	0.25	–\$60,000	\$50,000

$$EV(A_1) = 0.75(\$100{,}000) + 0.25(-\$60{,}000) = \$60{,}000$$

$$EV(A_2) = 0.75(\$50{,}000) + 0.25(\$50{,}000) = \$50{,}000$$

The optimal act is A_1, to apply for the patent while going ahead with the market development. The expected value with certain prediction is

$$EV(CP) = 0.75(\$100{,}000) + 0.25(\$50{,}000) = \$87{,}500$$

and the expected value of perfect information is

$$\begin{aligned} EV(PI) &= EV(CP) - EV(A_{\text{opt}}) \\ &= \$87{,}500 - \$60{,}000 = \$27{,}500 \end{aligned}$$

For a firm in different circumstances from Mercury Products, the same market opportunity thus leads to the opposite action, A_1 instead of A_2. A large firm with greater financial resources might well have a utility function that can be measured by money gains and losses. Maximizing expected money value would be appropriate because the firm would be well prepared through diversification of projects to take gambles that had positive expected money values even though there was some probability of substantial loss.

Exercises

19-7 *Utility functions.* Consider individuals A and B with the following utilities for selected money gains and losses.

	–\$1000	–\$500	–\$100	\$0	\$100	\$500	\$1000
A	–100	–80	–46	0	46	80	100
B	–100	–12	–1	0	1	12	100

a. Show whether A and whether B would prefer a certain gain of \$500 to a 50-50 chance of \$0 or \$1000.

b. Show whether A and whether B would prefer a certain loss of \$100 to a 10-90 chance of losing \$1000 or nothing.

c. Show whether A and whether B would prefer a certain gain of $100 to a 55-45 chance on a $1000 gain or a $1000 loss.
d. Plot the utilities versus dollar gain and loss for A and B, and sketch a smooth curve through each set of points.

19-8 *Attitudes toward risk.* Sketch the general shapes of the utility functions implied by the following attitudes.
a. A person who says, "I just do not take chances. If I can buy insurance against a loss, I do it; I almost never gamble for gains either, because the gamble never seems to be worth taking."
b. A person who says, "I believe in taking a chance on a large loss as well as a large gain. I would rather spend $100 on state lottery tickets than on disability insurance."

19-9 *The utility of grades.* Suppose you are about to take a course that you consider reasonably well suited to your abilities and interests. Final course grades will be A, B, C, D, or E. Let the utility of a C be zero and the utility of an A be 100.
a. What complementary probabilities of A and C would make you indifferent to receiving a B for sure? Find your utility for a B.
b. What complementary probabilities of B and D would make you indifferent to receiving a C for sure? Find your utility for a D.
c. Plot your utilities against equally spaced letter grades. Do you think the curve reflects your feelings about the desirability of different grades? Explain.

19-10 *Determining utility.* An entrepreneur is considering a business proposition that could lose $20,000 or gain $40,000. The entrepreneur would not consider the proposition at all unless the probability of gain exceeded 0.50. If there were another venture that yielded a sure $20,000, the probability of success in the risky venture would have to be 0.8 before the entrepreneur would prefer the gamble.
a. Let $U(\$0) = 0$ and $U(\$40{,}000) = 1000$ units, and determine the entrepreneur's utility for +$20,000 and −$20,000.
b. Plot the entrepreneur's utility curve and describe the kind of behavior represented.

19.3 Decision Trees

In Chapter 4 we used probability trees in organizing and solving probability problems. In a similar way, decision trees can be used to solve decision problems, especially when the problem involves a sequence of decisions. Consider the following decision problem.

Example

A construction firm has an opportunity to bid on a shopping center project, which is being divided into two phases. Bids are due on the land-preparation phase this month and on the construction phase in three months. Firms may submit bids on either phase. The firm figures that the cost of preparing a bid on the land-preparation phase is $50,000 and the cost of preparing a bid on the construction phase is $100,000. The firm can make $100,000 on the land-preparation project and

$500,000 on the construction project, exclusive of the bidding costs. If the company does not take on the construction project, they will accept a subcontracting job on a state highway project that will provide a profit of $200,000.

Company officials assess the probability of a successful bid on the land-preparation phase as 0.5. The probability of success on the construction phase is assessed at 0.5 if the firm has been awarded the land-preparation project and 0.3 otherwise. What action should the construction firm take regarding this opportunity?

Figure 19-3 presents the decision tree for this problem. The decision tree follows the time sequence of action alternatives and event-states. The firm must decide now whether to submit a bid on the land-preparation phase. The outcome of this bid, if submitted, will be known before the firm has to decide whether to bid on the construction phase. In any event, the firm must then make a decision about the construction bid, and sometime later will learn whether the bid, if submitted, was successful.

There are two kinds of branches on the decision tree. Branches stemming from the squares are decision branches. At each square, the decision maker has a choice of which branch (act) to take. The second kind of branch stems from a circle, and represents occurrences over which the decision maker has no control. Each of these sets of branches represent uncertain events, or states. In Figure 19-3 we have entered the given probabilities for these uncertain events.

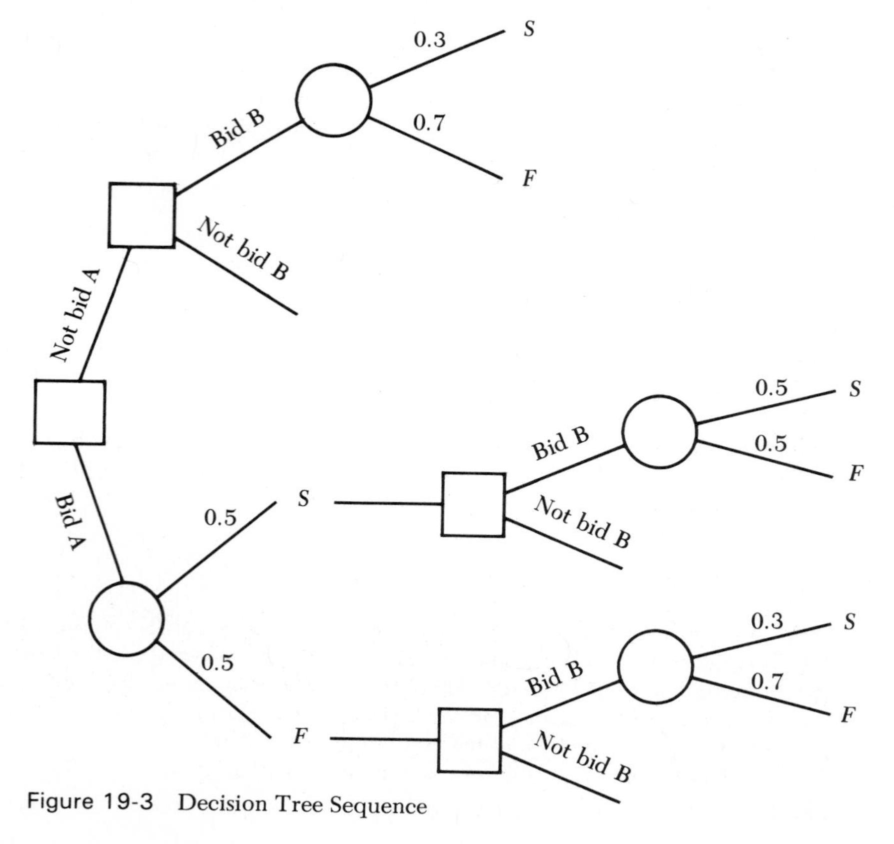

Figure 19-3 Decision Tree Sequence

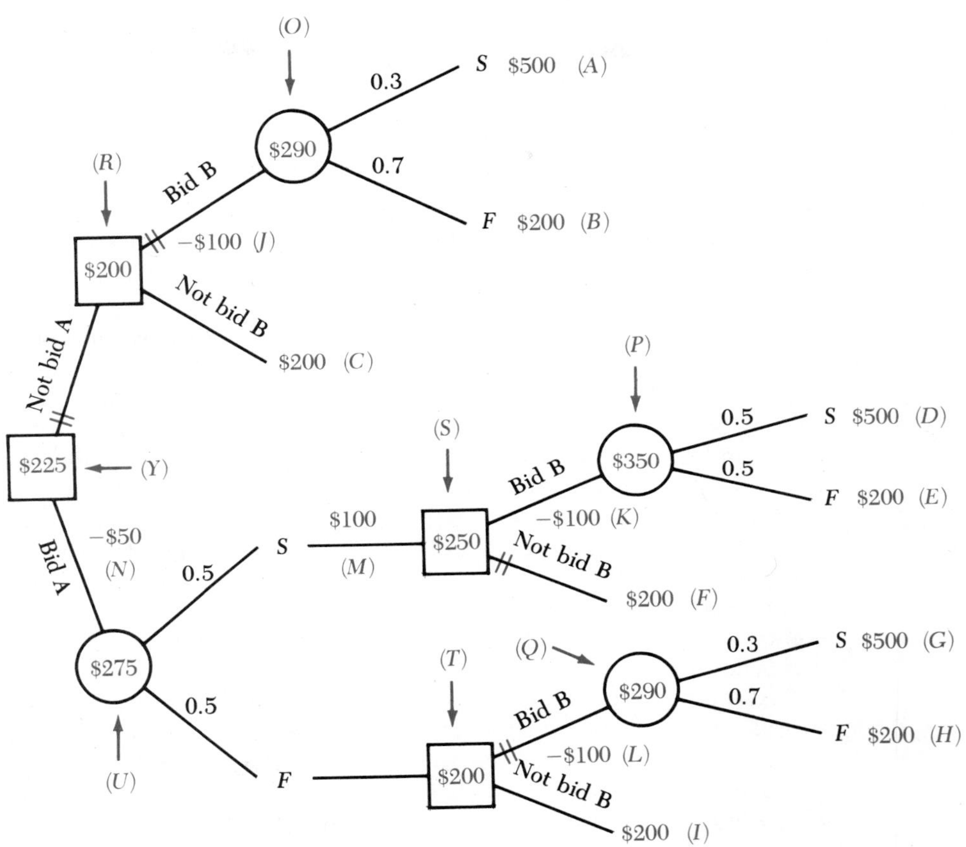

Figure 19-4 Solved Decision Tree

Solving the Decision Tree

Figure 19-4 represents the solved decision tree. The letters in parentheses would not normally appear on the tree. We have added them so that we can refer to the sequence of entries and operations by which the tree is solved.

Payoffs The solution of a decision tree proceeds from right to left, or backward in time. Generally there will be payoffs associated with the terminal events (states) at the extreme right. These payoffs, A to I in our example, are entered first.

Frequently, other payoffs or costs are associated with actions or intermediate outcomes elsewhere on the decision tree. In our example, bidding on B (the construction phase) involves a cost (in thousands) of $100, so this negative payoff is entered wherever the tree depicts the act *bid B* (J, K, L). Other intermediate payoffs entered are M, the $100 (thousand) profit should the firm be successful in bidding on the land-preparation phase, and N, the $50 (thousand) cost of bidding on the construction phase.

Chance Forks Working backward to the left, we shall first encounter chance forks (circled). We can start with any one of these forks. In Figure 19-4, we started with the fork at O. The general procedure is to enter in the circle the expected value of the chance event, or gamble, represented by the branches stemming from the fork. Therefore at O, we record the expected value

$$0.3(500) + 0.7(200) = 290 \text{ (thousands of dollars)}$$

In the lower sets of branches, we see two chance forks with which we can work. So we take the expected values of those gambles and enter the results in the circles P and Q.

Decision Forks Proceeding farther to the left in the upper set of branches, we now encounter a decision (boxed) fork. At any such fork, we have to decide which branch stemming from the fork is the best action. For the case in point, R, we see that not bidding on B (the construction phase) has a certain payoff (in thousands) of \$200, while bidding on B has an expected payoff of \$290, which we figured above, less the cost of \$100 to prepare the bid for the construction phase. The payoff of \$200 from not bidding on B (the payoff from the highway subcontracting job) is better than the \$290 − \$100 = \$190 expected payoff from bidding on the construction phase. Consequently we block the inferior action(s) with slashed lines and enter the payoff from the preferred action in the box at the decision fork. This means that if the sequence of events should bring us to that fork, we shall take the action (travel down the branch) that is not blocked.

At this point in solving our decision tree, we cannot proceed farther to the left on the upper set of branches. We must work back now along the lower sets of branches. There we encounter the decision forks at S and T. At S, we see that bidding on the construction phase is better than not bidding and we place the value of bidding (\$250) in the box and block the branch for the action *not bid B*. At T, we see that not bidding is the better action and enter the \$200 in the box there, at the same time blocking the branch for the action *bid B*.

Finishing Up The solution of the decision tree continues by working backward (to the left) and performing the procedure appropriate to the forks encountered. In Figure 19-4, we are now at U, where we see that bidding on the land-preparation phase amounts to a gamble on receiving \$250 + \$100 = \$350 if the bid is successful and \$200 if the bid is unsuccessful (and the firm turns to the highway subcontract). The expected value of the gamble is \$275, which is entered in the circle at the chance fork U. At V, we now compare traveling down the action branch *bid A* to the branch *not bid A*. We see that \$275 less the cost of bidding on the land-preparation phase (\$50) is better than the \$200 value for not bidding. Therefore we bring \$275 − \$50 = \$225 back and enter that amount in the box at V and block the inferior action *not bid A*.

The decision tree is now solved. Reading from left to right reveals the optimal course of action for the firm. The optimal course is to bid on the land-preparation phase. If the bid is successful, the firm will then bid on the construction phase. If the land-preparation bid is not successful, the firm will not bid on the construction phase. The expected return from following this optimal course of action is \$225 (thousand).

19.4 Opportunity Costs

In working with decision problems, we have worked with payoffs. Ordinary costs, such as the cost of preparing a bid, were regarded as negative payoffs. In this section we transform a payoff table into a table of *opportunity costs*. Opportunity costs provide some important insights into decision problems. Further, the use of opportunity costs in some decision problems enables us to use some short cuts in solving them.

Recall the payoff table for the *Recworld* newsstand problem. There the operator was considering the number of copies to stock each week. The cost to the newsstand was $2 a copy and the selling price was $5 a copy. Demand each week varied from 4 to 7 copies according to a probability distribution determined on the basis of past sales. We worked out a payoff table for the alternative acts of stocking 4, 5, 6, and 7 copies. That payoff table was as follows:

States	*Acts: Stock*			
(number demanded)	*4*	*5*	*6*	*7*
4	$12	$10	$8	$6
5	$12	$15	$13	$11
6	$12	$15	$18	$16
7	$12	$15	$18	$21

If 7 copies were stocked and 5 demanded, the newsstand would receive $25 from selling 5 copies and would pay $14 for 7 copies, so the return would be $25 – $14 = $11. The unsold copies are a complete loss.

Consider how you, as the newsstand operator, might feel if a particular state happened and you had earlier made the decision to stock a particular number of copies. For example, consider the state *4 copies are demanded.* If this happens and you had decided to stock 4 copies for the week, you would be content. If you had decided to stock 5 copies, however, you would wish you had been smart (or lucky) enough to stock 4 copies, because if you had, you would have made $12 instead of $10. In a sense, you would *regret* your decision to stock 4 copies, and the measure of your regret would be $2. Similarly, if you stock 6 copies and only 4 were demanded, your regret about your decision would be $12 – $8 = $4. On the other hand, if you stocked 6 copies and 7 were demanded (and you made $18), you would wish you had stocked 8 copies becuase you could have made $21 by so doing. Your regret in that situation would be $21 – $18 = $3.

What we called *regret* above is also called *opportunity cost*. The term *opportunity cost* comes from economics, where we learn that the real costs of an alternative are the gains *foregone* by having to give up other alternatives. Among the opportunity costs of an education, for example, are the earnings that could be made if one were not in school. A less frequently used word for the same idea is opportunity *loss*.

As described above, the regret associated with the ith state and the jth act in a decision problem is

$$R_{ij} = (V_{max} \mid S_i) - V_{ij} \tag{19-4}$$

where V refers to entries in the payoff table, i indexes the states, and j indexes the acts. Using the number of copies for the subscripts i and j, some of the regret figures are

$$R_{45} = \$12 - \$10 = 2$$
$$R_{65} = \$18 - \$15 = 3$$
$$R_{77} = \$21 - \$21 = 0$$

The entire regret table for the *Recworld* newsstand problem is as follows.

States (number demanded)	*Acts: Stock*			
	4	*5*	*6*	*7*
4	\$0	\$2	\$4	\$6
5	\$3	\$0	\$2	\$4
6	\$6	\$3	\$0	\$2
7	\$9	\$6	\$3	\$0

When the decision criterion is expected value, a decision problem can be solved using either the regrets table or the original payoff table. The optimal act is the same. Recall that the probabilities were 0.40, 0.30, 0.20, and 0.10 for 4, 5, 6, and 7 demands, respectively. The expected regrets for each act are calculated below and paired with the expected payoffs, which were shown in Table 19-1.

	EV
$ER(A_4) = 0.40(0) + 0.30(\$3) + 0.20(\$6) + 0.10(\$9) = \$3$	\$12
$ER(A_5) = 0.40(\$2) + 0.30(0) + 0.20(\$3) + 0.10(\$6) = \$2$	\$13
$ER(A_6) = 0.40(\$4) + 0.30(\$2) + 0.20(0) + 0.10(\$3) = \$2\frac{1}{2}$	\$12½
$ER(A_7) = 0.40(\$6) + 0.30(\$4) + 0.20(\$2) + 0.10(0) = \$4$	\$11

The criterion to minimize expected regret leads to the same optimal act, *stock* 5, as does the criterion to maximize expected payoff. Further, the differences in the values of the acts are the same for expected regret as for expected payoff. For example, stocking 4 copies a week is \$1 worse than stocking 5 copies, and stocking 7 copies a week is \$2 worse than stocking 5 copies.

An important result that arises in using regrets is that the expected regret of the optimal act is also the expected value of perfect information.

$$EV(PI) = ER(A_{opt}) \tag{19-5}$$

In our example,

$$EV(PI) = ER(A_5)$$
$$= \$2$$

This calls our attention to a foregone opportunity associated even with this best act. That opportunity is to obtain perfect information, which would allow the newsstand operator to switch to other acts (stock levels) when these acts would be better than stocking 5 copies. What the operator stands to gain by perfect information about weekly demand before it occurs is the opportunity foregone by using the existing probability distribution of demand to select an optimal act. Of course, the *cost* of perfect information may or may not exceed its value.

A Marginal Approach

A look at the regrets table for the *Recworld* newsstand problem reveals a feature that leads to a short cut for solving a class of decision problems. The feature is that for every copy *over*stocked in a week, there is an opportunity loss of \$2, and for every copy *under*stocked in a week, there is an opportunity loss of \$3. The opportunity loss of \$2 is the cost that could have been avoided if an unsold copy had not been stocked and the opportunity loss of \$3 is the amount that could have been made if there had been a copy in stock to meet an otherwise unmet demand. Using Table 19-3, we shall examine these opportunity losses more closely.

Table 19-3 Expected Net Gain from Stocking Additional Copies of *Recworld*

Demand x	$P(x)$	$P(X \geq x)$	$P(X < x)$	*Expected Net Gain from Stocking xth Copy*
4	0.40	1.00	0.00	1.0(\$3) – 0.0(\$2) = \$3.00
5	0.30	0.60	0.40	0.6(\$3) – 0.4(\$2) = \$1.00
6	0.20	0.30	0.70	0.3(\$3) – 0.7(\$2) = –\$0.50
7	0.10	0.10	0.90	0.1(\$3) – 0.9(\$2) = –\$1.50

What is the value of stocking the fourth copy? If demand equals or exceeds 4 copies, the operator will sell the fourth copy. We note that $P(X \geq 4) = 1.0$, a certainty. The operator is sure to sell the fourth copy and make the \$3 profit on that copy. There is no chance that demand will be less than 4 copies and leave the fourth copy unsold, which would cause a \$2 loss on the copy. The expected (in fact certain) net gain from stocking the fourth copy is \$3.

Now ask if it is worthwhile to stock a fifth copy. From $P(X \geq 5) = 0.60$, we see there is a 60% chance of selling the fifth copy and gaining the \$3 profit. Correspondingly, there is a 0.40 probability of not selling the fifth copy and incurring a loss of \$2 on that copy. Stocking the fifth copy is worthwhile because the expected net gain from stocking it is \$1.00, a positive amount. However, we see that the expected net gain from stocking the sixth copy is negative. The sixth copy should not be stocked and the optimal stock level is, therefore, 5 copies.

This kind of problem is called a *linear-loss* problem. Opportunity losses here are a linear function of the number of copies understocked ($3 times that number) and the number of copies overstocked ($2 times that number).

We can formulate a general rule for finding the optimal value of a decision variable in a linear-loss problem. The decision variable in the *Recworld* newsstand problem is weekly demand. We are seeking to find the optimal level of demand to use for a weekly stock order. Let the decision variable be x and let c_u and c_o be the opportunity losses from *under*estimating and from *over*estimating the decision variable by one unit, respectively. Given these symbols, we first calculate the criterion probability, P_c.

$$P_c = \frac{c_o}{c_u + c_o} \tag{19-6}$$

> The optimal value of the decision variable in a linear-loss problem will be the largest value of x for which $P(X \geq x) \geq P_c$.

In our example,

$$P_c = \frac{c_o}{c_o + c_u} = \frac{\$2}{\$2 + \$3} = 0.40$$

Now, we need only examine the cumulative *equal to or greater than* probability column in Table 19-3. Reading down the column, we find that the last cumulative probability that equals or exceeds $0.40 = P_c$ is 0.60. This cumulative probability is associated with a demand level of $x = 5$. Therefore 5 copies is the optimal weekly stock level.

Exercises

19-11 *Zoning application.* A realtor is considering two investment opportunities on which outlays would be the same. The realtor is certain that property A would bring a profit of $100,000 in two years. Over the same period, there is an opportunity to turn a profit of $50,000 on property B as is, but if rezoning can be obtained, the profit would be $100,000 more. Costs of $10,000 would be incurred in preparing and presenting the case to the zoning review board. The realtor assesses the probability of a successful appeal for rezoning at 0.40.

a. Present the realtor's problem as a decision tree.
b. Solve the decision tree.

19-12 *Crashing a project.* It is early March and the vice-president of Planimetrics, Inc., is concerned about the schedule for completing an environmental impact study to be delivered to a client by May 15. Given current budgeted workloads on the project, the firm will receive a net profit of $20,000 on the project if the study is delivered on time. If the report is delivered late, a penalty of $5000 will be assessed. The project can be "crashed" now—that is, additional personnel and overtime hours can be assigned to the project. This would cost $2000. By crashing now, the vice-president estimates the probability of being on time in April and completing the project on time in May as 0.8 and the probability of being behind schedule on April 15 as 0.2. If the project is behind schedule on April 15, it can be crashed again at a cost of $2000, but the probability is only 0.5 that this measure would lead to delivering the study on time in May.

Alternatively, the vice-president can elect to do nothing about the project schedule now. Under this option, the assessed probability of being on schedule on April 15 and completing the project on time in May is 0.5. If the project is behind schedule on April 15, it can be crashed then, with the same costs and outcomes as given above. If the project is behind and not crashed in April, it is certain to be delivered late.

a. Construct a decision tree for the vice-president's problem.
b. Should the project be "crashed" now, or should the vice-president wait until April 15 and reevaluate progress then?

19-13 *Make or buy decision.* Tomorrow Enterprises has obtained the rights to market a distinctive golfing shirt bearing the name of a player who has recently won several major tour events. Current demand for the shirts is uncertain because of the distinctive nature of the shirts and the unknown appeal of the new name. Later demand additionally depends on whether the player continues to be a winner. The firm can either make the shirts or buy them from others for its own private label. Generally, if demand is high it pays the firm to make the shirts, and if demand is low it pays to buy them from others. Tomorrow Enterprises evaluates the probability of high, medium, and low current demand as 0.4, 0.4, and 0.2, respectively. Given high current demand, the probabilities that later demand are high or low are 0.7 and 0.3, respectively, while these probabilities given medium current demand are 0.5. If current demand is low, it is a certainty that later demand will be low.

Tomorrow Enterprises can decide now to make the shirts to meet both current and later demand. Alternatively, they can buy the shirts in anticipation of current demand and decide whether to continue to buy or make the shirts themselves to meet later demand. Payoffs, in thousands of dollars, given various acts and states are as follows.

	Current Demand			*Later Demand*	
	High	*Medium*	*Low*	*High*	*Low*
Make	500	300	100	400	100
Buy	400	300	200		

The payoffs from later demand are in addition to the payoffs from current demand. If the firm buys to meet current demand, the following are additional payoffs for the act-state combinations for later demand.

	Later Demand	
	High	*Low*
Buy	300	200
Make	350	100

a. Construct a decision tree.
b. What is the optimal decision or decision sequence, and what is the expected value of the optimal decision(s)?

19-14 *New market opportunity.* A manufacturer of hand calculators has an opportunity to enter a Latin American market. Management evaluated three courses of action against demand in the new market as follows.

Demand Level	*Sell Through Agents*	*Hire Foreign Sales Force*	*Build Foreign Plant*
Low	\$2,000,000	–\$2,000,000	–\$4,000,000
Medium	\$3,000,000	\$4,000,000	\$0
High	\$4,000,000	\$6,000,000	\$8,000,000

The probabilities for low, medium, and high demand are evaluated at 0.10, 0.50, and 0.40, respectively.

a. Find the expected value of the alternative acts and the expected value of perfect information.
b. Transform the payoff table into a table of opportunity losses.
c. Find the expected opportunity loss for each act.
d. Is the expected opportunity loss for the optimal act equal to the $EV(PI)$ in (a)?

19-15 *Book rights.* A publisher owns the rights to the story of a team of balloonists who are planning a trans-Pacific flight. The publisher values the rights at \$1,000,000 should the team undertake the flight and succeed, \$200,000 should the team undertake the flight and fail, and \$0 should the team not undertake the flight. The publisher has an offer of \$200,000 for the rights.

a. Set up the opportunity loss table for the acts *sell the rights* and *do not sell the rights*.
b. Suppose the probability that the flight will be undertaken is 0.5, and the probability of success for the flight is 0.6. Find the expected value of perfect information by working with the opportunity losses.
c. If the publisher knows that the flight will be undertaken, is uncertainty about the success of the flight of concern in the decision about selling the rights? Explain.

19-16 *Inventorying cakes.* A bakery makes and decorates an elaborate birthday cake at a cost of \$4 and sells the cakes for \$10. There is no market for day-old birthday cakes, so any cakes not sold the day they are made are given away to clubs. Daily demand for the cakes is as follows.

Demand	0	1	2	3	4
Probability	0.20	0.30	0.20	0.20	0.10

a. Determine the payoff table for the acts *stock 0, 1, 2, 3,* and *4* cakes.
b. Find the expected value of each act.
c. Find the opportunity loss table and the expected opportunity loss for each act.
d. Find the optimal act using the marginal loss approach.

19-17 *Back-up equipment.* A crew of salvage divers is being sent to dive at a site for a period. Diving equipment is checked out each morning and, if found faulty, a replacement suit is used while the original equipment is repaired during the day. On any given day the following probability distribution for the number of replacement suits required is applicable.

Number	0	1	2	3	4
Probability	0.37	0.37	0.18	0.06	0.02

The cost of renting and maintaining replacement suits is $100 a day and the estimated loss from down-time should a crew member not be able to dive for lack of a suit is $1500.

a. Construct the opportunity loss table for the acts of providing 0, 1, 2, 3, and 4 replacement suits.
b. Find the expected opportunity loss for each of the acts.
c. Find the optimal act using the marginal loss approach. Is the optimal act the same as the optimal act from (b)?

Summary

Decision making involves a set of *uncertain states*, a set of *alternative acts*, and a set of *payoffs* for alternative acts taken in conjunction with the possible states. When a particular act is not the best under all possible states, the *Bayesian decision criterion* assesses a probability distribution on the states and selects the act with the largest expected payoff. The cost of operating under uncertainty, also called the *expected value of perfect information*, is the difference between the expected payoff with certain (or perfect) prediction and the expected payoff of the best act under the Bayesian decision criterion.

The appropriate payoffs in many decision problems are money gains and losses. In these situations, the marginal utility of additional units of monetary gain to the decision maker will be constant. When this is not so, a utility curve of the decision maker can be determined. Money gains can then be translated into utility units and the decision problem solved in terms of expected utility.

A *decision tree* is a useful tool in solving Bayesian decision problems. Alternative acts, chance events and their probabilities, and payoffs are arranged chronologically on the decision tree. Then the optimal act can be determined by a process of working backward along the tree.

Finally, the important concept of *opportunity losses* (or *regrets*) was introduced. Minimizing expected opportunity loss in a decision problem is identical to maximizing expected payoff. An important result, however, is that the expected opportunity loss of the optimal act is also the expected value of perfect information. Opportunity losses can also be used to effect a short cut to the solution of a common class of decision problems called *linear-loss problems*.

Supplementary Exercises

19-18 *Olympic rings.* A jeweler has designed an Olympic ring containing five gemstones set in the Olympic insignia, which will be placed on a consignment basis for sale in selected outlets. The rings cost the manufacturer $200 each and bring a profit of $300 each when sold. Rings not sold are returned to the manufacturer. At a cost of $100 each, unsold Olympic rings can be reworked into a conventional setting for the gemstones and sold by the manufacturer for $250 each. The manufacturer estimates that demand for the Olympic rings at a particular outlet will be as follows.

Demand	5	6	7	8	9	10
Probability	0.1	0.2	0.3	0.2	0.1	0.1

a. Prepare a table showing the expected payoff for each act-state combination.
b. From your table in (a), construct the table of opportunity losses.
c. Calculate the expected payoff and the expected opportunity loss for the acts to consign 7, 8, and 9 rings to the outlet.
d. Using the marginal opportunity loss approach, find the marginal gain (or loss) from (i) consigning the eighth ring and (ii) consigning the ninth ring to the outlet.

19-19 *Publication venture.* A trade association with 1000 members is considering publishing a quarterly newsletter on activities in the trade. Financial success of the newsletter depends on the proportion of members who will subscribe. Probabilities and payoffs are as follows.

Proportion Subscribing	*Probability*	*Return to Trade Association*
0.30	0.1	–$20,000
0.40	0.3	–$5,000
0.50	0.4	$10,000
0.60	0.2	$30,000

a. What is the expected return to the association?
b. Find the expected value with certain prediction.
c. Find the expected opportunity loss of publishing the newsletter.
d. From your answers above, show two ways of expressing the expected value of perfect information.

19-20 *Dividing the cache.* A trapper operating in a remote area gathered $10,000 worth of furs during a season. Winter was approaching, and the furs would have to be cached until spring, when the trapper could return with pack animals to bring the furs out. The trapper evaluated the probability at 0.10 that a cache would be found and the furs stolen. If two or more caches were used, the outcomes were considered independent. Should the trapper have used one cache or divided the furs equally between two caches? Solve for each case below.

a. If the utility of dollar gains is proportional to the amount of gain.
b. If the utility of dollar gains is proportional to the square root of dollar gain.
c. If the utility of dollar gains is proportional to the square of dollar gain.
d. Do the results of (a), (b), and (c) suggest any generalization about the relation between spreading of risk and the shape of a utility function?

19-21 *Utility curves.* Consider three industrial competitors. The utility curves for gains and losses for the firms are as follows.

Alpha	$U(\text{gains}) = 40X - 2(X)^2$	$U(\text{losses}) = 40X - 2(X)^2$
Beta	$U(\text{gains}) = 40X + 2(X)^2$	$U(\text{losses}) = 40X - 2(X)^2$
Omega	$U(\text{gains}) = 40X + 4(X)^2$	$U(\text{losses}) = 40X - 4(X)^2$

a. Sketch graphs of each utility function by finding utilities at $X = 1, 2, 3,$ and 4 (gains) and $X = -1, -2, -3,$ and -4 (losses).
b. Compare the attitudes of Alpha and Beta with respect to different kinds of risks.
c. Compare the attitudes of Beta and Omega with respect to different risky propositions.

19-22 *Land purchase option.* Profitline Investments, Inc., has an opportunity to purchase a parcel of land for $220,000. If a proposed highway project is approved and the highway alignment intersects the property, the property will be worth $400,000. Otherwise the property will be worth $100,000. The probability that the project will be approved is 0.4, and the probability that the alignment will intersect the property, given that the project is approved, is 0.7.

a. What is the expected value of the property?
b. What is the expected value of an option to purchase the property for $220,000 that could be exercised after it was known whether the project was approved but would not remain open until the time that the exact highway alignment was known?

19-23 *Stocking a cattle ranch.** In a normal year in the foothills of California, 3000 acres of range land will produce sufficient feed to carry a beef herd of 360 cows, which will be sold at an eight-month weight of 500 pounds. In a poor year, supplemental feed has to be purchased to carry such a herd, and in a good year there is some extra income from selling excess feed. Income projections for three herd sizes under three different range conditions were worked out as follows.

Range Condition	*Size of Herd*		
	280 cows	*360 cows*	*415 cows*
Poor	–$6000	–$10,000	–$12,500
Normal	$1000	$6,000	$5,500
Good	$2000	$7,500	$10,000

*Adapted from Albert N. Halter and Gerald W. Dean, *Decisions Under Uncertainty with Research Applications*, Cincinnati, Ohio: South-Western Publishing Company, 1971.

In the past 44 years, the U.S. Crop and Livestock Reporting Service rated range conditions *poor* in 7 years, *normal* in 22 years, and *good* in 15 years.

a. Convert the payoff table into an opportunity-loss table.
b. Find the size of herd that minimizes the expected opportunity loss.
c. If the herd sizes given were the only alternatives, what is the expected value to a rancher with 3000 acres of a perfect forecast of range conditions prior to stocking the ranch for a season?

19-24 *Inventor's decisions.* An inventor is trying to decide whether to invest $15,000 required to develop a new process for treating an industrial waste. There is a 60-40 chance of success. If the process is successful, the inventor can apply for a patent at a cost of $5000, with a 50-50 chance that the patent would be approved. Regardless of patent application or approval, the inventor would have to decide whether to market the process directly or to sell distribution rights to another firm. Profits (excluding patent cost) from these alternatives depend on market acceptance and patent status.

Market Acceptance	*Probability*	*Application*		*No Application*
		Approved	*Denied*	
If inventor markets:				
Substantial	0.3	$60,000	$28,000	$44,000
Moderate	0.4	$40,000	$22,000	$32,000
Light	0.3	$24,000	$12,000	$20,000
If inventor sells rights:		$40,000	$25,000	$30,000

a. Set up the decision tree for the inventor.
b. Solve the decision tree to find the optimal decision sequence and its expected value.

20

DECISION MAKING WITH SAMPLE INFORMATION

The significance of information in decision making is that it can change our beliefs about the occurrence of states. If the probabilities that we hold about the occurrence of states change, then we may want to change our choice of acts.

Example

The traditional route for a seaborne invasion of Europe from the British Isles was via Pas-de-Calais at the narrowest point of the English Channel. The Allies knew in World War II that this was the route that would be given the highest probability by the Germans and hence would be most heavily defended. In planning the Normandy invasion in 1944, every effort was made by the Allied high command to reinforce that belief and to maintain secrecy about the real point of attack. A fake military base, airfields, and port were created in southeastern England to carry out the deception.*

The principle underlying revision of probabilities is based on the idea of *conditional probability*. In the above example, it was important to the Allies that Hitler increase his probability assignment of a Calais invasion point upward given the "evidence" that was created just for this purpose.

This chapter continues our study of Bayesian decision making. The important added element is the use of sample information. Sample information is used to revise, or change, the probabilities of states. Changes in the probabilities of states can lead to changes in the optimal act. We must first understand the process of revision of probabilities.

In Chapter 19, we studied the concept of the expected value of perfect information. Of course, *perfect* information is rarely available in practice. More often sample information, which is by nature imperfect, is all that is available to improve our knowledge about states. In the second section of this chapter, we see how the value of sample information can be determined before a sample is taken. To be worthwhile in a decision making situation, the expected value of sample information must exceed its cost. The value of sample information is therefore an important consideration in business decisions.

Later in the chapter, we extend the ideas of Bayesian decisions to the important case of population means. Here alternative acts have payoffs that depend on the value of the population mean that in fact exists, or occurs. We discuss revising our belief about the population mean given the information of a random sample of observations from the population. Bayesian decision making when the decision parameter is a population mean is then compared with decisions based on classical inference procedures studied earlier in the text.

20.1 Revision of Probabilities

The table below gives the numbers of conventional and Federal Housing Administration (FHA) mortgages (in thousands) granted for new, private, single-family housing units in the United States in 1977, classified by size of mortgage.

State (type of mortgage)	*Event or Evidence*		*Total*
	Under $35,000	*$35,000 and Over*	
Conventional	51	542	593
FHA	50	117	167
Total	101	659	760

*See Ken Follett, *Eye of the Needle* (New York: The New American Library, Inc., 1978).

We shall consider the labels *State* and *Event or Evidence* shortly. For the moment, we observe that the table is an ordinary cross-classification table like those encountered in Chapter 4. There we learned that a marginal probability, such as the probability of a conventional mortgage, is constructed from the appropriate marginal totals. Specifically, $P(\text{conventional}) = 593/760 = 0.780$. Also, we can construct conditional probabilities by using the inner parts of the table. The probability of a conventional mortgage, given that the mortgage is under \$35,000, is $51/101 = 0.505$.

Consider these probabilities as measures of belief. The first says that if a mortgage were randomly selected from the entire population of 760,000 mortgages, our belief that the mortgage is a conventional mortgage is 0.780. The second says that if we know the mortgage was selected from the 101,000 mortgages for under \$35,000, our belief that the mortgage is conventional is 0.505.

Cross-classification tables involve two kinds of events. In our case, they are types and sizes of the mortgages. We have designated type of mortgage as *states* and size of mortgage as *evidence*. In this context, the probability of a conventional mortgage is the probability of a particular state (of the mortgage). The measure of our belief in this state is 0.780. However, if we come into possession of information that the mortgage was for under \$35,000, we have *evidence* that changes our belief about the state of the mortgage. The measure of our belief that the mortgage is conventional is now 0.505. The first probability is called a *prior probability* of a state. The second is called a *revised probability* of a state. We revised our earlier belief on the basis of information, or evidence, that was observed.

Bayes' Theorem

A formal rule for revising probabilities is *Bayes' theorem*.

> If an observed event or evidence (E) can result from several states, $S_1, S_2, \ldots, S_n$, the probability that the evidence resulted from S_i is the ratio of the joint probability of S_i and the evidence to the marginal probability of the evidence.
>
> $$P(S_i \mid E) = \frac{P(S_i) * P(E \mid S_i)}{P(E)} \qquad \textbf{(20-1)}$$

In Bayes' theorem, $P(S_i)$ is the prior probability of a state and $P(S_i \mid E)$ is the revised probability of the state. Thus Bayes' theorem is an equation for revising probabilities of states in the light of new evidence. In our mortgage example, we have the following.

$$P(S_i \mid E) = P(\text{Conventional} \mid \text{Under } \$35{,}000)$$

$$P(S_i) = P(\text{Conventional}) = 593/760 = 0.780$$

$$P(E \mid S_i) = P(\text{Under } \$35{,}000 \mid \text{Conventional}) = 51/593 = 0.086$$

$$P(E) = P(\text{Under } \$35{,}000) = 101/760 = 0.133$$

Substituting these in Bayes' theorem, we have

$$P(\text{Conventional} \mid \text{Under } \$35{,}000) = \frac{0.780(0.086)}{0.133} = 0.505$$

which is the same as the earlier result read directly from the table.

Likelihood

Some insight into Bayes' theorem is provided by the concept of *likelihood*. The second element in the numerator of Bayes' theorem is the probability of observed evidence given a state, $P(E \mid S_i)$. Probabilities of this type are called likelihoods. In the mortgage example, 8.6% of the conventional mortgages were for under $35,000 and 29.9% of the FHA mortgages were for under $35,000. We say that the likelihood of encountering a mortgage under $35,000 is much smaller for conventional than for FHA mortgages. The event, or evidence, of a mortgage under $35,000 is more *likely* given an FHA than a conventional mortgage.

A more extensive example will underscore the role of Bayes' theorem in combining prior probabilities and likelihoods.

Example

A manufacturer receives shipments of a component in boxes of 100 components each. The current belief about the percentage of defective parts in a box is represented by the following probability distribution.

Percent defective	0	1	2	3	4	5	Total
Probability	.20	.30	.25	.15	.08	.02	1.00

The *state* of a shipment that is of concern is the percentage of defectives. A box is received today and a sample of four components is taken, with replacement, from the box. One defective is found. What is the revised probability distribution of the percentage of defective parts in today's box?

Before answering let us consider the prior probability distribution and the evidence informally. In the prior distribution, probability is concentrated at 1 and 2 percent defective. One out of four sample components were found defective, a sample percentage of 25. The evidence should increase our belief in higher percentages defective and decrease our belief in smaller percentages defective. Because it is impossible to observe a defective part from a box with no defectives, our belief in 0.0 percent defective should go to zero.

Application of Bayes' theorem will adjust the probabilities formally. Table 20-1 organizes the situation. The prior distribution for percentage defective (states) is repeated in the column headed $P(S_i)$. Then, under the heading *likelihood*, we must find the probability of the event we observed (one defective in four) *given* each of the possible states for percentage defective. These will be recognized to be binomial probabilities. For instance, the probability of one defective in four *given* that the true percentage defective is one percent is

Table 20-1 Columnar Setup for Revising Probabilities

State S_i	*Prior Probability* $P(S_i)$	*Likelihood* $P(E \mid S_i)$	$P(S_i) * P(E \mid S_i)$	*Revised Probability* $P(S_i \mid E)$
0	0.20	0.0000	0.0000	0.00
1	0.30	0.0388	0.0116	0.19
2	0.25	0.0753	0.0188	0.31
3	0.15	0.1095	0.0164	0.27
4	0.08	0.1416	0.0113	0.18
5	0.02	0.1715	0.0034	0.05
Total	1.00		0.0615	1.00

$$P(R = 1 \mid n = 4, \pi = 0.01)$$

which is

$$4(0.01)(0.99)^3 = 0.0388$$

The probabilities of the evidence *given* other states are similarly determined. In those *likelihoods* the evidence stays the same. The manufacturer *did observe* one defective and that is the evidence we are taking into account. In terms of the binomial, r stays the same but π changes. The last likelihood in the table is

$$P(R = 1 \mid n = 4, \pi = 0.05)$$

which is

$$4(.05)(.95)^3 = 0.1715$$

We see that one defective in four is much more likely to occur from a 5 percent defective population than from a 1 percent defective population.

In the next column, the prior probabilities and the likelihoods are combined by multiplication. Each of these products is the numerator in the application of Bayes' theorem to a particular state. The denominator in the application of Bayes' theorem for each state is the marginal probability of the evidence of one defective. Adding the joint probabilities in the product column for each of the ways of observing one defective gives this marginal probability. Then, the revised probability that the percentage of defective products in the given box is 2% ($S = 2$), for example, is

$$P(S = 2 \mid r = 1) = \frac{P(S = 2) * P(r = 1 \mid S = 2)}{P(r = 1)}$$

$$= \frac{0.25(0.0753)}{0.0615} = \frac{0.0188}{0.0615} = 0.31$$

Figure 20-1 shows the graph of the elements in Bayes' theorem for this example. Comparing the revised probability with the prior distribution, we see that the belief

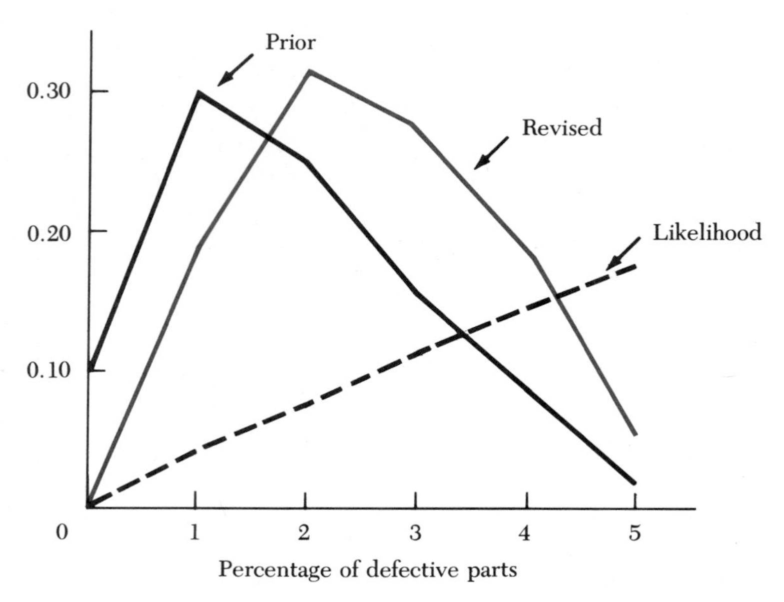

Figure 20-1 Prior Probability, Likelihood, and Revised Probability

that the percentage defective is zero or 1 has decreased, and the belief about higher percentages defective has increased. This is what we expected to happen when we considered that 25% (one in four) of the components sampled were defective. The reason the revised probabilities are not *more* different from the prior probabilities is that the sample size was small and hence the weight of the evidence is not overwhelming.

Reevaluating a Course of Action

At the beginning of this section, we emphasized that a change in belief can lead to a change in the best course of action. To illustrate this in detail, suppose our manufacturer receiving boxes of components is faced with the following alternatives concerning each box of 100 components.

Act A_1: Test all the components prior to assembly, and repair those found defective. The cost of testing is $100, and the cost to repair each component found defective is $40.

Act A_2: Ship the box on to the assembly operation. The defective components will be found upon testing the completed assembly. However, at this point it will cost $80 per defective component to dismantle the assembly, repair the component, and reinstall it.

Table 20-2 shows the costs for each act as a function of the percentage (which is also the number) of defective components in a box. Also shown are the opportunity losses for each act. For example, if 2% of the components in a box (2 components) are defective, under act A_1 a cost of $100 + 2($40) = $180 would be incurred. If act A_2 were taken under the same circumstance, a cost of 2($80) = $160 would be incurred. Act A_2, to ship the components on to assembly, is optimal for this state, so there is no opportunity loss connected with this state-act combination. The opportunity loss connected with act A_1 and a 2% defective box is $180 – $160 = $20.

Table 20-2 Costs and Opportunity Losses for Components Problem

Percent (and Number) of Defective Components	*Costs*		*Opportunity Losses*	
	A_1:*Test*	A_2:*Ship*	A_1:*Test*	A_2:*Ship*
0	$100	$ 0	$100	$ 0
1	140	80	60	0
2	180	160	20	0
3	220	240	0	20
4	260	320	0	60
5	300	400	0	100

If we calculate the expected opportunity loss (regret) for each act, we can then select the act that minimizes expected opportunity loss. Suppose the prior probabilities in Table 20-1 represent past experience with shipments provided by a particular supplier. For act A_1, to test all the components, we have the expected regrets

$$ER(A_1) = 0.20(\$100) + 0.30(\$60) + 0.25(\$20) = \$43$$

and for act A_2, to ship the box on to assembly, we have

$$ER(A_2) = 0.15(\$20) + 0.08(\$60) + 0.02(\$100) = \$9.80$$

The optimal act is to ship each box on to the assembly operation.

As we learned earlier, the expected value of perfect information, $EV(PI)$, is the expected regret of the optimal act. Thus the $EV(PI)$ here is $ER(A_2)$ = $9.80.

Let us return to the point in our example when the manufacturer took a sample of four components from the current shipment and observed one defective. The probabilities of states revised for this occurrence are given in Table 20-1. Using these revised probabilities the expected regrets are

$$\mathrm{ER}(A_1) = 0.0(\$100) + 0.19(\$60) + 0.31(\$20) = \$17.60$$
$$ER(A_2) = 0.27(\$20) + 0.18(\$60) + 0.05(\$100) = \$21.20$$

The optimal act is now A_1, to test and repair all defective components before shipping the box on to the assembly operation. The expected value of perfect information is now $17.60.

Exercises

20-1 *Preference for insurance plans.* In Exercise 19-2, probabilities that customers preferred one of two types of insurance plans were given as

Plan 1	0.6
Plan 2	0.4

a. Assume these probabilities are based on a census of customers, and that it is also known that 10% of the customers preferring plan 1 and 70% of the customers preferring plan 2 are under 30 years of age. Construct the joint probability table for plan preference and age group, and revise the preference probabilities for the event that a customer is under 30 years of age.

b. In Exercise 19-2, a payoff table for expected commissions per customer prospect was given, as follows.

Plan Preferred	*Present First*	
	Plan 1	*Plan 2*
Plan 1	\$40	\$20
Plan 2	\$30	\$40

What is the optimal act if it is known that a customer prospect is under 30 years of age?

20-2 *Revision given sample information.* Three defectives are observed in a random sample of six products from a large lot.

a. Revise the following prior distribution of states (proportion defective) for the sample evidence.

x	0.1	0.3	0.5	0.7	0.9
$P(x)$	0.2	0.2	0.2	0.2	0.2

b. Revise the following prior distribution of states for the same sample evidence.

x	0.1	0.3	0.5	0.7	0.9
$P(x)$	0.7	0.2	0.1	—	—

c. Describe the effect of the sample evidence in changing the prior belief in each of the above instances.

20-3 *Predicting stock price changes.* A stockbroker kept a record of six-month stock price predictions made by a charting service, with the results as shown in the following table.

Predicted Movement	*Actual Movement*			*Total*
	Down	*Unchanged*	*Up*	
Down	40	10	10	60
Unchanged	20	20	10	50
Up	10	10	30	50
Total	70	40	50	160

Use the data to construct

a. The probability of a price decrease, given that the service predicts a decrease.

b. The probability that the service predicted an increase when a price actually went up.

c. The broker holds the following belief about the future price movement of certain stocks before she notices the charting service's prediction given below.

Belief	*Stock A*	*Stock B*	*Stock C*
Down	$^2/_6$	$^4/_6$	$^3/_6$
Unchanged	$^2/_6$	$^1/_6$	$^2/_6$
Up	$^2/_6$	$^1/_6$	$^1/_6$
Prediction	Down	Up	Down

Assume the past performance of the charting service's predictions is relevant and revise the stockbroker's beliefs in light of the evidence of the service's predictions. (Give your answers as fractions.)

20-4 *Evidence and uncertainty.* In Exercise 20-2, does the evidence decrease uncertainty in both (a) and (b)? Does the information about the service's prediction decrease uncertainty about the price movement of each of the stocks in Exercise 20-3?

20-5 *Marginal probability of evidence.* Prior experience with shipments of four experimental components led a research and development department to establish the following prior probability distribution for the number defective in a current shipment of four components.

Number defective	0	1	2	3	4
Probability	6/10	3/10	1/10	—	—

One component from the shipment is to be tested.

a. What will the revised probabilities be if the component tested is defective?

b. What will the revised probabilities be if the component tested is not defective?

c. What is the marginal probability of observing a defective component? Of observing a good component?

20-6 *Diagnosing trouble.* Of the failures of a certain engine, 90% are caused by failure of the fuel injection system and 10% are caused by failure of the ignition system. If the fuel injection system fails, there is a 10% chance of fire; there is a 30% chance of fire if the ignition system fails. In the most recent failure there was a fire.

a. What is the probability that the fuel injection system failed?
b. Fill in the appropriate probabilities on a tree diagram that starts with the causes of failure and contains branches for the evidence of fire and no fire.

20-7 *Incoming lot quality.* The quality of large lots of containers received by a packing company from a manufacturer is a Bernoulli process with 0.8 probability of a 1% defective lot and 0.2 probability of a 2% defective lot. From a lot just received, a random sample of 20 containers reveals no defectives. What is the revised probability that there are only 1% defectives in the lot?

20-8 *Franchise profitability.* A fast-food franchise operation uses a two-stage procedure for establishing new units. In the first stage, cities, towns, and neighborhoods in large cities are selected using data on population and number of competing outlets. Following this, an analysis may be made of buying power and buying habits of potential customers for a unit in a particular location. This more detailed market analysis is designed to estimate the annual gross sales of the prospective unit. Since the franchise is a standardized operation, gross sales is a major determinant of net profit (payoff) from the franchise unit.

In one case, the first stage analysis has established the following.

Gross Sales	*Probability*	*Net Profit*
Under $200,000	0.5	–$50,000
$200,000–$300,000	0.3	$150,000
Over $300,000	0.2	$250,000

a. What is the expected net profit from establishing the unit?
b. What is the expected value of perfect information?
c. After completing the buying power study, the marketing research department figured that the probability of the survey outcome given that gross sales would be under $200,000 was 0.08, and the probability of the same evidence given true gross sales of $200,000 to $300,000 was 0.05, and given gross sales of over $300,000 was 0.02. What is the expected net profit from the unit after revising the state probabilities in the light of the buying power survey?
d. What is the expected value of perfect information at this point?

20.2 The Expected Value of Sample Information

The expected value of perfect information tells us what we can expect to gain in a decision situation by completely removing uncertainty about the best act to take. In some cases, it may be a sufficient guide in deciding whether to seek information before acting. In many cases, however, prospective information may be sufficiently imperfect that it is important to take that fact into account. As we shall see, figuring the expected value of sample information involves the revision of probabilities. Instead of revising probabilities just for information received (as in the preceding section), we must revise prior probabilities for information that a prospective sample *might* yield. We will introduce the expected value of sample information with an imaginary conversation between an art dealer and an acquaintance who is a professor of fine arts.

Transcription

Dealer: Hello, Leslie. B.J. Alcott here. I'm up in Newport. A local person named Kim Fogg claims to have a genuine Gresham. Wants $1500 for it. I've got a customer I know would pay $3500.

Prof: So, what's the problem? Sounds like a good deal to me.

Dealer: Well, as you know, Gresham painted in the same style as Fowler and neither of them signed their work. I know it's one or the other. If I knew it was a Fowler I'd offer $400 and sell it in New York for $1000.

Prof: But you said Fogg wanted $1500.

Dealer: Fogg is a little eccentric, but no fool. Kim knows who painted it. I'm sure of that. And that's a good price for a Fowler. But if it's a Gresham and I offer $400, that'll be the end of it. I'll be sent packing. It's all right for Fogg to try to take advantage of me. But you know these people. You can't go around insulting their intelligence.

Prof: I get the picture. You know you'd have to get the painting certified by Schaffer in New York to get $3500 for it from your customer. If you pay $1500 for it as a Gresham and it isn't, you're going to be out $500.

Dealer: Right. That's why I called you. Schaffer would do this one for free, but I'm up here and have to make a decision. You're supposed to be an expert on these early colonial period artists. Why don't you come up here and help me out? Take a look at the painting and give me your best judgment.

Prof: I might do that. You know, I've made these judgments before. They're not perfect. You still have to get the certification to know for sure. Those people have the lab equipment for analyzing the paints and so on. Anyway, if they say it's a Gresham, it is as far as the market is concerned. But I've looked at 200 paintings involving these artists; 60 that turned out to be Greshams and 140 Fowlers. That corresponds pretty closely to their comparative outputs I would think, so that's the odds I'd give on your situation without even coming up.

Dealer: Why do I want you, then?

Prof: Don't interrupt me. I'm coming to that. The point is, I was right on 40 of the Greshams and 130 of the Fowlers. We're not infallible, you know, but we do our best. I'll come up if you want. Favor to an old friend and all that! No fee. . . just expenses. . . say $150.

Dealer: I knew you'd come through, Leslie. Let me call you back.

In this situation, B.J. has two alternatives that can be taken now. Even if Leslie is not invited up to look at the painting, the phone call has served to provide prior probabilities for the true authorship of the painting. The decision problem facing B.J., if no further information is sought, is as follows.

State S	*Probability* *P*(S)	*Action*	
		Pay $1500 A_G	*Offer $400* A_F
S_G Gresham	60/200	$2000	0
S_F Fowler	140/200	–$500	$600

The expected values of these acts are

$$EV(A_G) = 0.30(\$2000) + 0.70(-\$500) = \$250$$

$$EV(A_F) = 0.30(\$0) + 0.70(\$600) = \$420$$

The best act without any further information is to assume the painting is a Fowler and offer Fogg $400. Either negotiations will be terminated by Fogg (S_G) or the offer will be accepted and the painting resold in New York for a $600 profit ($S_F$).

B.J. could also find the expected value of perfect information from

$$EV(CP) = 0.30(\$2000) + 0.70(\$600) = \$1020$$

$$\begin{aligned} EV(PI) &= EV(CP) - EV(A_{\text{opt}}) \\ &= \$1020 - \$420 \\ &= \$600 \end{aligned}$$

At this point we know that no information could be worth more than $600. But the problem revolves around the worth of Leslie's expertise.

Here it is useful to turn to a decision tree. The tree is shown in Figure 20-2. With the decision tree, we shall work out the expected value of the act of inviting Leslie to make a judgment and acting on that judgment. The value of this prospective information will be the difference between the expected value of obtaining and acting on the information and the expected value of the best act that can be taken without it. The decision tree follows the prospective time sequence of events and acts. These are as follows.

1. Leslie renders a judgment. This will be either E_G or E_F, the evidence (or information) predicts the painting is a Gresham or a Fowler.
2. The decision maker, B.J., considers the alternative acts. Logically these are A_G and A_F for *each* of the possible judgments. We are quite sure that the best act will be to follow the expert's judgment, but we shall let that be shown by the decision tree.
3. For each information-act combination, one of the possible states will eventually prove to be true and the payoff for that act-state combination realized.

The decision tree shown in Figure 20-2 has the final act-state payoffs entered. Note that these are the same act-state payoffs that were considered earlier. What

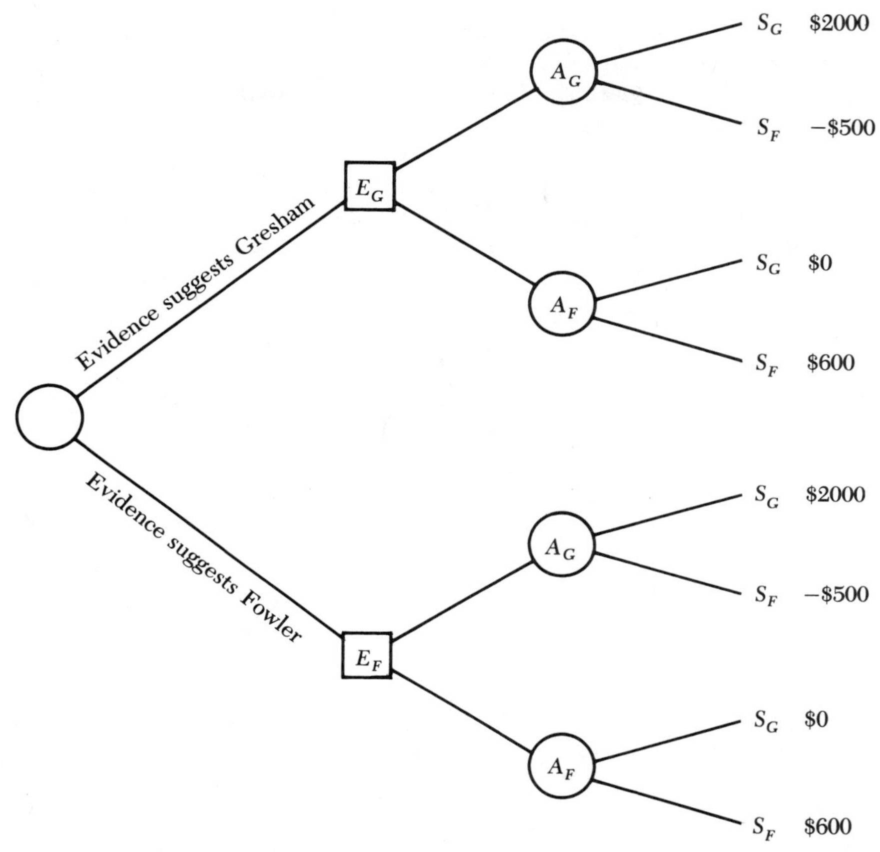

Figure 20-2 Decision Tree for Imperfect Information

remains is to enter the appropriate probabilities and to work back through the decision tree to find the expected value of the optimal acts, given the possible information outcomes.

The following is Leslie's record as a predictor of Greshams and Fowlers.

State	*Prediction*		*Total*
	E_G	E_F	
S_G	40	20	60
S_F	10	130	140
Total	50	150	200

Now look at the decision tree in Figure 20-2 and consider what probability belongs on the upper right-hand branch. That branch represents the condition that the painting turns out to be a Gresham after B.J. *acts* as if it is *following Leslie's judgment that it is*. The branch calls for the probability that the true state is a Gresham *given* that Leslie says it is. This is $P(S_G \mid E_G)$ from the table just presented, which is 40/50. It is the probability of a particular state *given* particular evidence. In other situations these kinds of probabilities might have to be arrived at through Bayes' theorem, but Leslie provided the relevant record in a form that does not require use of the theorem. The rest of the probabilities of states given information outcomes needed for the final branches are entered in Figure 20-3 in fractional form so that they can be traced back to the table of Leslie's performance record.

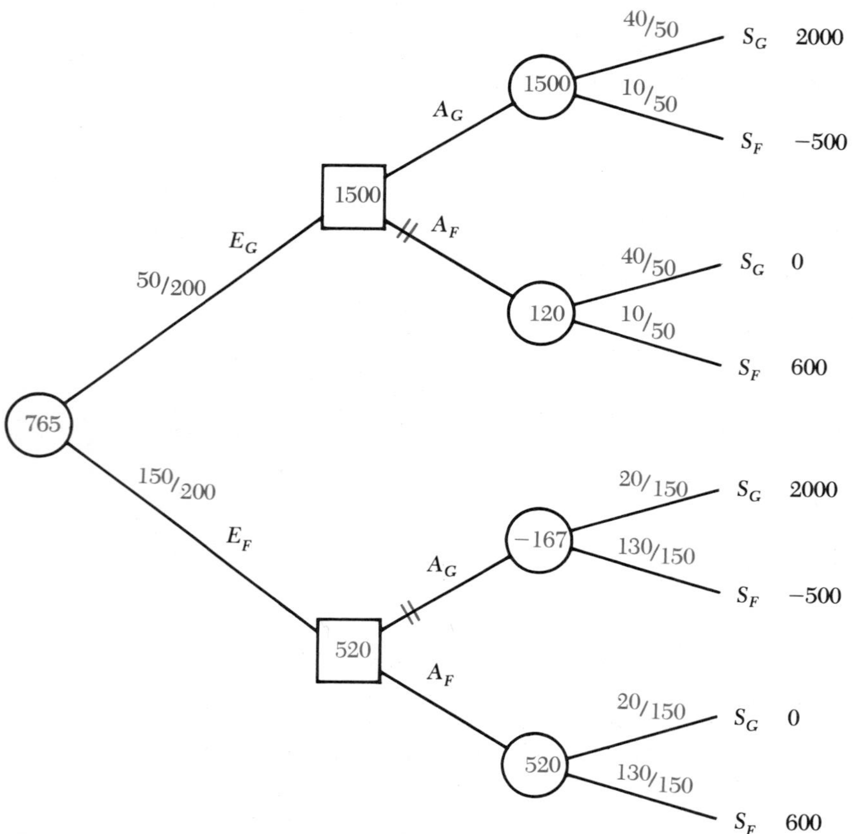

Figure 20-3 Solved Decision Tree for Imperfect Information

The probabilities that belong on the initial branches of the decision tree are the *marginal* probabilities of the possible *information outcomes*. The marginal probability that Leslie will predict the painting to be a Gresham is 50/200, and the probability is 150/200 for a prediction of a Fowler.

The decision tree is solved in Figure 20-3. Not surprisingly, the optimal act given E_G is A_G, and the optimal act given E_F is A_F. Thus inviting Leslie to make a judgment on the painting's artist can be seen to be a gamble with a 50/200 probability of E_G with a consequent expected payoff of \$1500 and a $^{150}/_{200}$ probability of E_F with a consequent expected payoff of \$520. This gamble has an expected value of \$765, which is the expected payoff of acting on the information that would be provided by Leslie. We use the symbols $EV(IA)$ for the *expected value* of the *information acts*. This terminology emphasizes that actions will follow the gathering of information. The expected value of the information, which will be called the *expected value* of *sample information* $EV(SI)$, is

$$EV(SI) = EV(IA) - EV(A_{\text{opt}}) \tag{20-2}$$

Here, as with the expected value of perfect information, $EV(A_{\text{opt}})$ is the expected value of the optimal act based on the prior probabilities, that is, without new information. For B.J.'s decision problem, $EV(A_{\text{opt}})$ was \$420 and

$$\begin{aligned} EV(SI) &= \$765 - \$420 \\ &= \$345 \end{aligned}$$

Leslie's expenses will be \$150, so the prospective value of the information is considerably greater in this instance than its cost. B.J. is now ready to make that return call.

A Calculation Method

If the number of information outcomes, possible acts, or states are large, the decision-tree method becomes rather cumbersome. Also, calculating the revised probabilities of states for each possible information outcome becomes tedious. The strength of the decision-tree approach is in showing the underlying logic. In this section, we shall illustrate a calculation method for finding the optimal acts conditional on sample information and the expected value of those information acts, $EV(IA)$, that can be carried out by hand for modest problems or put on the computer for larger problems. We shall first use it for the artist problem.

The inputs to the calculation are two tables. The first of these is the payoff table for act-state combinations arranged so the acts are the rows and the states are the columns. Call this the V_{ji} table. For the artist problem,

$$V_{ji} = \begin{pmatrix} \$2000 & -\$500 \\ \$0 & \$600 \end{pmatrix}$$

The second table is the *joint probabilities* of states and information outcomes, arranged so the states are the rows and the information outcomes are the columns. For the artist problem, this table is

$$P_{ik} = \begin{pmatrix} ^{40}/_{200} & ^{20}/_{200} \\ ^{10}/_{200} & ^{130}/_{200} \end{pmatrix}$$

Now define

$$v_{jk} = \sum_i [V_{ji} * P_{ik}] \tag{20-3}$$

From the two tables above, we calculate a table of values of v_{jk}. For example,

$$v_{11} = \sum_i [V_{1i} * P_{i1}]$$

$$= \$2000(^{40}/_{200}) + (-\$500)(^{10}/_{200}) = \$375$$

The rest of the v_{jk}'s are

$$v_{12} = \$2000(^{20}/_{200}) + (-\$500)(^{130}/_{200}) = -\$125$$

$$v_{21} = \$0(^{40}/_{200}) + \$600(^{10}/_{200}) = \$30$$

$$v_{22} = \$0(^{20}/_{200}) + \$600(^{130}/_{200}) = \$390$$

So the table of v_{jk} values is as follows.

Acts (A_j)	*Information* (E_k)	
	1 Gresham	*2 Fowler*
1 Act as if Gresham	\$375	–\$125
2 Act as if Fowler	\$30	\$390

The optimal act given an information outcome will be indicated by the act that has the maximum entry in the column corresponding to the information outcome.

$$v_{opt} \mid E_k = v_{max} \mid E_k \tag{20-4}$$

The expected value of the information acts will be the sum of the column maximum values that identified the optimal acts for each information outcome.

$$EV(IA) = \sum_k (v_{max} \mid E_k) \tag{20-5}$$

From the above table,

$$EV(IA) = \$375 + \$390 = \$765$$

which is the same result as from the decision tree.

A further convenience of this calculation method is that the expected values of the alternative acts in the absence of sampling can be determined from the table of v_{jk}. These are the expected values of the alternative acts using only the prior information, or probabilities. They are obtained by adding the v_{jk} entries for each row. In the present example,

$$EV(A_1) = \$375 + (-\$125) = \$250$$

$$EV(A_2) = \$30 + \$390 = \$420$$

With the expected value of the information acts and the expected value of the optimal act using only the prior probabilities, both available from the table of v_{jk}, we can then determine the expected value of sample information.

Exercises

20-9 *Customizing the sales presentation.* Suppose the agent in the insurance example of Exercise 20-1 uses the age of a prospect as a prediction of whether the prospect will prefer plan 1 or plan 2.

a. Draw a decision tree showing the possible information outcomes, alternative acts, and preference states.
b. Enter appropriate probabilities and payoffs and solve the decision tree. (*Note:* It will be helpful to refer to Exercises 19-2 and 20-1.)
c. What is the expected value of the information acts, $EV(IA)$?
d. What is the expected value of sample information, $EV(SI)$?

20-10 *Industrial intelligence.* Exercise 19-1 involved alternative acts that could be taken by a soft drink bottler in the face of uncertainty concerning the packaging strategy that would be adopted by a major competitor. Payoffs (in ten thousands of dollars) and probabilities were as follows.

Competitor's Action		*Bottler's Strategy*		
S	$P(S)$	A_1 (*lift-top*)	A_2 (*screw-top*)	A_3 (*standard*)
S_1 Standard	5/10	14	13	16
S_2 Screw-top	3/10	9	8	6
S_3 Lift-top	2/10	6	8	5

For an expenditure of $5000, the bottler believes that information about the competitor's action can be obtained that is 80% reliable. This means that given any true state (competitor's act) the probability is 8/10 that the information would predict the true state and 1/10 that the information would predict each of the incorrect states.

a. Calculate the revised probabilities for the competitor's acts given each of the intelligence predictions, carrying out joint probabilities and revised probabilities as fractions.
b. Construct and solve a decision tree for the information strategy. Carry out the work in fractions and leave the $5000 cost to be considered later.
c. What is the expected value of the (sample) information? How much better is the information strategy than the optimal act without seeking the information?

20-11 *Alternative calculation.*

a. Set up the V_{ji} (acts by states) payoff table for the bottler's problem in the preceding exercise.
b. Set up the P_{ik} (joint probabilities by states and information outcomes) table and carry out the sums of products calculations to find the elements in the v_{jk} (acts by information outcomes) table.
c. Identify the optimal act given each information outcome from the v_{jk} table and find the expected value of the information acts, $EV(IA)$. Is this the same result as in (b)?

20-12 *Range conditions forecast.** The ranch operator in Exercise 19-23 can obtain a forecast of range conditions during the season prior to stocking the ranch for the season. Over the past 44 seasons, these forecasts have been as follows.

Actual Conditions	*Forecast Conditions*			
	Poor	*Normal*	*Good*	*Total*
Poor	5	2	0	7
Normal	5	14	3	22
Good	1	5	9	15
Total	11	21	12	44

a. Use the payoff table for the alternative herd sizes in Exercise 19-23 and the joint probability distribution of states and information outcomes to obtain the v_{jk} table for forecast conditions versus herd size.

b. From the table of v_{jk}, determine the optimal herd size given each forecast, the expected value of these optimal acts, the expected value of the alternative herd sizes in the absence of a forecast, and the expected value (incremental) of the forecast.

20.3 Bayesian Statistics for Population Means

Previous sections on decision making have emphasized the elements in decision problems. These were seen in Chapter 19 to be the act-states payoff table and the probability distribution of states. In Section 20.1, we saw how new evidence, or information, is used to revise the probabilities of states. We shall now turn our attention to the situation in which the parameter that affects payoffs given any act is the population mean. In this section, we look at how the information of a random sample is used to revise the prior probabilities for a population mean. In the next section, we shall study how belief about the population mean can be used to select the optimal act in a decision making situation.

Revision Given a Single Observation

We begin with an example involving a single observation.

Example

The fills dispensed by a flour-bagging machine are normally distributed with a standard deviation of 2.0 lb. Yesterday morning the machine filled 600 bags with a true mean setting of 102.0 lb and the bags were stacked at one end of a storage shed.

*Adapted from Albert N. Halter and Gerald W. Dean, *Decisions Under Uncertainty with Research Applications* (Cincinnati, Ohio: South-Western Publishing Company, 1971).

Yesterday afternoon the machine filled 400 bags, which were stacked at the other end of the shed. It was discovered late in the day that in the afternoon the setting of the machine had been accidentally changed to 100.0 lb. This morning the shed supervisor finds a bag lying between the two stacks. Quickly weighing it, the supervisor finds the content weight to be 99 lb (to the nearest pound), that is, a true weight between 98.5 and 99.5 lb. What is the probability that the bag came from yesterday morning's production?

This is a problem involving revision of probabilities. Before weighing the bag, the probability that the bag came from yesterday morning's stack is $600/(600 + 400) = 0.60$. But now we must take account of the evidence of the weight of the bag. We need to know the probability of a bag weight between 98.5 and 99.5 lb if the bag came from a normal population with a mean of 102.0 and a standard deviation of 2.0 (yesterday morning's) on the one hand, and the probability of the same outcome if the observation were drawn from a normal population with a mean of 100.0 and a standard deviation of 2.0 (yesterday afternoon's population). We can find these answers with the normal table of areas.

Given $\mu = 102.0$:

$$z = \frac{99.5 - 102.0}{2.0} = -1.25 \qquad P(Z \leq -1.25) = 0.1056$$

$$z = \frac{98.5 - 102.0}{2.0} = -1.75 \qquad P(Z \leq -1.75) = \underline{0.0401}$$

$$0.0655$$

Given $\mu = 100.0$:

$$z = \frac{99.5 - 100.0}{2.0} = -0.25 \qquad P(Z \leq -0.25) = 0.4013$$

$$z = \frac{98.5 - 100.0}{2.0} = -0.75 \qquad P(Z \leq -0.75) = \underline{0.2266}$$

$$0.1747$$

Now we have the probabilities of the evidence given each state and we can proceed with the revision of probabilities.

States (S)	P(S)	P(E | S)	P(S) * P(E | S)	P(S | E)
$\mu = 102.0$	0.60	0.0655	0.0393	0.36
$\mu = 100.0$	0.40	0.1747	0.0699	0.64
			0.1092	1.00

We find the evidence considerably changes the belief about the states. The revised probability that the errant bag came from yesterday morning's output ($\mu = 102.0$) is only 0.36, as compared with the probability of 0.60 in the absence of the new information.

Revision Given a Sample Mean

Now suppose there were four bags found of unknown origin. The superintendent is told that the night custodian pulled them all from the same stack, but cannot remember which. The mean weight of the four bags is 100.8 to the nearest tenth of a pound. What are the revised probabilities for the origin of the bags? The procedure is similar, but this time we must find the *probability of a sample mean* (between 100.75 and 100.85), given each of the possible parent populations. The standard error of the mean of a random sample of size $n = 4$ from either population is $2.0/\sqrt{4} = 1.0$ (ignoring the finite population multiplier).

The z-values and resulting probabilities are as follows.

Given $\mu = 102.0$:

$$z = \frac{100.85 - 102.0}{1.0} = -1.15 \qquad P(Z \leq -1.15) = 0.1251$$

$$z = \frac{100.75 - 102.0}{1.0} = -1.25 \qquad P(Z \leq -1.25) = \underline{0.1056}$$

$$0.0195$$

Given $\mu = 100.0$:

$$z = \frac{100.85 - 100.0}{1.0} = 0.85 \qquad P(Z \leq 0.85) = 0.8023$$

$$z = \frac{100.75 - 100.0}{1.0} = 0.75 \qquad P(Z \leq 0.75) = \underline{0.7734}$$

$$0.0289$$

And now we carry out the revision.

States (S)	P(S)	P(E | S)	P(S) * P(E | S)	P(S | E)
$\mu = 102.0$	0.60	0.0195	0.01170	0.50
$\mu = 100.0$	0.40	0.0289	0.01156	0.50
			0.02326	1.00

It is important to appreciate that the probability of the sample evidence given each state is determined from the sampling distribution of the mean *conditional* on that state. Also, it is important to see that the standard deviation of the sampling distributions (in both instances, $\sigma_{\bar{X}} = 1.0$) is crucial in determining these probabilities.

Prior Odds, Likelihood Ratio, and Revised Odds

In the preceding examples, there were only two possible states. In such cases, it is common to think in terms of odds in favor of one state over the other. For example, the prior odds favoring $\mu = 102.0$ over $\mu = 100.0$ in both examples was 6 to 4. In the two-state case, Bayes' theorem can be phrased in terms of odds. The theorem is simply that the revised odds are the product of the prior odds and the likelihood ratio. In the example involving the mean weight of four bags, the ratio of the likelihood of the sample mean given $\mu = 102.0$ to the likelihood given $\mu = 100.0$ is 0.0195/0.0289. Multiplying the prior odds by the likelihood ratio gives

$$\left(\frac{6}{4}\right)\left(\frac{0.0195}{0.0289}\right) = \frac{0.1170}{0.1156} \simeq \frac{1}{1}$$

The revised odds are very close to even that the bags came from the population with a mean of 102.0 as opposed to the population with a mean of 100.0.

Revision of a Normal Prior for a Population Mean

We now give an example in which the statistical elements parallel those in the previous example, except that the prior distribution of states is continuous rather than discrete.

Example

An experimental automobile has just been completed. It has been designed for gasoline economy; the belief held by the development team about the average miles per gallon (mpg) that the car would attain in city driving can be expressed by a normal probability distribution with a mean of 100 mpg and a standard deviation of 3.0 mpg. The car is supplied with a carefully measured gallon of gasoline and driven under city driving conditions until the gasoline runs out. This is repeated 75 times. The mean of the 75 trials is 104.0 mi and the standard deviation is 15 mi. Assuming that the evidence constitutes a random sample, what is the revised belief about the long-run average mileage in city driving?

Notice that the true long-run mean, μ, is viewed as a random variable. Before the experimental runs, the probability distribution of this random variable has

$$E(\mu) = 100.0$$

$$\sigma_\mu = 3.0$$

The sample provides new evidence about the long-run population mean. The evidence is a sample mean of 104.0 mi for 75 random observations. We now want to revise the prior probability distribution of states (population mean) in the light of this evidence. First, we need to define two quantities.

$$I_P = \frac{1}{\sigma_\mu^2} \tag{20-6}$$

$$I_S = \frac{1}{s_{\bar{X}}^2} \tag{20-7}$$

where I_P can be thought of as the information content of the prior and I_S as the information content of the sample evidence. Note that the smaller the variances in Equations 20-6 and 20-7, the greater the information content. Our objective is to find the mean and standard deviation of the revised belief about the population mean, μ. Under the conditions of a normal prior distribution and approximately normal sampling distribution of the (sample) mean, the revised probability distribution of the population mean is normal with

$$E'(\mu) = \frac{I_P * E(\mu) + I_S * \bar{x}}{I_P + I_S} \tag{20-8}$$

$$(\sigma'_\mu)^2 = \frac{1}{I_p + I_S} \tag{20-9}$$

Here, the primes distinguish the revised parameters from the prior parameters.

In the formula for the revised expectation of the population mean, $E'(\mu)$, it can be seen that the revised mean is a weighted average of the mean (expectation) of the prior distribution and the sample mean. The weights are the information content values, which are the reciprocals of the variance of the prior distribution and the variance of the sampling distribution of the mean, respectively. The prior expectation and the sample mean are being combined with weights that reflect the degree of certainty about the population mean implicit in the two values. This is the idea of information content. For example, a prior distribution with a standard deviation of 5.0 mpg would have less information content than the development team's prior distribution and a sample mean of 100 trials would have more information content than the sample mean of 75 trials.

The results of our present example are

$$I_P = \frac{1}{\sigma^2_\mu} = \frac{1}{(3)^2} = \frac{1}{9}$$

$$I_S = \frac{1}{s^2_{\bar{X}}} = \frac{1}{s^2/n} = \frac{1}{15^2/75} = \frac{1}{3}$$

$$E'(\mu) = \frac{(1/9)(100.0) + (1/3)(104.0)}{(1/9) + (1/3)} = 103.0$$

$$(\sigma'_\mu)^2 = \frac{1}{(1/9) + (1/3)} = \frac{1}{4/9}$$

$$= 2.25$$

The denominator of the formula for the variance of the revised distribution, that is, the reciprocal of the revised variance, can be thought of as the information content of the revised distribution. It is the sum of the information contents of the prior distribution and the sample mean. Notice that the information content of the prior distribution in our example is $1/9$, the information content of the sample mean is $3/9$, and the information content of the revised distribution is the sum of these, or $4/9$. The revised variance is the reciprocal of $4/9$, or 2.25, so the standard deviation of the revised distribution is

$$\sigma'_\mu = \sqrt{2.25}$$

$$= 1.5$$

The team's revised belief about the long-run average mileage for the experimental car is a normal distribution with a mean of 103.0 and a standard deviation of 1.5. This is a considerable change from their prior belief. For example, before the experimental runs, the team held a 0.50 probability that the long-run average mileage would exceed 100.0 mpg. Now, after the runs, the probability that the average exceeds 100 mpg is about 0.975. Figure 20-4 shows these probabilities from the prior and the revised distributions.

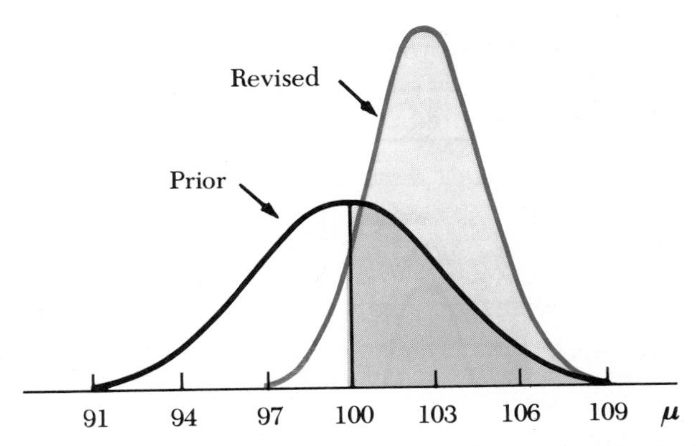

Figure 20-4 Prior and Revised Distributions for Population Mean

20.4 Decision Making Involving Means

In the previous section, we were concerned with revising the probability distribution of a population mean on the basis of sample information. In this section, we look at decisions in which the population mean is the critical parameter. This means that the payoffs or cost consequences of alternative acts will depend on the true value of the population mean. Since a probability distribution will have been assessed for the population mean—either a prior or a revised distribution (if sample data have been collected)—the optimal act can be determined.

Inferences about means—hypothesis testing and estimation—are central topics in classical statistics. However, two important elements not a part of classical procedures are added when decision making involving means is viewed from the Bayesian perspective:

1. The true mean is viewed as a random variable. The probability distribution assessed for the true mean may be a prior or a revised one. Subjective beliefs may be incorporated in and expressed by the probabilities assigned.
2. Cost consequences of alternative acts are explicitly included in the analysis.

The Bayesian approach represents an alternative to basing a decision involving a population mean on classical inference procedures. Therefore we shall at times in this section use the language of statistical inference in discussing the Bayesian procedures and results. This will enable us to compare the procedures at certain points.

Two Possible Values

The problem of the unidentified flour bags in the preceding section is an example in which there are two possible values for the population mean. Either the bags came from yesterday morning's output ($\mu = 102.0$) or from yesterday afternoon's output ($\mu = 100.0$). Suppose the seriousness of decision errors is represented by the following opportunity costs (regrets) measured in utility terms.

True State	*Decision*	
	$\mu = 102.0$	$\mu = 100.0$
$\mu = 102.0$	0	400
$\mu = 100.0$	100	0

What this means is that concluding that the sample came from a population with a mean of 100.0 when the parent population mean is really 102.0 is four times as serious an error as concluding that the parent population mean is 102.0 when it is really 100.0. It is more important that yesterday morning's good production run not be shipped out incomplete than it is that the substandard run not be missing four bags. We are assuming that such considerations have been duly weighed in arriving at the opportunity cost numbers.

Once the consequences have been expressed numerically, all that remains is to work out the expected opportunity loss of the alternative acts. In the present case, the revised probabilities—after observing the mean weight of the four bags—were

$$P(\mu = 102.0) = 0.50 \qquad P(\mu = 100.0) = 0.50$$

The expected opportunity losses then are

Act 1: Conclude $\mu = 102.0$ and stack the bags with yesterday morning's output.

$$ER(A_1) = 0.50(0) + 0.50(100) = 50$$

Act 2: Conclude $\mu = 100.0$ and stack the bags with yesterday afternoon's output.

$$ER(A_2) = 0.50(400) + 0.50(0) = 200$$

The optimal act is clearly A_1, despite the fact that the probabilities do not favor yesterday morning's output as the source of the unidentified bags. A little thought will reveal that, with the cost consequences given, the bags would be stacked with yesterday morning's output as long as the probability exceeded 0.2 that the population ($\mu = 102.0$) produced the sample. Let us look at the situation in terms of hypothesis testing.

True State	*Decision*	
	Accept H_0	*Accept H_A*
H_0 true ($\mu = 102.0$)		Type I error $c = 400$
H_A true ($\mu = 100.0$)	Type II error $c = 100$	

Since a type I error is four times as costly as a type II error in this situation, we would not accept any gamble where the risk of that graver error is any more than 1/4 the risk of the less serious error. We avoid type I errors by accepting the null hypothesis. We shall risk the more serious type I error only if the probability of error in rejecting the null hypothesis is less than 1/4 the probability of error in accepting the null hypothesis.

From these considerations, we can formulate a Bayesian decision rule for the two-alternative case. Where the alternatives are H_0 and H_A, the optimal decision is to accept H_0 as long as the probability that H_0 is true exceeds the critical probability

$$P_c = \frac{c_{\mathrm{II}}}{c_{\mathrm{I}} + c_{\mathrm{II}}} \tag{20-10}$$

where c_{I} and c_{II} are the opportunity losses from commission of type I and type II errors, respectively.

In the current example,

$$P_c = \frac{100}{400 + 100} = 0.2$$

Since the probability that H_0 ($\mu = 102.0$) is true is 0.50, which exceeds the critical probability, the optimal decision is to accept H_0.

In classical hypothesis testing, we do not deal with probabilities that the hypothesis is true, or explicitly with cost consequences of type I and type II errors. Nevertheless, some comments can be made about classical hypothesis testing from a Bayesian viewpoint.

1. If the null and alternative hypotheses are regarded in some sense as equally likely, then classical hypothesis tests using modest to small sample sizes and small allowable type I error risks, in which it is relatively difficult to reject the null hypothesis, implicitly assume that the consequences of type I error are more serious than type II error.

2. Alternatively, if the consequences of type I and type II errors are regarded as comparable in the situation described above, classical hypothesis testing makes sense from a Bayesian viewpoint if there is an implicit high prior probability that the null hypothesis is true.

It should be noted that the circumstances described do generally correspond to the way practicing classical statisticians operate. The null hypothesis is chosen to represent a belief that is sensible to hold unless there is substantial evidence that it is false. There is perhaps less difference in practice between "classicists" and "Bayesians" than in the languages they use to explain what they do.

Bayesian Estimation

We treated the preceding example, in which there were two possible values for the true mean and alternative acts to accept the one or the other, in the language of hypothesis testing. It could just as well have been considered as an estimation problem. There are two possible estimates of the mean, namely the two possible true values. The optimal estimate is the one with the smaller expected opportunity loss. In the example, estimating the true mean to be 102.0 has a smaller expected opportunity loss than the alternative estimate.

In the language of estimation, the rule given in connection with Formula 20-10 can be stated as follows.

> Where c_u is the opportunity loss of an underestimate and c_o is the opportunity loss of an overestimate, the optimal estimate of the mean is the value at which the $c_u/(c_u + c_o)$ cumulative lower-tail probability is reached.
>
> $$\hat{\mu}_{\text{opt}} = \mu_{c_u/(c_u + c_o)} \qquad \textbf{(20-11)}$$

Recalling the example, the revised probabilities and resulting cumulative lower-tail probability distribution of the mean are

Mean (μ)	$P(\mu)$	*Cum P*
100.0	0.50	0.50
102.0	0.50	1.00

The opportunity cost of an underestimate of the mean is 400, the regret of concluding the mean is 100.0 when it is really 102.0, and the opportunity cost of an overestimate is 100. The ratio $c_u/(c_u + c_o)$ is $400/(400 + 100) = 0.80$. It is clear that this cumulative probability is reached at $\mu = 102.0$.

Viewing this kind of problem as one of estimation is probably simpler than applying the language of hypothesis testing. We do not have to remember which is the null and which is the alternative hypothesis and which error is the type I and which is the type II error. If the cost consequences of an underestimate are larger than the cost of an overestimate, the critical ratio will be high; otherwise it will be low.

In the example of the gasoline mileage of the experimental car, we found a revised probability distribution of the long-run average miles per gallon in city driving. The distribution was normal with a mean of 103.0 and a standard deviation of 1.5 mpg. Here the state variable is continuous. The optimal point estimate of the true mean will depend on costs of underestimating and overestimating. In this situation, however, the consequence of errors of estimate may depend on the sizes of the errors. A large overestimate is likely to be more costly than a small overestimate, and a large underestimate is more costly than a small underestimate. A situation where these costs would be explicit is the following.

Example

A printer is approached by a professional association with 20,000 members. From time to time, the association offers members, through mailings, the opportunity to buy items useful to members at the same time that they promote the organization. The printer has been asked about printing up packets of personal note cards. The outside of the cards would feature the emblem of the association, and they would be advertised to members at \$2.25 a packet. The printer's cost would be \$2.00 a packet based on quantities anticipated by the organization, but the organization will pay only for the packets actually ordered by the members. They do offer to conduct a trial mailing at their expense based on a random sample of 100 members. This is done, and a revised distribution of the average number of packets ordered per member is found to be normal with a mean of 5.2 and a standard deviation of 0.6 packets. What is the optimal number of packets for the printer to print?

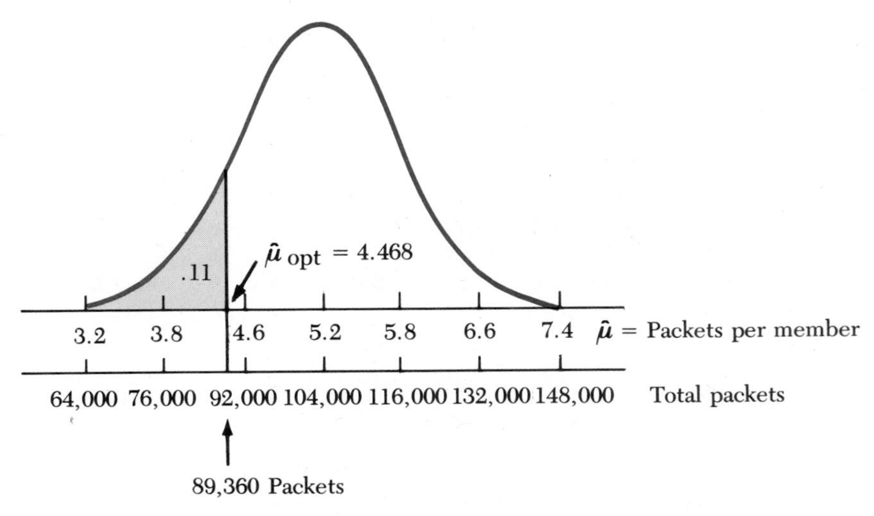

Figure 20-5 Optimal Estimate of μ and Optimal Stock Level Problem

You may recognize this kind of problem as the continuous equivalent of the newsstand optimal order problem with which we began Chapter 19. Figure 20-5 shows the normal revised distribution of average sales per member. Since total sales to members is 20,000 times sales per member, we show another scale for total sales under the mean sales scale. In terms of the marginal approach to the optimal stock level problem discussed in Section 19.4, the printer would print and hold in stock a number such that the probability of selling the last unit stocked would be

$$\frac{c_o}{c_o + c_u} = \frac{\$2.00}{\$2.00 + \$0.25} = 0.89$$

This calls for a low stock level because the gain from selling an additional packet is small compared to the loss from having it remain unsold. In terms of estimates, the cost of overestimating demand rises eight times as fast with the amount of error as does the cost of underestimating. The optimal estimate of the average order per customer to use to determine number of packets to print will be determined by the same rule as in Equation 20-11, where c_o and c_u are the costs per unit of overestimate and underestimate, respectively. The optimal estimate will lie at the

$$\frac{c_u}{c_o + c_u} = \frac{\$0.25}{\$2.00 + \$0.25} = 0.11$$

level of the lower-tail cumulative probability distribution of the true mean. For a normal distribution, this level is at $z = -1.22$, and

$$\begin{aligned}\hat{\mu}_{\text{opt}} &= E'(\mu) + (-1.22)\sigma'_{\mu} \\ &= 5.2 - 1.22(0.6) \\ &= 4.468\end{aligned}$$

which corresponds to 4.468(20,000) = 89,360 packets.

Exercises

20-13 *Predicting sales success.* In the past few years, an insurance company has given its new sales agents a sales aptitude test. Agents who proved successful during their first year with the firm had a mean score of 56 on the test and agents who did not prove successful during the first year had a mean score of 49. The distribution of scores for both groups was normal with a standard deviation of 5 points. A recently employed agent scored 53 on the test.

a. What is the probability of a score of 53 (to the nearest point) if the new agent belongs to the successful population?

b. What is the probability of a score of 53 if the new agent belongs to the unsuccessful population?

c. The personnel officer who interviewed the agent gave a subjective probability that the agent would succeed in the first year as 0.70. Revise this subjective probability for the evidence provided by the aptitude test.

20-14 *Credit-scoring systems.** A finance company that purchases sales contracts on mobile homes developed a scoring system for rating the safety of contracts. The scoring system took account of items such as the age of the creditor, time on present job, amount of down payment, number of children, and so on. Analysis was made of a large number of successful and unsuccessful (home was repossessed) loans. The mean score for the successful loans was 31 and the mean score for the bad loans was 27. Both populations of scores were normally distributed with a standard deviation of 5 points.

a. What is the probability of a score of 26 from the successful loan group?
b. What is the probability of a score of 26 from the unsuccessful loan group?
c. During a period when 80% of all loans made are successful, what is the probability that an application with a score of 26 will be a successful loan?

20-15 *Distinguishing product quality.* In the manufacture of a pin for a certain gear assembly, the standard for mean breaking strength of the pins is 20 lb, and it is important that mean breaking strength not be 19 lb or less. Assume that the standard deviation for a large lot of pins produced is 1.0 lb and that the prior odds in favor of a mean of 20 lb as against a mean of 19 lb are four to one.

a. If a sample of 4 pins yields a mean breaking strength of 19.3 lb (to the nearest tenth), what are the revised odds in favor of a mean of 20 as against 19 lb?
b. If a sample of 16 pins yields a mean breaking strength of 19.3 lb, what are the revised odds for a mean of 20 as against 19 lb? (Use nearest tenth.)

20-16 *Test marketing.* A manufacturer of home products sold through company representatives who call door-to-door in their neighborhoods has developed a new food blender. Based on similar product introductions in the past, the company has assessed the probability distribution of first-month sales per representative as normal with a mean of 21.0 and a standard deviation of 2.0. To gain further evidence, 36 representatives are randomly selected to test-market the new blender in their areas for a month. The average for the sample of 36 was 22.4 sales and the standard deviation was 6.0 sales.

a. Before the test marketing, what was the probability that sales per month per representative were fewer than 17.5?
b. Find the mean and standard deviation of probability distribution of sales per month per representative after the results of the test marketing are in.
c. Based on the revised distribution, what is the probability that mean first month sales are fewer than 20.0?

20-17 *Silkworm yields.* Silkworms spin cocoons containing between 500 and 1300 yd of fiber. In a particular silkworm operation, the operator's prior (to harvest) distribution of mean yield from cocoons is normally distributed with a mean of 800 and a standard deviation of 30 yd. At the beginning of the harvest, a random sample of 49 cocoons yields a mean of 830 yd and a standard deviation of 140 yd of fiber.

a. Find the mean and standard deviation of the revised distribution of mean yield.

---

*Adapted from James H. Myers and Edward W. Forgy, "The Development of Numerical Credit Evaluation Systems," *Journal of the American Statistical Association*, September 1963, pp. 799–806.

b. Use the revised distribution to find the probability that mean yield will exceed 800 yd.
c. Assuming that cocoon yields are normally distributed, estimate the probability that a single cocoon will yield over 800 yd of silk fiber.
d. Would it matter in your answer to (b) whether single cocoon yields were normally distributed? Explain.

20-18 *Credit-granting decision.* Failure to grant a loan that would be successful involves foregone interest income and granting a loan that proves unsuccessful involves bad-debt loss. Should credit be granted on a loan application with a score of 26 in Exercise 20-14 in each case?
a. If, in the period involved, the potential bad-debt loss is three times the potential interest income?
b. If, in the period involved, the potential bad-debt loss is two times the potential interest income?

20-19 *Product quality decision.* If the quality control department in Exercise 20-15 concludes that the mean breaking strength is 20 when it is really 19, the opportunity loss will be 1.5 times the opportunity loss from concluding that the mean is 19 when it is really 20. What is the optimal decision given the sample result in each case?
a. Part (a). b. Part (b).

20-20 *Optimal production run.* The home products manufacturer in Exercise 20-16 has 1000 sales representatives. Results of the test marketing are to be used to establish the size of production run to support first month sales of the new blender. The cost accounting department estimates that the opportunity cost of understocking per unit is six times the opportunity cost of overstocking per unit. What is the optimal size of production run to support first month sales?

20-21 *Estimating mean thread yield.* Silk thread is produced by winding together the fibers from several silkworm cocoons. The machine time necessary for this operation is directly proportional to the product of mean fiber yield per cocoon and the number of cocoons to be harvested. Assume the number of cocoons to be harvested is known and that the marginal loss per unit overestimate of machine time is 2.0 times the marginal loss per unit underestimate.
a. What is the optimal estimate of mean yield to use in estimating machine time prior to sampling the current harvest in Exercise 20-17?
b. What is the optimal estimate of mean yield given the results of the sample of 49 cocoons from the current harvest?

Summary

New information in a decision-making situation can change our belief about alternative states and thus change the expected values of the alternative acts. *Bayes' theorem* is used to reevaluate probabilities of states in the light of new information. Once probabilities of states have been revised, alternative acts and the expected value of perfect information can be reevaluated.

The act of gathering new information before making a decision is itself an act with economic value that can be determined. A *decision tree* is constructed to incorporate the possible information outcomes, alternative acts, revised probabilities of states given each information outcome, and the act-state payoffs. Solving this decision tree yields the optimal acts conditional on each information outcome and the expected value of the information acts. The expected value of sample information is the excess of this value over the expected value of the optimal act in the absence of gathering new information. A tabular calculation method that produces these same values was also introduced.

When the decision parameter is a population mean, new information will take the form of a sample mean from the population. Revision of prior probabilities for the population mean was examined for two situations. In the first of these, the possible population means are discrete and limited, and in the second the possible values are continuous. Formulas for revising a normal prior distribution were given for the latter case.

Bayesian decision making involving means was examined for the discrete and continuous cases. In both of these cases, the problem reduces to optimal estimation of a mean in the light of costs of underestimation and overestimation. A solution method for the linear opportunity cost case was developed and illustrated.

Supplementary Exercises

20-22 *Mining processes.* Ace Mining Company uses two processes to recover minerals from ore. Process A requires more capital and will be more profitable than process B if the mineral content of the ore is high. The larger investment for process A will lead to a larger loss if the mineral content of the ore is low. Once installed, either process will be continued until the ore is exhausted because variable returns exceed variable costs on both processes even for low-grade ore. A cost analysis and preliminary consideration of ore quality for an ore body under consideration have produced the following.

Ore Quality		*Payoff (millions of dollars)*	
	P	*Process A*	*Process B*
Low	4/19	−4	−0.3
Medium	3/10	−1	0
High	3/10	4	2.5

a. Ace can of course choose not to mine the ore body at all. Find the best course of action given the above information and an assumption that the company's utility function is measured by money return.

b. What is the expected value of perfect information?

20-23 *Extensive ore tests.* For \$100,000, Ace Mining Company (Exercise 20-22) can conduct extensive tests on ore quality. Experience with these tests on 50 ore bodies subsequently mined has enabled Ace to construct the following table.

Test Result	*Actual Quality*			
	Low	*Medium*	*High*	*Total*
Low	15	3	2	20
Medium	5	10	5	20
High	0	2	8	10
Total	20	15	15	50

a. Set up a decision tree for the alternative of conducting extensive tests before deciding whether to mine with process A or process B or not to mine at all.
b. Solve the decision tree to find the expected value of the information acts, $EV(IA)$.
c. What is the expected value of conducting the extensive tests, $EV(SI)$?

20-24 *Ace Mining, continued.* Ace has another ore body under consideration. The payoff table is the same as in Exercise 20-22, but preliminary study has established that $P(\text{Low}) = 0.2$, $P(\text{Medium}) = 0.3$, and $P(\text{High}) = 0.5$.
a. Using conditional probabilities of test results given states determined from the table in Exercise 20-23, find the joint probabilities of states and information, P_{ik}.
b. Determine the table for v_{jk}. (Do not include the $100,000 cost of conducting extensive tests.)
c. From the table of v_{jk}, find the optimal act without extensive testing and its expected value.
d. From the table of v_{jk}, determine the optimal acts conditional on test outcomes and the expected value of the optimal acts, $EV(IA)$. Should extensive tests be conducted?

20-25 *Predicting employee turnover.* The personnel department of an organization employing a large professional staff has for several years administered a standard personality inventory instrument to newly employed staff members. Recently, a study was made of differences between employees leaving the company within two years and those remaining longer than two years. It was found that a score based on measures of responsibility, socialization, and tolerance distinguished maximally between the groups. The mean score for the "stayers" was 55, and the mean score for the "movers" was 45, and both distributions were normal. The standard deviation for both groups was 10. If 70% of new employees remain over two years, find each probability.
a. That a new employee with a score of 43 will stay.
b. That a new employee with a score of 53 will stay.

20-26 *Cash requirements.* A firm estimates its needs for cash at the end of the next quarter as a normal distribution with a mean of $650,000 and a standard deviation of $30,000. The firm considers the marginal consequences of underestimating cash needs to be four times as serious as the consequences of overestimating cash needs. What cash balance should be planned for the end of the next quarter?

20-27 *Tire lifetimes.* Premier Rent-a-Car Agency was assured by the manufacturer that the mean useful lifetime of the brand of tires used on its fleet was normal with a mean of 50 and a standard deviation of 3.0 thousand miles. Premier accepted this assessment as a prior distribution, but checked a random sample of 100 of the tires in use on its vehicles. The sample yielded a mean lifetime of 54 and a standard deviation of 10 thousand miles.

a. Revise the mean and standard deviation of the prior distribution for the sample results.
b. What is the probability that the mean useful lifetime of the brand of tire used exceeds 55 thousand miles?
c. Find the 5th and 95th percentiles of the revised probability distribution of mean useful lifetime.

20-28 *Continuous demand distribution.* A natural health food store sells Supermilk Punch, made from fresh milk and three natural additives. Each batch of punch is made to meet demand for 2 days because health regulations do not permit any longer storage time. The cost of the punch is $3 per gallon and it is sold for $8 a gallon. Two-day demand is estimated to be normally distributed with an expected demand of 30 gal and a standard deviation of 5 gal. What is the optimal amount of Supermilk Punch to make every 2 days?

APPENDIX A

REFERENCE TABLES

$P(R = r \mid n, \pi)$

n	r	.05	.10	.15	.20	.25	.30	.35	.40	.45	.50	.55	.60	.65	.70	.75	.80	.85	.90	.95
1	0	.9500	.9000	.8500	.8000	.7500	.7000	.6500	.6000	.5500	.5000	.4500	.4000	.3500	.3000	.2500	.2000	.1500	.1000	.0500
	1	.0500	.1000	.1500	.2000	.2500	.3000	.3500	.4000	.4500	.5000	.5500	.6000	.6500	.7000	.7500	.8000	.8500	9000	.9500
2	0	.9025	.8100	.7225	.6400	.5625	.4900	.4225	.3600	.3025	.2500	.2025	.1600	.1225	.0900	.0625	.0400	.0225	.0100	.0025
	1	.0950	.1800	.2550	.3200	.3750	.4200	.4550	.4800	.4950	.5000	.4950	.4800	.4550	.4200	.3750	.3200	.2550	.1800	.0950
	2	.0025	.0100	.0225	.0400	.0625	.0900	.1225	.1600	.2025	.2500	.3025	.3600	.4225	.4900	.5625	.6400	.7225	.8100	.9025
3	0	.8574	.7290	.6141	.5120	.4219	.3430	.2746	.2160	.1664	.1250	.0911	.0640	.0429	.0270	.0156	.0080	.0034	.0010	.0001
	1	.1354	.2430	.3251	.3840	.4219	.4410	.4436	.4320	.4084	.3750	.3341	.2880	.2389	.1890	.1406	.0960	.0574	.0270	.0071
	2	.0071	.0270	.0574	.0960	.1406	.1890	.2389	.2880	.3341	.3750	.4084	.4320	.4436	.4410	.4219	.3840	.3251	.2430	.1354
	3	.0001	.0010	.0034	.0080	.0156	.0270	.0429	.0640	.0911	.1250	.1664	.2160	.2746	.3430	.4219	.5120	.6141	.7290	.8574
4	0	.8145	.6561	.5220	.4096	.3164	.2401	.1785	.1296	.0915	.0625	.0410	.0256	.0150	.0081	.0039	.0016	.0005	.0001	.0000
	1	.1715	.2916	.3685	.4096	.4219	.4116	.3845	.3456	.2995	.2500	.2005	.1536	.1115	.0756	.0469	.0256	.0115	.0036	.0005
	2	.0135	.0486	.0975	.1536	.2109	.2646	.3105	.3456	.3675	.3750	.3675	.3456	.3105	.2646	.2109	.1536	.0975	.0486	.0135
	3	.0005	.0036	.0115	.0256	.0469	.0756	.1115	.1536	.2005	.2500	.2995	.3456	.3845	.4116	.4219	.4096	.3865	.2916	.1715
	4	.0000	.0001	.0005	.0016	.0039	.0081	.0150	.0256	.0410	.0625	.0915	.1296	.1785	.2401	.3164	.4096	.5220	.6561	.8145
5	0	.7738	.5905	.4437	.3277	.2373	.1681	.1160	.0778	.0503	.0312	.0185	.0102	.0053	.0024	.0010	.0003	.0001	.0000	.0000
	1	.2036	.3280	.3915	.4096	.3955	.3602	.3124	.2592	.2059	.1562	.1128	.0768	.0488	.0284	.0146	.0064	.0022	.0004	.0000
	2	.0214	.0729	.1382	.2048	.2637	.3087	.3364	.3456	.3369	.3125	.2757	.2304	.1811	.1323	.0879	.0512	.0244	.0081	.0011
	3	.0011	.0081	.0244	.0512	.0879	.1323	.1811	.2304	.2757	.3125	.3369	.3456	.3364	.3087	.2637	.2048	.1382	.0729	.0214
	4	.0000	.0004	.0022	.0064	.0146	.0284	.0488	.0768	.1128	.1562	.2059	.2592	.3124	.3602	.3955	.4096	.3915	.3280	.2036
	5	.0000	.0000	.0001	.0003	.0010	.0024	.0053	.0102	.0185	.0312	.0503	.0778	.1160	.1681	.2373	.3277	.4437	.5905	.7738
6	0	.7351	.5314	.3771	.2621	.1780	.1176	.0754	.0467	.0277	.0156	.0083	.0041	.0018	.0007	.0002	.0001	.0000	.0000	.0000
	1	.2321	.3543	.3993	.3932	.3560	.3025	.2437	.1866	.1359	.0938	.0609	.0369	.0205	.0102	.0044	.0015	.0004	.0001	.0000
	2	.0305	.0984	.1762	.2458	.2966	.3241	.3280	.3110	.2780	.2344	.1861	.1382	.0951	.0595	.0330	.0154	.0055	.0012	.0001
	3	.0021	.0146	.0415	.0819	.1318	.1852	.2355	.2765	.3032	.3125	.3032	.2765	.2355	.1852	.1318	.0819	.0415	.0146	.0021
	4	.0001	.0012	.0055	.0154	.0330	.0595	.0951	.1382	.1861	.2344	.2780	.3110	.3280	.3241	.2966	.2458	.1762	.0984	.0305
	5	.0000	.0001	.0004	.0015	.0044	.0102	.0205	.0369	.0609	.0938	.1359	.1866	.2437	.3025	.3560	.3932	.3993	.3543	.2321
	6	.0000	.0000	.0000	.0001	.0002	.0007	.0018	.0041	.0083	.0156	.0277	.0467	.0754	.1176	.1780	.2621	.3771	.5314	.7351
7	0	.6983	.4783	.3206	.2097	.1335	.0824	.0490	.0280	.0152	.0078	.0037	.0016	.0006	.0002	.0001	.0000	.0000	.0000	.0000
	1	.2573	.3720	.3960	.3670	.3115	.2471	.1848	.1306	.0872	.0547	.0320	.0172	.0084	.0036	.0013	.0004	.0001	.0000	.0000
	2	.0406	.1240	.2097	.2753	.3115	.3177	.2985	.2613	.2140	.1641	.1172	.0774	.0466	.0250	.0115	.0043	.0012	.0002	.0000
	3	.0036	.0230	.0617	.1147	.1730	.2269	.2679	.2903	.2918	.2734	.2388	.1935	.1442	.0972	.0577	.0287	.0109	.0026	.0002
	4	.0002	.0026	.0109	.0287	.0577	.0972	.1442	.1935	.2388	.2734	.2918	.2903	.2679	.2269	.1730	.1147	.0617	.0230	.0036
	5	.0000	.0002	.0012	.0043	.0115	.0250	.0466	.0774	.1172	.1641	.2140	.2613	.2985	.3177	.3115	.2753	.2097	.1240	.0406
	6	.0000	.0000	.0001	.0004	.0013	.0036	.0084	.0172	.0320	.0547	.0872	.1306	.1848	.2471	.3115	.3670	.3960	.3720	.2573
	7	.0000	.0000	.0000	.0000	.0001	.0002	.0006	.0016	.0037	.0078	.0152	.0280	.0490	.0824	.1335	.2097	.3206	.4783	.6983

Table A-1 Binomial Distributions

$P(R = r \mid n, \pi)$

n	r	.05	.10	.15	.20	.25	.30	.35	.40	.45	.50	.55	.60	.65	.70	.75	.80	.85	.90	.95
8	0	.6634	.4305	.2725	.1678	.1001	.0576	.0319	.0168	.0084	.0039	.0017	.0007	.0002	.0001	.0000	.0000	.0000	.0000	.0000
	1	.2793	.3826	.3847	.3355	.2670	.1977	.1373	.0896	.0548	.0312	.0164	.0079	.0033	.0012	.0004	.0001	.0000	.0000	.0000
	2	.0515	.1488	.2376	.2936	.3115	.2965	.2587	.2090	.1569	.1094	.0703	.0413	.0217	.0100	.0038	.0011	.0002	.0000	.0000
	3	.0054	.0331	.0839	.1468	.2076	.2541	.2786	.2787	.2568	.2188	.1719	.1239	.0808	.0467	.0231	.0092	.0026	.0004	.0000
	4	.0004	.0046	.0185	.0459	.0865	.1361	.1875	.2322	.2627	.2734	.2627	.2322	.1875	.1361	.0865	.0459	.0185	.0046	.0004
	5	.0000	.0004	.0026	.0092	.0231	.0467	.0808	.1239	.1719	.2188	.2568	.2787	.2786	.2541	.2076	.1468	.0839	.0331	.0054
	6	.0000	.0000	.0002	.0011	.0038	.0100	.0217	.0413	.0703	.1094	.1569	.2090	.2587	.2965	.3115	.2936	.2376	.1488	.0515
	7	.0000	.0000	.0000	.0001	.0004	.0012	.0033	.0079	.0164	.0312	.0548	.0896	.1373	.1977	.2670	.3355	.3847	.3826	.2793
	8	.0000	.0000	.0000	.0000	.0000	.0001	.0002	.0007	.0017	.0039	.0084	.0168	.0319	.0576	.1001	.1678	.2725	.4305	.6634
9	0	.6302	.3874	.2316	.1342	.0751	.0404	.0207	.0101	.0046	.0020	.0008	.0003	.0001	.0000	.0000	.0000	.0000	.0000	.0000
	1	.2985	.3874	.3679	.3020	.2253	.1556	.1004	.0605	.0339	.0176	.0083	.0035	.0013	.0004	.0001	.0000	.0000	.0000	.0000
	2	.0629	.1722	.2597	.3020	.3003	.2668	.2162	.1612	.1110	.0703	.0407	.0212	.0098	.0039	.0012	.0003	.0000	.0000	.0000
	3	.0077	.0446	.1069	.1762	.2336	.2668	.2716	.2508	.2119	.1641	.1160	.0743	.0424	.0210	.0087	.0028	.0006	.0001	.0000
	4	.0006	.0074	.0283	.0661	.1168	.1715	.2194	.2508	.2600	.2461	.2128	.1672	.1181	.0735	.0389	.0165	.0050	.0008	.0000
	5	.0000	.0008	.0050	.0165	.0389	.0735	.1181	.1672	.2128	.2461	.2600	.2508	.2194	.1715	.1168	.0661	.0283	.0074	.0006
	6	.0000	.0001	.0006	.0028	.0087	.0210	.0424	.0743	.1160	.1641	.2119	.2508	.2716	.2668	.2336	.1762	.1069	.0446	.0077
	7	.0000	.0000	.0000	.0003	.0012	.0039	.0098	.0212	.0407	.0703	.1110	.1612	.2162	.2668	.3003	.3020	.2597	.1722	.0629
	8	.0000	.0000	.0000	.0000	.0001	.0004	.0013	.0035	.0083	.0176	.0339	.0605	.1004	.1556	.2253	.3020	.3679	.3874	.2985
	9	.0000	.0000	.0000	.0000	.0000	.0000	.0001	.0003	.0008	.0020	.0046	.0101	.0207	.0404	.0751	.1342	.2316	.3874	.6302
10	0	.5987	.3487	.1969	.1074	.0563	.0282	.0135	.0060	.0025	.0010	.0003	.0001	.0000	.0000	.0000	.0000	.0000	.0000	.0000
	1	.3151	.3874	.3474	.2684	.1877	.1211	.0725	.0403	.0207	.0098	.0042	.0016	.0005	.0001	.0000	.0000	.0000	.0000	.0000
	2	.0746	.1937	.2759	.3020	.2816	.2335	.1757	.1209	.0763	.0439	.0229	.0106	.0043	.0014	.0004	.0001	.0000	.0000	.0000
	3	.0105	.0574	.1298	.2013	.2503	.2668	.2522	.2150	.1665	.1172	.0746	.0425	.0212	.0090	.0031	.0008	.0001	.0000	.0000
	4	.0010	.0112	.0401	.0881	.1460	.2001	.2377	.2508	.2384	.2051	.1596	.1115	.0689	.0368	.0162	.0055	.0012	.0001	.0000
	5	.0001	.0015	.0085	.0264	.0584	.1029	.1536	.2007	.2340	.2461	.2340	.2007	.1536	.1029	.0584	.0264	.0085	.0015	.0001
	6	.0000	.0001	.0012	.0055	.0162	.0368	.0689	.1115	.1596	.2051	.2384	.2508	.2377	.2001	.1460	.0881	.0401	.0112	.0010
	7	.0000	.0000	.0001	.0008	.0031	.0090	.0212	.0425	.0746	.1172	.1665	.2150	.2522	.2668	.2503	.2013	.1298	.0574	.0105
	8	.0000	.0000	.0000	.0001	.0004	.0014	.0043	.0106	.0229	.0439	.0763	.1209	.1757	.2335	.2816	.3020	.2759	.1937	.0746
	9	.0000	.0000	.0000	.0000	.0000	.0001	.0005	.0016	.0042	.0098	.0207	.0403	.0725	.1211	.1877	.2684	.3474	.3874	.3151
	10	.0000	.0000	.0000	.0000	.0000	.0000	.0000	.0001	.0003	.0010	.0025	.0060	.0135	.0282	.0563	.1074	.1969	.3487	.5987
11	0	.5688	.3138	.1673	.0859	.0422	.0198	.0088	.0036	.0014	.0005	.0002	.0000	.0000	.0000	.0000	.0000	.0000	.0000	.0000
	1	.3293	.3835	.3248	.2362	.1549	.0932	.0518	.0266	.0125	.0054	.0021	.0007	.0002	.0000	.0000	.0000	.0000	.0000	.0000
	2	.0867	.2131	.2866	.2953	.2581	.1998	.1395	.0887	.0513	.0269	.0126	.0052	.0018	.0005	.0001	.0000	.0000	.0000	.0000
	3	.0137	.0710	.1517	.2215	.2581	.2568	.2254	.1774	.1259	.0806	.0462	.0234	.0102	.0037	.0011	.0002	.0000	.0000	.0000
	4	.0014	.0158	.0536	.1107	.1721	.2201	.2428	.2365	.2060	.1611	.1128	.0701	.0379	.0173	.0064	.0017	.0003	.0000	.0000
	5	.0001	.0025	.0132	.0388	.0803	.1321	.1830	.2207	.2360	.2256	.1931	.1471	.0985	.0566	.0268	.0097	.0023	.0003	.0000
	6	.0000	.0003	.0023	.0097	.0268	.0566	.0985	.1471	.1931	.2256	.2360	.2207	.1830	.1321	.0803	.0388	.0132	.0025	.0001
	7	.0000	.0000	.0003	.0017	.0064	.0173	.0379	.0701	.1128	.1611	.2060	.2365	.2428	.2201	.1721	.1107	.0536	.0158	.0014
	8	.0000	.0000	.0000	.0002	.0011	.0037	.0102	.0234	.0462	.0806	.1259	.1774	.2254	.2568	.2581	.2215	.1517	.0710	.0137
	9	.0000	.0000	.0000	.0000	.0001	.0005	.0018	.0052	.0126	.0269	.0513	.0887	.1395	.1998	.2581	.2953	.2866	.2131	.0867
	10	.0000	.0000	.0000	.0000	.0000	.0000	.0002	.0007	.0021	.0054	.0125	.0266	.0518	.0932	.1549	.2362	.3248	.3835	.3293
	11	.0000	.0000	.0000	.0000	.0000	.0000	.0000	.0000	.0002	.0005	.0014	.0036	.0088	.0198	.0422	.0859	.1673	.3138	.5688

Table A-1 Binomial Distributions (*continued*)

$P(R = r \mid n, \pi)$

n	r	π = .05	.10	.15	.20	.25	.30	.35	.40	.45	.50	.55	.60	.65	.70	.75	.80	.85	.90	.95
12	0	.5404	.2824	.1422	.0687	.0317	.0138	.0057	.0022	.0008	.0002	.0001	.0000	.0000	.0000	.0000	.0000	.0000	.0000	.0000
	1	.3413	.3766	.3012	.2062	.1267	.0712	.0368	.0174	.0075	.0029	.0010	.0003	.0001	.0000	.0000	.0000	.0000	.0000	.0000
	2	.0988	.2301	.2924	.2835	.2323	.1678	.1088	.0639	.0339	.0161	.0068	.0025	.0008	.0002	.0000	.0000	.0000	.0000	.0000
	3	.0173	.0852	.1720	.2362	.2581	.2397	.1954	.1419	.0923	.0537	.0277	.0125	.0048	.0015	.0004	.0001	.0000	.0000	.0000
	4	.0021	.0213	.0683	.1329	.1936	.2311	.2367	.2128	.1700	.1208	.0762	.0420	.0199	.0078	.0024	.0005	.0001	.0000	.0000
	5	.0002	.0038	.0193	.0532	.1032	.1585	.2039	.2270	.2225	.1934	.1489	.1009	.0591	.0291	.0115	.0033	.0006	.0000	.0000
	6	.0000	.0005	.0040	.0155	.0401	.0792	.1281	.1766	.2124	.2256	.2124	.1766	.1281	.0792	.0401	.0155	.0040	.0005	.0000
	7	.0000	.0000	.0006	.0033	.0115	.0291	.0591	.1009	.1489	.1934	.2225	.2270	.2039	.1585	.1032	.0532	.0193	.0038	.0002
	8	.0000	.0000	.0001	.0005	.0024	.0078	.0199	.0420	.0762	.1208	.1700	.2128	.2367	.2311	.1936	.1329	.0683	.0213	.0021
	9	.0000	.0000	.0000	.0001	.0004	.0015	.0048	.0125	.0277	.0537	.0923	.1419	.1954	.2397	.2581	.2362	.1720	.0852	.0173
	10	.0000	.0000	.0000	.0000	.0000	.0002	.0008	.0025	.0068	.0161	.0339	.0639	.1088	.1678	.2323	.2835	.2924	.2301	.0988
	11	.0000	.0000	.0000	.0000	.0000	.0000	.0001	.0003	.0010	.0029	.0075	.0174	.0368	.0712	.1267	.2062	.3012	.3766	.3413
	12	.0000	.0000	.0000	.0000	.0000	.0000	.0000	.0000	.0001	.0002	.0008	.0022	.0057	.0138	.0317	.0687	.1422	.2824	.5404
13	0	.5133	.2542	.1209	.0550	.0238	.0097	.0037	.0013	.0004	.0001	.0000	.0000	.0000	.0000	.0000	.0000	.0000	.0000	.0000
	1	.3512	.3672	.2774	.1787	.1029	.0540	.0259	.0113	.0045	.0016	.0005	.0001	.0000	.0000	.0000	.0000	.0000	.0000	.0000
	2	.1109	.2448	.2937	.2680	.2059	.1388	.0836	.0453	.0220	.0095	.0036	.0012	.0003	.0001	.0000	.0000	.0000	.0000	.0000
	3	.0214	.0997	.1900	.2457	.2517	.2181	.1651	.1107	.0660	.0349	.0162	.0065	.0022	.0006	.0001	.0000	.0000	.0000	.0000
	4	.0028	.0277	.0838	.1535	.2097	.2337	.2222	.1845	.1350	.0873	.0495	.0243	.0101	.0034	.0009	.0001	.0000	.0000	.0000
	5	.0003	.0055	.0266	.0691	.1258	.1803	.2154	.2214	.1989	.1571	.1089	.0656	.0336	.0142	.0047	.0011	.0001	.0000	.0000
	6	.0000	.0008	.0063	.0230	.0559	.1030	.1546	.1968	.2169	.2095	.1775	.1312	.0833	.0442	.0186	.0058	.0011	.0001	.0000
	7	.0000	.0001	.0011	.0058	.0186	.0442	.0833	.1312	.1775	.2095	.2169	.1968	.1546	.1030	.0559	.0230	.0063	.0008	.0000
	8	.0000	.0000	.0001	.0011	.0047	.0142	.0336	.0656	.1089	.1571	.1989	.2214	.2154	.1803	.1258	.0691	.0266	.0055	.0003
	9	.0000	.0000	.0000	.0001	.0009	.0034	.0101	.0243	.0495	.0873	.1350	.1845	.2222	.2337	.2097	.1535	.0838	.0277	.0028
	10	.0000	.0000	.0000	.0000	.0001	.0006	.0022	.0065	.0162	.0349	.0660	.1107	.1651	.2181	.2517	.2457	.1900	.0997	.0214
	11	.0000	.0000	.0000	.0000	.0000	.0001	.0003	.0012	.0036	.0095	.0220	.0453	.0836	.1388	.2059	.2680	.2937	.2448	.1109
	12	.0000	.0000	.0000	.0000	.0000	.0000	.0000	.0001	.0005	.0016	.0045	.0113	.0259	.0540	.1029	.1787	.2774	.3672	.3512
	13	.0000	.0000	.0000	.0000	.0000	.0000	.0000	.0000	.0000	.0001	.0004	.0013	.0037	.0097	.0238	.0550	.1209	.2542	.5133
14	0	.4877	.2288	.1028	.0440	.0178	.0068	.0024	.0008	.0002	.0001	.0000	.0000	.0000	.0000	.0000	.0000	.0000	.0000	.0000
	1	.3593	.3559	.2539	.1539	.0832	.0407	.0181	.0073	.0027	.0009	.0002	.0001	.0000	.0000	.0000	.0000	.0000	.0000	.0000
	2	.1229	.2570	.2912	.2501	.1802	.1134	.0634	.0317	.0141	.0056	.0019	.0005	.0001	.0000	.0000	.0000	.0000	.0000	.0000
	3	.0259	.1142	.2056	.2501	.2402	.1943	.1366	.0845	.0462	.0222	.0093	.0033	.0010	.0002	.0000	.0000	.0000	.0000	.0000
	4	.0037	.0349	.0998	.1720	.2202	.2290	.2022	.1549	.1040	.0611	.0312	.0136	.0049	.0014	.0003	.0000	.0000	.0000	.0000
	5	.0004	.0078	.0352	.0860	.1468	.1963	.2178	.2066	.1701	.1222	.0762	.0408	.0183	.0066	.0018	.0003	.0000	.0000	.0000
	6	.0000	.0013	.0093	.0322	.0734	.1262	.1759	.2066	.2088	.1833	.1398	.0918	.0510	.0232	.0082	.0020	.0003	.0000	.0000
	7	.0000	.0002	.0019	.0092	.0280	.0618	.1082	.1574	.1952	.2095	.1952	.1574	.1082	.0618	.0280	.0092	.0019	.0002	.0000
	8	.0000	.0000	.0003	.0020	.0082	.0232	.0510	.0918	.1398	.1833	.2088	.2066	.1759	.1262	.0734	.0322	.0093	.0013	.0000
	9	.0000	.0000	.0000	.0003	.0018	.0066	.0183	.0408	.0762	.1222	.1701	.2066	.2178	.1963	.1468	.0860	.0352	.0078	.0004
	10	.0000	.0000	.0000	.0000	.0003	.0014	.0049	.0136	.0312	.0611	.1040	.1549	.2022	.2290	.2202	.1720	.0998	.0349	.0037
	11	.0000	.0000	.0000	.0000	.0000	.0002	.0010	.0033	.0093	.0222	.0462	.0845	.1366	.1943	.2402	.2501	.2056	.1142	.0259
	12	.0000	.0000	.0000	.0000	.0000	.0000	.0001	.0005	.0019	.0056	.0141	.0317	.0634	.1134	.1802	.2501	.2912	.2570	.1229
	13	.0000	.0000	.0000	.0000	.0000	.0000	.0000	.0001	.0002	.0009	.0027	.0073	.0181	.0407	.0832	.1539	.2539	.3559	.3593
	14	.0000	.0000	.0000	.0000	.0000	.0000	.0000	.0000	.0000	.0001	.0002	.0008	.0024	.0068	.0178	.0440	.1028	.2288	.4877

Table A-1 Binomial Distributions (*continued*)

$P(R = r \mid n, \pi)$

n	r	.05	.10	.15	.20	.25	.30	.35	.40	.45	.50	.55	.60	.65	.70	.75	.80	.85	.90	.95
15	0	.4633	.2059	.0874	.0352	.0134	.0047	.0016	.0005	.0001	.0000	.0000	.0000	.0000	.0000	.0000	.0000	.0000	.0000	.0000
	1	.3658	.3432	.2312	.1319	.0668	.0305	.0126	.0047	.0016	.0005	.0001	.0000	.0000	.0000	.0000	.0000	.0000	.0000	.0000
	2	.1348	.2669	.2856	.2309	.1559	.0916	.0476	.0219	.0090	.0032	.0010	.0003	.0001	.0000	.0000	.0000	.0000	.0000	.0000
	3	.0307	.1285	.2184	.2501	.2252	.1700	.1110	.0634	.0318	.0139	.0052	.0016	.0004	.0001	.0000	.0000	.0000	.0000	.0000
	4	.0049	.0428	.1156	.1876	.2252	.2186	.1792	.1268	.0780	.0417	.0191	.0074	.0024	.0006	.0001	.0000	.0000	.0000	.0000
	5	.0006	.0105	.0449	.1032	.1651	.2061	.2123	.1859	.1404	.0916	.0515	.0245	.0096	.0030	.0007	.0001	.0000	.0000	.0000
	6	.0000	.0019	.0132	.0430	.0917	.1472	.1906	.2066	.1914	.1527	.1048	.0612	.0298	.0116	.0034	.0007	.0001	.0000	.0000
	7	.0000	.0003	.0030	.0138	.0393	.0811	.1319	.1771	.2013	.1964	.1647	.1181	.0710	.0348	.0131	.0035	.0005	.0000	.0000
	8	.0000	.0000	.0005	.0035	.0131	.0348	.0710	.1181	.1647	.1964	.2013	.1771	.1319	.0811	.0393	.0138	.0030	.0003	.0000
	9	.0000	.0000	.0001	.0007	.0034	.0116	.0298	.0612	.1048	.1527	.1914	.2066	.1906	.1472	.0917	.0430	.0132	.0019	.0000
	10	.0000	.0000	.0000	.0001	.0007	.0030	.0096	.0245	.0515	.0916	.1404	.1859	.2123	.2061	.1651	.1032	.0449	.0105	.0006
	11	.0000	.0000	.0000	.0000	.0001	.0006	.0024	.0074	.0191	.0417	.0780	.1268	.1792	.2186	.2252	.1876	.1156	.0428	.0049
	12	.0000	.0000	.0000	.0000	.0000	.0001	.0004	.0016	.0052	.0139	.0318	.0634	.1110	.1700	.2252	.2501	.2184	.1285	.0307
	13	.0000	.0000	.0000	.0000	.0000	.0000	.0001	.0003	.0010	.0032	.0090	.0219	.0476	.0916	.1559	.2309	.2856	.2669	.1348
	14	.0000	.0000	.0000	.0000	.0000	.0000	.0000	.0000	.0001	.0005	.0016	.0047	.0126	.0305	.0668	.1319	.2312	.3432	.3658
	15	.0000	.0000	.0000	.0000	.0000	.0000	.0000	.0000	.0000	.0000	.0001	.0005	.0016	.0047	.0134	.0352	.0874	.2059	.4633
16	0	.4401	.1853	.0743	.0281	.0100	.0033	.0010	.0003	.0001	.0000	.0000	.0000	.0000	.0000	.0000	.0000	.0000	.0000	.0000
	1	.3706	.3294	.2097	.1126	.0535	.0228	.0087	.0030	.0009	.0002	.0001	.0000	.0000	.0000	.0000	.0000	.0000	.0000	.0000
	2	.1463	.2745	.2775	.2111	.1336	.0732	.0353	.0150	.0056	.0018	.0005	.0001	.0000	.0000	.0000	.0000	.0000	.0000	.0000
	3	.0359	.1423	.2285	.2463	.2079	.1465	.0888	.0468	.0215	.0085	.0029	.0008	.0002	.0000	.0000	.0000	.0000	.0000	.0000
	4	.0061	.0514	.1311	.2001	.2252	.2040	.1553	.1014	.0572	.0278	.0115	.0040	.0011	.0002	.0000	.0000	.0000	.0000	.0000
	5	.0008	.0137	.0555	.1201	.1802	.2099	.2008	.1623	.1123	.0667	.0337	.0142	.0049	.0013	.0002	.0000	.0000	.0000	.0000
	6	.0001	.0028	.0180	.0550	.1101	.1649	.1982	.1983	.1684	.1222	.0755	.0392	.0167	.0056	.0014	.0002	.0000	.0000	.0000
	7	.0000	.0004	.0045	.0197	.0524	.1010	.1524	.1889	.1969	.1746	.1318	.0840	.0442	.0185	.0058	.0012	.0001	.0000	.0000
	8	.0000	.0001	.0009	.0055	.0197	.0487	.0923	.1417	.1812	.1964	.1812	.1417	.0923	.0487	.0197	.0055	.0009	.0001	.0000
	9	.0000	.0000	.0001	.0012	.0058	.0185	.0442	.0840	.1318	.1746	.1969	.1889	.1524	.1010	.0524	.0197	.0045	.0004	.0000
	10	.0000	.0000	.0000	.0002	.0014	.0056	.0167	.0392	.0755	.1222	.1684	.1983	.1982	.1649	.1101	.0550	.0180	.0028	.0001
	11	.0000	.0000	.0000	.0000	.0002	.0013	.0049	.0142	.0337	.0667	.1123	.1623	.2008	.2099	.1802	.1201	.0555	.0137	.0008
	12	.0000	.0000	.0000	.0000	.0000	.0002	.0011	.0040	.0115	.0278	.0572	.1014	.1553	.2040	.2252	.2001	.1311	.0514	.0061
	13	.0000	.0000	.0000	.0000	.0000	.0000	.0002	.0008	.0029	.0085	.0215	.0468	.0888	.1465	.2079	.2463	.2285	.1423	.0359
	14	.0000	.0000	.0000	.0000	.0000	.0000	.0000	.0001	.0005	.0018	.0056	.0150	.0353	.0732	.1336	.2111	.2775	.2745	.1463
	15	.0000	.0000	.0000	.0000	.0000	.0000	.0000	.0000	.0001	.0002	.0009	.0030	.0087	.0228	.0535	.1126	.2097	.3294	.3706
	16	.0000	.0000	.0000	.0000	.0000	.0000	.0000	.0000	.0000	.0000	.0001	.0003	.0010	.0033	.0100	.0281	.0743	.1853	.4401

Table A-1 Binomial Distributions (*continued*)

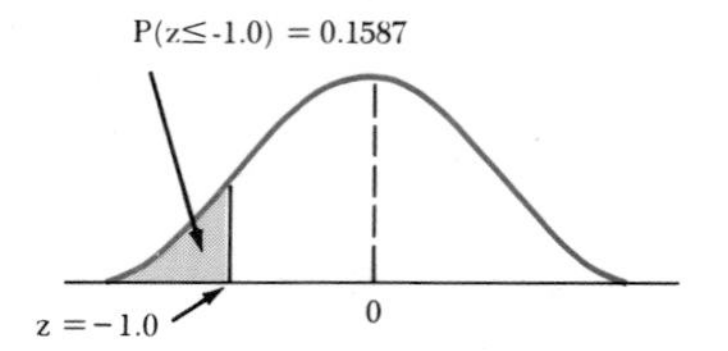

$P(Z \leq z)$

z	P	z	P	z	P
−3.09	.0010	−2.59	.0048	−2.09	.0183
−3.08	.0010	−2.58	.0049	−2.08	.0188
−3.07	.0011	−2.57	.0051	−2.07	.0192
−3.06	.0011	−2.56	.0052	−2.06	.0197
−3.05	.0011	−2.55	.0054	−2.05	.0202
−3.04	.0012	−2.54	.0055	−2.04	.0207
−3.03	.0012	−2.53	.0057	−2.03	.0212
−3.02	.0013	−2.52	.0059	−2.02	.0217
−3.01	.0013	−2.51	.0060	−2.01	.0222
−3.00	.0013	−2.50	.0062	−2.00	.0228
−2.99	.0014	−2.49	.0064	−1.99	.0233
−2.98	.0014	−2.48	.0066	−1.98	.0239
−2.97	.0015	−2.47	.0068	−1.97	.0244
−2.96	.0015	−2.46	.0069	−1.96	.0250
−2.95	.0016	−2.45	.0071	−1.95	.0256
−2.94	.0016	−2.44	.0073	−1.94	.0262
−2.93	.0017	−2.43	.0075	−1.93	.0268
−2.92	.0018	−2.42	.0078	−1.92	.0274
−2.91	.0018	−2.41	.0080	−1.91	.0281
−2.90	.0019	−2.40	.0082	−1.90	.0287
−2.89	.0019	−2.39	.0084	−1.89	.0294
−2.88	.0020	−2.38	.0087	−1.88	.0301
−2.87	.0021	−2.37	.0089	−1.87	.0307
−2.86	.0021	−2.36	.0091	−1.86	.0314
−2.85	.0022	−2.35	.0094	−1.85	.0322
−2.84	.0023	−2.34	.0096	−1.84	.0329
−2.83	.0023	−2.33	.0099	−1.83	.0336
−2.82	.0024	−2.32	.0102	−1.82	.0344
−2.81	.0025	−2.31	.0104	−1.81	.0351
−2.80	.0026	−2.30	.0107	−1.80	.0359
−2.79	.0026	−2.29	.0110	−1.79	.0367
−2.78	.0027	−2.28	.0113	−1.78	.0375
−2.77	.0028	−2.27	.0116	−1.77	.0384
−2.76	.0029	−2.26	.0119	−1.76	.0392
−2.75	.0030	−2.25	.0122	−1.75	.0401
−2.74	.0031	−2.24	.0125	−1.74	.0409
−2.73	.0032	−2.23	.0129	−1.73	.0418
−2.72	.0033	−2.22	.0132	−1.72	.0427
−2.71	.0034	−2.21	.0136	−1.71	.0436
−2.70	.0035	−2.20	.0139	−1.70	.0446
−2.69	.0036	−2.19	.0143	−1.69	.0455
−2.68	.0037	−2.18	.0146	−1.68	.0465
−2.67	.0038	−2.17	.0150	−1.67	.0475
−2.66	.0039	−2.16	.0154	−1.66	.0485
−2.65	.0040	−2.15	.0158	−1.65	.0495
−2.64	.0041	−2.14	.0162	−1.64	.0505
−2.63	.0043	−2.13	.0166	−1.63	.0516
−2.62	.0044	−2.12	.0170	−1.62	.0526
−2.61	.0045	−2.11	.0174	−1.61	.0537
−2.60	.0047	−2.10	.0179	−1.60	.0548

Table A-2 Standard Normal Distribution

$P(Z \leq z)$

z	P	z	P	z	P
−1.59	.0559	−1.00	.1587	−0.41	.3409
−1.58	.0571	−0.99	.1611	−0.40	.3446
−1.57	.0582	−0.98	.1635	−0.39	.3483
−1.56	.0594	−0.97	.1660	−0.38	.3520
−1.55	.0606	−0.96	.1685	−0.37	.3557
−1.54	.0618	−0.95	.1711	−0.36	.3594
−1.53	.0630	−0.94	.1736	−0.35	.3632
−1.52	.0643	−0.93	.1762	−0.34	.3669
−1.51	.0655	−0.92	.1788	−0.33	.3707
−1.50	.0668	−0.91	.1814	−0.32	.3745
−1.49	.0681	−0.90	.1841	−0.31	.3783
−1.48	.0694	−0.89	.1867	−0.30	.3821
−1.47	.0708	−0.88	.1894	−0.29	.3859
−1.46	.0721	−0.87	.1922	−0.28	.3897
−1.45	.0735	−0.86	.1949	−0.27	.3936
−1.44	.0749	−0.85	.1977	−0.26	.3974
−1.43	.0764	−0.84	.2005	−0.25	.4013
−1.42	.0778	−0.83	.2033	−0.24	.4052
−1.41	.0793	−0.82	.2061	−0.23	.4090
−1.40	.0808	−0.81	.2090	−0.22	.4129
−1.39	.0823	−0.80	.2119	−0.21	.4168
−1.38	.0838	−0.79	.2148	−0.20	.4207
−1.37	.0853	−0.78	.2177	−0.19	.4247
−1.36	.0869	−0.77	.2207	−0.18	.4286
−1.35	.0885	−0.76	.2236	−0.17	.4325
−1.34	.0901	−0.75	.2266	−0.16	.4364
−1.33	.0918	−0.74	.2297	−0.15	.4404
−1.32	.0934	−0.73	.2327	−0.14	.4443
−1.31	.0951	−0.72	.2358	−0.13	.4483
−1.30	.0968	−0.71	.2389	−0.12	.4522
−1.29	.0985	−0.70	.2420	−0.11	.4562
−1.28	.1003	−0.69	.2451	−0.10	.4602
−1.27	.1020	−0.68	.2483	−0.09	.4641
−1.26	.1038	−0.67	.2514	−0.08	.4681
−1.25	.1057	−0.66	.2546	−0.07	.4721
−1.24	.1075	−0.65	.2578	−0.06	.4761
−1.23	.1093	−0.64	.2611	−0.05	.4801
−1.22	.1112	−0.63	.2643	−0.04	.4840
−1.21	.1131	−0.62	.2676	−0.03	.4880
−1.20	.1151	−0.61	.2709	−0.02	.4920
−1.19	.1170	−0.60	.2743	−0.01	.4960
−1.18	.1190	−0.59	.2776	−0.00	.5000
−1.17	.1210	−0.58	.2810	0.01	.5040
−1.16	.1230	−0.57	.2843	0.02	.5080
−1.15	.1251	−0.56	.2877	0.03	.5120
−1.14	.1271	−0.55	.2912	0.04	.5160
−1.13	.1292	−0.54	.2946	0.05	.5199
−1.12	.1314	−0.53	.2981	0.06	.5239
−1.11	.1335	−0.52	.3015	0.07	.5279
−1.10	.1357	−0.51	.3050	0.08	.5319
−1.09	.1379	−0.50	.3085	0.09	.5359
−1.08	.1401	−0.49	.3121	0.10	.5398
−1.07	.1423	−0.48	.3156	0.11	.5438
−1.06	.1446	−0.47	.3192	0.12	.5478
−1.05	.1469	−0.46	.3228	0.13	.5517
−1.04	.1492	−0.45	.3264	0.14	.5557
−1.03	.1515	−0.44	.3300	0.15	.5596
−1.02	.1539	−0.43	.3336	0.16	.5636
−1.01	.1562	−0.42	.3372	0.17	.5675

Table A-2 Standard Normal Distribution (*continued*)

$P(Z \leq z)$

z	P	z	P	z	P
0.18	.5714	0.77	.7793	1.36	.9131
0.19	.5753	0.78	.7823	1.37	.9147
0.20	.5793	0.79	.7852	1.38	.9162
0.21	.5832	0.80	.7881	1.39	.9177
0.22	.5871	0.81	.7910	1.40	.9192
0.23	.5910	0.82	.7939	1.41	.9207
0.24	.5948	0.83	.7967	1.42	.9222
0.25	.5987	0.84	.7995	1.43	.9236
0.26	.6026	0.85	.8023	1.44	.9251
0.27	.6064	0.86	.8051	1.45	.9265
0.28	.6103	0.87	.8078	1.46	.9279
0.29	.6141	0.88	.8106	1.47	.9292
0.30	.6179	0.89	.8133	1.48	.9306
0.31	.6217	0.90	.8159	1.49	.9319
0.32	.6255	0.91	.8186	1.50	.9332
0.33	.6293	0.92	.8212	1.51	.9345
0.34	.6331	0.93	.8238	1.52	.9357
0.35	.6368	0.94	.8264	1.53	.9370
0.36	.6406	0.95	.8289	1.54	.9382
0.37	.6443	0.96	.8315	1.55	.9394
0.38	.6480	0.97	.8340	1.56	.9406
0.39	.6517	0.98	.8365	1.57	.9418
0.40	.6554	0.99	.8389	1.58	.9429
0.41	.6591	1.00	.8413	1.59	.9441
0.42	.6628	1.01	.8438	1.60	.9452
0.43	.6664	1.02	.8461	1.61	.9463
0.44	.6700	1.03	.8485	1.62	.9474
0.45	.6736	1.04	.8508	1.63	.9484
0.46	.6772	1.05	.8531	1.64	.9495
0.47	.6808	1.06	.8554	1.65	.9505
0.48	.6844	1.07	.8577	1.66	.9515
0.49	.6879	1.08	.8599	1.67	.9525
0.50	.6915	1.09	.8621	1.68	.9535
0.51	.6950	1.10	.8643	1.69	.9545
0.52	.6985	1.11	.8665	1.70	.9554
0.53	.7019	1.12	.8686	1.71	.9564
0.54	.7054	1.13	.8708	1.72	.9573
0.55	.7088	1.14	.8729	1.73	.9582
0.56	.7123	1.15	.8749	1.74	.9591
0.57	.7157	1.16	.8770	1.75	.9599
0.58	.7190	1.17	.8790	1.76	.9608
0.59	.7224	1.18	.8810	1.77	.9616
0.60	.7257	1.19	.8830	1.78	.9625
0.61	.7291	1.20	.8849	1.79	.9633
0.62	.7324	1.21	.8869	1.80	.9641
0.63	.7357	1.22	.8888	1.81	.9649
0.64	.7389	1.23	.8907	1.82	.9656
0.65	.7422	1.24	.8925	1.83	.9664
0.66	.7454	1.25	.8943	1.84	.9671
0.67	.7486	1.26	.8962	1.85	.9678
0.68	.7517	1.27	.8980	1.86	.9686
0.69	.7549	1.28	.8997	1.87	.9693
0.70	.7580	1.29	.9015	1.88	.9699
0.71	.7611	1.30	.9032	1.89	.9706
0.72	.7642	1.31	.9049	1.90	.9713
0.73	.7673	1.32	.9066	1.91	.9719
0.74	.7703	1.33	.9082	1.92	.9726
0.75	.7734	1.34	.9099	1.93	.9732
0.76	.7764	1.35	.9115	1.94	.9738

Table A-2 Standard Normal Distribution (*continued*)

$P(Z \leq z)$

z	P	z	P	z	P
1.95	.9744	2.34	.9904	2.72	.9967
1.96	.9750	2.35	.9906	2.73	.9968
1.97	.9756	2.36	.9909	2.74	.9969
1.98	.9761	2.37	.9911	2.75	.9970
1.99	.9767	2.38	.9913	2.76	.9971
2.00	.9772	2.39	.9916	2.77	.9972
2.01	.9778	2.40	.9918	2.78	.9973
2.02	.9783	2.41	.9920	2.79	.9974
2.03	.9788	2.42	.9922	2.80	.9974
2.04	.9793	2.43	.9925	2.81	.9975
2.05	.9798	2.44	.9927	2.82	.9976
2.06	.9803	2.45	.9929	2.83	.9977
2.07	.9808	2.46	.9931	2.84	.9977
2.08	.9812	2.47	.9932	2.85	.9978
2.09	.9817	2.48	.9934	2.86	.9979
2.10	.9821	2.49	.9936	2.87	.9979
2.11	.9826	2.50	.9938	2.88	.9980
2.12	.9830	2.51	.9940	2.89	.9981
2.13	.9834	2.52	.9941	2.90	.9981
2.14	.9838	2.53	.9943	2.91	.9982
2.15	.9842	2.54	.9945	2.92	.9982
2.16	.9846	2.55	.9946	2.93	.9983
2.17	.9850	2.56	.9948	2.94	.9984
2.18	.9854	2.57	.9949	2.95	.9984
2.19	.9857	2.58	.9951	2.96	.9985
2.20	.9861	2.59	.9952	2.97	.9985
2.21	.9864	2.60	.9953	2.98	.9986
2.22	.9868	2.61	.9955	2.99	.9986
2.23	.9871	2.62	.9956	3.00	.9987
2.24	.9875	2.63	.9957	3.01	.9987
2.25	.9878	2.64	.9959	3.02	.9987
2.26	.9881	2.65	.9960	3.03	.9988
2.27	.9884	2.66	.9961	3.04	.9988
2.28	.9887	2.67	.9962	3.05	.9989
2.29	.9890	2.68	.9963	3.06	.9989
2.30	.9893	2.69	.9964	3.07	.9989
2.31	.9896	2.70	.9965	3.08	.9990
2.32	.9898	2.71	.9966	3.09	.9990
2.33	.9901				

Table A-2 Standard Normal Distribution (*continued*)

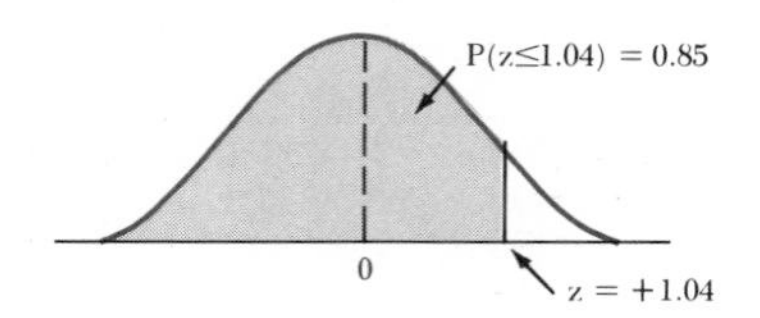

$P(Z \leq z)$

P	z	P	z	P	z
.001	−3.09	.051	−1.64	.110	−1.23
.002	−2.88	.052	−1.63	.120	−1.18
.003	−2.75	.053	−1.62	.130	−1.13
.004	−2.65	.054	−1.61	.140	−1.08
.005	−2.58	.055	−1.60	.150	−1.04
.006	−2.51	.056	−1.59	.160	−0.99
.007	−2.46	.057	−1.58	.170	−0.95
.008	−2.41	.058	−1.57	.180	−0.92
.009	−2.37	.059	−1.56	.190	−0.88
.010	−2.33	.060	−1.56	.200	−0.84
.011	−2.29	.061	−1.55	.210	−0.81
.012	−2.26	.062	−1.54	.220	−0.77
.013	−2.23	.063	−1.53	.230	−0.74
.014	−2.20	.064	−1.52	.240	−0.71
.015	−2.17	.065	−1.51	.250	−0.67
.016	−2.14	.066	−1.51	.260	−0.64
.017	−2.12	.067	−1.50	.270	−0.61
.018	−2.10	.068	−1.49	.280	−0.58
.019	−2.08	.069	−1.48	.290	−0.55
.020	−2.05	.070	−1.48	.300	−0.52
.021	−2.03	.071	−1.47	.310	−0.50
.022	−2.01	.072	−1.46	.320	−0.47
.023	−2.00	.073	−1.45	.330	−0.44
.024	−1.98	.074	−1.45	.340	−0.41
.025	−1.96	.075	−1.44	.350	−0.38
.026	−1.94	.076	−1.43	.360	−0.36
.027	−1.93	.077	−1.43	.370	−0.33
.028	−1.91	.078	−1.42	.380	−0.31
.029	−1.90	.079	−1.41	.390	−0.28
.030	−1.88	.080	−1.41	.400	−0.25
.031	−1.87	.081	−1.40	.410	−0.23
.032	−1.85	.082	−1.39	.420	−0.20
.033	−1.84	.083	−1.39	.430	−0.18
.034	−1.83	.084	−1.38	.440	−0.15
.035	−1.81	.085	−1.37	.450	−0.13
.036	−1.80	.086	−1.37	.460	−0.10
.037	−1.79	.087	−1.36	.470	−0.08
.038	−1.77	.088	−1.35	.480	−0.05
.039	−1.76	.089	−1.35	.490	−0.03
.040	−1.75	.090	−1.34	.500	−0.00
.041	−1.74	.091	−1.33	.510	0.02
.042	−1.73	.092	−1.33	.520	0.05
.043	−1.72	.093	−1.32	.530	0.08
.044	−1.71	.094	−1.32	.540	0.10
.045	−1.70	.095	−1.31	.550	0.13
.046	−1.69	.096	−1.30	.560	0.15
.047	−1.68	.097	−1.30	.570	0.18
.048	−1.66	.098	−1.29	.580	0.20
.049	−1.65	.099	−1.29	.590	0.23
.050	−1.65	.100	−1.28	.600	0.25

Table A-3 Inverted Standard Normal Distribution

$P(Z \leq z)$

P	z	P	z	P	z
.610	0.28	.915	1.37	.958	1.73
.620	0.31	.916	1.38	.959	1.74
.630	0.33	.917	1.39	.960	1.75
.640	0.36	.918	1.39	.961	1.76
.650	0.38	.919	1.40	.962	1.77
.660	0.41	.920	1.41	.963	1.79
.670	0.44	.921	1.41	.964	1.80
.680	0.47	.922	1.42	.965	1.81
.690	0.50	.923	1.43	.966	1.83
.700	0.52	.924	1.43	.967	1.84
.710	0.55	.925	1.44	.968	1.85
.720	0.58	.926	1.45	.969	1.87
.730	0.61	.927	1.45	.970	1.88
.740	0.64	.928	1.46	.971	1.90
.750	0.67	.929	1.47	.972	1.91
.760	0.71	.930	1.48	.973	1.93
.770	0.74	.931	1.48	.974	1.94
.780	0.77	.932	1.49	.975	1.96
.790	0.81	.933	1.50	.976	1.98
.800	0.84	.934	1.51	.977	2.00
.810	0.88	.935	1.51	.978	2.01
.820	0.92	.936	1.52	.979	2.03
.830	0.95	.937	1.53	.980	2.05
.840	0.99	.938	1.54	.981	2.08
.850	1.04	.939	1.55	.982	2.10
.860	1.08	.940	1.56	.983	2.12
.870	1.13	.941	1.56	.984	2.14
.880	1.18	.942	1.57	.985	2.17
.890	1.23	.943	1.58	.986	2.20
.900	1.28	.944	1.59	.987	2.23
.901	1.29	.945	1.60	.988	2.26
.902	1.29	.946	1.61	.989	2.29
.903	1.30	.947	1.62	.990	2.33
.904	1.30	.948	1.63	.991	2.37
.905	1.31	.949	1.64	.992	2.41
.906	1.32	.950	1.65	.993	2.46
.907	1.32	.951	1.65	.994	2.51
.908	1.33	.952	1.66	.995	2.58
.909	1.33	.953	1.68	.996	2.65
.910	1.34	.954	1.69	.997	2.75
.911	1.35	.955	1.70	.998	2.88
.912	1.35	.956	1.71	.999	3.09
.913	1.36	.957	1.72		
.914	1.37				

Table A-3 Inverted Standard Normal Distribution (*continued*)

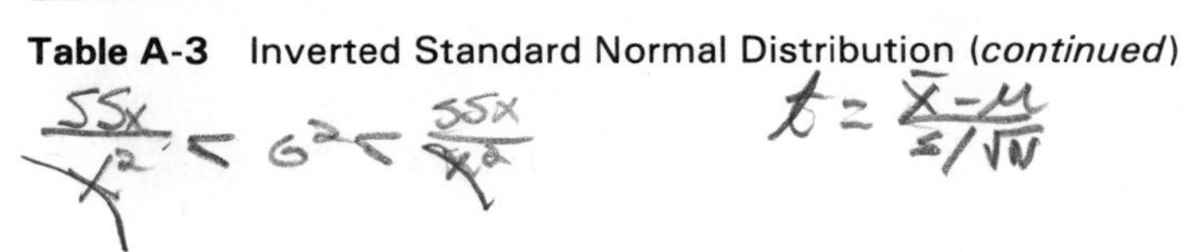

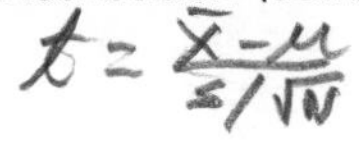

t distribution for 2 df

$P(t \leq 1.8856) = 0.90$

0

$t = +1.8856$

df	$t_{0.75}$	$t_{0.90}$	$t_{0.95}$	$t_{0.975}$	$t_{0.99}$	$t_{0.995}$
1	1.0000	3.0777	6.3138	12.7062	31.8207	63.6574
2	0.8165	1.8856	2.9200	4.3027	6.9646	9.9248
3	0.7649	1.6377	2.3534	3.1824	4.5407	5.8409
4	0.7407	1.5332	2.1318	2.7764	3.7469	4.6041
5	0.7267	1.4759	2.0150	2.5706	3.3649	4.0322
6	0.7176	1.4398	1.9432	2.4469	3.1427	3.7074
7	0.7111	1.4149	1.8946	2.3646	2.9980	3.4995
8	0.7064	1.3968	1.8595	2.3060	2.8965	3.3554
9	0.7027	1.3830	1.8331	2.2622	2.8214	3.2498
10	0.6998	1.3722	1.8125	2.2281	2.7638	3.1693
11	0.6974	1.3634	1.7959	2.2010	2.7181	3.1058
12	0.6955	1.3562	1.7823	2.1788	2.6810	3.0545
13	0.6938	1.3502	1.7709	2.1604	2.6503	3.0123
14	0.6924	1.3450	1.7613	2.1448	2.6245	2.9768
15	0.6912	1.3406	1.7531	2.1315	2.6025	2.9467
16	0.6901	1.3368	1.7459	2.1199	2.5835	2.9208
17	0.6892	1.3334	1.7396	2.1098	2.5669	2.8982
18	0.6884	1.3304	1.7341	2.1009	2.5524	2.8784
19	0.6876	1.3277	1.7291	2.0930	2.5395	2.8609
20	0.6870	1.3253	1.7247	2.0860	2.5280	2.8453
21	0.6864	1.3232	1.7207	2.0796	2.5177	2.8314
22	0.6858	1.3212	1.7171	2.0739	2.5083	2.8188
23	0.6853	1.3195	1.7139	2.0687	2.4999	2.8073
24	0.6848	1.3178	1.7109	2.0639	2.4922	2.7969
25	0.6844	1.3163	1.7081	2.0595	2.4851	2.7874
26	0.6840	1.3150	1.7056	2.0555	2.4786	2.7787
27	0.6837	1.3137	1.7033	2.0518	2.4727	2.7707
28	0.6834	1.3125	1.7011	2.0484	2.4671	2.7633
29	0.6830	1.3114	1.6991	2.0452	2.4620	2.7564
30	0.6828	1.3104	1.6973	2.0423	2.4573	2.7500
31	0.6825	1.3095	1.6955	2.0395	2.4528	2.7440
32	0.6822	1.3086	1.6939	2.0369	2.4487	2.7385
33	0.6820	1.3077	1.6924	2.0345	2.4448	2.7333
34	0.6818	1.3070	1.6909	2.0322	2.4411	2.7284
35	0.6816	1.3062	1.6896	2.0301	2.4377	2.7238
36	0.6814	1.3055	1.6883	2.0281	2.4345	2.7195
37	0.6812	1.3049	1.6871	2.0262	2.4314	2.7154
38	0.6810	1.3042	1.6860	2.0244	2.4286	2.7116
39	0.6808	1.3036	1.6849	2.0227	2.4258	2.7079
40	0.6807	1.3031	1.6839	2.0211	2.4233	2.7045
41	0.6805	1.3025	1.6829	2.0195	2.4208	2.7012
42	0.6804	1.3020	1.6820	2.0181	2.4185	2.6981
43	0.6802	1.3016	1.6811	2.0167	2.4163	2.6951
44	0.6801	1.3011	1.6802	2.0154	2.4141	2.6923
45	0.6800	1.3006	1.6794	2.0141	2.4121	2.6896

Table A-4 Student *t* Distributions

$s^2 = \frac{\Sigma(x-\bar{x})^2}{n-1}$

df	$t_{0.75}$	$t_{0.90}$	$t_{0.95}$	$t_{0.975}$	$t_{0.99}$	$t_{0.995}$
46	0.6799	1.3002	1.6787	2.0129	2.4102	2.6870
47	0.6797	1.2998	1.6779	2.0117	2.4083	2.6846
48	0.6796	1.2994	1.6772	2.0106	2.4066	2.6822
49	0.6795	1.2991	1.6766	2.0096	2.4049	2.6800
50	0.6794	1.2987	1.6759	2.0086	2.4033	2.6778
51	0.6793	1.2984	1.6753	2.0076	2.4017	2.6757
52	0.6792	1.2980	1.6747	2.0066	2.4002	2.6737
53	0.6791	1.2977	1.6741	2.0057	2.3988	2.6718
54	0.6791	1.2974	1.6736	2.0049	2.3974	2.6700
55	0.6790	1.2971	1.6730	2.0040	2.3961	2.6682
56	0.6789	1.2969	1.6725	2.0032	2.3948	2.6665
57	0.6788	1.2966	1.6720	2.0025	2.3936	2.6649
58	0.6787	1.2963	1.6716	2.0017	2.3924	2.6633
59	0.6787	1.2961	1.6711	2.0010	2.3912	2.6618
60	0.6786	1.2958	1.6706	2.0003	2.3901	2.6603
61	0.6785	1.2956	1.6702	1.9996	2.3890	2.6589
62	0.6785	1.2954	1.6698	1.9990	2.3880	2.6575
63	0.6784	1.2951	1.6694	1.9983	2.3870	2.6561
64	0.6783	1.2949	1.6690	1.9977	2.3860	2.6549
65	0.6783	1.2947	1.6686	1.9971	2.3851	2.6536
66	0.6782	1.2945	1.6683	1.9966	2.3842	2.6524
67	0.6782	1.2943	1.6679	1.9960	2.3833	2.6512
68	0.6781	1.2941	1.6676	1.9955	2.3824	2.6501
69	0.6781	1.2939	1.6672	1.9949	2.3816	2.6490
70	0.6780	1.2938	1.6669	1.9944	2.3808	2.6479
71	0.6780	1.2936	1.6666	1.9939	2.3800	2.6469
72	0.6779	1.2934	1.6663	1.9935	2.3793	2.6459
73	0.6779	1.2933	1.6660	1.9930	2.3785	2.6449
74	0.6778	1.2931	1.6657	1.9925	2.3778	2.6439
75	0.6778	1.2929	1.6654	1.9921	2.3771	2.6430
76	0.6777	1.2928	1.6652	1.9917	2.3764	2.6421
77	0.6777	1.2926	1.6649	1.9913	2.3758	2.6412
78	0.6776	1.2925	1.6646	1.9908	2.3751	2.6403
79	0.6776	1.2924	1.6644	1.9905	2.3745	2.6395
80	0.6776	1.2922	1.6641	1.9901	2.3739	2.6387
81	0.6775	1.2921	1.6639	1.9897	2.3733	2.6379
82	0.6775	1.2920	1.6636	1.9893	2.3727	2.6371
83	0.6775	1.2918	1.6634	1.9890	2.3721	2.6364
84	0.6774	1.2917	1.6632	1.9886	2.3716	2.6356
85	0.6774	1.2916	1.6630	1.9883	2.3710	2.6349
86	0.6774	1.2915	1.6628	1.9879	2.3705	2.6342
87	0.6773	1.2914	1.6626	1.9876	2.3700	2.6335
88	0.6773	1.2912	1.6624	1.9873	2.3695	2.6329
89	0.6773	1.2911	1.6622	1.9870	2.3690	2.6322
90	0.6772	1.2910	1.6620	1.9867	2.3685	2.6316

Table A-4 Student t Distributions (*continued*)

$x \pm t(s/\sqrt{n}) = 2$ answers

df	$t_{0.75}$	$t_{0.90}$	$t_{0.95}$	$t_{0.975}$	$t_{0.99}$	$t_{0.995}$
91	0.6772	1.2909	1.6618	1.9864	2.3680	2.6309
92	0.6772	1.2908	1.6616	1.9861	2.3676	2.6303
93	0.6771	1.2907	1.6614	1.9858	2.3671	2.6297
94	0.6771	1.2906	1.6612	1.9855	2.3667	2.6291
95	0.6771	1.2905	1.6611	1.9853	2.3662	2.6286
96	0.6771	1.2904	1.6609	1.9850	2.3658	2.6280
97	0.6770	1.2903	1.6607	1.9847	2.3654	2.6275
98	0.6770	1.2902	1.6606	1.9845	2.3650	2.6269
99	0.6770	1.2902	1.6604	1.9842	2.3646	2.6264
100	0.6770	1.2901	1.6602	1.9840	2.3642	2.6259
102	0.6769	1.2899	1.6599	1.9835	2.3635	2.6249
104	0.6769	1.2897	1.6596	1.9830	2.3627	2.6239
106	0.6768	1.2896	1.6594	1.9826	2.3620	2.6230
108	0.6768	1.2894	1.6591	1.9822	2.3614	2.6221
110	0.6767	1.2893	1.6588	1.9818	2.3607	2.6213
112	0.6767	1.2892	1.6586	1.9814	2.3601	2.6204
114	0.6766	1.2890	1.6583	1.9810	2.3595	2.6196
116	0.6766	1.2889	1.6581	1.9806	2.3589	2.6189
118	0.6766	1.2888	1.6579	1.9803	2.3584	2.6181
120	0.6765	1.2886	1.6577	1.9799	2.3578	2.6174
122	0.6765	1.2885	1.6574	1.9796	2.3573	2.6167
124	0.6765	1.2884	1.6572	1.9793	2.3568	2.6161
126	0.6764	1.2883	1.6570	1.9790	2.3563	2.6154
128	0.6764	1.2882	1.6568	1.9787	2.3558	2.6148
130	0.6764	1.2881	1.6567	1.9784	2.3554	2.6142
132	0.6764	1.2880	1.6565	1.9781	2.3549	2.6136
134	0.6763	1.2879	1.6563	1.9778	2.3545	2.6130
136	0.6763	1.2878	1.6561	1.9776	2.3541	2.6125
138	0.6763	1.2877	1.6560	1.9773	2.3537	2.6119
140	0.6762	1.2876	1.6558	1.9771	2.3533	2.6114
142	0.6762	1.2875	1.6557	1.9768	2.3529	2.6109
144	0.6762	1.2875	1.6555	1.9766	2.3525	2.6104
146	0.6762	1.2874	1.6554	1.9763	2.3522	2.6099
148	0.6762	1.2873	1.6552	1.9761	2.3518	2.6095
150	0.6761	1.2872	1.6551	1.9759	2.3515	2.6090
200	0.6757	1.2858	1.6525	1.9719	2.3451	2.6006
300	0.6753	1.2844	1.6499	1.9679	2.3388	2.5923
400	0.6751	1.2837	1.6487	1.9659	2.3357	2.5882
500	0.6750	1.2832	1.6479	1.9647	2.3338	2.5857
600	0.6749	1.2830	1.6474	1.9639	2.3326	2.5840
700	0.6748	1.2828	1.6470	1.9634	2.3317	2.5829
800	0.6748	1.2826	1.6468	1.9629	2.3310	2.5820
900	0.6748	1.2825	1.6465	1.9626	2.3305	2.5813
1000	0.6747	1.2824	1.6464	1.9623	2.3301	2.5808
∞	0.6745	1.2816	1.6449	1.9600	2.3263	2.5758

Table A-4 Student *t* Distributions (*continued*)

D. B. Owen, *Handbook of Statistical Tables* (Addison-Wesley Publishing Co., Inc., 1962, pp. 28–30). U.S. Department of Energy. Reprinted with permission of the publisher.

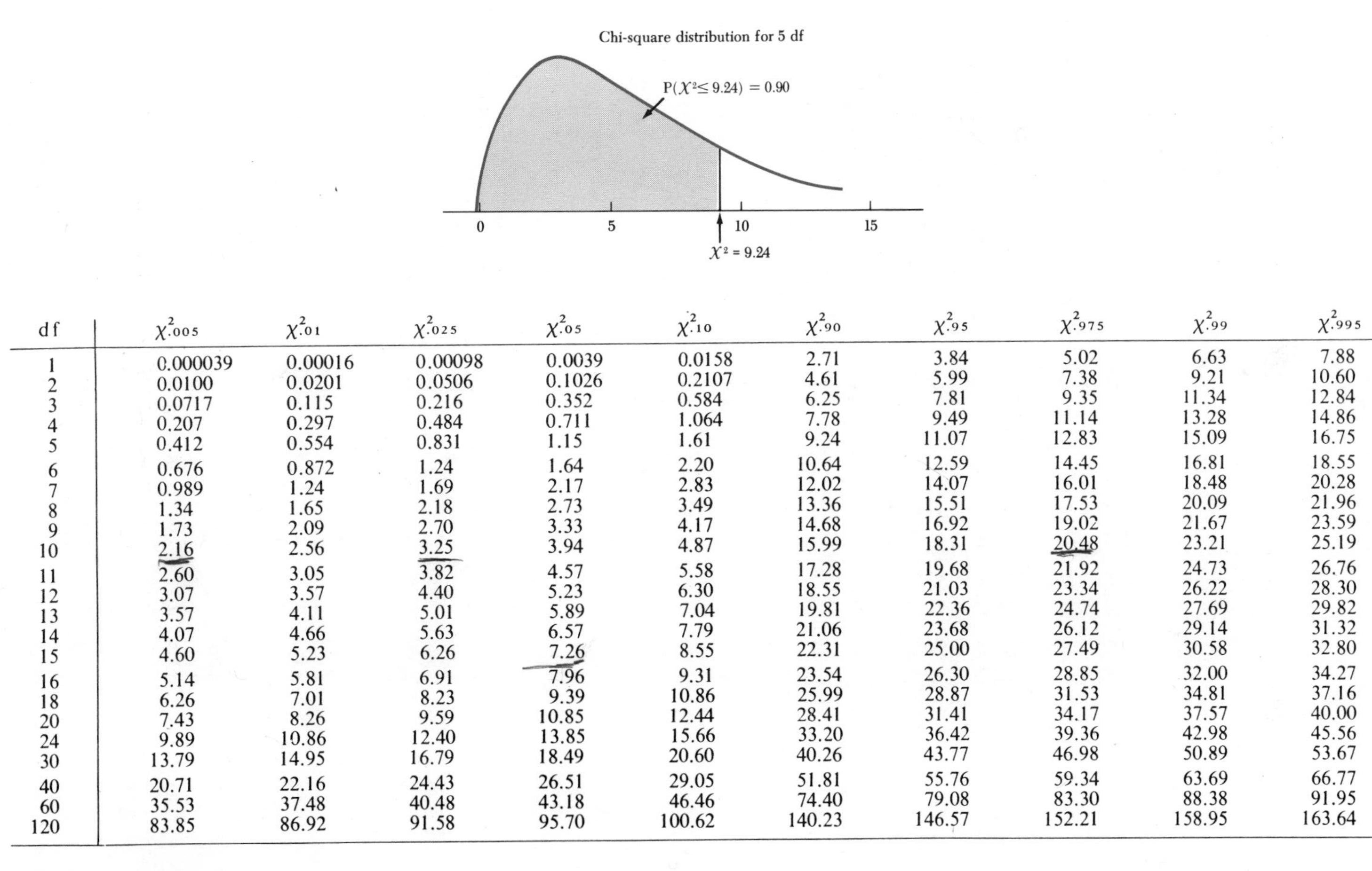

df	$\chi^2_{.005}$	$\chi^2_{.01}$	$\chi^2_{.025}$	$\chi^2_{.05}$	$\chi^2_{.10}$	$\chi^2_{.90}$	$\chi^2_{.95}$	$\chi^2_{.975}$	$\chi^2_{.99}$	$\chi^2_{.995}$
1	0.000039	0.00016	0.00098	0.0039	0.0158	2.71	3.84	5.02	6.63	7.88
2	0.0100	0.0201	0.0506	0.1026	0.2107	4.61	5.99	7.38	9.21	10.60
3	0.0717	0.115	0.216	0.352	0.584	6.25	7.81	9.35	11.34	12.84
4	0.207	0.297	0.484	0.711	1.064	7.78	9.49	11.14	13.28	14.86
5	0.412	0.554	0.831	1.15	1.61	9.24	11.07	12.83	15.09	16.75
6	0.676	0.872	1.24	1.64	2.20	10.64	12.59	14.45	16.81	18.55
7	0.989	1.24	1.69	2.17	2.83	12.02	14.07	16.01	18.48	20.28
8	1.34	1.65	2.18	2.73	3.49	13.36	15.51	17.53	20.09	21.96
9	1.73	2.09	2.70	3.33	4.17	14.68	16.92	19.02	21.67	23.59
10	2.16	2.56	3.25	3.94	4.87	15.99	18.31	20.48	23.21	25.19
11	2.60	3.05	3.82	4.57	5.58	17.28	19.68	21.92	24.73	26.76
12	3.07	3.57	4.40	5.23	6.30	18.55	21.03	23.34	26.22	28.30
13	3.57	4.11	5.01	5.89	7.04	19.81	22.36	24.74	27.69	29.82
14	4.07	4.66	5.63	6.57	7.79	21.06	23.68	26.12	29.14	31.32
15	4.60	5.23	6.26	7.26	8.55	22.31	25.00	27.49	30.58	32.80
16	5.14	5.81	6.91	7.96	9.31	23.54	26.30	28.85	32.00	34.27
18	6.26	7.01	8.23	9.39	10.86	25.99	28.87	31.53	34.81	37.16
20	7.43	8.26	9.59	10.85	12.44	28.41	31.41	34.17	37.57	40.00
24	9.89	10.86	12.40	13.85	15.66	33.20	36.42	39.36	42.98	45.56
30	13.79	14.95	16.79	18.49	20.60	40.26	43.77	46.98	50.89	53.67
40	20.71	22.16	24.43	26.51	29.05	51.81	55.76	59.34	63.69	66.77
60	35.53	37.48	40.48	43.18	46.46	74.40	79.08	83.30	88.38	91.95
120	83.85	86.92	91.58	95.70	100.62	140.23	146.57	152.21	158.95	163.64

Table A-5 Chi-Square Distributions

$\chi^2_{stat} = \frac{s^2(n-1)}{\sigma^2}$

F distribution for 50 and 5 df

P(F≤ 4.44) = 0.95

Values of *F* $F = 4.44$

Right tail of the distribution for $P = 0.05$ (lightface type), 0.01 (boldface type)

m_2 = Degrees of Freedom for Denominator	m_1 = Degrees of Freedom for Numerator											
m_2	1	2	3	4	5	6	7	8	9	10	11	12
1	161	200	216	225	230	234	237	239	241	242	243	244
	4052	**4999**	**5403**	**5625**	**5764**	**5859**	**5928**	**5981**	**6022**	**6056**	**6082**	**6106**
2	18.51	19.00	19.16	19.25	19.30	19.33	19.36	19.37	19.38	19.39	19.40	19.41
	98.49	**99.01**	**99.17**	**99.25**	**99.30**	**99.33**	**99.34**	**99.36**	**99.38**	**99.40**	**99.41**	**99.42**
3	10.13	9.55	9.28	9.12	9.01	8.94	8.88	8.84	8.81	8.78	8.76	8.74
	34.12	**30.81**	**29.46**	**28.71**	**28.24**	**27.91**	**27.67**	**27.49**	**27.34**	**27.23**	**27.13**	**27.05**
4	7.71	6.94	6.59	6.39	6.26	6.16	6.09	6.04	6.00	5.96	5.93	5.91
	21.20	**18.00**	**16.69**	**15.98**	**15.52**	**15.21**	**14.98**	**14.80**	**14.66**	**14.54**	**14.45**	**14.37**
5	6.61	5.79	5.41	5.19	5.05	4.95	4.88	4.82	4.78	4.74	4.70	4.68
	16.26	**13.27**	**12.06**	**11.39**	**10.97**	**10.67**	**10.45**	**10.27**	**10.15**	**10.05**	**9.96**	**9.89**
6	5.99	5.14	4.76	4.53	4.39	4.28	4.21	4.15	4.10	4.06	4.03	4.00
	13.74	**10.92**	**9.78**	**9.15**	**8.75**	**8.47**	**8.26**	**8.10**	**7.98**	**7.87**	**7.79**	**7.72**
7	5.59	4.74	4.35	4.12	3.97	3.87	3.79	3.73	3.68	3.63	3.60	3.57
	12.25	**9.55**	**8.45**	**7.85**	**7.46**	**7.19**	**7.00**	**6.84**	**6.71**	**6.62**	**6.54**	**6.47**
8	5.32	4.46	4.07	3.84	3.69	3.58	3.50	3.44	3.39	3.34	3.31	3.28
	11.26	**8.65**	**7.59**	**7.01**	**6.63**	**6.37**	**6.19**	**6.03**	**5.91**	**5.82**	**5.74**	**5.67**
9	5.12	4.26	3.86	3.63	3.48	3.37	3.29	3.23	3.18	3.13	3.10	3.07
	10.56	**8.02**	**6.99**	**6.42**	**6.06**	**5.80**	**5.62**	**5.47**	**5.35**	**5.26**	**5.18**	**5.11**
10	4.96	4.10	3.71	3.48	3.33	3.22	3.14	3.07	3.02	2.97	2.94	2.91
	10.04	**7.56**	**6.55**	**5.99**	**5.64**	**5.39**	**5.21**	**5.06**	**4.95**	**4.85**	**4.78**	**4.71**
11	4.84	3.98	3.59	3.36	3.20	3.09	3.01	2.95	2.90	2.86	2.82	2.79
	9.65	**7.20**	**6.22**	**5.67**	**5.32**	**5.07**	**4.88**	**4.74**	**4.63**	**4.54**	**4.46**	**4.40**
12	4.75	3.88	3.49	3.26	3.11	3.00	2.92	2.85	2.80	2.76	2.72	2.69
	9.33	**6.93**	**5.95**	**5.41**	**5.06**	**4.82**	**4.65**	**4.50**	**4.39**	**4.30**	**4.22**	**4.16**
13	4.67	3.80	3.41	3.18	3.02	2.92	2.84	2.77	2.72	2.67	2.63	2.60
	9.07	**6.70**	**5.74**	**5.20**	**4.86**	**4.62**	**4.44**	**4.30**	**4.19**	**4.10**	**4.02**	**3.96**
14	4.60	3.74	3.34	3.11	2.96	2.85	2.77	2.70	2.65	2.60	2.56	2.53
	8.86	**6.51**	**5.56**	**5.03**	**4.69**	**4.46**	**4.28**	**4.14**	**4.03**	**3.94**	**3.86**	**3.80**
15	4.54	3.68	3.29	3.06	2.90	2.79	2.70	2.64	2.59	2.55	2.51	2.48
	8.68	**6.36**	**5.42**	**4.89**	**4.56**	**4.32**	**4.14**	**4.00**	**3.89**	**3.80**	**3.73**	**3.67**
16	4.49	3.63	3.24	3.01	2.85	2.74	2.66	2.59	2.54	2.49	2.45	2.42
	8.53	**6.23**	**5.29**	**4.77**	**4.44**	**4.20**	**4.03**	**3.89**	**3.78**	**3.69**	**3.61**	**3.55**
17	4.45	3.59	3.20	2.96	2.81	2.70	2.62	2.55	2.50	2.45	2.41	2.38
	8.40	**6.11**	**5.18**	**4.67**	**4.34**	**4.10**	**3.93**	**3.79**	**3.68**	**3.59**	**3.52**	**3.45**
18	4.41	3.55	3.16	2.93	2.77	2.66	2.58	2.51	2.46	2.41	2.37	2.34
	8.28	**6.01**	**5.09**	**4.58**	**4.25**	**4.01**	**3.85**	**3.71**	**3.60**	**3.51**	**3.44**	**3.37**
19	4.38	3.52	3.13	2.90	2.74	2.63	2.55	2.48	2.43	2.38	2.34	2.31
	8.18	**5.93**	**5.01**	**4.50**	**4.17**	**3.94**	**3.77**	**3.63**	**3.52**	**3.43**	**3.36**	**3.30**
20	4.35	3.49	3.10	2.87	2.71	2.60	2.52	2.45	2.40	2.35	2.31	2.28
	8.10	**5.85**	**4.94**	**4.43**	**4.10**	**3.87**	**3.71**	**3.56**	**3.45**	**3.37**	**3.30**	**3.23**
21	4.32	3.47	3.07	2.84	2.68	2.57	2.49	2.42	2.37	2.32	2.28	2.25
	8.02	**5.78**	**4.87**	**4.37**	**4.04**	**3.81**	**3.65**	**3.51**	**3.40**	**3.31**	**3.24**	**3.17**
22	4.30	3.44	3.05	2.82	2.66	2.55	2.47	2.40	2.35	2.30	2.26	2.23
	7.94	**5.72**	**4.82**	**4.31**	**3.99**	**3.76**	**3.59**	**3.45**	**3.35**	**3.26**	**3.18**	**3.12**
23	4.28	3.42	3.03	2.80	2.64	2.53	2.45	2.38	2.32	2.28	2.24	2.20
	7.88	**5.66**	**4.76**	**4.26**	**3.94**	**3.71**	**3.54**	**3.41**	**3.30**	**3.21**	**3.14**	**3.07**
24	4.26	3.40	3.01	2.78	2.62	2.51	2.43	2.36	2.30	2.26	2.22	2.18
	7.82	**5.61**	**4.72**	**4.22**	**3.90**	**3.67**	**3.50**	**3.36**	**3.25**	**3.17**	**3.09**	**3.03**
25	4.24	3.38	2.99	2.76	2.60	2.49	2.41	2.34	2.28	2.24	2.20	2.16
	7.77	**5.57**	**4.68**	**4.18**	**3.86**	**3.63**	**3.46**	**3.32**	**3.21**	**3.13**	**3.05**	**2.99**
26	4.22	3.37	2.98	2.74	2.59	2.47	2.39	2.32	2.27	2.22	2.18	2.15
	7.72	**5.53**	**4.64**	**4.14**	**3.82**	**3.59**	**3.42**	**3.29**	**3.17**	**3.09**	**3.02**	**2.96**

Table A-6 *F* Distributions

m_1 = Degrees of Freedom for Numerator												
14	16	20	24	30	40	50	75	100	200	500	∞	m_2
245	246	248	249	250	251	252	253	253	254	254	254	1
6142	**6169**	**6208**	**6234**	**6258**	**6286**	**6302**	**6323**	**6334**	**6352**	**6361**	**6366**	
19.42	19.43	19.44	19.45	19.46	19.47	19.47	19.48	19.49	19.49	19.50	19.50	2
99.43	**99.44**	**99.45**	**99.46**	**99.47**	**99.48**	**99.48**	**99.49**	**99.49**	**99.49**	**99.50**	**99.50**	
8.71	8.69	8.66	8.64	8.62	8.60	8.58	8.57	8.56	8.54	8.54	8.53	3
26.92	**26.83**	**26.69**	**26.60**	**26.50**	**26.41**	**26.35**	**26.27**	**26.23**	**26.18**	**26.14**	**26.12**	
5.87	5.84	5.80	5.77	5.74	5.71	5.70	5.68	5.66	5.65	5.64	5.63	4
14.24	**14.15**	**14.02**	**13.93**	**13.83**	**13.74**	**13.69**	**13.61**	**13.57**	**13.52**	**13.48**	**13.46**	
4.64	4.60	4.56	4.53	4.50	4.46	4.44	4.42	4.40	4.38	4.37	4.36	5
9.77	**9.68**	**9.55**	**9.47**	**9.38**	**9.29**	**9.24**	**9.17**	**9.13**	**9.07**	**9.04**	**9.02**	
3.96	3.92	3.87	3.84	3.81	3.77	3.75	3.72	3.71	3.69	3.68	3.67	6
7.60	**7.52**	**7.39**	**7.31**	**7.23**	**7.14**	**7.09**	**7.02**	**6.99**	**6.94**	**6.90**	**6.88**	
3.52	3.49	3.44	3.41	3.38	3.34	3.32	3.29	3.28	3.25	3.24	3.23	7
6.35	**6.27**	**6.15**	**6.07**	**5.98**	**5.90**	**5.85**	**5.78**	**5.75**	**5.70**	**5.67**	**5.65**	
3.23	3.20	3.15	3.12	3.08	3.05	3.03	3.00	2.98	2.96	2.94	2.93	8
5.56	**5.48**	**5.36**	**5.28**	**5.20**	**5.11**	**5.06**	**5.00**	**4.96**	**4.91**	**4.88**	**4.86**	
3.02	2.98	2.93	2.90	2.86	2.82	2.80	2.77	2.76	2.73	2.72	2.71	9
5.00	**4.92**	**4.80**	**4.73**	**4.64**	**4.56**	**4.51**	**4.45**	**4.41**	**4.36**	**4.33**	**4.31**	
2.86	2.82	2.77	2.74	2.70	2.67	2.64	2.61	2.59	2.56	2.55	2.54	10
4.60	**4.52**	**4.41**	**4.33**	**4.25**	**4.17**	**4.12**	**4.05**	**4.01**	**3.96**	**3.93**	**3.91**	
2.74	2.70	2.65	2.61	2.57	2.53	2.50	2.47	2.45	2.42	2.41	2.40	11
4.29	**4.21**	**4.10**	**4.02**	**3.94**	**3.86**	**3.80**	**3.74**	**3.70**	**3.66**	**3.62**	**3.60**	
2.64	2.60	2.54	2.50	2.46	2.42	2.40	2.36	2.35	2.32	2.31	2.30	12
4.05	**3.98**	**3.86**	**3.78**	**3.70**	**3.61**	**3.56**	**3.49**	**3.46**	**3.41**	**3.38**	**3.36**	
2.55	2.51	2.46	2.42	2.38	2.34	2.32	2.28	2.26	2.24	2.22	2.21	13
3.85	**3.78**	**3.67**	**3.59**	**3.51**	**3.42**	**3.37**	**3.30**	**3.27**	**3.21**	**3.18**	**3.16**	
2.48	2.44	2.39	2.35	2.31	2.27	2.24	2.21	2.19	2.16	2.14	2.13	14
3.70	**3.62**	**3.51**	**3.43**	**3.34**	**3.26**	**3.21**	**3.14**	**3.11**	**3.06**	**3.02**	**3.00**	
2.43	2.39	2.33	2.29	2.25	2.21	2.18	2.15	2.12	2.10	2.08	2.07	15
3.56	**3.48**	**3.36**	**3.29**	**3.20**	**3.12**	**3.07**	**3.00**	**2.97**	**2.92**	**2.89**	**2.87**	
2.37	2.33	2.28	2.24	2.20	2.16	2.13	2.09	2.07	2.04	2.02	2.01	16
3.45	**3.37**	**3.25**	**3.18**	**3.10**	**3.01**	**2.96**	**2.89**	**2.86**	**2.80**	**2.77**	**2.75**	
2.33	2.29	2.23	2.19	2.15	2.11	2.08	2.04	2.02	1.99	1.97	1.96	17
3.35	**3.27**	**3.16**	**3.08**	**3.00**	**2.92**	**2.86**	**2.79**	**2.76**	**2.70**	**2.67**	**2.65**	
2.29	2.25	2.19	2.15	2.11	2.07	2.04	2.00	1.98	1.95	1.93	1.92	18
3.27	**3.19**	**3.07**	**3.00**	**2.91**	**2.83**	**2.78**	**2.71**	**2.68**	**2.62**	**2.59**	**2.57**	
2.26	2.21	2.15	2.11	2.07	2.02	2.00	1.96	1.94	1.91	1.90	1.88	19
3.19	**3.12**	**3.00**	**2.92**	**2.84**	**2.76**	**2.70**	**2.63**	**2.60**	**2.54**	**2.51**	**2.49**	
2.23	2.18	2.12	2.08	2.04	1.99	1.96	1.92	1.90	1.87	1.85	1.84	20
3.13	**3.05**	**2.94**	**2.86**	**2.77**	**2.69**	**2.63**	**2.56**	**2.53**	**2.47**	**2.44**	**2.42**	
2.20	2.15	2.09	2.05	2.00	1.96	1.93	1.89	1.87	1.84	1.82	1.81	21
3.07	**2.99**	**2.88**	**2.80**	**2.72**	**2.63**	**2.58**	**2.51**	**2.47**	**2.42**	**2.38**	**2.36**	
2.18	2.13	2.07	2.03	1.98	1.93	1.91	1.87	1.84	1.81	1.80	1.78	22
3.02	**2.94**	**2.83**	**2.75**	**2.67**	**2.58**	**2.53**	**2.46**	**2.42**	**2.37**	**2.33**	**2.31**	
2.14	2.10	2.04	2.00	1.96	1.91	1.88	1.84	1.82	1.79	1.77	1.76	23
2.97	**2.89**	**2.78**	**2.70**	**2.62**	**2.53**	**2.48**	**2.41**	**2.37**	**2.32**	**2.28**	**2.26**	
2.13	2.09	2.02	1.98	1.94	1.89	1.86	1.82	1.80	1.76	1.74	1.73	24
2.93	**2.85**	**2.74**	**2.66**	**2.58**	**2.49**	**2.44**	**2.36**	**2.33**	**2.27**	**2.23**	**2.21**	
2.11	2.06	2.00	1.96	1.92	1.87	1.84	1.80	1.77	1.74	1.72	1.71	25
2.89	**2.81**	**2.70**	**2.62**	**2.54**	**2.45**	**2.40**	**2.32**	**2.29**	**2.23**	**2.19**	**2.17**	
2.10	2.05	1.99	1.95	1.90	1.85	1.82	1.78	1.76	1.72	1.70	1.69	26
2.86	**2.77**	**2.66**	**2.58**	**2.50**	**2.41**	**2.36**	**2.28**	**2.25**	**2.19**	**2.15**	**2.13**	

m_2 = Degrees of Freedom for Denominator

Table A-6 *F* Distributions (*continued*)

m_2 = Degrees of Freedom for Denominator; m_1 = Degrees of Freedom for Numerator

m_2	1	2	3	4	5	6	7	8	9	10	11	12
27	4.21 **7.68**	3.35 **5.49**	2.96 **4.60**	2.73 **4.11**	2.57 **3.79**	2.46 **3.56**	2.37 **3.39**	2.30 **3.26**	2.25 **3.14**	2.20 **3.06**	2.16 **2.98**	2.13 **2.93**
28	4.20 **7.64**	3.34 **5.45**	2.95 **4.57**	2.71 **4.07**	2.56 **3.76**	2.44 **3.53**	2.36 **3.36**	2.29 **3.23**	2.24 **3.11**	2.19 **3.03**	2.15 **2.95**	2.12 **2.90**
29	4.18 **7.60**	3.33 **5.42**	2.93 **4.54**	2.70 **4.04**	2.54 **3.73**	2.43 **3.50**	2.35 **3.33**	2.28 **3.20**	2.22 **3.08**	2.18 **3.00**	2.14 **2.92**	2.10 **2.87**
30	4.17 **7.56**	3.32 **5.39**	2.92 **4.51**	2.69 **4.02**	2.53 **3.70**	2.42 **3.47**	2.34 **3.30**	2.27 **3.17**	2.21 **3.06**	2.16 **2.98**	2.12 **2.90**	2.09 **2.84**
32	4.15 **7.50**	3.30 **5.34**	2.90 **4.46**	2.67 **3.97**	2.51 **3.66**	2.40 **3.42**	2.32 **3.25**	2.25 **3.12**	2.19 **3.01**	2.14 **2.94**	2.10 **2.86**	2.07 **2.80**
34	4.13 **7.44**	3.28 **5.29**	2.88 **4.42**	2.65 **3.93**	2.49 **3.61**	2.38 **3.38**	2.30 **3.21**	2.23 **3.08**	2.17 **2.97**	2.12 **2.89**	2.08 **2.82**	2.05 **2.76**
36	4.11 **7.39**	3.26 **5.25**	2.86 **4.38**	2.63 **3.89**	2.48 **3.58**	2.36 **3.35**	2.28 **3.18**	2.21 **3.04**	2.15 **2.94**	2.10 **2.86**	2.06 **2.78**	2.03 **2.72**
38	4.10 **7.35**	3.25 **5.21**	2.85 **4.34**	2.62 **3.86**	2.46 **3.54**	2.35 **3.32**	2.26 **3.15**	2.19 **3.02**	2.14 **2.91**	2.09 **2.82**	2.05 **2.75**	2.02 **2.69**
40	4.08 **7.31**	3.23 **5.18**	2.84 **4.31**	2.61 **3.83**	2.45 **3.51**	2.34 **3.29**	2.25 **3.12**	2.18 **2.99**	2.12 **2.88**	2.07 **2.80**	2.04 **2.73**	2.00 **2.66**
42	4.07 **7.27**	3.22 **5.15**	2.83 **4.29**	2.59 **3.80**	2.44 **3.49**	2.32 **3.26**	2.24 **3.10**	2.17 **2.96**	2.11 **2.86**	2.06 **2.77**	2.02 **2.70**	1.99 **2.64**
44	4.06 **7.24**	3.21 **5.12**	2.82 **4.26**	2.58 **3.78**	2.43 **3.46**	2.31 **3.24**	2.23 **3.07**	2.16 **2.94**	2.10 **2.84**	2.05 **2.75**	2.01 **2.68**	1.98 **2.62**
46	4.05 **7.21**	3.20 **5.10**	2.81 **4.24**	2.57 **3.76**	2.42 **3.44**	2.30 **3.22**	2.22 **3.05**	2.14 **2.92**	2.09 **2.82**	2.04 **2.73**	2.00 **2.66**	1.97 **2.60**
48	4.04 **7.19**	3.19 **5.08**	2.80 **4.22**	2.56 **3.74**	2.41 **3.42**	2.30 **3.20**	2.21 **3.04**	2.14 **2.90**	2.08 **2.80**	2.03 **2.71**	1.99 **2.64**	1.96 **2.58**
50	4.03 **7.17**	3.18 **5.06**	2.79 **4.20**	2.56 **3.72**	2.40 **3.41**	2.29 **3.18**	2.20 **3.02**	2.13 **2.88**	2.07 **2.78**	2.02 **2.70**	1.98 **2.62**	1.95 **2.56**
55	4.02 **7.12**	3.17 **5.01**	2.78 **4.16**	2.54 **3.68**	2.38 **3.37**	2.27 **3.15**	2.18 **2.98**	2.11 **2.85**	2.05 **2.75**	2.00 **2.66**	1.97 **2.59**	1.93 **2.53**
60	4.00 **7.08**	3.15 **4.98**	2.76 **4.13**	2.52 **3.65**	2.37 **3.34**	2.25 **3.12**	2.17 **2.95**	2.10 **2.82**	2.04 **2.72**	1.99 **2.63**	1.95 **2.56**	1.92 **2.50**
65	3.99 **7.04**	3.14 **4.95**	2.75 **4.10**	2.51 **3.62**	2.36 **3.31**	2.24 **3.09**	2.15 **2.93**	2.08 **2.79**	2.02 **2.70**	1.98 **2.61**	1.94 **2.54**	1.90 **2.47**
70	3.98 **7.01**	3.13 **4.92**	2.74 **4.08**	2.50 **3.60**	2.35 **3.29**	2.23 **3.07**	2.14 **2.91**	2.07 **2.77**	2.01 **2.67**	1.97 **2.59**	1.93 **2.51**	1.89 **2.45**
80	3.96 **6.96**	3.11 **4.88**	2.72 **4.04**	2.48 **3.56**	2.33 **3.25**	2.21 **3.04**	2.12 **2.87**	2.05 **2.74**	1.99 **2.64**	1.95 **2.55**	1.91 **2.48**	1.88 **2.41**
100	3.94 **6.90**	3.09 **4.82**	2.70 **3.98**	2.46 **3.51**	2.30 **3.20**	2.19 **2.99**	2.10 **2.82**	2.03 **2.69**	1.97 **2.59**	1.92 **2.51**	1.88 **2.43**	1.85 **2.36**
125	3.92 **6.84**	3.07 **4.78**	2.68 **3.94**	2.44 **3.47**	2.29 **3.17**	2.17 **2.95**	2.08 **2.79**	2.01 **2.65**	1.95 **2.56**	1.90 **2.47**	1.86 **2.40**	1.83 **2.33**
150	3.91 **6.81**	3.06 **4.75**	2.67 **3.91**	2.43 **3.44**	2.27 **3.14**	2.16 **2.92**	2.07 **2.76**	2.00 **2.62**	1.94 **2.53**	1.89 **2.44**	1.85 **2.37**	1.82 **2.30**
200	3.89 **6.76**	3.04 **4.71**	2.65 **3.88**	2.41 **3.41**	2.26 **3.11**	2.14 **2.90**	2.05 **2.73**	1.98 **2.60**	1.92 **2.50**	1.87 **2.41**	1.83 **2.34**	1.80 **2.28**
400	3.86 **6.70**	3.02 **4.66**	2.62 **3.83**	2.39 **3.36**	2.23 **3.06**	2.12 **2.85**	2.03 **2.69**	1.96 **2.55**	1.90 **2.46**	1.85 **2.37**	1.81 **2.29**	1.78 **2.23**
1000	3.85 **6.66**	3.00 **4.62**	2.61 **3.80**	2.38 **3.34**	2.22 **3.04**	2.10 **2.82**	2.02 **2.66**	1.95 **2.53**	1.89 **2.43**	1.84 **2.34**	1.80 **2.26**	1.76 **2.20**
∞	3.84 **6.64**	2.99 **4.60**	2.60 **3.78**	2.37 **3.32**	2.21 **3.02**	2.09 **2.80**	2.01 **2.64**	1.94 **2.51**	1.88 **2.41**	1.83 **2.32**	1.79 **2.24**	1.75 **2.18**

Table A-6 *F* Distributions (*continued*)

m_1 = Degrees of Freedom for Numerator												m_2 = Degrees of Freedom for Denominator
14	16	20	24	30	40	50	75	100	200	500	∞	m_2
2.08	2.03	1.97	1.93	1.88	1.84	1.80	1.76	1.74	1.71	1.68	1.67	27
2.83	**2.74**	**2.63**	**2.55**	**2.47**	**2.38**	**2.33**	**2.25**	**2.21**	**2.16**	**2.12**	**2.10**	
2.06	2.02	1.96	1.91	1.87	1.81	1.78	1.75	1.72	1.69	1.67	1.65	28
2.80	**2.71**	**2.60**	**2.52**	**2.44**	**2.35**	**2.30**	**2.22**	**2.18**	**2.13**	**2.09**	**2.06**	
2.05	2.00	1.94	1.90	1.85	1.80	1.77	1.73	1.71	1.68	1.65	1.64	29
2.77	**2.68**	**2.57**	**2.49**	**2.41**	**2.32**	**2.27**	**2.19**	**2.15**	**2.10**	**2.06**	**2.03**	
2.04	1.99	1.93	1.89	1.84	1.79	1.76	1.72	1,69	1.66	1.64	1.62	30
2.74	**2.66**	**2.55**	**2.47**	**2.38**	**2.29**	**2.24**	**2.16**	**2.13**	**2.07**	**2.03**	**2.01**	
2.02	1.97	1.91	1.86	1.82	1.76	1.74	1.69	1.67	1.64	1.61	1.59	32
2.70	**2.62**	**2.51**	**2.42**	**2.34**	**2.25**	**2.20**	**2.12**	**2.08**	**2.02**	**1.98**	**1.96**	
2.00	1.95	1.89	1.84	1.80	1.74	1.71	1.67	1.64	1.61	1.59	1.57	34
2.66	**2.58**	**2.47**	**2.38**	**2.30**	**2.21**	**2.15**	**2.08**	**2.04**	**1.98**	**1.94**	**1.91**	
1.98	1.93	1.87	1.82	1.78	1.72	1.69	1.65	1.62	1.59	1.56	1.55	36
2.62	**2.54**	**2.43**	**2.35**	**2.26**	**2.17**	**2.12**	**2.04**	**2.00**	**1.94**	**1.90**	**1.87**	
1.96	1.92	1.85	1.80	1.76	1.71	1.67	1.63	1.60	1.57	1.54	1.53	38
2.59	**2.51**	**2.40**	**2.32**	**2.22**	**2.14**	**2.08**	**2.00**	**1.97**	**1.90**	**1.86**	**1.84**	
1.95	1.90	1.84	1.79	1.74	1.69	1.66	1.61	1.59	1.55	1.53	1.51	40
2.56	**2.49**	**2.37**	**2.29**	**2.20**	**2.11**	**2.05**	**1.97**	**1.94**	**1.88**	**1.84**	**1.81**	
1.94	1.89	1.82	1.78	1.73	1.68	1.64	1.60	1.57	1.54	1.51	1.49	42
2.54	**2.46**	**2.35**	**2.26**	**2.17**	**2.08**	**2.02**	**1.94**	**1.91**	**1.85**	**1.80**	**1.78**	
1.92	1.88	1.81	1.76	1.72	1.66	1.63	1.58	1.56	1.52	1.50	1.48	44
2.52	**2.44**	**2.32**	**2.24**	**2.15**	**2.06**	**2.00**	**1.92**	**1.88**	**1.82**	**1.78**	**1.75**	
1.91	1.87	1.80	1.75	1.71	1.65	1.62	1.57	1.54	1.51	1.48	1.46	46
2.50	**2.42**	**2.30**	**2.22**	**2.13**	**2.04**	**1.98**	**1.90**	**1.86**	**1.80**	**1.76**	**1.72**	
1.90	1.86	1.79	1.74	1.70	1.64	1.61	1.56	1.53	1.50	1.47	1.45	48
2.48	**2.40**	**2.28**	**2.20**	**2.11**	**2.02**	**1.96**	**1.88**	**1.84**	**1.78**	**1.73**	**1.70**	
1.90	1.85	1.78	1.74	1.69	1.63	1.60	1.55	1.52	1.48	1.46	1.44	50
2.46	**2.39**	**2.26**	**2.18**	**2.10**	**2.00**	**1.94**	**1.86**	**1.82**	**1.76**	**1.71**	**1.68**	
1.88	1.83	1.76	1.72	1.67	1.61	1.58	1.52	1.50	1.46	1.43	1.41	55
2.43	**2.35**	**2.23**	**2.15**	**2.06**	**1.96**	**1.90**	**1.82**	**1.78**	**1.71**	**1.66**	**1.64**	
1.86	1.81	1.75	1.70	1.65	1.59	1.56	1.50	1.48	1.44	1.41	1.39	60
2.40	**2.32**	**2.20**	**2.12**	**2.03**	**1.93**	**1.87**	**1.79**	**1.74**	**1.68**	**1.63**	**1.60**	
1.85	1.80	1.73	1.68	1.63	1.57	1.54	1.49	1.46	1.42	1.39	1.37	65
2.37	**2.30**	**2.18**	**2.09**	**2.00**	**1.90**	**1.84**	**1.76**	**1.71**	**1.64**	**1.60**	**1.56**	
1.84	1.79	1.72	1.67	1.62	1.56	1.53	1.47	1.45	1.40	1.37	1.35	70
2.35	**2.28**	**2.15**	**2.07**	**1.98**	**1.88**	**1.82**	**1.74**	**1.69**	**1.62**	**1.56**	**1.53**	
1.82	1.77	1.70	1.65	1.60	1.54	1.51	1.45	1.42	1.38	1.35	1.32	80
2.32	**2.24**	**2.11**	**2.03**	**1.94**	**1.84**	**1.78**	**1.70**	**1.65**	**1.57**	**1.52**	**1.49**	
1.79	1.75	1.68	1.63	1.57	1.51	1.48	1.42	1.39	1.34	1.30	1.28	100
2.26	**2.19**	**2.06**	**1.98**	**1.89**	**1.79**	**1.73**	**1.64**	**1.59**	**1.51**	**1.46**	**1.43**	
1.77	1.72	1.65	1.60	1.55	1.49	1.45	1.39	1.36	1.31	1.27	1.25	125
2.23	**2.15**	**2.03**	**1.94**	**1.85**	**1.75**	**1.68**	**1.59**	**1.54**	**1.46**	**1.40**	**1.37**	
1.76	1.71	1.64	1.59	1.54	1.47	1.44	1.37	1.34	1.29	1.25	1.22	150
2.20	**2.12**	**2.00**	**1.91**	**1.83**	**1.72**	**1.66**	**1.56**	**1.51**	**1.43**	**1.37**	**1.33**	
1.74	1.69	1.62	1.57	1.52	1.45	1.42	1.35	1.32	1.26	1.22	1.19	200
2.17	**2.09**	**1.97**	**1.88**	**1.79**	**1.69**	**1.62**	**1.53**	**1.48**	**1.39**	**1.33**	**1.28**	
1.72	1.67	1.60	1.54	1.49	1.42	1.38	1.32	1.28	1.22	1.16	1.13	400
2.12	**2.04**	**1.92**	**1.84**	**1.74**	**1.64**	**1.57**	**1.47**	**1.42**	**1.32**	**1.24**	**1.19**	
1.70	1.65	1.58	1.53	1.47	1.41	1.36	1.30	1.26	1.19	1.13	1.08	1000
2.09	**2.01**	**1.89**	**1.81**	**1.71**	**1.61**	**1.54**	**1.44**	**1.38**	**1.28**	**1.19**	**1.11**	
1.69	1.64	1.57	1.52	1.46	1.40	1.35	1.28	1.24	1.17	1.11	1.00	∞
2.07	**1.99**	**1.87**	**1.79**	**1.69**	**1.59**	**1.52**	**1.41**	**1.36**	**1.25**	**1.15**	**1.00**	

Table A-6 *F* Distributions (*continued*)

	μ				
r	.020	.040	.060	.080	.100
0	.980	.961	.942	.923	.905
1	.020	.038	.057	.074	.090
2	.000	.001	.002	.003	.005
Sum	1.000	1.000	1.000	1.000	1.000

r	.150	.200	.250	.300	.350
0	.861	.819	.779	.741	.705
1	.129	.164	.195	.222	.247
2	.010	.016	.024	.033	.043
3	.000	.001	.002	.003	.005
4	.000	.000	.000	.000	.000
5	.000	.000	.000	.000	.000
6	.000	.000	.000	.000	.000
Sum	1.000	1.000	1.000	1.000	1.000

					μ						
.400	.450	.500	.550	.600	.650	.700	.750	.800	.850	.900	.950
.670	.638	.607	.577	.549	.522	.497	.472	.449	.427	.407	.387
.268	.287	.303	.317	.329	.339	.348	.354	.359	.363	.366	.367
.054	.065	.076	.087	.099	.110	.122	.133	.144	.154	.165	.175
.007	.010	.013	.016	.020	.024	.028	.033	.038	.044	.049	.055
.001	.001	.002	.002	.003	.004	.005	.006	.008	.009	.011	.013
.000	.000	.000	.000	.000	.001	.001	.001	.001	.002	.002	.002
.000	.000	.000	.000	.000	.000	.000	.000	.000	.000	.000	.000
1.000	1.000	1.000	1.000	1.000	1.000	1.000	1.000	1.000	1.000	1.000	1.000

						μ					
r	1.00	1.10	1.20	1.30	1.40	1.50	1.60	1.70	1.80	1.90	2.00
0	.368	.333	.301	.273	.247	.223	.202	.183	.165	.150	.135
1	.368	.366	.361	.354	.345	.335	.323	.311	.298	.284	.271
2	.184	.201	.217	.230	.242	.251	.258	.264	.268	.270	.271
3	.061	.074	.087	.100	.113	.126	.138	.150	.161	.171	.180
4	.015	.020	.026	.032	.039	.047	.055	.064	.072	.081	.090
5	.003	.004	.006	.008	.011	.014	.018	.022	.026	.031	.036
6	.001	.001	.001	.002	.003	.004	.005	.006	.008	.010	.012
7	.000	.000	.000	.000	.001	.001	.001	.001	.002	.003	.003
8	.000	.000	.000	.000	.000	.000	.000	.000	.000	.001	.001
Sum	1.000	1.000	1.000	1.000	1.000	1.000	1.000	1.000	1.000	1.000	1.000

Table A-7 Poisson Distributions $P(R = r \mid \mu)$

	μ														
r	2.20	2.40	2.60	2.80	3.00	3.20	3.40	3.60	3.80	4.00	4.20	4.40	4.60	4.80	5.00
0	.111	.091	.074	.061	.050	.041	.033	.027	.022	.018	.015	.012	.010	.008	.007
1	.244	.218	.193	.170	.149	.130	.113	.098	.085	.073	.063	.054	.046	.040	.034
2	.268	.261	.251	.238	.224	.209	.193	.177	.162	.147	.132	.119	.106	.095	.084
3	.197	.209	.218	.222	.224	.223	.219	.212	.205	.195	.185	.174	.163	.152	.140
4	.108	.125	.141	.156	.168	.178	.186	.191	.194	.195	.194	.192	.188	.182	.175
5	.048	.060	.074	.087	.101	.114	.126	.138	.148	.156	.163	.169	.173	.175	.175
6	.017	.024	.032	.041	.050	.061	.072	.083	.094	.104	.114	.124	.132	.140	.146
7	.005	.008	.012	.016	.022	.028	.035	.042	.051	.060	.069	.078	.087	.096	.104
8	.002	.002	.004	.006	.008	.011	.015	.019	.024	.030	.036	.043	.050	.058	.065
9	.000	.001	.001	.002	.003	.004	.006	.008	.010	.013	.017	.021	.026	.031	.036
10	.000	.000	.000	.000	.001	.001	.002	.003	.004	.005	.007	.009	.012	.015	.018
11	.000	.000	.000	.000	.000	.000	.001	.001	.001	.002	.003	.004	.005	.006	.008
12	.000	.000	.000	.000	.000	.000	.000	.000	.000	.001	.001	.001	.002	.003	.003
13	.000	.000	.000	.000	.000	.000	.000	.000	.000	.000	.000	.000	.001	.001	.001
14	.000	.000	.000	.000	.000	.000	.000	.000	.000	.000	.000	.000	.000	.000	.000
Sum	1.000	1.000	1.000	1.000	1.000	1.000	1.000	1.000	1.000	1.000	1.000	1.000	1.000	1.000	1.000

	μ													
r	5.20	5.40	5.60	5.80	6.00	6.20	6.40	6.60	6.80	7.00	7.20	7.40	7.60	7.80
0	.006	.005	.004	.003	.002	.002	.002	.001	.001	.001	.001	.001	.001	.001
1	.029	.024	.021	.018	.015	.013	.011	.009	.008	.006	.005	.005	.004	.003
2	.075	.066	.058	.051	.045	.039	.034	.030	.026	.022	.019	.017	.014	.012
3	.129	.119	.108	.098	.089	.081	.073	.065	.058	.052	.046	.041	.037	.032
4	.168	.160	.152	.143	.134	.125	.116	.108	.099	.091	.084	.076	.070	.063
5	.175	.173	.170	.166	.161	.155	.149	.142	.135	.128	.120	.113	.106	.099
6	.151	.156	.158	.160	.161	.160	.159	.156	.153	.149	.144	.139	.134	.128
7	.113	.120	.127	.133	.138	.142	.145	.147	.149	.149	.149	.147	.145	.143
8	.073	.081	.089	.096	.103	.110	.116	.121	.126	.130	.134	.136	.138	.139
9	.042	.049	.055	.062	.069	.076	.082	.089	.095	.101	.107	.112	.117	.121
10	.022	.026	.031	.036	.041	.047	.053	.059	.065	.071	.077	.083	.089	.094
11	.010	.013	.016	.019	.023	.026	.031	.035	.040	.045	.050	.056	.061	.067
12	.005	.006	.007	.009	.011	.014	.016	.019	.023	.026	.030	.034	.039	.043
13	.002	.002	.003	.004	.005	.007	.008	.010	.012	.014	.017	.020	.023	.026
14	.001	.001	.001	.002	.002	.003	.004	.005	.006	.007	.009	.010	.012	.015
15	.000	.000	.000	.001	.001	.001	.002	.002	.003	.003	.004	.005	.006	.008
16	.000	.000	.000	.000	.000	.000	.001	.001	.001	.001	.002	.002	.003	.004
17	.000	.000	.000	.000	.000	.000	.000	.000	.000	.001	.001	.001	.001	.002
18	.000	.000	.000	.000	.000	.000	.000	.000	.000	.000	.000	.000	.001	.001
Sum	1.000	1.000	1.000	1.000	1.000	1.000	1.000	1.000	1.000	1.000	1.000	1.000	1.000	1.000

Table A-7 Poisson Distributions $P(R=r\,|\,\mu)$ *(continued)*

	μ										
r	8.0	8.5	9.0	9.5	10.0	10.5	11.0	11.5	12.0	12.5	13.0
0	.000	.000	.000	.000	.000	.000	.000	.000	.000	.000	.000
1	.003	.002	.001	.001	.000	.000	.000	.000	.000	.000	.000
2	.011	.007	.005	.003	.002	.002	.001	.001	.000	.000	.000
3	.029	.021	.015	.011	.008	.005	.004	.003	.002	.001	.001
4	.057	.044	.034	.025	.019	.014	.010	.007	.005	.004	.003
5	.092	.075	.061	.048	.038	.029	.022	.017	.013	.009	.007
6	.122	.107	.091	.076	.063	.051	.041	.033	.025	.020	.015
7	.140	.129	.117	.104	.090	.077	.065	.053	.044	.035	.028
8	.140	.138	.132	.123	.113	.101	.089	.077	.066	.055	.046
9	.124	.130	.132	.130	.125	.118	.109	.098	.087	.077	.066
10	.099	.110	.119	.124	.125	.124	.119	.113	.105	.096	.086
11	.072	.085	.097	.107	.114	.118	.119	.118	.114	.109	.101
12	.048	.060	.073	.084	.095	.103	.109	.113	.114	.113	.110
13	.030	.040	.050	.062	.073	.083	.093	.100	.106	.109	.110
14	.017	.024	.032	.042	.052	.063	.073	.082	.090	.097	.102
15	.009	.014	.019	.027	.035	.044	.053	.063	.072	.081	.088
16	.005	.007	.011	.016	.022	.029	.037	.045	.054	.063	.072
17	.002	.004	.006	.009	.013	.018	.024	.031	.038	.047	.055
18	.001	.002	.003	.005	.007	.010	.015	.020	.026	.032	.040
19	.000	.001	.001	.002	.004	.006	.008	.012	.016	.021	.027
20	.000	.000	.001	.001	.002	.003	.005	.007	.010	.013	.018
21	.000	.000	.000	.000	.001	.002	.002	.004	.006	.008	.011
22	.000	.000	.000	.000	.000	.001	.001	.002	.003	.004	.006
23	.000	.000	.000	.000	.000	.000	.001	.001	.002	.002	.004
24	.000	.000	.000	.000	.000	.000	.000	.000	.001	.001	.002
25	.000	.000	.000	.000	.000	.000	.000	.000	.000	.001	.001
26	.000	.000	.000	.000	.000	.000	.000	.000	.000	.000	.001
Sum	1.000	1.000	1.000	1.000	1.000	1.000	1.000	1.000	1.000	1.000	1.000

Table A-7 Poisson Distributions $P(R=r \mid \mu)$ *(continued)*

10 09 73 25 33	76 52 01 35 86	34 67 35 48 76	80 95 90 91 17	39 29 27 49 45
37 54 20 48 05	64 89 47 42 96	24 80 52 40 37	20 63 61 04 02	00 82 29 16 65
08 42 26 89 53	19 64 50 93 03	23 20 90 25 60	15 95 33 47 64	35 08 03 36 06
99 01 90 25 29	09 37 67 07 15	38 31 13 11 65	88 67 67 43 97	04 43 62 76 59
12 80 79 99 70	80 15 73 61 47	64 03 23 66 53	98 95 11 68 77	12 17 17 68 33
66 06 57 47 17	34 07 27 68 50	36 69 73 61 70	65 81 33 98 85	11 19 92 91 70
31 06 01 08 05	45 57 18 24 06	35 30 34 26 14	86 79 90 74 39	23 40 30 97 32
85 26 97 76 02	02 05 16 56 92	68 66 57 48 18	73 05 38 52 47	16 62 38 85 79
63 57 33 21 35	05 32 54 70 48	90 55 35 75 48	28 46 82 87 09	83 49 12 56 24
73 79 64 57 53	03 52 96 47 78	35 80 83 42 82	60 93 52 03 44	35 27 38 84 35
98 52 01 77 67	14 90 56 86 07	22 10 94 05 58	60 97 09 34 33	50 50 07 39 98
11 80 50 54 31	39 80 82 77 32	50 72 56 82 48	29 40 52 42 01	52 77 56 78 51
83 45 29 96 34	06 28 89 80 83	13 74 67 00 78	18 47 54 06 10	68 71 17 78 17
88 68 54 02 00	86 50 75 84 01	36 76 66 79 51	90 36 47 64 93	29 60 91 10 62
99 59 46 73 48	87 51 76 49 69	91 82 60 89 28	93 78 56 13 68	23 47 83 41 13
65 48 11 76 74	17 46 85 09 50	58 04 77 69 74	73 03 95 71 86	40 21 81 65 44
80 12 43 56 35	17 72 70 80 15	45 31 82 23 74	21 11 57 82 53	14 38 55 37 63
74 35 09 98 17	77 40 27 72 14	43 23 60 02 10	45 52 16 42 37	96 28 60 26 55
69 91 62 68 03	66 25 22 91 48	36 93 68 72 03	76 62 11 39 90	94 40 05 64 18
09 89 32 05 05	14 22 56 85 14	46 42 75 67 88	96 29 77 88 22	54 38 21 45 98
91 49 91 45 23	68 47 91 76 86	46 16 28 35 54	94 75 08 99 23	37 08 92 00 48
80 33 69 45 98	26 94 03 68 58	70 29 73 41 35	53 14 03 33 40	42 05 08 23 41
44 10 48 19 49	85 15 74 79 54	32 97 92 65 75	57 60 04 08 81	22 22 20 64 13
12 55 07 37 42	11 10 00 20 40	12 86 07 46 97	96 64 48 94 39	28 70 72 58 15
63 60 64 93 29	16 50 53 44 84	40 21 95 25 63	43 65 17 70 82	07 20 73 17 90
61 19 69 04 46	26 45 74 77 74	51 92 43 37 29	65 39 45 95 93	42 58 26 05 27
15 47 44 52 66	95 27 07 99 53	59 36 78 38 48	82 39 61 01 18	33 21 15 94 66
94 55 72 85 73	67 89 75 43 87	54 62 24 44 31	91 19 04 25 92	92 92 74 59 73
42 48 11 62 13	97 34 40 87 21	16 86 84 87 67	03 07 11 20 59	25 70 14 66 70
23 52 37 83 17	73 20 88 98 37	68 93 59 14 16	26 25 22 96 63	05 52 28 25 62
04 49 35 24 94	75 24 63 38 24	45 86 25 10 25	61 96 27 93 35	65 33 71 24 72
00 54 99 76 54	64 05 18 81 59	96 11 96 38 96	54 69 28 23 91	23 28 72 95 29
35 96 31 53 07	26 89 80 93 54	33 35 13 54 62	77 97 45 00 24	90 10 33 93 33
59 80 80 83 91	45 42 72 68 42	83 60 94 97 00	13 02 12 48 92	78 56 52 01 06
46 05 88 52 36	01 39 09 22 86	77 28 14 40 77	93 91 08 36 47	70 61 74 29 41
32 17 90 05 97	87 37 92 52 41	05 56 70 70 07	86 74 31 71 57	85 39 41 18 38
69 23 46 14 06	20 11 74 52 04	15 95 66 00 00	18 74 39 24 23	07 11 89 63 38
19 56 54 14 30	01 75 87 53 79	40 41 92 15 85	66 67 43 68 06	84 96 28 52 07
45 15 51 49 38	19 47 60 72 46	43 66 79 45 43	59 04 79 00 33	20 82 66 95 41
94 86 43 19 94	36 16 81 08 51	34 88 88 15 53	01 54 03 54 56	05 01 45 11 76
98 08 62 48 26	45 24 02 84 04	44 99 90 88 96	39 09 47 34 07	35 44 13 18 80
33 18 51 62 32	41 94 15 09 49	89 43 54 85 81	88 69 54 19 94	37 54 87 30 43
80 95 10 04 06	96 38 27 07 74	20 15 12 33 87	25 01 62 52 98	94 62 46 11 71
79 75 24 91 40	71 96 12 82 96	69 86 10 25 91	74 85 22 05 39	00 38 75 95 79
18 63 33 25 37	98 14 50 65 71	31 01 02 46 74	05 45 56 14 27	77 93 89 19 36
74 02 94 39 02	77 55 73 22 70	97 79 01 71 19	52 52 75 80 21	80 81 45 17 48
54 17 84 56 11	80 99 33 71 43	05 33 51 29 69	56 12 71 92 55	36 04 09 03 24
11 66 44 98 83	52 07 98 48 27	59 38 17 15 39	09 97 33 34 40	88 46 12 33 56
48 32 47 79 28	31 24 96 47 10	02 29 53 68 70	32 30 75 75 46	15 02 00 99 94
69 07 49 41 38	87 63 79 19 76	35 58 40 44 01	10 51 82 16 15	01 84 87 69 38

Table A-8 Random Numbers

09 18 82 00 97	32 82 53 95 27	22 08 63 04	83 38 98 73 74	64 27 85 80 44
90 04 58 54 97	51 98 15 06 54	94 93 88 19 97	91 87 07 61 50	68 47 66 46 59
73 18 95 02 07	47 67 72 62 69	62 29 06 44 64	27 12 46 70 18	41 36 18 27 60
75 76 87 64 90	20 97 18 17 49	90 42 91 22 72	95 37 50 58 71	93 82 34 31 78
54 01 64 40 56	66 28 13 10 03	00 68 22 73 98	20 71 45 32 95	07 70 61 78 13
08 35 86 99 10	78 54 24 27 85	13 66 15 88 73	04 61 89 75 53	31 22 30 84 20
28 30 60 32 64	81 33 31 05 91	40 51 00 78 93	32 60 46 04 75	94 11 90 18 40
53 84 08 62 33	81 59 41 36 28	51 21 59 02 90	28 46 66 87 95	77 76 22 07 91
91 75 75 37 41	61 61 36 22 69	50 26 39 02 12	55 78 17 65 14	83 48 34 70 55
89 41 59 26 94	00 38 75 83 91	12 60 71 76 46	48 94 97 23 06	94 54 13 74 08
77 51 30 38 20	86 83 42 99 01	68 41 48 27 74	51 90 81 39 80	72 89 35 55 07
19 50 23 71 74	69 97 92 02 88	55 21 02 97 73	74 28 77 52 51	65 34 46 74 15
21 81 85 93 13	93 27 88 17 57	04 68 67 31 56	07 08 28 50 46	31 85 33 84 52
51 47 46 64 99	68 10 72 36 21	94 04 99 13 45	42 83 60 91 91	08 00 74 54 49
99 55 96 83 31	62 53 52 41 70	69 77 71 28 30	74 81 97 81 42	43 86 07 28 34
33 71 34 80 07	93 58 47 28 69	51 92 66 47 21	58 30 32 98 22	93 17 49 39 72
85 27 48 68 93	11 30 32 92 70	28 83 43 41 37	73 51 59 04 00	71 14 84 36 43
84 13 38 96 40	44 03 55 21 66	73 85 27 00 91	61 22 26 05 61	62 32 71 84 23
56 73 21 62 34	17 39 59 61 31	10 12 39 16 22	85 49 65 75 60	81 60 41 88 80
65 13 85 68 06	87 64 88 52 61	34 31 36 58 61	45 87 52 10 69	85 64 44 72 77
38 00 10 21 76	81 71 91 17 11	71 60 29 29 37	74 21 96 40 49	65 58 44 96 98
37 40 29 63 97	01 30 47 75 86	56 27 11 00 86	47 32 46 26 05	40 03 03 74 38
97 12 54 03 48	87 08 33 14 17	21 81 53 92 50	75 23 76 20 47	15 50 12 95 78
21 82 64 11 34	47 14 33 40 72	64 63 88 59 02	49 13 90 64 41	03 85 65 45 52
73 13 54 27 42	95 71 90 90 35	85 79 47 42 96	08 78 98 81 56	64 69 11 92 02
07 63 87 79 29	03 06 11 80 72	96 20 74 41 56	23 82 19 95 38	04 71 36 69 94
60 52 88 34 41	07 95 41 98 14	59 17 52 06 95	05 53 35 21 39	61 21 20 64 55
83 59 63 56 55	06 95 89 29 83	05 12 80 97 19	77 43 35 37 83	92 30 15 04 98
10 85 06 27 46	99 59 91 05 07	13 49 90 63 19	53 07 57 18 39	06 41 01 93 62
39 82 09 89 52	43 62 26 31 47	64 42 18 08 14	43 80 00 93 51	31 02 47 31 67
59 58 00 64 78	75 56 97 88 00	88 83 55 44 86	23 76 80 61 56	04 11 10 84 08
38 50 80 73 41	23 79 34 87 63	90 82 29 70 22	17 71 90 42 07	95 95 44 99 53
30 69 27 06 68	94 68 81 61 27	56 19 68 00 91	82 06 76 34 00	05 46 26 92 00
65 44 39 56 59	18 28 82 74 37	49 63 22 40 41	08 33 76 56 76	96 29 99 08 36
27 26 75 02 64	13 19 27 22 94	07 47 74 46 06	17 98 54 89 11	97 34 13 03 58
91 30 70 69 91	19 07 22 42 10	36 69 95 37 28	28 82 53 57 93	28 97 66 62 52
68 43 49 46 88	84 47 31 36 22	62 12 69 84 08	12 84 38 25 90	09 81 59 31 46
48 90 81 58 77	54 74 52 45 91	35 70 00 47 54	83 82 45 26 92	54 13 05 51 60
06 91 34 51 97	42 67 27 86 01	11 88 30 95 28	63 01 19 89 01	14 97 44 03 44
10 45 51 60 19	14 21 03 37 12	91 34 23 78 21	88 32 58 08 51	43 66 77 08 83
12 88 39 73 43	65 02 76 11 84	04 28 50 13 92	17 97 41 50 77	90 71 22 67 69
21 77 83 09 76	38 80 73 69 61	31 64 94 20 96	63 28 10 20 23	08 81 64 74 49
19 52 35 95 15	65 12 25 96 59	86 28 36 82 58	69 57 21 37 98	16 43 59 15 29
67 24 55 26 70	35 58 31 65 63	79 24 68 66 86	76 46 33 42 22	26 65 59 08 02
60 58 44 73 77	07 50 03 79 92	45 13 42 65 29	26 76 08 36 37	41 32 64 43 44
53 85 34 13 77	36 06 69 48 50	58 83 87 38 59	49 36 47 33 31	96 24 04 36 42
24 63 73 87 36	74 38 48 93 42	52 62 30 79 92	12 36 91 86 01	03 74 28 38 73
83 08 01 24 51	38 99 22 28 15	07 75 95 17 77	97 37 72 75 85	51 97 23 78 67
16 44 42 43 34	36 15 19 90 73	27 49 37 09 39	85 13 03 25 52	54 84 65 47 59
60 79 01 81 57	57 17 86 57 62	11 16 17 85 76	45 81 95 29 79	65 13 00 48 60

Table A-8 Random Numbers (*continued*)

N	0	1	2	3	4	5	6	7	8	9
10	0000	0043	0086	0128	0170	0212	0253	0294	0334	0374
11	0414	0453	0492	0531	0569	0607	0645	0682	0719	0755
12	0792	0828	0864	0899	0934	0969	1004	1038	1072	1106
13	1139	1173	1206	1239	1271	1303	1335	1367	1399	1430
14	1461	1492	1523	1553	1584	1614	1644	1673	1703	1732
15	1761	1790	1818	1847	1875	1903	1931	1959	1987	2014
16	2041	2068	2095	2122	2148	2175	2201	2227	2253	2279
17	2304	2330	2355	2380	2405	2430	2455	2480	2504	2529
18	2553	2577	2601	2625	2648	2672	2695	2718	2742	2765
19	2788	2810	2833	2856	2878	2900	2923	2945	2967	2989
20	3010	3032	3054	3075	3096	3118	3139	3160	3181	3201
21	3222	3243	3263	3284	3304	3324	3345	3365	3385	3404
22	3424	3444	3464	3483	3502	3522	3541	3560	3579	3598
23	3617	3636	3655	3674	3692	3711	3729	3747	3766	3784
24	3802	3820	3838	3856	3874	3892	3909	3927	3945	3962
25	3979	3997	4014	4031	4048	4065	4082	4099	4116	4133
26	4150	4166	4183	4200	4216	4232	4249	4265	4281	4298
27	4314	4330	4346	4362	4378	4393	4409	4425	4440	4456
28	4472	4487	4502	4518	4533	4548	4564	4579	4594	4609
29	4624	4639	4654	4669	4683	4698	4713	4728	4742	4757
30	4771	4786	4800	4814	4829	4843	4857	4871	4886	4900
31	4914	4928	4942	4955	4969	4983	4997	5011	5024	5038
32	5051	5065	5079	5092	5105	5119	5132	5145	5159	5172
33	5185	5198	5211	5224	5237	5250	5263	5276	5289	5302
34	5315	5328	5340	5353	5366	5378	5391	5403	5416	5428
35	5441	5453	5465	5478	5490	5502	5514	5527	5539	5551
36	5563	5575	5587	5599	5611	5623	5635	5647	5658	5670
37	5682	5694	5705	5717	5729	5740	5752	5763	5775	5786
38	5798	5809	5821	5832	5843	5855	5866	5877	5888	5899
39	5911	5922	5933	5944	5955	5966	5977	5988	5999	6010
40	6021	6031	6042	6053	6064	6075	6085	6096	6107	6117
41	6128	6138	6149	6160	6170	6180	6191	6201	6212	6222
42	6232	6243	6253	6263	6274	6284	6294	6304	6314	6325
43	6335	6345	6355	6365	6375	6385	6395	6405	6415	6425
44	6435	6444	6454	6464	6474	6484	6493	6503	6513	6522
45	6532	6542	6551	6561	6571	6580	6590	6599	6609	6618
46	6628	6637	6646	6656	6665	6675	6684	6693	6702	6712
47	6721	6730	6739	6749	6758	6767	6776	6785	6794	6803
48	6812	6821	6830	6839	6848	6857	6866	6875	6884	6893
49	6902	6911	6920	6928	6937	6946	6955	6964	6972	6981
50	6990	6998	7007	7016	7024	7033	7042	7050	7059	7067
51	7076	7084	7093	7101	7110	7118	7126	7135	7143	7152
52	7160	7168	7177	7185	7193	7202	7210	7218	7226	7235
53	7243	7251	7259	7267	7275	7284	7292	7300	7308	7316
54	7324	7332	7340	7348	7356	7364	7372	7380	7388	7396
N	0	1	2	3	4	5	6	7	8	9

Table A-9 Mantissas of Common Logarithms (Base 10)

N	0	1	2	3	4	5	6	7	8	9
55	7404	7412	7419	7427	7435	7443	7451	7459	7466	7474
56	7482	7490	7497	7505	7513	7520	7528	7536	7543	7551
57	7559	7566	7574	7582	7589	7597	7604	7612	7619	7627
58	7634	7642	7649	7657	7664	7672	7679	7686	7694	7701
59	7709	7716	7723	7731	7738	7745	7752	7760	7767	7774
60	7782	7789	7796	7803	7810	7818	7825	7832	7839	7846
61	7853	7860	7868	7875	7882	7889	7896	7903	7910	7917
62	7924	7931	7938	7945	7952	7959	7966	7973	7980	7987
63	7993	8000	8007	8014	8021	8028	8035	8041	8048	8055
64	8062	8069	8075	8082	8089	8096	8102	8109	8116	8122
65	8129	8136	8142	8149	8156	8162	8169	8176	8182	8189
66	8195	8202	8209	8215	8222	8228	8235	8241	8248	8254
67	8261	8267	8274	8280	8287	8293	8299	8306	8312	8319
68	8325	8331	8338	8344	8351	8357	8363	8370	8376	8382
69	8388	8395	8401	8407	8414	8420	8426	8432	8439	8445
70	8451	8457	8463	8470	8476	8482	8488	8494	8500	8506
71	8513	8519	8525	8531	8537	8543	8549	8555	8561	8567
72	8573	8579	8585	8591	8597	8603	8609	8615	8621	8627
73	8633	8639	8645	8651	8657	8663	8669	8675	8681	8686
74	8692	8698	8704	8710	8716	8722	8727	8733	8739	8745
75	8751	8756	8762	8768	8774	8779	8785	8791	8797	8802
76	8808	8814	8820	8825	8831	8837	8842	8848	8854	8859
77	8865	8871	8876	8882	8887	8893	8899	8904	8910	8915
78	8921	8927	8932	8938	8943	8949	8954	8960	8965	8971
79	8976	8982	8987	8993	8998	9004	9009	9015	9020	9025
80	9031	9036	9042	9047	9053	9058	9063	9069	9074	9079
81	9085	9090	9096	9101	9106	9112	9117	9122	9128	9133
82	9138	9143	9149	9154	9159	9165	9170	9175	9180	9186
83	9191	9196	9201	9206	9212	9217	9222	9227	9232	9238
84	9243	9248	9253	9258	9263	9269	9274	9279	9284	9289
85	9294	9299	9304	9309	9315	9320	9325	9330	9335	9340
86	9345	9350	9355	9360	9365	9370	9375	9380	9385	9390
87	9395	9400	9405	9410	9415	9420	9425	9430	9435	9440
88	9445	9450	9455	9460	9465	9469	9474	9479	9484	9489
89	9494	9499	9504	9509	9513	9518	9523	9528	9533	9538
90	9542	9547	9552	9557	9562	9566	9571	9576	9581	9586
91	9590	9595	9600	9605	9609	9614	9619	9624	9628	9633
92	9638	9643	9647	9652	9657	9661	9666	9671	9675	9680
93	9685	9689	9694	9699	9703	9708	9713	9717	9722	9727
94	9731	9736	9741	9745	9750	9754	9759	9763	9768	9773
95	9777	9782	9786	9791	9795	9800	9805	9809	9814	9818
96	9823	9827	9832	9836	9841	9845	9850	9854	9859	9863
97	9868	9872	9877	9881	9886	9890	9894	9899	9903	9908
98	9912	9917	9921	9926	9930	9934	9939	9943	9948	9952
99	9956	9961	9965	9969	9974	9978	9983	9987	9991	9996
N	0	1	2	3	4	5	6	7	8	9

Table A-9 Mantissas of Common Logarithms (Base 10) (*continued*)

From Table 1, *Manual of Mathematics and Mechanics* by G. R. Clements and L. T. Wilson.

r	.00	.01	.02	.03	.04	.05	.06	.07	.08	.09
.0	.00000	.01000	.02000	.03001	.04002	.05004	.06007	.07012	.08017	.09024
.1	.10034	.11045	.12058	.13074	.14093	.15114	.16139	.17167	.18198	.19234
.2	.20273	.21317	.22366	.23419	.24477	.25541	.26611	.27686	.28768	.29857
.3	.30952	.32055	.33165	.34283	.35409	.36544	.37689	.38842	.40006	.41180
.4	.42365	.43561	.44769	.45990	.47223	.48470	.49731	.51007	.52298	.53606
.5	.54931	.56273	.57634	.59014	.60415	.61838	.63283	.64752	.66246	.67767
.6	.69315	.70892	.72500	.74142	.75817	.77530	.79281	.81074	.82911	.84795
.7	.86730	.88718	.90764	.92873	.95048	.97295	.99621	1.02033	1.04537	1.07143
.8	1.09861	1.12703	1.15682	1.18813	1.22117	1.25615	1.29334	1.33308	1.37577	1.42192
.9	1.47222	1.52752	1.58902	1.65839	1.73805	1.83178	1.94591	2.09229	2.29756	2.64665

Table A-10 Values of $w = \frac{1}{2} \ln \frac{1+r}{1-r}$

For negative values of r, put a minus sign in front of the tabled numbers.

From *Introduction to Statistical Analysis*, 3rd edition, by Wilfrid J. Dixon and Frank J. Massey, Jr.

APPENDIX B

DATA SETS FOR COMPUTER EXERCISES

Computer exercises refer to five data files: real estate, personnel, utilities, retailers, and baseball. The real estate file comes from a computer simulation. The others are actual data from a variety of sources. The personnel data are from a firm in which the identity of individuals is discussed. The salaries are linear combinations of actual salaries. The retailers and utilities files are for the *Fortune* top 50 organizations in 1982. (*Fortune*, July 13, 1983). The baseball data covers all the teams in the American and National Leagues for the period of 1969 to 1976. These data are aggregates for each team in each year.

These data files were selected to give students realistic experience with multivariate data. With the exception of the real estate data, the files contain anomalies (such as outliers) typically found in actual data files. Our intent is to provide data that can be used with a wide variety of computers and software packages. The real estate and baseball files contain only numeric variables. The other files contain both numeric and character variables.

Variable Definitions

Real Estate Data

Variable Name	*Definition and Code*
PRICE	Selling price in thousands of dollars
INCOME	Buyer's income in thousands of dollars
AGE	Age of buyer in years
RORNR	Whether or not the buyer lived in the community *before* this purchase (Yes = 0, No = 1)
OORN	Whether the purchased house was old (0) or new (1)

Personnel Data (Data values begin on page A-137.)

Variable Name	*Definition and Code*
ID	Employee identification number
DEPART	Department where employee works
HIRED	Year employee was hired
RETIRE	Year employee is scheduled to retire
COLLEGE	College identification code where employee received his degree
DEGREE	Graduate degree = G, undergraduate degree = U
YEAR	Year degree granted
SALARY	Employee's annual salary
LEVEL	Management position (LEVEL1 is entry level and ENTRY4 is top)
LEVELYR	Year employee's management position was attained
DEFER	Whether or not the employee participates in the tax deferral plan (Participates = Y, does not participate = N)

Utilities Data

Variable Name	*Definition and Code*
RANK82	Ranking of utility by total assets in 1982
RANK81	Ranking by total assets in 1981
STATE	State where the utility is located
ASSETS	Total assets in millions of dollars
REVENUES	Operating revenues in millions of dollars
INCOME	Net income in millions of dollars
EQUITY	Stockholders' equity in millions of dollars
EMPLOY	Number of employees
EPS82	Earnings per share in 1982
EPS81	Earnings per share in 1981
EPS72	Earnings per share in 1972
EPSGR	Earnings per share growth rate 1972–1982 in percent (the average annual growth, compounded)
RETINV82	Total return to investors for 1982 in percent
RETINVAV	Total return to investors for 1972–1982 (averaged and compounded) in percent
TYPE	Standard industrial classification of the utility

Retailers Data

Variable Name	*Definition and Code*
RANK82	Rank of retailer in 1982 by revenues (sales)
RANK81	Rank of retailer in 1981 by revenues (sales)
STATE	State where retailer is located
REVENUE	Sales in millions of dollars
ASSETS	Total assets in millions of dollars
INCOME	Net income in millions of dollars
EQUITY	Stockholders' equity in millions of dollars
EMPLOY	Number of employees
EPS82	Earnings per share in 1982
EPS81	Earnings per share in 1981
EPS72	Earnings per share in 1972
EPSGR	Earnings per share growth rate 1972–1982 in percent (the average annual growth compounded)
RETINV82	Total return to investors for 1982 in percent
RETINVAV	Total return to investors for 1972–1982 (averaged and compounded) in percent

Baseball Data (Data values begin on page A-151.)

Variable Name	*Definition and Code*
LEAGUE	National = 1, American = 2
DIVISION	East = 1, West = 2
TEAM	Team identification (see following table)
STAND	Team standing in division for the year
RUNS	Total runs scored by the team in the year
H2B	Total doubles hit by the team in the year
H3B	Total triples hit by the team in the year
HR	Total home runs hit by the team in the year
TBB	Total bases on balls by the team in the year
SB	Total sacrifice bunts by the team in the year
HP	Total team batters hit by pitchers in the year
SF	Total sacrifice flies hit by the team in the year
SH	Total sacrifice hits by the team in the year
GIDP	Total number of times the team grounded into a double play
CS	Total number of times team players were caught stealing
YEAR	Year in which the season was played
H1B	Total number of singles hit by the team in the year
OUT	Total number of outs by the team in the year

Note: The teams are identified by a combination of league, division, and team variables, in that order; thus 113 = CUBS and 213 = RED SOX.

NATIONAL LEAGUE		*AMERICAN LEAGUE*	
East	*West*	*East*	*West*
111 Phillies	121 Dodgers	211 Yankees	221 A's
112 Pirates	122 Reds	212 Orioles	222 Twins
113 Cubs	123 Astros	213 Red Sox	223 White Sox
114 Cards	124 Giants	214 Milwaukee Brewers	224 Rangers
115 Mets	125 Braves	215 Indians	225 Royals
116 Expos	126 Padres	216 Tigers	226 Angels

Except for 1969–1971:

214 Washington Senators
224 Seattle Pilots–Milwaukee Brewers

Table B-1 *Fortune* 50 Retailers Data

OBS	RANK82	RANK81	RETAILER	STATE	REVENUE	ASSETS	INCOME	EQUITY
1	1	1	SEARS ROEBUCK	IL	30020	36643	861	8812
2	2	2	SAFEWAY STORES	CA	17633	3891	160	1137
3	3	3	K MART	MI	16772	7344	262	2601
4	4	5	KROGER	OH	11902	2730	144	871
5	5	4	J C PENNEY	NY	11414	6682	392	3228
6	6	7	LUCKY STORES	CA	7973	1609	92	526
7	7	.	HOUSEHOLD INTERNATIONAL	IL	7768	8235	125	1332
8	8	9	FEDERATED DEPARTMENT STORES	OH	7699	4574	233	2105
9	9	8	AMERICAN STORES	UT	7508	1444	90	357
10	10	11	WINN-DIXIE STORES	FL	6764	1040	104	515
11	11	13	SOUTHLAND	TX	6757	1843	108	703
12	12	6	F W WOOLWORTH	NY	6590	2489	-353	999
13	13	10	GREAT ATLANTIC & PACIFIC TEA	NJ	6227	1142	-102	302
14	14	16	DAYTON HUDSON	MN	5661	2984	207	1349
15	15	12	MONTGOMERY WARD	IL	5584	4048	-75	1603
16	16	14	JEWEL COMPANIES	IL	5572	1475	88	576
17	17	.	BATUS	KY	5165	3745	214	1051
18	18	17	GRAND UNION	NJ	4137	756	25	212
19	19	18	ALBERTSON'S	ID	3940	831	58	340
20	20	19	MAY DEPARTMENT STORES	MO	3670	2625	142	1033
21	21	30	WAL-MART STORES	AR	3448	1187	124	488
22	22	24	MELVILLE	NY	3262	1247	142	701
23	23	30	SUPERMARKETS GENERAL	NJ	3247	693	38	206
24	24	23	ALLIED STORES	NY	3216	2283	91	868
25	25	25	ASSOCIATED DRY GOODS	NY	3189	1701	78	751
26	26	22	CARTER HAWLEY HALE STORES	CA	3055	1891	49	693
27	27	26	R H MACY	NY	2979	1619	136	786
28	28	27	DILLON COMPANIES	KA	2824	558	51	209
29	29	21	ARA SERVICES	PA	2806	1206	44	509
30	30	28	MCDONALD'S	IL	2715	3263	301	1529
31	31	.	WICKES COMPANIES	CA	3638	1899	-249	-230
32	32	32	MARRIOT	MD	2541	2063	94	516
33	33	.	PUBLIX SUPER MARKETS	FL	2507	481	47	293
34	34	29	RAPID-AMERICAN	NY	2401	1592	46	60
35	35	31	STOP & SHOP	MA	2342	618	35	181
36	36	34	ZAYRE	MA	2140	748	35	295
37	37	35	JACK ECKERD	FL	2080	754	71	530
38	38	36	WALGREEN	IL	2040	616	56	299
39	39	37	TANDY	TX	2033	1228	224	813
40	40	33	SIGMOR	TX	1888	632	-16	189
41	41	39	GIANT FOOD	MD	1684	392	17	133
42	42	.	BEST PRODUCTS	VA	1582	1059	29	322
43	43	41	REVCO D S	OH	1555	501	50	285
44	44	40	WALDBAUM	NY	1491	273	17	102
45	45	42	MERCANTILE STORES	DE	1429	780	70	458
46	46	.	BROWN GROUP	MO	1397	622	60	364
47	47	.	EVANS PRODUCTS	OR	1268	949	-65	219
48	48	47	U S SHOE	OH	1254	590	60	329
49	49	43	FIRST NATIONAL SUPERMARKETS	OH	1201	201	-5	37
50	50	50	SERVICE MERCHANDISE	KY	1195	586	32	205

Table B-1 *Fortune* 50 Retailers Data (continued)

OBS	RANK82	EMPLOY	EPS82	EPS81	EPS72	EPSGR	RETINV82	RETINVAV
1	1	403600	2.46	2.06	1.99	2.14	95.29	-1.70
2	2	156478	6.11	4.39	3.55	5.38	82.63	6.56
3	3	240000	2.06	1.75	1.00	7.49	49.95	-5.21
4	4	131962	4.84	4.51	0.69	21.51	58.22	20.67
5	5	173000	5.35	5.50	2.86	6.46	75.86	-1.95
6	6	64000	1.80	1.88	0.81	8.31	42.16	8.60
7	7	84000	2.06	2.66	2.42	-1.60	61.65	2.94
8	8	124600	4.79	5.33	2.46	6.89	36.95	2.71
9	9	62449	8.07	5.42	0.62	29.36	123.56	19.59
10	10	64500	4.18	3.67	1.49	10.87	53.58	8.62
11	11	49900	3.02	2.65	0.69	15.91	25.69	7.47
12	12	103600	-11.71	2.64	2.60	.	53.78	5.40
13	13	45000	-2.72	-1.35	0.59	.	113.00	-5.80
14	14	80000	4.29	3.62	0.85	17.57	87.50	19.93
15	15	83100	.	.	.	.	.	.
16	16	37500	6.82	8.03	2.67	9.83	38.41	9.38
17	17	64000	.	.	.	.	.	.
18	18	32000	.	.	.	.	.	.
19	19	32000	3.81	3.14	0.59	20.51	81.96	23.67
20	20	69300	4.87	4.31	2.09	8.83	95.64	8.92
21	21	46000	1.82	1.25	0.09	35.08	135.49	28.53
22	22	61844	5.46	5.26	1.18	16.56	92.82	12.38
23	23	31000	4.38	3.61	0.49	24.49	98.13	15.56
24	24	63000	4.41	4.38	1.64	10.40	52.82	13.42
25	25	61000	4.31	4.17	3.16	3.15	89.40	4.38
26	26	36000	1.55	1.55	1.87	-1.86	12.26	-4.22
27	27	49000	4.11	3.68	0.92	16.15	133.85	19.47
28	28	24700	2.76	2.28	0.47	19.37	63.10	17.06
29	29	115000	3.86	4.70	3.02	2.48	40.47	-6.24
30	30	140000	5.00	4.36	0.63	23.02	40.35	2.41
31	31	28000	-17.12	-17.47	1.42	.	-57.63	-11.12
32	32	109200	3.46	3.21	0.55	20.19	63.93	6.52
33	33	30137	9.43	7.68	.	.	.	.
34	34	40000	.	2.03	2.36	.	.	.
35	35	29000	6.87	6.06	1.62	15.54	152.83	21.84
36	36	34000	4.74	3.38	1.79	10.23	157.77	10.89
37	37	30600	1.90	2.19	0.73	10.04	4.84	1.08
38	38	31000	3.73	2.91	0.89	15.41	135.15	24.30
39	39	31000	2.17	1.65	0.09	37.47	50.37	36.65
40	40	4600	-0.96	1.90	0.41	.	14.74	19.30
41	41	14400	3.41	3.74	1.69	7.27	167.70	20.48
42	42	12800	2.45	2.50	0.25	25.64	38.74	2.82
43	43	20800	2.42	2.31	0.47	17.81	71.81	71.28
44	44	6500	6.29	3.86	0.74	23.86	136.69	18.16
45	45	19000	11.86	9.82	3.52	12.92	110.73	6.87
46	46	28000	5.55	5.05	2.00	10.75	103.56	17.47
47	47	13000	-5.64	1.03	1.71	.	-50.71	-4.43
48	48	25000	5.43	5.33	1.25	15.82	68.33	20.69
49	49	13000	-1.39	2.52	.	.	222.42	.
50	50	9963	2.40	1.82	0.12	34.93	139.79	26.14

Table B-2 *Fortune* 50 Utilities Data

OBS	RANK81	RANK82	UTILITY	STATE	ASSETS	REVENUES	INCOME	EQUITY
1	1	1	A T & T	NY	148186	65093	7279	62214
2	2	2	GTE	CN	22294	12066	836	5816
3	4	3	PACIFIC GAS & ELECTRIC	CA	13635	6785	810	5944
4	6	4	COMMONWEALTH EDISON	IL	12582	4130	607	4373
5	3	5	SOUTHERN CO	GA	12301	4927	472	3504
6	5	6	AMERICAN ELECTRIC POWER	OH	12224	4180	335	3463
7	8	7	MIDDLE SOUTH UTILITIES	LA	10365	2902	311	2482
8	7	8	SOUTHERN CALIFORNIA EDISON	CA	10158	4303	556	3864
9	10	9	TEXAS UTILITIES	TX	8021	3238	429	2810
10	11	10	PUBLIC SERVICE ELECTRIC & GAS	NJ	7907	3874	343	3081
11	9	11	CONSOLIDATED EDISON	NY	7872	5067	493	3989
12	14	12	DETROIT EDISON	MI	7636	2123	255	2311
13	13	13	CONSUMERS POWER	MI	7594	2731	281	2374
14	12	14	VIRGINIA ELECTRIC & POWER	VA	7359	2361	279	2535
15	16	15	PHILADELPHIA ELECTRIC	PA	7158	2645	336	2627
16	15	16	DUKE POWER	NC	7058	2244	398	2813
17	17	17	FLORIDA POWER & LIGHT	FL	6851	2941	301	2207
18	18	18	HOUSTON INDUSTRIES	TX	5672	3838	126	1827
19	20	19	PENNSYLVANIA POWER & LIGHT	PA	5531	1220	279	1875
20	25	20	LONG ISLAND LIGHTING	NY	5350	1634	310	2033
21	26	21	OHIO EDISON	OH	5247	1430	216	1801
22	19	22	GENERAL PUBLIC UTILITIES	NJ	5181	2406	38	1437
23	24	23	COLUMBIA GAS SYSTEM	DE	5155	5071	185	1458
24	22	24	CENTRAL & SOUTH WEST	TX	5135	2389	234	1659
25	20	25	PANHANDLE EASTERN	TX	5061	3391	220	1442
26	21	26	CAROLINA POWER & LIGHT	NC	4951	1538	227	1748
27	23	27	UNITED TELECOMMUNICATIONS	KA	4754	2419	202	1504
28	27	28	NIAGARA MOHAWK POWER	NY	4736	2394	269	1891
29	31	29	UNION ELECTRIC	MO	4631	1218	205	1602
30	29	30	TEXAS EASTERN	TX	4542	5654	161	1358
31	32	31	PACIFIC POWER & LIGHT	OR	4412	1406	197	1585
32	35	32	EL PASO	TX	4180	4364	59	541
33	33	33	NORTHEAST UTILITIES	CN	4029	1763	151	1160
34	28	34	AMERICAN NATURAL RESOURCES	MI	3980	3097	161	944
35	34	35	INTER NORTH	NB	3954	4159	135	1362
36	38	36	ARIZONA PUBLIC SERVICE	AZ	3889	1064	231	1611
37	39	37	TRANSCO ENERGY	TX	3866	3869	137	739
38	36	38	CONTINENTAL TELECOM	GA	3865	1817	150	1066
39	37	39	CLEVELAND ELECTRIC ILLUMINATING	OH	3843	1109	209	1227
40	41	40	PUBLIC SERVICE CO OF INDIANA	IN	3813	810	233	1498
41	40	41	GULF STATES UTILITIES	TX	3806	1307	166	1365
42	42	42	BALTIMORE GAS & ELECTRIC	MD	3513	1595	175	1438
43	44	43	NORTHERN INDIANA PUBLIC SERVICE	IN	3450	1733	103	1043
44	43	44	ALLEGHENY POWER SYSTEM	NY	3421	1752	161	1078
45	46	45	CONSOLIDATED NATURAL GAS	PA	3272	3211	160	1013
46	49	46	PACIFIC LIGHTING	CA	3257	4359	129	1013
47	45	47	NOTHERN STATES POWER	MN	3197	1561	158	1157
48	47	48	SONAT	AL	3139	2681	205	1135
49	48	49	ILLINOIS POWER	IL	3093	1107	156	1229
50	50	50	ENSEARCH	TX	3079	3788	157	981

Table B-2 *Fortune* 50 Utilities Data (continued)

OBS	RANK82	EMPLOY	EPS82	EPS81	EPS72	EPSGR	RETINV82	RETINVAV	TYPE
1	1	991596	8.40	8.55	4.34	6.83	10.25	9.01	COMUNICATIONS
2	2	196000	4.70	4.32	2.60	6.10	38.63	11.39	COMUNICATIONS
3	3	26023	4.91	3.41	3.02	4.98	47.53	8.25	ELECTRIC
4	4	17200	3.75	3.06	3.13	1.82	41.12	6.50	ELECTRIC
5	5	29882	2.38	1.81	1.88	2.39	44.06	8.47	ELECTRIC
6	6	24269	2.03	2.37	2.63	-2.56	23.13	5.44	ELECTRIC
7	7	13865	2.33	2.44	1.98	1.64	30.97	3.75	ELECTRIC
8	8	15797	5.13	4.93	2.55	7.24	33.69	11.69	ELECTRIC
9	9	16500	3.85	3.51	1.95	7.04	29.93	3.29	ELECTRIC
10	10	13118	3.24	2.63	2.44	2.88	43.23	10.74	ELECTRIC
11	11	22609	3.55	3.22	1.06	13.06	36.50	14.74	ELECTRIC
12	12	11208	1.75	2.02	2.09	-1.76	30.14	7.21	ELECTRIC
13	13	12936	3.16	3.12	2.72	1.51	28.84	7.61	ELECTRIC
14	14	12818	1.98	1.77	2.08	-0.49	35.32	5.79	ELECTRIC
15	15	10156	2.39	2.25	2.08	1.40	39.89	8.68	ELECTRIC
16	16	20174	3.59	3.19	1.69	7.83	23.58	9.27	ELECTRIC
17	17	12514	5.49	4.25	2.69	7.39	34.58	6.75	ELECTRIC
18	18	10769	1.62	3.14	2.07	-2.42	22.27	2.04	GAS
19	19	8208	3.25	3.40	2.48	3.05	36.06	8.33	ELECTRIC
20	20	6012	2.70	2.55	2.20	2.07	35.25	7.57	ELECTRIC
21	21	7885	2.13	2.30	1.91	1.10	35.57	6.10	ELECTRIC
22	22	12420	0.61	-0.26	2.21	-12.08	0.00	-4.01	ELECTRIC
23	23	12258	5.10	5.50	3.20	4.77	-1.21	7.33	GAS
24	24	8936	2.82	2.54	1.63	5.63	37.33	4.62	ELECTRIC
25	25	7300	5.36	6.52	2.89	6.37	22.31	7.55	GAS
26	26	8342	3.17	3.06	2.86	1.03	18.03	6.25	ELECTRIC
27	27	29309	2.49	3.47	1.49	5.27	3.86	6.75	COMUNICATIONS
28	28	10300	2.64	2.35	1.81	3.85	40.48	9.44	ELECTRIC
29	29	7294	2.17	1.90	1.35	4.86	40.96	8.47	ELECTRIC
30	30	6500	6.28	8.70	3.20	6.97	17.99	4.60	GAS
31	31	9252	3.03	2.82	2.18	3.35	30.29	7.42	ELECTRIC
32	32	8200	1.20	3.07	2.05	-5.21	-2.58	9.77	ELECTRIC
33	33	8692	1.76	1.29	1.57	1.15	46.95	8.24	ELECTRIC
34	34	11759	6.77	6.91	4.19	4.91	-13.19	6.17	GAS
35	35	10621	3.01	5.43	0.97	11.19	-7.61	15.43	GAS
36	36	7047	3.30	3.26	2.29	3.72	38.18	9.51	ELECTRIC
37	37	3542	5.26	5.41	2.01	10.10	-40.38	11.50	GAS
38	38	21698	2.32	2.34	1.60	3.79	11.46	3.68	COMUNICATIONS
39	39	5411	3.01	2.52	2.15	3.42	37.13	8.09	ELECTRIC
40	40	5351	4.55	3.42	2.07	8.19	36.27	7.10	ELECTRIC
41	41	4815	1.95	2.24	1.59	2.06	25.80	5.01	ELECTRIC
42	42	9118	4.07	3.70	2.83	3.70	36.27	9.37	ELECTRIC
43	43	6715	1.55	1.37	2.25	-3.66	28.42	2.20	ELECTRIC
44	44	5869	3.41	3.03	2.35	3.79	61.44	9.71	ELECTRIC
45	45	10099	3.90	3.72	1.65	8.98	-0.52	12.27	GAS
46	46	10099	4.57	4.52	2.23	7.44	14.48	10.85	GAS
47	47	7520	4.79	3.89	2.75	5.71	35.35	8.68	ELECTRIC
48	48	7000	5.05	4.24	1.18	15.65	-17.92	8.30	GAS
49	49	3993	3.04	2.84	2.38	2.48	27.16	6.29	ELECTRIC
50	50	21078	2.70	3.55	1.12	9.20	-18.41	6.89	GAS

Table B-3 Housing Sales Data

OBS	PRICE	INCOME	AGE	RORNR	OORN
1	88	39	28	0	1
2	91	34	28	0	1
3	66	34	28	1	1
4	78	14	29	0	1
5	83	28	35	0	1
6	63	25	29	0	0
7	104	57	31	1	0
8	116	52	32	1	1
9	56	13	33	0	0
10	98	31	48	1	0
11	84	29	37	1	0
12	83	36	32	0	0
13	84	45	33	1	1
14	42	7	28	0	1
15	76	24	29	1	0
16	64	53	32	0	1
17	70	20	27	1	1
18	63	36	30	1	0
19	70	15	32	0	1
20	97	55	27	1	1
21	50	15	27	0	1
22	84	32	31	0	1
23	91	32	29	0	1
24	91	41	54	1	1
25	59	11	30	0	0
26	52	13	29	0	1
27	87	45	26	1	1
28	63	28	47	1	1
29	67	13	28	0	1
30	29	3	29	0	1
31	36	15	28	1	1
32	59	25	26	0	0
33	77	25	30	0	1
34	35	17	32	0	0
35	113	62	29	1	1
36	60	38	32	1	0
37	77	32	36	1	0
38	87	50	41	0	0
39	83	59	28	1	0
40	92	42	47	1	1
41	56	18	32	0	0
42	75	50	30	0	1
43	98	29	27	0	1
44	90	56	47	0	1
45	67	42	21	0	0
46	113	48	38	1	1
47	97	63	28	1	1
48	101	55	57	1	1
49	59	17	28	1	1
50	167	120	32	1	1

OBS	PRICE	INCOME	AGE	RORNR	OORN
51	53	22	27	1	1
52	76	18	32	1	1
53	94	36	36	0	1
54	108	57	27	1	1
55	80	38	28	0	1
56	73	39	33	0	1
57	62	11	27	1	1
58	81	62	24	0	1
59	88	48	52	0	0
60	61	45	57	0	1
61	130	73	38	0	1
62	102	60	31	1	1
63	46	18	28	0	1
64	125	90	29	1	1
65	94	45	39	1	0
66	46	13	32	0	0
67	141	102	54	1	0
68	63	32	35	0	1
69	78	21	34	0	1
70	77	41	33	0	0
71	62	17	25	0	1
72	76	11	30	0	0
73	109	52	37	1	0
74	83	45	47	1	0
75	62	11	30	0	1
76	62	31	28	0	1
77	83	36	48	1	1
78	74	41	45	0	0
79	76	17	27	0	1
80	113	64	50	0	1
81	78	34	45	1	1
82	129	83	27	1	1
83	42	21	29	1	1
84	71	34	30	1	0
85	41	10	25	0	1
86	27	10	29	0	1
87	18	8	33	0	1
88	43	13	33	0	1
89	50	4	28	0	1
90	46	7	29	0	0
91	126	88	53	1	0
92	50	22	28	1	0
93	55	22	33	1	1
94	97	46	31	1	1
95	70	38	29	0	1
96	78	13	31	0	1
97	76	46	32	1	0
98	78	15	30	0	1
99	67	41	36	1	0
100	85	60	46	0	1

ANSWERS TO ODD-NUMBERED EXERCISES

Chapter 1 Statistics and Data

1-1 a. Elements: employees; characteristic: attitude about job performance; type of measuring scale: ordinal; variable: employee attitude about job performance, measured on a 5-point scale.
b. Discrete; only 5 values of the variable are possible.
c. The data are quantitative, because they represent the degree of agreement on a 5-point scale. Data will consist of a number between 1 and 5 for each employee. For a given use of the questionnaire the data will be cross-section data. Results at different times are a time series, which may display important shifts in attitude.

1-3 a. An example of a cross-section study is the comparison on a given day of the average price per share for stocks in the electronics industry with that for basic manufacturing. The average price-earnings ratio in the electronics industry compared with that in financial institutions is another example.
b. A plot of daily average prices quoted in each major branch of industry would show different trends in prices.

1-5 a. The inventory is a census of stock on hand at the time it is taken. It is a sample of annual inventories for the past ten years, for example.
b. These records constitute a census of prices of a 5-pound chuck roast in this town at the first of every month. This could also be a sample of prices for this item throughout the month, or a sample of prices for popular cuts of meat.
c. The temperatures recorded constitute a census for the time they were taken. Usually, however, they are viewed as a sample to be followed by other samples in an effort to detect any trends if illness is present or suspected.

1-7 Attitudes in the census may be distorted because of a sense of organizational coercion. Editing, coding, and tabulating results for the census will require considerable expense if accurate and timely results are to be obtained. Even a small percentage of nonresponse in the census would mean a relatively large absolute number of persons to be tracked down for completion of the questionnaire. The rate of processing errors per 100 interviews probably will be considerably higher with the census because the volume of work will prohibit the same intensity of checking that can be done in the survey.

1-9 The number of people in each urban concentration in the state is basic. In addition, describing the population in each place with respect to age, income, occupation, ages of children, number of agents currently in each market, and group insurance plan coverage are examples of relevant information. The census and insurance industry trade publications are two obvious secondary sources.

1-11 Intuitively, the average weight of several packages could be expected to fluctuate from the typical weight of all packages less than the weight of a single package.

1-13 a. An accident is an element to be observed. It is not a value of a variable so it cannot be an observation.
b. The number of accidents at a given intersection during a given time period is the value of a variable. Hence it is an observation.

c. Here weather is a variable measured on a 4-point nominal scale. For computer processing each of the four points would be assigned an arbitrary numerical value.

1-15 In small samples great care can be taken to get each observation and to make it correctly. Consequently, nonsampling errors can be kept low. In surveys with small samples, however, sampling errors tend to be larger than in those with large samples, because small samples have information on a smaller portion of the population and tend to be less correctly representative, that is, they tend to vary farther from population values of variables.

Chapter 2 Organizing and Presenting Data

2-1 A majority (55%) of voters interviewed with an opinion favor Walters.

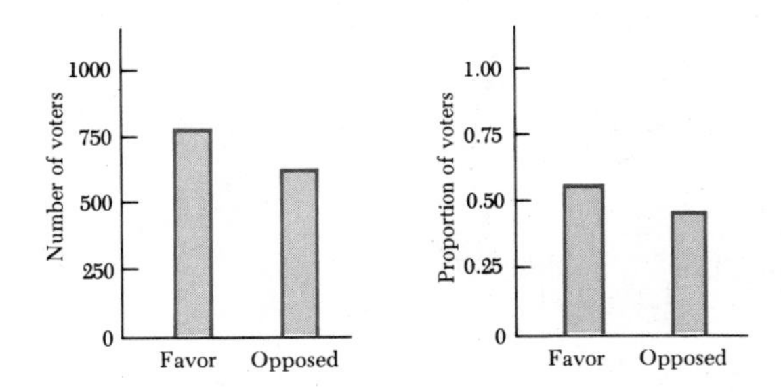

2-3 a.

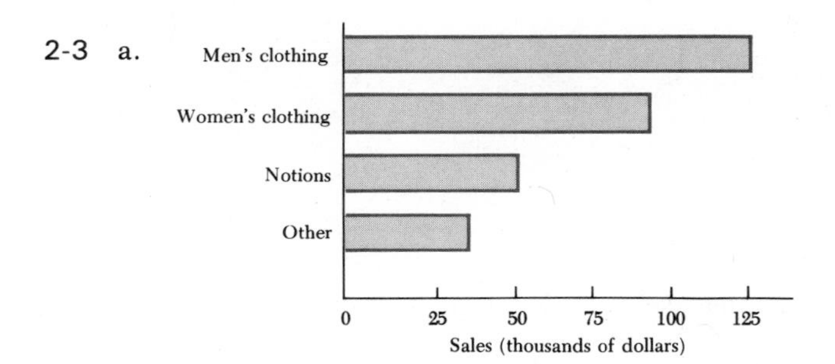

b. Men's clothing records the greatest sales with women's clothing a strong second. Notions contribute about 40 percent as much as men's clothing and other departments about 25 percent as much.

2-5 a.

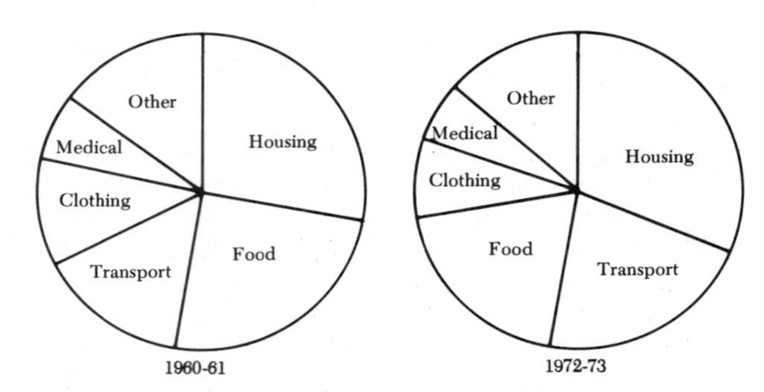

b. Transportation's share of consumption has increased substantially, as has housing's share. Food, clothing, and *other* items have decreased. Medical care is about the same share in the latter period as in the former.

2-7 a. The frequency table has been placed horizontally for convenience.

Number of Days	1	0	2	1	1	2	4	1	1	2	3	2	1
Parking Tickets	32	33	34	35	36	37	38	39	40	41	42	43	44

Number of Days	0	1	3	0	2	1	0	1	0	0	0	0	0	1
Parking Tickets	45	46	47	48	49	50	51	52	53	54	55	56	57	58

The frequency diagram has vertical lines proportional in height to the number of days in place of the numbers shown above.

b.

Number of Tickets	*Number of Days*
32-35	4
36-39	8
40-43	8
44-47	5
48-51	3
52-55	1
56-59	1
	30

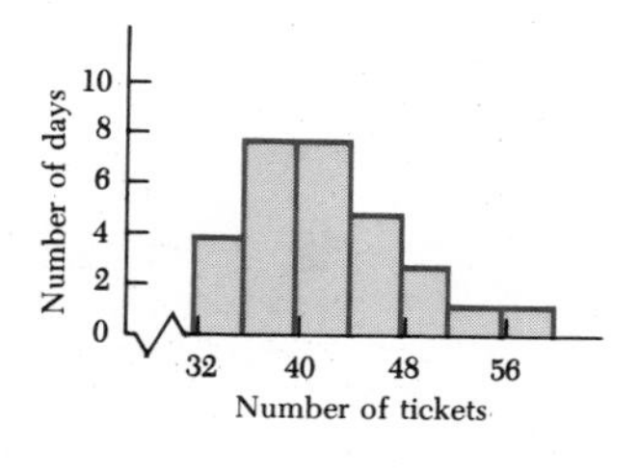

c. (b) is recommended because it brings out the pattern whereas (a) does not.

2-9 a. There are 120 observations. An average of 10 per class makes 12 classes.
b. The range is 13.11 − 12.84 = 0.27. The range divided by 12 is 0.0225. Rounding this result to the same number of decimal places as the data produces 0.02 inches as the estimate.

c.

Length of Stock (inches)	*Number of Pieces*
12.84-12.85	1
12.86-12.87	0
12.88-12.89	1
12.90-12.91	3
12.92-12.93	7
12.94-12.95	9
12.96-12.97	12
12.98-12.99	18
13.00-13.01	27
13.02-13.03	15
13.04-13.05	10
13.06-13.07	11
13.08-13.09	5
13.10-13.11	1
	120

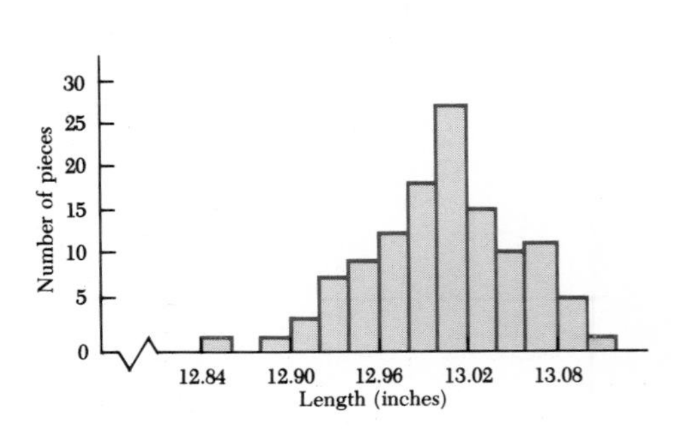

d.

Length of Stock (inches)	*Number of Pieces*
12.84-12.86	1
12.87-12.89	1
12.90-12.92	7
12.93-12.95	12
12.96-12.98	24
12.99-13.01	33
13.02-13.04	19
13.05-13.07	17
13.08-13.10	5
13.11-13.13	1
	120

This result is better than that in (c) because the sawtooth effect in the earlier result has been removed here.

2-11 a. Restricting ourselves to a choice between relative-frequency histograms and relative-frequency polygons, we must choose the latter if both are to be distinguishable on the same set of axes.

b. A much higher proportion of minority entrepreneurs are under 45 years of age. This is true for each 5-year age group as well as for the entire group less than 45 years old. The minority distribution is offset to the left of the nonminority distribution by about 5 years. Note the height adjustments in the first minority class and the last nonminority class.

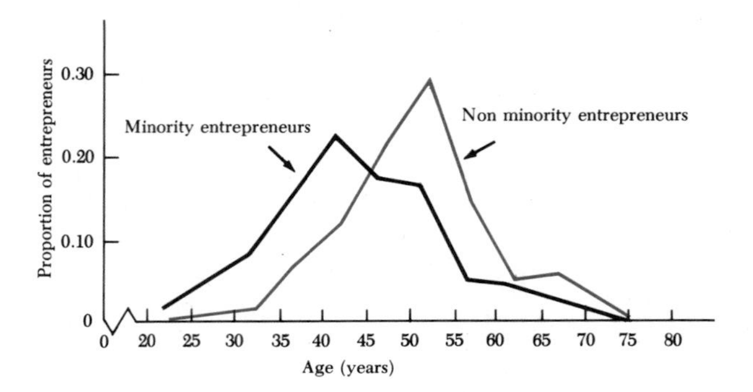

2-13 a. A relatively few high-speed drivers who never stop can be expected to produce a long tail to the right.

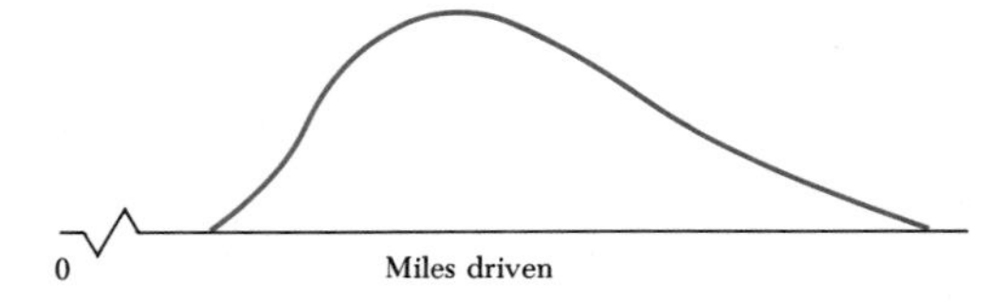

b. There are just a few young men who become bald. This produces a tail to the left. Mortality reduces the number of very old bald men.

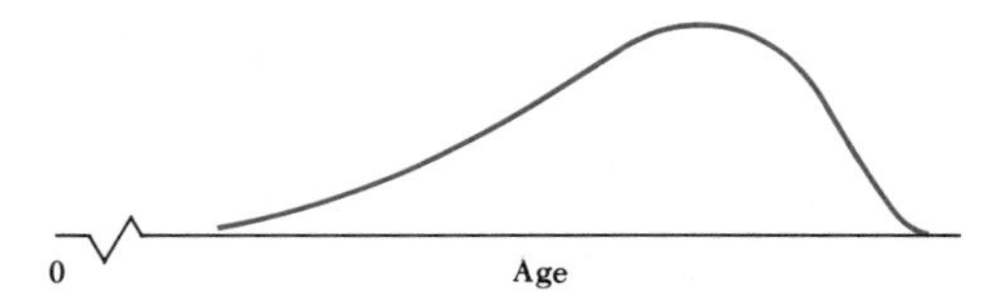

c. Heavy demand will force the machine to run out of stock and thereby put an upper limit on the amount of money. At the other end, holidays, inclement weather, and similar events can occasionally reduce demand below normal.

2-15

Tires per Month	*Cumulative Frequency*
0	11
1	19
2	25
3	29
4	33
5	35
6	36

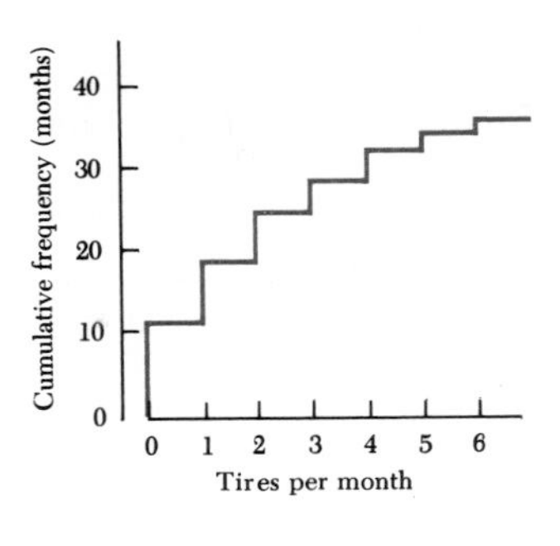

The dealer replaced 2 tires or less in 25 of the 36 months.

2-17 a.

Daily Sales	*Cumulative Frequency*
$ 90.00– 99.99	18
100.00–109.99	43
110.00–119.99	55
120.00–129.99	64
130.00–139.99	70
140.00–149.99	73
150.00–199.99	75

b.

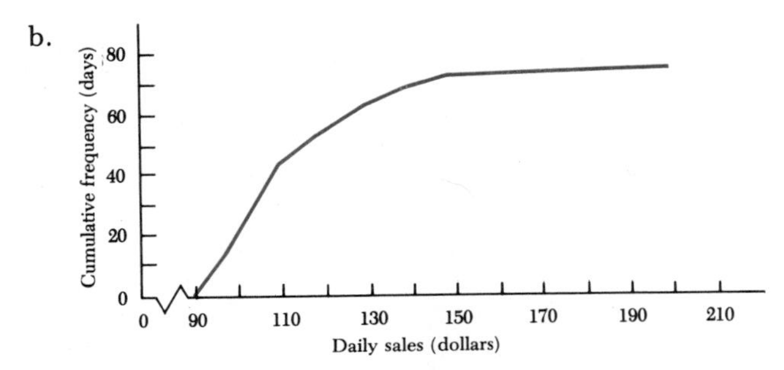

c. For 43 days during the 75-day period sales were less than $110.

2-19 The 42 observations range from 14 to 44. Stems of 1L, 1U, and so on will make a total of 7 stems with an average of 6 observations per stem, which should be adequate.

Unit: 1 travel ad; 1L | 4 represents 14 ads
1U | 5 represents 15 ads

Stem	Leaves
1L	4
1U	5899
2L	12
2U	6667888

Unit: 1 travel ad; 1L | 4 represents 14 ads
1U | 5 represents 15 ads

Stem	Leaves
3L	1233334
3U	555566799999
4L	013333444

The peak is in the upper 30's and negative skewness is present. The 7 observations from 14 through 22 may indicate a second class of magazines with a peak in the upper teens. The absence of any observations from 22 to 26 lends support to this possibility.

2-21

Unit: 1 dollar; 3 | 6 represents $36

Owner		*Renter*	
4		4	896
5	63	5	81015055
6	95383	6	1155184879
7	27915703826225758	7	923
8	07650032	8	1
9	211	9	

Those who own their homes typically spend more per week on groceries than do those who rent. In subsequent chapters we will learn how to determine whether this constitutes evidence for a real difference or should be considered only a sampling fluctuation.

2-23 The most important drop from 1971 to 1976 was marketing as a source for executives. Increases in finance and production as sources were primarily responsible for offsetting the drop in marketing.

Business Backgrounds of Executives

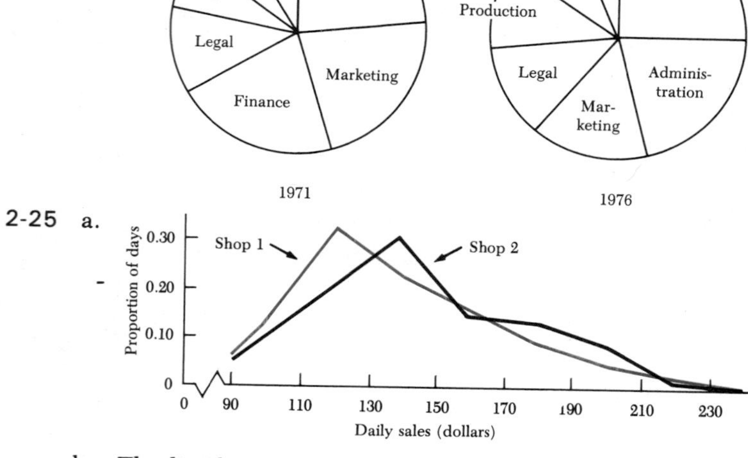

2-25 a.

b. The distribution for shop 2 is located somewhat to the right of the distribution for shop 1 and has a somewhat greater portion of its observations in the upper tail. Relative frequency polygons adjust for differences in number of observations whereas polygons do not.

c. It would not be easy, or sometimes even possible, to tell which shop was which in each class.

Chapter 3 Summary Descriptive Measures

3-1 Each weekly sales observation is the sum of the daily sales for the same week. The central location of the weekly distribution should be from 5 to 7 times that of the daily distribution, depending on how many days are included in a week. Dispersion of the weekly distribution should be greater. Both distributions can be expected to show considerable positive skewness.

3-3 a. \$192 satisfies the definition of the median.
b. \$186 is the mode.
c. Since the arithmetic mean is the sum of the observations in a data set divided by the number of observations in that set, it follows that the mean times the number of observations is equal to the sum of the observations.

3-5 a. $\bar{X} = \frac{32 + 41 + 35 + 38 + 27}{5} = \frac{173}{5} = 34.6$ passengers

b. $34.6(5) = 173 = 32 + 41 + 35 + 38 + 27$

3-7

X	f	fX
0	42	0
1	33	33
2	12	24
3	6	18
4	3	12
5	2	10
6	1	6
7	0	0
8	1	8
	100	111

Mean errors per page: $\bar{X} = \frac{111}{100} = 1.11$

3-9 a.

This Year			*Last Year*		
(X)	(f)	(fX)	(X)	(f)	(fX)
2	13	26	2	10	20
7	18	126	7	24	168
12	7	84	12	5	60
17	2	34	17	1	17
	40	270		40	265
$\bar{X} = 6.75$ listings/week			$\bar{X} = 6.625$ listings/week		

b. This year's listings are slightly greater than last year's, on the average.

3-11 In order of magnitude the list of observations is 0, 1, 2, 2, 3, 3, 4, 5, 8. The median observation is $(n + 1)/2 = (9 + 1)/2 = 5$. The fifth observation on the list is 3. Therefore, the median is 3 ads per page. The median proportion of area covered by ads per page would almost certainly be a better measure because a reader's impression probably is based more on space devoted to ads than on number of ads.

3-13

X	f	F
0	11	11
1	8	19
2	6	25
3	4	29
4	4	33
5	2	35
6	1	36

There are 36 observations, so the depth of the median is $(36 + 1)/2 = 18.5$ from either end. Beginning at the lower end, we see that the 18.5th case is in the second class. Hence, the median value is 1 tire replacement per month.

3-15 There are 158 observations in the distribution and $(n + 1)/2 = 79.5$ is the central observation. Thirty-eight observations precede the class 6000-8999 miles and there are 46 observations in this class. Hence, the central observation is in this class and it is the median class. The lower boundary of the class is 5999.5 miles and the upper boundary is 8999.5 miles. Hence,

$$Md = 5999.5 \times \frac{3000}{46}\left(\frac{158}{2} - 38\right) = 8673.4 \text{ miles}$$

3-17 The modal class runs from 150 to 250 hours because 91 is the maximum frequency in any class. The mode is 200 hours, the midvalue of the modal class. There is positive skewness with the longer tail extending upward through 550 hours. The median is in the class next above the modal class, that is, in the 250- to 350-hour class.

$$Md = 250 + \frac{100}{58}\left(\frac{260}{2} - 126\right) = 256.90$$

Since this is the middlemost value, one can argue that most people would consider it more representative than the mode. Positive skewness is present.

3-19 The two modes for the distribution in the second year suggest different accident populations—the "before" population with a mode of 3.3 and the "after" with a mode of 2.8. This could be substantiated by separating the data into before and after September sets and looking at the independent frequency distributions for each set. If this proves to be the case, the before and after distributions for the second year should be presented separately. Incidentally, with modes smaller than means, it appears that there is positive skewness in all these distributions. If so, medians probably would be the preferred measure of central location.

3-21

X	$X - \overline{X}$	X^2
$ 22,500	−19,400	506,250,000
78,300	36,400	6,130,890,000
47,700	5,800	2,275,290,000
32,100	− 9,800	1,030,410,000
28,900	−13,000	835,210,000
$209,500		10,778,050,000

$$\overline{X} = \frac{209{,}500}{5} = \$41{,}900$$

a. $$\text{Mean deviation} = \frac{\Sigma \mid X - \overline{X} \mid}{n} = \frac{(19{,}400 + \cdots + 13{,}000)}{5} = \$16{,}800$$

b. From Equation 3-12, working in thousands of dollars, the variance is

$$\frac{10{,}778.05 - (209.5^2/5)}{5 - 1} = 500$$

or 500,000,000. The square root is $22,361, which is the standard deviation.

3-23

X	f	fX	fX^2
0	8	0	0
1	8	8	8
2	8	16	32
3	3	9	27
4	0	0	0
5	2	10	50
6	1	6	36
	30	49	153

$$s^2 = \frac{153 - (49^2/30)}{30 - 1} = 2.5161$$

$$s = 1.59 \text{ flaws}$$

3-25

Age	X	f	fX	fX^2
25-29	27	4	108	2,916
30-34	32	6	192	6,144
35-39	37	6	222	8,214
40-44	42	2	84	3,528
45-49	47	2	94	4,418
		20	700	25,220

$$s^2 = \frac{25{,}220 - (700^2/20)}{20 - 1} = 37.89;$$

$$s = 6.16 \text{ months}$$

3-27 Since the stem and leaf diagram shows the data collection to be unimodal, a boxplot is appropriate. There are 80 observations, so the depth of the median is (80 + 1)/2, or 40.5, and the depth of the quartiles is (40 + 1)/2, or 20.5. The 20th case is 99.7 and the 21st is 99.8. Hence, $Q1$ is 99.75 lb. Similarly, the median is 100.0 lb and $Q3$ is 100.2 lb. The *IQR* is 100.2 − 99.75 = 0.45 and K = 1.5(0.45) = 0.675 lb. The *LIF* is 99.75 − 0.675 = 99.075, which makes 99.1 the lower adjacent value. The *UIF* is 100.2 + 0.675 = 100.875, which makes 100.7 the upper adjacent value. Given these adjacent values, we find no outliers. The median is 100.0, exactly what we expect. The boxplot below shows the box to be essentially symmetrical around 100.0 and the upper whisker to be very nearly the same length as the box. The lower whisker is somewhat longer than the box.

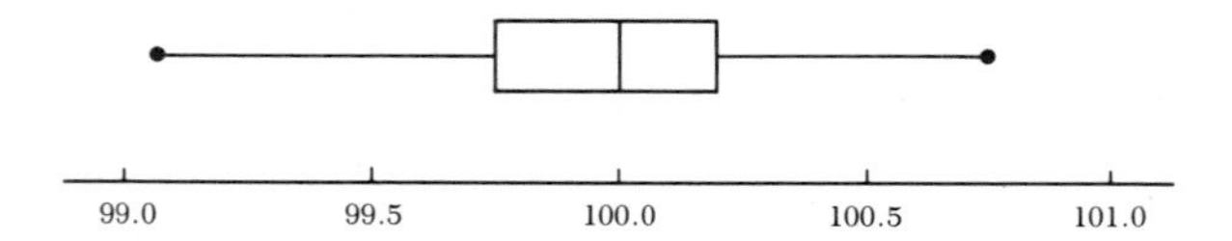

3-29 There are 120 observations in the set. This makes the median depth 60.5, and the value of the median is 13.00 inches. The quartile depth is (60 + 1)/2, or 30.5. The quartiles are 12.97 inches and 13.03 inches. The *IQR* is 0.06 and K is 0.09. The *LIF* is 12.97 − .09 = 12.88, which makes the lower adjacent value 12.89. The *UIF* is 13.12 and the *UAV* is 13.11. As the boxplot below illustrates, except for the outlier at 12.84, the conformance with expectations for a normal distribution is very high. The plot is symmetrical and the whiskers are very nearly the same length as the box.

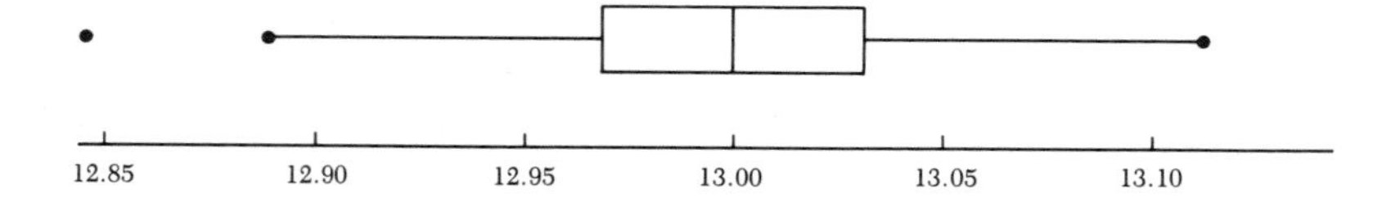

3-31 a. $n = 754$; $\Sigma X = 36{,}950$; $\overline{X} = 36{,}950/754 = 49.005$ years; (754 + 1)/2 = 377.5 is the median value. There are 363 executives 49 years old or less and 401 are age 50 or less. Hence the median is 50. The single-value class with maximum frequency is 54 years which has 46 executives. This means that 54 is the mode. With the mean less than the median and the median less than the mode, we have evidence for negative skewness.

b. We find from the data that $\Sigma X^2 = 1{,}853{,}162$. Then

$$s^2 = \frac{1{,}853{,}612 - 754(49.0053)^2}{753} = 56.9269;$$

$$s = 7.545$$

c.

Age	f	F	X	fX	fX^2
20-24	4	4	22.5	90.0	2,025.00
25-29	11	15	27.5	302.5	8,318.75
30-34	19	34	32.5	617.5	20,068.75
35-39	43	77	37.5	1,612.5	60,468.75
40-44	108	185	42.5	4,590.0	195,075.00
45-49	178	363	47.5	8,455.0	401,612.50
50-54	209	572	52.5	10,972.5	576,056.25
55-59	148	720	57.5	8,510.0	489,325.00
60-64	28	748	62.5	1,750.0	109,375.00
65-69	4	752	67.5	270.0	18,225.00
70-74	2	754	72.5	145.0	10,512.50
	754			37,315.0	1,891,062.50

$\overline{X} = 37{,}315/754 = 49.489$ years versus 49.005 for original observations.

Since $n/2$ is 377, the median class has a lower boundary of 50 years. Then

$$Md = 50 + \frac{5}{209}\left(\frac{754}{2} - 363\right) = 50.33 \text{ years}$$

as compared with a median of 50 years for the ungrouped data. The modal class runs from 50 to 55 years because 209 is the maximum number of observations in any class. The class midvalue is 52.5 years, the mode of the distribution. This compares with 54 years for the ungrouped data. The mode for the grouped data is preferable because it is far more likely to be a more stable estimate of the population mode.

d.

$$s^2 = \frac{1{,}891{,}062.5 - 754(49.4894)^2}{753} = 58.9179$$

$$s = 7.676 \text{ years}$$

These compare with 56.927 and 7.545 for the original observations.

3-33

Age	f	X	fX	fX^2
Under 25	8	20	160	3,200
25-34	11	30	330	9,900
35-44	23	40	920	36,800
45-54	16	50	800	40,000
55-64	9	60	540	32,400
65 and over	4	69	276	19,044
	71		3026	141,344

a.

$$\overline{X} = \frac{3026}{71} = 42.6197 \text{ years}$$

b.
$$s^2 = \frac{141{,}344 - 71(42.6197)^2}{70} = 176.8120;$$
$$s = 13.2971$$

c.
$$\bar{X} - s = 42.6197 - 13.2971 = 29.3226$$
$$\bar{X} + s = 55.9168$$

As a rough estimate we can say that 6 of the 11 observations in the 25-35 class are above 29, the lower limit of the range of interest. Then we have 6 observations from this class plus 23 and 16 from the next two classes plus an estimated 1 from the 55-65 class. This is a total of 46 observations of 71, or 64.8%, estimated to be in the designated range. This compares with the normal curve guideline of 68.3%.

3-35 The coefficient of variation for the first commodity is 0.47/3.02 = 0.1556, or 15.56%, and the coefficient of variation for the second commodity is 0.63/5.14 = 0.1226, or 12.26%. Relative to their means, the first commodity was more variable than the second.

3-37 For the verbal test

$$z = \frac{115 - 100}{10} = +1.5 \text{ standard deviations}$$

For the mathematical reasoning test

$$z = \frac{30 - 32}{6} = -0.333 \text{ standard deviations}$$

On the verbal test, the normal curve guideline indicates that the applicant scored better than 84% of those used to establish the mean and standard deviation (50 + (68.3/2)). Actually the performance is considerably better than 84% indicates, because this percentage is for only 1 standard deviation above the mean.

On the mathematical reasoning test, the standard score falls below the mean and is therefore below the median if the test scores are essentially symmetrical. Well over half of those who took the test scored better than this applicant.

Chapter 4 Probability

4-1 a. Relative frequency.
b. Degree of belief.
c. Equally likely events.
(b) could be established by relative frequency of past events, but this would not take into account relevant current information.

4-3 a. The percentage of employees who are college graduates.
b. The percentage of (college graduate) employees who are male.
c. The percentage of employees who are male college graduates.
d. The percentage of male employees who are college graduates.

4-5 a. $P(\text{Head}) = 50/100 = 1/2$
b. $P(\text{Rain}) = 1/(1 + 4) = 1/5$
c. $P(\text{Evert wins}) = 5/(5 + 3) = 5/8$

4-7 a. 3160/13433 = 0.235
b. (4526 + 3581)/13433 = 0.604
c. 3486/13433 = 0.260

4-9 a. 326/1574 = 0.207
b. 154/208 = 0.740
c. Age 41–50; Lt. Colonel
d. 112/478 = 0.234
e. 112/1574 = 0.071

4-11

	S_2	F_2	
S_1	0.08	0.12	0.20
F_1	0.08	0.72	0.80
	0.16	0.84	1.00

a. $P(S_2 \mid S_1) = P(S_2 \text{ and } S_1)/P(S_1) = 0.08/0.20 = 0.40$
b. $P(S_1 \mid S_2) = P(S_1 \text{ and } S_2)/P(S_2) = 0.08/0.16 = 0.50$

4-13 a. 0.10 + 0.12 = 0.22
b. 0.10 + 0.60 − 0.10(0.60) = 0.64
c. 0.10 + 0.60 − 0.08 = 0.62

4-15 0.30(1 − 0.60)(0.25) = 0.03

4-17 a.

	Shirts		
Shorts	*Yes*	*No*	*Total*
Yes	500	100	600
No	200	200	400
Total	700	300	1000

b. The marginal probability of interest in shirts is 700/1000 = 0.70. The probability of interest in shirts given interest in shorts is 0.83, so interest in shirts is not independent of interest in shorts. Under independence these probabilities would be equal. Alternatively, under independence the probability of interest in shirts *and* shorts would be 0.60(0.70) = 0.42. Since this probability is 0.50 for the boutique, interest in shirts is not independent of interest in shorts.

4-19 a. $P(\text{Tornado and Flood}) = P(\text{Tornado}) * P(\text{Flood}) = 0.001(0.005)$
$= 0.000005$ or 5 in a million
b. $P(\text{Tornado or Flood}) = P(\text{Tornado}) + P(\text{Flood}) - P(\text{Tornado and Flood})$
$= 0.001 + 0.005 - 0.000005 = 0.005995$
c. $P(\text{Tornado}' \text{ and Flood}') = (1 - 0.001)(1 - 0.005) = 0.994005$

4-21 a. Not independent; mutually exclusive.
b. No; if Jones swims laps on half of the days when he/she does not play tennis.

4-23 $P(A \text{ and } B) = P(A) * P(B \mid A) = 0.30(0.75) = 0.225$

4-25 $P(A) = P(A \mid B) * P(B) + P(A \mid B') * P(B')$
$= 0.60(0.70) \quad + 0.40(1 - 0.70) = 0.540$

4-27 a. $P(\text{Two sales})$
$= P(S_1 \text{ and } S_2 \text{ and } F_3) + P(S_1 \text{ and } F_2 \text{ and } S_3) + P(F_1 \text{ and } S_2 \text{ and } S_3)$
$= 0.20(0.20)(0.80) + 0.20(0.80)(0.20) + 0.80(0.20)(0.20)$
$= 0.032 + 0.032 + 0.032 = 0.096$

b. $P(\text{At least one sales in three})$
$= 1 - 0.80(0.80)(0.80) = 0.488$

4-29 Using S for success and F for failure to sell

$P(S_1 \text{ or } S_2 \text{ or } S_3) = 0.6 + 0.4(0.5) + 0.4(0.5)(0.4) = 0.88$

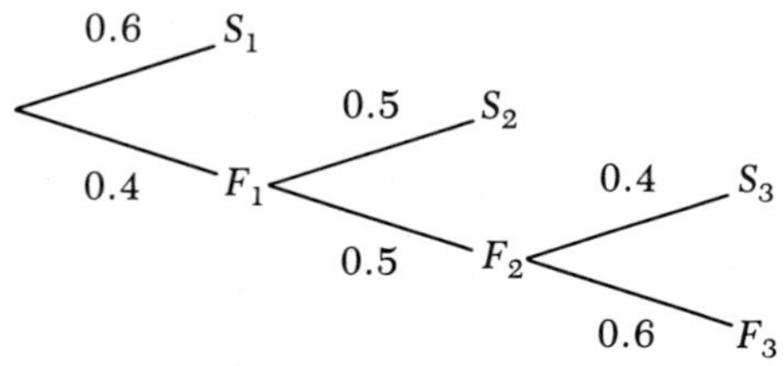

4-31 Relevant conditional and joint probabilities are

Actual (A)	Graded (G) 1	2	
1	0.90	0.10	1.00
2	0.20	0.80	1.00

Actual (A)	Graded (G) 1	2	
1	0.54	0.06	0.60
2	0.08	0.32	0.40
	0.62	0.38	1.00

a. $P(A_1 \text{ and } G_2) + P(A_2 \text{ and } G_1) = 0.06 + 0.08 = 0.14$
b. $P(A_1 \text{ and Misgraded})/P(\text{Misgraded}) = 0.06/0.14 = 0.429$
c. $P(A_1 \text{ and } G_2)/P(G_2) = 0.06/(0.06 + 0.32) = 0.158$

4-33 a. $(10 + 8)/28 = 0.64$
b. $5/23 = 0.22$
c. Assistant secretary; under secretary; deputy secretary.
d. Normal; Normal; Normal.

4-35 a. Needed here because A and B are not mutually exclusive.
b. Not needed here because being turned down by the personnel department and being turned down by an operating department are mutually exclusive ways of being turned down.

4-37 The following relevant tables can be constructed.

	Basketball Attend	Not	
Men	24	36	60
Women	16	24	40
	40	60	100

	Football Attend	Not	
Men	10	50	60
Women	10	30	40
	20	80	100

a. Yes, because $P(\text{Attend} \mid \text{Woman}) = P(\text{Attend} \mid \text{Man}) = 0.40$, or because the joint probabilities are the products of the marginal probabilities; for example, $P(\text{Attend and Woman}) = 0.40(0.60) = 0.24$.
b. No, because $P(\text{Attend} \mid \text{Woman}) = 10/60$ does not equal $P(\text{Attend}) = 20/100$, or because the joint probabilities are not the products of the marginal probabilities; for example, $P(\text{Attend and Woman}) = 10/100$ is not equal to $0.20(0.60)$.
c. Information is not given on the joint occurrence of football and basketball attendance.

4-39 a. $P(\text{Arrive late} \mid \text{Left on time}) = P(\text{Arrived late and Left on time})/P(\text{Left on time}) = 0.20/0.60 = 0.33$
b. $P(\text{Left on time} \mid \text{Arrived late}) = P(\text{Left on time and Arrived late})/P(\text{Arrived late}) = 0.20/0.50 = 0.40$
c. The joint probability table can be constructed.

	Arrive		
Leave	*On time*	*Late*	
On time	0.40	0.20	0.60
Late	0.10	0.30	0.40
	0.50	0.50	1.00

The events are not independent because $P(\text{Arrive late} \mid \text{Leave late}) = 0.30/0.40 = 0.75$ is not equal to the marginal probability of arriving late $= 0.50$.

4-41 $P(\text{Black or Walnut})$

$$= P(\text{Black}) + P(\text{Walnut}) - P(\text{Black and Walnut})$$
$$0.50 = 0.30 + P(\text{Walnut}) - 0.20$$
$$P(\text{Walnut}) = 0.50 + 0.20 - 0.30 = 0.40$$

4-43 The problem requires $P(\text{Good} \mid \text{Signal defective})$ which is equal to $P(\text{Good and Signal defective})/P(\text{Signal defective})$.

$$P(\text{Good and Signal defective}) = (1 - 0.02)(0.05) = 0.0490$$
$$P(\text{Defective and Signal defective}) = 0.02(0.99) = 0.0198$$
$$P(\text{Signal defective}) = 0.0688$$

Thus the required probability is $0.0490/0.0688 = 0.712$.

4-45 The joint probability table is

Embezzlement	*Discover (D)*		
Present (E)	*D*	*D′*	
E	0.63	0.07	0.70
E'	—	0.30	0.30
	0.63	0.37	1.00

$P(E \text{ and } D) = 0.70(0.90) = 0.63$
The problem requires $P(E' \mid D') = 0.30/0.37 = 0.81$

4-47 $P(A \text{ and } B) = P(A) * P(B)$ if independent. Events are not independent, because $0.30(0.21) = 0.063 \neq 0.095$.

4-49 a. $P(G \mid A) = P(G \text{ and } A)/P(A) = 0.20/0.30 = 0.67$
b. $P(A \mid G) = P(A \text{ and } G)/P(G) = 0.20/0.50 = 0.40$

4-51 $1 - [P(\text{Milk}) + P(\text{Eggs}) - P(\text{Milk and Eggs})] = 1 - [0.40 + 0.36 - 0.40(0.30)] = 0.36$

Chapter 5 Discrete Probability Distributions

5-1 a. Newborn infants
b. Weight and order of birth
c. Weight is continuous; order of birth is discrete

5-3 a. Automobile gasoline tanks
b. States
c. Service stations

5-5 a.

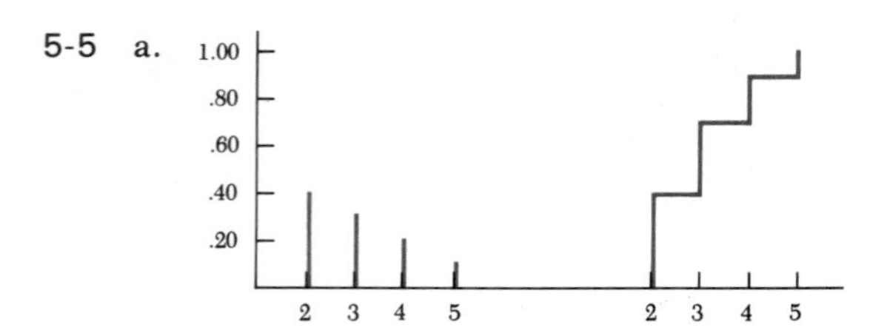

b. 0.70; 0.60; 0.40; 0.30

5-7

Week 1		Week 2	Prob
1/3 → 0	1/3	0	1/9
	1/3	1	1/9
	1/3	2	1/9
1/3 → 1	1/3	0	1/9
	1/3	1	1/9
	1/3	2	1/9
1/3 → 2	1/3	0	1/9
	1/3	1	1/9
	1/3	2	1/9

Sum	Prob
0	1/9
1	2/9
2	3/9
3	2/9
4	1/9

Total	*Probability*
0	1/9
1	2/9
2	3/9
3	4/9
4	5/9

5-9

First Draw	*Second Draw*		
	36	*42*	*48*
36	$(7/12)(6/11) = 42/132$	$(7/12)(4/11) = 28/132$	$(7/12)(1/11) = 7/132$
42	$(4/12)(7/11) = 28/132$	$(4/12)(3/11) = 12/132$	$(4/12)(1/11) = 4/132$
48	$(1/12)(7/11) = 7/132$	$(1/12)(4/11) = 4/132$	$(1/12)(0/11) = 0/132$

Sum	*P(Sum)*
72	42/132
78	56/132
84	26/132
90	8/132
Total	132/132

5-11 a. $\Sigma [x * P(x)] = 8.10$
b. As repeated selections of stocks are made, the average price-earnings ratio of the stocks selected will get closer and closer to 8.1.
c. $P(X \geq 9) + 0.30 + 0.10 = 0.40$
As more and more stocks are selected, the proportion having price-earnings ratios of 9 or more will approach 0.40.

5-13

y = Sum	$P(y)$	$y * P(y)$	$(y - \mu) * P(y)$
−190	0.10	−19	4410
−90	0.10	−9	1210
10	0.40	4	40
110	0.40	44	3240
	1.00	20	8900

$E(Y) = 20$; $\text{Var}(Y) = 8900$

5-15 a. In thousands of dollars the alternatives are

A		*B*	
x	$P(x)$	x	$P(x)$
0	0.90	0	0.70
200	0.10	50	0.30

A: $\mu = 200(0.10) = 20$; $\sigma^2 = (200)^2(0.10) - (20)^2 = 3600$; $\sigma = 60$
B: $\mu = 50(0.30) = 15$; $\sigma^2 = (50)^2(0.30) - (15)^2 = 525$; $\sigma = 22.9$

b. Alternative A is the riskier proposition. The chance of gain is smaller but the possible gain is larger.
c. Answers will reflect different subjective preferences for expected value as opposed to risk, or variance.

5-17 a. No, there is not a constant probability of success.
b. No, if they improve as expected the probability of success will not be constant.
c. Possibly, barring something systematic in the order of calls that is not mentioned in the exercise, such as calling on the easy firms first and the hard ones later.

5-19 a. $P(2) = 3(0.60)^2(0.40)^1 = 0.432$
b. $P(0) = 1(0.20)^0(0.80)^2 = 0.64$
c. $P(4) = 1(0.90)^4(0.10)^0 = 0.6561$

5-21 a. $P(R = 2 \mid n = 5, \pi = 0.80) = 0.0512$ (Table 1, Appendix A)
b. $P(F_1 \text{ and } F_2 \text{ and } F_3 \text{ and } S_4 \text{ and } S_5) = 0.2(0.2)(0.2)(0.8)(0.8) = 0.00512$
c. $P(R = 5 \mid n = 5, \pi = 0.80) = 0.3277$ (Table 1, Appendix A)

5-23 a. $\mu = \pi = 0.1$
$\sigma = \sqrt{\pi(1 - \pi)} = \sqrt{.09} = 0.3$
b. $n\pi = 9(0.1) = 0.9$
$\sigma = \sqrt{9(0.1)(0.9)} = 0.9$
c. $n\pi = 25(0.1) = 2.5$
$\sigma = \sqrt{2.5(.9)} = 1.5$

5-25 $\mu = n\pi$ $\sigma = \sqrt{n\pi(1 - \pi)}$
a. $\mu = 9(0.10) = 0.9$; $\sigma = \sqrt{9(0.10)(0.90)} = 0.9$
b. $\mu = 9(0.50) = 4.5$; $\sigma = \sqrt{9(0.50)(0.50)} = 1.5$
c. $\mu = 9(0.90) = 8.1$; $\sigma = \sqrt{9(0.90)(0.10)} = 0.9$

5-27

	Probability of Success (Bernoulli)		
	0.10	*0.20*	*0.50*
Bernoulli	0.30	0.40	0.50
$n = 4$	0.60	0.80	1.00
$n = 9$	0.90	1.20	1.50
$n = 16$	1.20	1.60	2.00

5-29 a. $(0.50 + 0.10)(0.50 + 0.10) = 0.36$
b. $0.50(0.50) + 0.50(0.10) + 0.10(0.50) = 0.35$
c. $0.10(0.10) = 0.01$

5-31 a. The distributions and their graphs are, in thousands of dollars

First offer					Alternative		
x	1	2	5	10	x	2	4
$P(x)$	0.60	0.10	0.20	0.10	$P(x)$	0.60	0.40

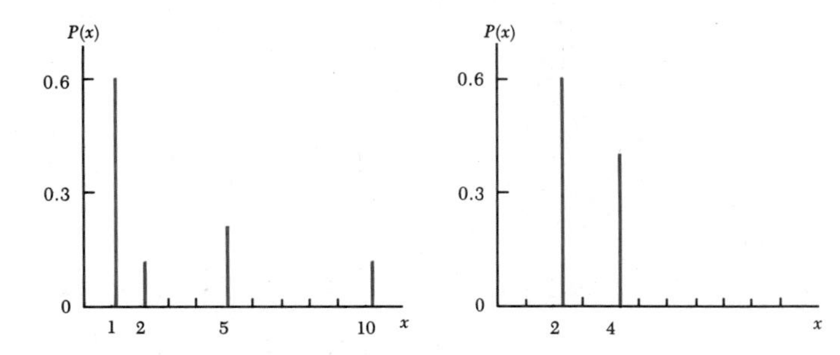

b. First offer:
$\mu = 1(0.60) + 2(0.10) + 5(0.20) + 10(0.10) = 2.8$
$\sigma^2 = 1(0.60) + 4(0.10) + 25(0.20) + 100(0.10) - (2.8)^2 = 8.16$
$\sigma = \sqrt{8.16} = 2.86$
Alternative:
$\mu = 2(0.6) + 4(0.4) = 2.8$
$\sigma^2 = 4(0.6) + 16(0.4) - (2.8)^2 = 0.96$
$\sigma = \sqrt{0.96} = 0.98$

c. Yes, the outcomes in the first offer are much more varied.

5-33 a.

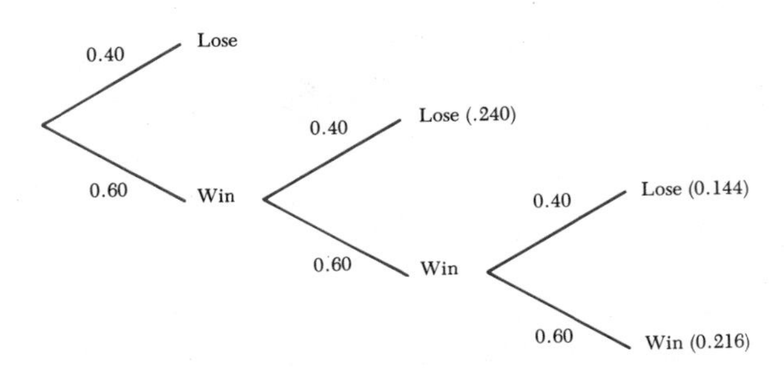

x	0	1	2	3
$P(x)$	0.400	0.240	0.144	0.216

b. $E(\$ \text{ prize}) = 5(0.400) + 10(0.240) + 20(0.144) + 50(0.216)$
$= 18.08$ or \$18,080

5-35 a. $P(R = 3 \mid n = 10, \pi = 0.20) = 0.2013$
b. $P(R = 3 \mid n = 10, \pi = 0.50) = 0.1172$

5-37 Adding probabilities for $r = 11$ and $r = 12$ for each value of π

π	0.50	0.55	0.60	0.65	0.70	0.75	0.80
	0.0031	0.0083	0.0196	0.0425	0.0850	0.1584	0.2749
	0.85	0.90	0.95	1.00			
	0.4434	0.6590	0.8817	1.00			

a.

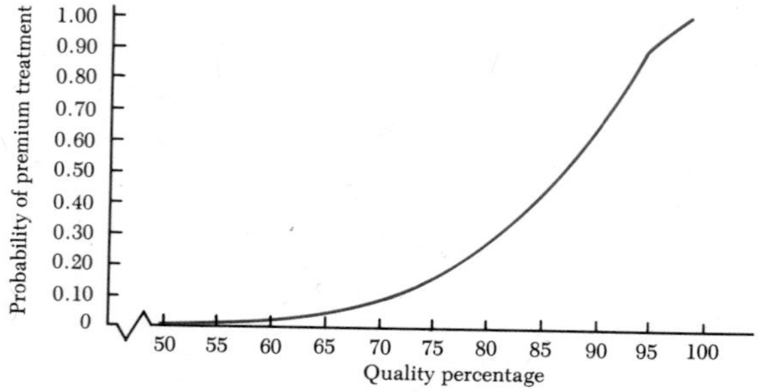

b. The cooperative could show the buyer that the chances of a 65%-quality lot being classed as premium is small, for example. What is not known is the relative frequency of 65%-quality lots, however.

5-39 $P(R = 2 \mid n = 14, \pi = 0.15) = 0.2912$
$P(R = 2 \mid n = 14, \pi = 0.05) = 0.1229$

5-41 a, b.

Number of Machines (n)	$P(R \geq 1 \mid \pi = 0.20) \times$	*(Number of Firms)* =	*(Expected Number Needing Service)*
1	0.2000	10	2.0
2	0.3600	40	14.4
3	0.4880	30	14.6
4	0.5904	20	11.8
5	0.6723	10	6.7
Total		110	49.5

c. From the above, the expected number of firms with a service need is 49.5 out of 110, or a probability of need of 49.5/110 = 0.45. If visits are restricted to firms with three or more machines, the probability of service need is (14.6 + 11.8 + 6.7) + (30 + 20 + 10) = 33.1/60 = 0.55.

5-43 a. $0.036 + 0.012 + 0.003 + 0.001 = 0.052$
b. $r = 6$, because $P(R > 6) = 0.004$ and $P(R > 5) = 0.016$

5-45 $P(R \leq 3 \mid \mu = 8.5) = 0.000 + 0.002 + 0.007 + 0.021 = 0.030$

5-47 a. $P(R > 3 \mid \mu = 1.0) = 0.015 + 0.003 + 0.001 = 0.019$
b. $P(R > 6 \mid \mu = 2.0) = 0.003 + 0.001 = 0.004$

5-49 a. Stock = 7 where $P(R \geq 7) = 0.034$
b. Stock = 11 where $P(R \geq 11) = 0.042$

c. Stock 15 for $\mu = 9$ where $P(R \geq 15) = 0.041$; Stock 19 for $\mu = 12$ where $P(R \geq 19) = 0.038$

Chapter 6 Continuous Random Variables and the Normal Distribution

6-1 a, b.

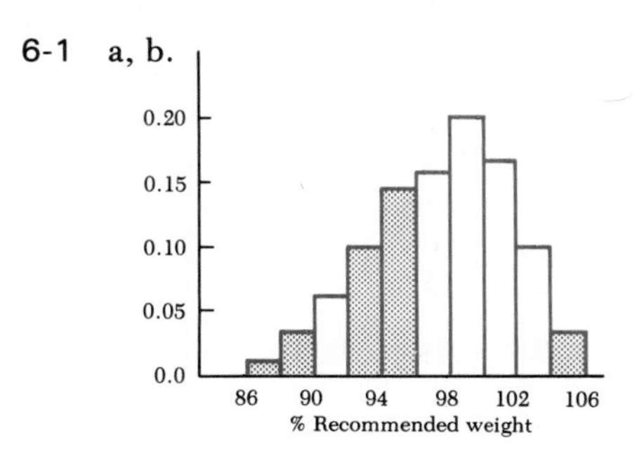

c, d.

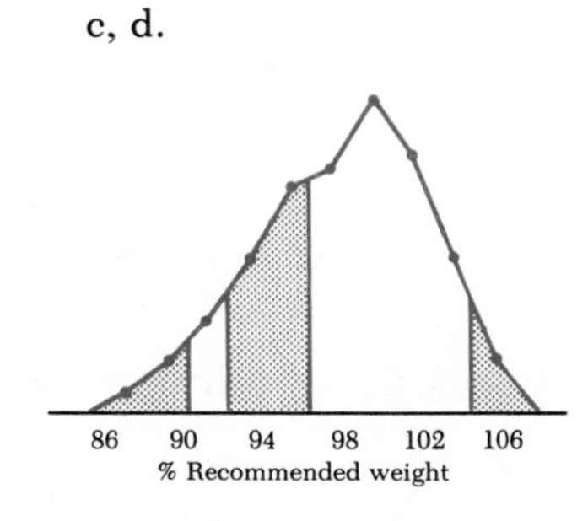

6-3 a.

Rate (%)	7.50–7.99	8.00–8.49	8.50–8.99	9.00–9.49	9.50–9.99	Total
P(Rate)	0.012	0.044	0.194	0.542	0.208	1.000

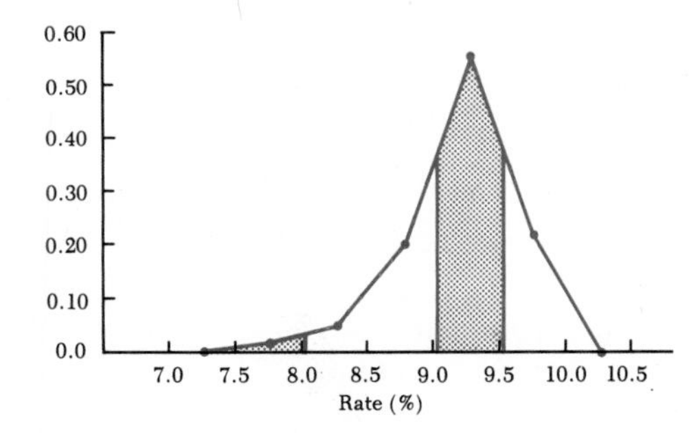

6-5 a. $P(X > 15)$
b. $P(0.0 \leq Z \leq 1.7)$
c. $x_{.15}$
d. $z_{.95}$

6-7 a. 0.6985
b. $1 - 0.9938 = 0.0062$
c. $0.8943 - 0.3085 = 0.5858$

6-9 a. 1.65; $P(Z < 1.65) = 0.9505$
b. 2.33; $P(Z < 2.33) = 0.9901$
c. 2.58; $P(Z < 2.58) = 0.9951$
d. 3.09; $P(Z < 3.09) = 0.9990$

6-11 a. $P[Z \leq (55 - 50)/2] = P(Z \leq 2.5)$
b. $P[(47 - 50)/2 \leq Z \leq (53 - 50)/2] = P(-1.5 \leq Z \leq 1.5)$
c. $P[Z \leq (46 - 50)/2 \text{ or } Z \geq (54 - 50)/2] = P(Z \leq -2.0 \text{ or } Z \geq 2.0)$

6-13 a. $P(Z \le -0.40) = 0.3446$
b. $P(-0.80 \le Z \le 0.40) = 0.6554 - 0.2119 = 0.4435$
c. $P(Z \le -1.50 \text{ or } Z > 1.50) = 0.0668 + 0.0668 = 0.1336$
d. $40 + (-1.04)(10) = 29.6$
e. $40 + 1.96(10)$; 20.4 to 59.6
f. $40 + 1.28(10)$; 27.2 to 52.8

6-15 a. $z = (610 - 600)/8 = 1.25$; $P(Z < 1.25) = 0.8943$
b. $z = (580 - 600)/8 = -2.50$; $P(Z < -2.50) = 0.0062$
c. $z = (590 - 600)/8 = -1.25$; $z = (610 - 600)/8 = 1.25$;
$P(-1.25 \le Z \le 1.25) = 0.8943 - 0.1057 = 0.7886$

6-17 a. $x_{0.80} = 600 + z_{0.80}(8) = 600 + (0.84)(8) = 606.72$
b. $x_{0.98} = 600 + z_{0.98}(8) = 600 + (2.05)(8) = 616.40$

6-19 a. $z = 5000/3600 = 1.39$; $P(-1.39 \le Z \le 1.39) = 0.9177 - 0.0823 = 0.8354$
b. $z_{0.25} = -0.67$; $z_{0.75} = 0.67$; $x_{0.25} = 20{,}000 + (-0.67)(3600) = 17{,}588$;
$z_{0.75} = 20{,}000 + (0.67)(3600) = 22{,}412$

6-21 $\mu = 25 - z_{0.01}(0.2) = 25 - (-2.33)(0.2) = 25.466$

6-23 a. $z = (0 - 0.65)/3.5 = -0.19$; $P(Z > -0.19) = 1 - 0.4247 = 0.5753$
b. $z = (5 - 0.65)/3.5 = 1.24$; $P(Z > 1.24) = 1 - 0.8925 = 0.1075$
c. $z_{0.95} = 1.65$; $x_{0.95} = 0.65 + (1.65)(3.5) = 6.43$

6-25 a. $\mu = 16(0.5) = 8$; $\sigma = \sqrt{16(0.5)(1 - 0.5)} = 2$; $z = (11.5 - 8)/2 = 1.75$;
$P(Z \ge 1.75) = 1 - 0.9599 = 0.0401$ versus 0.0383
b. $z = (5.5 - 8.0)/2 = -1.25$; $z = (6.5 - 8.0)/2 = -0.75$;
$P(-0.75 \le Z \le -1.25) = 0.2266 - 0.1057 = 0.1209$ versus 0.1222
c. $\mu = 14(0.6) = 8.4$; $\sigma = \sqrt{14(0.6)(1 - 0.6)} = 1.83$; $z = (6.5 - 8.4)/1.83 = -1.04$
$P(Z \le -1.04) = 0.1492$ versus 0.1501

6-27 a. $\mu = 36(0.20) = 7.2$; $\sigma = \sqrt{36(0.20)(0.80)} = 2.4$; $z = (9.5 - 7.2)/2.4 = 0.96$;
$P(Z \ge 0.96) = 1 - 0.8315 = 0.1685$
b. $\mu = 64(0.20) = 12.8$; $\sigma = \sqrt{64(0.20)(0.80)} = 3.2$; $z = (14.5 - 12.8)/3.2 = 0.53$;
$P(Z \ge 0.53) = 1 - 0.7019 = 0.2981$
c. $\mu = 100(0.20) = 20$; $\sigma = \sqrt{100(0.20)(0.80)} = 4.0$; $z = (24.5 - 20.0)/4.0 = 1.12$;
$P(Z \ge 1.12) = 1 - 0.8686 = 0.1314$

6-29 $\mu = 225(0.04) = 9.0$; $\sigma = \sqrt{225(0.04)(0.96)} = 2.94$;
$x_{0.05} = 9.0 + (-1.65)(2.94) = 13.85$; 15 defectives (14.5–15.5)

6-31 $\mu = 20(0.5) = 10$; $\sigma = \sqrt{20(0.5)(0.5)} = 2.236$; $z = (13.5 - 10)/2.236 = 1.57$;
$P(Z \ge 1.57) = 1 - 0.9418 = 0.0582$

6-33 a. $z_{0.80} = 0.84$; $x_{0.80} = 78 + 0.84(5) = 82.2$
b. $z_{0.90} = 1.28$; $x_{0.90} = 78 + 1.28(5) = 84.4$
c. $z_{0.99} = 2.33$; $x_{0.99} = 78 + 2.33(5) = 89.6$

6-35 a. $x_{0.01} = 20$; $\mu = x_{0.01} - (-2.33)\sigma = 20 + 2.33(0.8) = 21.86$
b. $x_{0.99} = 12$; $\mu = x_{0.99} - (2.33)\sigma = 12 - 2.33(0.8) = 10.14$

6-37 a. $\mu = 36(0.50) = 18$; $\sigma = \sqrt{36(0.50)(0.50)} = 3.0$; $z = (23.5 - 18)/3.0 = 1.83$;
$P(Z \ge 1.83) = 1 - 0.9664 = 0.0336$
b. $z = (24.5 - 18)/3.0 = 2.17$; $P(Z \ge 2.17) = 1 - 0.9850 = 0.0150$

c. 25 or more homemakers who prefer the "new and improved" polish.

6-39 $x_{0.99} = 50$; $\mu = 50 - x_{0.99}\sigma = 50 - 2.33(0.10) = 49.767$

6-41 a. $P[Z > (0 - 0.7)/0.4] = P(Z > -1.75) = 1 - 0.0401 = 0.9599$
b. $P[Z > (0 - 0.9)/1.4] = P(Z > -0.64) = 1 - 0.2611 = 0.7389$
c. $3.5 + (-1.65)(0.8) = \$2.18$ mil
d. The second has lower expectation but less risk of loss than the third

6-43 a. $E(X - Y) = 100{,}000 - 130{,}000 = -30{,}000$ (dollars);
$\text{Var}(X - Y) = (15{,}000)^2 + (8{,}000)^2 = 289(10)^6$;
$\sigma = 17{,}000$; $z = [-40{,}000 - (-30{,}000)]/17{,}000 = -0.59$; $P(Z \leq -0.59) = 0.2776$
b. $z = [-60{,}000 - (-10{,}000)]/17{,}000 = -1.76$; $P(Z \leq -1.76) = 0.0392$

6-45 a. $E(X + Y) = 1500 + 1000 = 2500$;
$\text{Var}(X + Y) = (100)^2 + (80)^2 = 16{,}400$; $\sigma = 128.1$;
$z = (2300 - 2500)/128.1 = -1.56$; $z = (2700 - 2500)/128.1 = 1.56$;
$P(-1.56 \leq Z \leq 1.56) = 0.9406 - 0.0594 = 0.8812$
b. $z_{0.025} = -1.96$; $z_{0.975} = 1.96$;
$(x + y)_{0.025} = 2500 - 1.96(128.1) = 2249$;
$(x + y)_{0.975} = 2500 + 1.96(128.1) = 2751$

6-47 a. Let y = sum; $E(Y) = 90$; $\sigma(Y) = \sqrt{8.0^2 + 6.0^2} = 10$;
$P[Z < (100 - 90)/10] = P(Z < 1.0) = 0.8413$
b. Let y = difference; $E(Y) = 10$; $\sigma(Y) = \sqrt{8.0^2 + 6.0^2} = 10$;
$P[Z < (0 - 10)/10] = P(Z < -1.0) = 0.1587$
c. $P[Z(A) < (40 - 50)/8 \text{ and } Z(B) < (30 - 40)/6] = P[Z(A) < -1.25 \text{ and } Z(B) < -1.67]$
$= 0.1057(0.0475) = 0.0501$

Chapter 7 Sampling Methods and Sampling Distributions

7-1 a. Easy, because the agents can be listed and numbered.
b. Difficult, because a list would not exist. Accidents reported to police or hospitals would be difficult to assemble and would not be a complete list anyway.
c. It would not be easy to mark and list or number all the mature trees in a 36-mile area—much less find the ones selected if that could be done.

7-3 For example, beginning with row (100-83) = 17 of Table A-8 of Appendix A:

Random number	80	12	43	56	35	17	72	70	80	15
Unit selected	—	12	43	56	35	17	—	—	—	15
Random number	45	31	82	23	74	21				
Unit selected	45	31	—	23	—	21				

7-5 See Exercise 7-2 for selection of dwelling units. The following is an example only.

Unit number	30	27	06	27	56	19	06	34	44	39
x	73.9	77.0	49.8	77.0	78.8	68.2	49.8	72.6	68.2	81.1

a. $\bar{x} = \Sigma x/n = 696.4/10 = 69.64$; $\sigma_{\bar{X}} = \sigma/\sqrt{n} = 11.72/\sqrt{10} = 3.71$;
$z = 2/3.71 = 0.54$; $P(-0.54 \leq Z \leq 0.54) = 0.7054 - 0.2946 = 0.4108$
b. $z = 5/3.71 = 1.35$; $P(-1.35 \leq Z \leq 1.35) = 0.9115 - 0.0885 = 0.8230$

7-7 See Exercise 7-5 for an example of sample observations.

a, b.

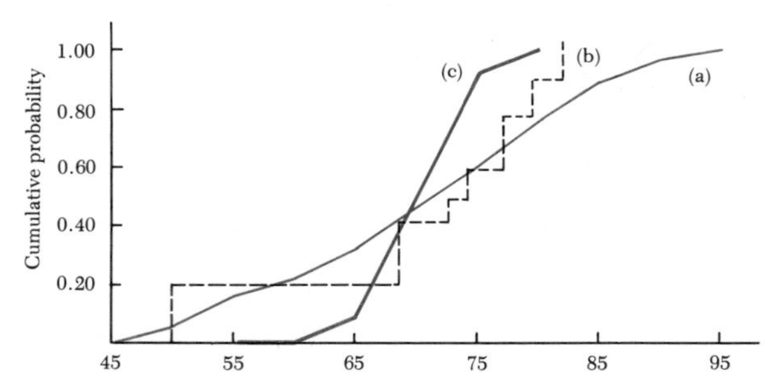

c. $z = (60 - 69.79)/3.71 = -2.64$; $P(Z \leq -2.64) = 0.0041$
$z = (65 - 69.79)/3.71 = -1.29$; $P(Z \leq -1.29) = 0.0985$
$z = (70 - 69.79)/3.71 = 0.06$; $P(Z \leq 0.06) = 0.5239$
$z = (75 - 69.79)/3.71 = 1.40$; $P(Z \leq 1.40) = 0.9192$
$z = (80 - 69.79)/3.71 = 2.75$; $P(Z \leq 2.75) = 0.9970$

d. (a) and (b), the population distribution and the distribution of sample observations.

7-9 a.

Samples	1,1	1,2	1,3	2,1	2,2	2,3	3,1	3,2	3,3
Means	1.0	1.5	2.0	1.5	2.0	2.5	2.0	2.5	3.0

$\bar{x}$	1.0	1.5	2.0	2.5	3.0
$P(\bar{x})$	1/9	2/9	3/9	2/9	1.9

b.

Sample	*Mean*	*Sample*	*Mean*	*Sample*	*Mean*
1,1,1	3/3	2,1,1	4/3	3,1,1	5/3
1,1,2	4/3	2,1,2	5/3	3,1,2	6/3
1,1,3	5/3	2,1,3	6/3	3,1,3	7/3
1,2,1	4/3	2,2,1	5/3	3,2,1	6/3
1,2,2	5/3	2,2,2	6/3	3,2,2	7/3
1,2,3	6/3	2,2,3	7/3	3,2,3	8/3
1,3,1	5/3	2,3,1	6/3	3,3,1	7/3
1,3,2	6/3	2,3,2	7/3	3,3,2	8/3
1,3,3	7/3	2,3,3	8/3	3,3,3	9/3

$\bar{x}$	3/3	4/3	5/3	6/3	7/3	8/3	9/3
$P(\bar{x})$	1/27	3/27	6/27	7/27	6/27	3/27	1/27

c. Despite the fact that the population values are equally probable, there is central clustering in the sampling distribution of the mean for $n = 2$. For $n = 3$ there is not only central clustering, but there is a suggestion of the normal distribution shape.

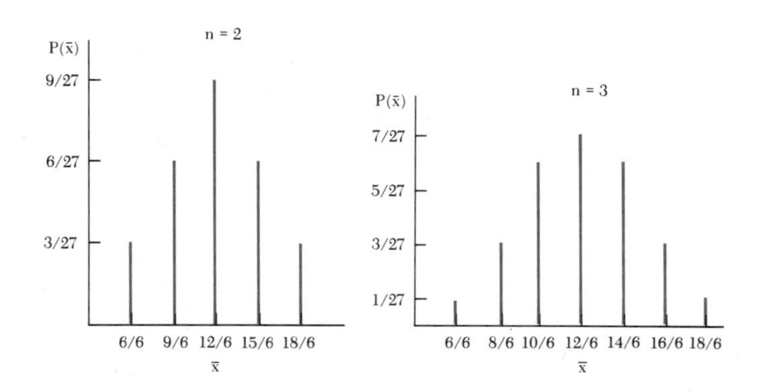

7-11 a. Confuses sampling distribution of the mean with the distribution of the sample observations.
b. Only in the limit, or as sample size increases.
c. The probability distribution of the sample mean will tend toward normality—not the distribution of the sample observations.
d. No, the central limit theorem assures us that if the sample is large enough, the sampling distribution of the mean will approximate the normal distribution even when the population is not normally distributed.

7-13 a. $\mu_{\bar{X}} = 80$; $\sigma_{\bar{X}} = 10/\sqrt{4} = 5$; $z = (87.5 - 80)/5 = 1.5$;
$P(Z \geq 1.5) = 1 - 0.9332 = 0.0668$
b. $z = (68.75 - 80)/5 = -2.25$; $P(Z \leq -2.25) = 0.0122$
c. $z = (75 - 80)/5 = -1.0$; $z = (87.5 - 80)/5 = 1.5$;
$P(-1.0 \leq Z \leq 1.5) = 0.9332 - 0.1587 = 0.7745$

7-15 $\mu_{\bar{X}} = 80$; $\sigma_{\bar{X}} = 5$
a. $\bar{x}_{0.10} = 80 + z_{0.10}(5) = 80 + (-1.28)(5) = 73.60$
b. $\bar{x}_{0.40} = 80 + z_{0.40}(5) = 80 + (-0.25)(5) = 78.75$
c. $\bar{x}_{0.70} = 80 + z_{0.70}(5) = 80 + (0.52)(5) = 82.60$

7-17 a. $\mu = 32.1$; $\sigma = 0.2$; $z = (32.0 - 32.1)/0.2 = -0.5$;
$P(Z \leq -0.5) = 0.3085$
b. $\mu_{\bar{X}} = 32.1$; $\sigma_{\bar{X}} = 0.2/\sqrt{4} = 0.1$; $z = (32.0 - 32.1)/0.1 = -1.0$;
$P(Z \leq -1.0) = 0.1587$
c. $\mu_{\bar{X}} = 32.1$; $\sigma_{\bar{X}} = 0.2/\sqrt{16} = 0.05$; $z = (32.0 - 32.1)/0.05 = -2.0$;
$P(Z \leq -2.0) = 0.0228$

7-19 a. $\mu_{\bar{X}} = 48{,}000$; $\sigma_{\bar{X}} = 2000/\sqrt{10} = 632.5$
b. $\bar{x}_{0.05} = 48{,}000 + (-1.65)(632.5) = 46{,}956$;
$\bar{x}_{0.95} = 48{,}000 + 1.65(632.5) = 49{,}044$

7-21 a. Yes, owing to the central limit theorem.
b. No, because the sample size may not be large enough to rely on the central limit theorem.

7-23 a. For $n = 4$, $\mu_{\bar{X}} = 3$, and $\sigma_{\bar{X}} = 3/\sqrt{4} = 1.5$;
for $n = 36$, $\mu_{\bar{X}} = 3$, and $\sigma_{\bar{X}} = 3/\sqrt{36} = 0.5$
b. Not for $n = 4$; yes, for $n = 36$.
c. For $n = 36$, $\bar{x}_{0.95} = 3 + 1.65(0.5) = 3.83$;
for $n = 81$, $\bar{x}_{0.95} = 3 + 1.65(3/\sqrt{81}) = 3.55$

7-25 a. $\mu = \pi = 0.20$; $\sigma = \sqrt{\pi(1-\pi)} = \sqrt{0.20(0.80)} = 0.40$
b. $\mu = n\pi = 625(0.20) = 125$; $\sigma = \sqrt{n\pi(1-\pi)} = \sqrt{625(0.20)(0.80)} = 10$
c. $\mu_P = \pi = 0.20$; $\sigma_P = \sqrt{\pi(1-\pi)/n} = \sqrt{0.20(0.80)/625} = 0.016$

7-27 a. Example, using sample of Exercise 7-4: $r = 10$; $\mu = n\pi = 12(0.68) = 8.16$; $\sigma = \sqrt{12(0.68)(0.32)} = 1.62$; $z = (10.5 - 8.16)/1.62 = 1.44$; $P(Z < -1.44 \text{ or } > 1.44) = 2(0.0749) = 0.1498$
b. $z = (9.5 - 8.16)/1.62 = 0.83$; $P(Z < -0.83 \text{ or } > 0.83) = 2(0.2033) = 0.4066$

7-29 a. $\mu_P = \pi = 0.10$; $\sigma_P = \sqrt{\pi(1-\pi)/n} = \sqrt{0.10(0.90)/100} = 0.03$
b. $z = (0.050 - 0.100)/0.03 = -1.67$; $P(Z \le -1.67) = 0.0475$

7-31 a. $z = 0.01/0.01 = 1.0$; $P(Z \le -1.0) + P(Z \ge 1.0) = 2(0.1587) = 0.3174$
b. $z = 0.01/0.008 = 1.25$; $P(Z \le -1.25) + P(Z \ge 1.25) = 2(0.1056) = 0.2112$
c. $z = 0.01/0.006 = 1.67$; $P(Z \le -1.67) + P(Z \ge 1.67) = 2(0.0475) = 0.0950$

7-33 For example, consider the days as being numbered sequentially from 01 to 84. Reading two-digit random numbers beginning with the fifty-first row of Table A-8 of Appendix A, we get

Random Number	x	*Random Number*	x	*Random Number*	x
09	6	22	4	85	—
18	7	08	3	80	10
82	4	63	12	44	7
00	—	04	Repeat	90	—
97	—	83	11	04	Repeat
32	6	38	6	58	9
82	Repeat	98	—	54	6
53	6	73	9	97	—
95	—	74	10	51	10
27	10	64	8	98	—
04	4	27	Repeat	15	5

$\bar{x} = \Sigma x/n = 153/21 = 7.29$

a. $\sigma_{\bar{X}} = (\sigma/\sqrt{n})\sqrt{(N-n)/(N-1)} = (3.02/\sqrt{21})\sqrt{(84-21)/(84-1)} = 0.574$
b. $z = (7.29 - 7.38)/0.574 = -0.16$; $P(Z \le -0.16) + P(Z \ge 0.16) = 0.4364 + 0.4364 = 0.8728$

7-35 a. Stratified random sampling.
b. For the examples worked, the stratified sample mean.
c. No, stratified random sampling is still a better (more reliable) procedure because the sample mean is *expected* to be closer, even though in a particular case it may not be.

7-37 a. No, there is no list of visitors from which to draw.
b. If only a few clusters and consequently many visitors per cluster are selected, the ranger might not have the time during a tour stay to arrange and conduct the

required interviews. With more clusters (tours) and fewer interviews per tour, the work could be accomplished more easily.

7-39 a. Blocks 1, 4, 5, and 8 of each eight blocks in a mile will border a major intersection and blocks 2, 3, 6, and 7 will not. Thus, half of the possible samples of every fourth block will contain retail concentrations and half will not. No sample of every fourth block will tend to be representative of the entire arterial.

b. Every fifth block tends to avoid the problem in (a) and over a long stretch of the arterial will prove representative.

c. This would probably be satisfactory.

7-41 $\mu_{\bar{X}} = 0.5833(\$75.90) + 0.4167(\$61.23) = \$69.79$

$$\sigma_{\bar{X}}^2 = \frac{0.5833(8.93)^2 + 0.4167(9.60)^2}{24} = 3.54;$$

$\sigma_{\bar{X}} = \$1.88$

Chapter 8 Estimation of Means and Proportions

8-1 a. A parameter is a summary descriptive measure for a statistical *population*; an estimate is a summary descriptive measure for a *sample*. Examples: μ and $\bar{x}$; σ and s.

b. An estimator is a random variable with a sampling distribution. An estimate is a particular numerical value of an estimator. For example, the sample mean with its sampling distribution is an estimator of the population mean, while the mean of a specific sample is an estimate.

8-3 The second estimator is biased, because the mean of its sampling distribution is less than the population variance.

8-5 a. Although the population of ages may not be normally distributed, for all practical purposes, the sample means of samples of 1200 will be. Hence, as an estimator of μ, the sample mean will be unbiased, consistent, and efficient. The sample mean in this case is 41,500/1200 = 34.58 years of age.

b. The sample proportion p is an unbiased, consistent, and efficient estimator of π. It is, therefore, the best choice in this case. For the sample given, p is 744/1200 = 0.62, or 62% of the sample prefers smaller cars.

8-7 a. Assuming the process is in a steady state, we can expect the population of diameters to be normally distributed. In this case, $\bar{X}$ is the preferred estimator of μ. The sum of the 5 observations is 15.037. Thus the sample mean is 15.037/5 = 3.0074 inches.

b. On the likely assumption that the population of diameters is normally distributed, the sample median is an unbiased estimator of the population mean. This estimator is, however, less efficient than the sample mean and should be rejected.

c. The unbiased estimator is s^2, the sum of squared deviations from the sample mean divided by one less than the number of observations in the sample. Hence, $s^2 = (\Sigma x^2 - n\bar{x}^2)/(n-1) = (45.222853 - 45.2222738)/4 = 0.0001448$. Then $s = \sqrt{0.0001448} = 0.012033$ inches.

d. The sample mean appears to be close enough to 3.00 inches for the difference to be only sampling fluctuation. But the sample variance and standard deviation appear to be too large for the difference from the desired variance to be only sampling fluctuation.

8-9 The applicable formula for this case is the interval between

$$\bar{x} \pm 2.58(\sigma/n^{1/2}) = 17.66 \pm 2.58(0.31/8^{1/2}) = 17.66 \pm 2.58\,(0.31/2.83) = 17.66 \pm 0.28$$

or 17.38 to 17.94 oz. The 99% confidence interval estimate of the mean fill weight is 17.38 to 17.94 oz.

8-11 a. The distribution has 0.025 less than −2.5706 and 0.025 greater than +2.5706, a total of 0.05 farther than 2.5706. Hence, 0.95 is within 2.5706 of 0.

b. The distribution has 0.025 greater than 2.5706, so 0.975 is less than +2.5706.

c. 0.50 lies above a t-value of 0.

8-13 Sum = 4845; sum of squares = 1,501,709; $n = 16$; sample mean = 302.81; $s = 48.0156$; $t_{15,.95} = 1.7531$

$$\bar{x} \pm t(s/n^{1/2}) = 302.81 \pm 1.7531(48.0156/4) = 302.81 \pm 21.04 \quad \text{or} \quad 282 \text{ to } 324$$

The 90% confidence interval estimate of the mean water usage for the neighborhood is from 282 to 324 gallons.

8-15 Sample mean = 1183 hr; $s = 97$ hr; $t_{99,.995} = 2.6264$; $n = 100$

$$\bar{x} \pm t(s/n^{1/2}) = 1183 \pm 2.6264(97/10) = 1183 \pm 25.5 \quad \text{or} \quad 1157.5 \text{ to } 1208.5 \text{ hr}$$

The 99% confidence interval estimate of the shipment's mean lifetime is 1157.5 to 1208.5 hr.

8-17 a. $\Sigma x = -5283$; $\Sigma x^2 = 4{,}967{,}297$; $\bar{x} = -5283/7 = -\$755$; $s^2 = (\Sigma x^2 - n\bar{x}^2)/(n-1) = 980{,}141/6 = 163{,}357$; $s = \$404$. For 6 degrees of freedom, t for the central 95% is ±2.447. The 0.95 confidence interval is

$$\bar{x} \pm t\frac{s}{\sqrt{n}} = -755 \pm 2.447\frac{404}{\sqrt{7}} = -755 \pm 374 \quad \text{or} \quad -\$1{,}129 \text{ to } -\$381$$

b. Since the confidence interval doesn't include 0, the bias is negative and between \$381 and \$1129 for the stipulated level of confidence. The standard deviation is estimated at \$404 and indicates very good efficiency. If an appraiser is known to be about \$750 low with little variation, that amount can be added to correct the bias. Lack of variability is usually more desirable than complete lack of bias.

8-19 $p = 0.54$; $z_{.95} = 1.65$. The interval lies between the values

$$p \pm z[p(1-p)/n]^{1/2} = 0.54 \pm 1.65[.54(.46)/175]^{1/2} = 0.54 \pm 1.65[0.0377] = 0.54 \pm 0.06 \quad \text{or} \quad 0.48 \text{ to } 0.60$$

The 90% confidence interval estimate of the proportion not in favor of raising taxes is 0.48 to 0.60.

8-21 $n = 36;\ x = 12;\ \alpha = 0.10;\ 1 - \alpha/2 = 0.95$

$$L = x/[x + (n - x + 1)F_1]$$

where $F_1 = F_{.95;2(n-x+1),2x} = F_{.95;2(36-12+1),24} = F_{.95;50,24} = 1.86$

Hence, $L = 12/[12 + (36 - 12 + 1)(1.86)] = 12/[12 + 46.5] = 12/58.5 = 0.2051$ or 0.21

$$U = (x + 1)F_2/[(x + 1)F_2 + (n - x)]$$

where $F_2 = F_{.95;2(x+1),2(n-x)} = F_{.95;26,48} = 1.73$, by interpolation

Hence, $U = 13(1.73)/[13(1.73) + 24] = 22.49/46.49 = 0.4838$ or 0.48

The 0.90 confidence interval estimate is from 21 to 48% of the employees, an interval that is not very informative because of the small sample.

8-23 $n = 100;\ r = np = 85;\ 1 - r = 15$; so the normal approximation doesn't apply. Use Equations 8-7 and 8-8 with $x = 85$.

$$F_1 = F_{.95;2(100-85+1),170} = F_{.95;32,170} = 1.51$$

$$L = 85/[85 + (100 - 85 + 1)1.51] = 0.779$$

$$F_2 = F_{.95;2(86),2(15)} = F_{.95;172,30} = 1.67$$

$$U = 86(1.67)/[86(1.67) + 15] = 0.905$$

The 90% confidence interval estimate of the proportion of the chemist's time being spent in chemistry is 0.779 to 0.905.

8-25 $\alpha = 0.01;\ 1 - \alpha/2 = 0.995;\ z_{1-\alpha/2} = 2.58;\ h = 0.6$ in.; $\sigma = 1.1$ in.

$$n = [z(\sigma)/h]^2 = [2.58(1.1)/0.6]^2 = 22.4 \quad \text{or} \quad 23 \text{ lengths}$$

8-27 $\alpha = 0.10;\ 1 - \alpha/2 = 0.95;\ z_{1-\alpha/2} = 1.65;\ h = 0.04$

$$n = [z^2\pi(1 - \pi)]/h^2 = [1.65^2(0.50)(0.50)]/0.04^2 = 425.4$$

or 426 stockholders, assuming this is less than 10% of all stockholders.

8-29 For 90% confidence, $z = 1.65$.

$$n = \frac{z^2\sigma^2}{h^2} = \frac{1.65^2(2920)^2}{125^2} = 1485.6 \quad \text{or} \quad 1486 \text{ households}$$

8-31 Because the sample size (10) is more than 10% of the population (60), Equation 8-11 applies: For $n = 10$, the degrees of freedom for t is 9, and $t_{.975} = 2.2622$. The estimate is

$$24.62 \pm 2.2622\sqrt{\left(\frac{0.38}{10}\right)\left(\frac{60 - 10}{60 - 1}\right)} = 24.62 \pm 0.25 \quad \text{or} \quad 24.36 \text{ to } 24.87 \text{ lb}$$

8-33 The sample size, 200, is more than 10% of the population, 900. Hence, Equation 8-12 is the one to use. From Table A-2, $z_{.95} = 1.65$. Furthermore, $p = 135/200 = 0.675$.

$$0.675 \pm 1.65\sqrt{\left(\frac{.675(1 - .675)}{200}\right)\left(\frac{900 - 200}{900 - 1}\right)} = 0.675 \pm 0.048 \quad \text{or} \quad 0.627 \text{ to } 0.723$$

8-35 a. For any given sample size, the mean of the sampling distribution of the biased estimator is not equal to the parameter being estimated, but the mean of the sampling distribution of the unbiased estimator is equal to the parameter. The standard deviation (standard error) of the efficient estimator is less than the standard error of the inefficient estimator.

b. As sample size increases, the sampling distribution of any consistent estimator (biased or unbiased, efficient or inefficient) becomes more tightly concentrated around the parameter and the parameter is the limiting value of the estimator as sample size increases without limit. The limiting value of an inconsistent estimator is not equal to the parameter.

8-37 Estimated mean tensile strength where $t_{.995,119} = 2.6177$ is

$$\bar{x} \pm t\frac{s}{\sqrt{n}} = 89.3 \pm 2.6177\frac{1.8}{\sqrt{120}} = 89.3 \pm 0.43 \quad \text{or} \quad 88.9 \text{ to } 89.7 \text{ pounds}$$

8-39 From the classical view of probability, 0.95 is a level of confidence; it is not a probability. *Before* an interval is constructed $\bar{X} \pm 1.96\,\sigma/\sqrt{n}$ is a random interval for which there is a probability of 0.95 of including the population mean. *After* the value of the sample mean and the limits of the interval are known, however, the interval either includes μ with probability 1 or does not include it with probability 0, even though we don't know which.

8-41 a. For the frequency distribution, $n = \Sigma f = 200$; $\Sigma fX = 417$; $\Sigma fX^2 = 1167$; $\bar{x} = 417/200 = 2.085$.

$$s = \sqrt{\frac{\Sigma fX^2 - n\bar{x}^2}{n-1}} = \sqrt{\frac{1167 - 869.445}{199}} = 1.223$$

$$\bar{x} \pm t\frac{s}{\sqrt{n}} = 2.085 \pm 1.6525\frac{1.223}{\sqrt{200}} = 2.085 \pm 0.143$$

or 1.942 to 2.228 vacancies per block.

b. For 5000 blocks, the 0.90 confidence interval is

$$5000(2.085 \pm 0.143) = 10{,}425 \pm 715$$

or 9710 to 11,140 vacancies in the city.

8-43 Assuming that $\pi = 0.10$,

$$n = \frac{1.65^2(.1)(.9)}{.05^2} = 98$$

but $n\pi = 0.1(98)$, or 9.8, which is well below the 50 required for the normal approximation. From a pocket calculator that computes terms of binomial distributions, trial and error shows that, for $n = 100$, $Pr(p \leq .05) = 0.06$ and $Pr(p \leq .15) = 0.96$. Hence, for a sample of 100, the probability of being within .05 of .10 is 0.90, as required.

Chapter 9 Hypothesis Tests for Means and Proportions

9-1
1. Hypothesis statement: H_0: $\mu = 24$ oz; H_A: $\mu \neq 24$ oz.
2. Test statistic: $\overline{X}$. Sampling distribution: Normal, with mean μ and standard deviation $\sigma/n^{1/2}$.
3. For the 0.05 significance level, $\alpha = 0.05$ and $z_{1-\alpha/2} = 1.96$. Accept if $\overline{X}$ lies in acceptance region:

$$24 \pm 1.96(0.73/36^{1/2}) = 24 \pm 0.24 \text{ oz} \quad \text{or} \quad 23.76 \text{ to } 24.24 \text{ oz}$$

4. The observed mean weight for a random sample of 36 bottles is 24.68 oz.
5. Reject the null hypothesis that the mean process weight is 24 oz. The process mean seems to be out of control on the high side.

9-3
1. Hypothesis statement: H_0: $\mu = 4.00$ in.; H_A: $\mu \neq 4.00$ in.
2. Test statistic: t. Sampling distribution: A t distribution with $n - 1 = 8$ degrees of freedom.
3. For $\alpha = 0.01$ and 8 degrees of freedom, t for a two-sided test is 3.3554. Accept the null hypothesis if the sample mean lies in the interval

$$4.000 \pm 3.3554(s/9^{1/2})$$

$$\text{where} \quad s^2 = \frac{\Sigma(x^2) - [\Sigma(x)^2/n]}{n-1} = \frac{147.7708 - (36.467^2/9)}{9-1} = 0.00132361$$

$$\text{and} \quad s = 0.03638$$

Hence, the acceptance interval is

$$4.000 \pm 3.3554(0.03638/3) \quad \text{or} \quad 3.9597 \text{ to } 4.0403 \text{ in.}$$

4. The sample mean is $36.467/9 = 4.052$ in.
5. Since the sample mean lies outside the acceptance region, reject the null hypothesis. The mean length of the shipment apparently is too great.

9-5
1. Hypothesis statement: H_0: $\mu = 4150$; H_A: $\mu \neq 4150$ hr.
2. Test statistic: t. Sampling distribution: A t distribution with 99 df.
3. Acceptance region: $t_{.975} = 1.9842$ for 99 df

$$4150 \pm 1.9842(90/100^{1/2}) = 4150 \pm 17.86 \quad \text{or} \quad 4132.14 \text{ to } 4167.86$$

4. The sample mean is 4170 hr.
5. Reject the null hypothesis, because the sample mean is greater than the upper limit of the acceptance region.

9-7
1. Hypothesis statement: H_0: $\mu \leq \$23.41$; H_A: $\mu > \$23.41$.
2. Test statistic: z. Sampling distribution: Although the population probably is positively skewed, 64 observations should be enough to make the distribution of the sample mean essentially normal.

3. For $\alpha = 0.05$ and an upper-tailed test, $z_{.95} = 1.65$. The acceptance region lies below the critical value:

$$\$23.41 + 1.65(\$9.60/64^{1/2}) = \$25.39$$

4. The sample mean is \$25.79.
5. Reject the null hypothesis of no increase over \$23.41. The mean donation has increased following the change.

9-9
1. Hypothesis statement: H_0: $\mu \geq 80$; H_A: $\mu < 80$.
2. Test statistic: t. Sampling distribution: A t distribution with 207 df.
3. For the 0.05 level of significance and with 207 df, the value of t for a one-sided test is 1.6525. The acceptance region for the null hypothesis is

$$80 - 1.6525(7.21/208^{1/2}) = 79.17 \text{ or greater}$$

4. The sample mean is 78.3.
5. The nationwide trial indicates that mean deliveries would be less than 80 per month.

9-11 A type I error will be made if (a) the true population mean is 80 per month or greater and (b) the sample mean is less than the critical value, 79.17. A type II error will be made if (a) the true population mean is anything less than 80 per month and (b) the sample mean is greater than 79.17.

9-13
1. Hypothesis statement: H_0: $\pi = 0.60$; H_A: $\pi \neq 0.60$.
2. Check for applicability of normal approximation:

$$n = 900; \quad n\pi = 0.6(900) = 540; \quad n(1 - \pi) = 0.4(900) = 360$$

Normal approximation applies for the binomial distribution.

3. With $\alpha = 0.01$, $z = 2.58$ for a two-tailed test. Critical values are

$$p = 0.60 \pm 2.58\left[\frac{0.60(0.40)}{900}\right]^{1/2} = 0.60 \pm 0.04 \quad \text{or} \quad 0.56 \text{ and } 0.64$$

The acceptance region for the null hypothesis is

$$0.56 \leq p \leq 0.64$$

4. In the survey, 519 of 900 supported the proposed tax. Hence,

$$p = 519/900 = 0.58$$

5. The observed sample proportion falls in the acceptance region. The evidence supports the candidate's claim.

9-15
1. Hypothesis statement: H_0: $\pi \geq 0.25$; H_A: $\pi < 0.25$.
2. Since $n < 100$, the normal approximation does not apply. Use Appendix Table A-1 to find the binomial distribution for $n = 16$ and $\pi = 0.25$.
3. The critical value of r is 0, because Table A-1 shows that 0.01 of the distribution is in the lower tail if this is the critical value. In terms of p, the critical value is 0/16, or 0 also. The acceptance region for H_0 is any value of r or p greater than 0.
4. For the sample, r is 2 and p is $1/8$.
5. Since this result falls in the acceptance region for H_0, the evidence indicates that 25% or more of all students had the correct concept.

9-17 a. 1. Hypothesis statement: H_0: $\pi \leq 0.7$; H_A: $\pi > 0.7$.
2. Since n is greater than 100 and both $n\pi$ and $n(1 - \pi)$ are greater than 10, the normal approximation to the binomial can be used for the sampling distribution of the proportion.
3. For a one-tailed test with $\alpha = 0.05$, $z_{.95} = 1.65$. The critical value is

$$p = \pi_0 + 1.65\left[\frac{0.07(0.93)}{200}\right]^{1/2} = 0.07 + 0.025 \quad \text{or} \quad 0.095$$

Accept the null hypothesis if the sample proportion is less than 0.095.

b. Because the observed result, 0.11, falls above the acceptance region, the evidence suggests that the proportion of substandard dwellings has increased significantly at the 5% level of significance.

9-19 a. From Table A-2, the area in the upper tail beyond $z = 2.95$ is $1 - .9984 = 0.0016$. Since this is a one-tailed test, $P = 0.0016$.

b. From Table A-2, $Pr(z < -1.25) = Pr(z > 1.25) = 1 - .8943 = 0.1057$. Because we are concerned with a two-tailed test, the P-value is $2(0.1057) = 0.2114$.

c. From Table A-4, $Pr(t > 0.6870) = 0.25$ for 20 df. For a one-tailed test, this is also the P-value.

d. From Table A-4, $Pr(t > 2.7045) = Pr(t < -2.7045) = 1 - .995 = 0.005$ for 40 df. Because this is a two-tailed test, the probability of getting a t value more extreme in either direction is $2(0.005) = 0.010$.

e. From Table A-4, $Pr(t > 4.333) < Pr(t > 3.0545)$, which is $1 - .995$, or 0.005, for 12 degrees of freedom. Hence, for a one-tailed test, the P-value is less than 0.005.

9-21 1. Hypothesis statement: H_0: $\mu = 24$; H_A: $\mu \neq 24$.
2. The test statistic is now z, which is distributed as a standard normal variable with mean 0 and standard deviation 1 (see Table A-2).
3. With a two-tailed test and a 0.05 level of significance, $P_{\text{criterion}}$ is 0.05 and $z_{.975} = 1.96$. Hence, the acceptance region is

$$-1.96 \leq z \leq +1.96$$

4. From the sample,

$$z_{\text{observed}} = \frac{24.68 - 24}{0.73/(36)^{1/2}} = 5.58$$

The P-value associated with 5.58 in a two-tailed test is the probability area below -5.58 and above $+5.58$. From Table A-2, $P < 0.001$.

5. Since the observed value of z lies outside the acceptance region, or, alternatively, since the observed value of P is less than 0.05, reject the null hypothesis of 24. The evidence is much stronger than is indicated by a 5% level of significance.

9-23 a. $\bar{x}_{\text{criterion}} = 8 + 2.33(0.44/\sqrt{16}) = 8.26$ grams

b.

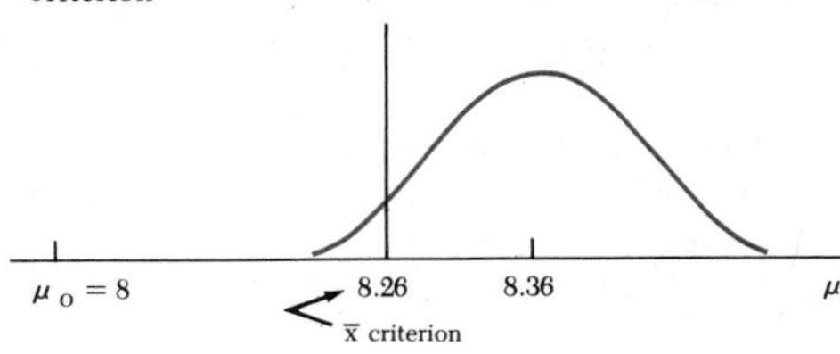

c. $z = (8.26 - 8.36)/0.11 = -0.91$; area in upper tail = 0.82, the power when $\mu = 8.36$.

9-25 a. $\bar{x}_{\text{criterion}} = \mu_0 - 1.65\sigma_{\bar{X}}$

$96.05 = 100 - 1.65\sigma_{\bar{X}}$

$\sigma_{\bar{X}} = 2.40$; $\sigma = 2.40\sqrt{25} = 12$ hours

b. A type I error is possible when the mean is 102. A type II error is possible for all other values of the mean because they are all less than $\mu_0 = 100$.

c.

μ	z	$1 - \beta$
99	1.229	0.1095
96	−0.021	0.5083
94	−0.8542	0.8035
92	−1.6875	0.9542
90	−2.5210	0.9941

$$z = \frac{\mu - 96.05}{2.40}$$

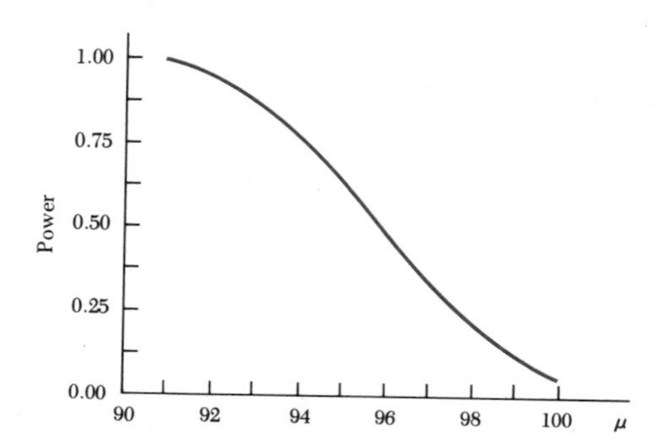

9-27 a. When performing at a population mean fill of 10 grains, sample means are normally distributed around 10 grains with a standard error of $0.2/\sqrt{16} = 0.05$ grains. The standard normal deviate for the lower critical value is

$$z_L = \frac{9.902 - 10}{0.05} = -1.96$$

and z_u is +1.96. Hence the probability of no interruption is $1 - \alpha = 0.95$ and the probability of erroneous interruption is $\alpha = 0.05$, and this is the probability of a type I error.

b. Given that $\mu = 9.95$, the sampling distribution of $\bar{X}$ is

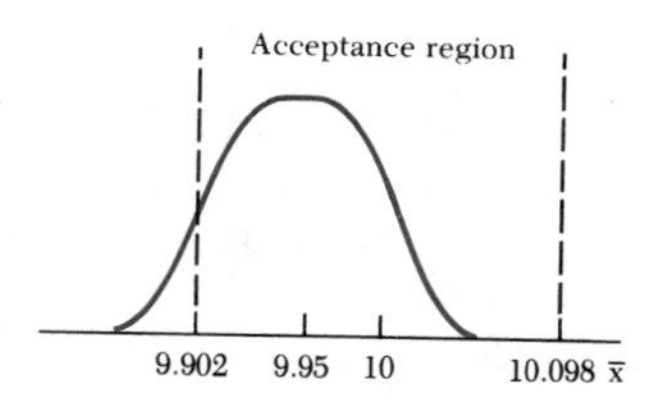

$$z_1 = \frac{9.902 - 9.95}{0.05} = -0.96; \quad z_2 = \frac{10.098 - 9.95}{0.05} = 2.96$$

so 0.33 is the area from $z = 0$ to z_1 and 0.4985 is the area from $z = 0$ to z_2. The area in the acceptance region (β) is, therefore, $0.33 + 0.4985 = 0.83$. Production will be continued erroneously under this condition. The probability of a correct decision, $1 - \beta$, is $1 - 0.83 = 0.17$.

State	Decision: Accept H_0	Decision: Reject H_0
H_0 true	Correct decision; probability: $1 - \alpha = 0.95$	Type I error; probability: $\alpha = 0.05$
H_0 false	Type II error; probability: $\beta = 0.83$	Correct decision; probability: $1 - \beta = 0.17$

9-29 Collect a random sample of the specified size and find the sample mean, $\bar{x}_{\text{observed}}$. Then find

$$z_{\text{observed}} = \frac{\bar{x}_{\text{observed}} - \mu_0}{\sigma/\sqrt{n}}$$

If z_{observed} is in the acceptance region ($z_{\text{observed}} \geq z_{\text{criterion}}$), accept the null hypothesis. If not, reject the null hypothesis. The level of significance is α, and the value of $z_{\text{criterion}}$ is found in Table 9-5.

9-31 As n increases, the power curve approaches a vertical line at $\mu_0 = 12$. For values of μ less than 12, the curve approaches a horizontal line at 1.0 on the power scale. The result is the power curve for an ideal lower-tailed test, because when μ is less than 12 the test is certain to reject H_0, as it should.

9-33
1. H_0: $\mu = 16$ feet; H_A: $\mu \neq 16$ feet
2. Population normal with $\sigma = 0.12$ foot. Test statistic: $z = (\bar{X} - \mu_0)/\sigma_{\bar{X}}$
3. Critical values: $z_L = -1.96$; $z_U = +1.96$
4. $n/N = 25/144 = 0.17$, so the finite population multiplier is required.

$$\sigma_{\bar{X}} = \frac{\sigma}{\sqrt{n}}\sqrt{\frac{N-n}{N-1}} = \frac{0.12}{\sqrt{25}}\sqrt{\frac{144-25}{144-1}} = 0.022 \text{ foot}$$

5. $$z_{\text{observed}} = \frac{16.06 - 16}{0.022} = +2.73$$

6. $z_{\text{observed}} > z_U$. Reject H_0. The mean length is significantly greater than 16 feet at the 0.05 significance level.

9-35 a.

$$n = \left[\frac{27(1.65 + 2.33)}{(435 - 420)}\right]^2 = 51.3 \quad \text{or} \quad 52 \text{ hours}$$

Sample observations are the numbers of assemblies produced in 52 different hours randomly selected following the changes.

b.

$$\bar{x}_{\text{criterion}} = 420 + 1.65(27)/\sqrt{52} \quad \text{or} \quad 426.18$$

Since management wants to detect an upward shift in the mean, an upper-tailed test is correct.

c. Find the mean number of assemblies per hour for the sample of 52 hours of production. If this value exceeds 426.16, reject H_0: $\mu \leq 420$. Otherwise do not reject H_0.

Chapter 10 Simple Linear Regression and Correlation

10-1 a. Two points on the line are (20, 0.6) and (60, 0.3).

$$b_1 = \frac{0.3 - 0.6}{60 - 20} = \frac{-0.3}{40} = -0.0075$$

$$Y_C = b_0 + b_1 X$$

$$0.6 = b_0 + (-0.0075)(20)$$

$$b_0 = 0.75$$

b. The percentage of those favoring lighter automobiles drops 0.75% for every year's increase in age.

c. No, the observations cover only persons from 20 to 60. It is to be expected that the proportion favoring lighter cars levels off above 60.

10-3 a. $Y_C = 0.266 - 0.011(5) = 0.211$; there is an estimated probability of 0.211 that a one-officer unit will make an arrest if the response time is 5 minutes.

b. $Y_C = 0.309 - 0.013(4) = 0.257$; there is an estimated probability of 0.257 that an arrest will be made if a two-officer unit responds after 4 minutes.

c. Although the two-officer response line drops more rapidly as response time increases, that line remains above the one-officer line for all response times under 10 minutes.

10-5 a.

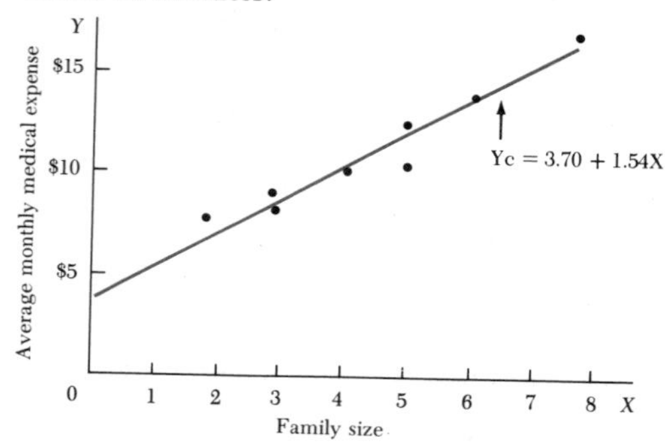

A straight line is satisfactory.

b.

X	Y	X^2	XY
3	8.5	9	25.50
2	7.5	4	15.00
3	8.0	9	24.00
5	12.0	25	60.00
4	9.5	16	38.00
5	10.0	25	50.00
6	13.0	36	78.00
8	16.5	64	132.00
36	85.0	188	422.50

$$b_1 = \frac{8(422.50) - (36)(85.0)}{8(188) - (36)^2} = 1.53846$$

$$b_0 = \frac{85}{8} - 1.53846\left(\frac{36}{8}\right) = 3.70192$$

$$Y_C = 3.70 + 1.54X$$

c. $Y_C = 3.70 + 1.54(4) = \$9.86$ per month

10-7 a. The dependent variable is the time to load the truck, in hours, and the independent variable is the estimated weight. This follows the useful time sequence for bidding on jobs and estimating costs.

b.
$$b_1 = \frac{10(66) - 21(28)}{10(52) - 21^2} = 0.91139$$

$$b_0 = \frac{28}{10} - 0.91139\left(\frac{21}{10}\right) = 0.88608$$

$$Y_C = 0.89 + 0.911X$$

c. The dimensions on b_1 are hours/ton. Hence, it takes an estimated 0.911 hours to load each ton.

10-9 a. $\Sigma(Y^2) = 968$. See the solution to Exercise 10-5 for other sums, slope, and intercept. Using Equation 10-7:

$$\text{SSTO} = 968 - (85^2/8) = 64.875$$
$$\text{SSR} = 3.70192(85) + 1.53846(422.5) - (85^2/8) = 61.53755$$

b. $r^2 = 61.53755/64.875 = 0.9486$; 94.9% of the variation in family medical expenses is attributable to the number of people in the family.

c. $r = +0.974$. The plus sign corresponds to the sign of b_1. The sign tells us the slope is positive, that expense increases with the number of family members.

10-11 a. $\Sigma(Y^2) = 327$. From the solution to Exercise 10-6, $\Sigma(X) = 50$; $\Sigma(X^2) = 318$; $n = 9$; $\Sigma(Y) = 51$; $\Sigma(XY) = 314$; $b_0 = 1.43094$; $b_1 = 0.76243$.

$$\text{SSTO} = 327 - (51^2/9) = 38$$
$$\text{SSR} = 1.43094(51) + 0.76243(314) - (51^2/9) = 23.38$$
$$\text{SSE} = 38 - 23.38 = 14.62$$
$$r^2 = 23.38/38 = 0.6153; \quad r = 0.7844$$

61.5% of the variability in performance rating is associated with scores on the test. The correlation coefficient indicates that the relationship is positive.

b. If the test with r^2 equal to 0.85 used the same performance rating procedure, then that test accounted for 72.25% of the variability in scores, while the current test accounted for only 61.5%.

10-13 a. From the solutions to the previous exercises,

$$r^2 = 0.95 \quad n = 8$$

$$\text{Then} \quad F = \frac{0.95(8-2)}{(1-0.95)} = 114 \quad \text{for the sample.}$$

From Table A-6, $F_{0.99,1,6} = 13.74$. Reject the hypothesis of no linear relationship in the population.

b. Because the hypothesis test supports the existence of a relationship in the population, it is logical to use the sample result for making estimates concerning that relationship. To find the sample standard error of estimate, we need the following: $\Sigma(Y^2) = 968$; $\Sigma(Y) = 85$; $\Sigma(XY) = 422.5$; $b_0 = 3.70192$; $b_1 = 1.53846$. Then,

$$s_{Y \cdot X} = \left(\frac{968 - 3.70(85) - 1.54(422.5)}{8 - 2}\right)^{1/2} = \left(\frac{2.85}{6}\right)^{1/2} = \$0.689$$

10-15 a. From solutions to the previous exercises,

$$r^2 = 0.6153 \quad n = 9$$

$$\text{Hence,} \quad F = \frac{0.6153(9 - 2)}{(1 - 0.6153)} = 11.2$$

From Table A-6, $F_{0.95,1,7} = 5.59$, so we can reject the null hypothesis.

b. Yes; because the null hypothesis was rejected, we can assume there is a linear relationship in the population from which the sample is assumed to have come.

c. From Equation 10-11, the sample standard error of estimate is

$$s_{Y \cdot X} = \left(\frac{327 - 1.43094(51) - 0.76243(314)}{9 - 2}\right)^{1/2}$$

$$= \left(\frac{14.619}{7}\right)^{1/2} = 1.45 \quad \text{performance points}$$

10-17 a.

$$b_1 = \frac{13(3255) - 65(535)}{13(453) - 65^2} = 4.53125$$

$$b_0 = \frac{535}{13} - 4.53125\left(\frac{65}{13}\right) = 18.4976$$

$$Y_C = 18.50 + 4.53X$$

$$Y_C = 18.50 + 4.53(7) = 50.2, \text{ and so forth}$$

b. Using Equation 10-11,

$$s_{Y \cdot X} = \sqrt{\frac{25{,}063 - 18.4976(535) - 4.53125(3255)}{13 - 2}} = 6.16$$

Ten of the 13 observations, or 76.9%, are within 1 standard error of the regression line, and 100% are within 2 standard errors. The normally distributed equivalents are 68.27% and 95.45%. The data were generated from the correct model so these results are an example of random fluctuations from expectation. The model is

$$\mu_{Y \cdot X} = 20 + 4X, \ \sigma_{Y \cdot X} = 5$$

c.

$$s_Y = \sqrt{\frac{25{,}063 - 13(535/13)^2}{13 - 1}} = 15.93$$

The estimate of the standard error of estimate is only 6.16, which is less than half the sample standard deviation of the Y values. Linear regression appears to have made a significant reduction in Y variability.

10-19 Null hypothesis: $B_1 = 0$

Test statistic: $t = \frac{b_1 - B_1}{s_{b_1}} = \frac{b_1}{s_{b_1}}$ with $n - 2 = 11$ degrees of freedom

Acceptance region: $-3.1058 < t_{\text{sample}} < 3.1058$

From Equation 10-14 and the solution to Exercise 10-9

$$s_{b_1} = \frac{6.161223}{\sqrt{453 - 13(5^2)}} = 0.54458$$

From Equation 10-12

$$t_{\text{sample}} = \frac{4.53125}{0.54458} = 8.32 \quad \text{Reject } H_0.$$

The sample evidence lends strong support to the conclusion that there is a linear relationship between number of ads and potential customers generated.

Assuming the 13 weeks were selected randomly from the 52 weeks in the past year, the population data are the 52 possible pairs of ad spots and potential customers, with ads in the range 0 to 10.

10-21 Equation 10-14 is the general expression for the desired confidence interval estimate. From the solutions to Exercises 10-9 and 10-11, $b_1 = 4.53125$ and $s_{b_1} = 0.54458$. In addition we need $t_{0.975}$ for 11 degrees of freedom; this is 2.2010. The 0.95 confidence interval estimate of B_1 is $4.53125 \pm 2.201(0.54458) = 4.53 \pm 1.20$, or $3.33 \leq B \leq 5.73$.

The slope of the population regression line is estimated to be between +3.33 and +5.73 with 95% confidence. This is an estimate of the number of potential customers per spot ad. Since the 0.95 interval estimate does not include 0, the null hypothesis $B_1 = 0$ can be rejected at the 0.05 level of significance. The acceptance region cannot include b_1 if the estimate does not include $B_1 = 0$.

10-23 a.

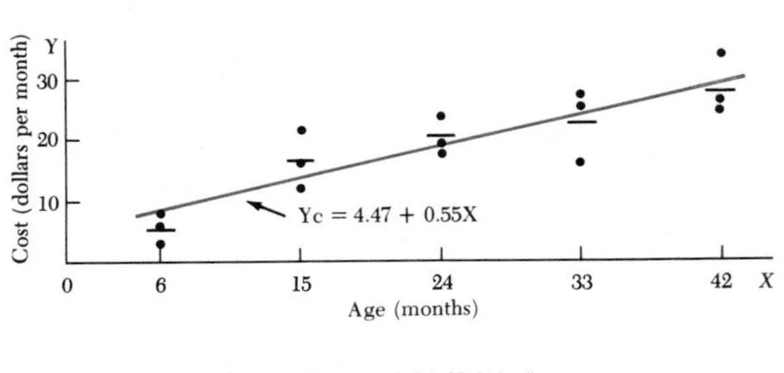

$$b_1 = \frac{15(7711.83) - 360(265.5)}{15(11{,}070) - 360^2} = 0.55137$$

$$b_0 = \frac{265.5 - 0.55137(360)}{15} = 4.46712$$

The straight line appears to be an adequate model.

b.

Age	*Mean Cost*
6	5.06
15	15.81
24	19.01
33	21.68
42	26.94

Although none of the class means lie directly on the sample regression line, there is no marked departure of the means from the line.

c. The point estimate is found from the regression equation:

$$\overline{Y}_C = 4.4671 + 0.5514(36) = \$24.32 \text{ per month}$$

$$t_{0.975,13} = 2.1604$$

$$s^2_{Y \cdot X} = \frac{5676.23 - 4.47(265.5) - 0.5514(7711.83)}{15 - 2} = 18.3186; \quad s_{Y \cdot X} = 4.28$$

$$s_{\overline{Y}_C} = 4.28\sqrt{\frac{1}{15} + \frac{(36 - 24)^2}{11{,}070 - 15(24)^2}} = 1.5188$$

The 0.95 interval estimate is

$$24.32 \pm 2.1604(1.5188) = 24.32 \pm 3.28$$

or 21.04 to 27.60 dollars per month.

10-25 a. As found in Exercise 10-12, $s_{Y \cdot X} = 0.0825$ cents per premium dollar. We can expect about two-thirds of the net cost data to fall this close to the relationship with age, and about 95% to be within 2 standard errors, or 0.165 cents.

b.

$$\begin{aligned} Y_C &= 1.3274 - 0.0109(10) = 1.2180245 \\ &= 1.3274 - 0.0109(20) = 1.108600141 \\ &= 1.3274 - 0.0109(36) = 0.9335211023 \end{aligned}$$

For $X = 10$

$$s_{\overline{Y}_C} = 0.0825\sqrt{\frac{1}{16} + \frac{(10 - 41.0625)^2}{32{,}367 - (657^2/16)}} = 0.040568$$

$$t_{0.975,14} = 2.1448$$

The 0.95 confidence interval estimate of $\mu_{Y \cdot X}$ is 1.218 ± .087, or 1.131 to 1.305 dollars. For $X = 20$

$$s_{\overline{Y}_C} = 0.0825\sqrt{\frac{1}{16} + \frac{(20 - 41.0625)^2}{32{,}367 - 16(41.0625)^2}} = 0.0314244$$

The 0.95 confidence interval estimate of $\mu_{Y \cdot X}$ for $X = 20$ is 1.1086 ± 0.0674, or 1.0412 to 1.176 dollars. For $X = 36$

$s_{\overline{Y}_C} = 0.02141$; the 0.95 confidence interval is 0.93352 ± 0.0459, or 0.8876 to 0.9794.

With 95% confidence, the true cost lies not more than 0.087 cents from the estimate for $X = 10$ and is within 6.8 cents for the other two ages.

10-27

$$r = \frac{15(7711.83) - (360)(265.5)}{\sqrt{[15(11{,}070) - 360^2][15(5676.2346) - 265.5^2]}} = +0.8696108092$$

Since r is positive, the relationship is direct and this checks with the positive value for b_1 in Exercise 10-23. $r^2 = 0.7562$ and r^2 is always nonnegative. Hence, the direction of the relationship between the variables cannot be inferred from it.

10-29 a.

$$b_1 = \frac{12(68{,}098.4) - 14.3(45{,}355)}{12(21.99) - 14.3^2} = 2838.93$$

$$b_0 = \frac{45{,}355 - 2838.93(14.3)}{12} = 396.52$$

$Y_C = 396.52 + 2838.93X$ where b_0 is the (meaningless) breaking strength of a cable with 0 diameter and where b_1 is the increase in pounds of breaking strength per inch increase in the diameter of the cable.

b.

$$s^2_{Y \cdot X} = \frac{211{,}461{,}083 - 396.52(45{,}355) - 2838.93(68{,}098.4)}{10}$$

$$s_{Y \cdot X} = 122.49 \text{ pounds}$$

About two-thirds of the breaking strengths will be within 122.5 pounds of the linear relationship with diameter. About 95% will be within two standard errors, or 245 pounds.

c.

$$s_{b_1} = \frac{122.49}{\sqrt{21.99 - 12(1.1917)^2}} = 55.06;$$

$$t_{0.995,10} = 3.1693$$

$b_1 \pm ts_{b_1} = 2838.93 \pm 174.50$, or 2664.43 to 3013.44. This interval is many standard errors above 0. Hence, it strongly suggests a positive relationship between diameter and strength.

10-31 a.

$$b_1 = \frac{20(749{,}200) - 4680(2606)}{20(1{,}307{,}600) - 4680^2} = 0.656043$$

$$b_0 = \frac{2606 - 0.656(4680)}{20} = -23.21404$$

b.

$$r = \frac{20(749{,}200) - 4680(2606)}{\sqrt{[20(1{,}307{,}600) - 4680^2][20(450{,}160) - 2606^2]}} = +0.909321$$

$$r^2 = 0.826865$$

This is the same result as with the following.

$$r^2 = \frac{-23.214(2606) + 0.656(749{,}200) - 20(130.3)^2}{450{,}160 - 20(130.3)^2} = 0.826865$$

This looks as if a strong linear regression relationship should be present in the population.

10-33

$$s^2_{Y \cdot X} = \frac{450{,}160 - (-23.214)(2606) - 0.656(749{,}200)}{20 - 2}$$

$$s_{Y \cdot X} = 32.61598; \; s_{b_1} = \frac{32.61598}{\sqrt{1{,}307{,}600 - 20(234)^2}} = 0.070757$$

$$b_1 \pm t s_{b_1} = 0.656 \pm 2.8784(0.07076) = 0.656 \pm 0.2037$$

or 0.4524 to 0.8597. This indicates significant positive slope in the population. For a two-tailed hypothesis test of H_0: $B_1 = 0$

$$t_{\text{sample}} = \frac{0.656 - 0}{0.07076} = 9.27$$

which is well above $t_{0.995,18} = 2.8794$, the criterion. The interval estimate goes on to establish confidence that the population slope is positive as well as being significant.

Chapter 11 Inferences from Two Samples

11-1 Because the populations are normally distributed with variances known to be 25, Equations 11-5 and 11-4 apply.

$$\sigma_{\bar{X}_1 - \bar{X}_2} = 5\sqrt{\frac{1}{9} + \frac{1}{6}} = 2.635$$

$z = 1.96$ for a 0.95 confidence level.

The 0.95 confidence interval estimate of the increase in performance is

$$(88.5 - 80) \pm 1.96(2.635) = 8.5 \pm 5.16$$

or 3.33 to 13.67 points. Mean test performance has improved from 3.33 to 13.67 points, at the 0.95 level of confidence. This presumably reflects improved performance in computer operation.

11-3 Given normal populations with equal, unknown standard deviations, Equations 11-6 and 11-7 are applicable.

$$s_{\bar{X}_1 - \bar{X}_2} = \sqrt{\frac{8(0.73)^2 + 10(0.91)^2}{9 + 11 - 2}\left(\frac{1}{9} + \frac{1}{11}\right)} = 0.3752 \text{ pounds}$$

$$t_{0.995,18} = 2.8784$$

The confidence interval estimate is

$$(4.20 - 5.64) \pm 2.8784(0.3752) = -1.44 \pm 1.08$$

or −2.52 to −0.36 pounds. At the 0.99 confidence level, mean weight loss with program A is from 0.36 to 2.52 pounds greater in the first month than with program B.

11-5 For large samples from nonnormal populations with unknown variances, Equations 11-8 and 11-9 apply.

$$s_{\bar{X}_1 - \bar{X}_2} = \sqrt{\frac{27^2}{156} + \frac{31^2}{169}} = \$3.22$$

$$z_{0.995} = 2.58$$

The 0.99 confidence interval estimate is

$$(\$410 - \$380) \pm 2.58(\$3.22) = \$30 \pm \$8.31$$

or \$21.69 to \$38.31. At the 0.99 confidence level, mean weekly earnings for construction workers in the proposed urban center are from \$21.69 to \$38.31 less than in the firm's present location.

11-7 Let X_1 be the score before training and X_2 be the score after training. Then the hypothesis statement is

$$H_0\colon \mu_2 \leq \mu_1;\ H_A\colon \mu_2 > \mu_1$$

or, equivalently

$$H_0\colon \mu_2 - \mu_1 \leq 0;\ H_A\colon \mu_2 - \mu_1 > 0.$$

Because the populations are normal with known, equal variances, Equations 11-5 and 11-10 apply. The acceptance region for H_0 is $z \leq 2.33$.

$$\sigma_{\bar{X}_1 - \bar{X}_2} = 5\sqrt{\frac{1}{9} + \frac{1}{6}} = 2.635$$

$$z_{\text{observed}} = \frac{(88.5 - 80) - 0}{2.635} = 3.23$$

Reject H_0. At the 0.01 level, there is a significant increase in mean examination score.

11-9 The hypothesis statement is

$$H_0\colon \mu_1 - \mu_2 \leq 0;\ H_A\colon \mu_1 - \mu_2 > 0$$

Equations 11-6 and 11-11 are applicable. The acceptance region for H_0 is $t \leq 2.4163$, where t has $24 + 21 - 2 = 43$ degrees of freedom.

$$s_{\bar{X}_1 - \bar{X}_2} = \sqrt{\frac{23(0.22)^2 + 20(0.13)^2}{24 + 21 - 2}\left(\frac{1}{24} + \frac{1}{21}\right)} = 0.055$$

$$t_{\text{observed}} = \frac{(5.53 - 5.04) - (0)}{0.055} = 8.91$$

Reject H_0. The observed value of t lies well above the acceptance region. The evidence supports the claim.

11-11 The hypothesis statement is

$$H_0\colon \mu_2 \leq \mu_1;\ H_A\colon \mu_2 > \mu_1$$

where μ_2 is the mean at the new location. Equations 11-8 and 11-12 apply.

$$s_{\bar{X}_1 - \bar{X}_2} = \sqrt{\frac{5000^2}{150} + \frac{6000^2}{180}} = \$605.53$$

$$z_{\text{observed}} = \frac{(\$84{,}750 - \$82{,}500) - (0)}{\$605.53} = 3.72$$

Reject H_0. The mean price is significantly greater at the new location because $z_{\text{criterion}} = 1.65$.

11-13

After (X_1)	*Before* (X_2)	$d = X_1 - X_2$	$d - \bar{d}$	$(d - \bar{d})^2$
40	37	+3	0	0
40	41	−1	−4	16
32	28	+4	1	1
35	32	+3	0	0
41	38	+3	0	0
37	35	+2	−1	1
36	30	+6	3	9
38	34	+4	1	1
		24		28

$$\bar{d} = 24/8 = 3;\ s_d = \sqrt{28/7} = 2;\ t_{0.975,7} = 2.3646$$

The 0.95 confidence interval estimate is

$$3 \pm 2.3646(2/\sqrt{8}) = 3 \pm 1.67$$

or 1.33 to 4.67. The estimated increase in mean number of assemblies/hour is from 1.33 to 4.67 following the changes.

11-15 For the data, $\Sigma d = 34$ and $\Sigma d^2 = 280$, where d = programmed course score minus lecture course score.

1. Hypothesis statement: H_0: $\mu_d \le 0$; H_A: $\mu_d > 0$
2. Test statistic: t with 11 degrees of freedom where

$$t = \frac{\bar{d} - \mu_d}{s_d/\sqrt{n}}$$

3. For $\alpha = 0.05$, the acceptance region for H_0 is $t \le 1.7959$.
4. $$\bar{d} = 34/12 = 2.833$$

$$s_d = \sqrt{\frac{280 - (34^2/12)}{12 - 1}} = 4.086$$

$$t_{\text{observed}} = \frac{2.833 - 0}{4.086/\sqrt{12}} = 2.40$$

5. Reject H_0. The observed value of t is greater than the upper limit of the acceptance region. Programmed learning produced a greater mean examination score.

11-17 Let $p_1 = 233/523 = 0.4455$ and $p_2 = 90/266 = 0.3383$. From Equations 11-18

$$s^2_{P_1} = \frac{0.4455(0.5545)}{523 - 1} = 0.0004732$$

$$s^2_{P_2} = \frac{0.3383(0.6617)}{266 - 1} = 0.0008447$$

$$s_{P_1 - P_2} = \sqrt{0.0004732 + 0.0008447} = 0.0363$$

The 0.90 confidence interval estimate of the difference in population proportions is

$$(0.4455 - 0.3383) \pm 1.645(0.0363) = 0.1072 \pm 0.0597$$

From 4.75% to 16.69% more Volvo drivers wore seat belts than did Colt drivers.

11-19

	Northern Route	*Southern Route*
$\bar{x}$	13.5	16.3
n	60	40
s	3.6	2.9

$$s_{\bar{X}_1 - \bar{X}_2} = \sqrt{\frac{3.6^2}{60} + \frac{2.9^2}{40}} = 0.6529$$

The 0.95 confidence interval estimate is

$$(13.5 - 16.3) \pm 1.96(0.6529) = -2.8 \pm 1.28$$

or −4.08 to −1.52. At the 0.95 confidence level, the mean weight on the northern route is from 1.52 to 4.08 tons less than on the southern route.

11-21 H_0: $\mu_d = 0$; H_A: $\mu_d \neq 0$

Acceptance region: $-1.8946 \leq t_{\text{observed}} \leq 1.8946$

where $t_{0.95,7} = 1.8946$

$$\Sigma d = 50; \quad \Sigma d^2 = 920; \quad \bar{d} = 6.25;$$

$$s_d = \sqrt{\frac{920 - (50^2/8)}{8 - 1}} = 9.316; \quad s_{\bar{d}} = \frac{9.316}{\sqrt{8}} = 3.294;$$

$$t_{\text{observed}} = 6.25/3.294 = 1.897$$

Reject H_0. The sample evidence supports the conclusion that program I results in greater weight gain.

Chapter 12 Analysis of Frequencies

12-1 a. \$6.57
b. \$14.00
c. \$29.14

12-3 a.

x	4	5	6	7	*Total*
f_o	12	16	15	24	
f_e	8.4	16.8	20.9	20.9	
$(f_o - f_e)^2/f_e$	1.54	0.04	1.67	0.46	$3.71 = \chi^2(df = 3)$

b. $\chi^2_{0.95}(df = 3) = 7.81$; the hypothesis cannot be rejected.

12-5 a.

Self $(130 + 132 + 101)/1687 = 0.215$
Friends $(213 + 272 + 232)/1687 = 0.425$
Relatives $(182 + 193 + 232)/1687 = 0.360$

b.

	1964	*1970*	*1974*
Self	113	128	121
Friends	223	254	240
Relatives	189	215	203

c. $\chi^2(df = 4) = 14.63$; $\chi^2_{0.95}(df = 4) = 9.49$; reject the hypothesis of no association and conclude that shopping-center companion preferences are not the same in the several years.

12-7 a.

	f_e			$(f_e - f_o)^2/f_e$		
	Office	*Type*	*Disc.*	*Office*	*Type*	*Disc.*
Gift	71	131	105	0.35	4.77	8.57
Personal	127	233	188	0.07	0.27	0.13
Business	55	100	81	1.16	2.89	7.72

$\chi^2(df = 4) = \Sigma[(f_o - f_e)^2/f_e] = 25.93$; $\chi^2_{0.99}(df = 4) = 13.28$; we conclude that association exists.

b.

	Office	*Type*	*Disc.*
Gift	0.261	0.228	0.361
Personal	0.490	0.519	0.489
Business	0.249	0.252	0.150
Total	1.000	0.999	1.000

12-9 a.

	f_e Yes	f_e No	$(f_o - f_e)^2/f_e$ Yes	$(f_o - f_e)^2/f_e$ No
50–59	34.7	17.3	0.15	0.31
60–69	144.7	72.3	0.65	1.30
70–79	48.0	24.0	3.00	6.00
80–89	16.0	8.0	2.25	4.50

$\chi^2(df = 4) = \Sigma[(f_o - f_e)^2/f_e] = 18.16$

b. $\chi^2_{0.95}(df = 4) = 9.49$; no.

12-11

	f_e				$(f_o - f_e)^2/f_e$			
	1	*2*	*3*	*4*	*1*	*2*	*3*	*4*
Prefer	240	240	240	240	0.27	1.67	0.60	1.35
Not prefer	160	160	160	160	0.40	2.50	0.90	2.02

$\chi^2(df = 4) = 9.71$; $\chi^2_{0.95}(df = 4) = 9.49$; reject the hypothesis.

12-13 a. $(\Sigma fx/\Sigma f)/n = (94/36)/4 = 0.65$

b.

x	*0*	*1*	*2*	*3*	*4*
f_o	4	3	7	11	11
f_e	0.5	4.0	11.2	13.8	6.4

c. $\chi^2(df = 2) = [(7 - 4.5)^2/4.5] + [(7 - 11.2)^2/11.2] + [(11 - 13.8)^2/13.8] + [(11 - 6.4)^2/6.4] = 6.84$;

$\chi^2_{0.95}(df = 2) = 5.99$; reject the hypothesis of an underlying binomial distribution.

12-15 a.

	f_o	z_{upper}	Normal Probability: *Cumulative*	Normal Probability: *Interval*	f_e	$(f_o - f_e)^2/f_e$
Under 25	4	−3.18 }	0.0059	0.0059	4.4 }	4.58
25–30	11	−2.52 }				
30–35	19	−1.86	0.0314	0.0255	19.2 }	
35–40	43	−1.19	0.1170	0.0856	64.5	7.17
40–45	108	−0.53	0.2981	0.1811	136.5	5.95
45–50	178	0.13	0.5517	0.2536	191.2	0.91
50–55	209	0.80	0.7881	0.2364	178.2	5.32
55–60	148	1.46	0.9279	0.1398	105.4	17.22
60–65	28	2.12	0.9830	0.0551	41.5	4.39
65–70	4	2.78	0.9973	0.0143	10.8 }	3.61
Over 70	2	∞	1.0000	0.0027	2.0 }	
	754					49.15

b. $\chi^2(df = 5) = 49.15$; $\chi^2_{0.95}(df = 5) = 11.07$; reject the hypothesis of normality.
c. The distribution is skewed to the left and more peaked than a normal distribution.

12-17 a. All expected frequencies are equal to 20. City A: $\chi^2 = 0.7$; city B: $\chi^2 = 3.7$; city C: $\chi^2 = 3.9$; $\chi^2_{0.95}(df = 2) = 5.99$; for each city, the hypothesis of equal preference is accepted.
b. Marginal preferences of 69, 60, and 51 lead to expected frequencies for each city of 23, 20, and 17; $\chi^2(df = 4) = 5.61$; $\chi^2_{0.95}(df = 4) = 9.49$; accept the hypothesis of independence.

12-21 a.

			Normal Probability			
	f_o	z_{upper}	*Cumulative*	*Interval*	f_e	$(f_o - f_e)^2/f_e$
Under 99.30	2	−2.16	0.0154	0.0154	1.2	
99.30– 99.50	3	−1.56	0.0594	0.0440	3.5	0.00
99.50– 99.70	8	−0.97	0.1660	0.1066	8.5	
99.70– 99.90	12	−0.38	0.3520	0.1860	14.9	0.56
99.90–100.10	21	0.21	0.5832	0.2312	18.5	0.34
100.10–100.30	18	0.81	0.7910	0.2078	16.6	0.12
100.30–100.50	9	1.40	0.9192	0.1282	10.3	0.16
100.50–100.70	6	1.99	0.9767	0.0575	4.6	0.04
Over 100.70	1		1.0000	0.0233	1.9	
	80					1.22

b. $\chi^2(df = 3) = 1.22$; $\chi^2_{0.95}(df = 3) = 7.81$; accept the hypothesis of normality.

12-23

Acreage	f_e	χ^2
Less than 50	.35(406) = 142.10	7.25
50–179	.43(406) = 174.58	.41
180–499	.18(406) = 73.08	1.94
500 & over	.04(406) = 16.24	8.52
		18.12

$\chi^2_{0.95,3} = 7.81$; reject H_0 and conclude that the method was biased.

12-25

	Under 5	*5–7.5*	*7.5–9.99*	*10–14.99*	*15–19.99*	*20–24.99*	*25 and over*	*Total*
Form 1	22	8	31	50	33	10	23	177
Form 2	18	12	17	48	43	23	28	189
Total	40	20	48	98	76	33	51	366

12-25 (cont.)

	Under 5	5–7.5	7.5–9.99	10–14.99	15–19.99	20–24.99	25 and over	Total
Expected								
Form 1	19.3	9.7	23.2	47.4	36.8	16.0	24.7	177
Form 2	20.7	10.3	24.8	50.6	39.2	17.0	26.3	189
χ^2								
Form 1	.36	.29	2.61	.14	.38	2.23	.11	6.13
Form 2	.34	.27	2.45	.13	.36	2.08	.11	5.74
							$\chi^2 =$	11.87

$\chi^2_{0.95,5} = 11.07$, and we can conclude that the forms do not elicit the same information.

Chapter 13 Analysis of Variance

13-1 a. $F_{0.99,10,20} = 3.37$. Hence 99% of the distribution lies below 3.37.

b. $F_{0.99,10,20} = 3.37$, as in part (a). Hence, if 99% of the distribution lies below 3.37, 1% must lie above that value.

c. $F_{0.99,20,10} = 4.41$, which differs from 3.37 as found in part (a). In general $F_{1-\alpha,a,b} \neq F_{1-\alpha,b,a}$.

13-3 a. The mean for observer A is $(51 + 36 + 35 + 62)/4 = 46$. Similarly, the means for observers B and C are 39 and 38. The grand mean is $(46 + 39 + 38)/3 = 41$.

b. Between-means estimate

$$= \frac{4[(46-41)^2 + (39-41)^2 + (38-41)^2]}{(3-1)} = 76$$

Within-columns estimate

$$= \frac{(51-46)^2 + \cdots + (50-39)^2 + \cdots + (34-38)^2}{3(4-1)} = \frac{1196}{9} = 132.89$$

c. $F_{\text{sample},2,9} = 76/132.89 = 0.5719$
$F_{\text{criterion}} = F_{0.95,2,9} = 4.26$

With no difference between population means, F_{sample} should be about 1. The actual result strongly suggests there is no significant difference between these means. This is the desired result. Results for different observers observing the same events should be consistent.

d. In each period, the actual number of cars being observed is identical for each observer. The top three rows look like random fluctuations caused by slight miscounts. But 62 is way out of line in the bottom row. Perhaps the observer transposed the numbers and meant to record 26.

13-5 a.

	A	B	C	
n_j	4	4	4	$n = 12$
C_j	184	156	152	$T = 492$
ΣX^2	8966	6418	6136	$\Sigma\Sigma X^2 = 21{,}520$

b.

$$\text{SST} = 21{,}520 - \frac{492^2}{12} = 1348$$

$$\text{SSA} = \left(\frac{184^2}{4} + \frac{156^2}{4} + \frac{152^2}{4}\right) - \frac{492^2}{12} = 20{,}324 - 20{,}172 = 152$$

$$\text{SSE} = 1348 - 152 = 1196$$

Source	*SS*	*df*	*MS*	
Observer	152	2	76	$F = 0.5719$
Residual	1196	9	132.89	
Total	1348	11		

$F_{0.95,2,9} = 4.26$

c. There is no evidence for a significant difference in the means of the observer population.

d. MSA = 76, the same as the between-means estimate found earlier. Similarly, MSE = 132.89 is the same as the previous within-columns estimate. Consequently, the same result is attained for F_{sample}.

13-7 a.

	A	B	C	
n_j	3	5	4	$n = 12$
C_j	225	449	334	$T = 1008$
ΣX^2	16,973	40,485	28,028	$\Sigma\Sigma X^2 = 85{,}486$

b.

$$\text{SST} = 85{,}486 - \frac{1008^2}{12} = 85{,}486 - 84{,}672 = 814$$

$$\text{SSA} = \frac{225^2}{3} + \frac{449^2}{5} + \frac{334^2}{4} - 84{,}672 = 85{,}084.2 - 84{,}672 = 412.2$$

$$\text{SSE} = 814 - 412.2 = 401.8$$

Source	*SS*	*df*	*MS*	
Calculator	412.2	2	206.1	$F_{\text{sample}} = 4.62$
Residual	401.8	9	44.64	
Total	814	11		

$F_{0.95,2,9} = 4.26$ $F_{0.99,2,9} = 8.02$

c. The three population mean scores are significantly different at the 0.05 level. The sample means are A = 75, B = 89.8, C = 83.5. Assuming more points means greater overall effectiveness, it appears that B may be better than the other two. A confidence interval estimate of the difference is required, however.

13-9 a. $K_1 = 70.6 - 49.6 = 21.0$
$K_2 = 49.6 - 39.8 = 9.8$
$K_3 = 39.8 - 34.4 = 5.4$
$K_4 = 49.6 - 34.4 = 15.2$

See the solution for Exercise 13-6 for means and other required inputs.

b. MSE = 142.15; $s^2(\overline{X}_j) = 142.15/5 = 28.43$ for every treatment mean because of equal sample sizes.

$$s(K_1) = s(K_2) = s(K_3) = \sqrt{142.15\left[\frac{(+1)^2}{5} + \frac{(-1)^2}{5}\right]} = \sqrt{56.86} = 7.541$$

c. $T^2 = (4 - 1)(2.46) = 7.38;\ T = 2.717$
$T \cdot s(K) = 2.717(7.541) = 20.5$

The intervals, in the above order, are

21.0 ± 20.5
9.8 ± 20.5
5.4 ± 20.5
15.2 ± 20.5

K_1 has the only interval which does not include 0. Hence, brand A population mean preference is significantly greater than that of any other brand at the 0.90 level.

d. No pair of brands B, C, and D have significantly different means. Therefore, they can be treated as a single group in the contrast.

$$K_5 = 70.6 - \frac{49.6 + 39.8 + 34.4}{3} = 29.33$$

$$s^2(K_5) = \frac{142.15}{5}\left[1^2 + \left(\frac{-1}{3}\right)^2 + \left(\frac{-1}{3}\right)^2 + \left(\frac{-1}{3}\right)^2\right] = 37.9066$$

$$s(K_5) = 6.1568;\ T \cdot s(K_5) = 16.7258$$

Interval for K_5: 29.33 ± 16.73, which does not include 0.

e. Brands B, C, and D constitute one group and brand A constitutes another. A is superior. The confidence level is 0.90 for this composite statement.

13-11 Let the three systems be A, B, and C; then the cross classified table with empty cells is

a.

	A	*B*	*C*
With young children			
Without young children			

b. Four subjects are required for each of 6 cells. Hence, a total of 24 subjects are needed — 12 with young children and 12 without.

13-13 a.

$$\text{SST} = (72.03 + 77.87 + 35.69) - \frac{(29.25 + 39.10 + 19.63)^2}{49} = 27.6296$$

$$\text{SSA} = \frac{29.25^2}{13} + \frac{39.10^2}{23} + \frac{19.63^2}{13} - \frac{87.98^2}{49} = 3.9548$$

$$\text{SSE} = 27.6296 - 3.9548 = 23.6748$$

b.

	SS	df	MS
Product	3.95	2	1.98
Residual	23.67	46	0.515
Total	27.63	48	

$F_{\text{sample}} = 3.84$

$F_{0.95,2,46} = 3.20$ $F_{0.99,2,46} = 5.10$

c. With estimating procedures which produce estimates so low that actual costs run 1.5 to 2.25 times as much as the estimates, the firm could very well be suffering an actual net loss on some or all of these new products.

13-15 Refer to the solution for Exercise 13-13 to get the necessary means and mean square.

a. $K_1 = 2.25 - 1.70 = 0.55$

$K_2 = 1.70 - 1.51 = 0.19$

$$K_3 = 2.25 - \frac{1.70 + 1.51}{2} = 0.645$$

b. $T^2 = (3-1)F_{0.90,2,46} = 2(2.42); T = 2.2$

MSE = 0.51467

$$s^2(K_1) = (+1)^2 \frac{0.51467}{13} + (-1)^2 \frac{0.51467}{23} = 0.06197;$$

$$s(K_1) = 0.2489;\ T \cdot s(K_1) = 0.5476$$

Interval for K_1: 0.55 ± 0.5476 which does not include 0.

$$s^2(K_2) = (1)^2 \frac{0.51467}{23} + (-1)^2 \frac{0.51467}{13} = 0.06197;$$

$$s(K_2) = 0.2489;\ T \cdot s(K_2) = 0.5476$$

Interval for K_2: 0.19 ± 0.5476 which includes 0.

$$s^2(K_3) = (1)^2 \frac{0.51467}{13} + \left(-\frac{1}{2}\right)^2 \frac{0.51467}{23} + \left(-\frac{1}{2}\right)^2 \frac{0.51467}{13} = 0.05508;$$

$$s(K_3) = 0.23469;\ T \cdot s(K_3) = 0.5163$$

Interval for K_3: 0.645 ± 0.5163 which does not include 0.

c. Product A, chemical entities, with a mean of 2.25 is significantly worse than B and is significantly worse than the common group composed of B and C. The confidence level for this composite statement is 0.90. If reviews of estimating procedures are to be undertaken, the fact that these procedures are worse for chemical entities may be important.

13-17 Refer to the solution for Exercise 13-7 to get the necessary treatment means and mean square.

a. $K_1 = \overline{X}_B - \overline{X}_A = 89.8 - 75 = 14.8$

$K_2 = 89.8 - 83.5 = 6.3$

$K_3 = 83.5 - 75 = 8.5$

$K_4 = \dfrac{89.8 + 83.5}{2} - 75 = 11.65$

b. $T^2 = (3 - 1)F_{0.90,2,9} = 2(3.00) = 6.00;\ T = 2.449$

MSE = 44.64

$$s^2(K_1) = (-1)^2 \frac{\text{MSE}}{3} + (+1)^2 \frac{\text{MSE}}{5};$$

$$s(K_1) = 4.88;\ T \cdot s(K) = 11.95$$

Interval for K_1: 14.8 ± 11.95 which does not include 0.

$$s^2(K_2) = (+1)^2 \frac{\text{MSE}}{5} + (-1)^2 \frac{\text{MSE}}{4};$$

$$s(K_2) = 4.48;\ T \cdot s(K) = 10.976$$

Interval for K_2: 6.3 ± 10.976 which includes 0.

$$s^2(K_3) = (-1)^2 \frac{\text{MSE}}{3} + (1)^2 \frac{\text{MSE}}{4};$$

$$s(K_3) = 5.10;\ T \cdot s(K) = 12.497$$

Interval for K_3: 8.5 ± 12.50 which includes 0.

$$s^2(K_4) = (-1)^2 \frac{\text{MSE}}{3} + \left(\frac{1}{2}\right)^2 \frac{\text{MSE}}{5} + \left(\frac{1}{2}\right)^2 \frac{\text{MSE}}{4}$$

$$s(K_4) = 4.46;\ T \cdot s(K) = 10.925$$

Interval for K_4: 11.65 ± 10.92 which does not include 0.

c. The composite statement is: Calculator B performs significantly better than A, but not significantly better than C. As a group, calculators B and C perform significantly better than A. The confidence level for the composite statement is 0.90. No additional contrasts would appear to contribute additional information.

13-19 The between-means estimate of the population variance includes the within-populations variability and it may also include the variability between population means. The denominator of the F ratio, the within-columns estimate of population variance, cannot include any variability for a possible difference in population means between the columns. Hence, the existence of true between-columns differences can be expected to produce only significantly large F values for the sample results.

Chapter 14 Multiple Regression

14-1 a. 4.817 is the intercept, b_0. It is the value of Y when all X variables are set equal to 0. $b_1 = 3.5$ cubic feet per crate, given shipments of equal weight. $b_2 = 5.297$ cubic feet per hundred pounds, given shipments containing the same number of crates.

b.
$$r^2_{Y\cdot 12} = \frac{4.817(1187) + 3.5(17{,}111) + 5.297(15{,}294) - (1187^2/12)}{150{,}101 - (1187^2/12)} = 0.9956$$

Over 99% of the original variation in space required is accounted for by this relationship. It appears that it may be useful, pending the outcome of later significance tests.

c.
$$Y_C = 4.817 + 3.5(19) + 5.297(16) = 156.069 \text{ cubic feet}$$

14-3 a. The accompanying scattergrams show no evidence of curvilinearity.

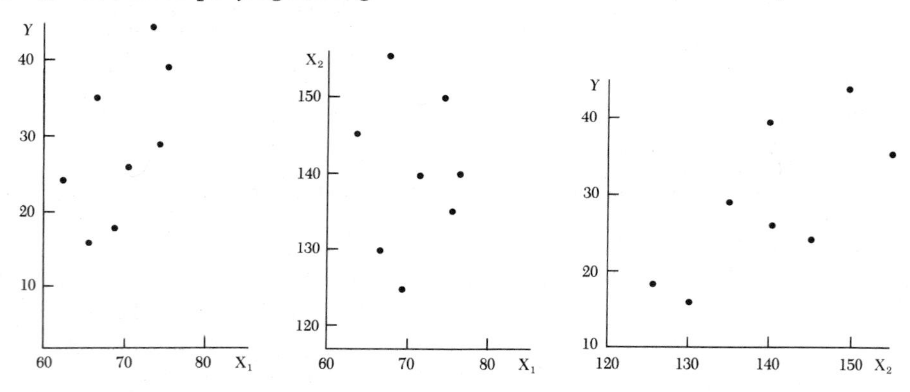

b. The normal equations are

$$\begin{aligned}
231 &= 8b_0 + 561b_1 + 1{,}120b_2 \\
16{,}936 &= 561b_0 + 39{,}493b_1 + 78{,}530b_2 \\
32{,}850 &= 1120b_0 + 78{,}530b_1 + 157{,}500b_2
\end{aligned}$$

From these, $b_0 = -169.6543$, $b_1 = 1.3384$, and $b_2 = 0.7477$.

c.
$$r^2_{Y\cdot 12} = \frac{-169.654(231) + 1.338(16{,}396) + 0.748(32{,}850) - (231^2/8)}{7355 - (231^2/8)} = 0.942$$

Hence, 94.2% of the variation in electricity consumption is attributable to this regression relationship.

14-5 a. As shown, the residual scattergrams suggest that the model is suitable.

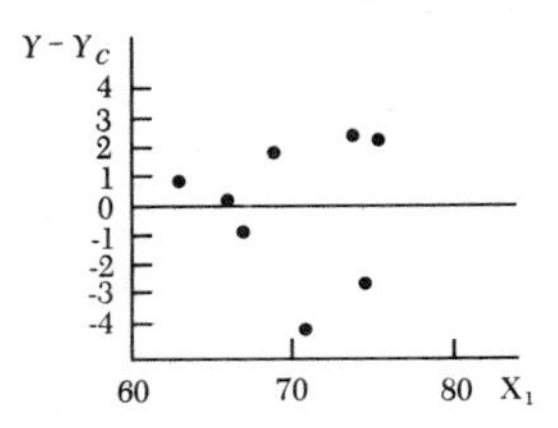

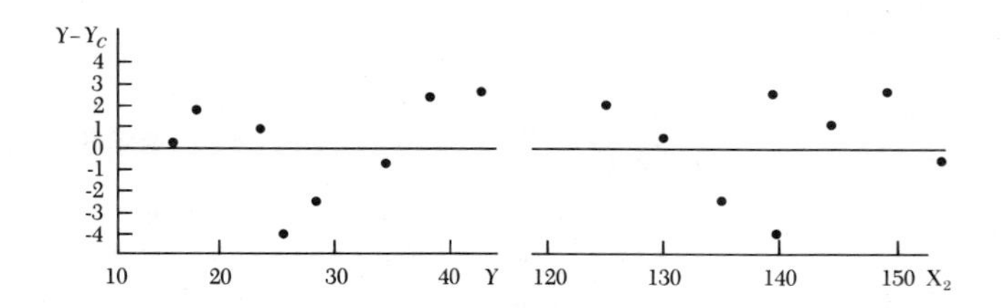

b. $\Sigma(Y - Y_C)^2 = 39.726518$

$s^2_{Y\cdot 12} = 39.726518/(8 - 3) = 7.9453$

$s_{Y\cdot 12} = 2.8187$ or 2.82 hundreds of KWH

c. In the list of residuals shown in the exercise, only one lies more than 2.82 from 0. The percentage is 100(1/8) = 12.5%, certainly not unusual.

14-7 First, the coefficient of multiple correlation is never negative. Second, it is the *square* of the coefficient of multiple correlation, the coefficient of multiple determination, which is the ratio of variation explained by regression to total Y-variation. A correct statement is: The coefficient of multiple correlation is 0.73. Hence, 53.29% of variation in family income is explained by the regression relationship.

14-9 a.

	SS	DF	MS	F STAT
REGRESSION	645.1467	2	322.5734	40.597
RESIDUAL	39.7283	5	7.9457	
TOTAL	684.8750	7		

b. H_0: $B_1 = B_2 = 0$. There is no regression of the form in the model present in the population.

c. From the F table, 0.01 of the distribution lies above an F-value of 13.27 under H_0. The observed value, 40.597, lies in the rejection region for the null hypothesis. Conclusion: There is evidence of linear regression in the population.

14-11 a. For $X_1 = 20$ and $X_2 = 18$, the regression point estimate is

$$Y_C = 4.817 + 3.5(20) + 5.297(18) = 170.163$$

which differs slightly from the computer output because of rounding.

b. The 0.99 confidence interval estimate of the mean response is

$170.1573 \pm 3.25(2.1187)$ or 170.1573 ± 6.8858

The mean space required for shipments consisting of 20 crates weighing 1800 pounds is from 163.27 to 177.04 cubic feet with 0.99 confidence.

c. An individual response is the space required for a single shipment consisting of 20 crates that weigh 1800 pounds. Single shipment space requirements would vary more than the mean space required for all such shipments.

14-13 a. For b_1, $t = 1.338361/0.228087 = 5.87$
For b_2, $t = 0.747691/0.106591 = 7.01$

b. H_0: $B_1 = 0$; H_A: $B_1 \neq 0$. The test statistic is t with $8 - 3 = 5$ degrees of freedom. At the 0.01 level, the acceptance region is $-4.032 \leq t \leq 4.032$ for the null hypothesis. The observed value of t is 5.87. Reject H_0; X_1 makes a significant marginal contribution to regression beyond that made by X_2.

For the test on B_2, the only difference is that the observed value of t is 7.01, which also leads to rejection of H_0 and the conclusion that X_2 makes a significant marginal contribution beyond that made by X_1.

c. Multicollinearity between X_1 and X_2 is low ($r_{12} = -0.031$). The observed value of F, 40.6, is significant at the 0.01 level, so it can be concluded that a regression relationship of the form indicated exists, and that both X_1 and X_2 make important contributions to the relationship. In the sample, X_1 explains about 36% of Y variance and X_2 explains an additional 54%, for a total of about 90%. ($0.609^2 = 0.37$, $0.737^2 = 0.54$). From Exercise 14-3, $r^2_{Y \cdot 12} = 0.94$, not far from the rough estimate just made.

14-15 For five classifications, four indicator variables are needed. Let them be X_1, X_2, X_3, and X_4. One possible coding scheme for these four variables is:

	X_1	X_2	X_3	X_4
Apprentice	1	0	0	0
Journeyman	0	1	0	0
Craftsman	0	0	1	0
Foreman	0	0	0	1
Management	0	0	0	0

14-17 a. and b.

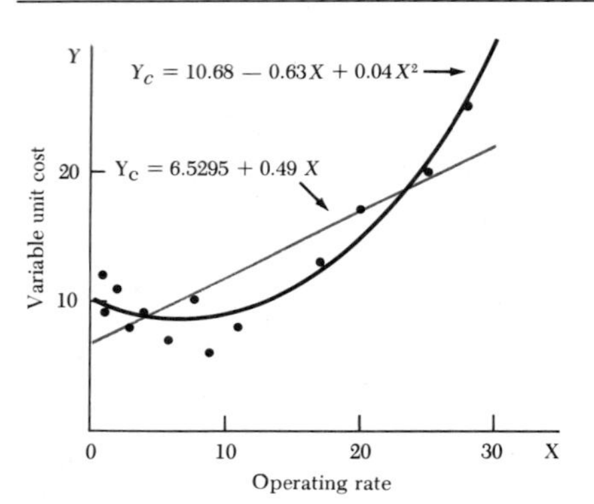

c. The second equation, the parabola, appears to fit the data much better than the straight line. This appearance is shown to be justified by the goodness-of-fit information in the next exercise.

14-19 The 120 sales districts are the units of observation. Populations of districts are included in the regression relationship. Since advertising within districts is proportional to population within the districts, advertising and population will have a correlation coefficient of 1, or very nearly so. Both variables are measuring the same characteristic in different units. Adding advertising to the relationship contributes no new information and will not explain any more of the variance in the dependent variable.

14-21 Equation 14-9 is

$$\Sigma(Y - \bar{Y})^2 = \Sigma(Y_C - \bar{Y})^2 + \Sigma(Y - Y_C)^2$$

By Equation 14-10 this is

$$SSTO = SSR + SSE$$

Then it follows that

$$MSE = \frac{SSE}{n-k} = \frac{\Sigma(Y - Y_C)^2}{n-k}$$

But the last quantity on the right is $s^2_{Y \cdot 12}$ and its square root is given by Equation 14-5. Hence, the square root of MSE is always the standard error of estimate in this type of analysis.

14-23 Because $r_{12} = 0$ in this case,

$$0.73632 + 0.21246 = 0.94878$$

$$r^2_{Y1} \quad + \quad r^2_{Y2} \quad = \quad r^2_{Y \cdot 12}$$

The separate contributions of the variables X_1 and X_2 can be assessed. The first contributes 0.73632 to the coefficient of determination and the second contributes 0.21246.

Chapter 15 Two-Way Analysis of Variance

15-1 a. This is a two-factor (age and sex), completely randomized (one student selected at random from all those with a given set of factor levels in the population) experiment without replication (only one student per combination of factor levels).

b. To change from a completely randomized to a randomized block experiment, the experimenter could select students from only two lists, male and female, and then block on age. For example, three males and three females could be randomly selected from each list. They could be arranged in "youngest," "next youngest," and "oldest" blocks after being selected.

c. In a completely randomized experiment, the population of units of observation is divided into subpopulations each of which satisfies one of the combinations of factor levels. The required number of units for the experiment is then selected by random sampling from each of these subpopulations. In a randomized block experiment, the population is not sorted into subpopulations on the factor that is controlled by blocking. Instead, units selected are arranged on this factor after selection.

d. Both row and column means may be used to estimate population means for a completely randomized two-factor experiment. The means for the blocked factor cannot be so used in a randomized block experiment.

e. Blocking may be necessary either because the population cannot be sorted into subpopulations or because the units of observation have already been selected when the decision to control on additional factors is made or for some similar reason.

15-3

A	B	C	D	E	Totals
9	11	12	10	12	54
6	7	10	9	9	41
2	5	8	6	6	27
17	23	30	25	27	122
81	121	144	100	144	590
36	49	100	81	81	347
4	25	64	36	36	165
121	195	308	217	261	1102

SST = $1102 - (122^2/15) = 1102 - 992.27 = 109.73$.

SSA = $(1/3)(17^2 + 23^2 + 30^2 + 25^2 + 27^2) - 992.27 = 31.73$.

SSB = $(1/5)(54^2 + 41^2 + 27^2) - 992.27 = 72.93$.

SSE = $109.73 - 31.73 - 72.93 = 5.07$.

Source	*SS*	*df*	*MS*	*F Ratio*
Modification	31.73	4	7.93	12.52
Operator	72.93	2		
Residual	5.07	8	0.63	
Total	109.73	14		

Criterion: $F_{0.01,4,8} = 7.01$.

Modifications have significantly different means.

15-5 From the solution to Exercise 15-2, SST = 308.00 and SSA = 183.33.

For a one-factor experiment, SSE = $308.00 - 183.33 = 124.67$.

Source	*SS*	*df*	*MS*	*F Ratio*
Training	183.33	3	61.11	3.92
Residual	124.67	8	15.58	
Total	308.00	11		

Criterion: $F_{0.05,3,8} = 4.07$.

Training is not significant at the 0.05 level, whereas it was significant at the 0.01 level previously. Blocking on the screening test greatly increased the sensitivity by making MSE only 5.03 earlier, whereas it is 15.58 here.

15-7 a.

	Squares of X(ijk)				*Cell, Column, and Row Sums*			
	1	2	3	*Total*	*1*	2	3	*Total*
	36	16	64	116				
I	81	36	100	217	18	12	24	54
	9	4	36	49				
	169	49	225	443				
II	121	36	121	278	36	18	36	90
	144	25	100	269				
				1372	54	30	60	144

$C = T^2/n = (144)^2/18 = 1152$

$\text{SST} = 1372 - 1152 = 220$

$\text{SSK} = [(36)^2 + \cdots + (60)^2]/3 - 1152 = 1320 - 1152 = 168$

$\text{SSA} = [(54)^2 + (30)^2 + (60)^2]/6 - 1152 = 1236 - 1152 = 84$

$\text{SSB} = [(54)^2 + (90)^2]/9 - 1152 = 1224 - 1152 = 72$

$\text{SSI} = \text{SSK} - \text{SSA} - \text{SSB} = 168 - 84 - 72 = 12$

Source	*SS*	*df*	*MS*	*F*
Program	84	2	42	9.7
Terminal	72	1	72	16.6
Interaction	12	2	6	1.39
Error	52	12	4.33	
Total	220	17		

Program and terminal effects are significant at the 0.01 level. The interaction is not significant at the 0.05 level.

b. Row and column means below produce the deviations given from the grand mean. The grand mean plus row and column deviation produce the point estimates shown for each treatment combination.

	Program 1	*Program 2*	*Program 3*	*Mean*	*Deviation*
Term 1	7	3	8	6	−2
Term 2	11	7	12	10	2
Mean	9	5	10	8	
Deviation	1	−3	2		

15-9 Letting X(1) and X(2) = 1 stand for calculators B and C, respectively, the input data are

```
0 0 83    1 0 96    0 1 86
0 0 70    1 0 90    0 1 90
0 0 72    1 0 80    0 1 74
          1 0 95    0 1 84
          1 0 88
```

a.
```
VARIABLE  REG. COEF.  STD.ERROR COEF.  COMPUTED T  BETA COEF.
    1     -30.79990      7.54060        -4.08454    -0.76399
    2     -20.99988      7.54059        -2.78491    -0.52090
    3     -36.19986      7.54059        -4.80067    -0.89793

INTERCEPT                70.59987

MULTIPLE CORRELATION      0.79172
STD. ERROR OF ESTIMATE   11.92272

           ANALYSIS OF VARIANCE FOR THE REGRESSION
    SOURCE OF VARI.       D.F.  SUM OF SQ.  MEAN SQ.  F VALUE
ATTRIBUTABLE TO REG.       3     3820.369   1273.456   8.958   13-11  13-12
DEVIATION FROM REG.       16     2274.420    142.151
     TOTAL                19     6094.789                      13-10
```

b. Treatment means:

Calculator A: Intercept = 75.0
Calculator B: 75.0 + 14.8 = 89.8
Calculator C: 75.0 + 8.5 = 83.5

Yes, these are the correct treatment means.

15-11 Using X(1) and X(2) = 1 for operators II and III, and X(3), X(4), X(5), and X(6) for modifications B, C, D, and E, respectively, the input data are

Operator		*0 0 0 0*	*1 0 0 0*	*0 1 0 0*	*0 0 1 0*	*0 0 0 1*
I	0 0	9	11	12	10	12
II	1 0	6	7	10	9	9
III	0 1	2	5	8	6	6

Computer regression output is as follows:

```
VARIABLE  REG. COEF.  STD.ERROR COEF.  COMPUTED T  BETA COEF.
   1      -2.60000       1.10755        -2.34753    -0.45315
   2      -5.40001       1.10755        -4.87564    -0.94116

INTERCEPT               10.80000

MULTIPLE CORRELATION     0.81526
STD. ERROR OF ESTIMATE   1.75119

           ANALYSIS OF VARIANCE FOR THE REGRESSION
  SOURCE OF VARI.        D.F.   SUM OF SQ.   MEAN SQ.    F VALUE
ATTRIBUTABLE TO REG.      2       72.933      36.467     11.891
DEVIATION FROM REG.      12       36.800       3.067
    TOTAL                14      109.733
```

```
VARIABLE   REG. COEF.   STD.ERROR COEF.   COMPUTED T   BETA COEF.
    3        2.00000        2.28035          0.87706      0.29578
    4        4.33333        2.28035          1.90029      0.64085
    5        2.66667        2.28035          1.16941      0.39437
    6        3.33334        2.28035          1.46176      0.49296

INTERCEPT                   5.66666

MULTIPLE CORRELATION        0.53776
STD. ERROR OF ESTIMATE      2.79285

          ANALYSIS OF VARIANCE FOR THE REGRESSION
  SOURCE OF VARI.          D.F.    SUM OF SQ.    MEAN SQ.    F VALUE
ATTRIBUTABLE TO REG.        4        31.733        7.933       1.017
DEVIATION FROM REG.        10        78.000        7.800
     TOTAL                 14       109.733

VARIABLE   REG. COEF.   STD.ERROR COEF.   COMPUTED T   BETA COEF.
    1       -2.60000        0.50332         -5.16565     -0.45315
    2       -5.40000        0.50332        -10.72868     -0.94116
    3        2.00000        0.64979          3.07792      0.29578
    4        4.33333        0.64979          6.66883      0.64085
    5        2.66666        0.64979          4.10389      0.39437
    6        3.33333        0.64979          5.12987      0.49296

INTERCEPT                   8.33333

MULTIPLE CORRELATION        0.97664
STD. ERROR OF ESTIMATE      0.79582

          ANALYSIS OF VARIANCE FOR THE REGRESSION
  SOURCE OF VARI.          D.F.    SUM OF SQ.    MEAN SQ.    F VALUE
ATTRIBUTABLE TO REG.        6       104.667       17.444      27.544
DEVIATION FROM REG.         8         5.067        0.633
     TOTAL                 14       109.733
```

Treatment means:

Operators	*Modifications*
I: Intercept = 12.0	A: Intercept = 5.67
II: 10.8 − 2.6 = 8.2	B: 5.67 + 2.00 = 7.67
III: 10.8 − 5.4 = 5.4	C: 5.67 + 4.33 = 10.00
	D: 5.67 + 2.67 = 8.34
	E: 5.67 + 3.33 = 9.00

15-13 Using X(1) = 1 for seed B, X(2) = 1 for feed B, and X(3) = X(1) * X(2), the independent variable coding is

	Feed A	*Feed B*
Seed A	0 0 0	1 0 0
Seed B	0 1 0	1 1 1

a. The regression output is

```
VARIABLE REG. COEF.  STD.ERROR COEF.  COMPUTED T  BETA COEF.
    1       4.12500       1.78723        2.30804     0.52500

INTERCEPT                  21.00000

MULTIPLE CORRELATION        0.52500
STD. ERROR OF ESTIMATE      3.57446

          ANALYSIS OF VARIANCE FOR THE REGRESSION
  SOURCE OF VARI.        D.F.   SUM OF SQ.    MEAN SQ.    F VALUE
ATTRIBUTABLE TO REG.       1       68.062       68.062      5.327
DEVIATION FROM REG.       14      178.875       12.777
     TOTAL                15      246.937
```

```
VARIABLE REG. COEF.  STD.ERROR COEF.  COMPUTED T  BETA COEF.
    1        4.12500      1.10343        3.73836     0.52500
    2        5.37500      1.10343        4.87120     0.68409

INTERCEPT                  18.31250

MULTIPLE CORRELATION        0.86233
STD. ERROR OF ESTIMATE      2.20685

          ANALYSIS OF VARIANCE FOR THE REGRESSION
  SOURCE OF VARI.        D.F.   SUM OF SQ.    MEAN SQ.    F VALUE
ATTRIBUTABLE TO REG.       2      183.625       91.812     18.852
DEVIATION FROM REG.       13       63.312        4.870
     TOTAL                15      246.937
```

```
VARIABLE REG. COEF.  STD.ERROR COEF.  COMPUTED T  BETA COEF.
    1       2.75000       1.52411        1.80433     0.35000
    2       4.00000       1.52411        2.62448     0.50909
    3       2.75000       2.15542        1.27585     0.30311

INTERCEPT                  19.00000

MULTIPLE CORRELATION        0.87991
STD. ERROR OF ESTIMATE      2.15542

          ANALYSIS OF VARIANCE FOR THE REGRESSION
  SOURCE OF VARI.        D.F.   SUM OF SQ.    MEAN SQ.    F VALUE
ATTRIBUTABLE TO REG.       3      191.187       63.729     13.717
DEVIATION FROM REG.       12       55.750        4.646
     TOTAL                15      246.937
```

The ANOVA is

Source	*SS*	*df*	*MS*	*F*
Seeds	68.06	1	68.06	14.6
Feeds	183.62 − 68.06 = 115.56	1	115.56	24.9
Interaction	191.19 − 68.06 − 115.56 = 7.57	1	7.57	1.63
Error	55.75	12	4.65	
Total	246.94	15		

$F(.05,1,12) = 4.75$. Thus, the main effects of seeds and of feeds are each significant, the interaction is not.

b. The means are estimated from the regression with only main (additive) effects.

Seed A, feed A:	Intercept = 18.3125
Seed B, feed A:	18.3125 + 4.125 = 22.4375
Seed A, feed B:	18.3125 + 5.375 = 23.6875
Seed B, feed B:	18.3175 + 4.125 + 5.375 = 27.8125

$\sqrt{\text{MSE}/4}$ from this model is $\sqrt{4.870/4} = 1.1034$. The half-width of the 95% confidence interval for each mean above is

$$2.1604(1.1034) = 2.3838$$

For any one interval, the confidence is 0.95.

15-15 Using X(1) = 1 for terminal II, X(2) and X(3) = 1 for program 2 and program 3, respectively, X(4) = X(1) * X(2), and X(5) = X(1) * X(3), the independent variable coding is

	Program		
Terminal	*1*	*2*	*3*
I	0 0 0 0 0	0 1 0 0 0	0 0 1 1 0
II	1 0 0 0 0	1 1 0 1 0	1 0 1 0 1

The computer regression output is

```
VARIABLE   REG. COEF.   STD.ERROR COEF.   COMPUTED T   BETA COEF.
   1         4.00000         1.43372         2.78994     0.57208

INTERCEPT                    6.00000

MULTIPLE CORRELATION         0.57208
STD. ERROR OF ESTIMATE       3.04138

             ANALYSIS OF VARIANCE FOR THE REGRESSION
  SOURCE OF VARI.        D.F.    SUM OF SQ.    MEAN SQ.     F VALUE
ATTRIBUTABLE TO REG.       1        72.000      72.000       7.784
DEVIATION FROM REG.       16       148.000       9.250
     TOTAL                17       220.000

VARIABLE   REG. COEF.   STD.ERROR COEF.   COMPUTED T   BETA COEF.
   1         4.00000         1.00791         3.96863     0.57208
   2        -4.00000         1.23443        -3.24037    -0.53936
   3         1.00001         1.23443         0.81010     0.13484

INTERCEPT                    6.99999

MULTIPLE CORRELATION         0.84208
STD. ERROR OF ESTIMATE       2.13809
```

```
            ANALYSIS OF VARIANCE FOR THE REGRESSION
     SOURCE OF VARI.          D.F.    SUM OF SQ.     MEAN SQ.     F VALUE
  ATTRIBUTABLE TO REG.         3       156.000        52.000       11.375
  DEVIATION FROM REG.         14        64.000         4.571
        TOTAL                 17       220.000
```

```
VARIABLE   REG. COEF.   STD.ERROR COEF.   COMPUTED T   BETA COEF.
    1        6.00002       1.69968          3.53010      0.85812
    2       -1.99998       1.69967         -1.17669     -0.26968
    3        6.00005       3.39935          1.76506      0.80905
    4       -4.00003       2.40371         -1.66411     -0.53936
    5       -6.00006       4.16335         -1.44116     -0.63961

INTERCEPT                  5.99999

MULTIPLE CORRELATION       0.87386
STD. ERROR OF ESTIMATE     2.08166
```

```
            ANALYSIS OF VARIANCE FOR THE REGRESSION
     SOURCE OF VARI.          D.F.    SUM OF SQ.     MEAN SQ.     F VALUE
  ATTRIBUTABLE TO REG.         5       168.000        33.600        7.754
  DEVIATION FROM REG.         12        52.000         4.333
        TOTAL                 17       220.000
```

The ANOVA is

Source	*SS*	*df*	*MS*	*F*
Terminals	72.0	1	72.0	16.6
Programs	156 − 72 = 84.0	2	42.0	9.7
Interaction	168 − 72 − 84 = 12.0	2	6.0	1.39
Error	52.0	12	4.33	
Total	220	17		

15-17 a. The analysis can be done using Equation 15-1 or with indicator variables for display locations (3) and sales volume classes (3). The multiple regression for the six indicator variables and for one of the sets of three variables are needed. The ANOVA table is

Source	*SS*	*df*	*MS*	*F*
Locations	319.25	3	106.42	4.57
Sales volume	242.75	3		
Error	209.75	9	23.31	
Total	771.75	15		

The F-value is significant at the 0.05 level.

b. The error variance would have been $(242.75 + 209.75)/12 = 37.71$, and the F-value would have been $106.42/37.71 = 2.82$ with 3 and 12 degrees of freedom, which is not significant at 0.05. We would have been unable to conclude that display locations made a difference.

15-19 a. Let A represent the methods and B the teams.

$$SST = 2213 - 42849/18 = 832.5$$
$$SS(AB) = 8935/3 - 2380.5 = 597.83$$
$$SS(A) = 15539/6 - 2380.5 = 209.33$$
$$SS(B) = 23805/9 - 2380.5 = 264.50$$
$$SS(I) = SS(AB) - SS(A) - SS(B) = 124.00$$

The ANOVA table is

Source	*SS*	*df*	*MS*	*F*
Methods	209.33	2	104.67	5.35
Teams	264.50	1	264.50	13.52
Interaction	124.00	2	62.00	3.17
Error	234.67	12	19.56	
Total	832.50	17		

The interaction is insignificant at $\alpha = 0.05$, teams are significant at 0.01, and methods are significant at 0.05.

b. Only methods and residual error would have been separated.

Methods: $209.33/2 = 104.67$
Residual: $623.34/15 = 41.56$

$$F = 104.67/41.56 = 2.52 \text{ (insignificant at 0.05)}$$

We would not have declared a difference in methods.

Chapter 16 Nonparametric Statistics

16-1 A one-tailed test is required by the consumer group. The prescribed ranking rule should produce high rank numbers for brand B if it really has lower amounts of abrasive material. The test is, therefore, upper tailed. Table 16-1 shows that $P(W \geq 19) = 0.072$ under H_0. Hence, $\alpha = 0.072$ for an acceptance region of $W \leq 18$ for H_0.

The ranks of the observations are

Brand A 1 2 3 5 6

Brand B 4 7 8 and $W = 4 + 7 + 8 = 19$

The observed value of W, 19, is not in the acceptance region; reject H_0. At the 0.072 level of significance, brand B contains less abrasive than brand A.

16-3 H_0: Populations have same location; H_A: Distribution B is greater than distribution A.

Test statistic: z as defined in Equation 16-3.

Acceptance region: Since industry B will get high rank numbers if its expenditures are greater than industry A, accept H_0 if $z_{\text{observed}} \leq +1.96$ for the 0.025 significance level. Calculation of z_{observed}: From Equations 16-1 and 16-2

$$\mu_W = \frac{12(29)}{2} = 174$$

$$\sigma_W = \sqrt{\frac{12(16)(29)}{12}} = 21.54$$

From the data for industry B:

$$W = 8.5 + 13 + 14 + 15 + 16 + 17 + 19.5 + 21 + 22 + 25 + 27 + 28 = 226$$

$$z_{\text{observed}} = \frac{(226 - 174)}{21.54} = 2.41$$

Conclusion: Reject H_0. At the 0.025 level, industry B has greater expenditures.

16-5 High ranks are assigned to fast delivery times. Therefore, if H_A is true, Nationwide can be expected to have a greater value of W than μ_W (the expected value when H_0 is true). This calls for an upper-tailed test. Because the test statistic is z and α is 0.05, the acceptance region for H_0 is $z \leq +1.645$.

The ranks for the smaller sample, Nationwide, are the terms in

$$W = 21 + 20 + 19 + 17 + 13 + 11 + 10 + 7 + 4 + 3 = 125$$

$\mu_W = 115$ and $\sigma_W = 15.166$, as before. Hence,

$$z_{\text{observed}} = \frac{(125 - 115)}{15.166} = +0.66$$

which means that H_0 must be accepted. The reversal requires sign changes for the critical and observed values of the test statistic, but leaves their absolute values unchanged. When ranking is reversed, a lower-tailed test must be changed to the equivalent upper-tailed test, and vice versa.

16-7 There are $k = 4$ treatments and at least five observations per treatment. Hence, under the null hypothesis of no differences among the four schools, the test statistic H will be distributed as chi-square with 3 df. For the 0.05 level of significance, the acceptance region for the null hypothesis is $H_{\text{observed}} \leq 7.81$.

The rank table and calculations for the test statistic are:

	School				
	A	B	C	D	
	16	24	1	2.5	
	13	17	2.5	20	
	12	14	19	6	
	9	11	22	8	
	10	7	15	18	
	21		4	23	
				5	
T_j	81	73	63.5	82.5	
n_j	6	5	6	7	$n = 24$
T_j^2/n_j	1093.5	1065.8	672.0	972.3	$S = 3803.6$

$$H_{\text{observed}} = \frac{12(3803.6)}{24(25)} - 3(25) = 1.07$$

Accept the null hypothesis. There is no evidence here to support the conclusion that there is any difference among the schools with respect to player effectiveness.

16-9 Values of T are 42, 88, and 80. These produce the same values for S and H_{observed} as before.

16-11 For 12 matched pairs, the test statistic, z, is very nearly normally distributed with mean 0 and standard deviation 1. The acceptance region for H_0 is $-1.96 \le z \le +1.96$, at the 0.05 level of significance. In the order given in the exercise, the absolute differences are:

1.2 2.1 1.7 1.3 0.6 1.8 0.4 1.6 0.2 0.1 0.3 0.5

The signed ranks are:

+7 +12 +10 −8 +6 +11 +4 −9 −2 +1 +3 +5

The sum of the signed ranks is $v = +40$. Further,

$$\mu_V = 0 \quad \text{and} \quad \sigma_V = \sqrt{12(13)(25)/6} = 25.5.$$

Then $z_{\text{observed}} = (40 - 0)/25.5 = +1.57$. Accept H_0; there is no evidence to support a difference in mileages for the two fuels.

16-13 For 14 matched pairs we can assume that z is distributed as a standard normal variate under the null hypothesis of no difference in productivity. For a two-tailed test at the 0.10 level of significance, the acceptance region for H_0 is $-1.645 \le z \le +1.645$. In the order given in the exercise the absolute differences are:

2.4 5.5 1.6 0.1 5.1 5.0 0.5 0.7 0.2 3.0 5.9 2.2 0.7 0.3

The signed ranks are:

9 −13 7 1 −12 −11 4 −5.5 2 10 −14 −8 5.5 −3

The sum of signed ranks is $v = -28$. Furthermore,

$$\mu_V = 0 \quad \text{and} \quad \sigma_V = \sqrt{14(15)(29)/6} = 31.859.$$

$$z_{\text{observed}} = \frac{(-28 - 0)}{31.859} = -0.88$$

Accept the null hypothesis; there is no evidence to support a difference in productivity.

16-15 Since there are 11 plus signs and 14 minus signs, the normal approximation applies if H_0 is true. At the 0.05 level, the acceptance region for H_0 is $-1.96 \le z \le +1.96$. For the sequence, $r = 8$, $n_1 = 11$, $n_2 = 14$,

$$\mu_R = \frac{25 + 2(11)(14)}{25} = 13.32$$

and

$$\sigma_R = \sqrt{\frac{2(11)(14)[2(11)(14) - 25]}{25^2(25 - 1)}} = 2.41$$

Then

$$z_{\text{observed}} = \frac{(8 - 13.32)}{2.41} = -2.21$$

Reject H_0. There appears to be too much clustering to support random fluctuations.

16-17 Let $n_1 = 10$ be the number of A's and $n_2 = 12$ be the number of B's. Low breaking-strengths for B will place many B's at the left of the sequence and tend to produce too few runs to support the null hypothesis of no difference in breaking-strengths for A and B. With the normal assumption, the acceptance region for H_0 is $z \geq -1.645$. From the data, $r = 6$,

$$\mu_R = \frac{22 + 2(120)}{22} = 11.909$$

$$\sigma_R = \sqrt{\frac{240(240 - 22)}{22^2(21)}} = 2.27$$

and

$$z_{\text{observed}} = \frac{(6 - 11.909)}{2.27} = -2.6$$

Reject H_0; the evidence supports lower breaking-strengths for B.

16-19 a. Ranks from greatest to least are:

Housewives:	8	9	3	10	11	7	6	4	2	1	5	12
Students:	3	7	10	2	11	12	5	1	8	4	6	9
Values of d are:	5	2	−7	8	0	−5	1	3	−6	−3	−1	3

$$r_s = 1 - \frac{6(232)}{12(144 - 1)} = +0.189$$

Since the sign of r_s is positive, there is a tendency toward consistency in the ratings for the two groups. The tendency may not be significant, however.

b. Acceptance region: for 10 df and a 0.05 level of significance, $-2.228 \leq t \leq +2.228$.

$$t_{\text{observed}} = 0.189\sqrt{\frac{12 - 2}{1 - 0.189^2}} = 0.61$$

Accept H_0; we conclude there is no evidence to support a relationship, either positive or negative, between the two series of ratings.

16-21 For 12 df and $\alpha = 0.05$, the acceptance region for the null hypothesis of no relationship is $-2.179 \leq t \leq 2.179$. The observed value of r_s is found as follows:

Rating Rank	*Price Rank*	d	d^2	*Rating Rank*	*Price Rank*	d	d^2
1	1	0	0	8	14	−6	36
2	3	−1	1	9	11	−2	4
3	8	−5	25	10	9	1	1
4	5	−1	1	11	6	5	25
5	3	2	4	12	7	5	25
6	13	−7	49	13	10	3	9
7	12	−5	25	14	3	11	121

$$n^2 = 196 \quad \Sigma d^2 = 326$$

$$r_s = 1 - \frac{6(326)}{14(195)} = +0.2835$$

$$t = 0.2835\sqrt{\frac{14-2}{1-0.2835^2}} = +1.024$$

Accept H_0; there is no evidence of association between rating and price.

16-23 a. Each observation was randomly and independently selected. Hence there are no matched pairs.

b. The rank sum test applies. In order of magnitude the two samples and the rank numbers are:

Process A		*Process B*	
119.9	7	104.6	1
124.7	9	106.0	2
125.1	10	107.7	3
129.2	12	111.8	4
135.8	14	113.8	5
138.5	16	119.6	6
141.6	18	122.5	8
144.8	19	127.1	11
148.9	21	131.7	13
151.9	22	136.3	15
174.6	23	139.9	17
185.6	24	148.2	20
		$W =$	105

Acceptance region: $z \geq -1.645$

$$\mu_W = \frac{12(24+1)}{2} = 150$$

$$\sigma_W = \sqrt{\frac{12(12)(25)}{12}} = 17.32$$

$$z_{\text{observed}} = \frac{105-150}{17.32} = -2.598$$

Reject H_0; process B produces shorter lifetimes.

16-25 a. The sequence of process symbols is

BBBBBB A B AA B A B A B A B AA B AAAA

With a predominance of B's first and A's last, a significant amount of clustering would be evidence of shorter lifetimes from process B.

$$\mu_R = \frac{24 + 2(12)(12)}{24} = 13$$

$$\sigma_R = \sqrt{\frac{2(12)(12)(288 - 24)}{576(23)}} = 2.40$$

$$z_{\text{observed}} = \frac{14 - 13}{2.40} = 0.42$$

For the lower-tailed test, accept if $z \geq -1.645$. Accept H_0.

b. The result here is not significant, but it was in Exercise 16-23. The runs test is much less sensitive than the rank sum test.

Chapter 17 Time Series and Index Numbers

17-1 a. Trend
b. Irregular
c. Seasonal
d. Cyclical

17-3 a.

	x	y	xy	T
1973	−7	7.3	−51.1	7.42
1974	−5	8.7	−43.5	8.56
1975	−3	9.8	−29.4	9.69
1976	−1	10.8	−10.8	10.82
1977	1	11.9	11.9	11.95
1978	3	13	39	13.09
1979	5	14.1	70.5	14.22
1980	7	15.5	108.5	15.35
Sum		91.1	95.1	

$a = \Sigma y/n = 91.1/8 = 11.3875$

$b = \Sigma xy/\Sigma x^2 = 95.1/168 = .5660714$

$x = 0$ in 1976½ and is in ½-year units

b.

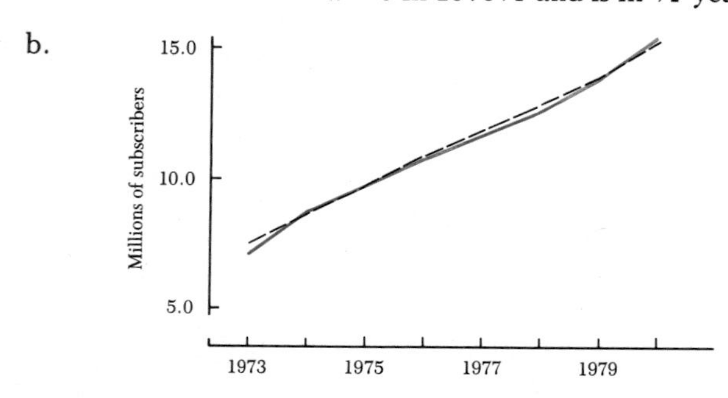

c. The trend appears to fit the series well.

17-5 a.

	1955	1960	1965	1970	1975	Total
y	377	393	479	708	1036	
$\log y$	2.58	2.59	2.68	2.85	3.02	13.72
x	−2	−1	0	1	2	
$x * \log y$	−5.16	−2.59	0	2.85	6.04	1.14
$\log T$	2.52	2.63	2.744	2.858	2.972	
T	328	427	554	721	938	

$\log a = 13.72/5 = 2.744$; $\log b = 1.14/10 = 0.114$

$\log T = 2.744 + 0.114x$

$x = 0$ in 1965 and is in 5-year units

b.

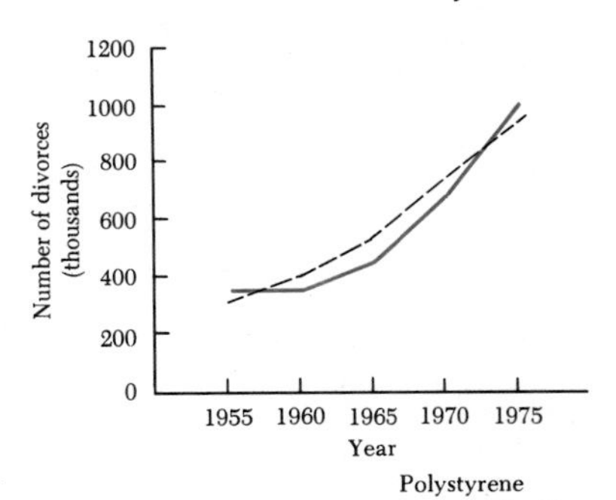

17-7 a.

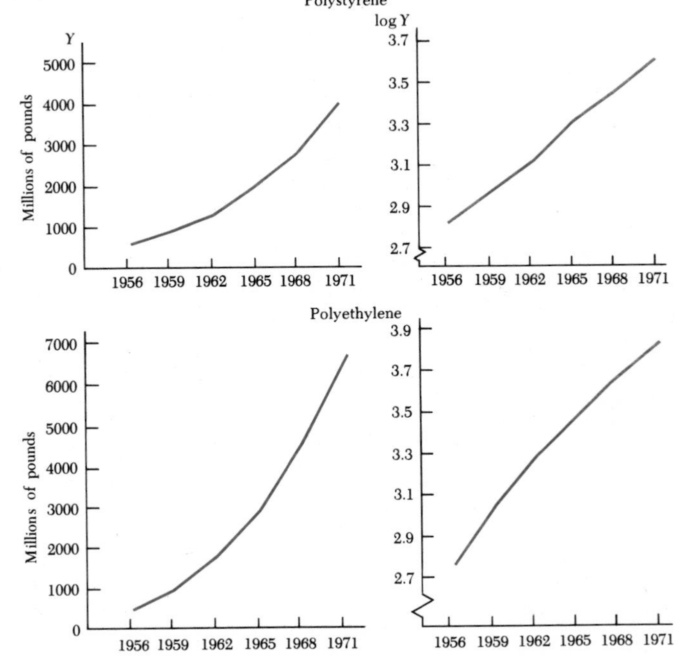

b. Polystyrene growth could be well described by an exponential trend because the log plot appears linear. Polyethylene growth does not follow either a straight-line arithmetic or a straight-line logarithmic (exponential) growth.

17-9 a.

	Week 1		*Week 2*		*Week 3*		*Week 4*	
	Total	*Moving Average*	*Total*	*Moving Average*	*Total*	*Moving Average*	*Total*	*Moving Average*
M			153	30.6	175	35.0	201	40.2
T			157	31.4	176	35.2	210	42.0
W	141	28.2	160	32.0	181	36.2	213	42.6
T	144	28.8	162	32.4	190	38.0		
F	149	29.8	168	33.6	194	38.8		

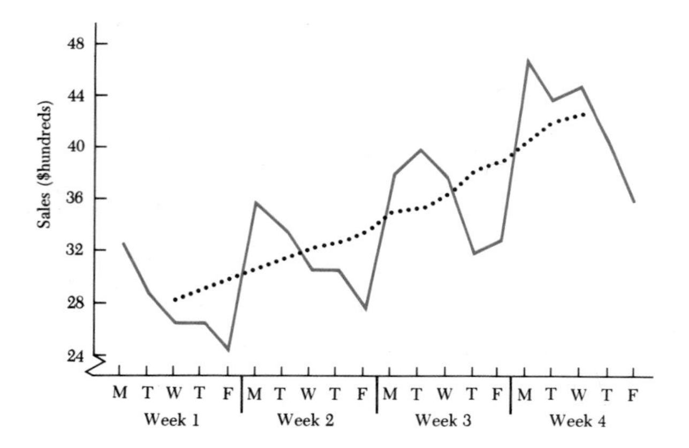

b. The five-day moving average smooths the series completely, indicating that the five-day cycle is a major element in the variability of the series.

c. A three-day moving average would fail to smooth out the five-day cycle.

17-11 a, c.

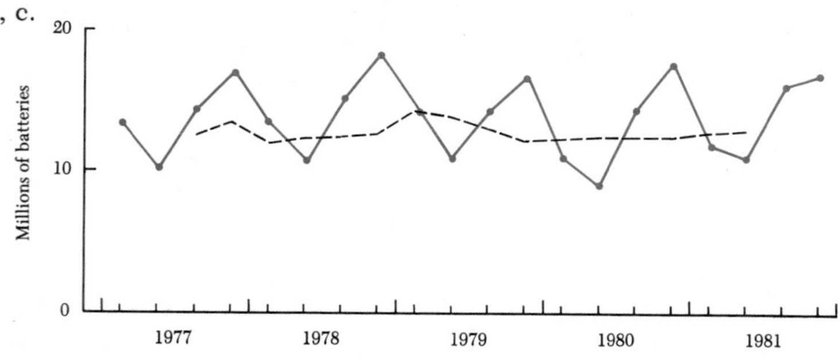

b.

	y	*4-term Total*	*Centered MA*
1977–1	13.1		
–2	10.0		
–3	14.3	54.5	13.60
–4	17.1	54.3	13.64
1978–1	12.9	54.8	13.79
–2	10.5	55.5	13.98
–3	15.0	56.3	14.18
–4	17.9	57.1	14.28
1979–1	13.7	57.1	14.13
–2	10.5	55.9	13.70
–3	13.8	53.7	13.00
–4	15.7	50.3	12.36
1980–1	10.3	48.6	12.15
–2	8.8	48.6	12.33
–3	13.8	50.0	12.68
–4	17.1	51.4	13.08
1981–1	11.7	53.2	13.50
–2	10.6	54.8	13.55
–3	15.4	53.6	
–4	15.9		

d. There is a weak upward trend with a cyclical peak in the first quarter of 1979 and a trough in late 1979 to early 1980.

17-13 a.

$$PI = \frac{\Sigma(p_1 q)}{\Sigma(p_0 q)}(100) = \frac{\$2.54(5500) + \$3.56(1800)}{\$1.33(5500) + \$1.33(1800)}(100) = 209.9$$

b.

	$p_0 q$	$w(\%)$
Corn	\$7315	75.34
Wheat	2394	24.66
	\$9709	100.00

$$\begin{aligned} PI = \Sigma[w(p_1/p_0)] &= 75.34(2.54/133) + 24.66(3.56/1.33) \\ &= 75.34(1.910) + 24.66(2.677) = 209.9 \end{aligned}$$

c. No, because the farmer's production is heavily concentrated on wheat, while corn is most important in the index calculated.

17-15 a. The index would assume, because of the fixed 1975 weights, that such substitution did not take place. Therefore the price index would overstate the effect of the price changes on consumers.

b. By the same sort of reasoning as in (a), the fixed-weight index would understate the effects of price increases on consumers.

17-17 a.

Year	Durables Current	Durables Constant	Durables Deflator	Nondurables Current	Nondurables Constant	Nondurables Deflator
1972	111	111	100	301	301	100
1974	123	112	110	373	303	123
1976	157	127	124	442	323	137
1978	199	146	136	530	346	153
1980	212	136	156	676	358	189

b. The price increases for nondurables are greater than for durables.

17-19 a, b, c.

Year	Y	x	xY	T
1959	703	−2	−1406	652.2
1964	1209	−1	−1209	1277.0
1969	1876	0	0	1901.8
1974	2579	1	2579	2526.6
1979	3142	2	6284	3151.4
Sum	9509		6248	

$a = 9509/5 = 1901.8$
$b = 6248/10 = 624.8$

Year	$\log Y$	x	$x(\log Y)$	$\log T$	T
1959	2.847	−2	−5.694	2.89636	787.7
1964	3.0824	−1	−3.0824	3.05931	1146.3
1969	3.2732	0	0	3.22226	1668.2
1974	3.4115	1	3.4115	3.38521	2427.8
1979	3.4972	2	6.9944	3.54816	3533.1
Sum	16.1113		1.6295		

$\log a = 16.1113/5 = 3.22226$
$\log b = 1.6295/10 = .16295$

d.

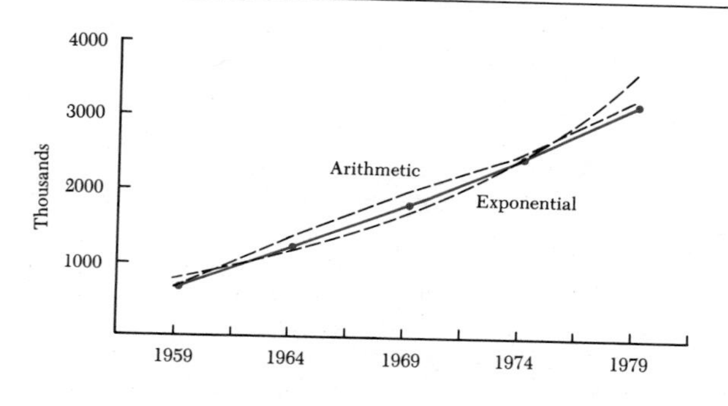

The arithmetic trend fits better, especially for the later years.

17-21 a, b.

Year	*y*	*x*	*xy*	*T*	%*T*	*95* | *100* | *110* |
1968	1051.8	−7	−7362.6	1033.9	101.7	*************
1969	1078.8	−6	−6472.8	1069.7	100.9	***********
1970	1075.3	−5	−5376.5	1105.5	97.3	****
1971	1107.5	−4	−4430.0	1141.3	97.0	****
1972	1185.9	−3	−3557.7	1177.2	100.7	***********
1973	1254.3	−2	−2508.6	1213.0	103.4	****************
1974	1246.3	−1	−1246.3	1248.8	99.8	*********
1975	1231.6	0	0	1284.7	95.9	*
1976	1298.2	1	1298.2	1320.5	98.3	******
1977	1369.7	2	2739.4	1356.3	101.0	************
1978	1438.6	3	4315.8	1392.2	103.3	****************
1979	1479.9	4	5919.6	1428.0	103.6	*****************
1980	1474.0	5	7370.0	1463.8	100.7	***********
1981	1502.6	6	9015.6	1499.7	100.2	**********
1982	1475.5	7	10328.5	1535.5	96.1	**
	$a =$ 1284.7		7092.6			
	$b =$ 35.830					

c. GNP peaks in 1973 and 1979, and reaches cyclical lows in 1970–71 and 1975.

17-23 a.

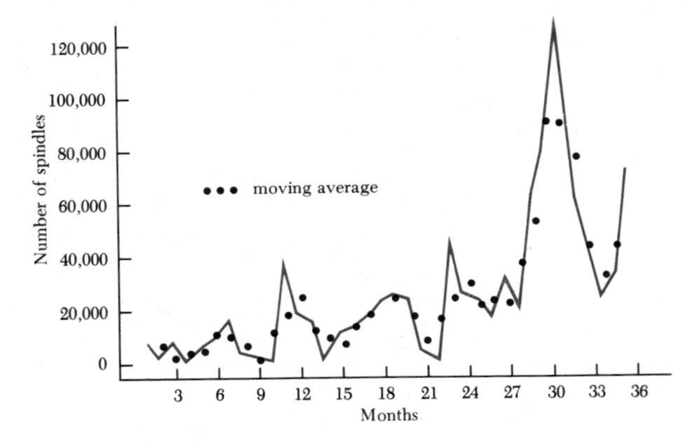

b. Moving average (reading across):

—	7242	4450	5883	5983	11003
10345	8040	2870	13978	19483	25283
13566	10600	9101	14268	18152	22235
24130	18189	9906	17163	24629	31660
22207	23683	22937	38114	54563	90372
90171	77895	43666	33832	43140	—

c. The moving average should bring out all elements of change, except month-to-month random changes, better than the original data. In the first 24 months there is a seasonal pattern in which replacement demand is up in the last few months of the year. In the third year, the demand level increases and there is a spurt in demand in the middle of the year.

17-25 a.

1965	492/0.964 = 510
1970	634/1.100 = 576
1975	1039/1.715 = 606
1980	1846/2.748 = 672

The adjusted series represents the value of shipments at 1967 prices.

b. Shipments valued at 1970 prices.

17-27 a. (76.1/63.9)($13,500) = $16,078 = cost of 1958 house in 1968; $2578 of the $4700 increase in price is attributable to cost; the remainder of the price change is attributable to quantity—the average 1968 house is larger than the average 1958 house.

b. (176.0/76.1)($18,200) = $42,097; of the $33,800 increase in home price, $23,897 is attributable to increase in cost and the remainder is attributable to quantity changes in the average house.

17-29 a, c.

	1963	*1966*	*1969*	*1972*	*1975*	*1978*	*Total*
y	99.1	111.4	124.9	139.1	157.2	175.8	
$\log y$	1.9961	2.0469	2.0966	2.1433	2.1965	2.245	12.7244
x	−5	−3	−1	1	3	5	
$x(\log y)$	−9.9805	−6.1407	−2.0966	2.1433	6.5895	11.225	1.74
$\log T$	1.9962	2.0460	2.0958	2.1456	2.1954	2.2452	

$$\log a = \Sigma \log y/n = 12.7244/6 = 2.1207$$
$$\log b = \Sigma x (\log y)/\Sigma x^2 = 1.74/70 = 0.0249$$

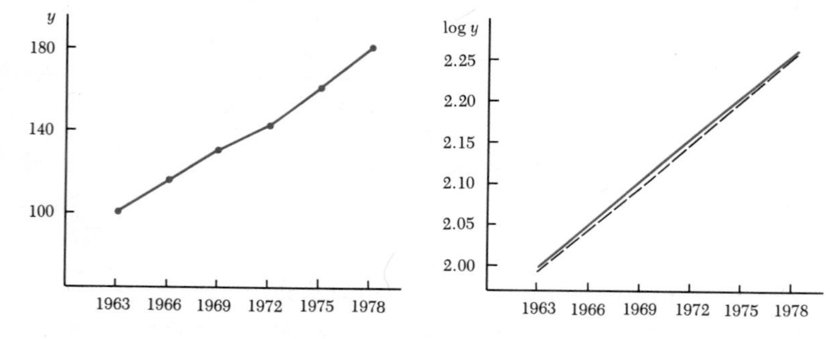

b, d. The logarithmic trend fits very closely.

Chapter 18 Time Series and Forecasting

18-1 a.

Year	Quarter 1	2	3	4	
1971			103.4	99.1	
1972	94.1	102.2	104.7	98.8	
1973	95.0	102.1	105.6	100.8	
1974	90.7	102.3	104.6	100.8	
1975	92.1	102.9	104.1	98.7	
1976	94.6	102.8			
Average	93.3	102.5	104.7	99.6	400.1
b. Seasonal Index	93.3	102.5	104.6	99.6	400.0

The individual y/MA ratios for each quarter show little variation; the seasonal pattern is quite stable.

18-3 a.

Year	Quarter 1	2	3	4
1974	22.0	20.5	19.4	18.9
1975	16.1	17.6	22.5	23.8
1976	24.2	23.9	25.2	24.0

b. During 1975-1976 there seems to be a strong recovery from a cyclical low, or an irregular upward change.

18-5 Adjusted series:

Year	1	2	3	4
1974	586	599	602	614
1975	604	615	613	618
1976	640	639	635	653

a. The difference is accounted for by irregular variation.
b. Only the adjusted data, since there are no moving-average values for the last two quarters.

18-7 a.

1977–1	[1598.17 – 33.37(29)](1.313) = 827.7
–2	[1598.17 – 33.37(30)](1.632) = 974.5
–3	[1598.17 – 33.37(31)](0.467) = 263.2
–4	[1598.17 – 33.37(32)](0.588) = 311.8

b. No, the moving-average line shows a definite cyclical pattern.

c. The actual data show that sales of room air conditioners recovered strongly from the cyclical low in 1976 and were actually high cyclically in 1977. The yearly total for 1977 was 3269, compared to a total for the four quarterly-forecasts of 2377.

18-9 a.

Year	Y	x	xY
1965	927	–5	–4635
1968	1046	–3	–3138
1971	1118	–1	–1118
1974	1244	1	1244
1977	1369	3	4107
1980	1485	5	7425
Sum	7189		3885

$$a = 7189/6 = 1198.167$$
$$b = 3885/70 = 55.5$$

$x = 0$ in 1972½ and is in 1½-year units.

b. $\text{Log } T(1984) = 3.111 + .0582(7.67) = 3.557394.$
$T = 3609$ in 1984 dollars.
$T = 1198 + 55.5(7.67) = 1624$ in 1972 dollars.

18-11 a, b. First differences:

	Sales	*Investment*		*Sales*	*Investment*
1966	55	–4.7	1972	479	9.8
1967	–54	–1.3	1973	203	–2.3
1968	456	5.6	1974	–421	–14.7
1969	131	0.4	1975	–473	–6.2
1970	33	–2.8	1976	435	8.9
1971	316	11.8			

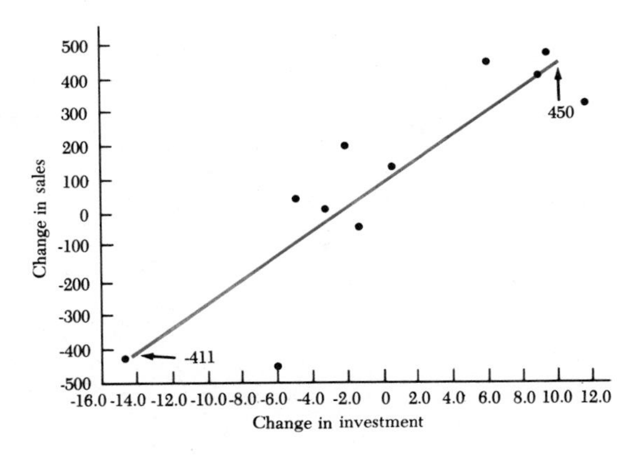

c. $d_x = 57.6 - 47.7 = 9.9$; $\hat{d}_y = 90.78 + 35.87(9.9) = 446$; estimated sales = 2515 + 446 = 2961

18-13 a. Differences:

Year	Dish.	Res. Inv.	Dur. Exp.
1974	–382	–14.7	–1.8
1975	–618	–6.2	10.7
1976	438	12.4	24.6
1977	216	9.5	21.4
1978	202	1.7	22.0
1979	–70	–3.3	13.2
1980	–750	–11.9	.9
1981	–254	–2.3	20.3
SS Dev.	1253627	631.84	703.52
S Prod. Dev.		24760.50	24106.42
Reg. Coeff.		39.188	34.265
Const. Term.		–79.752	–628.962
R Sq.		.774	.659
SE Est.		217.03	266.93

b. Residential investment produces a moderately higher coefficient of determination and consequently lower standard error. It is preferred on this account.

18-15 *Note:* In Exercise 18-15, $\hat{y}$ values are recorded to nearest integers, but carried fractionally in the exponential forecasting formula. RMSE values are based on the integer results.

	Product A			Product B		
Period	y	$w = 0.25$	$w = 0.75$	y	$w = 0.25$	$w = 0.75$
1	31			46		
2	39	31	31	49	46	46
3	37	33	37	52	47	50
4	24	34	37	49	48	52
5	24	32	27	54	48	50
6	25	30	25	50	50	53
7	45	29	25	65	50	51
8	36	33	40	51	54	61
9	36	33	37	46	53	54
10	36	34	36	51	51	48
11	45	35	36	70	51	50
12	20	37	43	50	56	65
13	52	33	26	77	54	54
14	64	38	45	84	60	71
15	47	44	59	62	66	81
16	45	45	50	55	65	67
17	63	45	46	78	63	58
18	53	50	59	78	66	73
19	61	50	54	91	69	77
20	45	53	59	70	75	87

		Product C			Product D	
Period	y	$w = 0.25$	$w = 0.75$	y	$w = 0.25$	$w = 0.75$
1	60			22		
2	81	60	60	11	22	22
3	68	65	76	19	19	14
4	78	66	70	24	19	18
5	55	69	76	5	20	22
6	62	65	60	32	17	9
7	74	64	62	42	20	26
8	64	66	71	37	26	38
9	40	66	66	43	29	37
10	40	59	46	21	32	42
11	41	55	42	35	29	26
12	36	51	41	30	31	33
13	41	47	37	18	31	31
14	45	46	40	41	27	21
15	30	46	44	37	31	36
16	34	42	33	34	32	37
17	39	40	34	45	33	35
18	42	40	38	58	36	42
19	39	40	41	35	41	54
20	46	40	39	39	40	40

	RMSE*		
Product	$w = 0.25$	$w = 0.50$	$w = 0.75$
A	11.6	11.7	12.6
B	12.2	12.1	12.9
C	12.1	10.8	10.9
D	11.8	11.9	12.7

* See Exercise 18-14 for $w = 0.50$.

18-17 a. New hires are seasonally high in the second and third quarters of the year and low in the first and fourth quarters.

b.

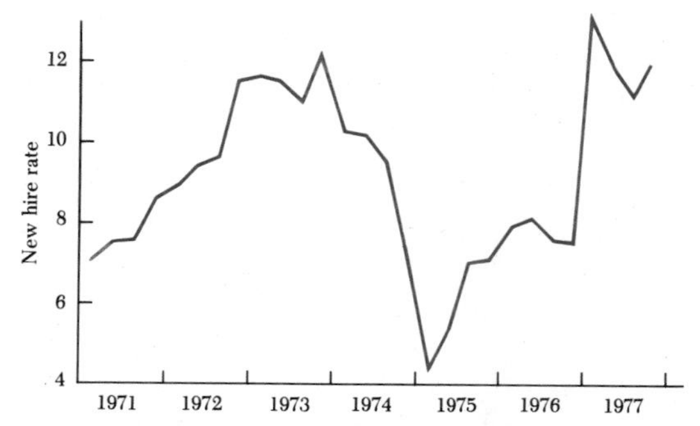

New hires reach a cyclical peak in 1973 and decline to a cyclical low in early 1975. Thereafter new hires recover through 1976 and 1977.

c.

$$1978\text{–}1 \quad [8.775 + 0.0303(28)](0.865) = 8.32$$
$$-2 \quad [8.775 + 0.0303(29)](1.128) = 10.89$$
$$-3 \quad [8.775 + 0.0303(30)](1.247) = 12.08$$
$$-4 \quad [8.775 + 0.0303(31)](0.760) = 7.38$$

The forecasts account for trend and seasonal only, and we know that cyclical variation has been an important element in the series.

18-19 a.

	Total	*MA*		*Total*	*MA*
Week 1—M			Week 3—M	175	35.0
T			T	176	35.2
W	141	28.2	W	181	36.2
T	144	28.8	T	190	38.0
F	149	29.8	F	194	38.8
Week 2—M	153	30.6	Week 4—M	201	40.2
T	157	31.4	T	210	42.0
W	160	32.0	W	213	42.6
T	162	32.4	T		
F	168	33.6	F		

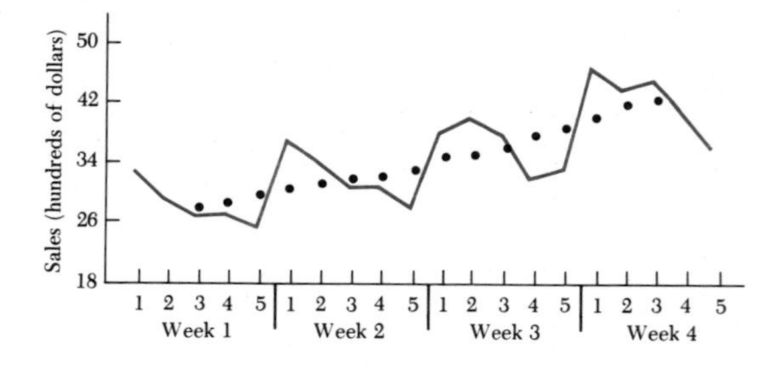

b. y/MA ratios:

Week	*Day* 1	2	3	4	5	*Total*
1				0.938	0.839	
2	1.176	1.083	0.969	0.957	0.833	
3	1.086	1.136	1.050	0.842	0.851	
4	1.169	1.048	1.056			
Average	1.144	1.089	1.025	0.912	0.841	5.011
Index	1.141	1.087	1.023	0.910	0.839	5.000

Monday's sales are typically about 14% above average daily sales for the week.

18-21 a.

	y	*MA*	*SI*	*y adj*		*y adjusted for seasonal* 550 | 900 | 1250 |
1977–1	900		121.23	742	1	*******
–2	1322		169.27	781	9	*********
–3	608	838.75	60.16	1011	7	******************
–4	439	935.38	49.34	890	7	*************
1978–1	1072	998.13	121.23	884	1	*************
–2	1923	997.50	169.27	1136	9	***********************
–3	509	1027.25	60.16	846	7	***********
–4	533	1019.88	49.34	1080	8	*********************
1979–1	1216	973.00	121.23	1003	1	******************
–2	1720	943.38	169.27	1016	9	******************
–3	337	905.63	60.16	560	7	*
–4	468	795.50	49.34	949	9	***************
1980–1	979	763.63	121.23	808	1	**********
–2	1076	806.00	169.27	636	9	***
–3	726	829.25	60.16	1208	8	**************************
–4	418	940.88	49.34	847	0	***********
1981–1	1215	982.13	121.23	1002	1	******************
–2	1733	932.38	169.27	1024	9	******************
–3	399		60.16	663	8	****
–4	347		49.34	703	1	******

b.

y/MA%	*Q1*	*Q2*	*Q3*	*Q4*	*Total*
1977			72.49	46.93	
1978	107.40	192.78	49.55	52.26	
1979	124.97	182.32	37.21	58.83	
1980	128.20	133.50	87.55	44.43	
1981	123.71	185.87			
Average	121.47	173.62	61.70	50.61	407.00
Index	118.99	170.63	60.64	49.74	400.00

c. Sales have recovered after the trend–cyclical decline of the earlier years, but lack any new trend or cycle.

18-23 a.

	d_y	d_x
1968	253	528
1969	184	80
1970	29	–139
1971	–35	–6
1972	73	245
1973	119	327
1974	221	140
1975	56	291
1976	251	434
1977	207	262
1978	555	603
Total	1863	2765

$$b = 263510/501364 = .5295$$

$$a = (1863 - .5295 * 2765)/11 = 40.80$$

$$\text{SST} = 265491; \quad \text{SSREG} = 140588$$

$$\text{SE EST} = 111.55; \quad \text{R SQ} = .5566$$

b. If regional income does not change, the expected change in originations is 41 thousand. For each change of $1 million in income, a further change of 530 originations is expected.

c. $$3372 + 40.80 + .5295(10000 - 9585) = 3633$$

d.

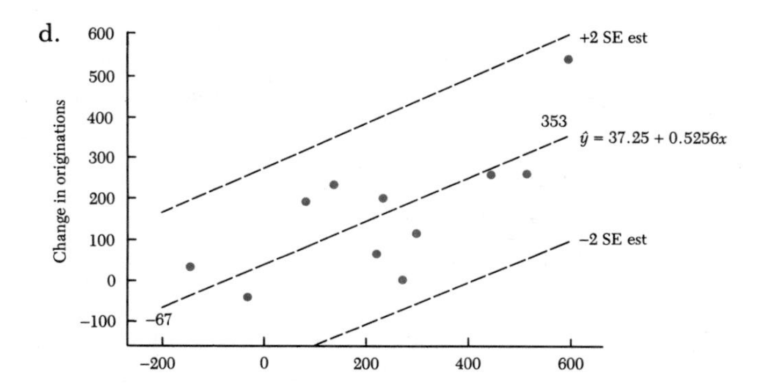

Chapter 19 Introduction to Decision Making

19-1 a.

$$EV(A_1) = 0.5(14) + 0.3(9) + 0.2(6) = 10.9$$
$$EV(A_2) = 0.5(13) + 0.3(8) + 0.2(8) = 10.5$$
$$EV(A_3) = 0.5(16) + 0.3(6) + 0.2(5) = 10.8$$

b.

$$EV(CP) = 0.5(16) + 0.3(9) + 0.2(8) = 12.3$$

c.

$$EV(PI) = EV(CP) - EV(A_{opt}) = 12.3 - 10.9 = 1.4$$

19-3 a.

State		*Act: Stock*		
Demand	*Probability*	*0*	*1*	*2*
0	0.80	—	−100	−200
1	0.15	—	1000	900
2	0.05	—	1000	2000

$$EV(\text{Stock1}) = 0.80(-100) + 0.15(1000) + 0.05(1000) = 120$$
$$EV(\text{Stock 2}) = 0.80(-200) + 0.15(900) + 0.05(2000) = 75$$

b. The optimal act is to stock one boat.

19-5

State Number Defective	*P(S)*	*Act*	
		Test None	*Test All*
0	0.6	0	−\$200
1	0.3	−\$500	−\$400
2	0.1	−\$1000	−\$600

a.

$$EV(\text{Test none}) = 0.6(\$0) + 0.3(-\$500) + 0.2(-\$1000) = -\$350$$
$$EV(\text{Test all}) = 0.6(-\$200) + 0.3(-\$400) + 0.1(-\$600) = -\$300$$

b. $$EV(CP) = 0.6(0) + 0.3(-\$400) + 0.1(-\$600) = -\$180$$
$$EV(PI) = EV(CP) - EV(A_{opt}) = -\$180 - (-\$300) = \$120$$

19-7 a. For A: $U(\$500) = 80$; $EU(\text{Gamble}) = 0.5 * U(\$0) + 0.5 * U(\$1000) = 0.5(0) + 0.5(100) = 50$; A prefers the certain gain.

For B: $U(\$500) = 12$; $EU(\text{Gamble}) = 0.5 * U(\$0) + 0.5 * U(\$1000) = 0.5(0) + 0.5(100) = 50$; B prefers the gamble.

b. For A: $U(-\$100) = -46$; $EU(\text{Gamble}) = 0.90(0) + 0.10(-100) = -10$; A prefers the gamble.

For B: $U(-\$100) = -1$; $EU(\text{Gamble}) = 0.90(0) + 0.10(-100) = -10$; B prefers the certain loss.

c. For A: $U(\$100) = 46$; $EU(\text{Gamble}) = 0.55(100) + 0.45(-100) = 10$; A prefers the certain gain.

For B: $U(\$100) = 1$; $EU(\text{Gamble}) = 0.55(100) + 0.45(-100) = 10$; B prefers the gamble.

d.

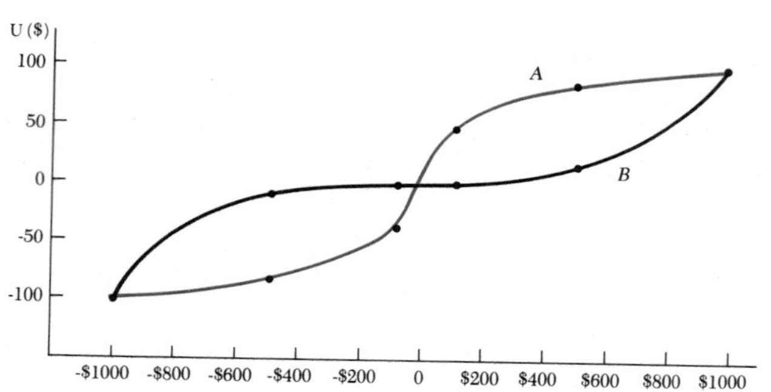

19-9 Example for a student who aspires to B grades:

a. 0.70 and 0.30; $U(B) = 0.70 * U(A) + 0.30 * U(C) = 0.70(100) + 0.30(0) = 70$

b. 0.90 and 0.10; $U(C) = 0.90 * U(B) + 0.10 * U(D)$;
$U(D) = U(C) - 0.90 * U(B) = 0 - 0.90(70) = -63$

c.

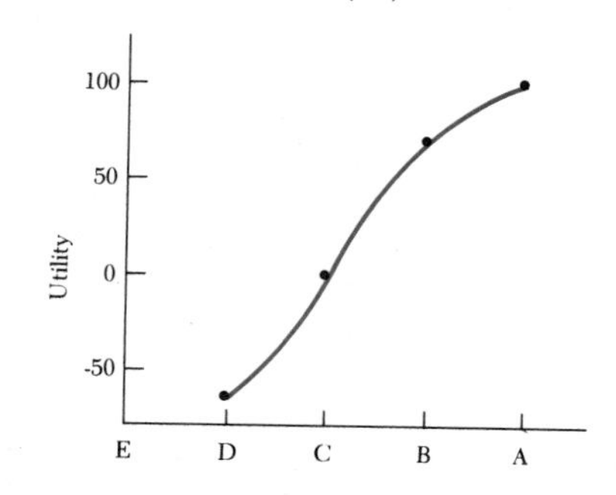

19-11 a.

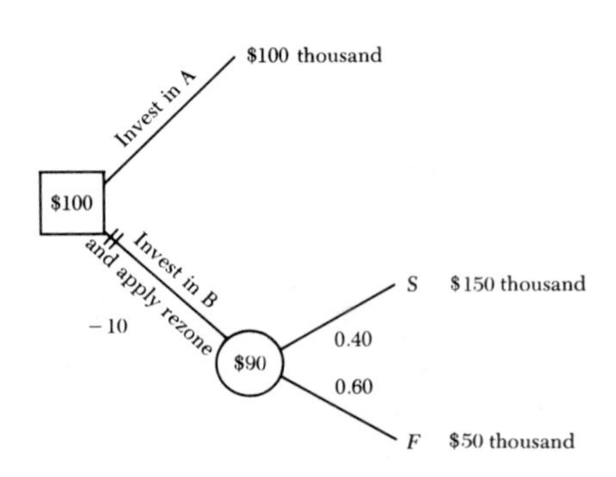

b. EV(Invest in B and apply rezone) = 0.40($150) + 0.60($50) − $10 = $80 (thousands)

19-13 a.

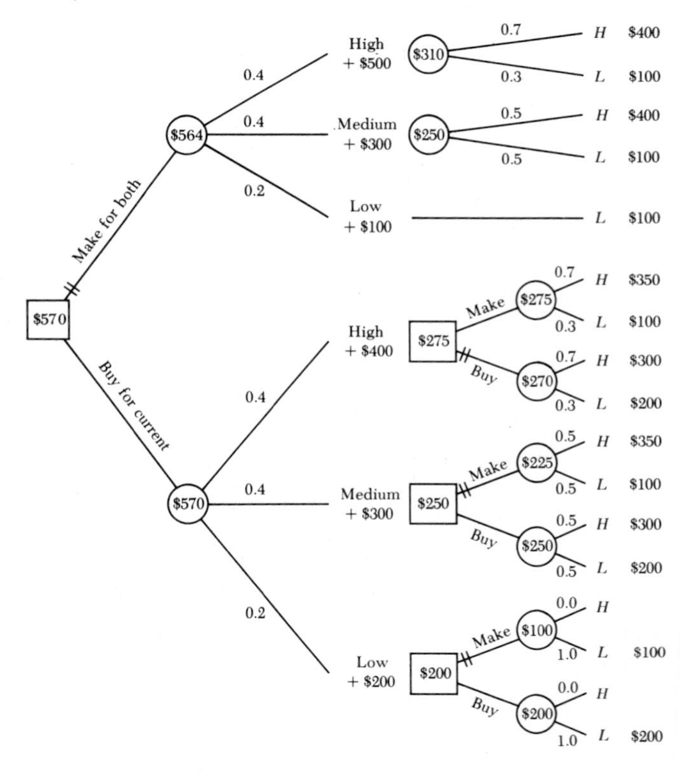

b. Buy for current demand; if later demand is high, make for future; if later demand is medium or low, buy for future. $EV(A_{opt})$ = $570.

19-15 a. (thousands of dollars)

State	Payoff Not Sell	Payoff Sell	Regrets Not Sell	Regrets Sell
Undertake — succeed (US)	1,000	200	0	800
Undertake — fail (UF)	200	200	0	0
Not undertake (NU)	0	200	200	0

b. $P(\text{U and S}) = 0.5(0.6) = 0.3$; $P(\text{U and F}) = 0.5 - 0.3 = 0.2$; $P(\text{NU}) = 1 - 0.5 = 0.5$; $ER(\text{Not sell}) = 0.5(200) = 40$; $ER(\text{Sell}) = 0.3(800) = 240$; $A_{\text{opt}} = \text{Not sell}$; $EV(PI) = ER(A_{\text{opt}}) = \$40$ (thousands)

c. No, because payoffs if the rights are not sold equal or exceed payoffs if the rights are sold whether or not the flight is a success.

19-17 a.

States		Opportunity Losses (dollars) if Provide				
Demand	Probability	0	1	2	3	4
0	0.37	0	100	200	300	400
1	0.37	1500	0	100	200	300
2	0.18	3000	1500	0	100	200
3	0.06	4500	3000	1500	0	100
4	0.02	6000	4500	3000	1500	0

b.	Expected Loss	1485	577	261	233	301
c.	Demand (d)	0	1	2	3	4
	$P(D \geq d)$	1.00	0.63	0.36	0.08	0.02

$P_C = 100/(100 + 1500) = 0.0625$; $d = 3$ is the highest d for which $P(D \geq d) \geq P_C$; therefore, the optimal act is to provide three replacement suits, which is the conclusion as in Part (b).

19-19 a. $EV(\text{Publish}) = 0.1(-\$20{,}000) + 0.3(-\$5000) + 0.4(\$10{,}000) + 0.2(\$30{,}000) = \$6500$

b. $EV(CP) = 0.4(\$0) + 0.4(\$10{,}000) + 0.2(\$30{,}000) = \$10{,}000$

c. $EOL(\text{Publish}) = 0.1(\$20{,}000) + 0.3(\$5000) = \$3500$

d. $EV(PI) = EV(CP) - EV(A_{\text{opt}}) = \$10{,}000 - \$6500 = \$3500$
$EV(PI) = EOL(A_{\text{opt}}) = EOL(\text{Publish}) = \$3500$

19-21 a.

Money (X)	*−4*	*−3*	*−2*	*−1*	*0*	*1*	*2*	*3*	*4*
Alpha	−192	−138	−88	−42	0	38	72	102	128
Beta	−192	−138	−88	−42	0	42	88	138	192
Omega	−224	−156	−96	−44	0	44	96	156	224

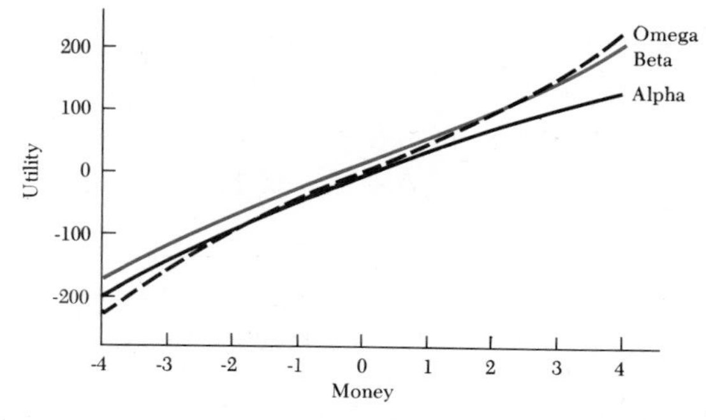

b. Alpha will insure against losses and will not prefer risky gains to receiving the fair value of the gamble for sure. Beta views losses the same way as Alpha, but will prefer risky gains to their sure equivalents. Alpha is risk-averse for both gains and losses and Beta is risk-averse for losses only.

c. Omega will pay more both to insure against loss and for an uncertain chance at gain than Beta.

19-23 a.

Range Condition	*Size of Herd* 280	*360*	*415*
Poor	0	\$4000	\$6500
Normal	\$5000	0	\$ 500
Good	\$8000	\$2500	0

b. $EOL(280) = (7/44)(\$0) + (22/44)(\$5000) + (15/44)(\$8000) = \$5227$
$EOL(360) = (7/44)(\$4000) + (22/44)(\$0) + (15/44)(\$2500) = \$1489$
$EOL(415) = (7/44)(\$6500) + (22/44)(\$500) + (15/44)(\$0) = \$1284$

The 415 cow herd minimizes expected opportunity loss.

c. $EV(PI) = EOL(415) = \$1284$

Chapter 20 Decision Making with Sample Information

20-1 a.

State Prefer	$P(S)$	< 30 $P(E_1 \mid S)$	≥ 30 $P(E_2 \mid S)$	Joint Probability E_1	E_2	$P(S \mid E_1)$
Plan 1	0.6	0.1	0.9	0.06	0.54	0.06/0.34 = 0.176
Plan 2	0.4	0.7	0.3	0.28	0.12	0.28/0.34 = 0.824
				0.34		

b. $EV(\text{Plan 1}) = 0.176(\$40) + 0.824(\$30) = \$31.76$
$EV(\text{Plan 2}) = 0.176(\$20) + 0.824(\$40) = \$36.48 = EV(A_{\text{opt}})$

20-3 a. 40/60 = 0.67
b. 30/50 = 0.60

c.

State	$P(S)$	$P(E \mid S)$	$P(S \text{ and } E)$	$P(S \mid E)$
Stock A: E = Predict down				
Down	2/6	40/70	80/420	80/143
Unchanged	2/6	10/40	35/420	35/143
Up	2/6	10/50	28/420	28/143
			143/420	
Stock B: E = Predict up				
Down	4/6	10/70	80/840	80/199
Unchanged	1/6	10/40	35/840	35/199
Up	1/6	30/50	84/840	84/199
			199/840	
Stock C: E = Predict down				
Down	3/6	40/70	240/840	240/338
Unchanged	2/6	10/40	70/840	70/338
Up	1/6	10/50	28/840	28/338
			338/840	

20-5

State: Number Defective	$P(S)$	E_1 = Defective $P(E_1 \mid S)$	E_2 = Good $P(E_2 \mid S)$	Joint Probability E_1	E_2	$P(S \mid E_1)$	$P(S \mid E_2)$
0	6/10	0/4	4/4	0/40	24/40	0/5	25/35
1	3/10	1/4	3/4	3/40	9/40	3/5	9/35
2	1/10	2/4	2/4	2/40	2/40	2/5	2/35
				5/40	35/40		

a. See column $P(S \mid E_1)$ in table above.
b. See column $P(S \mid E_2)$ in table above.
c. 5/40; 35/40

20-7

S	$P(S)$	$P(E \mid S)$	$P(S \text{ and } E)$	$P(S \mid E)$
1%	0.8	0.8179	0.6543	0.6543/0.7878 = 0.8305
2%	0.2	0.6676	0.1335	0.1335/0.7878 = 0.1695
			0.7878	

20-9 a, b.

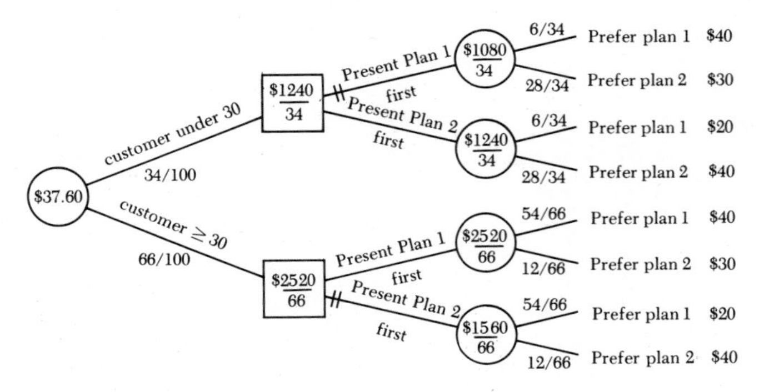

c. $EV(IA) = \$37.60$
d. $EV(SI) = EV(IA) - EV(A_{\text{opt}}) = \$37.60 - \$36.00 = \$1.60$

20-11 a.

$$V_{ji} = \text{Acts} \overset{\text{States}}{\begin{pmatrix} 14 & 9 & 6 \\ 13 & 8 & 8 \\ 16 & 6 & 5 \end{pmatrix}}$$

b.

$$P_{ik} = \text{States} \overset{\text{Information Outcomes}}{\begin{pmatrix} 40/100 & 5/100 & 5/100 \\ 3/100 & 24/100 & 3/100 \\ 2/100 & 2/100 & 16/100 \end{pmatrix}}$$

$$v_{jk} = \begin{pmatrix} 14 & 9 & 6 \\ 13 & 8 & 8 \\ 16 & 6 & 5 \end{pmatrix} \times \begin{pmatrix} 40/100 & 5/100 & 5/100 \\ 3/100 & 24/100 & 3/100 \\ 2/100 & 2/100 & 16/100 \end{pmatrix} = \begin{pmatrix} 5.99 & 2.98 & 1.93 \\ 5.60 & 2.73 & 2.17 \\ 6.68 & 2.34 & 1.78 \end{pmatrix}$$

c. Given I_1 A_3 is optimal; given I_2 A_1 is optimal; given I_3 A_2 is optimal; $EV(IA) = 6.68 + 2.98 + 2.17 = \$11.83$ (tens of thousands)

20-13 a. $\mu = 56$, $\sigma = 5$; for $x = 53$ use $52.5 \leq X \leq 53.5$; $z = (53.5 - 56)/5 = -0.5$; $z = (52.5 - 56)/5 = -0.7$; $P(-0.7 \leq Z \leq -0.5) = 0.3085 - 0.2420 = 0.0665$

b. $\mu = 53$, $\sigma = 5$; $z = (53.5 - 53)/5 = 0.10$; $z = (52.5 - 53)/5 = -0.10$; $P(-0.10 \leq Z \leq 0.10) = 0.5398 - 0.4602 = 0.0796$

c.

State (group)	*P(S)*	*P(E | S)*	*P(S and E)*	*P(S | E)*
Successful	0.70	0.0665	0.04655	0.661
Unsuccessful	0.30	0.0796	0.02388	0.339
			0.07043	

20-15 a. $\sigma_{\bar{X}} = 1/\sqrt{4} = 0.50$; for $P(19.25 \leq \bar{X} \leq 19.35)$ when $\mu = 20$; $z = (19.25 - 20)/0.50 = -1.5$ and $z = (19.35 - 20)/0.50 = -1.3$; $P(-1.5 \leq Z \leq -1.3) = 0.0968 - 0.0668 = 0.0300$.

When $\mu = 19$, $z = (19.25 - 19)/0.5 = 0.5$ and $z = (19.35 - 19)/0.5 = 0.70$; $P(0.5 \leq Z \leq 0.7) = 0.7580 - 0.6915 = 0.0665$.

Odds ($\mu = 20$ to $\mu = 19$) = $(4/1)(0.0300/0.0665) = 1.80$ to 1

b. $\sigma_{\bar{X}} = 1.0/\sqrt{16} = 0.25$; when $\mu = 20$, $z = (19.25 - 20)/0.25 = -3.0$ and $z = (19.35 - 20)/0.25 = -2.6$; $P(-3.0 \leq Z \leq -2.6) = 0.0047 - 0.0013 = 0.0034$

When $\mu = 19$, $z = (19.25 - 19)/0.25 = 1.0$ and $z = (19.35 - 19)/0.25 = 1.4$; $P(1.0 \leq Z \leq 1.4) = 0.9192 - 0.8413 = 0.0779$.

Odds ($\mu = 20$ to $\mu = 19$) = $(4/1)(0.0034/0.0779) = 0.175$ to 1

20-17 a. $I_P = (1/30)^2 = 1/900$, $I_S = 1/(140/\sqrt{49})^2 = 1/400$

$$E'(\mu) = \frac{(1/900)(800) + (1/400)(830)}{(1/900) + (1/400)} = 820.8$$

$\sigma_{\mu}'^2 = 1/[1/900) + (1/400)] = 276.9$; $\sigma_{\mu}' = \sqrt{276.9} = 16.64$

b. $z = (800 - 820.8)/16.64 = -1.25$; $P(Z \geq -1.25) = 1 - 0.1057 = 0.8943$

c. est $\sigma = s = 140$, est $\mu = E'(\mu) = 820.8$; $z = (800 - 820.8)/140 = -0.15$; $P(Z \geq -0.15) = 1 - 0.4404 = 0.5596$

20-19 a.

		Loss if conclude:	
State	*P(S)*	*(1)* $\mu = 19$	*(2)* $\mu = 20$
$\mu = 19$	.357	0	1.5
$\mu = 20$	.643	1	0

$P(\mu = 20) = 1.8/(1.8 + 1.0) = 0.643$; $EOL(1) = 0.643(1) = 0.643$ and $EOL(2) = 0.357(1.5) = 0.536$; conclude $\mu = 20$ is optimal.

b. $P(\mu = 20) = 0.175/(0.175 + 1.0) = 0.149$; $EOL(1) = 0.149(1) = 0.149$ and $EOL(2) = 0.851(1.5) = 1.277$; conclude $\mu = 19$ is optimal.

20-21 a. $c_u = 1.0$, $c_o = 2.0$; $c_u/(c_u + c_o) = 0.33$; $z_{0.33} = -0.44$; $\mu_{0.33} = 800 + (-0.44)(30) = 786.8$

b. $\mu'_{0.33} = 820.8 + (-0.44)(16.64) = 813.5$

20-23 a and b.

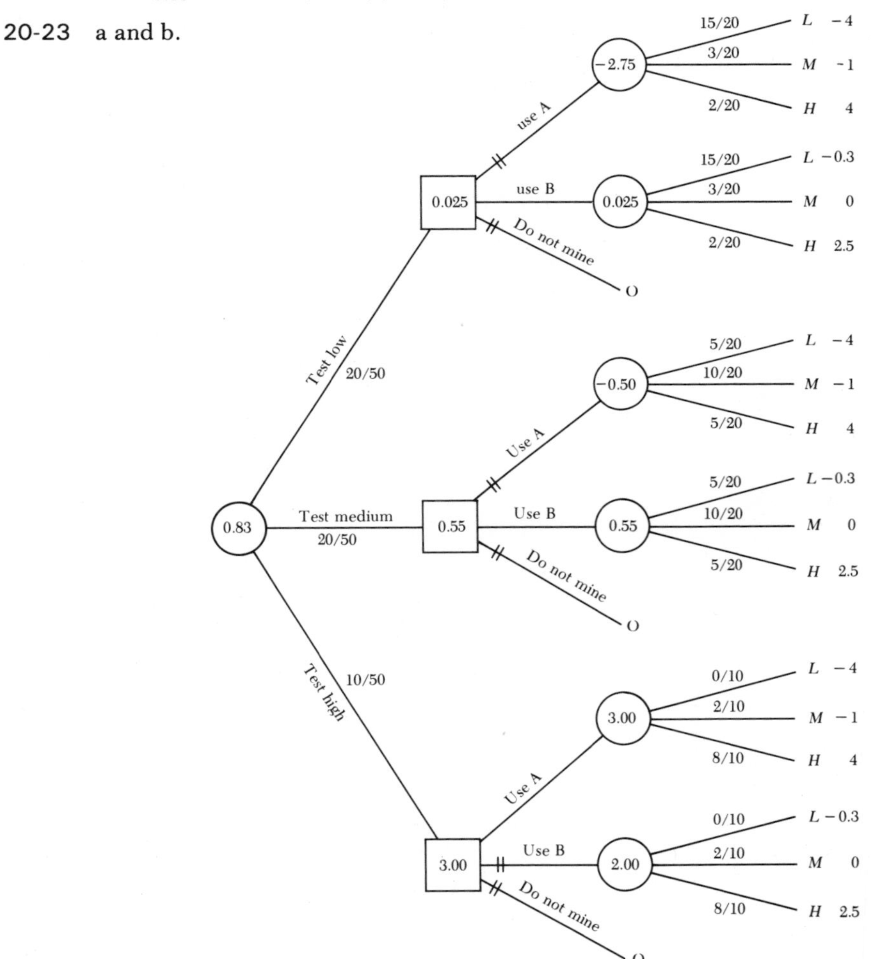

c. $EV(SI) = EV(IA) - EV(A_{\text{opt}}) = 0.83 - 0.63 = \$0.20$ (millions)

Since the test is worth \$200,000 and it costs \$100,000, the test should be conducted.

20-25 a. For $P(X = 43 \mid \text{Stayer})$; $z = (42.5 - 55)/10 = -1.25$ and $z = (43.5 - 55)/10 = -1.15$; $P(-1.25 \le Z \le -1.15) = 0.1251 - 0.1057 = 0.0194$.

For $P(X = 43 \mid \text{Mover})$; $z = (42.5 - 45)/10 = -0.25$ and
$z = (43.5 - 45)/10 = -0.15$; $P(-0.25 \leq Z \leq -0.15) = 0.4404 - 0.4013 = 0.0391$.

$P(\text{Stayer} \mid x = 43) = 0.70(0.0194)/[0.70(0.194) + 0.30(0.0391)] = 0.537$.

b. For $P(X = 53 \mid \text{Stayer})$; $z = (52.5 - 55)/10 = -0.25$ and
$z = (53.5 - 55)/10 = -0.15$; $P(-0.25 \leq Z \leq -0.15) = 0.0391$.

For $P(X = 53 \mid \text{Mover})$; $z = (52.5 - 45)/10 = 0.75$ and
$z = (53.5 - 45)/10 = 0.85$; $P(0.75 \leq Z \leq 0.85) = 0.8023 - 0.7734 = 0.0289$.

$P(\text{Stayer} \mid x = 43) = 0.70(0.0391)/[0.70(0.0391) + 0.30(0.0289)] = 0.759$.

20-27 $E(\mu) = 50$; $\sigma_\mu = 3.0$; $\bar{x} = 54$; $n = 100$; $s = 10$.

a. $I_P = 1/(3)^2 = 1/9$; $I_S = 1/(10/\sqrt{100})^2 = 1 = 9/9$

$$E'(\mu) = \frac{(1/9)(50) + (9/9)(54)}{(1/9) + (9/9)} = 53.6$$

$\sigma_\mu'^2 = 1/[(1/9) + (9/9)] = 0.90$; $\sigma'_\mu = 0.95$

b. $z = (55 - 53.6)/0.95 = 1.47$; $P(Z \geq 1.47) = 1 - 0.9292 = 0.0708$

c. $\mu_{0.95} = 53.6 + 1.65(0.95) = 55.17$;
$\mu_{0.05} = 53.6 - 1.65(0.95) = 52.03$

INDEX

OBS	RETIRE	HIRED	DEGREE	SALARY	ID	LEVEL	DEFER	LEVELYR	COLLEGE	YEAR	DEPART
1	1995	1969	G	$69,188	000-28-5703	LEVEL4	Y	1968	21988	1962	H
2	2006	1964	G	$57,141	000-30-5484	LEVEL4	Y	1975	18827	1964	A
3	2005	1972	G	$53,028	000-31-1884	LEVEL4	Y	1979	29428	1969	H
4	2021	1982	U	$35,224	000-32-9247	LEVEL1	N	1981	58568	1981	K
5	1993	1952	G	$67,092	000-48-5750	LEVEL4	Y	1964	18827	1951	I
6	2008	1976	G	$47,419	000-50-5172	LEVEL3	Y	1980	58542	1974	I
7	2001	1963	U	$57,324	000-50-7905	LEVEL4	Y	1978	27086	1956	I
8	2015	1975	G	$47,965	000-51-3191	LEVEL3	Y	1978	34710	1972	B
9	1995	1961	U	$56,863	000-68-4769	LEVEL3	N	1968	21988	1959	D
10	1994	1974	U	$54,813	000-89-2873	LEVEL2	N	1973	21344	1950	G
11	2001	1956	U	$57,140	001-10-5078	LEVEL3	N	1967	39560	1959	D
12	1991	1957	G	$68,831	001-28-5561	LEVEL4	Y	1962	15016	1953	D
13	2003	1961	U	$48,616	001-32-1596	LEVEL3	Y	1974	37275	1959	B
14	2014	1972	G	$55,731	001-33-6894	LEVEL3	N	1977	39560	1971	F
15	1991	1946	U	$73,300	001-48-9290	LEVEL4	N	1960	31236	1957	D
16	2010	1966	U	$48,492	001-52-1583	LEVEL2	N	1968	65828	1965	G
17	2010	1974	G	$48,717	001-90-5383	LEVEL3	N	1979	29626	1967	B
18	1996	1955	G	$62,796	002-08-9053	LEVEL3	Y	1962	28595	1966	A
19	2006	1958	U	$53,368	002-10-3988	LEVEL3	Y	1976	36986	1963	D
20	2017	1978	G	$49,609	002-12-8028	LEVEL3	Y	1981	26971	1977	C
21	2008	1970	G	$52,965	002-31-7375	LEVEL4	Y	1980	53684	1967	I
22	2023	1981	U	$50,054	002-52-1161	LEVEL2	Y	1980	28595	1977	C
23	2004	1972	G	$67,905	002-70-0337	LEVEL3	Y	1971	39378	1963	E
24	1991	1961	G	$84,564	002-88-0317	LEVEL4	Y	1960	33159	1949	J
25	1996	1958	G	$62,916	002-88-9283	LEVEL4	Y	1970	39534	1957	F
26	2005	1963	G	$69,458	002-90-1491	LEVEL4	Y	1974	18825	1962	E
27	2011	1968	G	$53,028	002-90-8506	LEVEL4	Y	1979	18728	1966	I
28	1999	1970	U	$53,473	003-09-2866	LEVEL4	N	1981	39534	1969	D
29	2011	1973	G	$47,537	003-10-8517	LEVEL3	Y	1977	31302	1974	J
30	1990	1959	U	$54,744	003-27-6321	LEVEL3	N	1968	39518	1958	D
31	1994	1960	G	$66,061	003-28-4744	LEVEL4	Y	1969	21344	1952	G
32	2001	1970	G	$58,727	003-29-6001	LEVEL4	N	1975	16287	1969	D
33	2004	1964	G	$56,506	003-30-0076	LEVEL3	Y	1969	69096	1966	A
34	2003	1975	G	$71,987	003-31-3767	LEVEL4	Y	1974	8961	1958	I
35	1999	1967	U	$78,716	003-49-2565	LEVEL4	Y	1966	35509	1960	E
36	2007	1965	G	$53,028	003-50-4935	LEVEL4	Y	1979	13612	1965	H
37	2010	1971	G	$47,629	003-50-8089	LEVEL3	Y	1977	18752	1970	D
38	2011	1974	G	$47,639	003-50-8782	LEVEL3	N	1978	39534	1973	J
39	2011	1975	U	$43,944	003-51-2068	LEVEL2	N	1975	39560	1974	K
40	2003	1971	G	$49,489	003-70-5038	LEVEL3	N	1977	51927	1970	H
41	2014	1979	G	$39,275	003-71-3032	LEVEL2	Y	1979	39534	1979	D
42	2002	1968	U	$49,094	004-09-6000	LEVEL3	N	1977	31236	1974	D
43	2003	1968	G	$46,040	004-10-1704	LEVEL2	Y	1971	31236	1973	J
44	2007	1963	U	$47,857	004-10-4686	LEVEL3	Y	1980	18752	1959	H
45	2011	1970	G	$52,964	004-10-7993	LEVEL4	Y	1980	18687	1969	J
46	2009	1969	G	$50,018	004-10-8900	LEVEL3	Y	1976	21988	1970	D
47	2008	1970	U	$54,406	004-10-8991	LEVEL2	Y	1969	31236	1968	D
48	2013	1972	G	$52,214	004-31-3805	LEVEL3	Y	1977	33876	1971	F
49	2021	1982	U	$55,681	004-32-8747	LEVEL2	Y	1981	13612	1979	C
50	2007	1968	G	$57,267	004-50-4751	LEVEL4	Y	1977	69096	1967	F

OBS	RETIRE	HIRED	DEGREE	SALARY	ID	LEVEL	DEFER	LEVELYR	COLLEGE	YEAR	DEPART
51	2016	1974	G	$51,958	004-71-5969	LEVEL3	Y	1976	39542	1973	J
52	2021	1977	U	$38,215	004-73-1901	LEVEL1	N	1976	31236	1976	K
53	1990	1970	G	$73,195	004-89-6443	LEVEL4	Y	1969	21988	1957	H
54	2012	1976	G	$56,227	004-90-8503	LEVEL3	Y	1979	39534	1976	C
55	2006	1967	G	$64,481	004-91-7706	LEVEL4	Y	1974	35509	1966	I
56	2007	1976	G	$59,250	004-92-1051	LEVEL4	Y	1975	35509	1966	I
57	2006	1970	G	$62,217	005-10-1783	LEVEL3	Y	1974	25935	1969	E
58	2011	1973	G	$57,064	005-10-9282	LEVEL4	Y	1978	26971	1967	B
59	1995	1962	U	$52,099	005-28-5061	LEVEL2	Y	1965	38075	1965	C
60	2009	1972	G	$59,558	005-30-4674	LEVEL3	Y	1974	18827	1971	C
61	1991	1956	U	$57,409	005-47-3548	LEVEL3	N	1962	31236	1973	C
62	1995	1954	U	$55,036	005-48-0282	LEVEL3	Y	1972	39560	1955	B
63	1993	1967	G	$65,179	005-48-0855	LEVEL4	Y	1973	48882	1962	H
64	2004	1970	G	$53,432	005-50-0116	LEVEL3	Y	1977	39534	1969	E
65	2012	1971	G	$51,794	005-51-2880	LEVEL4	Y	1981	39560	1969	B
66	2013	1969	G	$52,209	005-51-3043	LEVEL3	Y	1973	39542	1968	I
67	2013	1977	G	$50,393	005-51-3051	LEVEL3	Y	1976	35509	1969	I
68	1995	1949	G	$60,805	005-68-0783	LEVEL4	Y	1968	39518	1967	A
69	2010	1978	G	$67,013	005-70-8336	LEVEL4	Y	1979	25935	1968	C
70	2019	1977	G	$42,469	005-72-5779	LEVEL2	Y	1978	31195	1978	J
71	1995	1963	G	$79,228	005-90-0006	LEVEL4	Y	1962	31302	1951	H
72	2010	1981	G	$55,601	005-91-5906	LEVEL3	Y	1980	18752	1971	I
73	2023	1981	U	$41,463	005-92-0601	LEVEL2	Y	1980	18736	1980	F
74	2012	1969	G	$53,028	005-92-4170	LEVEL4	Y	1979	25848	1968	I
75	2023	1982	G	$42,161	005-93-3579	LEVEL2	Y	1981	69137	1981	B
76	1989	1949	G	$71,035	006-05-3917	LEVEL4	Y	1967	33159	1961	C
77	1984	1970	U	$50,424	006-08-4692	LEVEL2	Y	1969	31236	1965	D
78	2000	1962	G	$60,119	006-08-8274	LEVEL4	Y	1971	18728	1957	I
79	2000	1971	G	$47,287	006-09-1980	LEVEL3	N	1979	40237	1971	D
80	1990	1978	U	$74,265	006-10-0955	LEVEL4	N	1977	21988	1963	K
81	2013	1972	G	$52,334	006-11-3100	LEVEL3	Y	1976	58511	1972	A
82	2021	1978	U	$33,447	006-13-2950	LEVEL1	N	1977	39560	1977	A
83	2014	1972	G	$48,813	006-14-0771	LEVEL3	Y	1977	34710	1972	H
84	2004	1955	G	$74,398	006-27-6372	LEVEL4	Y	1964	18827	1953	H
85	2003	1966	U	$47,913	006-29-6772	LEVEL3	Y	1976	10535	1963	D
86	1987	1966	U	$56,192	006-46-8153	LEVEL4	Y	1980	.	.	J
87	1997	1967	U	$59,739	006-47-5924	LEVEL3	Y	1970	31236	1965	D
88	2009	1967	G	$58,423	006-50-4198	LEVEL4	Y	1975	31236	1964	B
89	2008	1972	G	$46,513	006-50-4882	LEVEL3	Y	1979	39534	1970	D
90	2012	1968	G	$52,322	006-50-9463	LEVEL3	Y	1975	31195	1971	J
91	1996	1966	G	$71,858	006-67-6778	LEVEL4	Y	1965	18752	1965	I
92	1992	1970	U	$47,285	006-69-3682	LEVEL2	Y	1975	.	.	A
93	2009	1967	G	$56,276	006-70-5731	LEVEL4	Y	1975	66652	1965	B
94	1990	1962	U	$55,761	006-85-8188	LEVEL4	N	1976	31236	1960	K
95	1987	1953	U	$57,148	006-86-7586	LEVEL4	Y	1975	29420	1951	D
96	1996	1968	U	$70,893	006-87-3087	LEVEL4	Y	1967	47405	1952	E
97	2005	1969	G	$66,356	006-89-7811	LEVEL4	Y	1973	15016	1968	C
98	2011	1969	G	$48,062	006-90-8276	LEVEL3	Y	1976	51513	1968	I
99	1988	1970	U	$64,246	007-05-4267	LEVEL4	N	1972	31236	1955	H
100	1993	1963	G	$57,892	007-08-0597	LEVEL4	Y	1976	39560	1964	B

OBS	RETIRED	HIRED	DEGREE	SALARY	ID	LEVEL	DEFER	LEVELYR	COLLEGE	YEAR	DEPART
101	1990	1966	G	$56,821	007-08-5416	LEVEL4	N	1976	39518	1968	D
102	1983	1943	G	$69,024	007-08-9016	LEVEL4	Y	1959	21344	1943	B
103	2000	1970	G	$57,971	007-08-9555	LEVEL4	Y	1976	13612	1963	J
104	2009	1966	G	$53,229	007-09-6759	LEVEL4	Y	1979	39518	1967	J
105	2005	1982	U	$42,161	007-09-6797	LEVEL2	Y	1981	58511	1973	K
106	2005	1970	G	$55,807	007-09-7147	LEVEL4	Y	1979	48692	1964	B
107	2005	1973	G	$47,629	007-09-7183	LEVEL3	N	1977	26971	1972	H
108	2009	1968	G	$43,263	007-10-1780	LEVEL2	N	1972	39518	1978	H
109	2019	1977	U	$42,373	007-10-4182	LEVEL2	N	1976	31236	1970	K
110	2000	1960	G	$56,466	007-10-5683	LEVEL3	Y	1966	28595	1961	J
111	1990	1954	G	$60,598	007-10-8780	LEVEL4	Y	1969	28133	1953	I
112	2018	1977	G	$52,045	007-11-2253	LEVEL2	Y	1976	31302	1975	I
113	2019	1980	U	$34,839	007-11-7093	LEVEL1	N	1979	39560	1977	K
114	2022	1980	G	$40,753	007-12-1619	LEVEL2	Y	1979	48692	1979	A
115	2023	1982	U	$33,712	007-12-3944	LEVEL1	N	1981	29420	1981	K
116	2016	1981	G	$48,147	007-12-4537	LEVEL2	Y	1980	39560	1976	I
117	1990	1962	U	$66,864	007-25-0425	LEVEL3	Y	1962	29420	1950	D
118	1990	1954	G	$63,596	007-26-7471	LEVEL4	Y	1969	39534	1954	I
119	1994	1962	U	$51,700	007-28-0649	LEVEL3	Y	1973	31236	1966	G
120	1992	1957	U	$57,782	007-28-0939	LEVEL3	N	1978	39560	1960	A
121	1994	1966	U	$60,123	007-28-1340	LEVEL3	Y	1967	31236	1958	G
122	1993	1954	G	$68,676	007-28-1636	LEVEL4	Y	1967	34355	1959	J
123	1993	1954	G	$65,835	007-28-3872	LEVEL4	Y	1969	18827	1953	B
124	1993	1962	U	$49,168	007-28-4434	LEVEL2	N	1963	40237	1961	K
125	1995	1973	U	$50,632	007-28-4486	LEVEL2	N	1972	37604	1958	D
126	1989	1975	U	$42,236	007-28-8711	LEVEL1	Y	1974	39560	1941	A
127	2002	1960	U	$50,417	007-29-1915	LEVEL2	N	1963	39560	1960	G
128	2003	1968	U	$51,047	007-29-2184	LEVEL3	Y	1974	39560	1959	A
129	2005	1979	G	$40,162	007-29-2629	LEVEL2	N	1978	31302	1975	H
130	2006	1971	G	$59,766	007-29-2839	LEVEL4	Y	1976	18827	1972	F
131	2004	1971	G	$49,625	007-29-3183	LEVEL2	Y	1971	69096	1971	J
132	2006	1966	U	$55,741	007-29-7192	LEVEL3	Y	1970	29420	1966	D
133	2007	1970	U	$53,091	007-29-7257	LEVEL4	Y	1980	35475	1960	D
134	2011	1972	G	$60,315	007-30-1373	LEVEL4	Y	1979	58485	1970	C
135	2013	1970	G	$58,751	007-30-4564	LEVEL4	Y	1981	39534	1971	C
136	2014	1972	U	$45,381	007-30-5100	LEVEL3	N	1979	39560	1971	D
137	2014	1970	U	$41,736	007-30-5703	LEVEL2	Y	1977	39560	1977	A
138	2015	1972	U	$43,473	007-30-5779	LEVEL2	N	1974	39560	1974	K
139	2019	1978	U	$39,385	007-30-8963	LEVEL1	Y	1977	39560	1973	D
140	1992	1964	U	$50,939	007-30-9563	LEVEL3	N	1972	69096	1946	D
141	2017	1974	U	$43,246	007-31-2801	LEVEL2	N	1975	31236	1972	K
142	2019	1978	G	$42,099	007-31-6338	LEVEL2	Y	1977	58526	1975	G
143	1997	1967	G	$66,563	007-31-6725	LEVEL4	Y	1971	363	1964	E
144	2021	1979	G	$43,730	007-34-5713	LEVEL2	Y	1978	10191	1977	B
145	2013	1981	G	$45,688	007-34-8031	LEVEL2	Y	1980	363	1978	I
146	1995	1952	G	$53,086	007-47-2164	LEVEL4	Y	1978	28133	1963	B
147	1995	1967	G	$72,482	007-47-2490	LEVEL4	Y	1971	41120	1966	E
148	1996	1965	G	$62,233	007-47-6306	LEVEL4	Y	1971	63427	1962	B
149	1990	1948	U	$60,914	007-48-0499	LEVEL3	Y	1957	29420	1947	C
150	1996	1957	U	$47,993	007-48-3873	LEVEL3	Y	1975	21344	1954	B

OBS	RETIRE	HIRED	DEGREE	SALARY	ID	LEVEL	DEFER	LEVELYR	COLLEGE	YEAR	DEPART
151	1996	1951	U	$61,902	007-48-5571	LEVEL4	Y	1979	39560	1950	E
152	1997	1962	U	$49,408	007-48-7912	LEVEL3	N	1978	39560	1961	G
153	1998	1960	U	$50,965	007-48-8802	LEVEL3	Y	1975	39560	1961	I
154	2003	1976	U	$40,612	007-48-9029	LEVEL2	N	1978	31195	1975	K
155	1987	1966	U	$58,453	007-48-9054	LEVEL3	N	1965	31236	1957	K
156	2001	1972	U	$59,012	007-48-9237	LEVEL3	N	1971	31236	1978	K
157	2000	1970	G	$44,472	007-48-9727	LEVEL2	Y	1977	31302	1977	J
158	2004	1977	U	$36,525	007-49-2803	LEVEL2	N	1980	39560	1976	K
159	2005	1974	U	$45,457	007-49-2988	LEVEL2	N	1975	31236	1960	K
160	2008	1964	U	$43,493	007-49-6805	LEVEL2	N	1970	39560	1964	K
161	2012	1972	G	$47,872	007-50-0491	LEVEL2	Y	1971	69187	1971	D
162	2011	1978	G	$45,945	007-50-0587	LEVEL2	N	1977	48882	1977	J
163	2011	1974	G	$47,091	007-50-0859	LEVEL3	N	1980	31236	1977	H
164	2011	1968	U	$46,202	007-50-1310	LEVEL2	Y	1971	39560	1972	I
165	1989	1942	U	$50,676	007-50-1534	LEVEL2	Y	1950	39560	1953	B
166	1994	1951	U	$49,278	007-50-1537	LEVEL3	N	1976	39560	1955	A
167	1990	1956	U	$55,685	007-50-1538	LEVEL3	Y	1964	21344	1953	B
168	1992	1955	G	$78,327	007-50-1582	LEVEL4	Y	1957	69096	1954	A
169	1993	1956	G	$64,381	007-50-1608	LEVEL4	Y	1966	40237	1955	B
170	2012	1971	G	$59,367	007-50-1615	LEVEL4	Y	1980	18728	1970	C
171	2002	1966	G	$55,580	007-51-3635	LEVEL4	Y	1975	363	1960	B
172	1997	1965	G	$58,230	007-51-3831	LEVEL4	Y	1980	363	1957	I
173	2020	1979	U	$37,293	007-51-6309	LEVEL2	N	1981	31195	1978	K
174	2021	1977	U	$38,173	007-51-7486	LEVEL2	N	1980	31236	1976	K
175	2001	1971	G	$61,416	007-52-4756	LEVEL4	Y	1978	363	1969	E
176	1991	1950	U	$61,456	007-66-2699	LEVEL4	Y	1973	39560	1954	B
177	1994	1957	G	$69,904	007-67-2536	LEVEL4	Y	1967	21344	1952	I
178	1995	1960	U	$50,241	007-67-2539	LEVEL3	Y	1976	31236	1976	D
179	1995	1966	U	$52,634	007-67-2685	LEVEL3	Y	1976	31236	1976	D
180	1997	1971	G	$68,169	007-67-2824	LEVEL4	Y	1970	28133	1955	B
181	1995	1959	U	$52,749	007-67-7844	LEVEL4	Y	1981	39560	1961	I
182	1996	1955	G	$67,624	007-68-0740	LEVEL4	Y	1965	28133	1954	J
183	2000	1966	G	$47,305	007-68-4441	LEVEL3	N	1979	31286	1979	H
184	2004	1960	U	$53,376	007-69-2509	LEVEL3	Y	1973	37275	1957	I
185	2010	1977	U	$47,257	007-69-6508	LEVEL3	Y	1981	52811	1964	C
186	2009	1964	G	$51,827	007-69-6655	LEVEL3	Y	1975	26327	1972	I
187	2009	1969	G	$55,058	007-69-7836	LEVEL4	Y	1978	39518	1967	I
188	2009	1971	G	$51,224	007-70-0488	LEVEL3	Y	1976	69121	1966	B
189	2019	1971	U	$42,479	007-70-1518	LEVEL2	N	1980	39560	1967	G
190	2012	1973	U	$50,657	007-70-1634	LEVEL3	Y	1976	29428	1970	D
191	2013	1978	U	$35,697	007-70-1810	LEVEL2	N	1981	39560	1977	K
192	2000	1961	G	$62,710	007-70-5054	LEVEL4	Y	1973	18752	1961	I
193	2015	1974	U	$42,302	007-70-8830	LEVEL2	N	1976	39560	1975	K
194	2017	1974	G	$61,902	007-70-9713	LEVEL3	Y	1977	39560	1972	E
195	2005	1970	G	$65,257	007-72-1273	LEVEL4	Y	1977	363	1962	F
196	2015	1972	G	$52,964	007-72-5724	LEVEL4	Y	1980	363	1968	H
197	1988	1946	U	$72,042	007-85-1185	LEVEL4	Y	1968	35623	1960	D
198	1988	1948	U	$54,504	007-85-1838	LEVEL3	Y	1961	39560	1948	J
199	1990	1950	U	$67,709	007-86-2300	LEVEL4	Y	1968	34710	1949	E
200	1998	1966	U	$51,284	007-87-1869	LEVEL3	Y	1974	16013	1950	H

OBS	RETIRE	HIRED	DEGREE	SALARY	ID	LEVEL	DEFER	LEVELYR	COLLEGE	YEAR	DEPART
201	1992	1969	G	$55,242	007-87-2013	LEVEL3	N	1974	39518	1966	D
202	1994	1968	G	$51,058	007-87-2441	LEVEL3	N	1973	31195	1965	D
203	2001	1967	G	$63,925	007-88-1479	LEVEL4	Y	1973	31236	1964	D
204	2002	1963	U	$64,921	007-88-8732	LEVEL3	Y	1967	25935	1964	E
205	1986	1947	G	$64,322	007-88-8838	LEVEL4	Y	1968	31302	1955	H
206	2003	1962	G	$59,198	007-89-2484	LEVEL4	Y	1973	39560	1962	F
207	2009	1972	G	$62,913	007-89-6645	LEVEL4	Y	1980	39560	1966	E
208	2010	1972	G	$52,710	007-89-7367	LEVEL4	Y	1981	29618	1969	G
209	1993	1960	U	$45,302	007-90-1383	LEVEL2	N	1967	39560	1948	G
210	2016	1978	G	$41,249	007-90-4752	LEVEL2	Y	1977	21344	1977	B
211	2019	1977	G	$44,587	007-90-8131	LEVEL3	Y	1981	39560	1977	B
212	2009	1968	G	$58,364	007-91-6855	LEVEL4	Y	1981	363	1963	I
213	2009	1975	G	$52,965	007-93-2175	LEVEL4	Y	1980	363	1970	I
214	1989	1982	U	$37,936	008-08-4514	LEVEL2	N	1981	24322	1981	D
215	1996	1965	G	$77,278	008-08-8365	LEVEL4	Y	1964	16287	1959	E
216	2008	1970	G	$61,777	008-10-9263	LEVEL3	Y	1974	25935	1967	E
217	2011	1971	G	$50,368	008-11-6566	LEVEL3	Y	1976	52553	1971	D
218	2015	1981	U	$44,984	008-12-0771	LEVEL2	Y	1980	29618	1980	F
219	2011	1970	G	$54,245	008-12-9602	LEVEL3	Y	1976	48692	1969	A
220	1991	1977	G	$73,497	008-29-6082	LEVEL4	Y	1976	39518	1965	E
221	2008	1970	G	$56,474	008-30-8196	LEVEL4	Y	1975	69096	1968	J
222	2008	1978	G	$53,734	008-50-8649	LEVEL3	Y	1977	16673	1967	B
223	2009	1974	G	$63,969	008-51-7686	LEVEL4	Y	1980	15016	1968	C
224	1997	1963	G	$62,227	008-68-9571	LEVEL4	Y	1969	48908	1956	D
225	2009	1970	G	$47,465	008-70-8982	LEVEL2	Y	1973	69096	1973	J
226	2009	1968	G	$54,152	008-71-3740	LEVEL4	Y	1977	18825	1964	H
227	1990	1950	U	$57,693	008-87-3058	LEVEL3	Y	1961	39560	1954	E
228	1990	1955	G	$73,974	008-87-3452	LEVEL4	Y	1964	26971	1951	I
229	1992	1958	G	$72,345	008-88-1191	LEVEL4	Y	1965	16673	1954	F
230	1997	1954	U	$55,689	008-88-9021	LEVEL4	Y	1974	39560	1957	A
231	1999	1970	G	$60,552	008-89-3191	LEVEL4	Y	1980	69096	1968	D
232	2007	1971	G	$53,028	008-90-8248	LEVEL4	Y	1979	69096	1966	I
233	2009	1968	G	$49,471	008-90-9818	LEVEL3	Y	1974	13612	1968	B
234	2011	1972	G	$64,975	008-91-3034	LEVEL4	Y	1981	39534	1971	C
235	2011	1975	G	$41,507	008-91-3477	LEVEL2	N	1976	16295	1976	D
236	2016	1973	G	$45,670	008-92-5408	LEVEL3	Y	1978	39534	1973	A
237	1999	1959	U	$54,704	009-10-5061	LEVEL4	Y	1980	51927	1965	A
238	2007	1964	G	$61,336	009-10-8632	LEVEL3	Y	1973	21344	1970	J
239	2001	1960	G	$66,640	009-29-6201	LEVEL4	Y	1970	363	1967	I
240	2010	1970	U	$54,967	009-30-5188	LEVEL3	Y	1975	37992	1969	D
241	2011	1970	G	$57,176	009-30-9394	LEVEL3	N	1980	39560	1967	I
242	1995	1960	G	$52,557	009-47-7157	LEVEL3	Y	1968	69153	1962	I
243	2006	1977	G	$72,570	009-50-4884	LEVEL4	Y	1976	363	1970	D
244	2010	1976	G	$45,116	009-50-8732	LEVEL2	N	1975	31302	1975	J
245	2012	1975	G	$61,911	009-51-2194	LEVEL3	Y	1977	26971	1969	E
246	2000	1959	G	$72,277	009-69-3337	LEVEL4	N	1970	26971	1961	H
247	2018	1981	G	$48,147	009-92-5128	LEVEL2	Y	1980	35509	1977	I
248	2000	1970	G	$56,461	010-29-6158	LEVEL4	Y	1975	21988	1962	J
249	2011	1970	G	$51,794	010-31-6777	LEVEL4	Y	1981	31302	1969	I
250	2017	1979	G	$41,611	010-32-5641	LEVEL2	Y	1978	18687	1977	H

OBS	RETIRE	HIRED	DEGREE	SALARY	ID	LEVEL	DEFER	LEVELYR	COLLEGE	YEAR	DEPART
251	1991	1967	G	$64,054	010-68-5026	LEVEL4	Y	1973	27086	1950	D
252	2009	1980	G	$40,162	010-71-3009	LEVEL2	N	1979	31302	1979	D
253	2015	1978	G	$45,612	010-72-0258	LEVEL3	Y	1979	21988	1973	J
254	2008	1964	U	$53,882	010-91-2932	LEVEL4	Y	1979	31236	1970	D
255	2012	1970	G	$45,046	011-31-6279	LEVEL2	Y	1974	48908	1974	H
256	2000	1969	G	$62,810	011-49-6744	LEVEL4	Y	1975	34710	1957	B
257	2013	1967	U	$50,816	011-71-6896	LEVEL3	Y	1977	48915	1966	H
258	2021	1976	U	$45,051	011-73-6324	LEVEL3	N	1980	21988	1975	D
259	1988	1961	U	$58,996	011-89-6759	LEVEL4	Y	1973	29585	1954	H
260	1997	1963	G	$58,211	012-30-0627	LEVEL4	N	1974	21988	1959	D
261	2012	1972	G	$53,027	012-51-7604	LEVEL4	Y	1979	28595	1971	D
262	2017	1974	G	$46,130	012-52-4943	LEVEL3	Y	1978	35509	1972	I
263	2019	1979	G	$46,224	012-53-2584	LEVEL2	Y	1978	28595	1979	J
264	2019	1977	G	$48,992	012-72-7917	LEVEL2	Y	1976	25848	1976	E
265	2012	1982	G	$57,653	012-91-7623	LEVEL3	Y	1981	21988	1972	I
266	2008	1969	U	$49,043	013-51-5986	LEVEL3	Y	1974	18752	1961	D
267	2012	1973	G	$57,603	013-71-2940	LEVEL4	Y	1981	18778	1972	E
268	1996	1962	G	$65,206	014-09-6130	LEVEL4	Y	1968	9719	1956	B
269	2002	1967	G	$63,286	014-09-6340	LEVEL4	Y	1971	28133	1963	F
270	2010	1977	G	$52,981	014-12-5555	LEVEL3	Y	1981	39518	1976	F
271	2001	1971	G	$54,022	014-31-6932	LEVEL3	Y	1975	21344	1966	E
272	2012	1970	G	$57,889	014-91-7118	LEVEL3	Y	1975	31302	1969	E
273	2017	1980	G	$50,809	015-32-8889	LEVEL2	Y	1979	39534	1977	E
274	1999	1968	G	$50,268	015-49-6683	LEVEL3	N	1977	58568	1967	H
275	2004	1971	G	$63,785	015-50-0543	LEVEL4	Y	1975	28595	1969	C
276	2007	1966	U	$45,261	015-71-2070	LEVEL2	N	1969	21988	1968	D
277	2020	1981	G	$43,801	015-73-6590	LEVEL2	Y	1980	39560	1980	F
278	2013	1968	U	$45,041	016-11-2827	LEVEL3	Y	1981	13612	1966	D
279	2006	1970	G	$61,914	016-90-9835	LEVEL4	Y	1980	29626	1963	E
280	2003	1969	G	$70,145	017-09-7067	LEVEL4	Y	1976	39518	1968	E
281	2010	1966	U	$48,112	017-32-5268	LEVEL3	Y	1975	22606	1963	D
282	2008	1971	G	$63,452	017-50-8058	LEVEL4	Y	1977	69187	1966	D
283	2004	1967	G	$60,826	017-70-5680	LEVEL4	Y	1973	26327	1966	J
284	1999	1976	G	$72,573	017-90-0444	LEVEL4	Y	1975	47405	1971	C
285	2018	1976	G	$55,785	017-92-5724	LEVEL3	Y	1978	39534	1975	C
286	2019	1980	G	$47,260	018-13-2538	LEVEL2	Y	1979	48710	1980	C
287	2001	1972	G	$76,467	018-32-4857	LEVEL4	Y	1971	21344	1967	C
288	2003	1964	G	$53,067	018-50-3973	LEVEL4	Y	1979	34710	1966	H
289	1990	1960	G	$66,232	018-65-4986	LEVEL4	Y	1960	16673	1958	E
290	2001	1957	G	$62,463	018-70-0591	LEVEL4	Y	1970	40237	1959	H
291	2007	1969	U	$48,567	019-10-8442	LEVEL3	Y	1976	26971	1968	D
292	2012	1979	U	$45,670	019-11-7433	LEVEL2	Y	1978	26327	1965	C
293	1999	1965	U	$50,931	019-49-1999	LEVEL2	N	1964	25773	1959	G
294	2007	1963	U	$51,955	019-50-9848	LEVEL3	N	1973	48915	1962	D
295	2011	1972	G	$51,794	019-51-3574	LEVEL4	Y	1981	53684	1971	D
296	1996	1964	G	$67,045	019-51-7419	LEVEL4	Y	1979	21344	1960	A
297	2017	1974	U	$43,889	019-52-0424	LEVEL2	N	1976	39560	1975	K
298	2007	1972	G	$48,933	019-52-8477	LEVEL3	Y	1976	27044	1971	I
299	2002	1966	G	$55,884	019-69-6270	LEVEL4	Y	1976	58511	1965	H
300	2013	1973	G	$56,357	019-91-6501	LEVEL3	Y	1977	39534	1975	C

OBS	RETIRE	HIRED	DEGREE	SALARY	ID	LEVEL	DEFER	LEVELYR	COLLEGE	YEAR	DEPART
301	1999	1968	G	$62,633	020-09-1890	LEVEL4	Y	1974	63427	1967	I
302	2003	1968	U	$79,968	020-10-5214	LEVEL4	Y	1967	35509	1960	E
303	2013	1970	G	$54,429	020-11-7061	LEVEL3	Y	1975	28595	1969	I
304	2012	1970	G	$47,431	020-11-7232	LEVEL2	N	1973	16287	1976	G
305	2002	1966	G	$48,786	020-29-6042	LEVEL3	Y	1974	18825	1968	H
306	2002	1965	G	$64,201	020-30-0040	LEVEL4	Y	1976	21988	1966	J
307	2010	1976	G	$55,923	020-30-9663	LEVEL3	N	1975	34710	1973	J
308	1997	1970	G	$59,680	020-68-9103	LEVEL3	Y	1969	40237	1953	I
309	2002	1965	G	$52,881	020-89-5860	LEVEL3	Y	1973	31236	1967	B
310	2006	1974	G	$48,580	020-90-8256	LEVEL3	N	1978	363	1967	B
311	2011	1972	G	$70,907	020-91-3513	LEVEL4	Y	1974	26327	1967	C
312	2015	1977	G	$41,372	021-32-0142	LEVEL2	N	1976	20857	1977	D
313	2021	1978	G	$48,464	021-93-2158	LEVEL2	Y	1977	16287	1979	C
314	2017	1973	G	$46,730	022-12-0218	LEVEL3	Y	1978	21344	1973	J
315	2021	1978	G	$41,494	022-13-3419	LEVEL2	N	1977	58956	1979	G
316	1995	1959	G	$75,115	022-28-4042	LEVEL4	Y	1966	18687	1957	J
317	1994	1964	G	$62,367	022-48-1663	LEVEL4	Y	1972	29626	1955	B
318	1985	1960	G	$71,851	023-30-8984	LEVEL4	Y	1959	58485	1941	I
319	2006	1978	G	$55,829	023-31-2137	LEVEL3	Y	1977	47300	1969	I
320	1999	1972	U	$53,667	023-48-8573	LEVEL3	Y	1976	21988	1955	J
321	2007	1968	G	$51,090	023-50-1641	LEVEL3	Y	1976	27044	1966	E
322	2011	1973	G	$59,335	023-51-2711	LEVEL4	Y	1978	16673	1967	G
323	2004	1965	G	$73,432	023-70-0355	LEVEL4	Y	1970	39518	1963	E
324	2007	1970	G	$58,781	023-72-8625	LEVEL4	Y	1975	26327	1968	I
325	1998	1969	G	$54,613	023-89-5878	LEVEL4	Y	1977	11418	1968	J
326	1986	1949	G	$74,664	024-05-9191	LEVEL4	Y	1959	21988	1958	J
327	1998	1966	G	$49,155	024-08-8321	LEVEL2	Y	1969	28595	1969	J
328	2002	1965	G	$53,028	024-09-6284	LEVEL4	Y	1979	58526	1967	I
329	2002	1963	G	$60,217	024-49-7618	LEVEL4	Y	1973	21344	1957	B
330	2002	1966	G	$54,620	024-50-4380	LEVEL4	Y	1977	53684	1965	I
331	2016	1981	G	$43,711	024-53-7031	LEVEL2	Y	1980	21988	1973	I
332	2014	1971	G	$63,574	024-71-6033	LEVEL4	Y	1981	39518	1970	E
333	1999	1965	G	$75,502	025-08-5221	LEVEL4	Y	1964	35509	1956	I
334	2011	1969	U	$42,297	025-11-1905	LEVEL2	N	1976	26971	1966	H
335	2021	1978	G	$54,561	025-14-1525	LEVEL3	Y	1981	21475	1977	E
336	2015	1972	G	$48,974	025-32-1134	LEVEL3	N	1976	29618	1973	H
337	2007	1974	U	$45,380	025-91-2253	LEVEL3	N	1979	39534	1970	A
338	2005	1981	G	$62,357	026-10-1552	LEVEL4	Y	1980	21988	1968	H
339	2020	1978	G	$45,761	026-32-9641	LEVEL2	Y	1977	11418	1976	H
340	2014	1973	G	$44,120	026-71-6148	LEVEL2	Y	1976	39560	1976	A
341	2005	1970	U	$49,633	026-90-5192	LEVEL3	Y	1976	35623	1972	D
342	2013	1968	U	$42,966	026-91-7557	LEVEL2	N	1972	39378	1967	H
343	2005	1962	G	$57,611	027-10-0227	LEVEL4	Y	1974	21344	1961	B
344	1999	1957	U	$50,873	027-29-7524	LEVEL2	Y	1956	22952	1955	I
345	2021	1980	G	$39,275	027-33-2063	LEVEL2	N	1979	48692	1979	H
346	2021	1982	G	$48,921	027-34-0847	LEVEL2	Y	1981	26971	1981	E
347	2003	1967	G	$66,437	027-50-8309	LEVEL4	Y	1969	28133	1960	J
348	1996	1970	U	$58,118	027-71-2045	LEVEL2	Y	1974	363	1949	D
349	2008	1978	U	$55,235	027-90-7914	LEVEL2	N	1977	29420	1976	K
350	2007	1968	G	$48,191	028-30-5237	LEVEL3	Y	1977	69080	1970	D

OBS	RETIRE	HIRED	DEGREE	SALARY	ID	LEVEL	DEFER	LEVELYR	COLLEGE	YEAR	DEPART
351	1993	1970	G	$64,240	028-31-6903	LEVEL4	Y	1976	363	1969	E
352	2013	1981	U	$39,914	028-31-7302	LEVEL1	Y	1980	58485	1975	J
353	2009	1976	G	$45,051	028-51-3001	LEVEL3	Y	1980	21988	1975	D
354	2019	1978	G	$41,229	028-52-9659	LEVEL2	Y	1977	31302	1977	D
355	1995	1958	G	$77,949	028-68-4166	LEVEL4	Y	1962	29626	1958	H
356	1990	1960	G	$92,011	028-87-1940	LEVEL4	Y	1959	29626	1956	E
357	1995	1959	G	$62,470	028-88-1041	LEVEL4	Y	1970	69096	1959	H
358	2017	1975	U	$53,682	028-92-0962	LEVEL3	Y	1979	48775	1978	C
359	2018	1981	G	$49,209	028-92-8480	LEVEL2	Y	1980	58956	1980	C
360	1993	1958	U	$51,058	029-08-1574	LEVEL3	Y	1973	16295	1951	D
361	2001	1967	U	$51,931	029-29-7828	LEVEL3	N	1975	26327	1962	D
362	2000	1970	G	$66,530	029-32-3927	LEVEL4	Y	1976	16673	1963	E
363	2017	1977	G	$45,740	029-32-4013	LEVEL2	Y	1976	15016	1975	B
364	2021	1982	G	$42,161	029-33-5870	LEVEL2	Y	1981	13612	1979	J
365	2021	1981	G	$51,744	029-52-9722	LEVEL2	Y	1980	52842	1980	E
366	2005	1967	G	$59,514	029-70-1697	LEVEL4	Y	1981	48733	1967	E
367	2009	1968	G	$53,454	029-70-4630	LEVEL4	Y	1978	19348	1968	H
368	2017	1977	G	$44,443	029-72-8829	LEVEL3	Y	1981	58568	1977	D
369	2001	1966	G	$52,525	029-90-1071	LEVEL4	Y	1981	37275	1965	A
370	2011	1972	G	$46,910	030-11-2747	LEVEL3	Y	1978	26327	1973	D
371	2013	1982	U	$37,092	030-19-5923	LEVEL2	N	1981	48733	1970	K
372	1995	1950	G	$76,744	030-27-5866	LEVEL4	Y	1962	26971	1956	E
373	2000	1963	G	$55,621	030-29-3003	LEVEL4	Y	1977	39560	1968	B
374	1999	1960	G	$59,943	030-89-2368	LEVEL4	Y	1973	28133	1959	I
375	1989	1966	U	$68,032	031-25-1796	LEVEL4	Y	1976	40237	1954	J
376	1987	1954	U	$59,281	031-25-4784	LEVEL3	Y	1956	21988	1946	D
377	1997	1981	G	$43,851	031-28-4828	LEVEL2	N	1980	18752	1969	H
378	2001	1982	U	$54,836	031-29-7601	LEVEL2	Y	1981	29428	1979	G
379	2006	1970	G	$47,882	031-30-1629	LEVEL3	Y	1978	18825	1972	D
380	2019	1978	G	$39,485	031-33-2738	LEVEL2	N	1977	47405	1976	D
381	2003	1965	G	$52,467	031-49-6771	LEVEL3	Y	1976	28133	1969	D
382	2013	1974	G	$47,120	031-71-6471	LEVEL3	Y	1978	31195	1974	J
383	2007	1966	G	$56,716	032-10-9554	LEVEL4	Y	1974	29626	1965	J
384	2019	1975	U	$48,724	032-32-5854	LEVEL3	Y	1980	26995	1974	F
385	2005	1971	G	$53,517	032-70-8596	LEVEL4	Y	1980	26971	1969	A
386	2016	1979	G	$44,454	032-92-0122	LEVEL2	Y	1978	39560	1975	I
387	2015	1974	G	$52,695	032-92-1157	LEVEL3	Y	1978	48684	1971	B
388	2005	1979	U	$42,533	032-93-7848	LEVEL2	N	1978	29428	1976	D
389	2003	1973	U	$60,094	033-49-2180	LEVEL4	Y	1975	18728	1965	D
390	2004	1966	U	$57,246	033-70-1151	LEVEL4	Y	1974	364	1960	D
391	2014	1975	U	$46,348	033-72-0688	LEVEL2	N	1974	22311	1973	D
392	2016	1974	G	$51,730	034-52-1507	LEVEL3	Y	1979	63709	1973	I
393	2012	1967	U	$48,736	035-11-7493	LEVEL3	N	1975	29428	1966	D
394	2018	1975	G	$47,068	035-32-8758	LEVEL3	Y	1978	34710	1973	J
395	2017	1978	U	$43,929	035-52-5342	LEVEL2	N	1977	52232	1970	K
396	1994	1954	G	$68,184	036-68-4896	LEVEL4	Y	1966	51927	1953	B
397	2009	1977	G	$55,127	036-70-5026	LEVEL4	Y	1978	58526	1968	J
398	1996	1952	U	$64,248	037-08-8971	LEVEL4	Y	1977	34710	1961	E
399	2007	1969	G	$64,731	037-12-0872	LEVEL4	Y	1979	18728	1967	E
400	2007	1977	G	$51,794	037-33-7330	LEVEL4	N	1981	363	1968	I

OBS	RETIRED	HIRED	DEGREE	SALARY	ID	LEVEL	DEFER	LEVELYR	COLLEGE	YEAR	DEPART
401	1992	1967	G	$64,802	037-48-0878	LEVEL4	Y	1970	26971	1954	A
402	1994	1966	U	$57,895	037-48-9127	LEVEL4	Y	1976	29634	1965	D
403	2002	1963	G	$61,833	037-91-2233	LEVEL4	Y	1974	363	1960	I
404	2000	1977	U	$47,860	038-29-6102	LEVEL3	N	1976	29420	1975	A
405	2007	1965	G	$52,602	038-31-3610	LEVEL3	Y	1973	26971	1963	I
406	1986	1947	U	$62,053	038-88-1601	LEVEL4	Y	1974	40237	1958	E
407	2012	1969	G	$49,844	039-11-3595	LEVEL3	Y	1979	15157	1968	D
408	2017	1978	U	$39,385	039-12-1836	LEVEL2	N	1977	18794	1973	D
409	2019	1979	U	$50,838	039-32-9663	LEVEL2	Y	1978	26971	1973	C
410	1995	1981	G	$73,156	039-48-0138	LEVEL4	Y	1980	21475	1960	E
411	2008	1980	U	$39,275	039-51-6553	LEVEL2	N	1979	29626	1966	K
412	2019	1980	G	$42,971	039-52-8122	LEVEL2	Y	1979	35509	1978	I
413	1989	1976	G	$77,873	039-70-1059	LEVEL4	Y	1975	27086	1949	B
414	1987	1965	U	$61,820	040-11-1957	LEVEL4	Y	1974	26971	1963	A
415	2021	1981	G	$40,759	040-13-3305	LEVEL2	N	1980	47300	1979	I
416	2016	1974	G	$54,062	040-31-7239	LEVEL3	Y	1977	26971	1973	C
417	2015	1973	G	$48,871	040-71-7293	LEVEL3	Y	1977	29618	1972	H
418	2014	1970	G	$47,497	040-71-7572	LEVEL3	N	1980	31302	1976	D
419	2006	1969	G	$48,208	040-90-5705	LEVEL3	Y	1975	28133	1971	D
420	2011	1974	G	$49,236	040-90-8839	LEVEL3	Y	1976	69121	1972	J
421	2002	1967	U	$44,299	040-91-2041	LEVEL2	N	1972	15001	1966	H
422	2014	1973	G	$50,820	041-51-7081	LEVEL3	Y	1977	21988	1970	I
423	1997	1961	G	$64,951	042-07-3155	LEVEL4	Y	1970	28266	1958	B
424	2012	1973	G	$45,464	042-11-2372	LEVEL3	Y	1981	46226	1973	B
425	2016	1973	U	$48,653	042-11-7329	LEVEL3	Y	1978	39560	1972	J
426	2002	1972	G	$50,219	042-29-5990	LEVEL3	Y	1977	21344	1966	A
427	1998	1966	G	$54,250	042-50-0100	LEVEL4	Y	1977	31302	1964	B
428	2012	1981	G	$44,984	042-53-3043	LEVEL2	N	1980	29626	1970	B
429	2002	1974	U	$53,001	043-10-0329	LEVEL3	Y	1973	58493	1976	B
430	2013	1972	U	$56,995	043-12-8274	LEVEL3	Y	1976	58511	1973	C
431	1992	1968	G	$78,940	043-13-3788	LEVEL4	Y	1967	363	1959	I
432	2021	1982	U	$36,669	043-14-4944	LEVEL2	N	1981	54426	1978	K
433	2002	1978	G	$45,945	043-30-8684	LEVEL2	N	1977	58511	1976	G
434	2014	1971	G	$53,824	043-33-6211	LEVEL3	Y	1976	16013	1971	E
435	2023	1982	G	$46,386	043-82-5823	LEVEL2	Y	1981	58511	1981	E
436	1993	1964	G	$59,531	043-87-6362	LEVEL4	Y	1971	51927	1958	B
437	1993	1961	G	$56,957	043-88-1536	LEVEL4	Y	1974	58511	1950	F
438	2010	1972	U	$69,272	043-92-1126	LEVEL4	Y	1975	47412	1969	C
439	2006	1968	G	$55,786	044-29-3019	LEVEL4	Y	1976	35509	1962	B
440	2013	1970	G	$54,353	044-50-4691	LEVEL4	Y	1980	58526	1969	I
441	2025	1979	U	$36,395	044-52-9237	LEVEL1	N	1978	48710	1978	A
442	2022	1982	G	$52,301	044-56-9176	LEVEL2	Y	1981	363	1977	E
443	2021	1982	G	$42,161	045-00-1101	LEVEL2	Y	1981	48692	1981	J
444	1987	1954	U	$58,029	045-67-3382	LEVEL4	Y	1981	26327	1949	E
445	1998	1971	G	$60,077	045-93-3728	LEVEL4	N	1977	51927	1956	J
446	1984	1971	G	$73,097	046-14-9366	LEVEL4	Y	1971	21988	1967	J
447	1985	1963	G	$55,840	046-65-8528	LEVEL3	Y	1979	18728	1962	I
448	2003	1977	U	$52,964	046-95-2115	LEVEL4	Y	1980	51407	1976	D
449	1987	1967	U	$47,154	047-14-3900	LEVEL3	N	1977	39560	1966	H
450	2012	1974	U	$50,285	047-59-6157	LEVEL3	N	1981	58526	1965	K

OBS	RETIRE	HIRED	DEGREE	SALARY	ID	LEVEL	DEFER	LEVELYR	COLLEGE	YEAR	DEPART
451	2005	1970	G	$63,454	047-94-8271	LEVEL3	Y	1969	63427	1965	E
452	2006	1970	G	$58,948	047-97-3246	LEVEL3	Y	1976	16287	1968	C
453	2008	1976	G	$61,128	048-35-6704	LEVEL4	Y	1978	47300	1968	I
454	2017	1982	G	$53,146	048-85-5209	LEVEL3	Y	1981	9719	1978	E
455	2009	1965	G	$60,647	050-35-6319	LEVEL4	Y	1973	52232	1965	J
456	2012	1979	U	$41,979	051-76-4762	LEVEL2	Y	1978	18687	1977	J
457	2010	1975	U	$69,207	051-97-2818	LEVEL4	Y	1979	28553	1970	E
458	2004	1965	G	$77,707	052-14-1525	LEVEL4	Y	1967	52842	1959	E
459	1997	1968	G	$53,924	052-30-8035	LEVEL3	Y	1974	19075	1966	H
460	2003	1965	G	$52,996	052-33-2661	LEVEL3	Y	1971	51927	1966	H
461	2015	1976	G	$44,174	052-39-2805	LEVEL2	Y	1975	52553	1976	D
462	2005	1967	G	$55,022	052-57-6557	LEVEL4	Y	1977	52553	1961	I
463	2007	1969	G	$58,228	052-61-7241	LEVEL3	Y	1975	52553	1968	I
464	2017	1973	U	$44,852	052-62-4475	LEVEL2	N	1974	58485	1971	K
465	2012	1972	G	$59,489	053-37-3833	LEVEL4	Y	1981	18695	1971	C
466	2009	1974	G	$52,203	053-55-3455	LEVEL3	Y	1977	58485	1975	C
467	2017	1979	G	$43,149	053-59-7687	LEVEL2	Y	1978	21344	1978	A
468	2021	1979	G	$43,456	053-63-2535	LEVEL2	Y	1978	15157	1977	F
469	2015	1979	G	$44,931	054-14-1292	LEVEL2	N	1978	48759	1971	B
470	1988	1950	G	$74,769	054-27-5894	LEVEL4	Y	1955	18827	1960	B
471	2012	1974	U	$48,947	054-72-4830	LEVEL3	Y	1976	22606	1964	D
472	2001	1968	G	$48,676	055-32-0184	LEVEL3	Y	1975	15016	1967	B
473	2000	1968	G	$55,673	056-10-0371	LEVEL4	Y	1981	16287	1968	H
474	2007	1969	G	$51,331	056-30-9524	LEVEL3	Y	1973	19348	1964	I
475	1993	1960	U	$85,623	056-67-3076	LEVEL4	Y	1961	53684	1959	E
476	1994	1966	G	$53,131	056-68-9394	LEVEL3	Y	1968	28595	1961	H
477	1999	1958	U	$56,287	056-89-6576	LEVEL4	Y	1974	40237	1957	D
478	2025	1982	G	$46,386	056-98-8711	LEVEL2	Y	1981	18825	1980	E
479	2000	1961	G	$55,049	057-29-6019	LEVEL4	Y	1976	46226	1960	D
480	2000	1964	G	$54,330	057-30-5270	LEVEL3	Y	1973	16287	1966	D
481	2007	1969	U	$47,803	057-71-3517	LEVEL3	N	1977	16287	1966	D
482	2013	1977	G	$48,102	058-72-4150	LEVEL2	Y	1976	18752	1975	H
483	1990	1962	G	$81,620	058-85-1369	LEVEL4	Y	1961	16287	1956	C
484	1984	1956	U	$65,580	058-87-6128	LEVEL3	Y	1957	18736	1950	C
485	2002	1965	G	$52,536	058-90-5346	LEVEL3	N	1973	29634	1964	H
486	1998	1961	G	$63,832	059-10-1638	LEVEL4	Y	1976	21988	1959	I
487	1988	1966	G	$70,352	060-28-4113	LEVEL4	Y	1980	63427	1964	C
488	1988	1958	G	$72,193	060-70-0643	LEVEL4	Y	1957	16673	1949	F
489	2009	1974	G	$57,593	060-74-0145	LEVEL4	N	1979	31302	1977	K
490	2022	1980	U	$35,726	061-76-8426	LEVEL1	N	1979	48775	1979	K
491	2008	1976	G	$67,392	061-95-2147	LEVEL4	Y	1975	363	1964	F
492	2003	1963	G	$52,812	062-32-8048	LEVEL3	N	1970	26971	1969	K
493	1987	1962	U	$64,632	062-92-0318	LEVEL4	Y	1964	363	1953	D
494	1993	1961	U	$58,218	063-10-0472	LEVEL4	Y	1981	10874	1960	D
495	2011	1979	U	$46,224	063-52-9514	LEVEL2	Y	1978	13612	1967	D
496	2016	1977	G	$47,020	064-32-4879	LEVEL3	Y	1979	39534	1976	J
497	2005	1970	G	$61,914	064-90-4975	LEVEL4	Y	1974	68682	1963	B
498	2019	1977	G	$45,051	065-13-3413	LEVEL3	Y	1980	31236	1977	H
499	2014	1980	G	$53,471	065-32-1282	LEVEL3	Y	1979	13547	1971	E
500	2020	1980	U	$51,696	065-33-7310	LEVEL2	Y	1979	12361	1972	C

OBS	RETIRE	HIRED	DEGREE	SALARY	ID	LEVEL	DEFER	LEVELYR	COLLEGE	YEAR	DEPART
501	2001	1979	G	$49,584	065-49-6137	LEVEL3	Y	1981	39560	1978	E
502	2014	1966	G	$88,564	065-52-4676	LEVEL4	Y	1966	18752	1957	E
503	2006	1965	G	$56,283	065-70-9812	LEVEL4	Y	1977	69096	1962	I
504	1998	1961	G	$68,178	065-89-5941	LEVEL4	Y	1966	69096	1954	I
505	2007	1972	G	$62,915	065-90-8714	LEVEL4	Y	1977	18728	1962	I
506	2009	1969	G	$54,514	065-91-3493	LEVEL4	Y	1978	58526	1967	I
507	2009	1972	G	$59,585	066-11-6473	LEVEL4	Y	1981	48684	1973	C
508	2009	1970	G	$52,842	066-30-8759	LEVEL3	Y	1978	31195	1970	D
509	1991	1967	G	$63,315	066-70-4500	LEVEL4	Y	1977	18728	1964	C
510	2008	1972	G	$55,465	067-30-9719	LEVEL4	Y	1979	66652	1964	B
511	2017	1982	G	$62,441	068-54-8854	LEVEL3	Y	1981	16013	1975	C
512	2019	1982	G	$56,526	068-73-2865	LEVEL2	Y	1981	18728	1981	C
513	2006	1972	G	$53,106	069-10-9702	LEVEL3	N	1976	52553	1971	G
514	2006	1982	G	$66,666	069-12-1698	LEVEL4	Y	1981	29634	1973	C
515	2005	1967	G	$64,116	069-31-7349	LEVEL4	Y	1971	16013	1962	I
516	2010	1973	G	$48,741	069-50-0548	LEVEL3	Y	1978	67173	1969	B
517	2008	1972	G	$55,858	069-51-2660	LEVEL4	Y	1981	18728	1970	A
518	2013	1971	G	$50,740	070-31-7179	LEVEL3	Y	1979	35509	1973	J
519	2013	1978	G	$50,530	070-72-1376	LEVEL3	Y	1977	40237	1969	H
520	1993	1971	U	$78,228	070-89-6760	LEVEL4	Y	1970	48882	1952	D
521	2014	1974	G	$45,381	071-51-2936	LEVEL3	Y	1979	18728	1974	H
522	2006	1964	G	$55,719	071-70-3948	LEVEL4	Y	1975	31302	1964	I
523	1997	1966	G	$77,290	072-08-4473	LEVEL4	Y	1965	63427	1963	I
524	2020	1978	G	$42,008	072-76-4254	LEVEL2	Y	1977	15016	1977	A
525	2005	1981	G	$76,756	072-91-7166	LEVEL4	Y	1980	31286	1969	G
526	2017	1981	G	$43,294	073-37-6135	LEVEL2	N	1980	18752	1976	I
527	1998	1960	G	$64,257	073-50-1175	LEVEL4	Y	1969	63435	1957	H
528	2000	1960	G	$64,909	074-10-1527	LEVEL4	Y	1969	18752	1960	B
529	1989	1961	G	$68,145	074-27-7545	LEVEL4	Y	1964	29626	1954	J
530	2014	1971	G	$57,786	074-33-3659	LEVEL3	Y	1975	16673	1970	I
531	2021	1977	G	$42,161	074-36-1408	LEVEL2	Y	1978	39560	1976	J
532	2016	1977	U	$58,678	074-36-4302	LEVEL3	Y	1976	13612	1972	C
533	2020	1978	G	$44,443	074-75-7005	LEVEL3	Y	1981	48908	1977	B
534	1997	1977	U	$61,368	075-08-8297	LEVEL4	Y	1976	47405	1970	D
535	1998	1969	G	$47,137	075-10-0214	LEVEL2	N	1968	18752	1968	H
536	1991	1972	G	$53,070	075-12-8289	LEVEL3	N	1973	52553	1954	G
537	1995	1968	G	$65,636	075-50-9056	LEVEL4	Y	1971	18687	1964	J
538	2005	1963	U	$48,611	075-51-7362	LEVEL3	N	1976	18752	1959	H
539	2015	1973	U	$42,986	075-73-7465	LEVEL2	N	1975	39560	1981	D
540	2011	1977	G	$48,609	076-12-8512	LEVEL3	Y	1980	48900	1972	F
541	1987	1965	U	$55,583	076-86-7767	LEVEL4	Y	1978	15016	1956	D
542	2020	1980	G	$41,582	076-94-1394	LEVEL2	Y	1979	69145	1976	B
543	2003	1969	G	$57,119	076-94-4757	LEVEL4	Y	1974	15016	1966	A
544	2000	1968	U	$69,643	077-09-6114	LEVEL4	Y	1968	31302	1963	J
545	2014	1976	G	$44,370	077-33-6899	LEVEL2	Y	1975	15016	1976	D
546	2011	1973	G	$52,755	077-72-8583	LEVEL3	Y	1978	58511	1972	B
547	1991	1953	G	$65,308	078-28-8999	LEVEL4	Y	1964	18825	1950	E
548	1984	1946	G	$74,407	078-47-2508	LEVEL4	Y	1960	18825	1945	J
549	1995	1949	U	$61,495	079-48-5262	LEVEL4	Y	1974	37035	1965	E
550	2006	1971	G	$58,898	079-74-0831	LEVEL4	Y	1979	18827	1965	F

OBS	RETIRE	HIRED	DEGREE	SALARY	ID	LEVEL	DEFER	LEVELYR	COLLEGE	YEAR	DEPART
551	2014	1972	G	$46,130	079-93-7504	LEVEL3	Y	1978	18778	1975	H
552	2008	1969	G	$56,939	080-34-4897	LEVEL4	Y	1976	69096	1967	I
553	2006	1980	G	$60,956	081-80-1767	LEVEL3	Y	1979	16287	1973	C
554	2013	1975	G	$46,770	082-18-5187	LEVEL3	Y	1978	16013	1969	J
555	2007	1970	G	$50,786	082-77-2824	LEVEL2	Y	1969	21344	1966	A
556	2011	1968	G	$49,163	082-98-4737	LEVEL3	Y	1974	18728	1969	D
557	2005	1972	U	$53,028	085-76-8194	LEVEL4	Y	1979	37629	1969	D
558	2004	1966	G	$60,174	086-39-3038	LEVEL4	Y	1976	69030	1965	E
559	2002	1969	G	$76,519	087-33-6505	LEVEL4	Y	1968	68682	1963	B
560	2014	1975	U	$51,330	087-80-0429	LEVEL3	N	1979	64097	1974	C
561	1990	1950	G	$62,267	088-87-6279	LEVEL4	Y	1976	18728	1959	J
562	2013	1981	G	$51,744	088-93-3301	LEVEL2	Y	1980	65828	1980	E
563	2014	1974	G	$47,250	089-53-6321	LEVEL3	Y	1977	18825	1971	I
564	2021	1981	G	$48,364	090-02-0816	LEVEL2	Y	1980	66454	1980	E
565	2007	1970	G	$55,327	090-21-5920	LEVEL3	Y	1976	29618	1970	E
566	2018	1982	G	$47,231	090-50-7740	LEVEL2	Y	1981	16673	1981	E
567	1996	1980	G	$64,737	091-30-0416	LEVEL4	Y	1979	48733	1951	B
568	1998	1972	G	$70,615	092-12-8097	LEVEL4	Y	1971	69187	1958	B
569	1995	1970	U	$49,853	092-35-7516	LEVEL3	N	1977	48775	1959	K
570	2005	1971	G	$75,095	092-94-5390	LEVEL4	Y	1973	35509	1964	I
571	2009	1969	G	$49,865	092-96-0963	LEVEL3	Y	1975	18827	1971	H
572	2002	1967	G	$48,132	093-62-0474	LEVEL3	Y	1977	66454	1969	D
573	2011	1972	G	$49,805	093-93-6786	LEVEL3	Y	1977	67173	1971	G
574	2019	1980	G	$45,541	093-96-4778	LEVEL2	N	1981	69104	1981	G
575	2002	1978	G	$46,491	093-97-2513	LEVEL2	N	1977	69096	1973	G
576	2002	1970	G	$70,753	094-30-4786	LEVEL4	Y	1974	69096	1960	I
577	2015	1969	G	$46,010	094-34-8753	LEVEL2	N	1972	39518	1977	G
578	1996	1969	U	$73,954	095-08-8078	LEVEL4	Y	1969	69096	1949	A
579	1996	1965	G	$52,871	095-11-6250	LEVEL3	N	1968	69080	1959	D
580	2015	1972	G	$46,870	095-14-8507	LEVEL2	Y	1972	51927	1972	J
581	2007	1968	G	$52,019	095-32-0604	LEVEL3	Y	1973	67173	1967	D
582	2004	1967	G	$51,747	095-50-4911	LEVEL3	Y	1973	39518	1966	H
583	2010	1969	G	$47,899	095-74-7917	LEVEL3	Y	1981	13612	1972	H
584	2009	1982	G	$57,709	095-78-4226	LEVEL3	Y	1981	16013	1974	I
585	2011	1967	G	$48,666	095-93-6435	LEVEL3	Y	1976	21988	1973	J
586	1996	1975	G	$75,694	096-51-6405	LEVEL4	Y	1974	58511	1959	I
587	2015	1974	G	$49,035	096-55-7246	LEVEL3	N	1977	63427	1973	D
588	1991	1969	G	$56,712	096-65-0870	LEVEL4	Y	1975	58485	1953	B
589	2012	1981	G	$51,744	096-74-4185	LEVEL2	Y	1980	26971	1979	E
590	2012	1973	G	$48,024	096-76-4609	LEVEL3	Y	1977	18687	1973	D
591	1995	1971	G	$62,778	096-90-4537	LEVEL4	Y	1972	67173	1958	B
592	2013	1981	G	$62,357	096-95-3273	LEVEL4	Y	1980	47183	1970	I
593	1988	1950	G	$69,950	097-05-8886	LEVEL4	Y	1960	21344	1944	A
594	2000	1967	G	$58,500	097-09-2153	LEVEL3	Y	1966	69137	1963	I
595	2000	1972	G	$69,291	097-31-7298	LEVEL4	Y	1971	16013	1961	J
596	2001	1964	G	$77,104	097-70-8061	LEVEL4	Y	1969	16673	1963	E
597	2006	1966	G	$57,764	098-14-0730	LEVEL4	Y	1975	18728	1964	E
598	2018	1980	G	$41,582	098-15-7374	LEVEL2	N	1979	48801	1977	B
599	2003	1968	G	$47,914	099-12-0149	LEVEL2	Y	1968	16287	1968	J
600	2012	1969	G	$50,181	099-53-6992	LEVEL3	Y	1974	48908	1968	H

OBS	RETIRE	HIRED	DEGREE	SALARY	ID	LEVEL	DEFER	LEVELYR	COLLEGE	YEAR	DEPART
601	2009	1969	G	$48,425	099-92-1567	LEVEL2	Y	1971	58511	1971	J
602	2006	1974	G	$51,364	099-96-9802	LEVEL3	Y	1979	363	1966	I
603	2014	1975	U	$41,611	100-14-0471	LEVEL2	N	1977	69153	1974	G
604	1995	1961	G	$60,511	100-27-5911	LEVEL4	Y	1971	28133	1957	B
605	2012	1970	G	$53,028	100-33-7320	LEVEL4	Y	1979	35509	1970	I
606	2005	1972	U	$44,443	100-72-0285	LEVEL2	Y	1977	24059	1969	A
607	2017	1976	G	$43,323	101-17-5904	LEVEL2	Y	1975	69137	1974	I
608	1996	1948	G	$78,013	101-34-8090	LEVEL4	N	1963	31236	1954	H
609	2016	1978	G	$46,539	101-36-4282	LEVEL2	Y	1977	69080	1974	B
610	2009	1970	G	$53,390	101-75-6290	LEVEL3	Y	1974	69137	1970	J
611	2010	1971	G	$48,654	102-12-7873	LEVEL3	N	1975	13605	1970	H
612	1991	1958	G	$60,283	102-28-1440	LEVEL4	Y	1972	18752	1957	D
613	2008	1968	G	$52,443	102-50-9219	LEVEL3	Y	1975	69121	1969	I
614	1991	1971	G	$84,724	103-25-0414	LEVEL4	Y	1970	48733	1950	I
615	1992	1972	G	$71,405	103-48-8303	LEVEL4	Y	1971	18687	1970	H
616	2016	1982	G	$53,991	104-61-2190	LEVEL3	Y	1981	69137	1974	I
617	2019	1977	U	$48,056	104-92-9794	LEVEL3	Y	1976	48775	1967	D
618	2002	1972	G	$54,139	105-23-1933	LEVEL4	Y	1979	68682	1971	H
619	2011	1969	G	$49,689	105-35-2455	LEVEL3	Y	1976	31302	1969	I
620	2004	1979	G	$64,351	105-54-5272	LEVEL3	Y	1978	55327	1965	C
621	2019	1979	G	$39,765	105-61-2414	LEVEL2	Y	1978	18736	1977	J
622	2006	1977	G	$53,084	106-33-6748	LEVEL3	Y	1976	28133	1973	J
623	1997	1962	G	$55,342	106-49-2282	LEVEL4	Y	1980	16287	1956	D
624	2008	1973	G	$50,083	106-52-1723	LEVEL3	N	1980	21344	1972	A
625	2014	1976	G	$50,990	106-73-2721	LEVEL3	N	1979	58526	1978	K
626	2023	1981	G	$42,956	106-96-8250	LEVEL2	Y	1980	40237	1980	A
627	2011	1982	G	$72,581	107-32-8940	LEVEL4	Y	1981	48908	1969	E
628	2019	1979	U	$46,225	107-33-7695	LEVEL2	Y	1978	18695	1975	F
629	2012	1974	G	$50,174	107-52-3937	LEVEL3	Y	1981	48908	1973	A
630	2004	1972	G	$49,631	107-70-9068	LEVEL3	N	1976	48908	1971	B
631	1999	1966	G	$56,892	107-72-4562	LEVEL4	Y	1979	48908	1965	J
632	1996	1967	G	$48,935	107-88-0400	LEVEL3	Y	1974	13612	1967	H
633	2013	1974	G	$44,041	108-14-5730	LEVEL2	Y	1973	16013	1972	H
634	2016	1977	U	$54,191	109-18-1270	LEVEL3	Y	1980	48775	1976	C
635	2021	1982	G	$46,386	109-39-2676	LEVEL2	Y	1981	39560	1979	F
636	2013	1970	G	$59,608	109-55-7254	LEVEL4	Y	1976	46226	1969	I
637	2008	1981	G	$51,744	109-60-0207	LEVEL2	N	1980	48692	1979	E
638	2005	1981	G	$62,357	109-60-0208	LEVEL3	Y	1980	48692	1972	E
639	2000	1974	G	$53,683	109-72-1071	LEVEL3	Y	1973	48824	1965	G
640	1990	1952	G	$69,516	109-85-9070	LEVEL4	Y	1966	18728	1947	I
641	2018	1977	G	$48,257	110-18-4627	LEVEL3	Y	1981	48692	1975	C
642	2019	1980	U	$36,613	110-35-3678	LEVEL1	N	1979	39560	1979	K
643	2011	1971	U	$64,923	110-60-8560	LEVEL4	Y	1974	363	1967	I
644	2001	1977	G	$58,752	110-92-1787	LEVEL4	Y	1980	35509	1965	I
645	2003	1971	G	$56,753	110-92-5221	LEVEL4	Y	1981	39518	1968	B
646	2011	1978	G	$39,485	110-95-3189	LEVEL2	Y	1977	69096	1977	D
647	2006	1974	G	$43,169	111-16-5042	LEVEL2	Y	1975	29634	1975	J
648	2021	1982	G	$42,358	111-20-0034	LEVEL2	Y	1981	18827	1981	A
649	2014	1975	G	$45,051	111-58-4441	LEVEL3	Y	1980	21344	1973	I
650	2023	1982	G	$37,936	111-59-6938	LEVEL2	Y	1981	11418	1980	J

OBS	RETIRE	HIRED	DEGREE	SALARY	ID	LEVEL	DEFER	LEVELYR	COLLEGE	YEAR	DEPART
651	2029	1982	G	$41,316	111-61-3046	LEVEL2	Y	1981	31302	1976	I
652	1934	1967	G	$61,456	111-78-1846	LEVEL4	Y	1978	47183	1964	I
653	1995	1965	G	$66,688	112-30-0874	LEVEL4	Y	1970	26971	1957	B
654	2018	1982	G	$62,441	112-39-3001	LEVEL3	N	1981	16013	1976	C
655	2011	1970	G	$53,326	112-55-6934	LEVEL4	Y	1979	48710	1967	B
656	2010	1979	G	$56,489	112-99-2904	LEVEL3	Y	1978	69656	1970	J
657	2012	1970	G	$50,211	113-36-0019	LEVEL3	N	1975	21344	1969	J
658	2001	1965	G	$62,034	113-72-8217	LEVEL4	Y	1972	48692	1963	A
659	2012	1971	G	$48,399	113-96-0403	LEVEL3	Y	1975	48759	1967	I
660	2006	1977	G	$44,443	114-17-3381	LEVEL3	Y	1981	13605	1975	H
661	2015	1979	G	$41,611	114-17-6456	LEVEL2	Y	1978	363	1971	B
662	2018	1982	G	$46,386	114-19-7030	LEVEL2	Y	1981	69153	1977	F
663	2011	1973	G	$53,027	114-19-7409	LEVEL4	Y	1979	48710	1970	A
664	2006	1974	U	$53,432	114-54-0846	LEVEL3	N	1973	48733	1972	G
665	2009	1981	G	$62,357	114-74-8168	LEVEL4	Y	1980	48908	1972	J
666	2016	1975	U	$41,842	114-76-4042	LEVEL2	Y	1974	48733	1970	D
667	2016	1977	G	$45,051	114-76-8033	LEVEL3	Y	1980	34710	1972	I
668	2012	1973	G	$50,064	115-32-5425	LEVEL3	Y	1981	31302	1973	J
669	2013	1971	G	$48,283	115-55-7560	LEVEL3	Y	1976	31286	1968	I
670	2016	1981	G	$48,147	115-57-7090	LEVEL2	N	1980	39560	1976	I
671	1994	1974	G	$52,964	115-72-1543	LEVEL4	N	1980	31302	1968	I
672	1999	1964	G	$66,181	144-45-5624	LEVEL4	Y	1968	48775	1959	B

OBS	LEAGUE	DIVISION	TEAM	STAND	RUNS	H2B	H3B	HR	TB	SB	HP	SF	SH	GIDP	CS	YEAR	H1B	OUT
1	1	1	1	5	645	227	35	137	549	73	26	36	61	119	49	69	905	3985
2	1	1	2	3	725	220	52	119	454	74	46	36	73	128	34	69	1166	3941
3	1	1	3	2	720	215	40	142	559	30	36	46	72	111	32	69	1003	4019
4	1	1	4	4	595	228	44	90	503	87	23	33	57	112	49	69	1041	4021
5	1	1	5	1	632	184	41	109	527	66	33	33	82	105	43	69	977	4011
6	1	1	6	6	582	202	33	125	529	52	38	30	57	132	52	69	940	3987
7	1	2	1	4	645	185	52	97	484	80	21	33	96	110	51	69	1071	4017
8	1	2	2	3	798	224	42	171	474	79	46	47	100	117	56	69	1121	3959
9	1	2	3	5	676	208	40	104	699	101	41	40	68	109	58	69	932	3955
10	1	2	4	2	713	187	28	136	711	71	66	42	82	121	32	69	974	4028
11	1	2	5	1	691	195	22	141	485	59	32	34	87	130	48	69	1053	3919
12	1	2	6	6	468	180	42	99	423	45	35	20	56	122	44	69	882	4032
13	2	1	1	5	562	210	44	94	565	119	14	40	63	105	74	69	899	3956
14	2	1	2	1	779	234	29	175	634	82	43	59	74	138	45	69	1027	3915
15	2	1	3	3	743	234	37	197	658	41	32	43	67	125	47	69	913	3988
16	2	1	5	6	573	173	24	119	535	85	24	43	47	147	37	69	956	3946
17	2	2	4	6	639	179	27	125	626	167	34	29	72	111	59	69	945	4057
18	2	1	6	2	701	188	29	182	578	35	30	43	63	113	28	69	917	4012
19	2	2	1	2	740	210	28	148	617	100	63	35	74	124	39	69	1014	4090
20	2	2	2	1	790	246	32	163	599	115	43	40	65	114	70	69	1079	4043
21	2	2	3	5	625	210	27	112	552	54	49	37	70	119	22	69	997	3985
22	2	1	4	4	694	171	40	148	630	52	32	36	51	158	40	69	1006	3924
23	2	2	5	4	586	179	32	98	522	129	43	38	57	125	70	69	1002	4026
24	2	2	6	3	528	151	29	88	516	54	32	41	75	105	39	69	953	3990
25	1	1	1	5	594	224	58	101	519	72	25	37	62	133	64	70	916	4024
26	1	1	2	1	729	235	70	130	444	66	44	53	53	117	34	70	1087	3998
27	1	1	3	2	806	228	44	179	607	39	20	36	75	110	16	70	973	3957
28	1	1	4	4	744	218	51	113	569	117	26	40	52	138	47	70	1115	4054
29	1	1	5	3	695	211	42	120	684	118	26	48	74	139	54	70	985	3946
30	1	1	6	6	687	211	35	136	659	65	39	35	107	106	45	70	902	4021
31	1	2	1	2	749	233	67	87	541	138	24	50	72	114	57	70	1128	3977
32	1	2	2	1	775	253	45	191	547	115	29	48	58	119	52	70	1009	3923
33	1	2	3	5	744	250	47	129	598	114	27	43	63	144	41	70	1020	3984
34	1	2	4	3	831	257	35	165	729	83	56	40	66	143	27	70	1003	3975
35	1	2	5	4	736	215	24	160	522	58	38	42	54	140	34	70	1096	3911
36	1	2	6	6	681	208	36	172	500	60	39	29	83	111	45	70	937	4030
37	2	1	1	2	680	208	41	111	588	105	25	46	60	115	61	70	1021	3996
38	2	1	2	1	792	213	25	179	717	84	44	46	64	110	39	70	1007	4011
39	2	1	3	3	786	252	28	203	594	50	40	47	34	137	48	70	967	3948
40	2	1	5	6	649	197	23	183	503	25	37	45	76	119	36	70	955	3986
41	2	2	4	5	613	202	24	126	592	91	36	32	115	130	73	70	953	3960
42	2	1	6	4	666	207	38	148	656	29	34	49	83	134	30	70	889	3961
43	2	2	1	3	678	208	24	171	584	131	36	36	73	121	68	70	935	3917
44	2	2	2	1	744	230	41	153	501	57	42	38	79	132	52	70	1014	3913
45	2	2	3	5	633	192	20	123	477	53	42	48	51	132	33	70	1059	3988
46	2	1	4	6	626	184	28	138	635	72	46	38	44	133	42	70	952	4025
47	2	2	5	4	611	202	41	97	514	97	21	27	63	126	53	70	1001	4036
48	2	2	6	2	631	197	40	114	447	69	29	37	69	118	27	70	1040	4023
49	1	1	1	6	558	209	35	123	499	63	34	45	55	98	39	71	922	4151

OBS	LEAGUE	DIVISION	TEAM	STAND	RUNS	H2B	H3B	HR	TB	SB	HP	SF	SH	GIDP	CS	YEAR	H1B	OUT
50	1	1	2	1	788	223	61	154	469	65	29	49	62	120	31	71	1117	3999
51	1	1	3	3	637	202	34	128	527	44	34	40	92	144	32	71	1037	3893
52	1	1	4	2	739	225	54	95	543	124	25	59	63	139	53	71	1168	3929
53	1	1	5	4	588	203	29	98	547	89	28	35	91	147	43	71	1035	3965
54	1	1	6	5	622	197	29	88	543	51	78	40	102	136	43	71	998	3887
55	1	2	1	2	663	213	38	95	489	76	21	50	72	140	40	71	1123	3914
56	1	2	2	4	586	203	28	138	438	59	27	25	69	138	33	71	937	3970
57	1	2	3	5	585	230	52	71	478	101	29	41	65	105	51	71	966	4068
58	1	2	4	1	706	224	36	140	654	101	37	37	69	126	36	71	948	3987
59	1	2	5	3	643	192	30	153	434	57	28	44	91	120	46	71	1059	4021
60	1	2	6	6	486	184	31	96	438	70	25	19	87	134	45	71	939	3982
61	2	1	1	4	648	195	43	97	581	75	37	56	77	140	55	71	1042	3896
62	2	1	2	1	742	207	25	158	672	66	46	37	85	125	38	71	992	3796
63	2	1	3	3	691	246	28	161	552	51	32	47	75	146	34	71	925	3895
64	2	1	5	5	543	200	20	109	467	57	31	29	67	127	37	71	974	4037
65	2	2	4	6	534	160	23	104	543	82	30	39	107	103	53	71	901	3894
66	2	1	6	2	701	214	38	179	540	35	55	37	62	129	43	71	968	3974
67	2	2	1	1	691	195	25	160	542	80	37	38	80	113	53	71	1003	3998
68	2	2	2	6	654	197	31	116	512	66	25	57	64	158	44	71	1062	3850
69	2	2	3	3	617	185	30	138	562	83	51	38	81	105	65	71	993	3931
70	2	1	4	5	537	189	30	86	575	68	28	35	58	127	45	71	914	3944
71	2	2	5	2	603	225	40	80	490	130	25	59	45	128	46	71	978	3844
72	2	2	6	4	511	213	18	96	441	72	29	31	83	141	34	71	944	4083
73	1	1	1	6	503	200	36	98	487	42	22	36	69	104	50	72	906	3904
74	1	1	2	1	691	251	47	110	404	49	25	50	52	110	30	72	1097	3875
75	1	1	3	2	685	206	40	133	565	69	28	43	67	126	47	72	967	3775
76	1	1	4	4	568	214	42	70	437	104	27	36	58	141	48	72	1057	3802
77	1	1	5	3	528	175	31	105	589	41	34	39	86	127	41	72	843	3854
78	1	1	6	5	513	156	22	91	474	68	49	30	108	96	66	72	936	3855
79	1	2	1	3	584	178	39	98	480	82	24	40	89	121	39	72	1034	3800
80	1	2	2	1	707	214	44	124	606	140	37	54	65	109	63	72	935	3815
81	1	2	3	2	708	233	38	134	524	111	32	51	62	99	56	72	954	3809
82	1	2	4	5	662	211	36	150	480	123	30	39	64	107	45	72	884	3857
83	1	2	5	4	628	186	17	144	532	47	35	38	55	131	35	72	1016	3784
84	1	2	6	6	488	168	38	102	407	78	15	25	90	97	46	72	873	3935
85	2	1	1	4	557	201	24	103	491	71	29	31	74	112	42	72	960	3768
86	2	1	2	3	519	193	29	100	507	78	36	40	65	113	41	72	831	3762
87	2	1	3	2	640	229	34	124	522	66	37	48	56	111	30	72	902	3808
88	2	1	5	5	472	187	18	91	420	49	26	33	83	112	53	72	924	3875
89	2	1	4	6	493	167	22	88	472	64	29	30	78	144	57	72	927	3776
90	2	1	6	1	558	179	32	122	483	17	31	34	74	125	21	72	873	3768
91	2	2	1	1	604	195	29	134	463	87	47	36	100	112	48	72	890	3840
92	2	2	2	3	537	182	31	93	478	53	35	28	73	129	41	72	971	3828
93	2	2	3	2	566	170	28	108	511	100	34	25	68	113	52	72	902	3762
94	2	2	4	6	461	166	17	56	503	126	30	34	84	100	73	72	853	3837
95	2	2	5	4	580	220	26	78	534	85	34	38	72	134	44	72	993	3716
96	2	2	6	5	454	171	26	78	358	57	25	25	66	125	37	72	974	3791
97	1	1	1	6	642	218	29	134	476	51	33	45	56	91	47	73	1000	4074
98	1	1	2	3	704	257	44	154	432	23	27	40	60	130	30	73	1010	4013

OBS	LEAGUE	DIVISION	TEAM	STAND	RUNS	H2B	H3B	HR	TB	SB	HP	SF	SH	GIDP	CS	YEAR	H1B	OUT
99	1	1	3	5	614	201	21	117	575	65	20	37	75	144	58	73	983	3897
100	1	1	4	2	643	240	35	75	531	100	29	50	89	149	46	73	1068	3911
101	1	1	5	1	608	198	24	85	540	27	23	36	108	147	22	73	1038	3965
102	1	1	6	4	668	190	23	125	695	77	44	28	115	144	68	73	1007	3880
103	1	2	1	2	675	219	29	110	497	109	28	57	81	123	50	73	1115	4008
104	1	2	2	1	741	232	34	137	639	148	31	51	78	118	55	73	995	3989
105	1	2	3	4	681	216	35	134	469	92	33	36	83	135	48	73	1006	4006
106	1	2	4	3	739	212	52	161	590	112	35	37	75	113	52	73	1027	3972
107	1	2	5	5	799	219	34	206	608	84	34	46	65	112	40	73	1038	4022
108	1	2	6	6	548	198	26	112	401	88	21	33	73	115	36	73	994	4012
109	2	1	1	4	641	212	17	131	489	47	22	34	27	142	43	73	1075	3915
110	2	1	2	1	754	229	48	119	648	146	43	49	58	126	64	73	1078	3937
111	2	1	3	2	738	235	30	147	581	114	31	44	54	143	45	73	1060	3898
112	2	1	5	6	680	205	29	158	471	60	32	43	40	125	68	73	1037	4038
113	2	1	4	5	708	229	40	145	563	110	42	35	61	114	66	73	985	4013
114	2	1	6	3	642	213	32	157	509	28	39	31	48	147	30	73	998	3961
115	2	2	1	1	758	216	28	147	595	128	35	53	67	118	57	73	1040	3958
116	2	2	2	3	738	240	44	120	598	87	34	40	34	150	46	73	1117	3954
117	2	2	3	5	652	228	38	111	537	83	32	42	49	146	73	73	1023	3929
118	2	2	4	6	619	195	29	110	503	91	27	50	45	132	53	73	1063	3959
119	2	2	5	2	755	239	40	114	644	105	26	47	49	134	69	73	1047	3934
120	2	2	6	4	629	183	29	93	509	59	34	45	60	131	47	73	1090	3979
121	1	1	1	3	676	233	50	95	469	115	26	44	84	112	58	74	1056	3948
122	1	1	2	1	751	238	46	114	514	55	38	53	54	140	31	74	1162	4002
123	1	1	3	6	669	221	42	110	621	78	29	40	30	121	73	74	1024	4056
124	1	1	4	2	677	216	46	83	531	172	44	55	68	142	62	74	1147	3986
125	1	1	5	5	572	183	22	96	597	43	29	45	87	148	23	74	985	4034
126	1	1	6	4	662	201	29	86	652	124	33	56	106	127	49	74	1039	3861
127	1	2	1	1	798	231	34	139	597	149	28	64	86	116	75	74	1107	3930
128	1	2	2	2	776	271	35	135	693	146	30	46	68	121	49	74	996	3977
129	1	2	3	4	653	222	41	110	471	108	29	48	83	124	65	74	1068	3924
130	1	2	4	5	634	228	38	93	548	107	30	24	75	113	51	74	1021	3989
131	1	2	5	3	661	202	37	120	571	72	24	42	109	124	44	74	1016	4034
132	1	2	6	6	541	196	27	99	564	85	20	35	83	106	45	74	917	4070
133	2	1	1	2	671	220	30	101	515	53	21	72	49	122	35	74	1100	3951
134	2	1	2	1	659	226	27	116	509	145	58	56	72	124	58	74	1049	3993
135	2	1	3	3	696	236	31	109	569	104	33	49	64	128	58	74	1073	3922
136	2	1	5	4	662	201	19	131	432	79	26	31	56	135	68	74	1044	3944
137	2	1	4	5	647	228	49	120	500	106	27	35	56	103	75	74	938	4034
138	2	1	6	6	620	200	35	131	436	67	24	37	41	132	38	74	1009	4061
139	2	2	1	1	689	205	37	132	568	164	38	51	60	108	93	74	941	3908
140	2	2	2	3	673	190	37	111	520	74	46	40	64	159	45	74	1192	3943
141	2	2	3	4	684	225	23	135	519	64	26	43	70	156	53	74	1109	3929
142	2	2	4	2	690	198	39	99	508	113	38	47	81	155	80	74	1146	3812
143	2	2	5	5	667	232	42	89	550	146	32	43	56	133	76	74	1085	4001
144	2	2	6	6	618	203	31	95	509	119	45	48	82	135	79	74	1043	3894
145	1	1	1	2	735	283	42	125	610	126	31	42	88	114	57	75	1056	3972
146	1	1	2	1	712	255	47	138	468	49	38	40	76	124	28	75	1004	3921
147	1	1	3	5	712	229	41	95	650	67	30	66	107	112	55	75	1054	3939
148	1	1	4	4	662	239	46	81	444	116	29	45	92	137	49	75	1161	3933
149	1	1	5	3	646	217	34	101	501	32	37	37	75	143	26	75	1078	4014
150	1	1	6	6	601	216	31	98	579	108	27	38	110	110	58	75	1001	4062
151	1	2	1	2	648	217	31	118	611	138	31	47	104	123	52	75	989	3975
152	1	2	2	1	840	278	37	124	691	168	35	45	66	122	36	75	1076	3944
153	1	2	3	6	664	218	54	84	523	133	32	44	97	118	62	75	1045	3996
154	1	2	4	3	659	235	45	84	604	99	22	51	62	139	47	75	1048	3896
155	1	2	5	5	583	179	28	107	543	55	18	35	72	139	38	75	1009	3962
156	1	2	6	4	552	215	22	78	506	85	37	46	133	128	50	75	1009	3977
157	2	1	1	3	681	230	39	110	486	102	30	53	54	142	59	75	1051	3843
158	2	1	2	2	682	224	33	124	580	104	38	46	73	114	55	75	1001	3978
159	2	1	3	1	796	284	44	134	565	66	34	53	75	137	58	75	1038	3811
160	2	1	5	4	688	201	25	153	525	106	24	40	64	145	89	75	1030	3850
161	2	1	4	5	675	242	34	146	553	65	28	49	73	131	64	75	921	3904
162	2	1	6	6	570	171	39	125	383	63	28	38	37	120	57	75	1003	3908
163	2	2	1	1	758	220	33	151	609	183	51	35	74	107	82	75	972	3932
164	2	2	2	4	724	215	28	121	563	81	40	46	62	127	48	75	1133	3890
165	2	2	3	5	655	209	38	94	611	101	40	52	50	150	54	75	1059	3940
166	2	2	4	3	714	208	17	134	613	102	25	41	64	125	62	75	1072	4043
167	2	2	5	2	710	263	58	118	591	155	29	51	68	109	75	75	992	3951
168	2	2	6	6	628	195	41	55	593	220	27	47	97	98	8	75	1033	3955
169	1	1	1	1	770	259	45	110	542	127	40	67	59	119	70	76	1091	3904
170	1	1	2	2	708	249	56	110	433	130	29	50	61	131	45	76	1084	3974
171	1	1	3	4	611	216	24	105	490	74	30	41	75	126	74	76	1041	4007
172	1	1	4	5	629	243	57	63	512	123	22	45	86	127	55	76	1069	3957
173	1	1	5	3	615	198	34	102	561	66	28	33	92	127	58	76	1000	3954
174	1	1	6	6	531	224	32	94	433	86	16	40	75	107	44	76	925	4046
175	1	2	1	2	608	200	34	91	486	144	29	47	91	129	55	76	1046	3972
176	1	2	2	1	857	271	63	141	681	210	28	60	67	103	57	76	1124	4000
177	1	2	3	3	625	195	50	66	530	150	21	39	57	127	57	76	1090	3936
178	1	2	4	4	595	211	37	85	518	88	25	48	80	122	55	76	1007	3990
179	1	2	5	6	620	170	30	82	589	74	19	47	107	131	61	76	1027	3905
180	1	2	6	5	570	216	37	64	488	92	23	42	125	119	46	76	1010	3923
181	2	1	1	1	730	231	36	120	470	163	35	46	50	94	65	76	1109	3965
182	2	1	2	2	619	213	28	119	519	150	23	35	57	119	61	76	966	4012
183	2	1	3	3	716	257	53	134	500	95	29	59	55	127	70	76	1004	3936
184	2	1	5	4	615	189	38	85	479	75	11	60	67	143	69	76	1111	3846
185	2	1	4	6	570	170	38	88	511	62	23	48	78	112	61	76	1030	3958
186	2	1	6	5	609	207	38	101	450	107	31	50	46	157	59	76	1055	3883
187	2	2	1	2	686	208	33	113	592	341	45	58	58	91	23	76	965	3943
188	2	2	2	3	743	222	51	81	550	146	41	49	93	121	75	76	1172	3927
189	2	2	3	6	586	209	46	73	471	120	34	55	79	104	53	76	1082	4018
190	2	2	4	5	616	213	26	80	568	87	29	45	72	141	45	76	1071	4024
191	2	2	5	1	713	259	57	65	484	218	31	71	71	103	6	76	1109	3947
192	2	2	6	4	550	210	23	63	534	126	42	48	92	130	80	76	969	3990

Student t Distributions

df	$t_{0.75}$	$t_{0.90}$	$t_{0.95}$	$t_{0.975}$	$t_{0.99}$	$t_{0.995}$
1	1.0000	3.0777	6.3138	12.7062	31.8207	63.6574
2	0.8165	1.8856	2.9200	4.3027	6.9646	9.9248
3	0.7649	1.6377	2.3534	3.1824	4.5407	5.8409
4	0.7407	1.5332	2.1318	2.7764	3.7469	4.6041
5	0.7267	1.4759	2.0150	2.5706	3.3649	4.0322
6	0.7176	1.4398	1.9432	2.4469	3.1427	3.7074
7	0.7111	1.4149	1.8946	2.3646	2.9980	3.4995
8	0.7064	1.3968	1.8595	2.3060	2.8965	3.3554
9	0.7027	1.3830	1.8331	2.2622	2.8214	3.2498
10	0.6998	1.3722	1.8125	2.2281	2.7638	3.1693
11	0.6974	1.3634	1.7959	2.2010	2.7181	3.1058
12	0.6955	1.3562	1.7823	2.1788	2.6810	3.0545
13	0.6938	1.3502	1.7709	2.1604	2.6503	3.0123
14	0.6924	1.3450	1.7613	2.1448	2.6245	2.9768
15	0.6912	1.3406	1.7531	2.1315	2.6025	2.9467
16	0.6901	1.3368	1.7459	2.1199	2.5835	2.9208
17	0.6892	1.3334	1.7396	2.1098	2.5669	2.8982
18	0.6884	1.3304	1.7341	2.1009	2.5524	2.8784
19	0.6876	1.3277	1.7291	2.0930	2.5395	2.8609
20	0.6870	1.3253	1.7247	2.0860	2.5280	2.8453
21	0.6864	1.3232	1.7207	2.0796	2.5177	2.8314
22	0.6858	1.3212	1.7171	2.0739	2.5083	2.8188
23	0.6853	1.3195	1.7139	2.0687	2.4999	2.8073
24	0.6848	1.3178	1.7109	2.0639	2.4922	2.7969
25	0.6844	1.3163	1.7081	2.0595	2.4851	2.7874
26	0.6840	1.3150	1.7056	2.0555	2.4786	2.7787
27	0.6837	1.3137	1.7033	2.0518	2.4727	2.7707
28	0.6834	1.3125	1.7011	2.0484	2.4671	2.7633
29	0.6830	1.3114	1.6991	2.0452	2.4620	2.7564
30	0.6828	1.3104	1.6973	2.0423	2.4573	2.7500
31	0.6825	1.3095	1.6955	2.0395	2.4528	2.7440
32	0.6822	1.3086	1.6939	2.0369	2.4487	2.7385
33	0.6820	1.3077	1.6924	2.0345	2.4448	2.7333
34	0.6818	1.3070	1.6909	2.0322	2.4411	2.7284
35	0.6816	1.3062	1.6896	2.0301	2.4377	2.7238
36	0.6814	1.3055	1.6883	2.0281	2.4345	2.7195
37	0.6812	1.3049	1.6871	2.0262	2.4314	2.7154
38	0.6810	1.3042	1.6860	2.0244	2.4286	2.7116
39	0.6808	1.3036	1.6849	2.0227	2.4258	2.7079
40	0.6807	1.3031	1.6839	2.0211	2.4233	2.7045
41	0.6805	1.3025	1.6829	2.0195	2.4208	2.7012
42	0.6804	1.3020	1.6820	2.0181	2.4185	2.6981
43	0.6802	1.3016	1.6811	2.0167	2.4163	2.6951
44	0.6801	1.3011	1.6802	2.0154	2.4141	2.6923
45	0.6800	1.3006	1.6794	2.0141	2.4121	2.6896
46	0.6799	1.3002	1.6787	2.0129	2.4102	2.6870
47	0.6797	1.2998	1.6779	2.0117	2.4083	2.6846
48	0.6796	1.2994	1.6772	2.0106	2.4066	2.6822
49	0.6795	1.2991	1.6766	2.0096	2.4049	2.6800
50	0.6794	1.2987	1.6759	2.0086	2.4033	2.6778
51	0.6793	1.2984	1.6753	2.0076	2.4017	2.6757
52	0.6792	1.2980	1.6747	2.0066	2.4002	2.6737
53	0.6791	1.2977	1.6741	2.0057	2.3988	2.6718
54	0.6791	1.2974	1.6736	2.0049	2.3974	2.6700
55	0.6790	1.2971	1.6730	2.0040	2.3961	2.6682
56	0.6789	1.2969	1.6725	2.0032	2.3948	2.6665
57	0.6788	1.2966	1.6720	2.0025	2.3936	2.6649
58	0.6787	1.2963	1.6716	2.0017	2.3924	2.6633
59	0.6787	1.2961	1.6711	2.0010	2.3912	2.6618
60	0.6786	1.2958	1.6706	2.0003	2.3901	2.6603
61	0.6785	1.2956	1.6702	1.9996	2.3890	2.6589
62	0.6785	1.2954	1.6698	1.9990	2.3880	2.6575
63	0.6784	1.2951	1.6694	1.9983	2.3870	2.6561
64	0.6783	1.2949	1.6690	1.9977	2.3860	2.6549
65	0.6783	1.2947	1.6686	1.9971	2.3851	2.6536
66	0.6782	1.2945	1.6683	1.9966	2.3842	2.6524
67	0.6782	1.2943	1.6679	1.9960	2.3833	2.6512
68	0.6781	1.2941	1.6676	1.9955	2.3824	2.6501
69	0.6781	1.2939	1.6672	1.9949	2.3816	2.6490
70	0.6780	1.2938	1.6669	1.9944	2.3808	2.6479